Methods in Molecular Biology

Series Editor
John M. Walker
School of Life and Medical Sciences
University of Hertfordshire
Hatfield, Hertfordshire, UK

For further volumes:
http://www.springer.com/series/7651

For over 35 years, biological scientists have come to rely on the research protocols and methodologies in the critically acclaimed *Methods in Molecular Biology* series. The series was the first to introduce the step-by-step protocols approach that has become the standard in all biomedical protocol publishing. Each protocol is provided in readily-reproducible step-by step fashion, opening with an introductory overview, a list of the materials and reagents needed to complete the experiment, and followed by a detailed procedure that is supported with a helpful notes section offering tips and tricks of the trade as well as troubleshooting advice. These hallmark features were introduced by series editor Dr. John Walker and constitute the key ingredient in each and every volume of the *Methods in Molecular Biology* series. Tested and trusted, comprehensive and reliable, all protocols from the series are indexed in PubMed.

Ctenophores

Methods and Protocols

Edited by

Leonid L. Moroz

Department of Neuroscience and McKnight Brain Institute, University of Florida, Gainesville, FL, USA;
Whitney Laboratory for Marine Biosciences University of Florida, St. Augustine, FL, USA

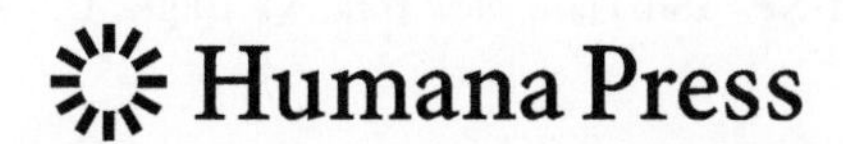

Editor
Leonid L. Moroz
Department of Neuroscience
and McKnight Brain Institute
University of Florida
Gainesville, FL, USA

Whitney Laboratory for Marine Biosciences
University of Florida
St. Augustine, FL, USA

ISSN 1064-3745 ISSN 1940-6029 (electronic)
Methods in Molecular Biology
ISBN 978-1-0716-3641-1 ISBN 978-1-0716-3642-8 (eBook)
https://doi.org/10.1007/978-1-0716-3642-8

© Springer Science+Business Media, LLC, part of Springer Nature 2024
This work is subject to copyright. All rights are reserved by the Publisher, whether the whole or part of the material is concerned, specifically the rights of translation, reprinting, reuse of illustrations, recitation, broadcasting, reproduction on microfilms or in any other physical way, and transmission or information storage and retrieval, electronic adaptation, computer software, or by similar or dissimilar methodology now known or hereafter developed.
The use of general descriptive names, registered names, trademarks, service marks, etc. in this publication does not imply, even in the absence of a specific statement, that such names are exempt from the relevant protective laws and regulations and therefore free for general use.
The publisher, the authors, and the editors are safe to assume that the advice and information in this book are believed to be true and accurate at the date of publication. Neither the publisher nor the authors or the editors give a warranty, expressed or implied, with respect to the material contained herein or for any errors or omissions that may have been made. The publisher remains neutral with regard to jurisdictional claims in published maps and institutional affiliations.

This Humana imprint is published by the registered company Springer Science+Business Media, LLC, part of Springer Nature.
The registered company address is: 1 New York Plaza, New York, NY 10004, U.S.A.

Paper in this product is recyclable.

Preface

Practical Guides for Enigmatic Phyla: "Aliens" of the Sea

With advances in modern biomedical sciences and the advent of the single-cell and genomic revolutions, humankind challenges nature and struggles for global planetary health. And understandably, most public focuses and research efforts rotate around human and clinical work.

One can ask a simple question: why should we care about some "minor" groups of organisms or cryptic, occasionally microscopic, creatures living between grains of sand or in plankton or the depth of oceans? In brief, the answer is conceptually the same as if we are looking for aliens in space. We wonder, we are curious, and we dream. On a deeper level, as in science fiction, we consciously or unconsciously look for alternatives, for conceptually new ways to make a brain [1], an organism [2, 3], and a life itself [4, 5].

In these quests, we nearly forget about virtually unexplored, and even unheard, "aliens" living in the sea. And there are many of them. Most of these "aliens," even professional biologists today, refer to as enigmatic groups [6] because so little is known about them. We know more about rocks on the moon than creatures living in our oceans. The animal kingdom has about 33 phyla (the major taxonomical units), but about two dozen are virtually unexplored. As stated by Sydney Brenner in the introduction to his Nobel Lecture: "It is . . . about *how the great diversity of the living world can both inspire and serve innovation in biological research*" [7].

Ctenophora or comb jellies [8, 9] and Placozoa [10, 11] are two little investigated animal phyla [6], essential to understanding fundamental principles of animal evolution and organization [6, 12] (Fig. 1).

Today, more scientists and laboratories are studying a model nematode worm than the numbers of cells in *Caenorhabditis elegans*, with hundreds of books and more than 37,000 scientific papers since 1974 when Sydney Brenner introduced this microscopic worm into experimental genetics and development [13]. His original question is still applicable today: "How genes might specify the complex structures found in higher organisms is a major [still] unsolved problem of biology" [7]. And greater biodiversity outcomes are needed to be explored as the wisdom of living word, with trillion experiments performed by *Mother Nature*, and accumulated over 3.5 billion years of biological evolution. Extant biodiversity, especially oceanic biotas, are unique, fragile 'gifts' to us, the buffer for planetary health in changing environments and the unreplaceable herritage of humankind.

There are more than 120,000 papers about the tiny fruit fly *Drosophila melanogaster*. In contrast, there are only about 700 publications about *all* extant ctenophores and ~300 articles about *Trichoplax*, the representative species of placozoan, even though these groups have 150 years of scientific history. Why is that?

To begin with, these creatures (i.e., comb jellies and placozoans) remained enigmatic primarily because they were less accessible to experimental scientists. But they are not rare, spreading for thousands of miles over the world's oceans. Admittedly, representatives of both phyla are more fragile than *Drosophila* and *C. elegans*, but some are still amenable to culture and experimental manipulations [14], as summarized in this book. The primary barrier is philosophical and sometimes psychological.

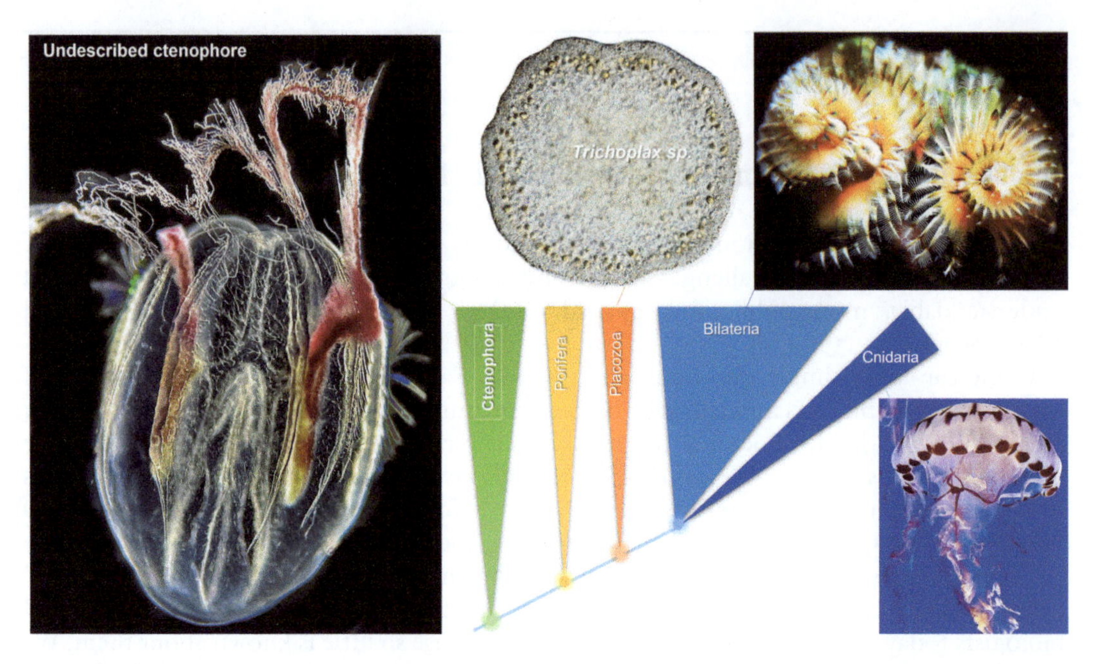

Fig. 1 Ctenophores, Placozoans, and Novel Animal Phylogeny suggest extensive convergent and parallel evolution of many traits in Metazoa and independent origins of animal complexities. The figure shows undescribed ctenophore species (Friday Harbor, WA, USA) and the placozoan *Trichoplax* sp. (Hawaii, also known as H2 haplotype without a formal description of this species)

These two groups were historically related to cnidarians (i.e., polyps, jellyfishes, or their kins, formally known as Coelenterata) or derived from simplified cnidarians' larval-like stages (e.g., *Trichoplax*). But a more in-depth view of their molecular makeups or genomes reveals a remarkably different design to their body plans and signaling, from the genome to behavior with single-cell resolution. Representatives of these animal lineages have developed many unique traits, including multiple examples of convergent evolution [15, 16].

Most intriguing, ctenophores belong to the earliest branching metazoan lineage, sister to all animals [17–23]. In other words, they are surviving descendants of the most ancestral animal branch (Fig. 1). As a result, the comb jellies independently developed a complexity comparable to many bilaterians with elaborated neuromuscular, digestive, locomotory, and sensory systems. Ctenophores might have preserved or developed alternative designs of neurons, neural circuits, muscles, intercellular communication systems, tissues and organs [17, 24, 25].

In contrast, placozoans are more derived, forming the sister group to the Cnidaria +Bilateria clade (see also [21]), but apparently preserved the ancestral, simpler body plan without symmetry and without organs (Fig. 1). In fact, placozoans are morphologically the simplest known free-living animals, maintaining many unique ancient traits (including complex modes of cilia-controlled locomotion), and with the potential to understand the integrative mechanisms at all levels of biological organization, from genome to behaviors.

No comprehensive reference source reflects the emerging approaches and methods inspired by recent research using ctenophores or placozoans. Thus, this volume presents a collection of introductory and methodological chapters, which facilitate studying these enigmatic species and can be applied to other non-traditional reference species focusing across non-bilaterian lineages.

Here, practical protocols are integrated with introductory chapters about general biology [26, 27], summarizing these enigmatic organisms' diversity, evolution, genomics, development, and neurobiology. Special chapters outline strategies and applications for culturing [28, 29], microscopy [30, 31], physiology [32, 33], electrical synapses [34] and bioluminescence [35, 36], molecular [37–39] and single-cell biology [40], transcription factors [41] as well as gene [39] and protein expression analyses [42]. The book also contains bioinformatics and computational chapters [43] covering peptidomics [44, 45] and epigenomics [46]. Finally, we also included an illustrative guide for various ctenophore lineages, with the taxonomy of all recognized, species and genera of this phylum [47], for the first time over 100 years. We did not include sections about the development, which is summarized elsewhere [48].

The described approaches would interest comparative and experimental biologists and interdisciplinary scientists aiming to decipher fundamental principles of animal organization, origin, and evolution of biological complexity in general, opening novel avenues for future synthetic biology [3, 49].

St. Augustine, FL, USA *Leonid L. Moroz*

References

1. Moroz LL, Romanova DY (2022) Alternative neural systems: what is a neuron? (Ctenophores, sponges and placozoans). Front Cell Dev Biol 10:1071961
2. Venetz JE, Del Medico L, Wolfle A, Schachle P, Bucher Y, Appert D et al (2019) Chemical synthesis rewriting of a bacterial genome to achieve design flexibility and biological functionality. Proc Natl Acad Sci U S A 116:8070–8079
3. Fredens J, Wang K, de la Torre D, Funke LFH, Robertson WE, Christova Y et al (2019) Total synthesis of *Escherichia coli* with a recoded genome. Nature 569:514–518
4. Gibson DG, Glass JI, Lartigue C, Noskov VN, Chuang RY, Algire MA et al (2010) Creation of a bacterial cell controlled by a chemically synthesized genome. Science 329:52–56
5. Malyshev DA, Dhami K, Lavergne T, Chen T, Dai N, Foster JM et al (2014) A semi-synthetic organism with an expanded genetic alphabet. Nature 509:385–388
6. Brusca RC, Giribet G, Moore W (2022) Invertebrates, 4th edn. Sinauer Associates of Oxford University Press, Sunderland/New York
7. Brenner S (2002) Nature's gift to science. Nobel Lecture NobelPrizeorg
8. Hernandez-Nicaise M-L (1991) Ctenophora. In: Harrison FW, Westfall JA (eds) Microscopic anatomy of invertebrates: Placozoa, Porifera, Cnidaria, and Ctenophora. Wiley, New York, pp 359–418
9. Tamm SL (1982) Ctenophora. In: Electrical conduction and behavior in "simple" invertebrates. Clarendon Press, Oxford, pp 266–358
10. Grell KG, Ruthmann A (1991) Placozoa. In: Harrison FW (ed) Microscopic anatomy of invertebrates. Wiley-Liss, New York, pp 13–27
11. Tessler M, Neumann JS, Kamm K, Osigus H-J, Eshel G, Narechania A et al (2022) Phylogenomics and the first higher taxonomy of Placozoa, an ancient and enigmatic animal phylum. Front Ecol Evol 10:10
12. Nielsen C (2019) Early animal evolution: a morphologist's view. R Soc Open Sci 6:190638
13. Brenner S (1974) The genetics of *Caenorhabditis elegans*. Genetics 77:71–94
14. Presnell JS, Bubel M, Knowles T, Patry W, Browne WE (2022) Multigenerational laboratory culture of pelagic ctenophores and CRISPR-Cas9 genome editing in the lobate *Mnemiopsis leidyi*. Nat Protoc 17: 1868–1900
15. Moroz LL (2015) Convergent evolution of neural systems in ctenophores. J Exp Biol 218:598–611
16. Dunn CW, Leys SP, Haddock SH (2015) The hidden biology of sponges and ctenophores. Trends Ecol Evol 30:282–291

17. Moroz LL, Kocot KM, Citarella MR, Dosung S, Norekian TP, Povolotskaya IS et al (2014) The ctenophore genome and the evolutionary origins of neural systems. Nature 510:109–114
18. Li Y, Shen XX, Evans B, Dunn CW, Rokas A (2021) Rooting the animal tree of life. Mol Biol Evol 38: 4322–4333
19. Ryan JF, Pang K, Schnitzler CE, Nguyen AD, Moreland RT, Simmons DK et al (2013) The genome of the ctenophore *Mnemiopsis leidyi* and its implications for cell type evolution. Science 342:1242592
20. Whelan NV, Halanych KM (2023) Available data do not rule out Ctenophora as the sister group to all other Metazoa. Nat Commun 14:711
21. Schultz DT, Haddock SHD, Bredeson JV, Green RE, Simakov O, Rokhsar DS (2023) Ancient gene linkages support ctenophores as sister to other animals. Nature 618:110–117
22. Whelan NV, Kocot KM, Moroz LL, Halanych KM (2015) Error, signal, and the placement of Ctenophora sister to all other animals. Proc Natl Acad Sci U S A 112:5773–5778
23. Whelan NV, Kocot KM, Moroz TP, Mukherjee K, Williams P, Paulay G et al (2017) Ctenophore relationships and their placement as the sister group to all other animals. Nat Ecol Evol 1:1737–1746
24. Moroz LL, Kohn AB (2016) Independent origins of neurons and synapses: insights from ctenophores. Philos Trans R Soc Lond Ser B Biol Sci 371:20150041
25. Moroz LL, Romanova DY, Kohn AB (2021) Neural versus alternative integrative systems: molecular insights into origins of neurotransmitters. Philos Trans R Soc Lond Ser B Biol Sci 376:20190762
26. Moroz LL (2024) Brief history of Ctenophora. In: Moroz LL (ed) Ctenophores. Methods in molecular biology. Humana, New York. p 1–26
27. Romanova DY, Moroz LL (2024) Brief history of Placozoa. In: Moroz LL (ed) Ctenophores. Methods in molecular biology. Humana, New York. p 103–122
28. Angel J-JS, Nordmann E-L, Sturm D, Sachkova M, Pang K, Burkhardt P (2024) Stable laboratory culture system for the ctenophore *Mnemiopsis leidyi*. In: Moroz LL (ed) Ctenophores. Methods in molecular biology. Humana, New York. p 123–145
29. Romanova DY, Varoqueaux F, Eitel M, Yoshida MA, Nikitin MA, Moroz LL (2024) Long-term culturing of placozoans (*Trichoplax and Hoilungia*). In: Moroz LL (ed) Ctenophores. Methods in molecular biology. Humana, New York. p 509–529
30. Norekian TP, Moroz LL (2024) Illustrated Neuroanatomy of Ctenophores: Immunohistochemistry. In: Moroz LL (ed) Ctenophores. Methods in molecular biology. Humana, New York. p 147–161
31. Norekian TP, Moroz LL (2024) Scanning electron microscopy of ctenophores: illustrative atlas. In: Moroz LL (ed) Ctenophores. Methods in molecular biology. Humana, New York. pp 163–184
32. Norekian T, Moroz LL (2024) Recording cilia activity in ctenophores. In: Moroz LL (ed) Ctenophores. Methods in molecular biology. Humana, New York. pp 307-313
33. Meech RW, Bilbaut A, Hernandez-Nicaise M-L (2024) Electrophysiology of ctenophore smooth muscle. In: Moroz LL (ed) Ctenophores. Methods in molecular biology. Humana, New York. pp 315–359
34. Kohn AB, Moroz LL (2024) Gap junctions in Ctenophora. In: Moroz LL (ed) Ctenophores. Methods in molecular biology. Humana, New York. pp 361–381
35. Markova SV, Vysotski ES (2024) Functional screening of cDNA expression library for novel ctenophore photoproteins. In: Moroz LL (ed) Ctenophores. Methods in molecular biology. Humana, New York. pp 289–306
36. Burakova LP, Markova SV, Malikova NP, Vysotski ES (2024) Expression, purification, and determination of sensitivity to calcium ions of ctenophore photoproteins. In: Moroz LL (ed) Ctenophores. Methods in molecular biology. Humana, New York. pp 269–287
37. Moraga DAA, Kohn AB, Bobkova Y, Panayotova NG, Moroz LL (2024) DNA isolation long-read genomic sequencing in ctenophores. In: Moroz LL (ed) Ctenophores. Methods in molecular biology. Humana, New York. pp 185–200
38. Kohn AB, Bobkova Y, Moroz LL (2024) RNA isolation from ctenophores. In: Moroz LL (ed) Ctenophores. Methods in molecular biology. Humana, New York. pp 201–214
39. Kohn AB, Bobkova Y, Moroz LL (2024) Gene expression patterns in the ctenophore *Pleurobrachia bachei*: in situ hybridization. In: Moroz LL (ed) Ctenophores. Methods in molecular biology. Humana, New York. pp 215–237

40. Li Y, Sun C, Romanova DY, Wu DO, Fang R, Moroz LL (2024) Analysis and visualization of single-cell sequencing data with Scanpy and MetaCell: a tutorial. In: Moroz LL (ed) Ctenophores. Methods in molecular biology. Humana, New York. pp 383–445
41. Mukherjee K, Moroz LL (2024) Parallel evolution of transcription factors in basal metazoans. In: Moroz LL (ed) Ctenophores. Methods in molecular biology. Humana, New York. pp 491–508
42. Mayer ML (2024) Expression and functional analysis of ctenophore glutamate receptor genes. In: Moroz LL (ed) Ctenophores. Methods in molecular biology. Humana, New York. pp 259–268
43. Hsiao LD, Moroz LL, Chalasani SH, Edsinger E (2024) Ocean to tree: leveraging single-molecule RNA-Seq to repair genome gene models and improve phylogenomic analysis of gene and species evolution. In: Moroz LL (ed) Ctenophores. Methods in molecular biology. Humana, New York. pp 461–490
44. Romanova DY, Moroz LL (2024) Bioinformatic prohormone discovery in basal metazoans: insights from *Trichoplax*. In: Moroz LL (ed) Ctenophores. Methods in molecular biology. Humana, New York
45. Muthye V, Lavrov D (2024) Characterization of the mitochondrial proteome in the ctenophore *Mnemiopsis leidyi* using MitoPredictor. In: Moroz LL (ed) Ctenophores. Methods in molecular biology. Humana, New York. pp 239–257
46. Dabe EC, Kohn A, Moroz LL (2024) DNA methylation in ctenophores. In: Moroz LL (ed) Ctenophores. Methods in molecular biology. Humana, New York. pp 447–459
47. Moroz LL, Collins R, Paulay G, (2024) Ctenophora: illustrated guide and taxonomy. In: Moroz LL (ed) Ctenophores. Methods in molecular biology. Humana, New York. pp 27–101
48. Martindale MQ, Henry JQ (1997) The Ctenophora. Sinauer, Sunderland
49. Kriegman S, Blackiston D, Levin M, Bongard J (2021) Kinematic self-replication in reconfigurable organisms. Proc Natl Acad Sci U S A 118:e2112672118

Contents

Contributors

YELENA BOBKOVA • *Whitney Laboratory for Marine Bioscience, University of Florida, St. Augustine, FL, USA*

LYUDMILA P. BURAKOVA • *Photobiology Laboratory, Institute of Biophysics SB RAS, Federal Research Center "Krasnoyarsk Science Center SB RAS", Krasnoyarsk, Russia*

PAWEL BURKHARDT • *Michael Sars Centre, University of Bergen, Bergen, Norway*

SREEKANTH H. CHALASANI • *Molecular Neurobiology Laboratory, Salk Institute for Biological Study, La Jolla, CA, USA*

RICHARD COLLINS • *Florida Museum of Natural History, Gainesville, FL, USA*

EMILY C. DABE • *Whitney Laboratory for Marine Biosciences, University of Florida, St. Augustine, FL, USA; Department of Neuroscience, McKnight Brain Institute, University of Florida, Gainesville, FL, USA*

LOLA CHENXI DENG • *Molecular Neurobiology Laboratory, Salk Institute for Biological Study, La Jolla, CA, USA*

ERIC EDSINGER • *Molecular Neurobiology Laboratory, Salk Institute for Biological Study, La Jolla, CA, USA*

MICHAEL EITEL • *Department of Earth and Environmental Sciences Palaeontology & Geobiology, LMU München, Munich, Germany*

RUOGU FANG • *Department of Electrical and Computer Engineering, Herbert Wertheim College of Engineering, University of Florida, Gainesville, FL, USA; J. Crayton Pruitt Family Department of Biomedical Engineering, Herbert Wertheim College of Engineering, University of Florida, Gainesville, FL, USA; Center for Cognitive Aging and Memory, McKnight Brain Institute and NSF Center for Big Learning, University of Florida, Gainesville, FL, USA*

MARI-LUZ HERNANDEZ-NICAISE • *Nice, France*

JAN HSIAO • *Molecular Neurobiology Laboratory, Salk Institute for Biological Study, La Jolla, CA, USA*

ANDREA B. KOHN • *Whitney Laboratory for Marine Biosciences, University of Florida, St. Augustine, FL, USA*

DENNIS V. LAVROV • *Department of Ecology, Evolution and Organismal Biology, Iowa State University, Ames, IA, USA*

YANJUN LI • *Department of Medicinal Chemistry, College of Pharmacy, University of Florida, Gainesville, FL, USA; Center for Natural Products, Drug Discovery and Development, University of Florida, Gainesville, FL, USA*

NATALIA P. MALIKOVA • *Photobiology Laboratory, Institute of Biophysics SB RAS, Federal Research Center "Krasnoyarsk Science Center SB RAS", Krasnoyarsk, Russia*

SVETLANA V. MARKOVA • *Photobiology Laboratory, Institute of Biophysics SB RAS, Federal Research Center "Krasnoyarsk Science Center SB RAS", Krasnoyarsk, Russia*

MARK L. MAYER • *National Institute of Neurological Disorders and Stroke, National Institutes of Health, Bethesda, MD, USA*

ROBERT W. MEECH • *School of Physiology, Pharmacology & Neuroscience, University of Bristol, University Walk, Bristol, UK*

DAVID MORAGA AMADOR • *Interdisciplinary Center for Biotechnology (ICBR), University of Florida, Gainesville, FL, USA*

LEONID L. MOROZ • *Department of Neuroscience, McKnight Brain Institute, University of Florida, Gainesville, FL, USA; Whitney Laboratory for Marine Bioscience, University of Florida, St. Augustine, FL, USA*
KRISHANU MUKHERJEE • *Whitney Laboratory for Marine Bioscience, University of Florida, St. Augustine, FL, USA*
VIRAJ MUTHYE • *Department of Ecology, Evolution and Organismal Biology, Iowa State University, Ames, IA, USA*
MIKHAIL A. NIKITIN • *Belozersky Institute for Physico-Chemical Biology, Lomonosov Moscow State University, Moscow, Russian Federation; Kharkevich Institute for Information Transmission Problems, RAS, Moscow, Russian Federation; Institute of Higher Nervous Activity and Neurophysiology of RAS, Moscow, Russian Federation*
EVA-LENA NORDMANN • *Michael Sars Centre, University of Bergen, Bergen, Norway*
TIGRAN P. NOREKIAN • *Whitney Laboratory for Marine Bioscience, University of Florida, St. Augustine, FL, USA*
NEDKA G. PANAYOTOVA • *Interdisciplinary Center for Biotechnology (ICBR), University of Florida, Gainesville, FL, USA*
KEVIN PANG • *Michael Sars Centre, University of Bergen, Bergen, Norway*
GUSTAV PAULAY • *Florida Museum of Natural History, Gainesville, FL, USA*
DARIA Y. ROMANOVA • *Institute of Higher Nervous Activity and Neurophysiology of RAS, Moscow, Russia*
MARIA SACHKOVA • *Michael Sars Centre, University of Bergen, Bergen, Norway*
JOAN J. SOTO-ANGEL • *Michael Sars Centre, University of Bergen, Bergen, Norway*
DANIELA STURM • *University of Plymouth, Plymouth, UK*
CHAOYUE SUN • *Department of Electrical and Computer Engineering, Herbert Wertheim College of Engineering, University of Florida, Gainesville, FL, USA*
FRÉDÉRIQUE VAROQUEAUX • *Department of Fundamental Neurosciences, University of Lausanne, Lausanne, Switzerland*
EUGENE S. VYSOTSKI • *Photobiology Laboratory, Institute of Biophysics SB RAS, Federal Research Center "Krasnoyarsk Science Center SB RAS", Krasnoyarsk, Russia*
DAPENG O. WU • *Department of Computer Science, City University of Hong Kong, Hong Kong, China*
MASA-AKI YOSHIDA • *Marine Biological Science Section, Education and Research Center for Biological Resources, Faculty of Life and Environmental Science, Shimane University, Okinoshima, Oki, Shimane, Japan*

Chapter 1

Brief History of Ctenophora

Leonid L. Moroz

Abstract

Ctenophores are the descendants of the earliest surviving lineage of ancestral metazoans, predating the branch leading to sponges (Ctenophore-first phylogeny). Emerging genomic, ultrastructural, cellular, and systemic data indicate that virtually every aspect of ctenophore biology as well as ctenophore development are remarkably different from what is described in representatives of other 32 animal phyla. The outcome of this reconstruction is that most system-level components associated with the ctenophore organization result from convergent evolution. In other words, the ctenophore lineage independently evolved as high animal complexities with the astonishing diversity of cell types and structures as bilaterians and cnidarians. Specifically, neurons, synapses, muscles, mesoderm, through gut, sensory, and integrative systems evolved independently in Ctenophora. Rapid parallel evolution of complex traits is associated with a broad spectrum of unique ctenophore-specific molecular innovations, including alternative toolkits for making an animal. However, the systematic studies of ctenophores are in their infancy, and deciphering their remarkable morphological and functional diversity is one of the hot topics in biological research, with many anticipated surprises.

Key words Ctenophora, Placozoa, Porifera, *Pleurobrachia*, *Mnemiopsis*, Neurons, Muscles, Development, Cell-type evolution, Phylogeny

1 Ctenophores as the Sister Lineage to All Other Animal Phyla

Ctenophores or comb jellies are true wonders of nature! They are the most unusual animals in the marine realm, both from structural and molecular standpoints. "Although it is easy in a given case to determine whether or not a particular animal is a ctenophore, it is equally difficult to establish how closely or distantly ctenophores are related to other forms of animals."—this Krumbach's note (1925) and the challenge [1] was reconfirmed by the leading experts at the beginning of the twenty-first century, with no morphological evidence that could link the phylum Ctenophora to any other extant phylum [2, 3]. This hundred-year enigma started to be uncovered only recently.

Arguably ctenophores are the descendants of the earliest surviving lineage of ancestral metazoans [4–8], predating the branch

Leonid L. Moroz (ed.), *Ctenophores: Methods and Protocols*, Methods in Molecular Biology, vol. 2757,
https://doi.org/10.1007/978-1-0716-3642-8_1, © Springer Science+Business Media, LLC, part of Springer Nature 2024

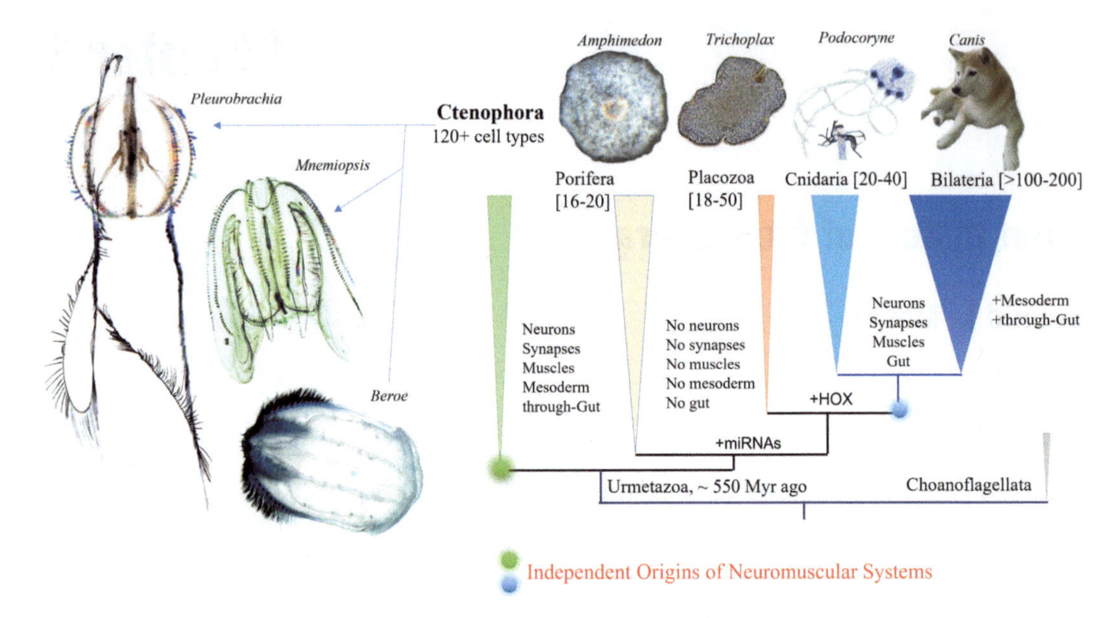

Fig. 1 Relationships among five basal metazoan clades with Choanoflagellata as the sister group to Metazoa. Three species (*Pleurobrachia bachei*, *Mnemiopsis leidyi*, and *Beroe* sp. from Antarctica) illustrate the phylum of Ctenophora as the descendents of the earliest branching animal lineage. The most recent comparative analyses suggest independent origins and convergent evolution of neurons, synapses, muscles, mesoderm, and through-gut in Metazoa (see text for details). Possible origins of microRNA and HOX gene cluster are indicated. Numbers under each lineage are the author's estimates of the diversity of cell types in basal metazoan clades

leading to sponges (Fig. 1). As a result, virtually every aspect of ctenophore biology, the systemic and molecular organization, as well as ctenophore development are remarkably different from what is described in other representatives of 32 animal phyla. In this respect, comb jellies are indeed *"aliens" of the sea*.

Ctenophores are exclusively marine species—from the surface to the record depth of 10,040 meters [9]. Most of the ctenophores, especially in deep habitats, are bioluminescent. The functional role of bioluminescence is unknown, but it is mediated by a distinct group of photoproteins [10–23] unrelated to the famous green fluorescent protein family.

These beautiful "*aliens of the sea*" (sometimes reaching 1.5 m—*Cestum*) can be easily recognized on a calm day in seawater [24] across the globe, from polar to tropical habitats [25]. Any curious observer can find ctenophores without difficulties (Fig. 2). Ctenophores are unmistakably distinguished from the canonical jellies (which belong to another phylum Cnidaria) by the presence of brightly iridescent [26] fused cilia assembled in eight comb rows [27, 28], hence, the name cteno-phora—comb bearers (Ancient Greek: κτείς *(kteis)* "comb" and φέρω *(pherō)* "to carry"). Fused locomotory cilia are the largest in the animal kingdom and are used to glide animals in the water with minimal disturbance, often as

Fig. 2 Diversity of ctenophore species. (1) Benthic ctenophores (Platyctenida). (2) Tentaculate ctenophores (Cydippida). (3) Atentaculate Beroida or Nuda (*Beroe*). (4) Lobata (*Bolinopsis* and *Mnemiopsis*). (5) Lobata: *Ocyropsis*. (6) *Labatolampea*

stealth predators [29, 30]. Such a mode of locomotion separates *comb* jellies from true jellyfishes that are moved by muscular jet-type propulsions. Most ctenophores are holopelagic, but some are creeping (Platyctenida) and even sessile (*Tjalfiella tristoma*, *Lyrocteis imperatoris*).

The first ctenophore drawing (*Bolinopsis* and *Mertensia*) was provided by a ship doctor Martens in 1671, in the vicinity of Spitzbergen [31]. The relationships of comb jellies with other organisms were unclear. The phylum Ctenophora was formally established in 1889 by Hatschek as a separate group distinct in their organization from cnidarians. However, until recently, their affinity with cnidarians was considered, forming a clade coelenterates. All current phylogenomic reconstructions reject this association.

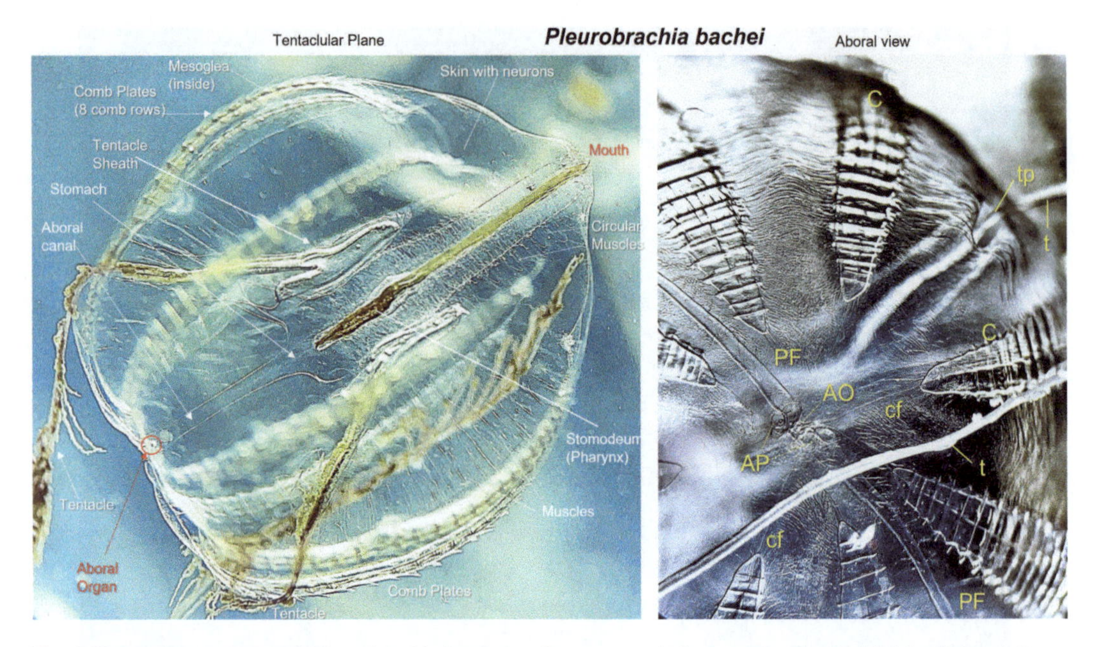

Fig. 3 Illustrative anatomy of *Pleurobrachia bachei* as the representative species for Cyddipida. Abbreviations: *AO* the aboral organ, *AP* anal pores, *C* comb plates, cf ciliated furrows, *PF* polar fields, *t* tentacles, *tp* tentacle pocket

Fig. 4 Two symmetry plans in ctenophores: tentacle and sagittal/esophageal axes (*Pleurobrachia bachei*). Abbreviations: *AO* the aboral organ, *PF* polar fields, *t* tentacles

Four ctenophore genomes have been sequenced, annotated, and published: two closely related cydippid species, *Pleurobrachia bachei* [5] (Figs. 3 and 4) and *Hormiphora californensis* [32], and

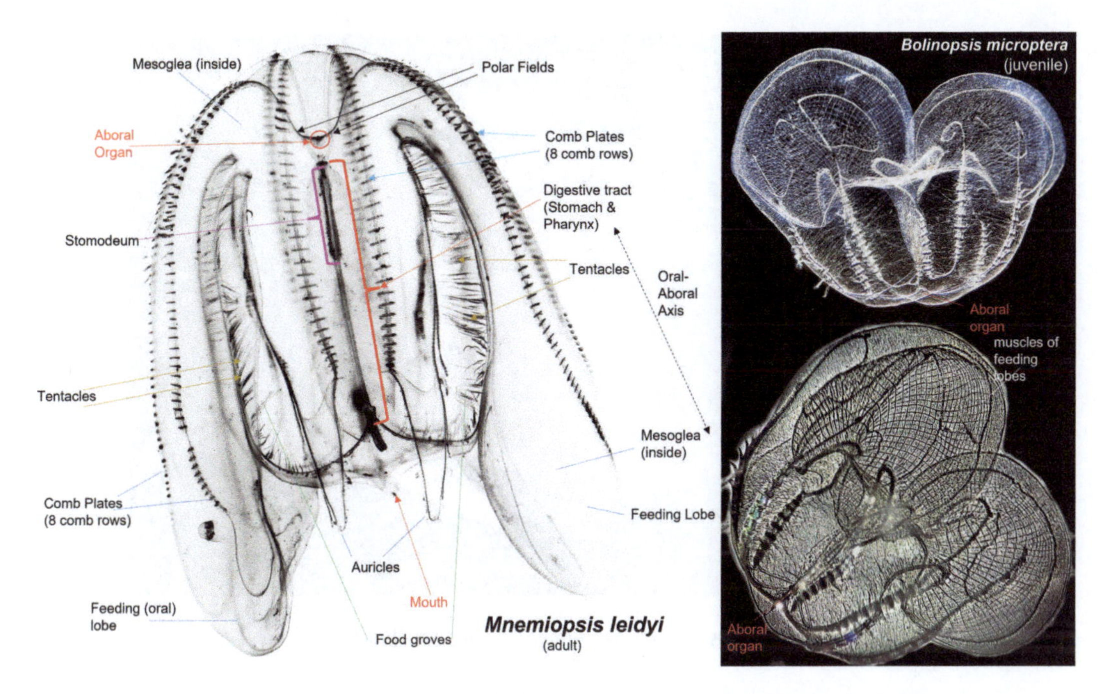

Fig. 5 Illustrative anatomy of *Mnemiopsis and Bolinopsis* as representative species for Lobata

two closely related lobates (Fig. 5), *Mnemiopsis leidyi* [33] and *Bolinopsis microptera* [6]. Three of them (*Hormiphora*, *Pleurobrachia,* and *Bolinopsis*) have chromosome-scale resolution [6, 34] with about 13 chromosomes, suggesting that a common $n = 13$ karyotype is ancestral to this cydippid-lobate group. These sequenced genomes are quite small, with estimated 1C sizes of 100–254 Mbp. Two additional genomes from atentaculate ctenophores (*Beroe forskalii* and *B. ovata*) were recently sequenced and deposited to NCBI (Bioprojects: PRJNA421807, PRJEB23672). The representatives of Beroida are active swimmers (Fig. 6) and often prey on other ctenophores (such as *Bolinopsis*, Fig. 7) and diverse pelagic invertebrates.

The sequencing of these ctenophore genomes and functional/developmental data provided convincing arguments that the ctenophores form the first branch of the animal tree of life, sister to the rest of all metazoans (Figs. 1 and 8). This conclusion is based on two compelling lines of evidence. First, integrative, interdisciplinary analysis of multiple traits and genes encoding neural, muscular, immune, mesoderm, and intracellular signaling components, combined with phylogenomics, revealed a reduced representation in each of these toolkits compared to sponges and the rest of metazoans [5]. This discovery led to the scenario that neurons, muscles, and mesoderm, systemic gut with two anuses, and sensory organs evolved more than once and independently in the ctenophores vs. Cnidaria+Bilateria clade [5, 35–37].

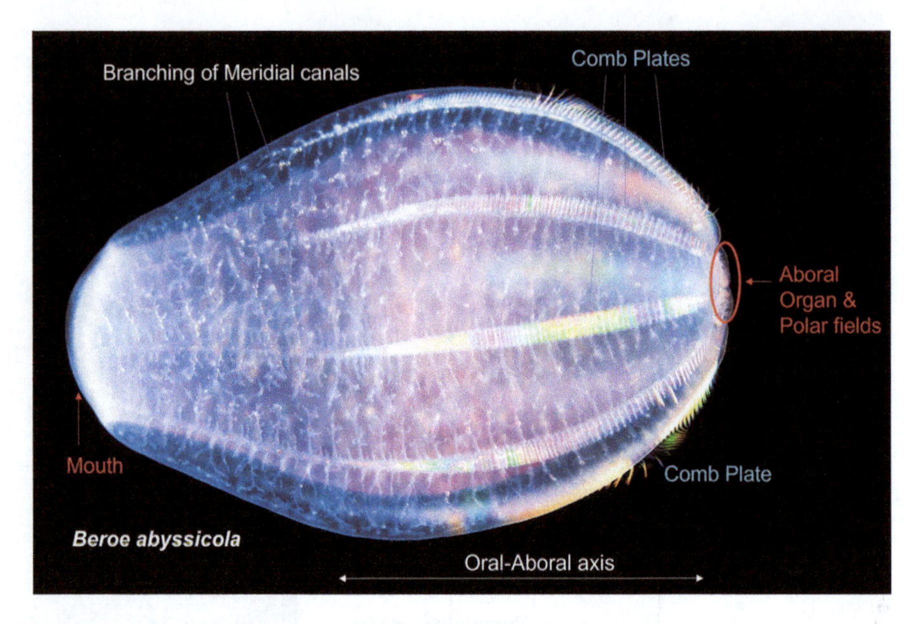

Fig. 6 Illustrative anatomy of *Beroe* as the representative species for Nuda

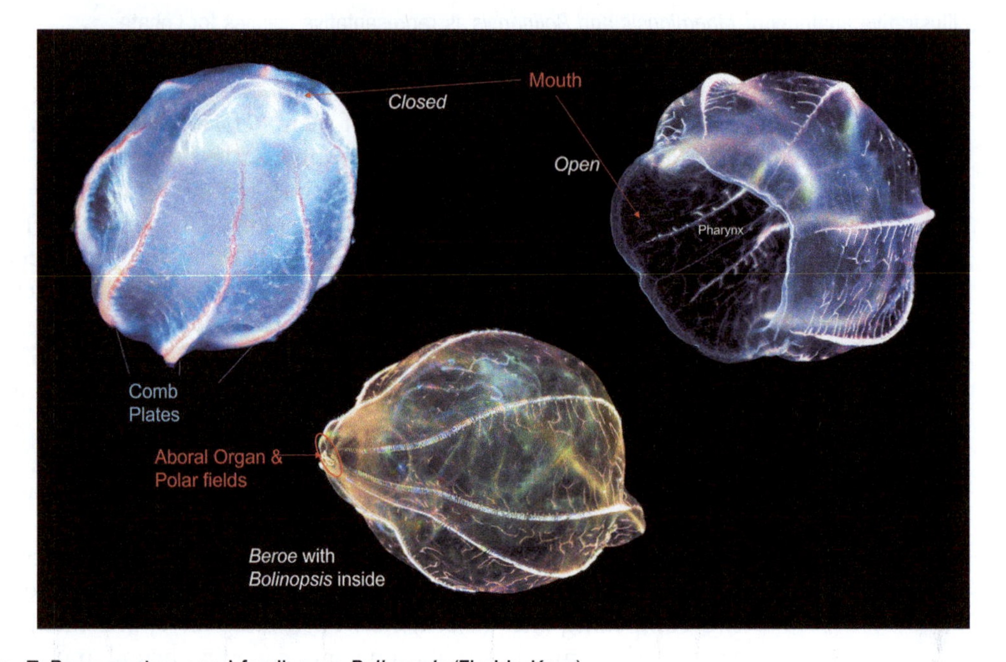

Fig. 7 *Beroe* anatomy and feeding on *Bolinopsis* (Florida Keys)

Second, the chromosome-level synteny analyses across Metazoa showed that ctenophores and unicellular eukaryotes share ancestral metazoan patterns, whereas sponges, bilaterians, and cnidarians share derived chromosomal rearrangements [6]. Schultz and colleagues pointed out: "the patterns of synteny shared by

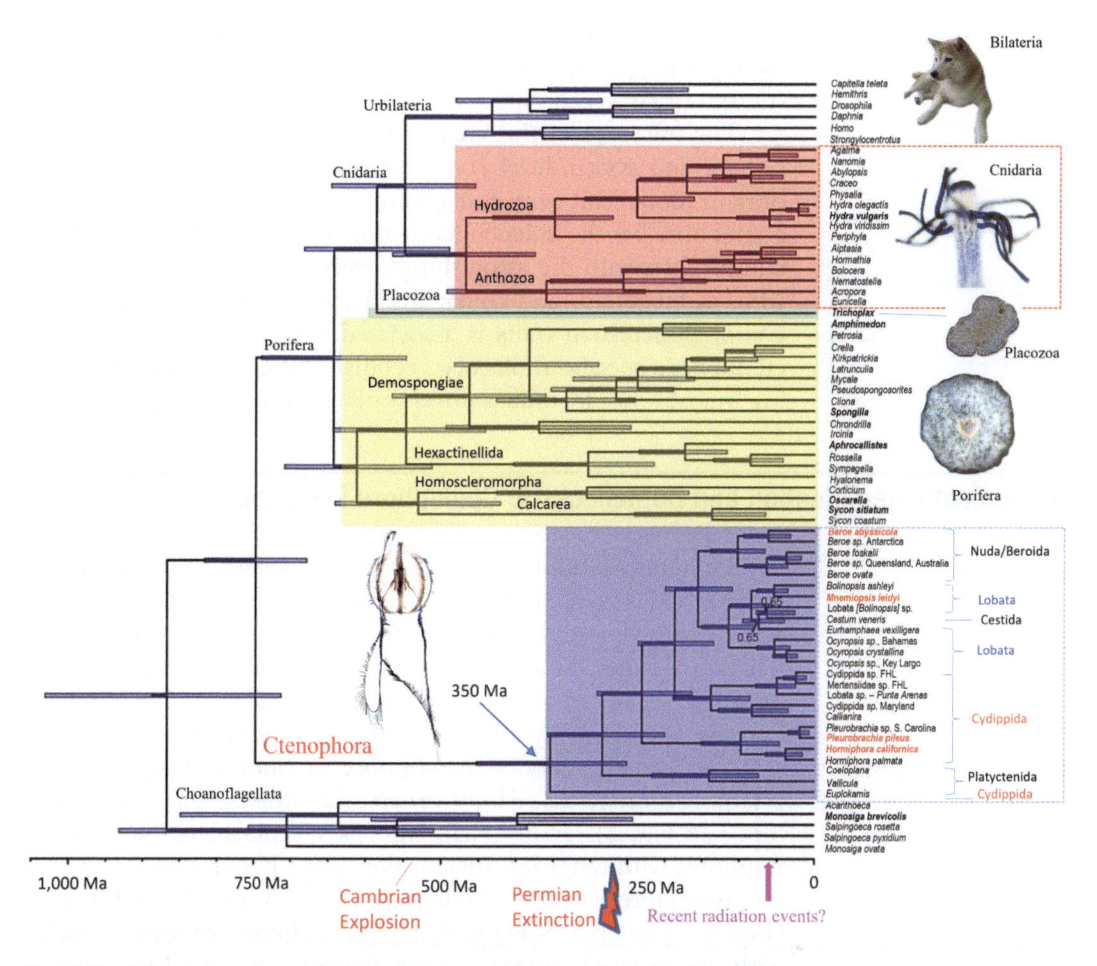

Fig. 8 Ctenophora as sister to the rest of Metazoa. The tree shows relationships among basal metazoan clades and within species of the phylum Ctenophora (Adapted and modified from Ref. [8]). Of note, this phylogeny does not support the classical ctenophore systematics and indicates the polyphyly of Lobata and Cydippida and the placement of Nuda/Beroida within Lobata

sponges, bilaterians, and cnidarians are the result of *rare and irreversible chromosome fusion-and-mixing events* that provide robust and unambiguous phylogenetic support for the ctenophore-sister hypothesis" [6]. More than 30 ctenophore transcriptomes were obtained in parallel, leading to the same conclusion and the ctenophore-first hypothesis [5, 7, 8] (Fig. 8).

Giant mitochondria [38] and compact mitochondrial genomes in ctenophores are also unique and highly derived due to their rapid evolutionary dynamics [39–47]. These findings prevent the use of mitogenomics for macrophylogeny. In contrast, mitogenomics is highly valuable for deciphering divergent evolution within the phylum [41, 42, 48, 49]. In addition, the diversity of mobile elements in ctenophores might support the origins of certain innovations and even facilitate transcription factors' evolution [50, 51];

many of transcription factor families (e.g. BHLH) resulted from ctenophore-specific difersification events, supporting complex tissue and organ specification.

The outcome of this ctenophore-first hypothesis is that most cellular and system-level components associated with the animal organization result from convergent evolution. In other words, the ctenophore lineage independently evolved such high level of animal complexities with the astonishing diversity of cell types and structures as bilaterians and cnidarians. Parallel and early evolution of complex metazoan traits is associated with a broad spectrum of ctenophore-specific molecular, cellular, developmental and feeding innovations, including novel toolkits for making an animal.

2 Recent Diversification and Bottlenecks in Ctenophore Evolution

Ctenophores are animals with exceptional rotational-type symmetry [52, 53] (Fig. 4), not recognized in other metazoans. There are 185 described species of Ctenophora (See Moroz, Collins, Paulay, Chapter 2, this book [198]), and likely this number could be doubled to incorporate recently discovered (but not formally described) and mostly unknown deep-water species.

The existing classical ctenophore taxonomy recognizes two established classes [2, 31], 9 orders, 32 families, and >50 genera (see also Fig. 2). Traditionally, the class Tentaculata includes ctenophores with tentacles, such as illustrated here representatives of the two largest orders: Cydippida (Figs. 3 and 4) and Lobata (Fig. 5). The class Nuda includes ctenophores without tentacles, with one order (Beroida) and two genera, *Neis* and *Beroe* (Figs. 6 and 7), which secondarily lost tentacles both in their larval and adult stages. The presence of tentacles in adults and larval ctenophores (cydippid larva) is likely the ancestral trait.

However, the emerging molecular phylogeny challenges the classical taxonomy [5, 7, 8], uncovering the polyphyly of Lobata and Cydippida. The parallel evolution of multiple traits (Figs. 8 and 9) includes two independent transitions to benthic lifestyles in Platyctenida or benthic ctenophores and *Lobatolampea*, respectively (Fig. 9, red arrows). Furthermore, the comparative phylogenomic analysis, using more than 30 ctenophore transcriptomes and molecular clock estimates, indicated that the ctenophore lineage went through a significant bottleneck about ~350–250 million years ago [8], with a possibility of the most recent diversification events that occurred around 100–60 million years ago (Fig. 8), which correlates with the Cretaceous–Tertiary (K–T) extinction at the end of the Mesozoic era, also ending the dinosaurs' epoch.

These evolutionary bottlenecks explain the loss of some distinctive features of ancient ctenophores found in fossils of about 20 species. Indeed, some Cambrian ctenophores possessed 16–80

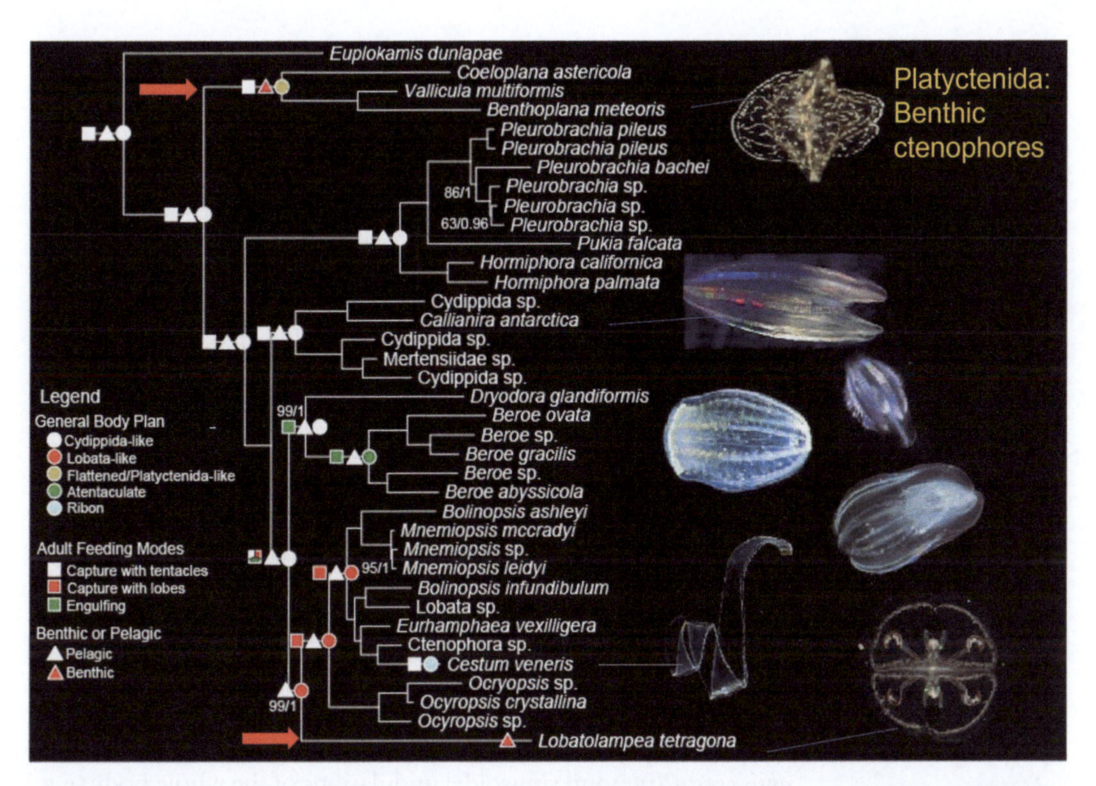

Fig. 9 Ctenophore phylogeny reveals parallel evolution of adaptive strategies in Ctenophora (Adapted and modified from Ref. [8]). Red arrows indicate two independent transitions from pelagic to a benthic lifestyle in Ctenophora

comb rows (vs. only eight comb rows in all extant ctenophores) [54]. There are also speculations that some ancestral ctenophores had sclerotized skeletons and could be secondarily sessile, forming a now-extinct clade Scleroctenophora [55]. Some Ediacaran fossils, such as *Eoandromeda*, were interpreted as an early stem-group ctenophore [56]. Zhao and colleagues also suggested that the earliest ctenophores were suspension feeders [57], implying that tentacles and predation occurred later. The earliest tentaculate ctenophores were found in the early Cambrian [58] and Devonian [59, 60]. Nevertheless, it isn't easy to reconstruct their history due to the poor preservation of ctenophores in fossil records.

3 Ctenophores as Predators

Ctenophores are carnivores (active or ambush predators), feeding on a broad range of animals [61–63]: from zooplanktons to other ctenophores (e.g., *Beroe*, Fig. 7; see also [64]), narcomedusae (e.g., *Haeckelia* [65–67], or larvaceans for *Dryodora* (see also [68–75]. As a result, ctenophores exhibit a remarkable diversity of behaviors [76–89], which are little investigated. Tentacles and their small branches (tentillae/tentilla) contain specialized sticky glue cells or colloblasts [90, 91], facilitating prey capture and performing other functions.

Ctenophores have highly elaborated digestive systems with well-developed tripartite **through-gut** [31]: mouth, pharynx, stomach, and a pair of anal pores with rhythmic contractions, often associated with defecation [92]. Such distinctive through-gut evolved in ctenophores independently from the rest of metazoans. Absorption of digested nutrients is transported to a branching gastro-endodermal canal system (meridional canals) and delivered to the rest of the body.

4 Ctenophore Life Is Based on Cilia and Alternative Neural Systems

It would be proper to say that virtually all ctenophore organization and their life is based on cilia [27, 93]. The diversity, complexity, and control of cilia in ctenophores are greater than that observed in other animals. In contrast to other animals, cilia, not muscles, are the primary effectors in many ctenophores. Muscles in ctenophores are usually involved in pray catching rather than in locomotion. Only a few species evolved muscular jet-like propulsion (e.g., *Ocyropsis crystalline*) and sinusoidal undulations of the whole body (e.g., *Cestum veneris*) during swimming escape responses. Some muscles are giant and well-characterized electrophysiologically [94–100]. These muscles control hydroskeleton tone, body shape, and feeding, which might be the original functions of muscle elements in animal ancestors.

Figure 10 illustrates cilia diversity in *Beroe abyssicola* with different types of cilia in the mouth (some serve as teeth for prey capture [101–103]) and body wall. At least six types of cilia [104] construct the aboral organ as a gravity center with dozens of living cells—**lithocytes** containing statolith [105–107]. Ciliated furrows are also efficient conductive pathways mediating various behaviors. There are multiple types of ciliated receptors formed by nonmotile cilia [102, 108–110].

The cilia are primarily used for locomotion with the unique ability to reverse cilia beating [111] and contain ctenophore-specific proteins CTENO64 and CTENO189, which are required for paddling of comb plates and locomotion of ctenophores [112] as well as reinforce the elastic connection among cilia to overcome the hydrodynamic drag of giant multiciliary plates [113].

A diverse spectrum of behaviors, ciliated and muscular locomotion, as well as feeding [30, 68, 93, 107], is controlled by quite complex neural systems, and, at least in part, it is coordinated by the aboral organ [107], an analog of the elementary brain.

The study of the neural organization of ctenophores was started in 1880s by R. Hertwig [114] as a logical expansion of

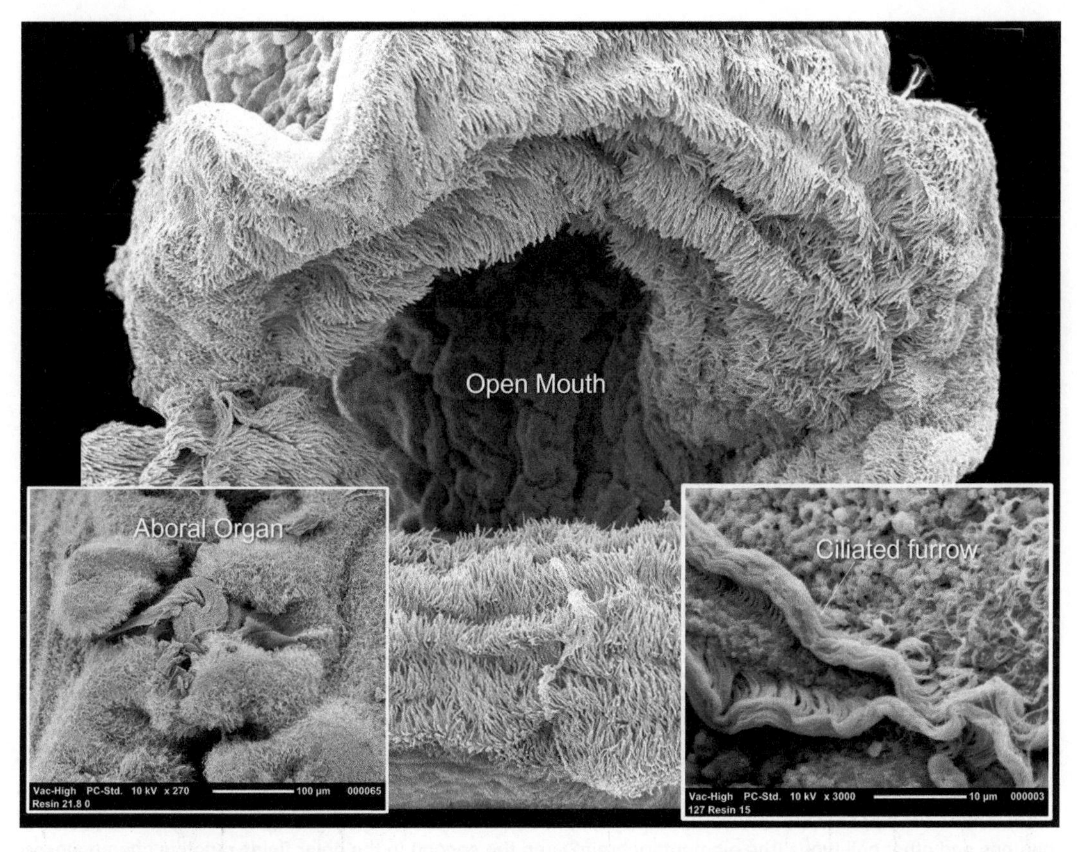

Fig. 10 Scanning electron microscopy of the mouth, aboral organ, and ciliated furrows of *Beroe abbysicola*. (See details in Refs. [102, 103])

similar studies on cnidarians by Hertwig's brothers [115–117]. This fundamental work led to the most well-known hypothesis of nervous system evolution [118, 119]. However, ctenophore neurons are elusive cells to stain with convenient histological dyes or bilaterian molecular markers due to the lack of pan-neuronal genes across Metazoa [120].

The overall microanatomy of neural systems is now described for 11 ctenophore species [27, 102, 108, 109, 121–128] and summarized in Fig. 11 [129]. About 10,000 neurons were counted in *Pleurobrachia bachei*, representing five distinct components: (i) the aboral organ, (ii) polar fields, (iii) conductive pathways, and (iv) subepithelial and (v) mesogleal nerve nets.

Integrative comparative analyses, including genomics, metabolomics, molecular mapping, and physiology, suggest that ctenophore neurons are remarkably different from all other studied neurons in Cnidaria and Bilateria, meaning, together with the current phylogenetic reconstruction, their independent origins and ongoing parallel evolution (summarized in [35–37, 130–133]. Recent volume electron microscopy reconstruction of

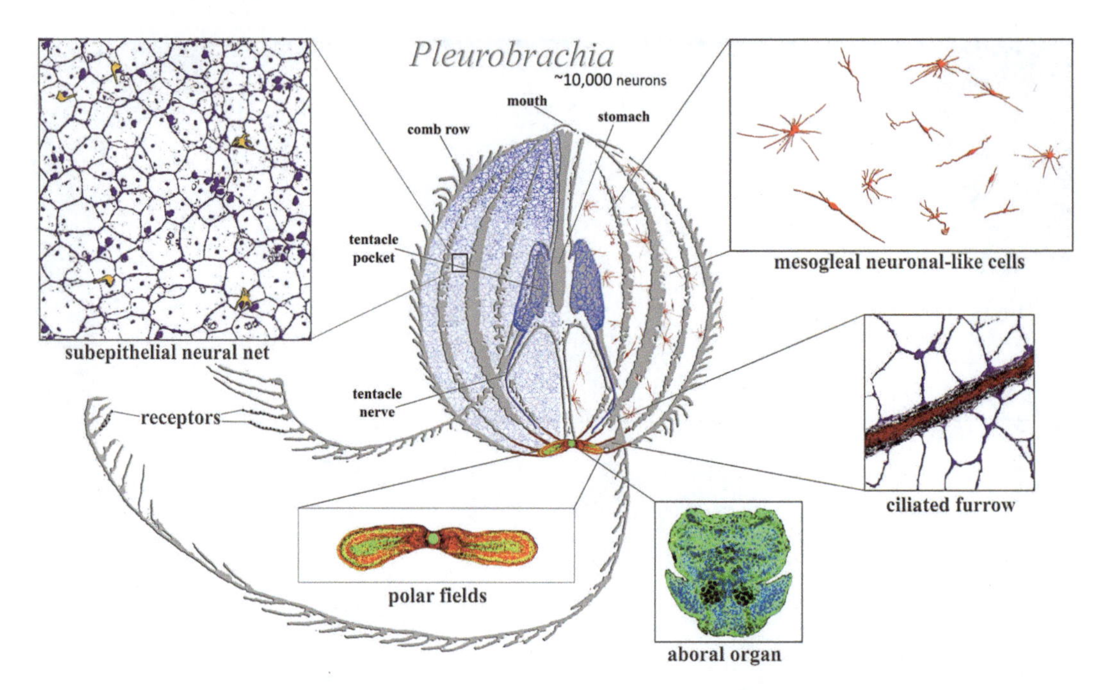

Fig. 11 Neural systems in ctenophores. The schematic diagram is based on the recent studies of several species [102, 108, 109, 127, 128, 196] with the cydippid *Pleurobrachia bachei* as a key reference model. Different colors indicate different cellular populations. Most neurons and receptors (yellow) are located within the subepithelial neural net in the skin (blue, magenta) and tentacle shields with two tentacular nerves (dark blue). There are two concentrations of neural elements: one in the aboral organ (green) with densely packed neurons and other cell types (the elementary brain?) and the second in the polar fields putative chemosensory structures (yellow/green, red marks phalloidin-labeled elements). The mesoglea has a diffuse population of neuron-like cells (red). Eight ciliated furrows (conductive ciliated cells—red lines) connect the aboral organ with comb plates. The ciliated furrows are closely associated with neural net elements (insert) and are possible under neuronal control. (Adapted from Ref. [129])

juvenile *Mnemiopsis* found that five neurons in the subepithelial network form the syncytium [134], which is likely a secondary adaptation for some neural elements. Still, most neurons and neuro-effector communications are chemical [197] with the distinct tripartite organization of ctenophore synapses, also known as "presynaptic triad." Each presumed presynaptic zone contains a three-layer complex of organelles: a single layer of synaptic vesicles lining the presynaptic membrane, a cistern of agranular endoplasmic reticulum just above the row of vesicles, followed by one or several mitochondria [27, 122, 125, 134–136].

The diversity of synaptic vesicles implies the variety of signal molecules and neurotransmitters—most of them are currently unknown. Gaseous nitric oxide (NO) was also implicated in intercellular signaling. However, nitric oxide synthase (NOS) was not detected in ctenophore neurons [137, 138]. Initial analysis of the *Pleurobrachia* genome and transcriptomes for dozen of related species, complemented by metabolomic and functional studies, indicated that the canonical bilaterian neurotransmitters such as serotonin, dopamine, octopamine, noradrenaline, adrenaline, histamine, and acetylcholine are absent in the ctenophores, and likely bilaterian innovations [37, 120, 139].

Glutamate was proposed as a candidate for neuromuscular transmission [5, 140] and small secretory peptides are major transmitters with about 100 of ctenophore-specific neuropeptides [5, 37]. The diversity and role of neuropeptides were subsequently validated in two other species *Mnemiopsis* [141] and *Bolinopsis* [142], confirming the hypothesis that the earliest transmitters can be secretory peptides [119] and neurons evolved from genealogically different secretory cell types [132]. Of note, none of the ctenophore neuropeptides had recognized homologs outside of this phylum, further supporting the hypothesis about the unique organization of ctenophore neural systems, their independent origins, and extensive parallel evolution.

5 Unique Ctenophore Development

Most ctenophores are direct developing, self-fertile hermaphrodites with a few exceptions, such as the presence of both sexes in *Ocyropsis* [143]. Gonads derive from the endoderm of meridional canals; one part represents the female and the second male gonads. Gametes are released through pores in the epidermis or through meridional canals and anal pores (personal observation in *Pleurobrachia bachei - see Fig 6, next Chapter*). Unlike other metazoans, **polyspermy** occurs in ctenophores such as *Beroe*. As many as 20 spermatozoa enter the egg, and the female pronucleus moves and "selects" a male pronucleus, and the position of the selection determines the position of the blastoporal pore [144–146]. Patterns of early development seemed to be shared across ctenophores and were observed for several decades of research, starting with classical pioneering work at the end of the nineteenth century [31, 147–161]. The latest progress is summarized in [162] using *Mnemiopsis leidyi* as a model. All available data indicate that ctenophore development distinctly differs from other basal metazoans (e.g., see Fig. 12 for *Pleurobrachia bachei*).

The early [147, 148] and controversial history of ctenophore embryology started with the pioneering work on biodiversity and the earliest developmental specification discovered in 1880s by C. Chun [149]. When C. Chun separated blastomers in two-cell embryos, he found that each half-embryo developed half of the adult structures in ctenophores, suggesting highly deterministic mechanisms even after the first division during the cleavage. G. Freeman showed that the oral-aboral axis is established at the time of the first cleavage that cleavage plays a causal role in

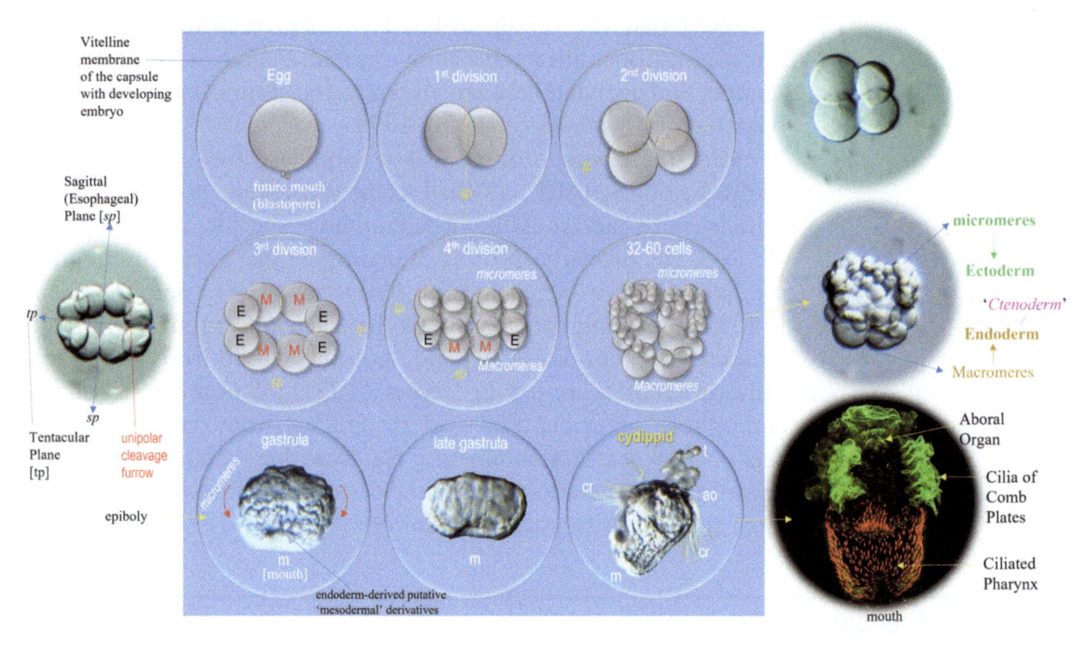

Fig. 12 Development in *Pleurobrachia bachei*. (Modified from Ref. [127]; see text for details)

setting up the axis and that comb plate-forming potential begins to be localized in the aboral region of the embryo at this time [163].

The first division starts with a characteristic unipolar cleavage furrow. Most cell fates are determined at the first cleavage stages and continue through 60-cell stages and gastrulation, as carefully characterized by microinjection and dye-tracing techniques [162, 164, 165]. Macromere lineages give rise to the endoderm and its derivatives (including endothelium of meridional canals, the mineral-containing lithocytes generated in the floor of the aboral organ). In contrast, aboral micromeres give rise to the ectoderm and its components (skin, comb rows, most of the aboral organ, tentacle epidermis with colloblasts, some neurons, and pharyngeal epithelium). Furthermore, in ctenophores, the epithelial might also be regulated differently than in bilaterians and cnidarians. Specifically, Par protein localization during the early development of *Mnemiopsis leidyi* suggests other modes of epithelial organization [166].

The most fascinating is the "mesoderm" development. According to the carefull work of E. Metschnikoff [150] in 1885, "ctenophores have a 'true' mesoderm of entodermal origin" [31] derived from small cells at their oral poles. These cells carried inward during the gastrulation process proliferate and "become the cells of the collenchyme, including muscle cells" [31].

Recent studies of Martindale and Henry on *Mnemiopsis* convincingly identified a distinct subset of macromer-derived "oral" micromeres, which subsequently move inside the embryo and differentiate into mesenchymal cells [162, 165]. The muscle cells are supposedly derived from a type of mesenchyme cell in the mesoglea; they are segregated early in embryonic development and, therefore, can be considered as "true" mesodermal derivatives (separate from epidermis and gastrodermis [167, 168]). Separate comparative analyses of *Pleurobrachia* [5] and *Mnemiopsis* [33] genomes revealed that ctenophores do not possess many canonical developmental regulatory genes required for bilaterian mesoderm specification. Moreover, these data and the ctenophore-sister phylogeny imply that muscles and mesoderm evolved independently in ctenophores. Thus, the ctenophore "mesoderm" might not be homologous to the bilaterian mesoderm as we know it today. As a result, the term "ctenoderm" was proposed to refer to cells residing in this layer [169].

Later, post-hatching development varies more than embryonic development, creating enormous diversity of ctenophore forms across the phylum. Lobate ctenophores are generally flattened in the tentacles plane, while Platyctenida are flattened in the aboral-oral direction.

For example, after hatching as a classical cydippid larva/or juvenile, tentacles are dramatically reduced in Lobata representatives and can even be lost in adult *Ocyropsis.* Representatives of the order Beroida lost their tentacles at all developmental stages and in adults.

In some benthic ctenophores Platyctenids, adults can also lose comb plates from their cydippid larvae. A fascinating case was discovered in the Greenland sessile *Tjalfiella tristoma*, which is viviparous; the young ctenophores grow in a womb [31, 170]. Finally, one species *Lampetia* has an undifferential larval stage that parasitizes salps [170]. This larval stage was initially not recognized as the same species and was called *Gastrodes.*

Does dissogeny exist in ctenophores? In *Mnemiopsis* (and possibly *Beroe*), *continuous* reproduction was reported from early juvenile animals to large mature adults [171]. These observations challenge the concept of **dissogeny** or the presence of *separate phases* of larval and adult reproduction (see also [172]). Edgar and colleagues suggested that "spawning at small body size should be considered the default, on-time developmental trajectory rather than precocious, stress-induced, or otherwise unusual for ctenophores. The ancestral ctenophore was likely a direct developer, consistent with the hypothesis that multiphasic life cycles were introduced after the divergence of the ctenophore lineage" [171]. Whether such an exceptional situation would be applied to other ctenophore species would be the subject of future research [172].

6 Ctenophores Are Kings of Regeneration

In contrast to highly deterministic "mosaic" development, many ctenophore species are capable of fast and efficient regeneration [173–181], the most characteristic for very fragile lobate ctenophores (Fig. 13), but also observed for tentacles and additional body parts (e.g., tentacles) in other lineages within Cydippida [182] and Platynectida. The creeping Platyctenida even can reproduce asexually from their fragments that could regenerate the whole animal with all organs [173, 176, 183, 184]. In contrast, Beroids have a minimal regeneration capability.

In *Bolinopsis* and *Mnemiopsis*, we noted the remarkable regeneration of the aboral organ, which takes 2.5–3.5 days at ambient temperatures, and restoration of observable behaviors within 5–6 days (n=45, author's observations). For example, I observed the regeneration of the aboral organ four times from the very same animal. After the first regeneration event, I fed animals following the recovery of their behaviors and repeated the procedure four times! Cellular, molecular, and genomic bases of such unique

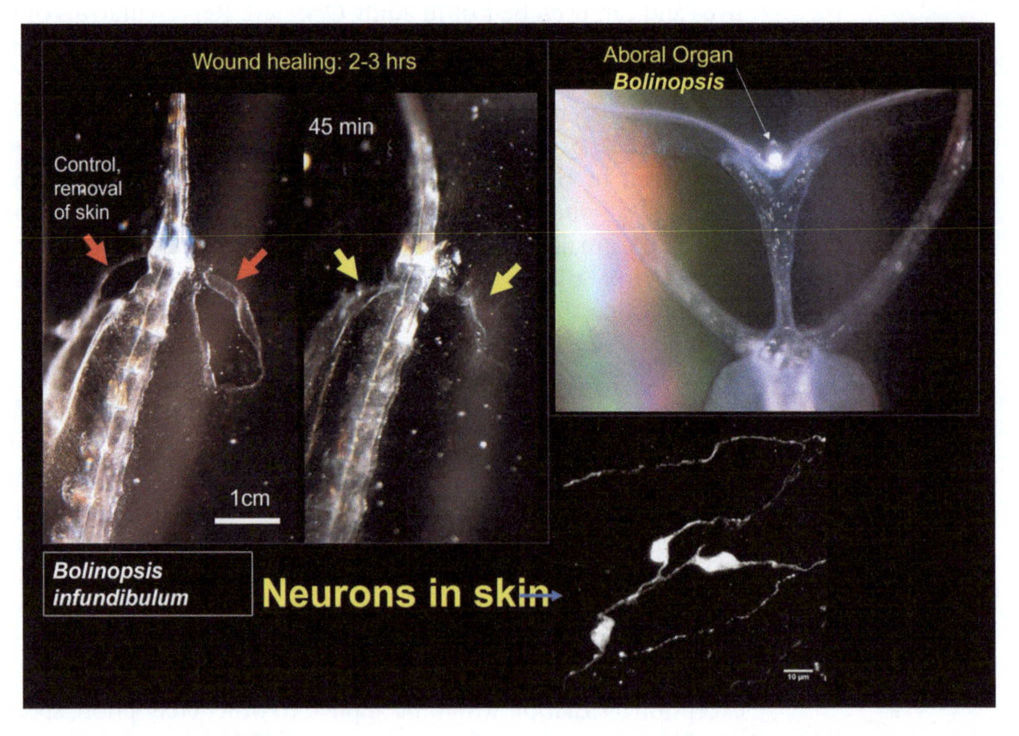

Fig. 13 Ctenophore regeneration. An illustrative example of wound healing in *Bolinopsis microptera*, where an experimental cut of the skin area induced its rapid closing within 1 h after the injury. The aboral organ in this species (shown on the right) can efficiently regenerate within 3 days (see text for details). The aboral organ's wound healing and regeneration are accompanied by notable reorganization of the subepithelial neural net (lower right)

regeneration capabilities are under intensive investigation [182, 185, 186] and can provide deep insights into the synthetic biology of the future.

7 Future Directions: Ctenophores as Key Reference Species: Culturing, Genomics, and Gene Editing

Systematic interdisciplinary studies of ctenophores are in their infancy, and deciphering the remarkable morphological and functional diversity is one of the hot topics in biological research over the following decades, with many anticipated surprises. Many of these surprises would be from examples of convergent evolution, including deciphering lineage-specific diversification across integrative systems and signaling in ctenophores (Fig. 14).

Several reasonably straightforward directions in the field are outlined below.

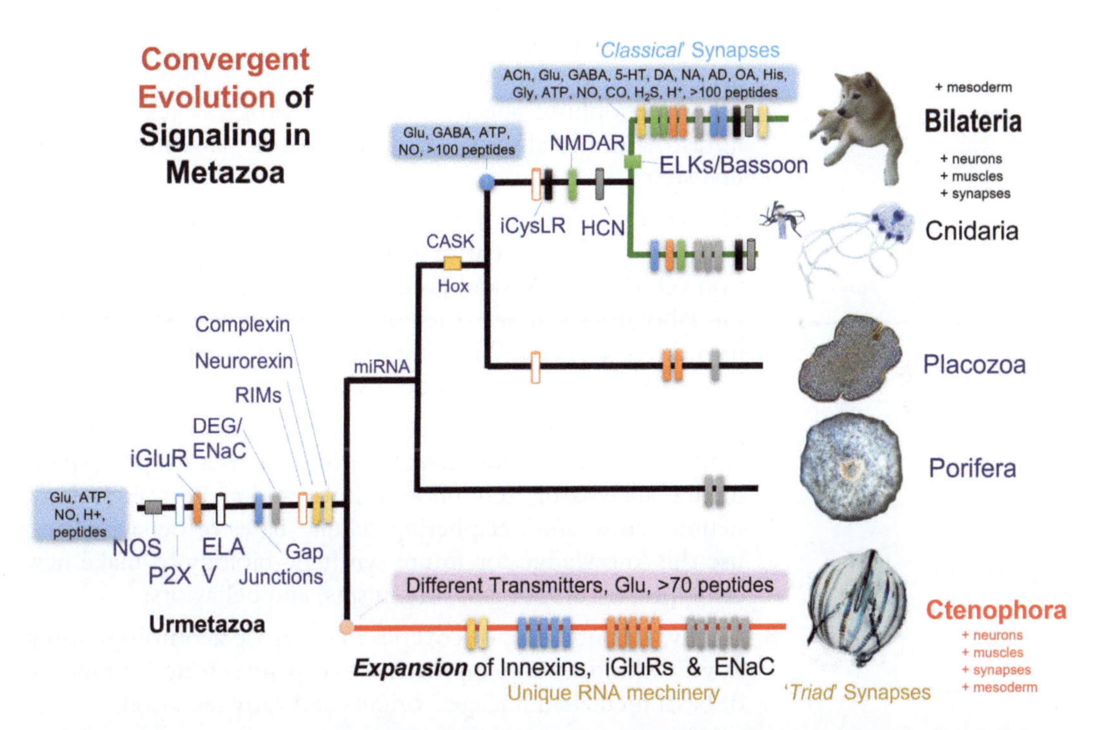

Fig. 14 Molecular innovations underlying the parallel evolution of neuromuscular organization and respective (neuro)transmitter systems in ctenophores vs. other basal metazoan lineages (Modified from Refs. [5, 37]). Bars indicate the presence or independent radiation of selected gene families (e.g., ionotropic glutamate receptors [iGluR], innexins [5, 199], acid-sensitive channels (ENA) in ctenophores and Cnidaria+Bilateria clades. Our model suggests that sponges and placozoans never developed "true" neural and muscular systems. However, both neurons and muscles independently evolved in common ancestors of the ctenophore vs. Cnidaria and Bilateria lineages with a distinct complement of signaling molecules and secretory peptides

1. Although most ctenophores cannot be routinely maintained in laboratory culture, we already see remarkable progress in this direction for some species [187–192], primarily using facilities of marine stations.
2. Ctenophore cells can be efficiently maintained in cell culture, enabling a diversity of experimental manipulations [95, 193, 194].
3. The remarkable breakthrough was a success in gene editing using CRISPR-cas9 technology in *Mnemiopsis* [188] and morpholinos in *Bolinopsis* [142].
4. Sequencing, chromosome-level, and functional annotation of genomes from dozens of diverse ctenophore species representing all families of the phylum is needed and will be achievable soon. This research will decipher ctenophore innovations and be a critical platform for virtually all directions in the field.
5. Nevertheless, most surprises are anticipated in the sea, from investigations of animals in their native habitats toward little explored functional biodiversity for these enigmatic species. This strategy would expand work from standard model organisms such as specialized and abundant *Mnemiopsis* to dozens of other ctenophore species. Here, the progress relies on the infrastructure of already established marine laboratories as the first step.
6. However, we expect the most discoveries by direct access to ctenophores in their native living habitats using remote operation vehicles (ROV) and even full-scale interdisciplinary floating laboratories at sea, such as the Ship-seq approach [195] introduced earlier and leading to the first systematic molecular access to more than 30 species [8].
7. Finally, we expect a shift from more traditional genomic or embryological/developmental approaches to a deeper experimental analysis of ctenophore cellular and system physiology, neuroscience, and deciphering cellular bases of behaviors and use this knowledge for future synthetic biology to make new cell types, tissues, organs, organisms, and behaviors.
8. Finally, we anticipate discoveries in (micro)paleontology using novel techniques and approaches to expand our understanding of basal metazoan lineages' origins and early radiation.

Acknowledgments

The author thanks the OGAP team led by Capt. Peter Molnar (vessel SAM), Tyler Meade, Matthew Stromberg as well as Mr. James F. Jacoby (vessel Miss Phebe II) and Mr. Steven Sablonski (vessel Copacetic) with help to collect ctenophores around the globe. This work was supported by the Human Frontiers Science Program (RGP0060/2017) and the National Science Foundation (IOS-1557923) grants to LLM. Research reported in this publication was also supported in part by the National Institute of Neurological Disorders and Stroke of the National Institutes of Health under award number R01NS114491 (to LLM). The content is solely the author's responsibility and does not necessarily represent the official views of the National Institutes of Health.

References

1. Krumbach T (1925) Erste und einzige Klasse der Actinaria Vierte Klasse des Stammes der Coelenterata. Ctenophora. In: Kukenthal W, Krumbach T (eds) Handbuch der Zoologie. de Gruyter, Berlin, pp 905–995
2. Harbison GR (1985) On the classification and evolution of the Ctenophora. In: Morris SC et al (eds) The origins and relationships of lower invertebrates. Clarendon Press, Oxford, pp 78–100
3. Podar M et al (2001) A molecular phylogenetic framework for the phylum Ctenophora using 18S rRNA genes. Mol Phylogenet Evol 21(2):218–230
4. Li Y et al (2021) Rooting the animal tree of life. Mol Biol Evol 38(10):4322–4333
5. Moroz LL et al (2014) The ctenophore genome and the evolutionary origins of neural systems. Nature 510(7503):109–114
6. Schultz DT et al (2023) Ancient gene linkages support ctenophores as sister to other animals. Nature 618(7963):110–117
7. Whelan NV et al (2015) Error, signal, and the placement of Ctenophora sister to all other animals. Proc Natl Acad Sci U S A 112(18): 5773–5778
8. Whelan NV et al (2017) Ctenophore relationships and their placement as the sister group to all other animals. Nat Ecol Evol 1(11): 1737–1746
9. Jamieson AJ, Lindsay DJ, Kitazato H (2023) Maximum depth extensions for hydrozoa, Tunicata and Ctenophora. Mar Biol 170(3): 33
10. Aghamaali MR et al (2011) Cloning, sequencing, expression and structural investigation of mnemiopsin from *Mnemiopsis leidyi*: an attempt toward understanding Ca2+–regulated photoproteins. Protein J 30(8): 566–574
11. Burakova LP, Kolmakova AA, Vysotski ES (2022) Recombinant light-sensitive photoprotein berovin from ctenophore *Beroe abyssicola*: bioluminescence and absorbance characteristics. Biochem Biophys Res Commun 624:23–27
12. Burakova LP et al (2021) Unexpected Coelenterazine degradation products of *Beroe abyssicola* photoprotein photoinactivation. Org Lett 23(17):6846–6849
13. Burakova LP, Vysotski ES (2019) Recombinant Ca(2+)-regulated photoproteins of ctenophores: current knowledge and application prospects. Appl Microbiol Biotechnol 103(15):5929–5946
14. Jafarian V et al (2011) A unique EF-hand motif in mnemiopsin photoprotein from *Mnemiopsis leidyi*: implication for its low calcium sensitivity. Biochem Biophys Res Commun 413(2):164–170
15. Markova SV et al (2012) The light-sensitive photoprotein berovin from the bioluminescent ctenophore *Beroe abyssicola*: a novel type of Ca(2+) -regulated photoprotein. FEBS J 279(5):856–870
16. Mohammadi Ghanbarlou R et al (2018) Molecular mechanisms governing the evolutionary conservation of Glycine in the 6 (th) position of loops IotaIotaIota and IotaV in photoprotein mnemiopsin 2. J Photochem Photobiol B 187:18–24
17. Molakarimi M et al (2019) Reaction mechanism of the bioluminescent protein mnemiopsin1 revealed by X-ray crystallography and QM/MM simulations. J Biol Chem 294(1): 20–27
18. Pashandi Z et al (2017) Photoinactivation related dynamics of ctenophore photoproteins: insights from molecular dynamics simulation under electric-field. Biochem Biophys Res Commun 490(2):265–270
19. Powers ML et al (2013) Expression and characterization of the calcium-activated photoprotein from the ctenophore *Bathocyroe fosteri*: insights into light-sensitive photoproteins. Biochem Biophys Res Commun 431(2):360–366

20. Schnitzler CE et al (2012) Genomic organization, evolution, and expression of photoprotein and opsin genes in *Mnemiopsis leidyi*: a new view of ctenophore photocytes. BMC Biol 10:107
21. Stepanyuk GA et al (2013) Spatial structure of the novel light-sensitive photoprotein berovin from the ctenophore *Beroe abyssicola* in the Ca (2+)-loaded apoprotein conformation state. Biochim Biophys Acta 1834(10):2139–2146
22. Ward WW, Seliger HH (1974) Properties of mnemiopsin and berovin, calcium-activated photoproteins from the ctenophores *Mnemiopsis* sp. and *Beroe ovata*. Biochemistry 13(7):1500–1510
23. Ward WW, Seliger HH (1974) Extraction and purification of calcium-activated photoproteins from the ctenophores *Mnemiopsis* sp. and *Beroe ovata*. Biochemistry 13(7): 1491–1499
24. Madin LP et al (2013) Scuba diving in blue water: a window on ecology and evolution in the epipelagic ocean. Research and Discoveries: The Revolution of Science Through Scuba
25. Harbison G, Madin L, Swanberg N (1978) On the natural history and distribution of oceanic ctenophores. Deep-Sea Res 25(3): 233–256
26. Welch V et al (2006) Optical properties of the iridescent organ of the comb-jellyfish Beroe cucumis (Ctenophora). Phys Rev E Stat Nonlinear Soft Matter Phys 73(4 Pt 1):041916
27. Hernandez-Nicaise M-L (1991) Ctenophora. In: Harrison FWFW, Westfall JA (eds) Microscopic anatomy of invertebrates: Placozoa, Porifera, Cnidaria, and Ctenophora. Wiley, New York, pp 359–418
28. Heimbichner Goebel WL et al (2020) Scaling of ctenes and consequences for swimming performance in the ctenophore *Pleurobrachia bachei*. Invertebr Biol 139(3):e12297
29. Colin SP et al (2010) Stealth predation and the predatory success of the invasive ctenophore *Mnemiopsis leidyi*. Proc Natl Acad Sci U S A 107(40):17223–17227
30. Gemmell BJ et al (2019) A ctenophore (comb jelly) employs vortex rebound dynamics and outperforms other gelatinous swimmers. R Soc Open Sci 6(3):181615
31. Hyman LH (1940) Invertebrates: protozoa through Ctenophora, vol 1. McGraw-Hill, New York/London, p 726
32. Schultz DT et al (2021) A chromosome-scale genome assembly and karyotype of the ctenophore *Hormiphora californensis*. G3 (Bethesda) 11(11)
33. Ryan JF et al (2013) The genome of the ctenophore *Mnemiopsis leidyi* and its implications for cell type evolution. Science 342(6164):1242592
34. Hoencamp C et al (2021) 3D genomics across the tree of life reveals condensin II as a determinant of architecture type. Science 372(6545):984–989
35. Moroz LL (2014) The genealogy of genealogy of neurons. Commun Integr Biol 7(6): e993269
36. Moroz LL (2015) Convergent evolution of neural systems in ctenophores. J Exp Biol 218(Pt 4):598–611
37. Moroz LL, Kohn AB (2016) Independent origins of neurons and synapses: insights from ctenophores. Philos Trans R Soc Lond Ser B Biol Sci 371(1685):20150041
38. Horridge GA (1964) The giant mitochondria of ctenophore comb plates. Q J Microsc Sci 105:301–310
39. Kohn AB et al (2012) Rapid evolution of the compact and unusual mitochondrial genome in the ctenophore, *Pleurobrachia bachei*. Mol Phylogenet Evol 63(1):203–207
40. Lavrov DV, Pett W (2016) Animal mitochondrial DNA as we do not know it: mt-genome organization and evolution in Nonbilaterian lineages. Genome Biol Evol 8(9):2896–2913
41. Arafat H et al (2018) Extensive mitochondrial gene rearrangements in Ctenophora: insights from benthic Platyctenida. BMC Evol Biol 18(1):65
42. Christianson LM et al (2022) Hidden diversity of Ctenophora revealed by new mitochondrial COI primers and sequences. Mol Ecol Resour 22(1):283–294
43. Formaggioni A, Luchetti A, Plazzi F (2021) Mitochondrial genomic landscape: a portrait of the mitochondrial genome 40 years after the first complete sequence. Life (Basel) 11(7)
44. Muthye V, Lavrov DV (2018) Characterization of mitochondrial proteomes of nonbilaterian animals. IUBMB Life 70(12): 1289–1301
45. Pett W et al (2011) Extreme mitochondrial evolution in the ctenophore *Mnemiopsis leidyi*: insight from mtDNA and the nuclear genome. Mitochondrial DNA 22(4): 130–142
46. Schultz DT et al (2020) Conserved novel ORFs in the mitochondrial genome of the ctenophore *Beroe forskalii*. PeerJ 8:e8356
47. Wang M, Cheng F (2019) The complete mitochondrial genome of the ctenophore *Beroe cucumis*, a mitochondrial genome

showing rapid evolutionary rates. Mitochondrial DNA B Resour 4(2):3774–3775
48. Alamaru A et al (2017) Molecular diversity of benthic ctenophores (Coeloplanidae). Sci Rep 7(1):6365
49. Schroeder A et al (2021) Suitability of a dual COI marker for marine zooplankton DNA metabarcoding. Mar Environ Res 170: 105444
50. Mukherjee K, Moroz LL (2023) Transposon-derived transcription factors across metazoans. Front Cell Dev Biol 11:1113046
51. Mukherjee K, Moroz LL (2023) Parallel evolution of transcription factors in basal metazoans. Ctenophores: Methods and Protocols, Methods Mol Biol, vol. 2757, https://doi.org/10.1007/978-1-0716-3642-8_20. This volume
52. Becklemishev VM (1964) Foundation for comprative anatomy of invertebrates, vol 1–2, 3rd edn. Nauka, Moscow. (in Russian)
53. Brusca RC, Giribet G, Moore W (2022) Invertebrates, 4th edn. Sinauer Associates of Oxford University Press, p 1104
54. Parry LA et al (2021) Cambrian comb jellies from Utah illuminate the early evolution of nervous and sensory systems in ctenophores. iScience 24(9):102943
55. Ou Q et al (2015) A vanished history of skeletonization in Cambrian comb jellies. Sci Adv 1(6):e1500092
56. Tang F et al (2011) *Eoandromeda* and the origin of Ctenophora. Evol Dev 13(5): 408–414
57. Zhao Y et al (2019) Cambrian sessile, suspension feeding stem-group ctenophores and evolution of the comb jelly body plan. Curr Biol 29(7):1112–1125 e2
58. Fu D et al (2019) The Qingjiang biota – a burgess shale-type fossil Lagerstatte from the early Cambrian of South China. Science 363(6433):1338–1342
59. Stanley GD, Sturmer W (1983) The first fossil ctenophore from the lower devonian of West Germany. Nature 303:518–520
60. Stanley GD, Stürmer W (1987) A new fossil ctenophore discovered by X-rays. Nature 328(6125):61–63
61. Haddock SH (2007) Comparative feeding behavior of planktonic ctenophores. Integr Comp Biol 47(6):847–853
62. Cordeiro M et al (2022) Oceanic lobate ctenophores possess feeding mechanics similar to the impactful coastal species *Mnemiopsis leidyi*. Limnol Oceanogr 67(12):2706–2717
63. Reeve MR, Walter MA (1979) Nutritional ecology of ctenophores – a review of recent research. In: Russell FS, Yonge M (eds) Advances in marine biology. Academic, pp 249–287
64. Swanberg N (1974) The feeding behavior of *Beroe ovata*. Mar Biol 24:69–76
65. Carre D, Carre C, Mills CE (1989) Novel cnidocysts of narcomedusae and a medusivorous ctenophore, and confirmation of kleptocnidism. Tissue Cell 21(5):723–734
66. Mills CE, Miller RL (1984) Ingestion of a medusa (*Aegina citrea*) by the nematocyst-containing ctenophore *Haeckelia rubra* (formerly *Euchlora rubra*): phylogenetic implications. Mar Biol 78(2):215–221
67. Carré C, Carré D (1980) Les cnidocystes du cténophore *Euchlora rubra* (Kölliker 1853). Cah Biol Mar 21:221–226
68. Potter B et al (2023) Quantifying the feeding behavior and trophic impact of a widespread oceanic ctenophore. Sci Rep 13(1):2292
69. Yip SY (1984) The feeding of *Pleurobrachia pileus* Müller (Ctenophora) from Galway Bay. Proc R Ir Acad Sect B Biol Geol Chem Sci 84B:109–122
70. Waggett R, Costello J (1999) Capture mechanisms used by the lobate ctenophore, *Mnemiopsis leidyi*, preying on the copepod *Acartia tonsa*. J Plankton Res 21(11): 2037–2052
71. Kremer P, Reeve MR, Syms MA (1986) The nutritional ecology of the ctenophore *Bolinopsis vitrea*: comparisons with *Mnemiopsis mccradyi* from the same region. J Plankton Res 8(6):1197–1208
72. Kremer P, Canino M, Gilmer R (1986) Metabolism of epipelagic tropical ctenophores. Mar Biol 90:403–412
73. Jaspers C et al (2018) Resilience in moving water: effects of turbulence on the predatory impact of the lobate ctenophore *Mnemiopsis leidyi*. Limnol Oceanogr 63(1):445–458
74. Buecher E, Gasser B (1998) Estimation of predatory impact of *Pleurobrachia rhodopis* (cydippid ctenophore) in the northwestern Mediterranean Sea: in situ observations and laboratory experiments. J Plankton Res 20(4): 631–651
75. Swanberg N, Båmstedt U (1989) The role of prey stratification in the predation pressure by the cydippid ctenophore *Mertensia ovum* in the Barents Sea. In: Coelenterate biology: recent research on Cnidaria and Ctenophora: proceedings of the fifth international conference on coelenterate biology, vol 1991. Springer

76. Townsend J et al (2020) Ink release and swimming behavior in the oceanic ctenophore *Eurhamphaea vexilligera*. Biol Bull 238(3): 206–213
77. Sutherland KR et al (2014) Ambient fluid motions influence swimming and feeding by the ctenophore *Mnemiopsis leidyi*. J Plankton Res 36(5):1310–1322
78. Matsumoto G, Harbison G (1993) In situ observations of foraging, feeding, and escape behavior in three orders of oceanic ctenophores: Lobata, Cestida, and Beroida. Mar Biol 117:279–287
79. Matsumoto G, Hamner W (1988) Modes of water manipulation by the lobate ctenophore *Leucothea* sp. Mar Biol 97:551–558
80. Matsumoto G (1991) Swimming movements of ctenophores, and the mechanics of propulsion by ctene rows. Hydrobiologia 216:319–325
81. Hamner W et al (1987) Ethological observations on foraging behavior of the ctenophore *Leucothea* sp. in the open sea 1. Limnol Oceanogr 32(3):645–652
82. Colin SP et al (2015) Elevating the predatory effect: sensory-scanning foraging strategy by the lobate ctenophore *Mnemiopsis leidyi*. Limnol Oceanogr 60(1):100–109
83. Swift HF et al (2009) Feeding behavior of the ctenophore *Thalassocalyce inconstans*: revision of anatomy of the order Thalassocalycida. Mar Biol 156(5):1049–1056
84. Decker MB, Breitburg DL, Purcell JE (2004) Effects of low dissolved oxygen on zooplankton predation by the ctenophore *Mnemiopsis leidyi*. Mar Ecol Prog Ser 280:163–172
85. Thuesen EV, Rutherford LD, Brommer PL (2005) The role of aerobic metabolism and intragel oxygen in hypoxia tolerance of three ctenophores: *Pleurobrachia bachei*, *Bolinopsis infundibulum* and *Mnemiopsis leidyi*. J Mar Biol Assoc U K 85(3):627–633
86. Båmstedt U, Martinussen MB (2015) Ecology and behavior of *Bolinopsis infundibulum* (Ctenophora; Lobata) in the Northeast Atlantic. Hydrobiologia 759:3–14
87. Titelman J et al (2012) Predator-induced vertical behavior of a ctenophore. Hydrobiologia 690:181–187
88. Falkenhaug T, Stabell OB (1996) Chemical ecology of predator-prey interactions in ctenophores. Mar Freshw Behav Phy 27(4): 249–260
89. Moss AG, Rapoza RC, Muellner L (2001) A novel cilia-based feature within the food grooves of the ctenophore *Mnemiopsis mccradyi* Mayer. Hydrobiologia 451:287–294
90. Leonardi ND, Thuesen EV, Haddock SHD (2020) A sticky thicket of glue cells: a comparative morphometric analysis of colloblasts in 20 species of comb jelly (phylum Ctenophora). Cienc Mar 46(4):211–225
91. Townsend J et al (2020) Colloblasts act as a biomechanical sensor for suitable prey in *Pleurobrachia*. bioRxiv:2020.06. 27.175059
92. Presnell JS et al (2016) The presence of a functionally tripartite through-gut in Ctenophora has implications for metazoan character trait evolution. Curr Biol 26(20):2814–2820
93. Tamm SL (2014) Cilia and the life of ctenophores. Invertebr Biol 133(1):1–46
94. Dubas F, Stein PG, Anderson PA (1988) Ionic currents of smooth muscle cells isolated from the ctenophore *Mnemiopsis*. Proc R Soc Lond B Biol Sci 233(1271):99–121
95. Stein PG, Anderson PA (1984) Maintenance of isolated smooth muscle cells of the ctenophore *Mnemiopsis*. J Exp Biol 110:329–334
96. Hernandez-Nicaise ML, Mackie G, Meech RW (1980) Giant smooth muscle cells of *Beroe*. J General Physiol 75:79–105
97. Bilbaut A et al (1988) Membrane currents that govern smooth muscle contraction in a ctenophore. Nature 331(6156):533–535
98. Anderson PAV (1984) The electrophysiology of single smooth muscle cells isolated from *Mnemiopsis*. J Compar Physiol B 154:257–268
99. Meech RW (2015) Electrogenesis in the lower Metazoa and implications for neuronal integration. J Exp Biol 218(Pt 4):537–550
100. Meech RW, Bilbaut A, Hernandez-Nicaise ML. Electrophysiology of ctenophore smooth muscle. In: Ctenophores: Methods and Protocols, Methods in Molecular Biology, vol. 2757, https://doi.org/10.1007/978-1-0716-3642-8_15 (in press)
101. Tamm S, Tamm S (1991) Macrociliary tooth patterns in beroid ctenophores. Biol Bull 181(2):355–356
102. Norekian TP, Moroz LL (2019) Neural system and receptor diversity in the ctenophore *Beroe abyssicola*. J Comp Neurol 527(12): 1986–2008
103. Norekian TP, Moroz LL (2023) Scanning electron microscopy of ctenophores: Illustrative atlas. in Ctenophores: Methods and Protocols, Methods in Molecular Biology, vol. 2757, https://doi.org/10.1007/978-1-0716-3642-8_6. This volume.
104. Jokura K, Inaba K (2020) Structural diversity and distribution of cilia in the apical sense organ of the ctenophore *Bolinopsis mikado*. Cytoskeleton (Hoboken) 77(10):442–455

105. Noda N, Tamm SL (2014) Lithocytes are transported along the ciliary surface to build the statolith of ctenophores. Curr Biol 24(19):R951–R952
106. Krisch B (1973) Über das Apikalorgan (statocyste) der ctenophore *Pleurobrachia pileus*. Z Zellforsh 142:241–262
107. Tamm SL (1982) Ctenophora. In: Electrical conduction and behavior in "simple" invertebrates. Clarendon Press, Oxford, pp 266–358
108. Norekian TP, Moroz LL (2019) Neuromuscular organization of the ctenophore *Pleurobrachia bachei*. J Comp Neurol 527(2): 406–436
109. Norekian TP, Moroz LL (2020) Comparative neuroanatomy of ctenophores: neural and muscular systems in *Euplokamis dunlapae* and related species. J Comp Neurol 528(3): 481–501
110. Tamm SL (1983) Motility and mechanosensitivity of macrocilia in the ctenophore *Beroe*. Nature 305(5933):430–433
111. Tamm SL, Tamm S (1981) Ciliary reversal without rotation of axonemal structures in ctenophore comb plates. J Cell Biol 89(3): 495–509
112. Jokura K et al (2019) CTENO64 is required for coordinated paddling of ciliary comb plate in ctenophores. Curr Biol 29(20): 3510–3516 e4
113. Jokura K et al (2022) Two distinct compartments of a ctenophore comb plate provide structural and functional integrity for the motility of giant multicilia. Curr Biol 32(23):5144–5152 e6
114. Hertwig R (1880) Ueber den Bau der Ctenophoren. Jenaische Z Naturwiss 14:393–457
115. Hertwig O, Hertwig R (1878) Das Nervensystem und die Sinnesorgane der Medusen (The nervous system and the sensory organs of the Medusa). Vogel, Leipzig, p 157
116. Hertwig O, Hertwig R (1879) Die Actinien anatomisch und histologisch mit besonderer Berucksichtigung des Nervenmuskelsystems untersucht. Jenaische Z. Naturwiss. 13:457–640
117. Hertwig O, Hertwig R (1880) Die Actinien anatomisch und histologisch mit besonderer Berucksichtigung des Nervenmuskelsystems untersucht. Jenaische Z. Naturwiss. 14:39–89
118. Parker GH (1919) The elementary nervous systems. Lippincott, Philadelphia, p 229
119. Moroz LL (2009) On the independent origins of complex brains and neurons. Brain Behav Evol 74(3):177–190
120. Moroz LL, Kohn AB (2015) Unbiased view of synaptic and neuronal gene complement in ctenophores: are there pan-neuronal and pan-synaptic genes across metazoa? Integr Comp Biol 55(6):1028–1049
121. Hernandez-Nicaise ML (1968) Specialized connexions between nerve cells and mesenchymal cells in ctenophores. Nature 217(5133):1075–1076
122. Hernandez-Nicaise ML (1973) The nervous system of ctenophores. III. Ultrastructure of synapses. J Neurocytol 2(3):249–263
123. Hernandez-Nicaise ML (1973) The nervous system of ctenophores. I. Structure and ultrastructure of the epithelial nerve-nets. Z Zellforsch Mikrosk Anat 137(2):223–250
124. Hernandez-Nicaise ML (1973) The nervous system of ctenophores. II. The nervous elements of the mesoglea of beroids and cydippids (author's transl). Z Zellforsch Mikrosk Anat 143(1):117–133
125. Hernandez-Nicaise ML (1974) Ultrastructural evidence for a sensory-motor neuron in Ctenophora. Tissue Cell 6(1):43–47
126. Jager M et al (2011) New insights on ctenophore neural anatomy: immunofluorescence study in *Pleurobrachia pileus* (Muller, 1776). J Exp Zool B Mol Dev Evol 316B(3): 171–187
127. Norekian TP, Moroz LL (2016) Development of neuromuscular organization in the ctenophore *Pleurobrachia bachei*. J Comp Neurol 524(1):136–151
128. Norekian TP, Moroz LL (2021) Development of the nervous system in the early hatching larvae of the ctenophore *Mnemiopsis leidyi*. J Morphol 282(10):1466–1477
129. Moroz LL, Romanova DY (2022) Alternative neural systems: what is a neuron? (ctenophores, sponges and placozoans). Front Cell Dev Biol 10:1071961
130. Moroz LL (2012) Phylogenomics meets neuroscience: how many times might complex brains have evolved? Acta Biol Hung 63 (Suppl 2):3–19
131. Moroz LL (2018) NeuroSystematics and periodic system of neurons: model vs reference species at single-cell resolution. ACS Chem Neurosci 9(8):1884–1903
132. Moroz LL (2021) Multiple origins of neurons from secretory cells. Front Cell Dev Biol 9:669087
133. Moroz LL, Romanova DY (2021) Selective advantages of synapses in evolution. Front Cell Dev Biol 9:726563
134. Burkhardt P et al (2023) Syncytial nerve net in a ctenophore adds insights on the evolution

of nervous systems. Science 380(6642): 293–297
135. Horridge GA (1965) Non-motile sensory cilia and neuromuscular junctions in a ctenophore independent effector organ. Proc R Soc Lond Biol 162:333–350
136. Horridge GA, Mackay B (1964) Neurociliary synapses in *Pleurobrachia* (Ctenophora). Q J Microsc Sci 105:163–174
137. Moroz LL, Mukherjee K, Romanova DY (2023) Nitric oxide signaling in ctenophores. Front Neurosci 17:1125433
138. Norekian TP, Moroz LL (2023) Recording cilia activity in ctenophores: effects of nitric oxide and low molecular weight transmitters. Front Neurosci 17:1125476
139. Moroz LL, Romanova DY, Kohn AB (1821) Neural versus alternative integrative systems: molecular insights into origins of neurotransmitters. Philos Trans R Soc Lond Ser B Biol Sci 2021(376):20190762
140. Moroz LL et al (2021) Evolution of glutamatergic signaling and synapses. Neuropharmacology 199:108740
141. Sachkova MY et al (2021) Neuropeptide repertoire and 3D anatomy of the ctenophore nervous system. Curr Biol 31(23): 5274–5285 e6
142. Hayakawa E et al (2022) Mass spectrometry of short peptides reveals common features of metazoan peptidergic neurons. Nat Ecol Evol 6(10):1438–1448
143. Harbison GR, Miller RL (1986) Not all ctenophores are hermaphrodites. Studies on the systematics, distribution, sexuality and development of two species of *Ocyropsis*. Mar Biol 90(3):413–424
144. Carre D, Sardet C (1984) Fertilization and early development in *Beroe ovata*. Dev Biol 105(1):188–195
145. Carre D, Rouviere C, Sardet C (1991) In vitro fertilization in ctenophores: sperm entry, mitosis, and the establishment of bilateral symmetry in *Beroe ovata*. Dev Biol 147(2):381–391
146. Sardet C, Carré D, Rouvière C (1990) Reproduction and development in ctenophores. Exp Embryol Aquat Plants Anim:83–94
147. Kowalevsky A (1866) Entwickelungsgeschichte der Rippenquallen, vol 10. Memoires L'Academie Imperial des Sciences de St.-Petersbourg, St.-Petersbourg
148. Agassiz A (1874) Embryology of the ctenophorae. Memoirs Am Acade Arts Sci 10(3): 357–398
149. Chun C (1880) Die Ctenophoren des Golfes von Neapel. Fauna Flora Neapel. Monogr:1–313
150. Metchnikoff E (1885) Vergleichend-embryologische Studien. 4. Über die Gastrulation und Mesodermbildung der Ctenophoren. 5 Über die Bildung der Wanderzellen bei Asterien und Echiniden. Z Wiss Zool 42: 648–673
151. Driesch H, Morgan TH (1895) Zur Analysis der ersten Entwickelungsstadien des Ctenophoreneies: I. Von der Entwickelung einzelner Ctenophorenblastomeren. Arch Mikrosk Anat 2:204–215
152. Ziegler HE (1898) Experimentelle Studien über die Zellteilung. Arch f Entwicklungsmech d Organismen 7:34–64
153. Fischel A (1903) Entwickelung und Organ-Differenzirung. Archiv für Entwicklungsmechanik der Organismen 15(4):679–750
154. Yatsu N (1911) Observations and experiments on the ctenophore egg: II. Notes on early cleavage stages and experiments on cleavage. 日本動物学彙報 7(5):333–346
155. Yatsu N (1912) Observations and experiments on the ctenophore egg. Annot Zool Japon 8:5
156. Reverberi G (1957) Mitochondrial and enzymatic segregation through the embryonic development in ctenophores
157. Reverberi G, Ortolani G (1965) The development of the ctenophores egg. Riv Biol 58: 113–137
158. Dunlap H (1966) Oogenesis in the Ctenophora. Ph.D. thesis, University of Washington, Seattle
159. Freeman GP, Reynolds GT (1973) The development of bioluminescence in the ctenophore *Mnemiopsis leidyi*. Dev Biol 31(1): 61–100
160. Dunlap H (1974) Ctenophora. In: Giese AC, Pearse JP (eds) Reproduction in marine invertebrates. Academic, New York, pp 201–265
161. Strathmann MF (2017) Reproduction and development of marine invertebrates of the northern Pacific coast: data and methods for the study of eggs, embryos, and larvae. University of Washington Press
162. Martindale MQ, Henry JQ (2015) Ctenophora. In: Wanninger A (ed) Evolutionary developmental biology of invertebrates 1: introduction, non-Bilateria, Acoelomorpha, Xenoturbellida, Chaetognatha. Springer, Vienna, pp 179–201
163. Freeman G (1977) The establishment of the oral-aboral axis in the ctenophore embryo, vol 42, p 237

164. Martindale MQ, Henry JQ (1997) Reassessing embryogenesis in the Ctenophora: the inductive role of e1 micromeres in organizing ctene row formation in the 'mosaic' embryo, *Mnemiopsis leidyi*. Development 124(10): 1999–2006
165. Martindale MQ, Henry JQ (1999) Intracellular fate mapping in a basal metazoan, the ctenophore *Mnemiopsis leidyi*, reveals the origins of mesoderm and the existence of indeterminate cell lineages. Dev Biol 214(2): 243–257
166. Salinas-Saavedra M, Martindale MQ (2020) Par protein localization during the early development of *Mnemiopsis leidyi* suggests different modes of epithelial organization in the metazoa. elife 9:9
167. Derelle R, Manuel M (2007) Ancient connection between NKL genes and the mesoderm? Insights from Tlx expression in a ctenophore. Dev Genes Evol 217(4):253–261
168. Burton PM (2008) Insights from diploblasts; the evolution of mesoderm and muscle. J Exp Zool B Mol Dev Evol 310(1):5–14
169. Giribet G, Edgecombe GD (2020) The invertebrate tree of life. Princeton University Press
170. Mortensen T (1912) Ctenophora Danish Ingolf-expedition. 5A(2):1–96
171. Edgar A, Ponciano JM, Martindale MQ (2022) Ctenophores are direct developers that reproduce continuously beginning very early after hatching. Proc Natl Acad Sci U S A 119(18):e2122052119
172. Soto-Angel JJ et al (2023) Are we there yet to eliminate the terms larva, metamorphosis, and dissogeny from the ctenophore literature? Proc Natl Acad Sci U S A 120(4): e2218317120
173. Mortensen T (1913) On regeneration in ctenophores. Vidensk Medd Fra Dan Nat Foren I Kjøbenhavn 66:45–51
174. Coonfield B (1936) Regeneration in *Mnemiopsis leidyi*. Agassiz The Biological Bulletin 71(3):421–428
175. Coonfield BR (1937) The regeneration of plate rows in *Mnemiopsis leidyi*, Agassiz. Proc Natl Acad Sci U S A 23(3):152–158
176. Freeman G (1967) Studies on regeneration in the creeping ctenophore, *Vallicula multiformis*. J Morphol 123(1):71–83
177. Korotkova GP, Pylilo IV (1970) Regenerative phenomena in Ctenophora larvae. Vestn Leningr Univ Biol 1:21–28
178. Henry JQ, Martindale MQ (2000) Regulation and regeneration in the ctenophore *Mnemiopsis leidyi*. Dev Biol 227(2):720–733
179. Tamm SL (2012) Regeneration of ciliary comb plates in the ctenophore *Mnemiopsis leidyi*. i. Morphology. J Morphol 273(1): 109–120
180. Komai T (1922) Studies on two aberrant ctenophores: *Coeloplana* and *Gastrodes*. The Author
181. Tanaka H (1931) Reorganization in regenerating pieces of *Coeloplana*. Kyoto Imp Univ Coll Sci Ser B 7(Pt.):5
182. Alie A et al (2011) Somatic stem cells express Piwi and Vasa genes in an adult ctenophore: ancient association of "germline genes" with stemness. Dev Biol 350(1):183–197
183. Tanaka, H Reorganization in regenerating pieces of *Coeloplana*, Memoirs Coll Sci Kyoto Imp Univ Ser B, 1932. 7(5): p. 223–246
184. Dawydoff C (1938) Multiplication asexuée chez les *Ctenoplana*. Acad Sci Paris Compt Rend 206:127–128
185. Ramon-Mateu J et al (2022) Studying Ctenophora WBR using *Mnemiopsis leidyi*. Methods Mol Biol 2450:95–119
186. Ramon-Mateu J et al (2019) Regeneration in the ctenophore *Mnemiopsis leidyi* occurs in the absence of a blastema, requires cell division, and is temporally separable from wound healing. BMC Biol 17(1):80
187. Angel J-JS et al (2023) Stable laboratory culture system for the ctenophore *Mnemiopsis leidyi*. In Ctenophores: Methods and Protocols, Methods Mol Biol, vol. 2757 (This volume)
188. Presnell JS et al (2022) Multigenerational laboratory culture of pelagic ctenophores and CRISPR-Cas9 genome editing in the lobate *Mnemiopsis leidyi*. Nat Protoc 17(8): 1868–1900
189. Patry WL et al (2020) Diffusion tubes: a method for the mass culture of ctenophores and other pelagic marine invertebrates. PeerJ 8:e8938
190. Bubel M, Knowles T, Patry WL (2019) Ctenophore culture at the Monterey Bay Aquarium
191. Baker LD, Reeve MR (1974) Laboratory culture of the lobate ctenophore *Mnemiopsis mccradyi* with notes on feeding and fecundity. Mar Biol 26:57–62
192. Courtney A, Merces GO, Pickering M (2020) Characterising the behaviour of the Ctenophore *Pleurobrachia pileus* in a Laboratory Aquaculture System. bioRxiv:2020.05.25.114744
193. Dieter AC, Vandepas LE, Browne WE (2022) Isolation and maintenance of in vitro cell

cultures from the ctenophore *Mnemiopsis leidyi*. Methods Mol Biol 2450:347–358
194. Vandepas LE et al (2017) Establishing and maintaining primary cell cultures derived from the ctenophore *Mnemiopsis leidyi*. J Exp Biol 220(Pt 7):1197–1201
195. Moroz LL (2015) Biodiversity meets neuroscience: from the sequencing ship (Ship-Seq) to deciphering parallel evolution of neural systems in Omic's era. Integr Comp Biol 55(6):1005–1017
196. Norekian TP, Moroz LL (2023) Illustrative neuroanatomy of ctenophores: Immunohistochemistry. In Ctenophores: Methods and Protocols, Methods Mol Biol, vol. 2757, https://doi.org/10.1007/978-1-0716-3642-8_5
197. Moroz LL (2023) Syncytial nets vs. chemical signaling: emerging properties of alternative integrative systems. Front Cell Dev Biol 11, 1320209. https://doi.org/10.3389/fcell.2023.1320209
198. Moroz LL, Collins R, Paulay G Ctenophora: Illustrated Guide and Taxonomy. In: Ctenophores: Methods and Protocols, Methods Mol Biol vol. 2757, https://doi.org/10.1007/978-1-0716-3642-8_2, this volume
199. Kohn AB, Moroz LL (2024) Gap Junctions in Ctenophora. In: Ctenophores: Methods and Protocols, Methods Mol Biol vol. 2757, https://doi.org/10.1007/978-1-0716-3642-8_16. This volume

Chapter 2

Ctenophora: Illustrated Guide and Taxonomy

Leonid L. Moroz, Richard Collins, and Gustav Paulay

Abstract

Ctenophores or comb jellies represent the first diverging lineage of extant animals – sister to all other Metazoa. As a result, they occupy a unique place in the biological sciences. Despite their importance, this diverse group of marine predators has remained relatively poorly known, with both the species and higher-level taxonomy of the phylum in need of attention. We present a checklist of the phylum based on a review of the current taxonomic literature and illustrate their diversity with images. The current classification presented remains substantially in conflict with recent phylogenetic results, and many of the taxa are not monophyletic or untested. This chapter summarizes the existing classification focusing on recognized families and genera with 185 currently accepted, extant species listed. We provide illustrative examples of ctenophore diversity covering all but one of the 33 families and 47 of the 48 genera, as well as about 25–30 undescribed species. We also list the 14 recognized ctenophore fossil species and note others that have been controversially attributed to the phylum. Analyses of unique ctenophore adaptations are critical to understanding early animal evolution and adaptive radiation of this clade of basal metazoans.

Key words Ctenophora, *Pleurobrachia*, *Mnemiopsis*, *Bolinopsis*, *Beroe*, *Cestum*, *Ocyropsis*, *Euplokamis*, Platyctenida, Benthic ctenophores, Feeding, Classification, Phylogeny, Fossils, Cambrian, Evolution

1 Introduction

Ctenophores or comb jellies (Figs. 1, 2, and 3) are the most likely candidates for the earliest-diverging lineage of animals – sister to the rest of Metazoa [1–6] (Fig. 4). As a result, they occupy a unique place in biological sciences, illuminating distinct paths of early animal evolution and origins of animal complexity [7–10].

Phylogenetic understanding of the relationships within the Ctenophora is still young but indicates that the traditional taxonomy will need to be substantially modified as it does not match recent molecular analyses. The two classes, most of the orders, and some of the families and genera for which phylogenetic data are available appear to be polyphyletic or nested within others [11–14] (Fig. 5), with multiple examples of gain and losses of particular traits [6, 15].

Leonid L. Moroz (ed.), *Ctenophores: Methods and Protocols*, Methods in Molecular Biology, vol. 2757, https://doi.org/10.1007/978-1-0716-3642-8_2, © Springer Science+Business Media, LLC, part of Springer Nature 2024

Fig. 1 Artistic Illustration of Ctenophore and Cnidarian diversity by Ernst Heinrich Philipp August Haeckel [86] (1899–1904; https:/commons.wikimedia.org/wiki/Kunstformen_der_Natur and https:/library.si.edu/digital-library/author/bibliographisches-institut-leipzig-germany). Left is the reproduction of plate 27 – Ctenophores: 1, 2 – *Haeckelia rubra*, Haeckeliidae; 3 – *Hormiphora foliosa* = *Hormiphora* sp., Cydippidae; 4 – *Callianira bialata*, Callianiridae; 5 – *Tinerfe cyanea*, Cydippidae; 6 – *Lampea pancerina*, Lampeidae

The species-level classification of ctenophores has not been reviewed in modern times, and the last genus-level overview was by Krumbach in 1925 [16], about 100 years ago. Dr. Claudia Mills (University of Washington, WA, USA) has initiated a summary of ctenophore taxa in 1998 and continues updating this ctenophore database [17] (https://faculty.washington.edu/cemills/Ctenolist.html). This became and was used as the basis for ctenophores in the World Register of Marine Species [WoRMS] (https://www.marinespecies.org/aphia.php?p=taxdetails&id=1248). In preparation for this chapter, GP reexamined the descriptions and synonymies of living ctenophore taxa from the primary literature and worked with Mills to reassess problematic ones. The results were used to update WoRMS and are presented below. Only published taxonomic information has been included. Numerous species and some higher taxa are poorly known, often based only on limited original descriptions, and should be considered *nomina dubia*.

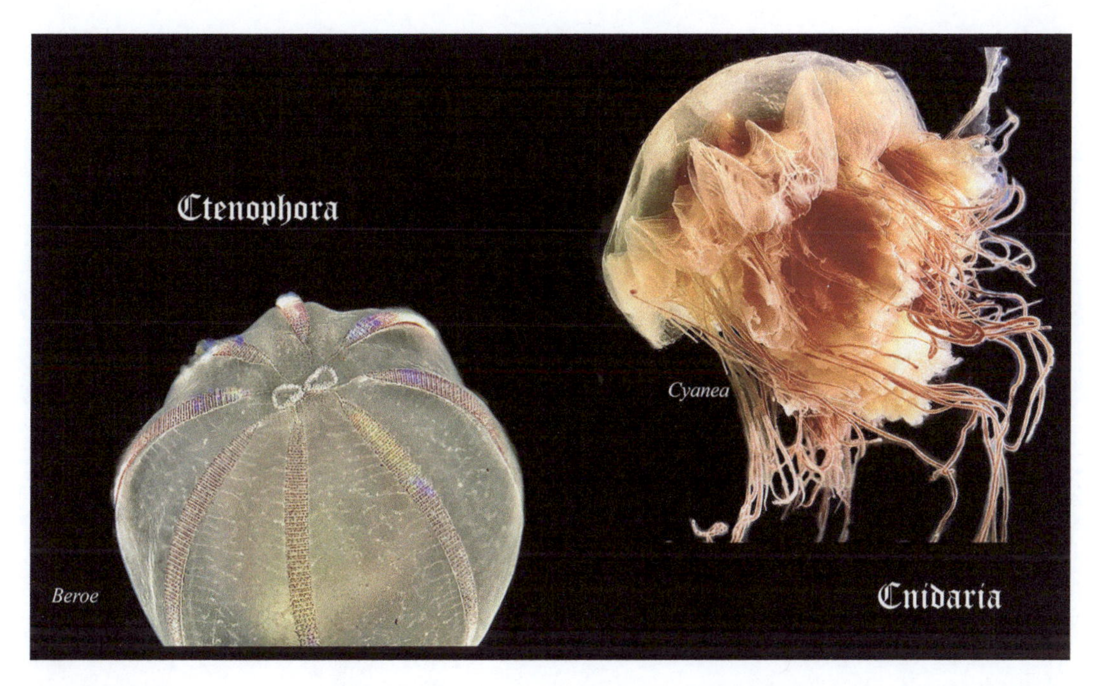

Fig. 2 Distinct body plans and locomotion modes in ctenophores vs. cnidarians. The left image is the top view of the comb jelly *Beroe abyssicola* with polar fields around the aboral organ and eight comb rows. The right image is the Lion's mane jellyfish, *Cyanea capillata* (Scyphozoa, Cnidaria). Both images are from Friday Harbor, WA, USA

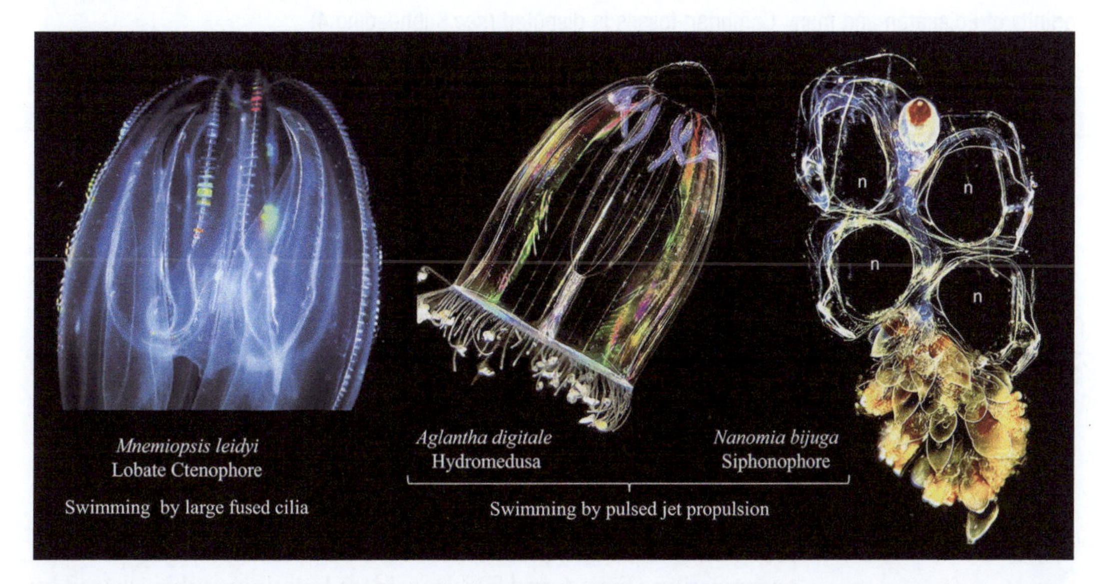

Fig. 3 Ctenophores primarily use their large ciliated plates for locomotion (e.g., the "sea walnut" *Mnemiopsis leidyi*, from Woods Hole, MA, USA). In contrast, cnidarians (e.g., *Aglantha* and *Nanomia*, both from Friday Harbor, WA, USA) primarily use muscular locomotion

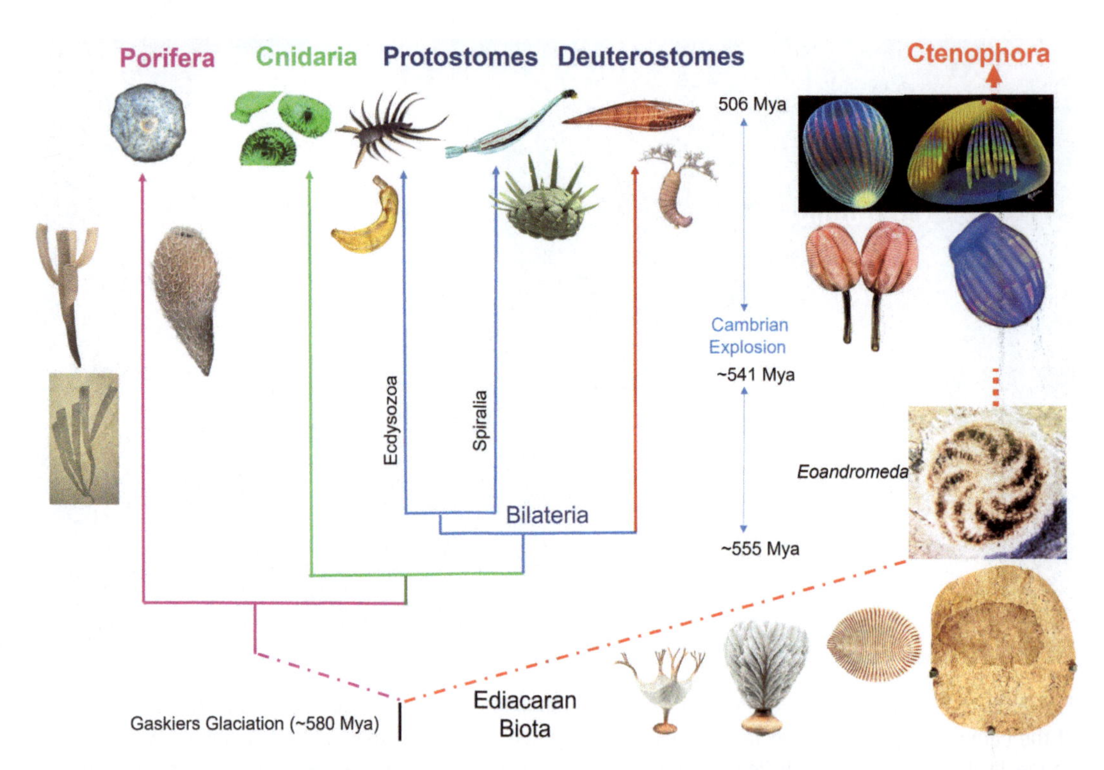

Fig. 4 Enigmatic origins of ctenophores as the sister lineage to all other Metazoa. Images of Cambrian animals are photos from reconstructions made by the Royal Ontario Museum© in Toronto, Canada. The ctenophore affinity of Ediacaran and many Cambrian fossils is disputed (*see* Subheading 4)

These are listed below, along with the well-known species, unless their status as *nomina dubia* has appeared in the literature and thus been captured in WoRMS. Synonymies of many species similarly remain contentious. The classification presented below reflects what is currently published and will undergo substantial changes as taxa are reassessed.

Traditionally [11, 13], two classes have been recognized in the phylum, but the paraphyly of the Tentaculata relative to the Nuda (ctenophores without tentacles) has long been suspected on morphological grounds [14] and confirmed by phylogenetics. Currently, nine orders are recognized (the two proposed by Ospovat [18] have not been assessed since their publication), with 33 families and 48 genera. Figure 5 summarizes relationships among selected ctenophore lineages, showing paraphyly of Cydippida and Lobata and placement of Nuda (Beroida). Cydippid-type organization (Figs. 6 and 7) is considered plesiomorphic, shared by many ctenophore lineages.

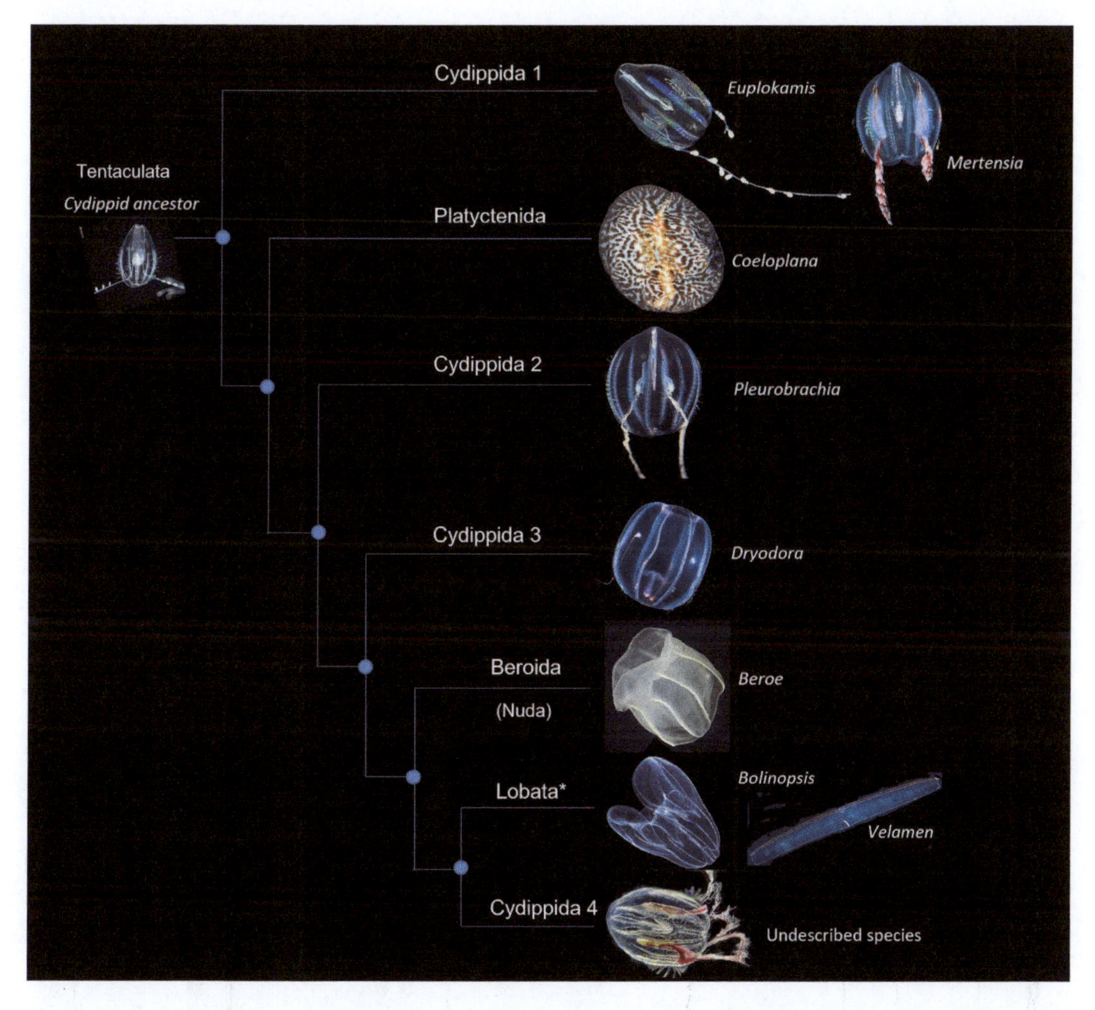

Fig. 5 Proposed phylogeny of several well-known ctenophore clades (based on [1, 6]). The revision of the higher-order taxonomy of ctenophores has not been systematically addressed, and we might suggest the name of the clade Euplokamida instead of Cydippida 1, as a tentative and representing one of the earliest branching lineages of ctenophores with unique adaptations in muscular and neural systems [65, 209]

2 List of Described Species and Illustrative Anatomy of the Phylum Ctenophora

Currently, 185 species of living ctenophores are recognized, although several dozen of these are too poorly known and should be considered *nomina dubia*. There are at least an additional 50 known but yet undescribed species and likely hundreds of ctenophore species (especially from deepwater habitats) awaiting discovery.

Below orders are arranged in order of decreasing species diversity: beginning with Cydippida with 62 described species (Figs. 6,

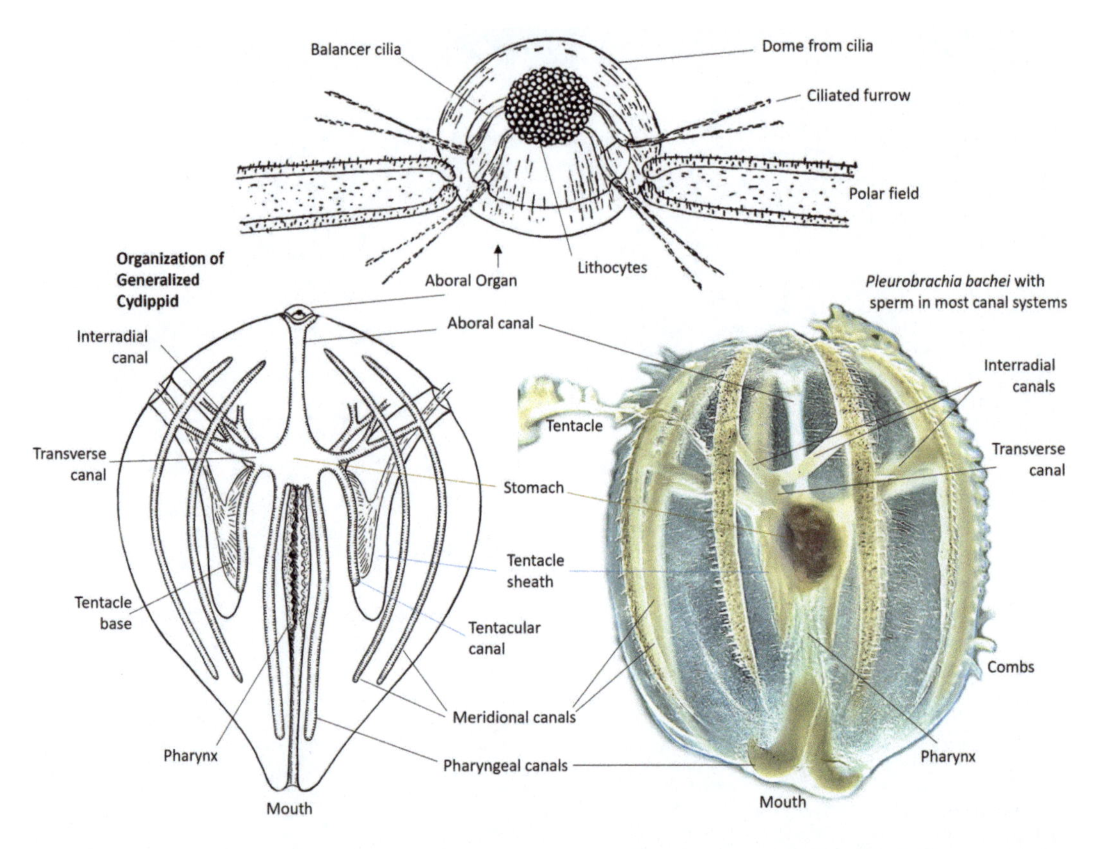

Fig. 6 General and microscopic anatomy of Cydippida with major organs and structures indicated (The left and top images are from [11]). The right image is *Pleurobrachia bachei*, with meridional canals filled with sperm. (The photo is taken in British Columbia, August 2023)

7, 8, 9, 10, 11, 12, 13, 14, 15, 16, 17, 18, 19, 20, 21, 22, 23, 24, 25, 26, 27, 28, 29, 30, 31, 32, 33, 34, 35, 36, 37, and 89); Platyctenida, 49 species (Figs. 38, 39, 40, 41, 42, 43, 44, 45, 46, 47, 48, 49, 50, 51, 52, 53, 54, 55, 56, 57, 58, 59, 60, 61, and 62); Lobata, 36 species (Figs. 63, 64, 65, 66, 67, 68, 69, 70, 71, 72, 73, 74, 75, 76, 77, 78, 79, 80, 81, 82, 83, 84, 85, 86, 87, 88, and 89); and Beroida, 30 species (Figs. 89, 90, 91, 92, 93, 94, 95, 96, 97, 98, 99, 100, and 101). Families are listed in alphabetical order. We also indicated the numbers of currently accepted species in each taxon.

Species with sequenced genomes are marked by #, while species with images in this chapter are bolded and referenced to appropriate figures.

For some taxa, we show multiple images to illustrate different aspects of morphology through ontogeny and by different types of illumination. We also provide illustrations of anatomy and selected ecological adaptations.

Several classical and recent manuscripts are useful in introducing details of ctenophore biology. These include monographic treatments from 1870 to 1940s focusing on ctenophore taxonomy and morphology with numerous illustrations [11, 19–27]. More recent reviews [28] summarize physiology [29–31], feeding [32], development [33, 34], and microscopic anatomy [35], as well as neurobiology [36–38]. Various aspects of ctenophore ecology and distribution can be found in the following publications: Harbison and colleagues [14, 39–41], Ospovat [18], Gershwin and colleagues about Australian ctenophores [42], as well as ctenophores in Antarctica [43], South Africa [44], New Zealand [45–47], Mexican seas [48], and many other regions worldwide [20, 24–26, 49–61].

3 Classification of Ctenophora

Phylum **Ctenophora** Eschscholtz, 1829 [62] (185 currently accepted species).

Marine, gelatinous, carnivorous animals with eight comb rows (or ctenes) formed by large fused cilia, biradial symmetry, true neurons, muscles, a gastrovascular system with canals in the mesoglea, an aboral organ with a statocyst that serves as an integrative center for locomotion, and other functions. As a result, the aboral organ is sometimes dubbed "an elementary brain" [1]. The aboral organ evolved independently from the analogous apical organs or sensory tufts in bilaterians and cnidarians, possesses a different organization [30, 31, 63–67], and should not be called "apical organ" in ctenophores.

Most ctenophores are simultaneous hermaphrodites with direct development and three embryonic layers (ectoderm, mesoderm, and endoderm). Most ctenophores are bioluminescent, and many that live in deep habitats are red-colored (Figs. 8, 9, 10, 35, and 75) to hide the animals in the darkness as red wavelengths are the first to be absorbed in water.

There are 199 species of ctenophores with valid names, including 185 extant and 14 fossils. Twelve other Cambrian and Ediacaran fossils were described as related to stem ctenophores, but their affinities are questionable and debated (*see* Subheading 4).

1. **"Cydippida"**: 13 families, 22 genera, 62 accepted species. A polyphyletic [1, 6] assemblage of pelagic ctenophores with well-developed tentacles that are retractible into specialized tentacle sheaths. Figures 6 and 7 provide overviews of typical representatives; Figs. 28 and 29 show some unusual feeding adaptations, and Figs. 7, 8, 9, 10, 11, 12, 13, 14, 15, 16, 17, 18, 19, 20, 21, 22, 23, 24, 25, 26, 27, 28, 29, 30, 31, 32, 33, 34, 35, 36, and 37 illustrate diversity.

1.1. Aulacoctenidae Lindsay & Miyake, 2007 (**1**) [68]
Aulacoctena Mortensen, 1913 (**1**)

Aulacoctena acuminata Mortensen, 1913 [69] (Fig. 9)

1.2. Bathyctenidae Mortensen, 1913 [70] (**2**)
Bathyctena Mortensen, 1912 [25] (**2**)

Bathyctena chuni (Moser, 1909) [71] (Fig. 10)

Bathyctena latipharyngea (Dawydoff, 1946) [72] (Fig. 11)

1.3. Callianiridae Eschscholtz, 1829 [62] (**4**)
Callianira Lamarck, 1816 (**4**) [73] (Fig. 12)

Callianira antarctica Chun, 1897 (Fig. 13) [74]

Callianira bialata Delle Chiaje, 1841 [75] (Fig. 1)

Callianira cristata Moser, 1909 [71]

Callianira ficalbi Curreri, 1900 [76]

1.4. Cryptocodidae Leloup, 1938 (**1**)
Cryptocoda Leloup, 1938 (**1**)

Cryptocoda gerlachi Leloup, 1938 [77] (Fig. 14)

1.5. Ctenellidae C. Carré & D. Carré, 1993 (**1**). Monotypic family without colloblasts but with labial suckers [78].
Ctenella C. Carré & D. Carré, 1993 (**1**)

Ctenella aurantia C. Carré & D. Carré, 1993 [78] (Fig. 15)

1.6. Cydippidae Gegenbaur, 1856 [79] (**28**)
Attenboroughctena Ceccolini & Cianferoni, 2020 [80] (**1**)

Attenboroughctena bicornis (C. Carré & D. Carré, 1991) [81] (Fig. 16)

Hormiphora L. Agassiz, 1860 (**16**) (Fig. 17) [82]

Hormiphora australis (Benham, 1907) [213]

Hormiphora californensis# (Torrey, 1904) [83]

Hormiphora cilensis (Ghigi, 1909) [84]

Hormiphora cucumis (Mertens, 1833) [85]

Hormiphora elliptica (Eschscholtz, 1829) [62]

Hormiphora foliosa Haeckel, 1904 (Fig. 1) [86]

Hormiphora hormiphora (Gegenbaur, 1856) [87] (Fig. 18)

Hormiphora luminosa Dawydoff, 1946 [72]

Hormiphora ochracea (A. Agassiz & Mayer, 1902) [88]

Hormiphora octoptera (Mertens, 1833) [85]

Hormiphora palmata Chun, 1898 [89]

Hormiphora piriformis Ghigi, 1909 [84]

Hormiphora polytrocha Dawydoff, 1946 [72]

Hormiphora punctata Moser, 1909 [71]

Hormiphora sibogae Moser, 1903 [90]

Hormiphora spatulata Chun, 1898 [89]

Minictena C. Carré & D. Carré, 1993 (**1**)

Minictena luteola C. Carré & D. Carré, 1993 [91, 92] (Fig. 19)

Pleurobrachia Fleming, 1822 (**8**) [93] (Figs. 6, 17, 20, 21, 22, and 23)

Pleurobrachia bachei# A. Agassiz, 1860 [82] (Figs. 6, 17, 20, 21, and 22)

Pleurobrachia brunnea Mayer, 1912 [24]

Pleurobrachia cyanea (Chun, 1889) [94]

Pleurobrachia globosa Moser, 1903 [90]

Pleurobrachia pigmentata Moser, 1903 [90]

Pleurobrachia pileus (O. F. Müller, 1776) [95] (Fig. 23)

Pleurobrachia rhodopis Chun, 1879 [74]

Pleurobrachia striata Moser, 1908 [96]

Tinerfe Chun, 1898 [89] (**1**)

Tinerfe cyanea (Chun, 1889) [94] (Fig. 24)

1.7. Dryodoridae Harbison, 1996 [41] (**1**)
Dryodora L. Agassiz, 1860 (**1**) [82]

Dryodora glandiformis (Mertens, 1833) [85] (Fig. 25).

1.8. Euplokamididae Mills, 1987 (**5**) [97]
Euplokamis Chun, 1879 (**5**) [74]

Euplokamis crinita (Moser, 1909) [71]

Euplokamis dunlapae Mills, 1987 (Figs. 7 and 26) [98]

Euplokamis evansae Gershwin, Zeidler & Davie, 2010 [42]

Euplokamis helicoides (Ralph & Kaberry, 1950) [46]

Euplokamis stationis Chun, 1879 [74]

1.9. Haeckeliidae Krumbach, 1925 [99] (**4**)
Haeckelia Carus, 1863 [100] (**4**)

Haeckelia beehleri (Mayer, 1912) [24] (Fig. 27)

Haeckelia bimaculata C. Carré & D. Carré, 1989 [101]

Haeckelia filigera (Chun, 1880) [20]

Haeckelia rubra (Kölliker, 1853) (Fig. 28) [102]

1.10. Lampeidae Stechow, 1921 (**5**) [103]
Lampea Stechow, 1921 (**5**) (Fig. 28)

Lampea dimidiata (Eschscholtz, 1829)

Lampea elongata (Quoy & Gaimard, 1833) [104]

Lampea komai (Dawydoff, 1937) [105]

Lampea lactea (Mayer, 1912) [24] (Fig. 29)

Lampea pancerina (Chun, 1879) [74] (Fig. 1)

1.11. Mertensiidae L. Agassiz, 1860 (**3**) [82]
Mertensia Lesson, 1829 (**2**) (Fig. 13)

Mertensia groenlandica Lesson, 1829 [106]

Mertensia ovum (Fabricius, 1780) [107, 108] (Figs. 7, 30, 31, and 32)

Charistephane Chun, 1879 (**1**)

Charistephane fugiens Chun, 1879 [74] (Fig. 33)

1.12. Pukiidae Gershwin, Zeidler & Davie, 2010 (**2**) [42]
Pukia Gershwin, Zeidler & Davie, 2010 (**2**)

Pukia falcata Gershwin, Zeidler & Davie, 2010 (Fig. 34)

Pukia ohtsukai Lindsay, 2017 [68] (Fig. 34)

1.13. Vampyroctenidae Townsend, Damian-Serrano & Whelan, 2020 (**1**)
Vampyroctena Townsend, Damian-Serrano & Whelan, 2020 (**1**)

Vampyroctena delmarvensis Townsend, Damian-Serrano & Whelan, 2020 [109] (Fig. 35)

Cydippida *incertae sedis* (of uncertain placement) (**4** species)

Duobrachium Ford, Bezio & Collins, 2020 (**1**)

Duobrachium sparksae Ford, Bezio & Collins, 2020 [56] (Fig. 36)

Paracelsia Dawydoff, 1946 (**1**)

Paracelsia quadriloba Dawydoff, 1946 [72] (Fig. 19)

Thoe Chun, 1879 (**1**)

Thoe paradoxa Chun, 1879 [74] (Fig. 37)

Tizardia Dawydoff, 1946 (**1**)

Tizardia phosphorea Dawydoff, 1946 [72] (Fig. 19)

2. **Order Platyctenida** Bourne 1900: Five families and six genera with 49 accepted species (Figs. 38, 39, 40, 41, 42, 43, 44, 45, 46, 47, 48, 49, 50, 51, 52, 53, 54, 55, 56, 57, 58, 59, 60, 61, and 62). A well-defined clade of benthic ctenophores, with two well-developed tentacles, differentiated aboral and oral sides, the latter a creeping sole derived from the stomodaeum. Comb rows lost in adults, except in the Ctenoplanidae; embryos brooded. The aboral organ can be reduced, and the statocyst is lost in *Savangia*. Most species epifaunal on invertebrates or algae [1, 6, 110, 111].

2.1. Coeloplanidae Willey, 1896 [112] (**33**)

Coeloplana Kowalevsky, 1880 [113] (**32**) (Figs. 38, 39, 40, 41, 42, 43, 44, 45, 46, 47, 48, 49, 50, 51, 52, 53, 54, 55, and 56).

Coeloplana agniae Dawydoff, 1930 [114]

Coeloplana anthostella Song & Hwang, 2010 [115]

Coeloplana astericola Mortensen, 1927 [116] (Fig. 39)

Coeloplana bannwarthii Krumbach, 1933 [117] (Fig. 40)

Coeloplana bocki Komai, 1920 [118]

Coeloplana duboscqui Dawydoff, 1930 [114]

Coeloplana echinicola Tanaka, 1932 [119]

Coeloplana fishelsoni Alamaru, Brokovich & Loya, 2016 [120] (Fig. 41)

Coeloplana gonoctena Krempf, 1921 [121]

Coeloplana huchonae Alamaru, Brokovich & Loya, 2016 [120]

Coeloplana indica Devanesen & Varadarajan, 1942 [122]

Coeloplana komaii Utinomi, 1963 [123]

Coeloplana krusadiensis Devanesen & Varadarajan, 1942 [122] (Fig. 42)

Coeloplana lineolata Fricke, 1970 [61]

Coeloplana loyai Alamaru & Brokovich, 2016 [120] (Figs. 43 and 44)

Coeloplana mellosa Gershwin, Zeidler & Davie, 2010 [42]

Coeloplana mesnili Dawydoff, 1938 [124]

Coeloplana meteoris Thiel, 1968 [125] (Figs. 45 and 46)

Coeloplana metschnikowii Kowalevsky, 1880 [113]

Coeloplana mitsukurii Abbott, 1902 [126] (Fig. 47)

Coeloplana perrieri Dawydoff, 1930 [114]

Coeloplana punctata Fricke, 1970 [61]

Coeloplana reichelti Gershwin, Zeidler & Davie, 2010 [42]

Coeloplana scaberiae Matsumoto & Gowlett-Holmes, 1996 [127] (Fig. 42)

Coeloplana sophiae Dawydoff, 1938 [124]

Coeloplana tattersalli Devanesen & Varadarajan, 1942 [122]

Coeloplana thomsoni Matsumoto, 1999 [128]

Coeloplana waltoni Glynn, Bayer & Renegar, 2014 [129]

Coeloplana weilli Dawydoff, 1938 [124]

Coeloplana willeyi Abbott, 1902 [126] (Fig. 48)

Coeloplana wuennenbergi Fricke, 1970 [61] (Fig. 49)

Coeloplana yulianicorum Alamaru, Brokovich & Loya, 2016 [120] (Fig. 50)

Vallicula Rankin, 1956 (**1**)

Vallicula multiformis Rankin, 1956 [130] (Fig. 57)

2.2. Ctenoplanidae Willey, 1896 [112] (**12**)
Ctenoplana Korotneff, 1886 (**12**) [131] (Figs. 38, 58, 59, and 60)

Ctenoplana agniae Dawydoff, 1929 [132]

Ctenoplana bengalensis Gnanamuthu & Nair, 1948 [133]

Ctenoplana caulleryi (Dawydoff, 1936) [134]

Ctenoplana duboscqui Dawydoff, 1929 [132]

Ctenoplana korotneffi Willey, 1896 [112]

Ctenoplana kowalevskii Korotneff, 1886 [131]

Ctenoplana maculomarginata Yosii, 1933 [135]

Ctenoplana maculosa Yosii, 1933 [135]

Ctenoplana neritica Fricke & Plante, 1971 [136]

Ctenoplana perrieri Dawydoff, 1936 [134]

Ctenoplana rosacea Willey, 1896 [112]

Ctenoplana yuri Dawydoff, 1929 [132]

2.3. Lyroctenidae Komai, 1942 (**2**) [137–140]
Lyrocteis Komai, 1941 (**2**) (Fig. 61)

Lyrocteis imperatoris Komai, 1941 (Fig. 62) [141]

Lyrocteis flavopallidus Robilliard & Dayton, 1972 [142]

2.4. Savangiidae Harbison & Madin, 1982 (**1**) [40].
Savangia Dawydoff, 1950 (**1**), no illustrations available; no specimens were recorded since the original description more than 70 years ago [143].

Savangia atentaculata Dawydoff, 1950 [143].

This truly enigmatic ctenophore species has only been found in the China Sea; it has six symmetrical aboral papillae beset with small tubercles; neither an aboral organ nor anal pores were reported. The oral part of the stomodaeum is permanently everted, forming a creeping sole as in other Platyctenida. *Savangia* is chocolate brown and can reach 2.5 cm [40, 143].

2.5. Tjalfiellidae Komai, 1922 (**1**)
Tjalfiella Mortensen, 1910 (**1**)

Tjalfiella tristoma Mortensen, 1910 [25] (Fig. 38)

3. **Order Lobata:** 9 families and 11 genera with 34 accepted species (Figs. 63, 64, 65, 66, 67, 68, 69, 70, 71, 72, 73, 74, 75, 76, 77, 78, 79, 80, 81, 82, 83, 84, 85, 86, 87, 88, and 89). A paraphyletic group of fragile pelagic ctenophores with reduced tentacles, body compressed in the tentacle plane, two large oral lobes and four auricles, and oral ends of gastrovascular canals anastomosing [1, 6].

3.1. Bathocyroidae Harbison & Madin, 1982 (**3**) [144]
Bathocyroe Madin & Harbison, 1978 (**3**)

Bathocyroe fosteri Madin & Harbison, 1978 [145] (Fig. 64)

Bathocyroe longigula Horita, Akiyama & Kubota, 2011 [146] (Figs. 64 and 65)

Bathocyroe paragaster (Ralph & Kaberry, 1950)

3.2. Bolinopsidae Bigelow, 1912 [147] (**12**)
Bolinopsis L. Agassiz, 1860 [82] (**10**)

Bolinopsis ashleyi Gershwin, Zeidler & Davie, 2010 [42] (Fig. 66)

Bolinopsis chuni (von Lendenfeld, 1884) [148]

Bolinopsis elegans (Mertens, 1833) [85]

Bolinopsis indosinensis Dawydoff, 1946 [72]

Bolinopsis infundibulum (O.F. Müller, 1776) [95] (Figs. 67 and 68)

Bolinopsis microptera# (A. Agassiz, 1865) [149] (Fig. 70)

Bolinopsis mikado (Moser, 1907) [150] (Fig. 69)

Bolinopsis ovalis (Bigelow, 1904) [151]

Bolinopsis rubripunctata Tokioka, 1964 [152]

Bolinopsis vitrea (L. Agassiz, 1860) [82] (Fig. 70)

Mnemiopsis L. Agassiz, 1860 [82] (**2**)

Mnemiopsis gardeni L. Agassiz, 1860 [82]

Mnemiopsis leidyi[#] A. Agassiz, 1865 [149] (Figs. 63 and 70)

3.3. Eurhamphaeidae L. Agassiz, 1860 [82] (**2**)
Deiopea Chun, 1879 [74] (**1**)

Deiopea kaloktenota Chun, 1879 (Fig. 71).

Eurhamphaea Gegenbaur, 1856 [87] (**1**)

Eurhamphaea vexilligera Gegenbaur, 1856 [153] (Figs. 72 and 73)

3.4. Kiyohimeidae Komai & Tokioka, 1940 [154] (**2**)
Kiyohimea Komai & Tokioka, 1940 (**2**)

Kiyohimea aurita Komai & Tokioka, 1940 [154] (Fig. 74)

Kiyohimea usagi Matsumoto & Robison, 1992 [155, 156] (Fig. 74)

Or ref. 85, Video: https://www.youtube.com/watch?app=desktop&v=WNX8xcAvSEY

3.5. Lampoctenidae Harbison, Matsumoto & Robison, 2001 (**1**)
Lampocteis Harbison, Matsumoto & Robison, 2001 (**1**) (Fig. 75)

Lampocteis cruentiventer Harbison, Matsumoto & Robison, 2001 [157] (Fig. 76)

3.6. Leucotheidae Krumbach, 1925 [16] (**6**)
Leucothea Mertens, 1833 [85] (**6**) (Fig. 77)

Leucothea filmersankeyi Gershwin, Zeidler & Davie, 2010 [42]

Leucothea grandiformis (Agassiz & Mayer, 1899) [158]

Leucothea japonica Komai, 1918 [159]

Leucothea multicornis (Quoy & Gaimard, 1824) [160] (Figs. 78, 79, and 80)

Leucothea ochracea Mayer, 1912 [24] (Fig. 81)

Leucothea pulchra Matsumoto, 1988 [161] (Fig. 77)

3.7. Lobatolampeidae Horita, 2000 (**1**)
Lobatolampea Horita, 2000 (**1**)

Lobatolampea tetragona Horita, 2000 [162, 163] (Fig. 82)

3.8. Ocyropsidae Krumbach, 1925 [99] (**5**). At least two species (*O. maculata* and *O. crystallina*) are dioecious rather than hermaphroditic [164].

Ocyropsis Mayer, 1912 [24] (**5**)

Ocyropsis crystallina (Rang, 1827) [165] (Figs. 83, 84, and 85)

Ocyropsis fusca (Rang, 1827) [165]

Ocyropsis maculata (Rang, 1827) [165] (Figs. 86 and 87)

Ocyropsis pteroessa Bigelow, 1904 [151]

Ocyropsis vance Gershwin, Zeidler & Davie, 2010 [42]

3.9. Pterygioctenidae Mills & Dubois, 2023 (**2**) [166]
Pterygiocteis Mills & Dubois, 2023 (**2**)

Pterygiocteis nigrolimbatus Mills & Dubois, 2023 [166] (Fig. 88)

Pterygiocteis pinnatus (Ralph & Kaberry, 1950) [46]

Lobata *incertae sedis.* The following species (*Axiotima gaedii*, Eschscholtz, 1825, 1829) has not been recognized after its original description [62, 167]; the original drawing is not interpretive.

4. **Order Beroida** Eschscholtz, 1825 (Class Nuda) [167]: One family, two genera with 30 accepted species (Figs. 89, 90, 91, 92, 93, 94, 95, 96, 97, 98, 99, 100, and 101). Muscular pelagic predators that typically feed on other ctenophores. There are no tentacles during any time in their development. The aboral organ is well-developed with numerous papillae. All species have a very wide mouth; the stomodeum (pharynx) occupies most of the body; meridional canals in most species have numerous side branches.

4.1. Beroidae Eschscholtz, 1825 (**30** species) [168]; (Figs. 89, 90, 91, 92, 93, 94, 95, 96, 97, 98, 99, and 100).
Beroe Müller, 1776 [95] (**29**)

Beroe abyssicola Mortensen, 1927 [116] (Figs. 90, 91, 92, 93, 94, and 95).

Beroe australis Agassiz & Mayer, 1899 [158]

Beroe baffini Kramp, 1942; the original description is not documented

Beroe basteri Lesson, 1829 [106]

Beroe beroe (Linnaeus, 1758) [169]

Beroe campana Komai, 1918 [159]

Beroe chiaji Lesson, 1836 [170]

Beroe compacta Moser, 1909 [71]

Beroe constricta Chamisso & Eysenhardt, 1821 [171]

Beroe cucumis Fabricius, 1780 [172] (Figs. 96, 97, and 98)

Beroe cyathina (A. Agassiz, 1860) [82]

Beroe fallax Lesson, 1836 [170]

Beroe flemingii (Eschscholtz, 1829) [62]

Beroe forskalii Milne Edwards, 1841 [173] (Fig. 99)

Beroe gracilis Künne, 1939 [174]

Beroe hyalina Moser, 1907 [150]

Beroe macrostomus Péron & Lesueur, 1807 [175]

Beroe mitraeformis Lesson, 1829 [106]

Beroe mitrata (Moser, 1907) [150] (Fig. 89)

Beroe ovata Bruguière, 1789 [176] (Fig. 100)

Beroe pandorina Moser, 1903 [90]

Beroe penicillata (Mertens, 1833) [85]

Beroe quoyi Lesson, 1836 [170]

Beroe ramosa Komai, 1921 [177]

Beroe roseus Quoy & Gaimard, 1824 [160]

Beroe rufescens (Eschscholtz, 1829) [62]

Beroe santonum Lesson, 1843 [178]

Beroe scoresbyi Lesson, 1836 [170]

Beroe shakespeari Benham, 1907 [47]

Neis Lesson, 1829 (**1**)

Neis cordigera Lesson, 1829 [106] (Fig. 101)

5. **Order Cestida (4)**. One family, two genera, four accepted species. These pelagic ctenophores are highly compressed in the tentacular plane with a long ribbon-like body (10–150 cm); four comb rows and the tentacles are reduced, but they have two rows of small tentilla along their oral margin.

5.1. Cestidae Gegenbaur, 1856 [79] (**4**)

Cestum Lesueur, 1813 [179] (**3**)

Cestum mertensii L. Agassiz, 1860 [82]

Cestum najadis Eschscholtz, 1829 [62]

Cestum veneris Lesueur, 1813 [179] (Fig. 102)

Velamen Krumbach, 1925 (**1**)

Velamen parallelum (Fol, 1869) [180] (Fig. 103)

6. **Order Cryptolobiferida** (**2**). Order with one family proposed by Ospovat [18] for two genera and species of poorly known tropical ctenophores; no recent photos.

6.1. Cryptolobatidae Ospovat, 1985 [18]

Cryptolobata Moser, 1909 (**1**)

Cryptolobata primitiva Moser, 1909 [71] (Fig. 104)

Lobocrypta Dawydoff, 1946 (**1**)

Lobocrypta annamita Dawydoff, 1946 [72] (Fig. 105a)

7. **Order Ganeshida**. One family and genus, two accepted species. Tropical ctenophores of intermediate form between cydippid and lobate body plan [14], with large mouth, body compressed in tentacular plane, tentacles with sheaths and tentilla, and circumoral canal. Known from earlier morphological observations; no recent photos.

   7.1. Ganeshidae Moser, 1907 [150] (**2**)

   Ganesha Moser, 1907 [150] (**2**)

   Ganesha annamita Dawydoff, 1946 [72] (Fig. 105d)

   Ganesha elegans (Moser, 1903) [90] (Fig. 105c)

8. **Order Cambojiida.** Monotypic order proposed by Ospovat for a poorly known tropical ctenophore known from two specimens [18]; no recent photos.

   8.1. Cambojiidae Ospovat, 1985 [18] (**1**)

   Cambodgia Dawydoff, 1946 (**1**)

   Cambodgia elegantissima Dawydoff, 1946 [72] (Fig. 105b)

9. **Order Thalassocalycida** (**1**). Monotypic. Extremely fragile pelagic ctenophores with medusa-bell shape and unique feeding strategy [181, 182]. Tentacles (without sheaths) are near the mouth.

   9.1. Thalassocalycidae Madin & Harbison, 1978 (**1**).

   Thalassocalyce Madin & Harbison, 1978 (**1**)

   Thalassocalyce inconstans Madin & Harbison, 1978 [181] (Fig. 106).

4 Fossil Ctenophora

Fragile ctenophores left a very limited fossil record, and the affinity of early fossils is often questionable. The middle Cambrian (~506–510 MA) *Fasciculus vesanus*, *Xanioascus canadensis*, *Ctenorhabdotus capulus* [183], *Ctenorhabdotus campanelliformis* and *Thalassostaphylos elegans* [184] from the Burgess Shale had multiple comb plates but no tentacles. The only known specimen of *Fasciculus vesanus* (114 mm) appears to have possessed ~80 comb rows and had bilaterial-like symmetry. *Xanioascus* and *Ctenorhabdotus* both had 24 comb rows (maximum sizes: 122 mm and 77 mm, respectively), and *Thalassostaphylos* had 16 rows [184] (but *see* [185] for evidence of tentacles in Cambrian ctenophores).

Class Scleroctenophora † (extinct) (**6**).

Armored, probably sessile, ctenophore-like animals from the early Cambrian Chengjiang biota (ca. 520 Ma) with oral-aboral axis, aboral sense organ, and eight ctene rows [186].

1. *Batofasciculus ramificans* Hou, Bergstöm, Wang, Feng & Chen, 1999† [187]. Chengjiang Biota, Lower Cambrian: Yunnan, China (see also [188] for the fossil affinity discussion).
2. *Galeactena hemispherica* Ou, Xiao, Han, Sun, Zhang, Zhang & Shu, 2015† Chengjiang Biota, Lower Cambrian: Yunnan, China
3. *Gemmactena actinala* Ou, Xiao, Han, Sun, Zhang, Zhang & Shu, 2015† Chengjiang Biota, Lower Cambrian: Yunnan, China
4. *Maotianoascus octonarius* Chen and Zhou, 1997†. Chengjiang Biota, Lower Cambrian: Yunnan, China [189]; see also [190] for possible embryos.
5. *Thaumactena ensis* Ou, Xiao, Han, Sun, Zhang, Zhang & Shu, 2015†. Chengjiang Biota, Lower Cambrian: Yunnan, China
6. *Trigoides aclis* Luo and Hu, 1999† [191] Chengjiang Biota, Early Cambrian: Yunnan, China (see also [192]).

Cydippida (**2**)

7. *Archeocydippida hunsrueckiana* Stanley & Stürmer, 1987† [193], Devonian: from Hunsrück Slate, Germany. The fossil with (two proposed tentacles) was reinterpreted as not a ctenophore [194] but then accepted [183] and again challenged [17].
8. *Paleoctenophora brasseli* Stanley & Stürmer, 1983† [195], from Devonian: Hunsrück Slate, Germany, also independently accepted as a ctenophore [183].

Others (**6**)

Family **Ctenorhabdotidae (6)** Conway Morris and Collins, 1996 [183], 24 comb rows.

9. ***Ctenorhabdotus capulus*** Conway Morris and Collins, 1996† [183]. Middle Cambrian: Burgess Shale, British Columbia, Canada (Fig. 107c)
10. *Ctenorhabdotus campanelliformis* Parry, Lerosey-Aubril, Weaver and Ortega-Hernandez, 2021† [184], Middle Cambrian: House Range of western Utah, USA

Family **Fasciculidae** Conway Morris and Collins, 1996 [183].

11. ***Fasciculus vesanus*** Simonetta and Della Cave, 1978† [183]. 80 comb rows. Middle Cambrian: Burgess Shale, British Columbia, Canada (Fig. 107b)

Family **Xanioascidae** Conway Morris and Collins, 1996 [183].

12. ***Xanioascus canadensis*** Conway Morris and Collins, 1996† [183]. 24 comb rows. Middle Cambrian: Burgess Shale, British Columbia, Canada (Fig. 107a)

Family is not defined for the species belows.

13. *Thalassostaphylos elegans* Parry, Lerosey-Aubril, Weaver and Ortega-Hernandez, 2021† [184]. Middle Cambrian: House Range of western Utah, USA; 16 comb rows.
14. *Daihuoides jakobvintheri* Klug, Kerr, Lee & Cloutier, 2021 † Later Devonian ctenophore-type organism; it is suggested to be a sister lineage to *Fasciculus* as a later surviving representative of stem-ctenophores [196].

Questionable Cambrian ctenophore-like fossils: (**8**)

15. ***Siphusauctum gregarium*** O'Brien and Caron, 2012† [197]. Middle Cambrian, Burgess Shale, ~510 million years old. A stalked, sessile fossil (*see also* Fig. 107d) with six radial parts with comb-like elements that surround an internal body cavity with a large stomach, conical median gut, and straight intestine. *The* animal was probably an active filter-feeder, with water passing through the calyx openings, capturing food particles with its comb-like elements. More than 1100 specimens were collected [197].
16. *Dinomischus isolatus*† – Proposed as a sessile stem-group ctenophore, Middle Cambrian, Burgess Shale [198].
17. *Dinomischus venustus*† – Proposed as a sessile stem-group ctenophore, Chengjiang Biota, Lower Cambrian [198].
18. *Daihua sanqiong*† – Proposed as a sessile stem-group ctenophore, Chengjiang Biota, Lower Cambrian, [198].
19. *Xianguangia sinica*† – Proposed as a sessile stem-group ctenophore, Chengjiang Biota, Lower Cambrian [198], with possible tentaculate-type structures [199].
20. *Sinoascus papillatus* Chen and Zhou, 1997†. Chengjiang Biota, Lower Cambrian: Yunnan, China. The fossil was initially interpreted as an early ctenophore, but this affinity is uncertain [188].
21. *Yunnanoascus haikouensis* Hu et al. 2007† [188]. Chengjiang Biota, Lower Cambrian: Yunnan, China. The fossil

was initially interpreted as a ctenophore [200], but later it was assigned to Cnidaria as a jellyfish with rhopalia [201].

22. *Stromatoveris psygmoglena* Shu, 2006† [202]. Chengjiang Biota, Lower Cambrian: Yunnan, China. The fossil has been interpreted as an early stem-group ctenophore; it can be a potential "link" between Ediacaran and Cambrian biotas [203]

Questionable Precambrian ctenophore fossils (4)

23. *Eoandromeda octobrachiata*† Ediacaran Biota, South China. The fossil has been interpreted as a ctenophore-like organism due to apparent spiral comb-like structures (4 pairs) and octoradial symmetry or as an early stem-group ctenophore [204].
24. *Dickinsonia* spp.† Ediacaran Biota. Interpreted as a possible ctenophore due to biradial symmetry [205].
25. *Rangea schneiderhoehni*, Gurich, 1930†. Dzik (2002) argues for possible ctenophore affinities of this Precambrian sea-pen-like organism [206].
26. *Namacalathus hermanastes*†† [207]; a biomineralized fossil similar to *Siphusauctum*, but also interpreted as lophotrochozoan [208].

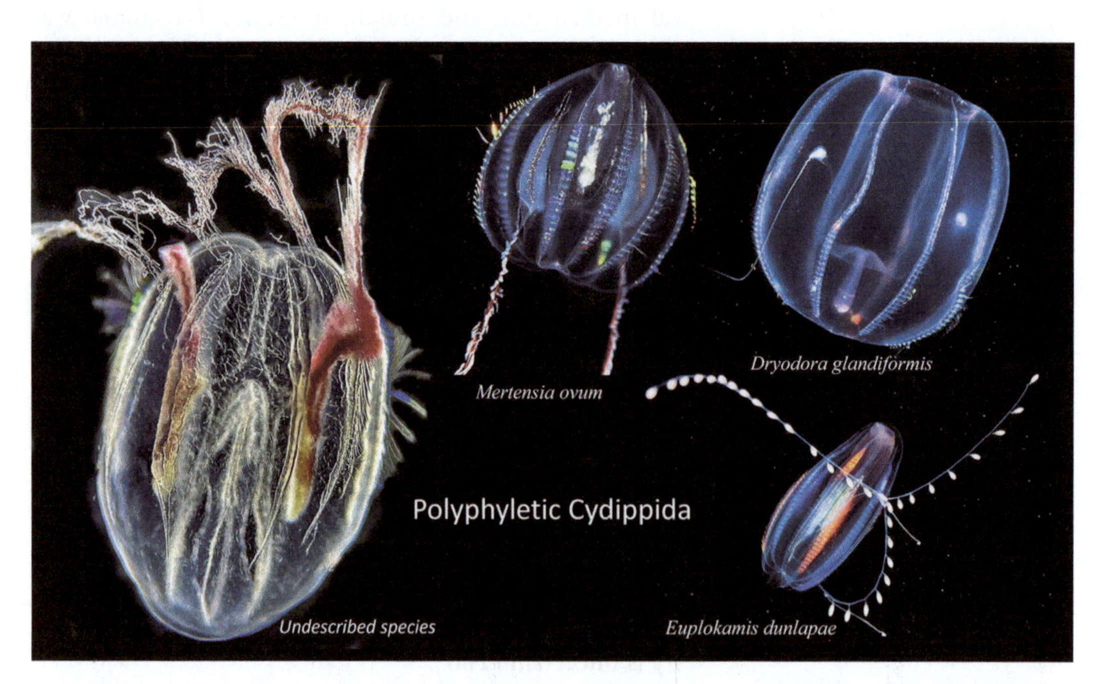

Fig. 7 Illustrated examples of different cydippid species. (The photo of an undescribed cydippid was taken at Friday Harbor Laboratories, Washington, USA. *Mertensia*, *Dryodora*, and *Euplokamis* are from Arctic photos by Alexander Semenov© (https:/www.flickr.com/photos/a_semenov/40592507363/in/photostream/))

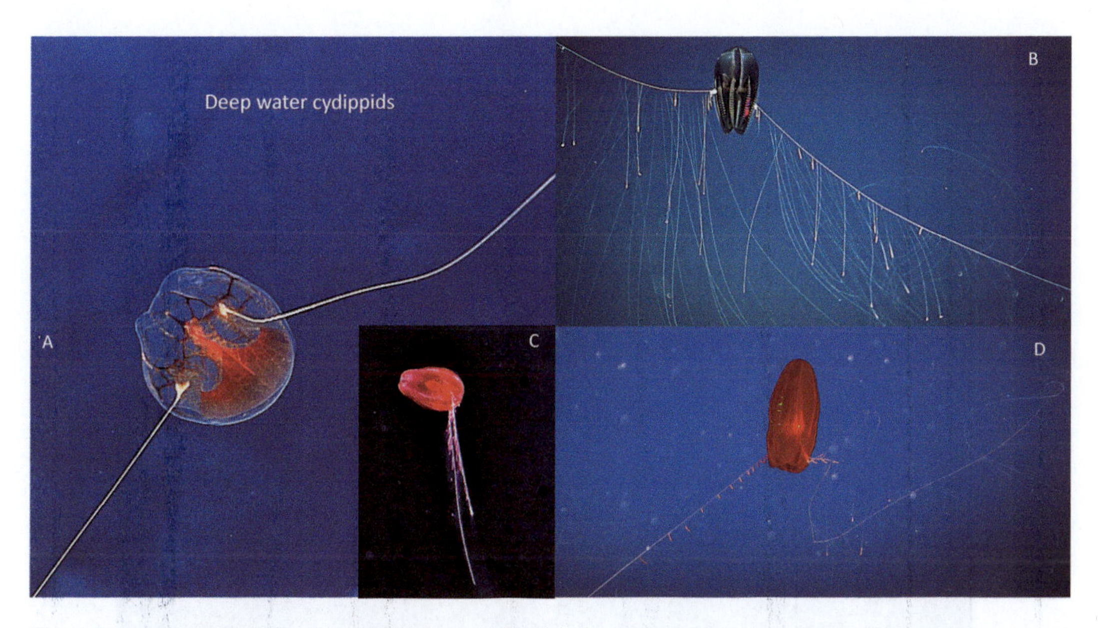

Fig. 8 Deepwater cydippids with red/black coloration. Their pigments hide the animals in the darkness as red light is the first wavelength to be absorbed in water from the visible spectrum. (**a**) A ctenophore with long tentacles (Aulacoctenidae) was collected at Southernmost Cone within the Pacific Remote Islands Marine National Monument. (Image courtesy of the NOAA Office of Ocean Exploration and Research, 2015 Hohonu Moana (https:/oceanexplorer.noaa.gov/okeanos/explorations/ex1504/logs/sept27/sept27.html; or https:/ www.flickr.com/photos/oceanexplorergov/22568727927/in/photostream/)). (**b**) This dark cydippid was observed ~1460 m deep in an area of the Western Gulf of Mexico. (Image courtesy of the NOAA Office of Ocean Exploration and Research, Gulf of Mexico, 2018 (https:/www.flickr.com/photos/oceanexplorergov/40 959988584/in/photostream/)). (**c**) Undescribed deepwater cydippid from the public exposition at the Monterey Bay Aquarium. (**d**) Undescribed cydippid. (Courtesy of the NOAA Office of Ocean Exploration and Research, Deep Connections 2019 (https:/www.flickr.com/photos/oceanexplorergov/48976460707/in/photostream/))

Fig. 9 Deep (**a**–**c**) *Aulacoctena acuminata*, its body can be up to 30 cm long with white tentacles; observed at depths of 500 to over 3000 m. (For *Aulacoctena* photos, credits go to Marsh Youngbluth/MAR-ECO, Census of Marine Life (https:/ocean.si.edu/ocean-life/invertebrates/red-mid-water-comb-jelly); (https:/oceanexplorer.noaa.gov/explorations/05arctic/logs/july19/media/aulacoctena.html); see also *video courtesy of The Hidden Ocean 2016: Chukchi Borderlands, Oceaneering-DSSI* (https:/oceanexplorer.noaa.gov/explorations/16arctic/logs/video/combjelly/combjelly_video.html)). However, these three ctenophores might also represent three different species

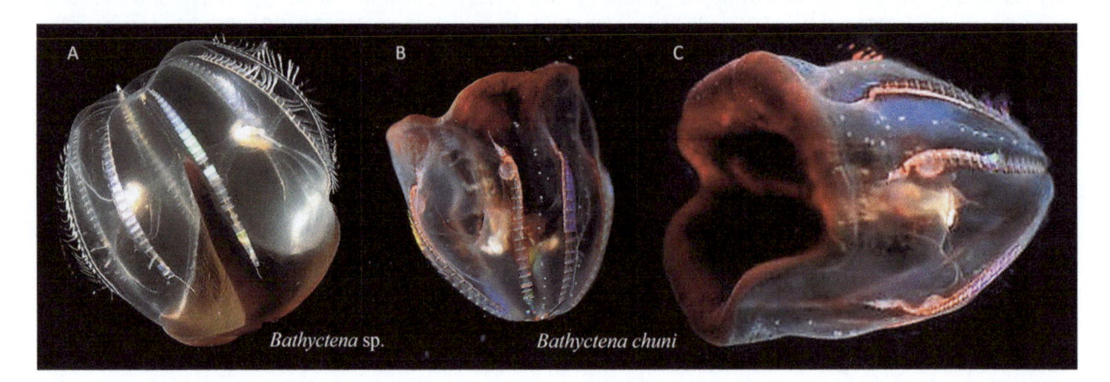

Fig. 10 (**a**) Deepwater *Bathyctena* sp. (?), Arctic Ocean, and (**b**, **c**) *Bathyctena chuni*. (Sources of images: (**a**) *Arctic Ocean diversity by* Russ Hopcroft – public domain: http:/www.arcodiv.org/watercolumn/ctenophores/Bathyctena.html; (**b**) public domain: https:/forums.unknownworlds.com/discussion/154142/arctic-jellyfish; (**c**) Photo: Steven Haddock© MBARI 2006. That animal was caught May 2006 at a depth of 697 meters off the California coast. https:/www.eurekalert.org/multimedia/929241. Photo Credit: Steve Haddock; see also [32]. *See additional images of Bathyctena chuni*, from [32], and undescribed cydippids (from [15]). Additional sources for deepwater ctenophore images: http:/www.coml.org/census-arts/the-deep/; https:/twitter.com/Emma_Hollen/status/965871500590043136/photo/1; https:/cflas.org/2015/05/13/comb-jellyfish/; https:/twitter.com/Emma_Hollen/status/965871500590043136; https:/imgur.com/gallery/BL8gS/comment/297275655, MBARI© and Academic Press© [15])

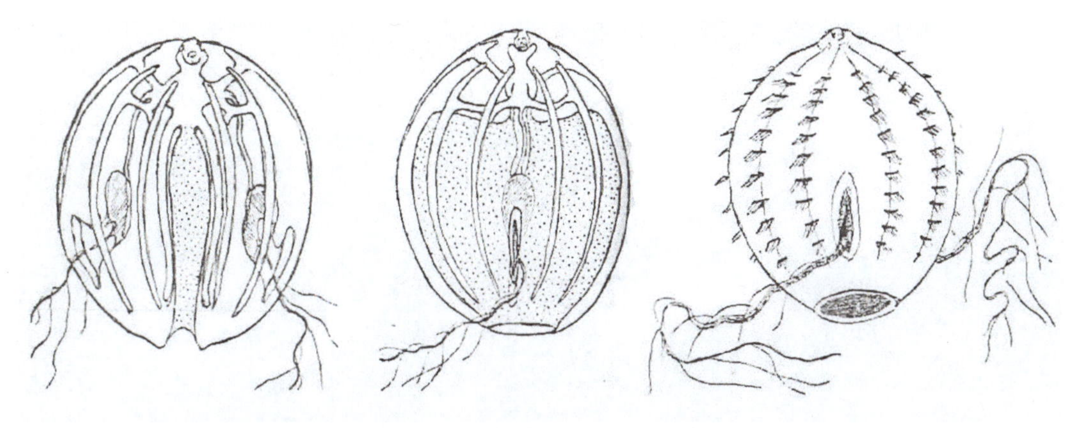

Fig. 11 Deepwater *Bathyctena latipharyngea* (Dawydoff, 1946). (from [72])

Fig. 12 *Callianira* spp. Florida, West Palm Beach, USA. Possible *Callianira cristata*

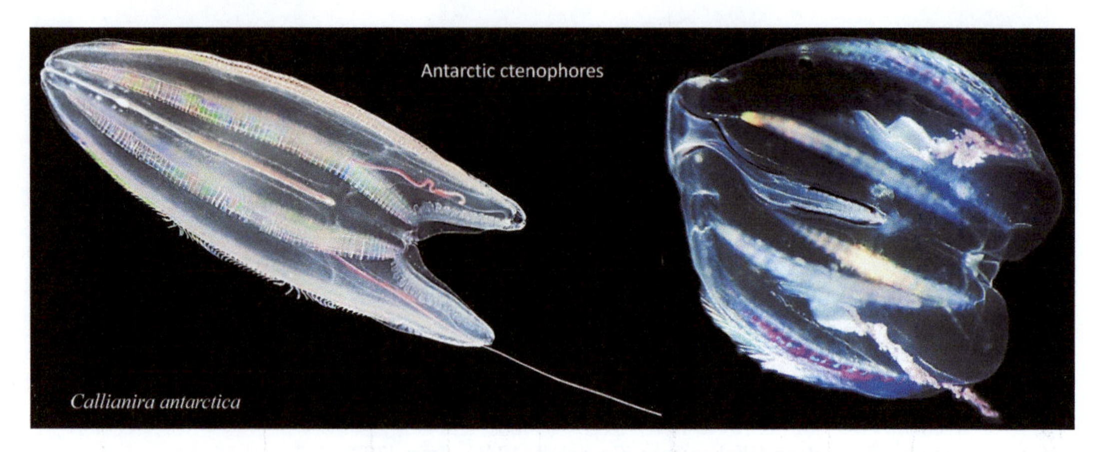

Fig. 13 *Callianira antarctica* and undescribed cydippid, Weddell Sea, Antarctica

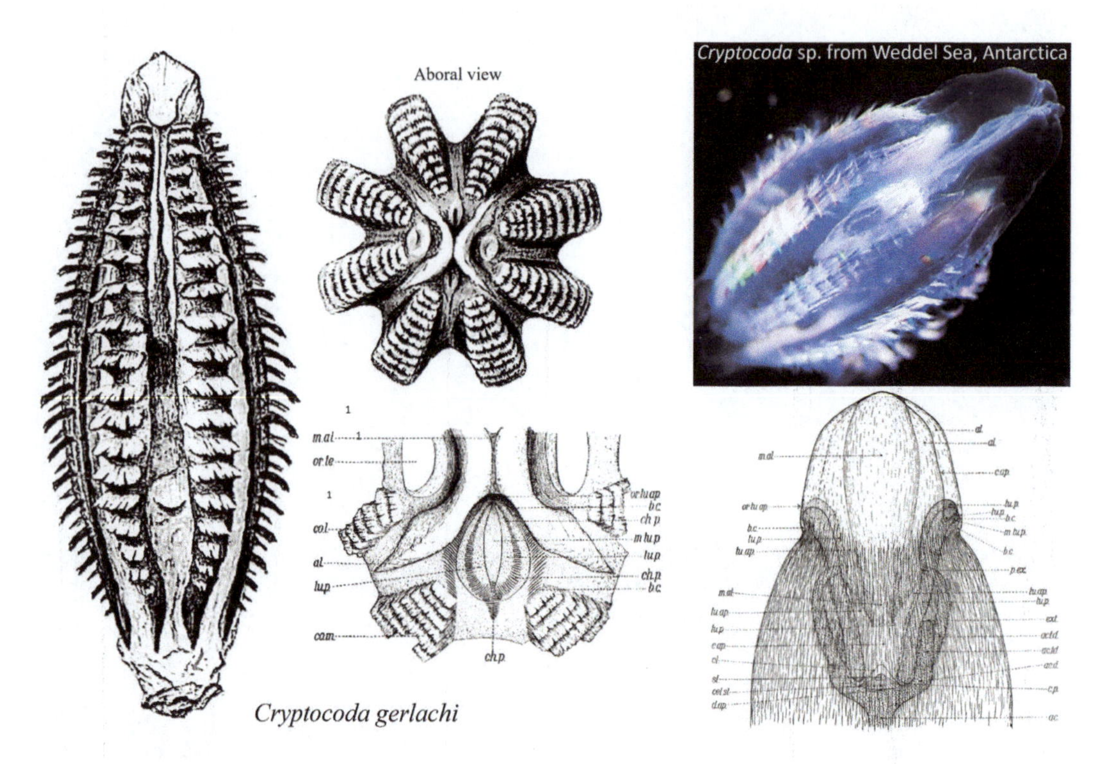

Fig. 14 *Cryptocoda gerlachi* Leloup, 1938. (from [77] with original abbreviations). The top right image might be *Cryptocoda* sp. from Weddel Sea, Antarctica

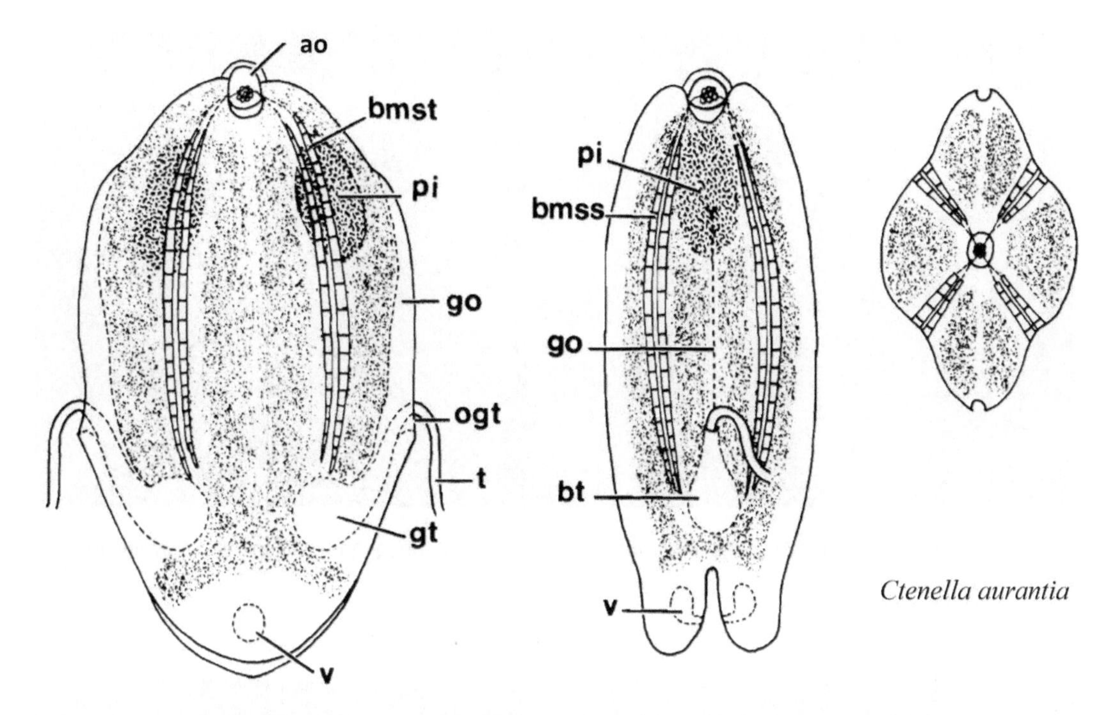

Fig. 15 *Ctenella aurantia* C. Carré & D. Carré, 1993. (from [78] with original abbreviations)

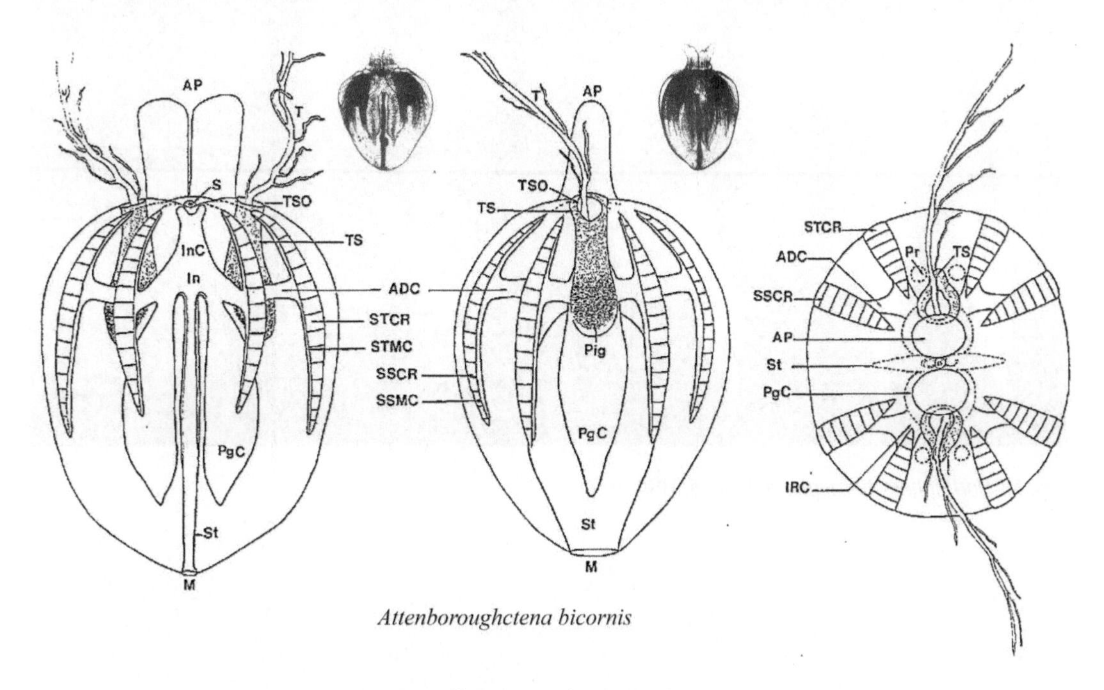

Fig. 16 *Attenboroughctena bicornis* (C. Carré & D. Carré, 1991). (from [81])

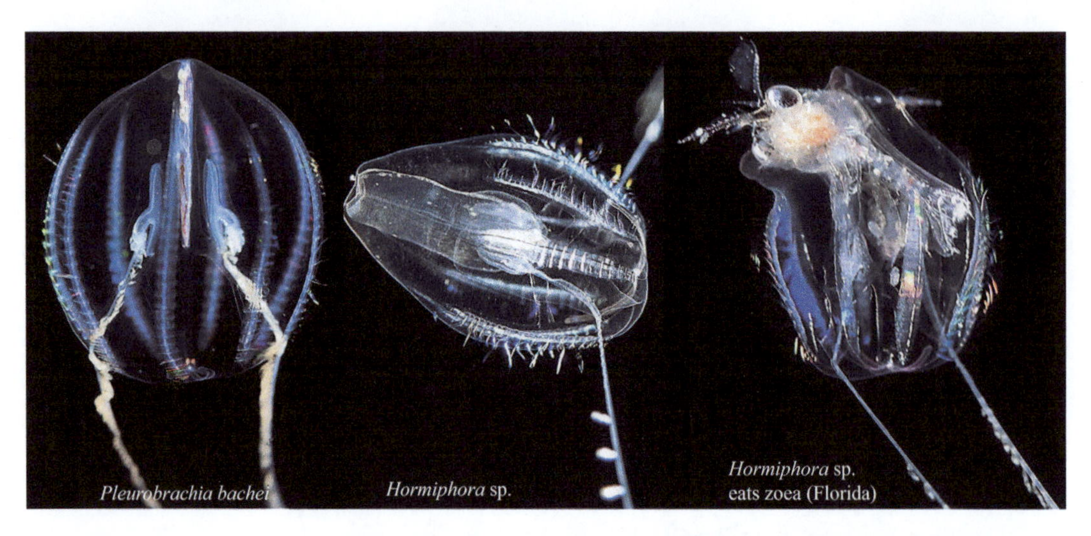

Fig. 17 A comparison of cydippids: *Pleurobrachia bachei*, San Juan Archipelago, vs. *Hormiphora* sp., Florida, USA

Fig. 18 *Hormiphora hormiphora*, Florida, USA

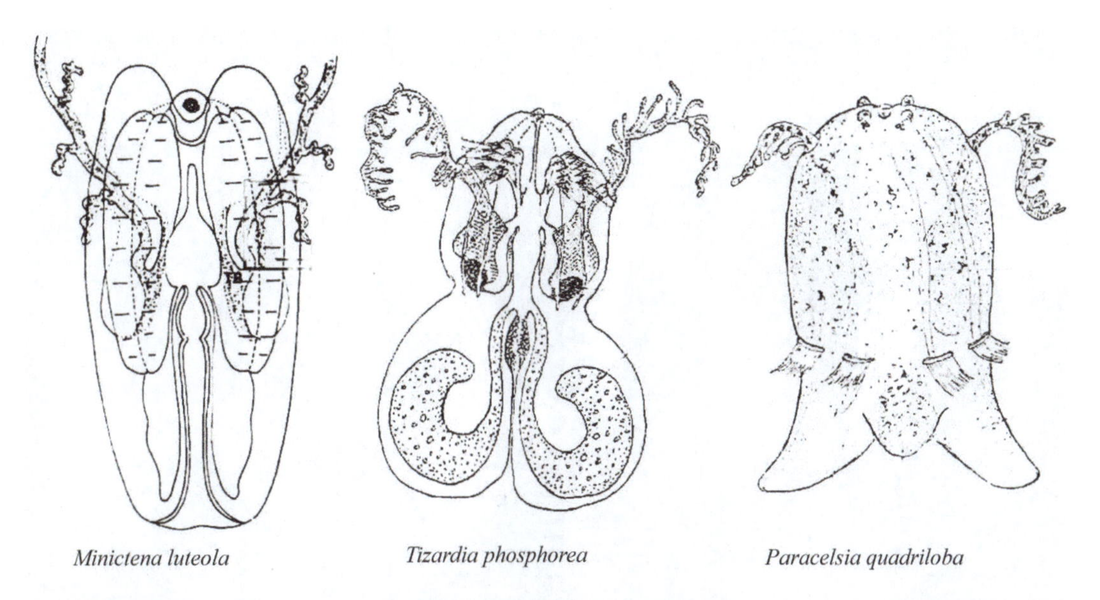

Fig. 19 *Minictena luteola* C. Carré & D. Carré, 1993. (from [91, 92]). *Tizardia phosphorea* Dawydoff, 1946. (Modified from [72]). *Paracelsia quadriloba* Dawydoff, 1946. (from [72])

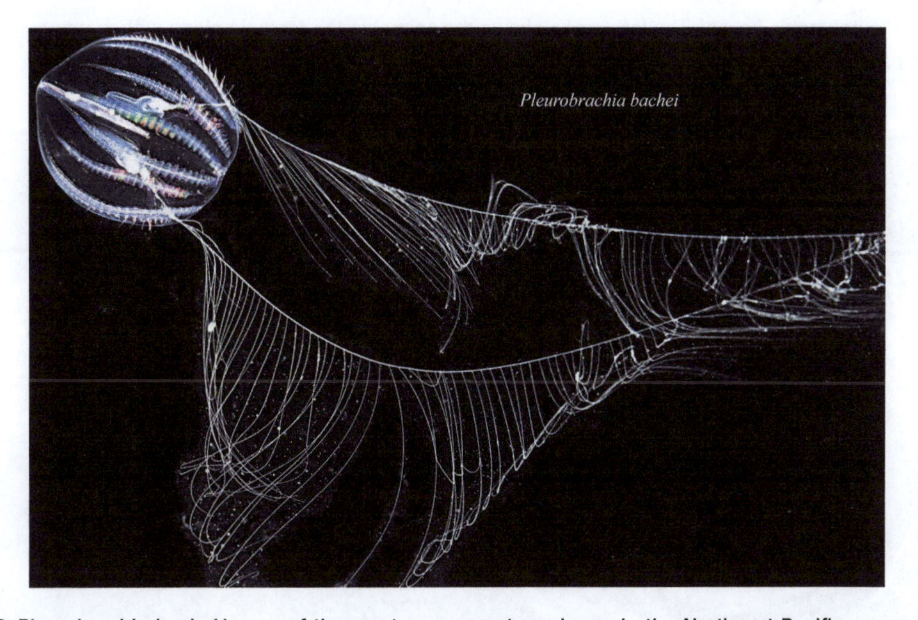

Fig. 20 *Pleurobrachia bachei* is one of the most common ctenophores in the Northeast Pacific

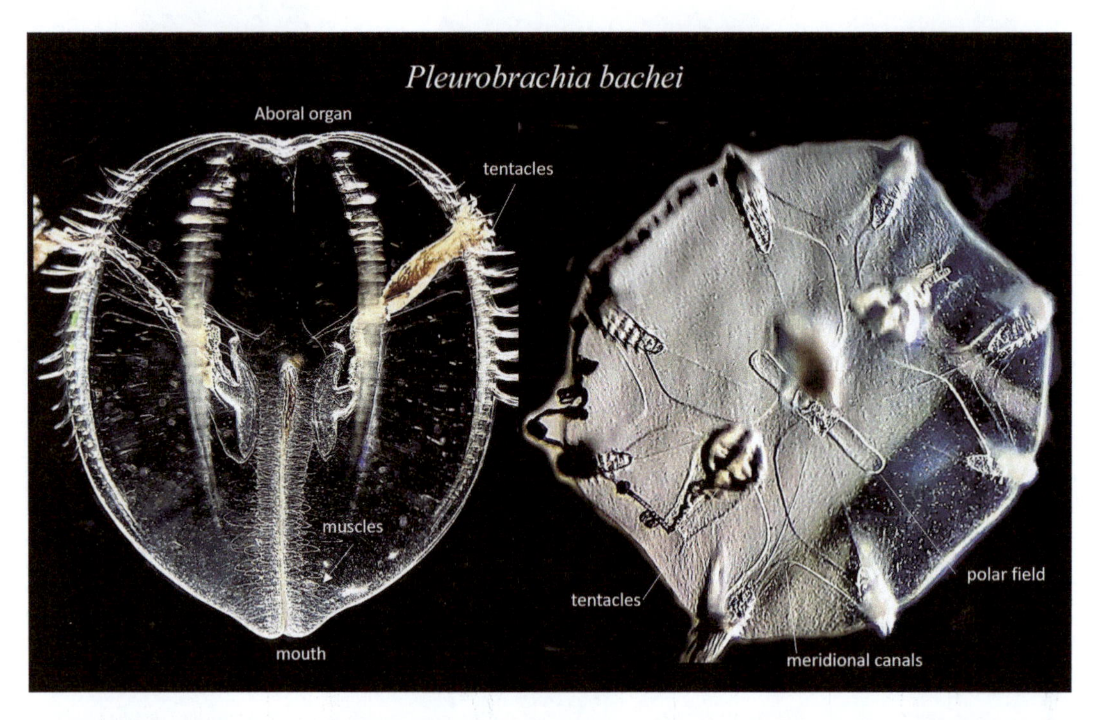

Fig. 21 *Pleurobrachia bachei.* The left image shows the animal in the tentacle plane with mouth and circular muscles around the pharynx. The right image is the aboral view of this transparent animal

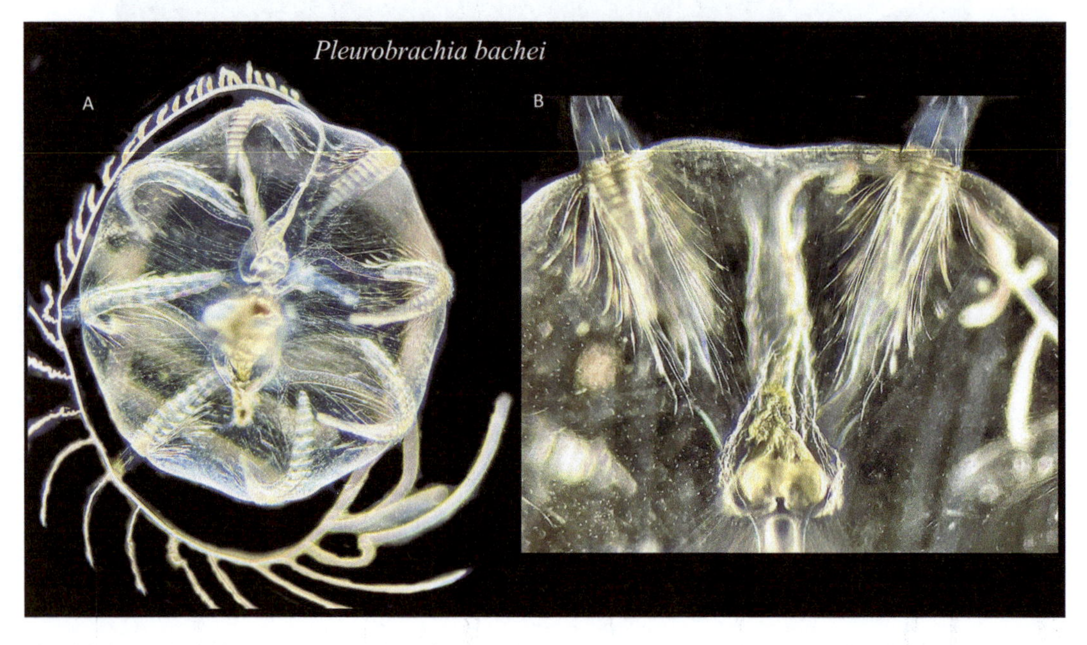

Fig. 22 *Pleurobrachia bachei.* (**a**) A view from the oral side shows tentacles with sheaths and meridional canals. (**b**) Side view of two comb plates with cilia and muscles. The bottom of the image shows a tentacle retracted in the tentacle pocket (sheath)

Fig. 23 *Pleurobrachia pileus*, Maine, USA

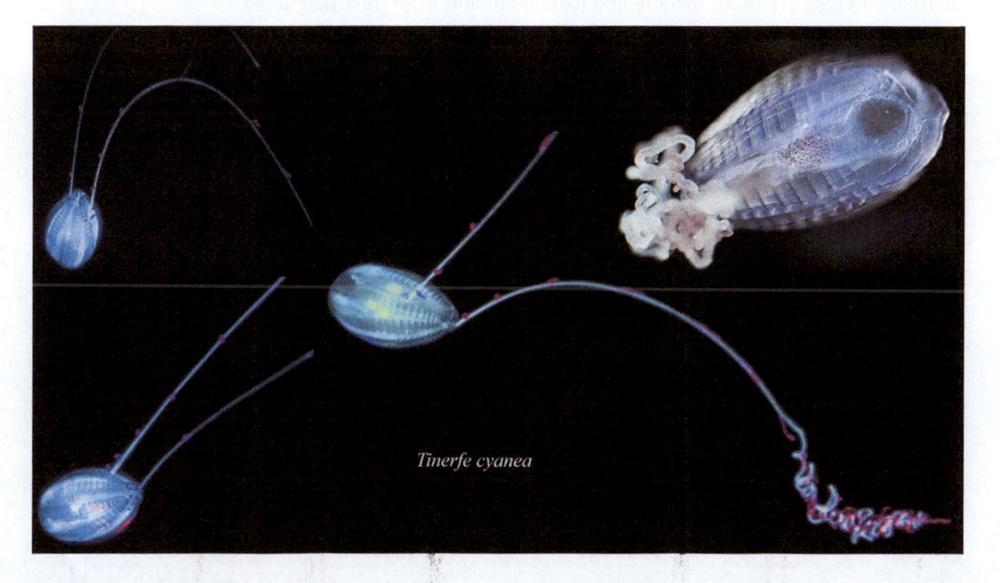

Fig. 24 *Tinerfe cyanea*, Florida, USA

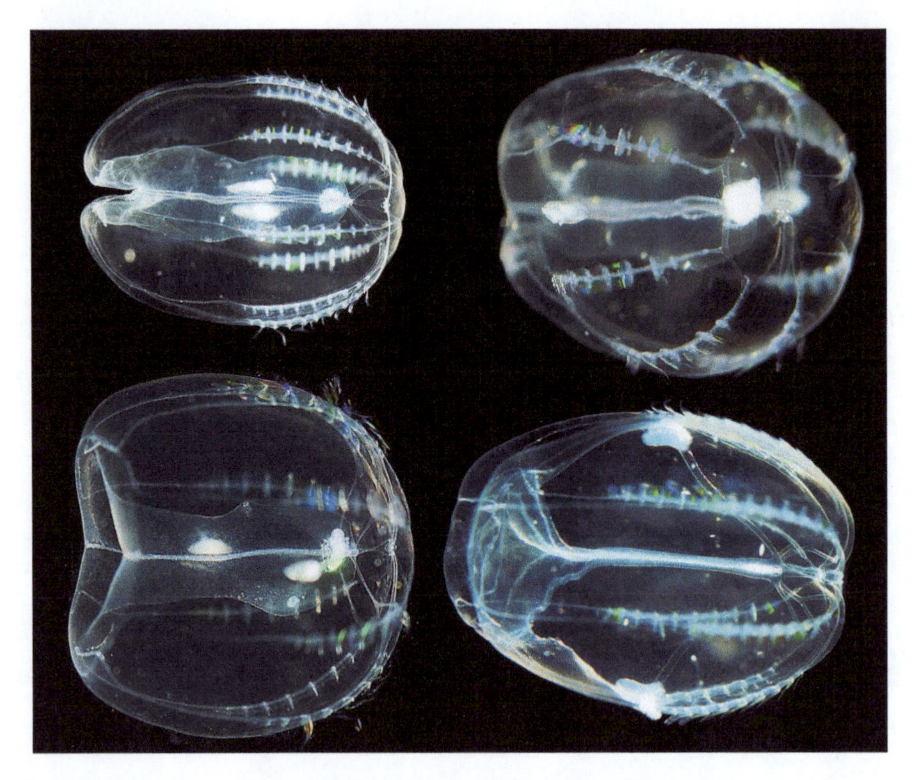

Fig. 25 These images are of an unidentified juvenile lobate from New Caledonia

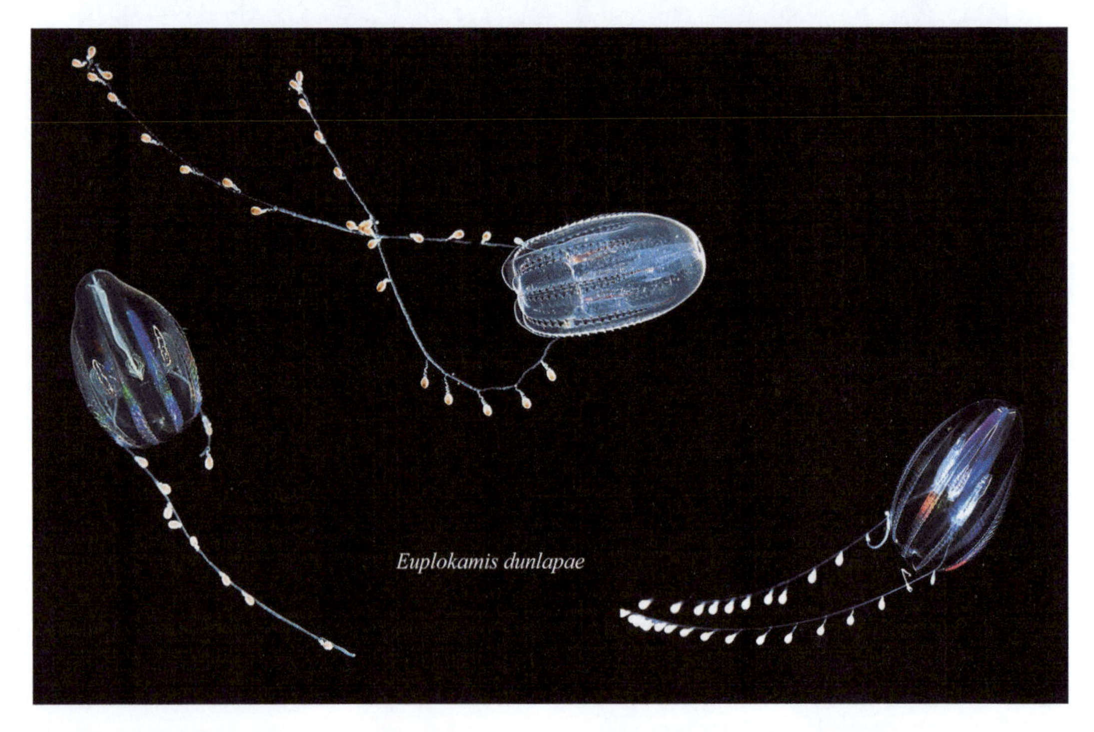

Fig. 26 *Euplokamis dunlapae* (Friday Harbor, Washington, USA, and Arctic Ocean) (Right Photo: Alexander Semenov©). *Euplokamis* also has unusual feeding because its rapidly coiling tentillae have unique striated muscles among ctenophores

Fig. 27 *Haeckelia beehleri*, Florida, USA

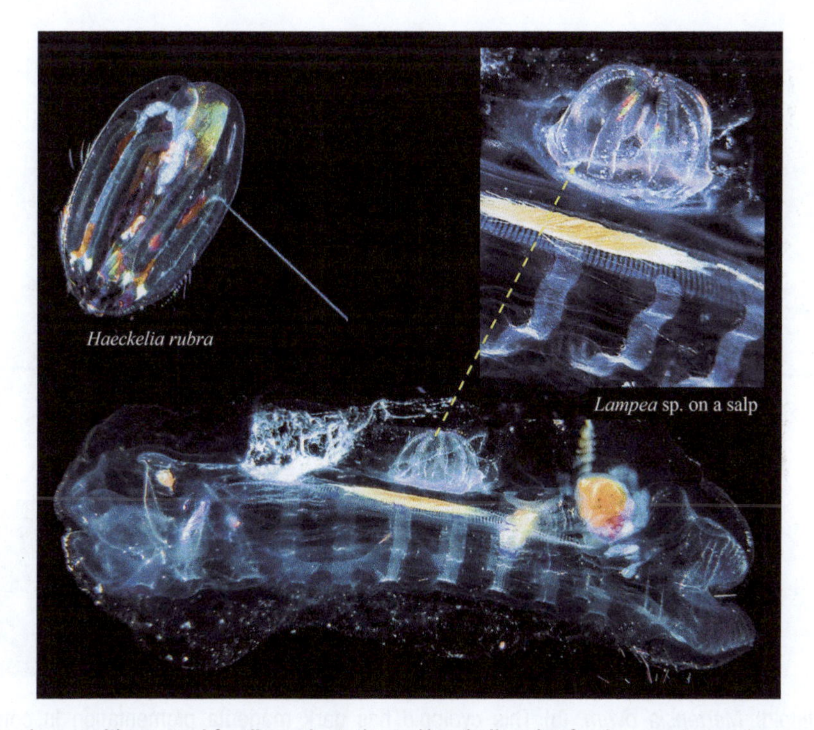

Fig. 28 Ctenophores with unusual feeding adaptations: *Haeckelia rubra* feeds on narcomedusae and is able to sequester their nematocysts for defense [102], and *Lampea* feeds on salps as a predator on small prey or as a parasite on larger individuals

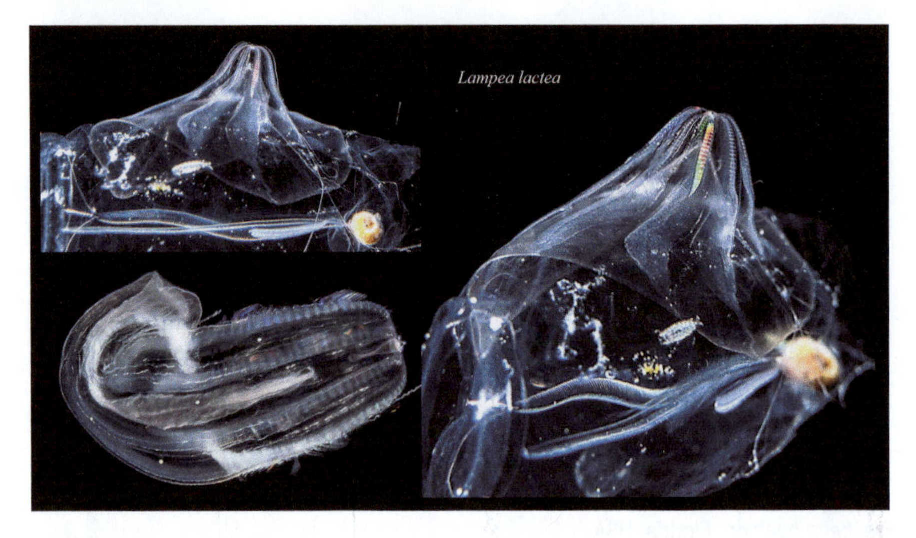

Fig. 29 *Lampea lactea*, Florida, USA

Fig. 30 Cydippid: *Mertensia ovum*. (**a**) This cydippid has dark magenta pigmentation in combs (**b**) and tentacles (**c**) [24]. (**a**) From the White Sea (Photo: Alexander Semenov©); (**b–d**) From Maine, USA

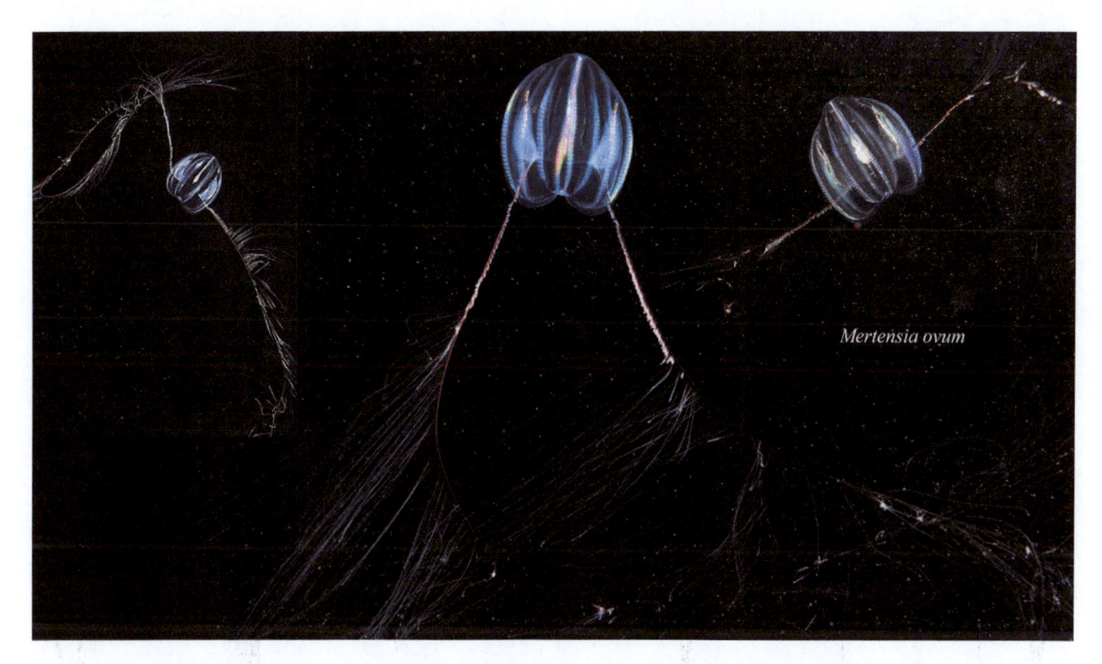

Fig. 31 Cydippid: *Mertensia ovum* from the White Sea. (Photo: Alexander Semenov©)

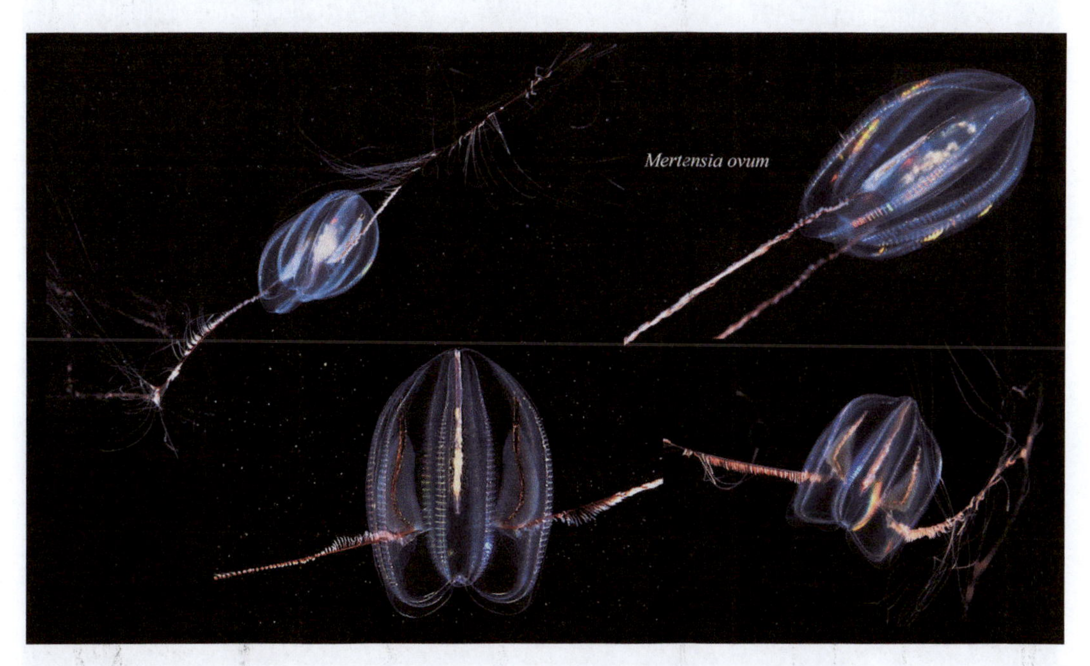

Fig. 32 Cydippid: *Mertensia ovum* from the White Sea. (Photo: Alexander Semenov©)

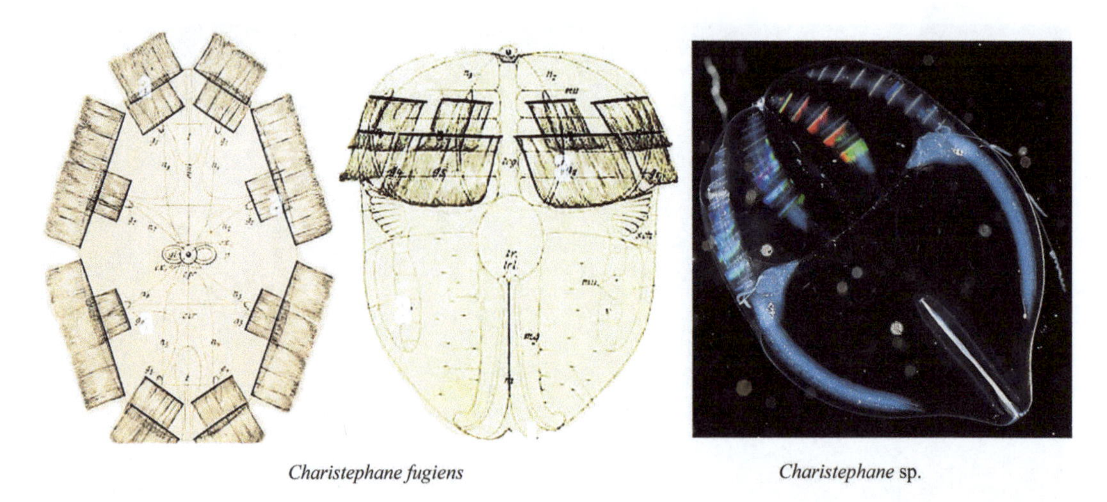

Fig. 33 *Charistephane fugiens* Chun, 1879. (from [74]. *Charistephane* sp. – https:/imgur.com/gallery/BL8gS/comment/297275655). The right image: Photo: Steven Haddock© MBARI 2006. The specimen was collected from 300 meters near Davidson Seamount, 100 km off the coast of California

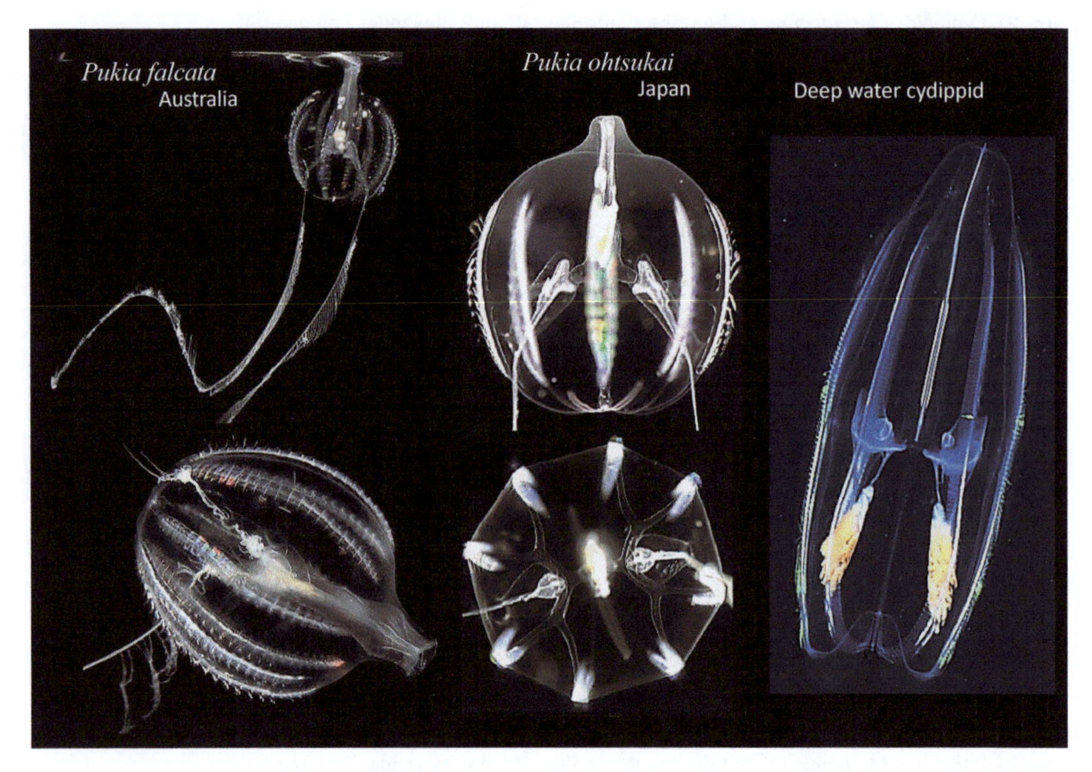

Fig. 34 *Pukia falcata*. (Photo: Denis Riek©; http:/www.roboastra.com/Cnidaria3/brac178.html; *Pukia ohtsukai* Lindsay, 2017; from [68], Okinawa, Japan. The right image is a deepwater cydippid at the Blake Escarpment, Atlantic Ocean, courtesy NOAA Ocean Exploration, Windows to the Deep 2019, https:/www.flickr.com/photos/oceanexplorergov/51816911735/in/photostream/)

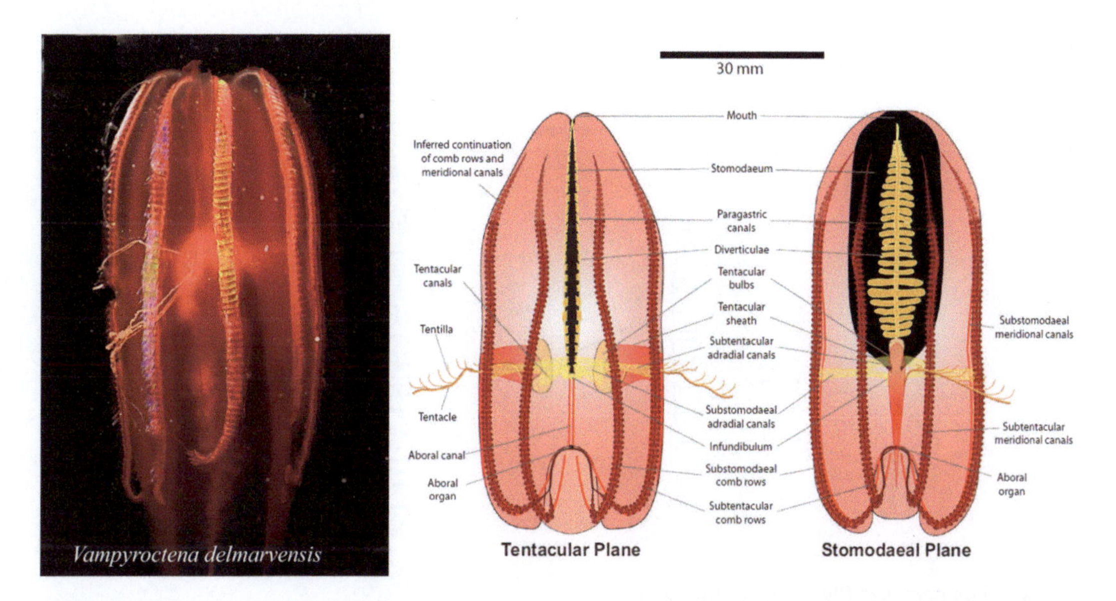

Fig. 35 Cydippida: *Vampyroctena delmarvensis*, from [109] (© Senckenberg Gesellschaft für Naturforschung 2020)

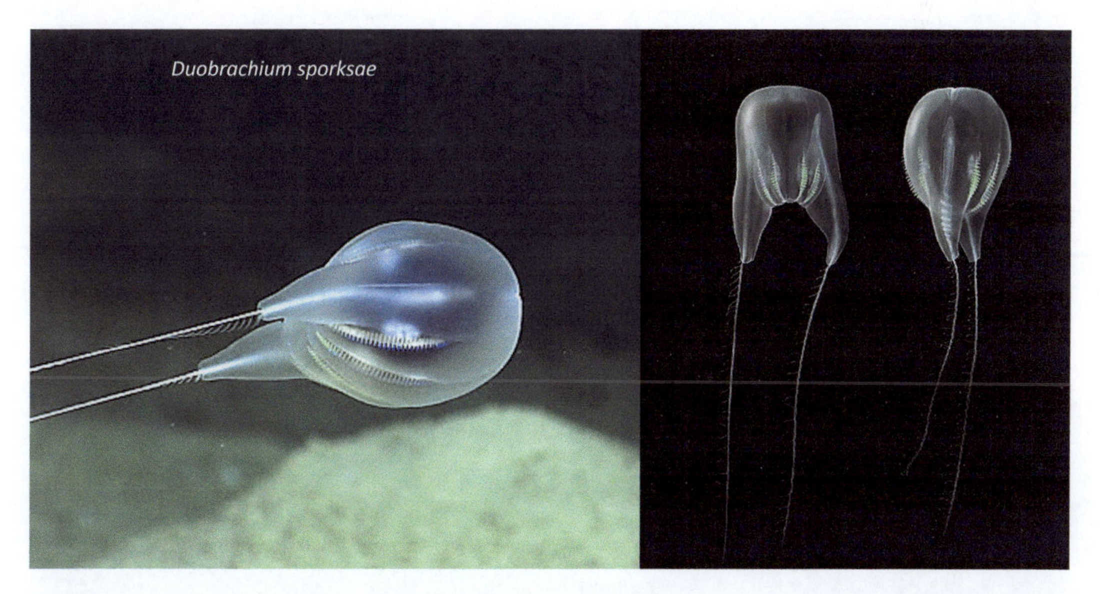

Fig. 36 The deepwater cydippid *Duobrachium sparksae* [56]. The species was found at a depth of approximately 3900 m, within meters of the seafloor in Guajataca Canyon, north-northwest of Puerto Rico (http:/ novataxa.blogspot.com/2020/11/duobrachium.html)

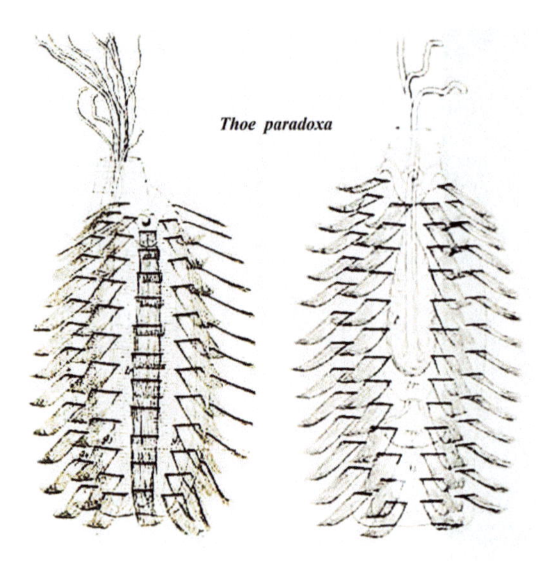

Fig. 37 *Thoe paradoxa* Chun, 1879. (from [74])

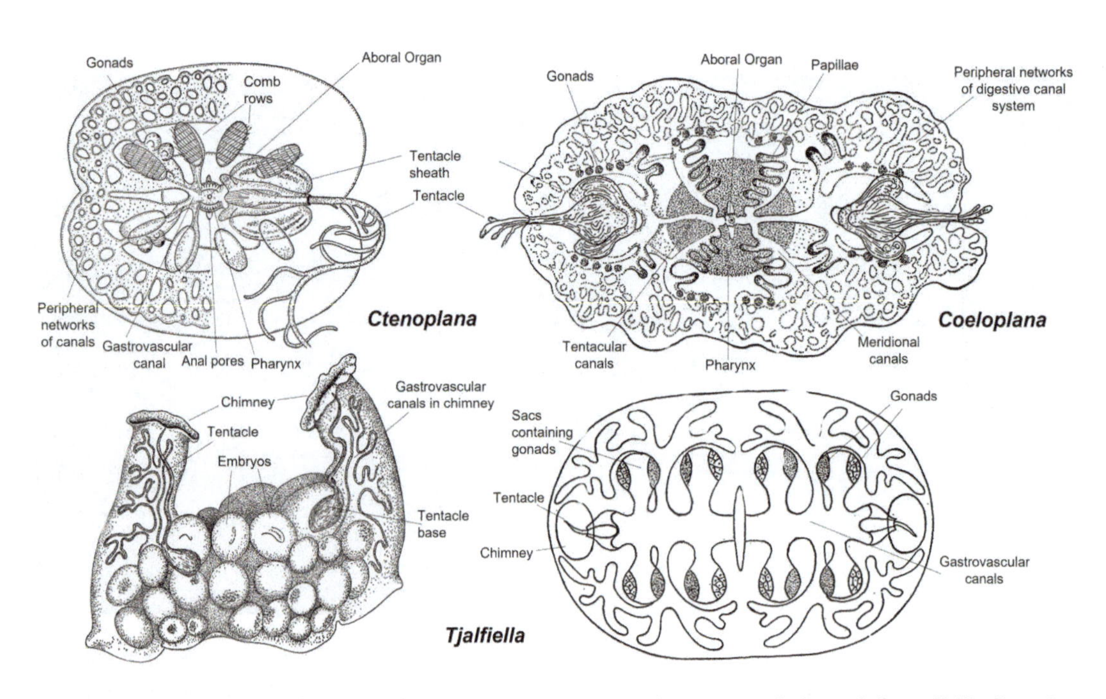

Fig. 38 General anatomy of Platyctenida with major organs and structures indicated (from [11]). See also images on the sea slug forum [210], and subsequent figures 26–47

a
b
Coeloplana astericola, on Echinaster luzonicus, Philippines
Coeloplana astericola, on Echinaster luzonicus, Okinawa

Fig. 39 *Coeloplana astericola*, on the sea star *Echinaster luzonicus* from the Philippines (**a**) and Okinawa (**b**)

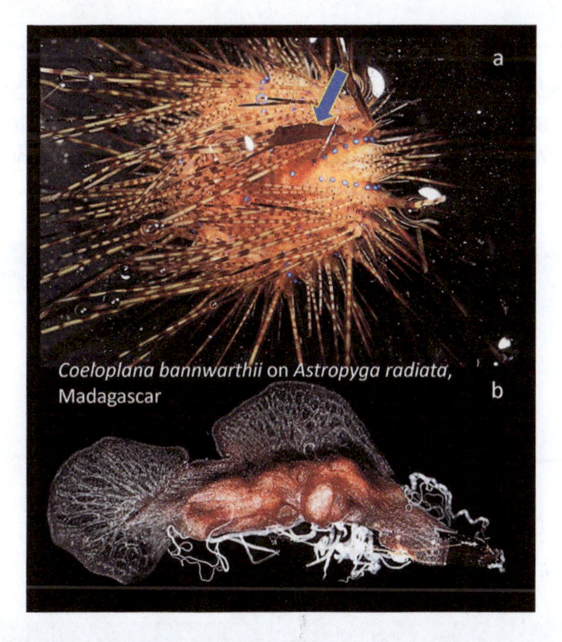

Fig. 40 *Coeloplana bannwarthii* on the sea urchin *Astropyga radiata*, Madagascar

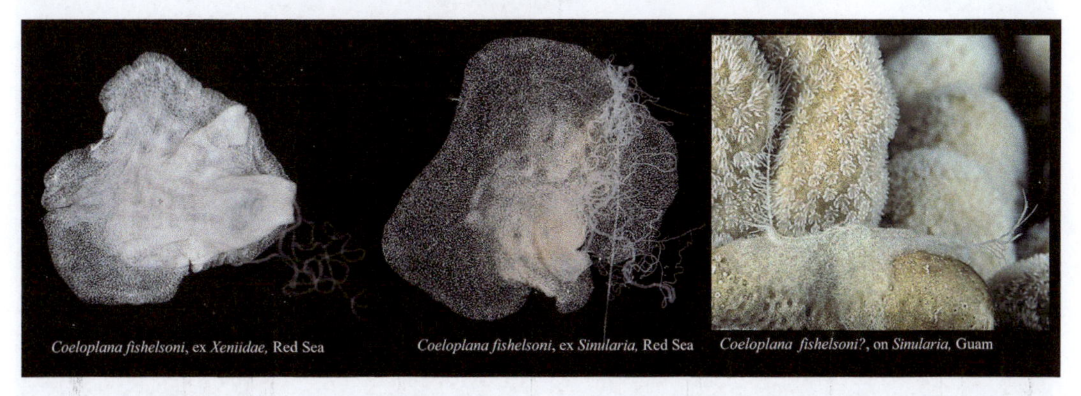

Coeloplana fishelsoni, ex Xeniidae, Red Sea
Coeloplana fishelsoni, ex Sinularia, Red Sea
Coeloplana fishelsoni?, on Sinularia, Guam

Fig. 41 *Coeloplana fishelsoni*, extracted from different species (soft corals *Xeniidae* and *Sinularia*, Red Sea), and in natural habitats, Guam (right image)

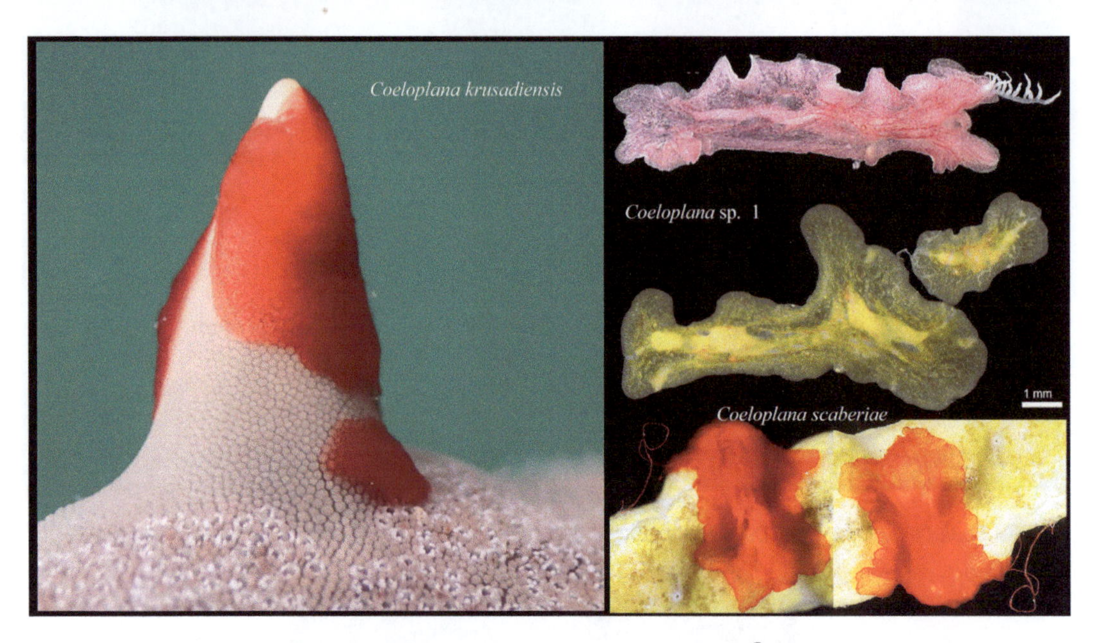

Fig. 42 *Coeloplana krusadiensis* from New Caledonia. (Photo: Claire Goiran© https:/guatemala.inaturalist.org/photos/249575709). Two color morphs of *Coeloplana* sp. 1 (undescribed species 1) from Oman [211]. *Coeloplana scaberiae* Matsumoto & Gowlett-Holmes, 1996 [127]. (Photo: Leon Altoff© – Barker Rocks, South Australia, https:/biocache.ala.org.au/occurrences/9428c840-bf78-4850-aa1c-4b6340c998bc)

Fig. 43 *Coeloplana loyai*, on the on the mushroom coral *Ctenactis echinata*, Red Sea. Bottom images are the same species collected from *Trachyphyllia geoffroyi* in New Caledonia

Fig. 44 *Coeloplana* cf. *loyai* from New Caledonia

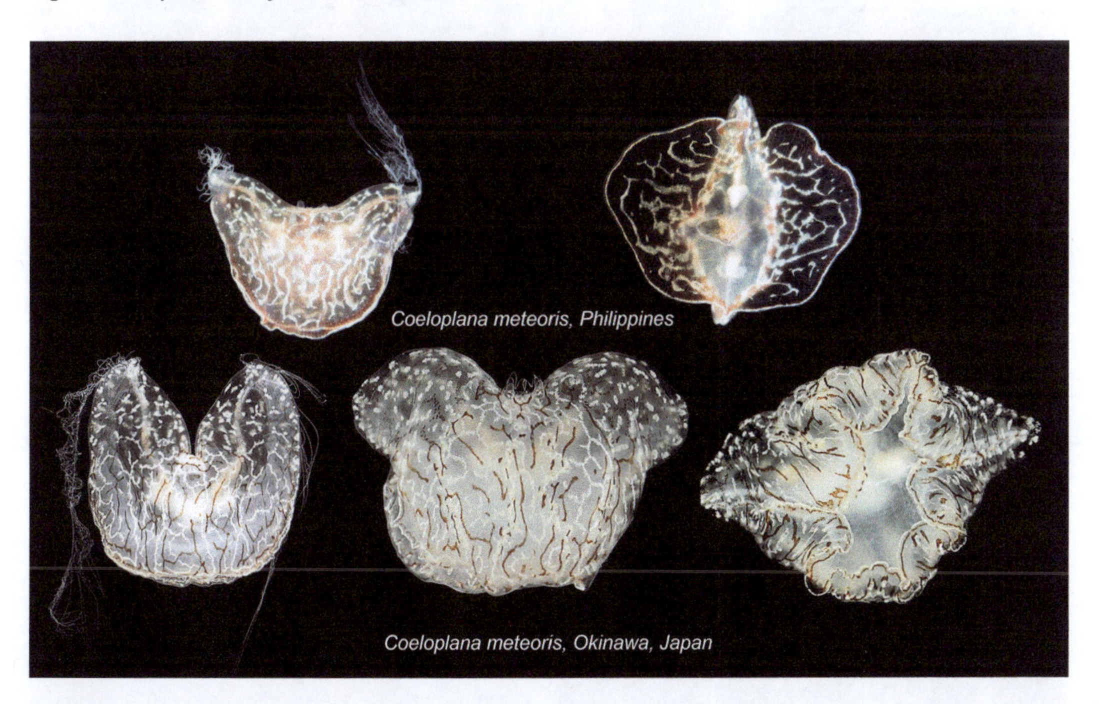

Fig. 45 *Coeloplana meteoris* from different locations

Fig. 46 The sea slug *Philinopsis ctenophoraphaga* feeding on the benthic ctenophore *Coenoplana meteoris*. (Images from [212]; http:/www.seaslugforum.net/find/15543) photos and description of event by Brian Francisco

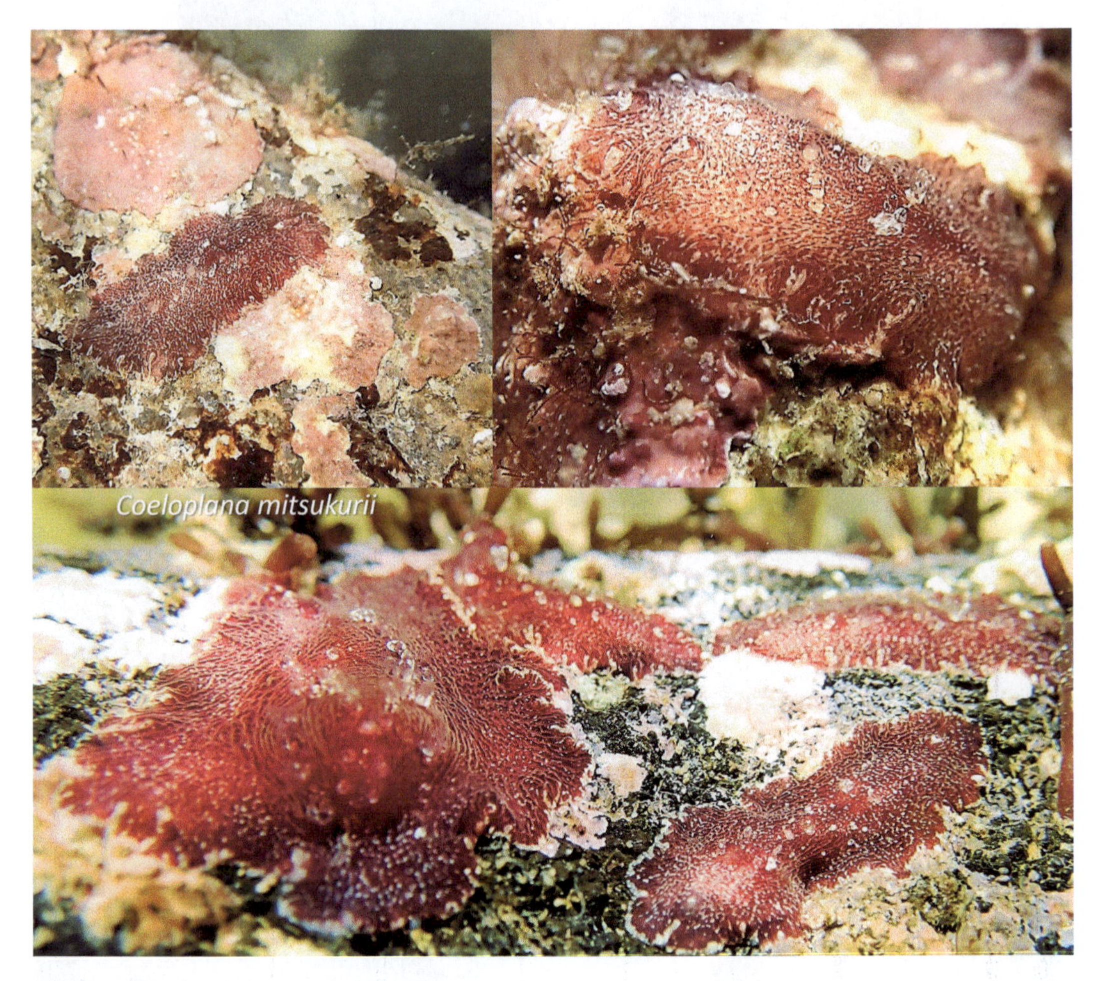

Fig. 47 *Coenoplana mitsukurii*, Oki Islands, Japan; photo: Masa-aki Yoshida. Photo by Masa-aki Yoshida, used with permission

Fig. 48 *Coeloplana willeyi*? extracted from the green alga *Halimeda*, Hawaii (left); *Coeloplana* sp. 2, extracted from the brown alga *Turbinaria*, Hawaii (right). Cory Pittman photos with permission

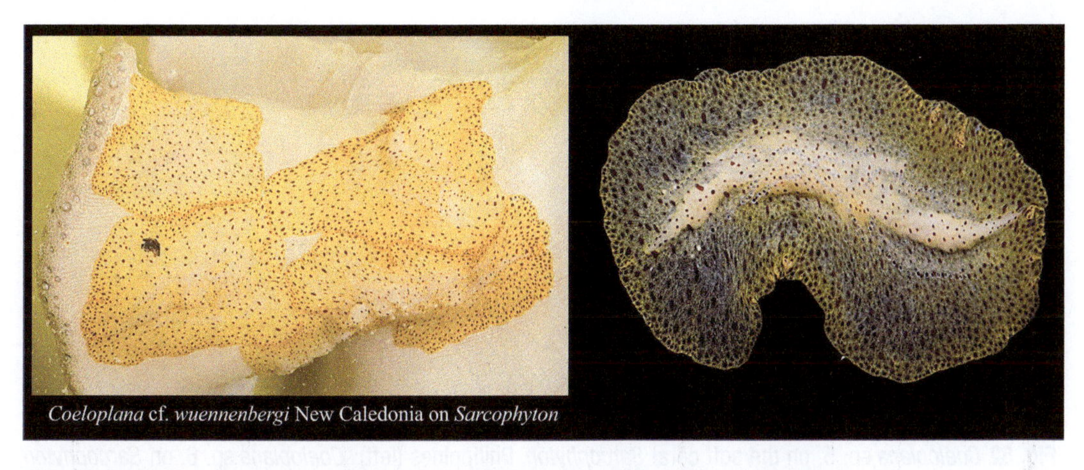

Fig. 49 *Coeloplana* cf. *wuennenbergi* on the soft coral *Sarcophyton*, New Caledonia

Fig. 50 (**a**) *Coeloplana* sp. 3, on the horned sea star *Protoreaster nodosus*, Philippines; (**b**) *Coeloplana yulianicorum*, on the soft coral *Sarcophyton*, Red Sea

Fig. 51 *Coeloplana* sp. 4, on the soft coral *Sarcophyton*, Philippines

Fig. 52 *Coeloplana* sp. 5, on the soft coral *Sarcophyton*, Philippines (left); *Coeloplana* sp. 6, on *Sarcophyton*, Philippines (right)

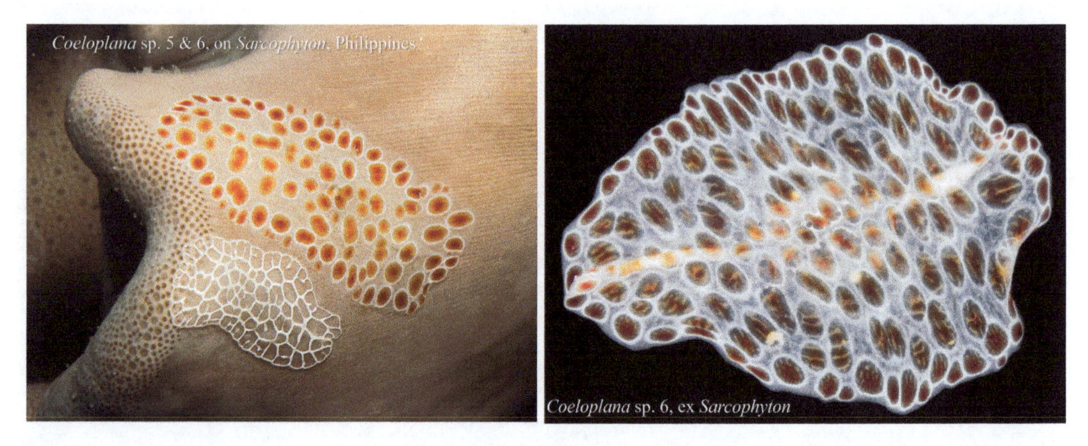

Fig. 53 *Coeloplana* sp. 5 and 6, on the soft coral *Sarcophyton*, Philippines (left); *Coeloplana* sp. 6, extracted from *Sarcophyton*

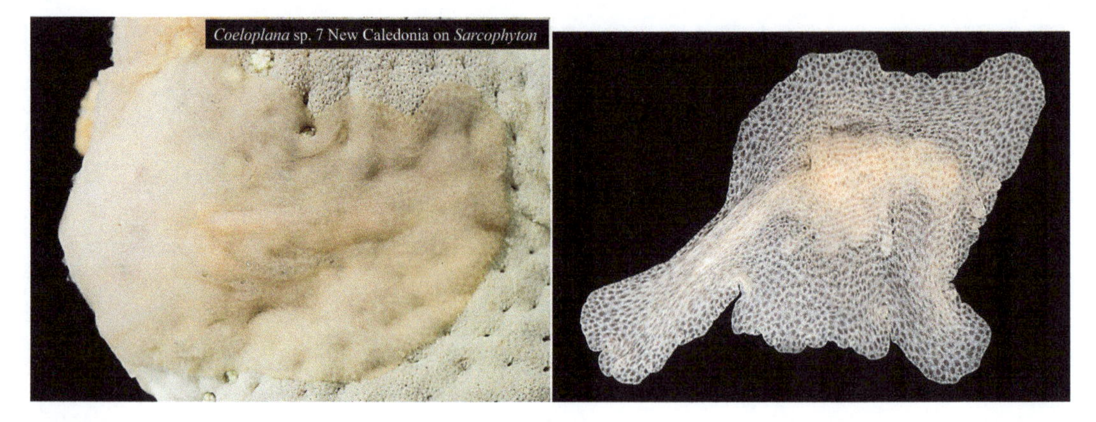

Fig. 54 *Coeloplana* sp. 7, New Caledonia, on the soft coral *Sarcophyton* (left), and the same species extracted (right)

Fig. 55 *Coeloplana* sp. 8, on the soft coral *Lobophytum*, Mascarene Islands

Fig. 56 *Coeloplana* sp. 9, extracted from the green alga *Halimeda*, Hawaii. *Coeloplana* sp. 10/9, extracted from the seaweed *Caulerpa*, Hawaii; Cory Pittman photos

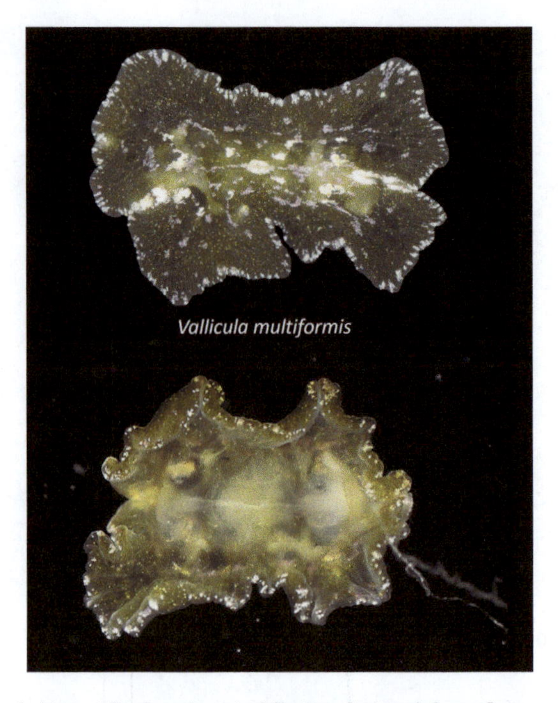

Fig. 57 *Vallicula multiformis* (dorsal [top] and ventral [bottom] views) from Oman

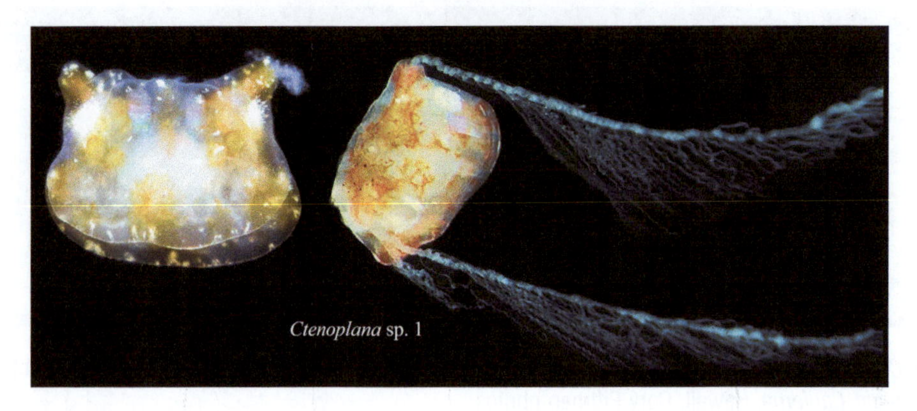

Fig. 58 *Ctenoplana* sp. 1 from New Caledonia

Fig. 59 *Ctenoplana* sp. 2 from New Caledonia

Fig. 60 *Ctenoplana* sp. 3 from New Caledonia

Fig. 61 The sessile platyctenid: *Lyrocteis* sp. (Image courtesy of the NOAA Office of Ocean Exploration and Research, 2016 Deepwater Exploration of the Marianas (https:/www.flickr.com/photos/oceanexplorergov/50218989191/in/photostream/))

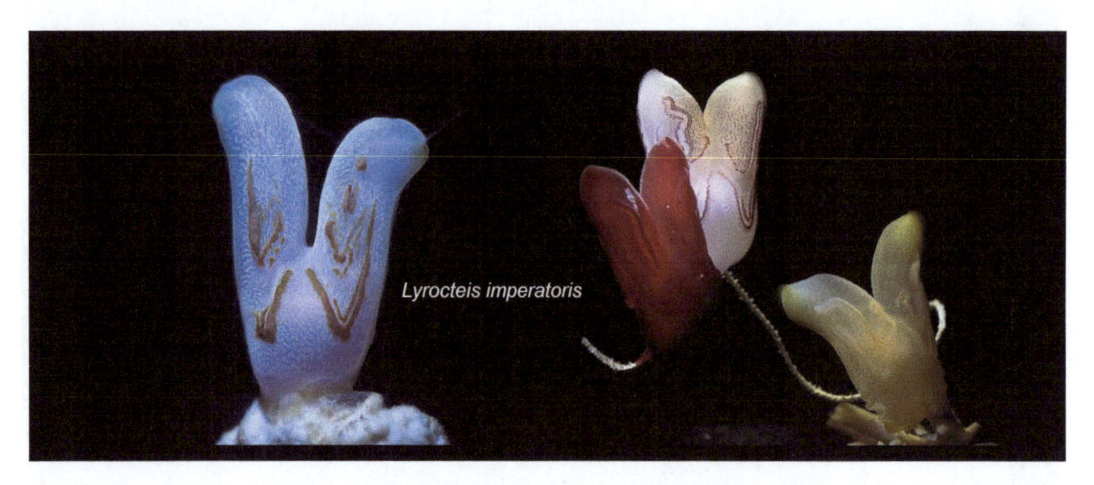

Fig. 62 The rare and beautiful sessile platyctenid: *Lyrocteis imperatoris* [141], Japan. (Image from https:/churaumi.okinawa/sp/en/fishbook/1459390728/). It attaches itself to rocks and gorgonian octocorals horny corals. This species varies in color, and there are many colorful individuals (© Okinawa Churaumi Aquarium)

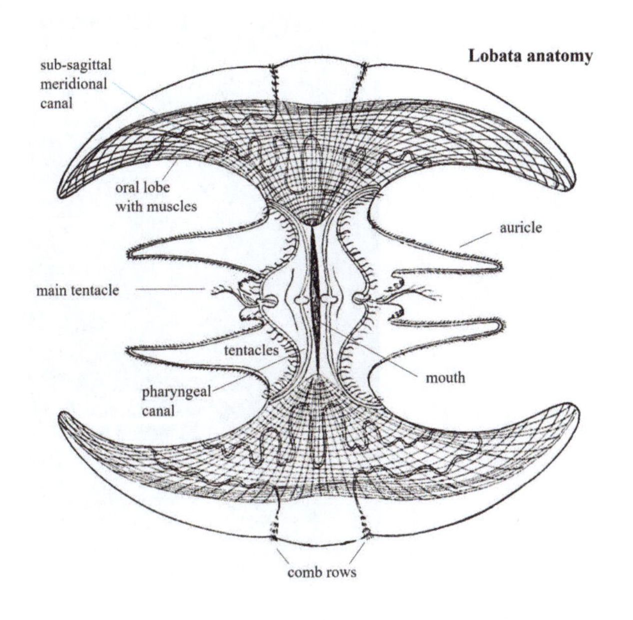

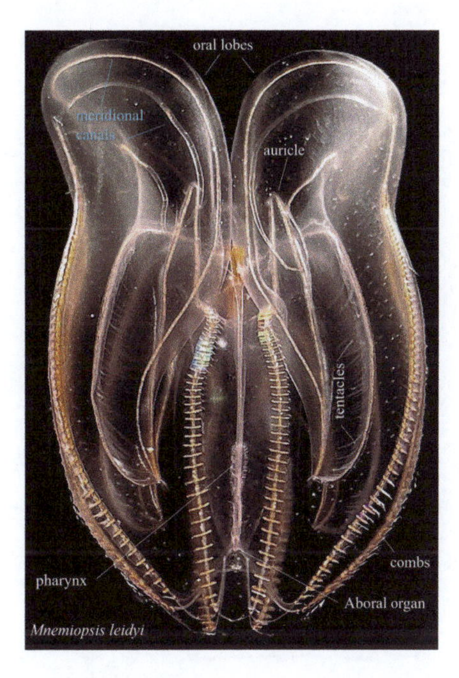

Fig. 63 General anatomy of Lobata with major organs and structures indicated (from [11]). The right image - *Mnemiopsis* from St. Augustine, Florida,. USA

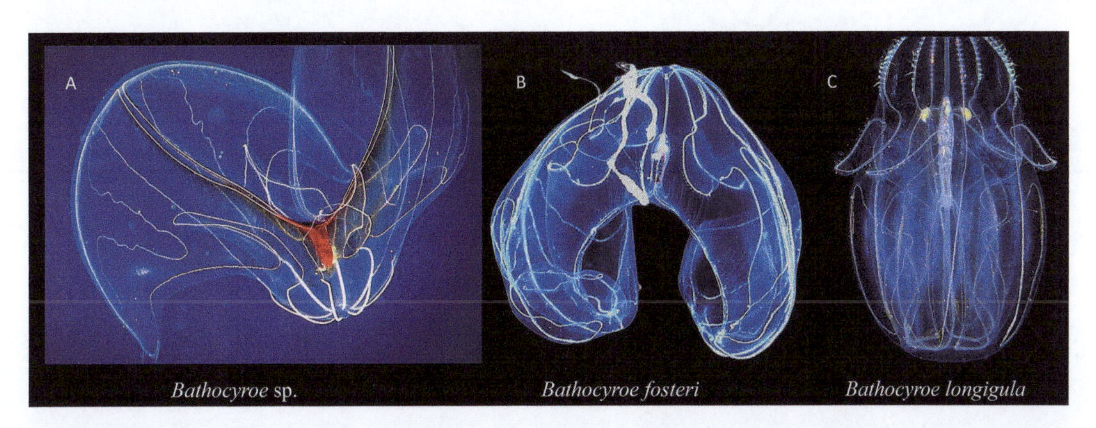

Fig. 64 (**a**) *Bathocyroe* sp. (https:/www.flickr.com/photos/oceanexplorergov/50091043518). (**b**) *Bathocyroe fosteri*. (Photo courtesy of Marsh Youngbluth; This image is in the public domain because it contains materials that originally came from the US National Oceanic and Atmospheric Administration, taken or made as part of an employee's official duties). (**c**) *Bathocyroe longigula*. (https:/www.facebook.com/blackwaterdive/posts/2854790224767581/)

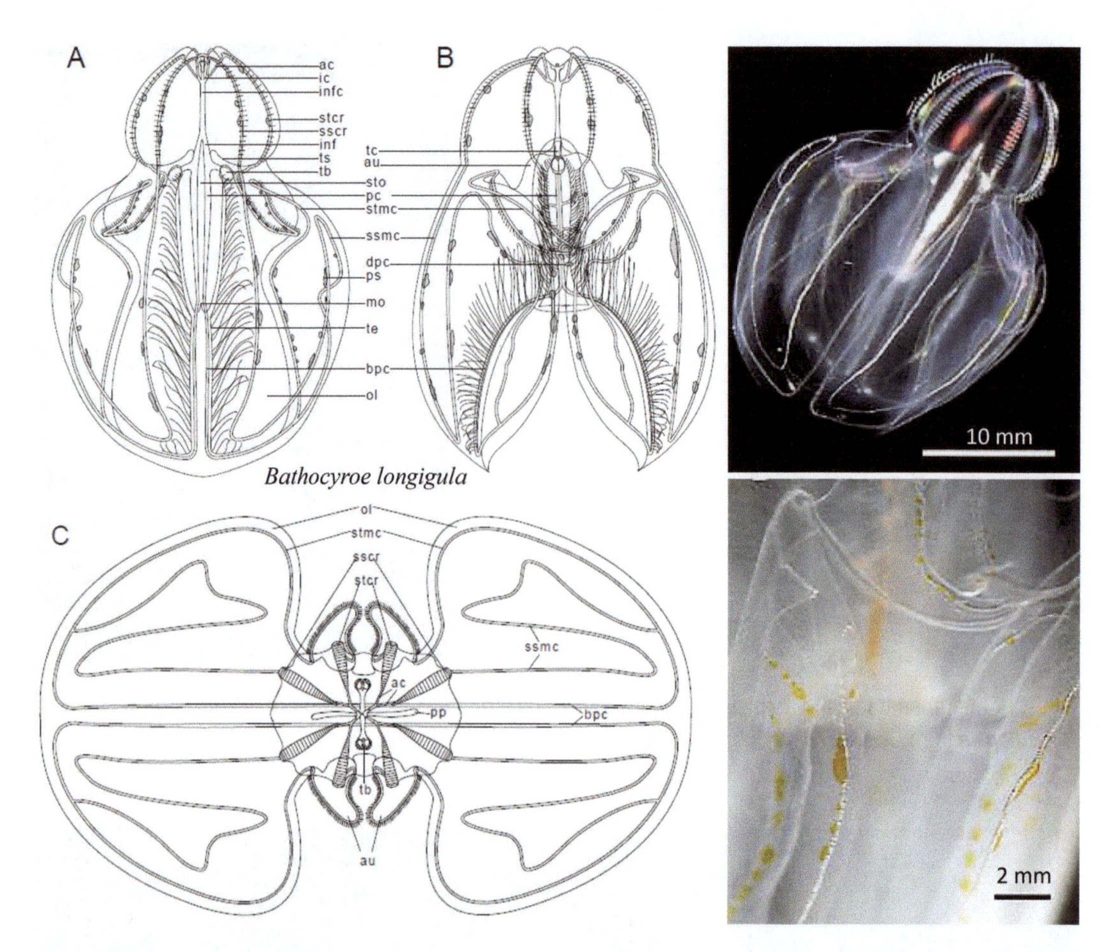

Fig. 65 *Bathocyroe longigula*, from [146], Japan. (**a**) The tentacular plane, (**b**) the stomodaeal plane, and (**c**) the aboral pole view. Abbreviations: ac, adradial canal; au, auricle; bpc, branch of paragastric canal; dpc, diverticulum of paragastric canal; ic, interradial canal; inf, infundibulum; infc, infundibular canal; mo, mouth opening; ol, oral lobe; pc, paragastric canal; ps, pigment spot; pp, pole plate; sscr, substomodaeal comb row; ssmc, substomodaeal meridional canal; stcr, subtentacular cmb row; stmc, subtentacular meridional canal; sto, stomodaeum; tb, tentacle bulb; tc, tentacular canal; te, tentacle; ts, tentacle sheath. Scale bar: 10 mm. (The right photos are the holotype image and the distribution of pigment spots (as in fig. 2 from [146]). Permission from the publisher)

Fig. 66 *Bolinopsis ashleyi*. The left image: Cairns Australia. (Photo: © djwitherall; (https:/guatemala.inaturalist.org/observations/183142120)). The central image: Sydney, Australia. (Photo: © Niki Hubbard, https:/guatemala.inaturalist.org/observations/142985877). The right image: Heron Island, Australia

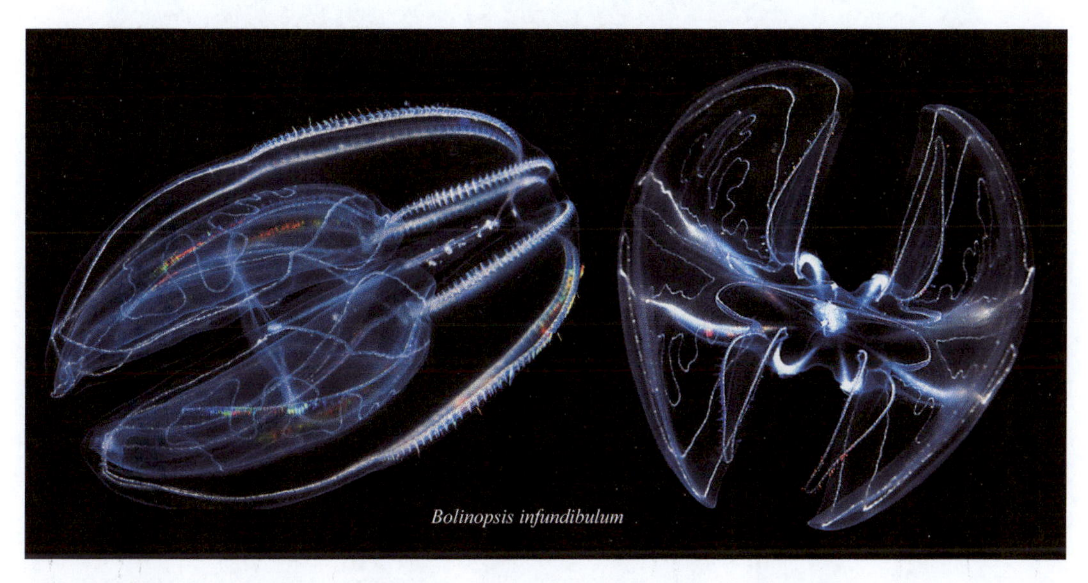

Fig. 67 Lobata: *Bolinopsis infundibulum* from White Sea. (Photo: Alexander Semenov©)

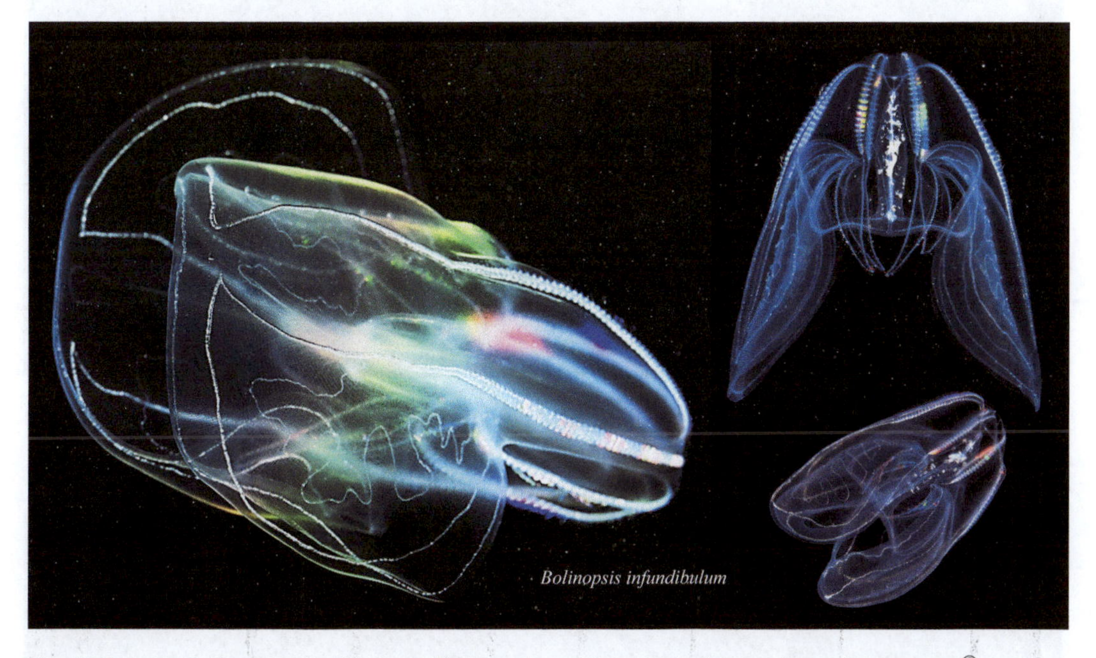

Fig. 68 Lobata: *Bolinopsis infundibulum*; different views (White Sea). (Photo: Alexander Semenov©)

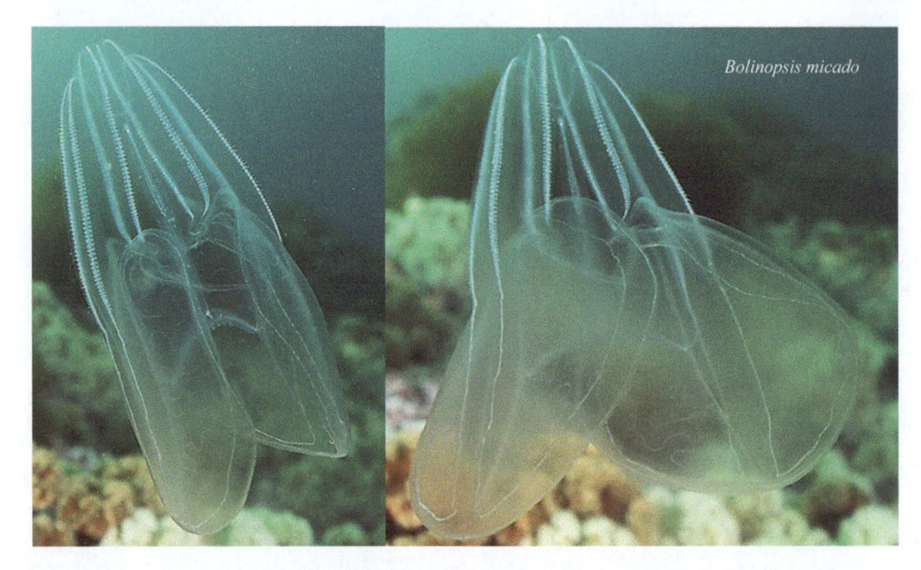

Fig. 69 Lobata: *Bolinopsis mikado* from Vityaz bay, Japan Sea. (Photo: Alexander Semenov©)

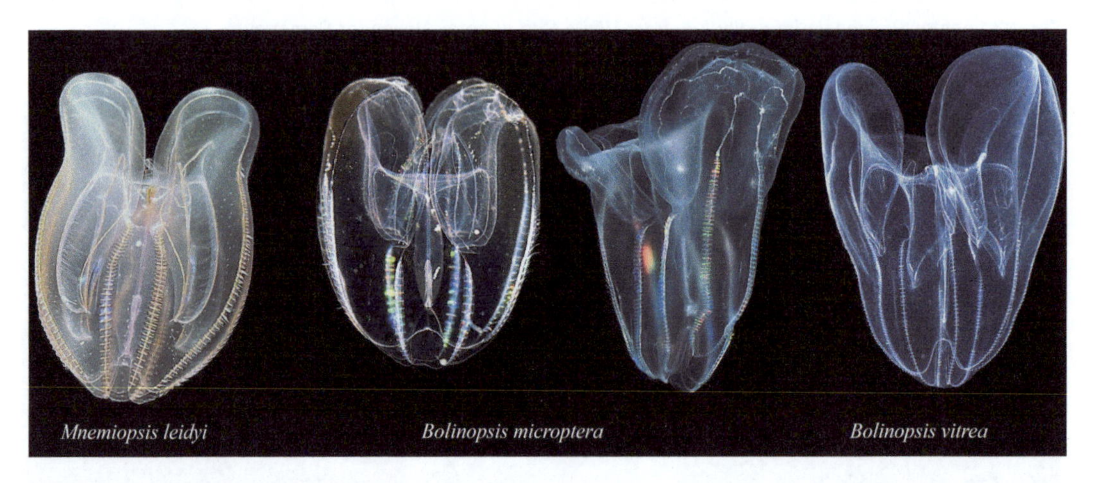

Fig. 70 Lobata: *Mnemiopsis leidyi* (Central Florida), *Bolinopsis microptera* (Friday Harbor, WA, USA), and *Bolinopsis vitrea* (South Florida)

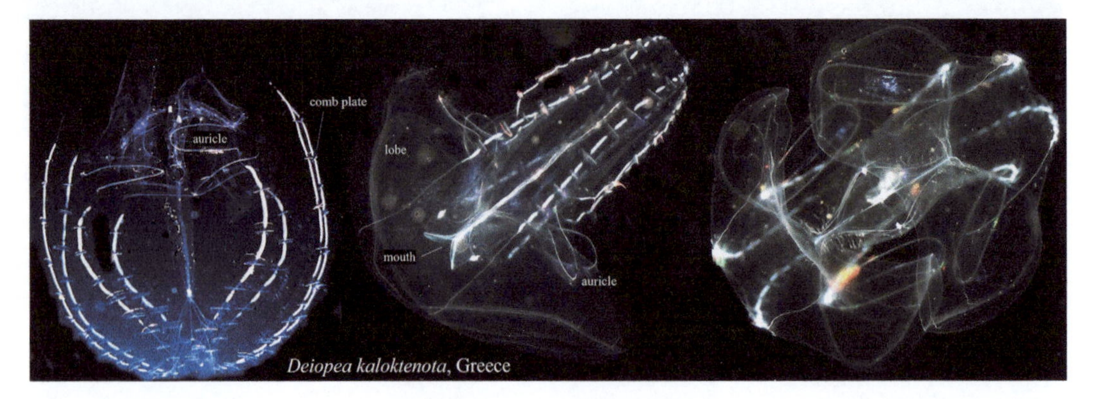

Fig. 71 *Deiopea kaloktenota*, Lobata, from Saronida, Attica, Greece. (Photos: iNaturalist Canada by Stergios Vasilis (https:/inaturalist.ca/taxa/949340-Deiopea-kaloktenota)). The left photo was taken by Stergios Vasilis. The original source is https:/www.inaturalist.org/observations/69830733. The other two shots are from https:/ www.inaturalist.org/taxa/949340-Deiopea-kaloktenota

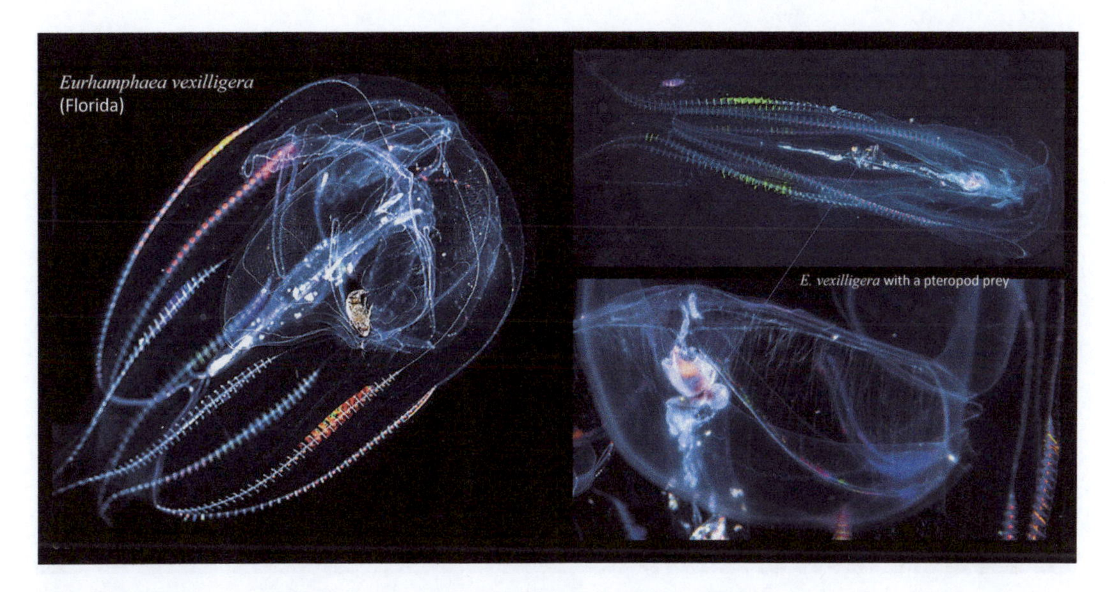

Fig. 72 *Eurhamphaea vexilligera* with crustacean and pteropod prey, Florida

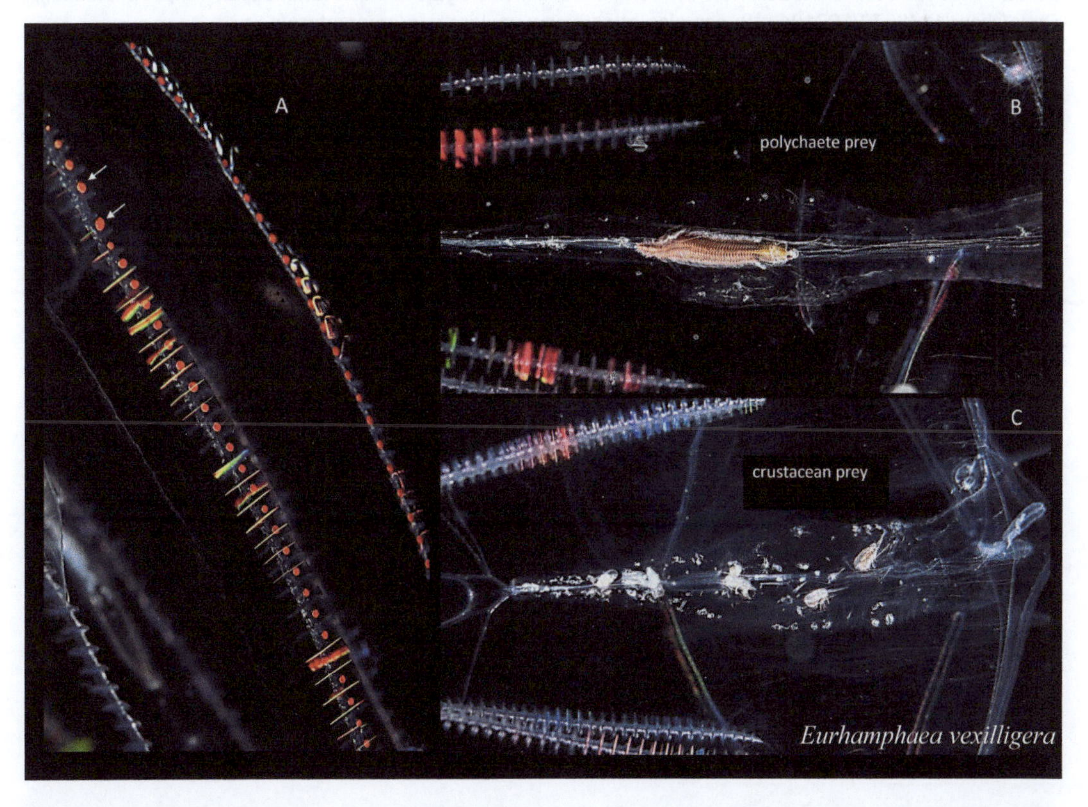

Fig. 73 *Eurhamphaea vexilligera* – ink release sites (arrows – (**a**), see also [153]) and with polychaete prey (**b**), crustacean prey (**c**)

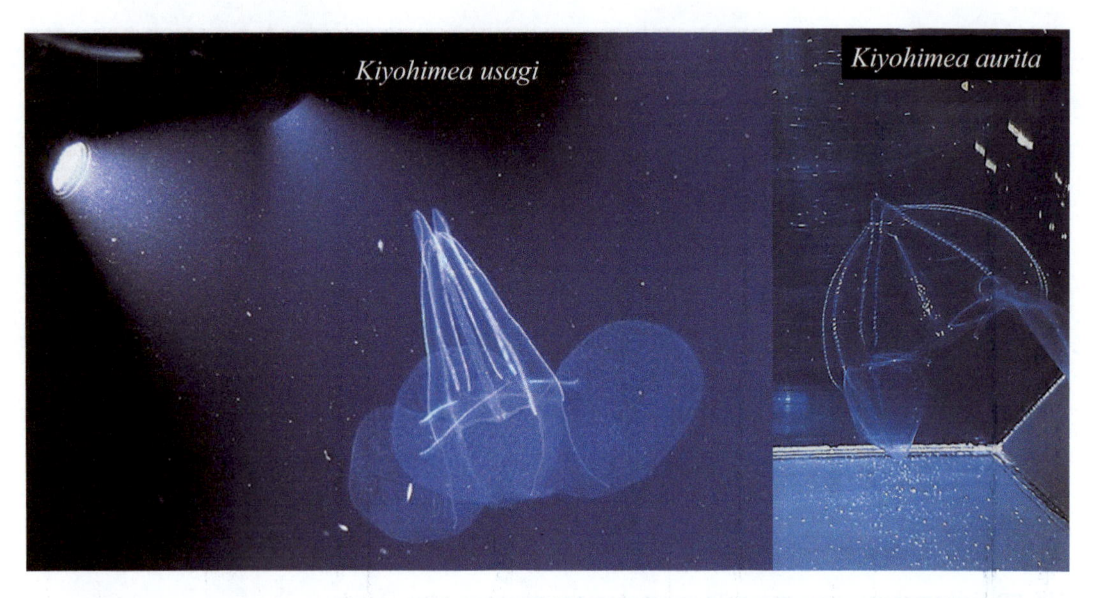

Fig. 74 *Kiyohimea usagi*. (Photo: Henk Jan T. Howing/GEOMAR, Cape Verde https:/twitter.com/GEOMAR_en/status/1140616744110874625/photo/2); *Kiyohimea aurita* Nagasaki Aquarium, Japan (https:/photozou.jp/photo/show/1143978/127124448; see also https:/www.umikirara.jp/en/creature/)

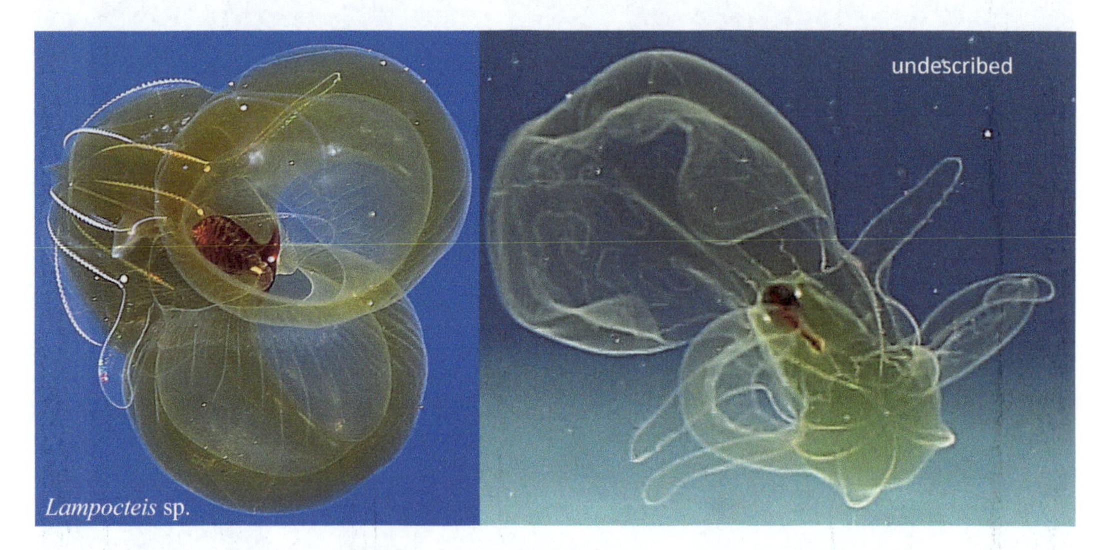

Fig. 75 *Lampocteis* sp. (Image courtesy of NOAA Ocean Exploration, Voyage to the Ridge 2022; https:/www.flickr.com/photos/oceanexplorergov/52648188412/in/photostream/) and undescribed deepwater lobate. (Image courtesy of the NOAA Office of Ocean Exploration and Research, Western Gulf of Mexico 2018; https:/www.flickr.com/photos/oceanexplorergov/51236540164/in/photostream/)

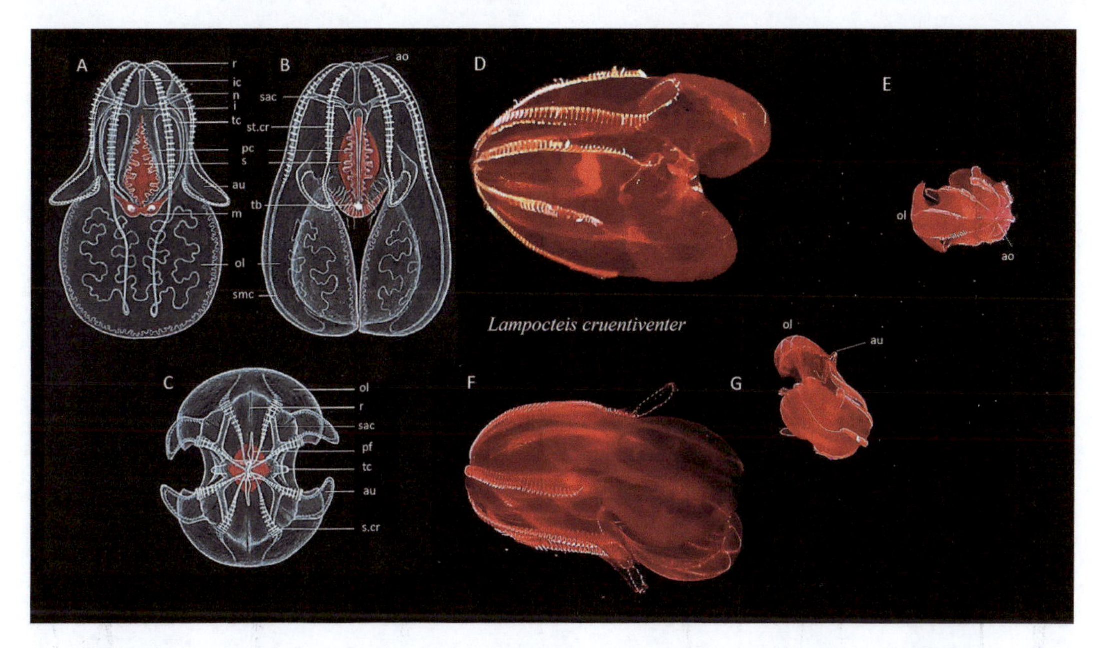

Fig. 76 Schematic anatomy of the bloodybelly comb jelly *Lampocteis cruentiventer*, (**a–c**) from [157]. This deepwater lobate (up to 15 cm) can be found at ~250–1500 m. The red pigmentation makes the ctenophore practically invisible to predators. It is also suggested such red color mask bioluminescence. The rest are the general views of living *Lampocteis*, from San Francisco area. Abbreviations: ao, aboral organ; au, auricles; i, infundibulum; ic, infundibular canal; m, mouth; n, notch; pc, paragastric canal; s, stomodaeum; sac, substomodaeal adradial canal; s.cr, substomodaeal comb row; smc, substomodaeal meridional canal; st.cr, subtentacular comb row; tb, tentacle bulb; tc, tentacular canal. (Credits: (**a**) [157] © Rosenstiel School of Marine, Atmospheric & Earth Science (see also images from Vincenzoxvivolo, CC BY-SA 4.0 https:/creativecommons.org/licenses/by-sa/4.0, via Wikimedia Commons); from https:/www.mbari.org/animal/bloody-belly-comb-jelly/. Additional images and videos of *L. cruentiventer* can be found at https:/www.mbari.org/animal/bloody-belly-comb-jelly/, see also [60])

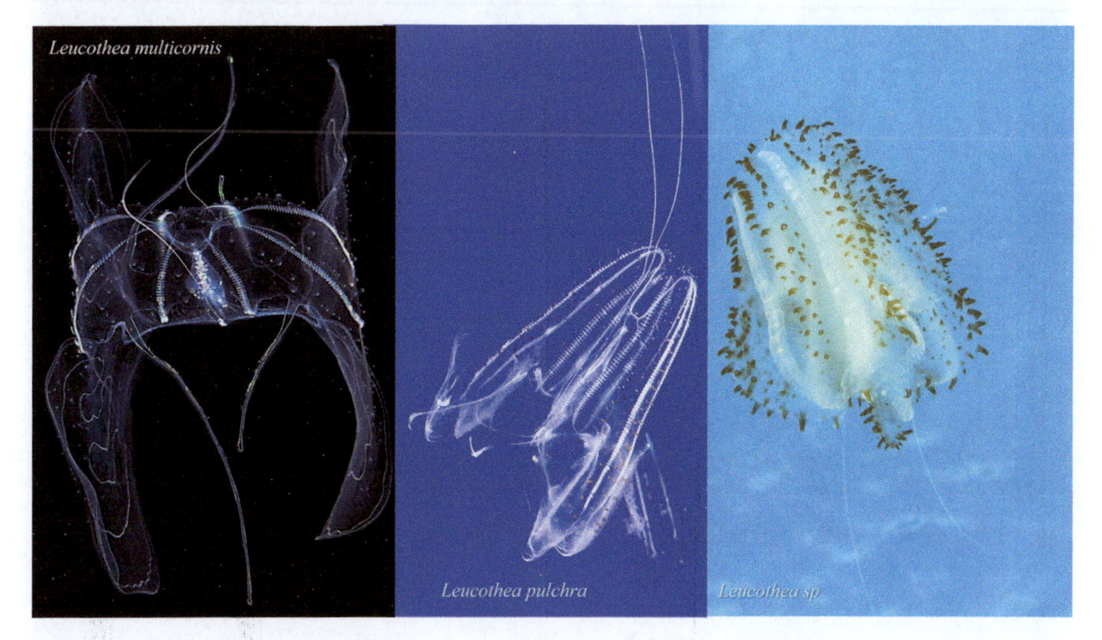

Fig. 77 *Leucothea multicornis*, Mediterranean Sea (Photo: Alexander Semenov©) and *Leucothea pulchra* (California), and *Leucothea* sp. The right image: Photo: Emily Hale©

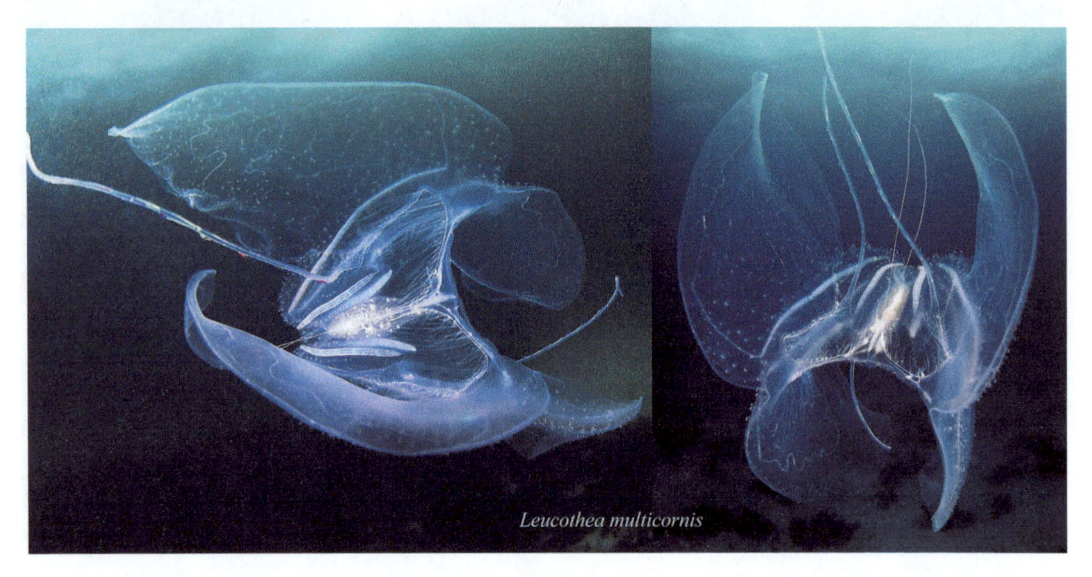

Fig. 78 *Leucothea multicornis*, Ponza Island, Italy, Tyrrhenian Sea. (Photo: Alexander Semenov©; https:/www.flickr.com/photos/a_semenov/15342819150/in/photostream/)

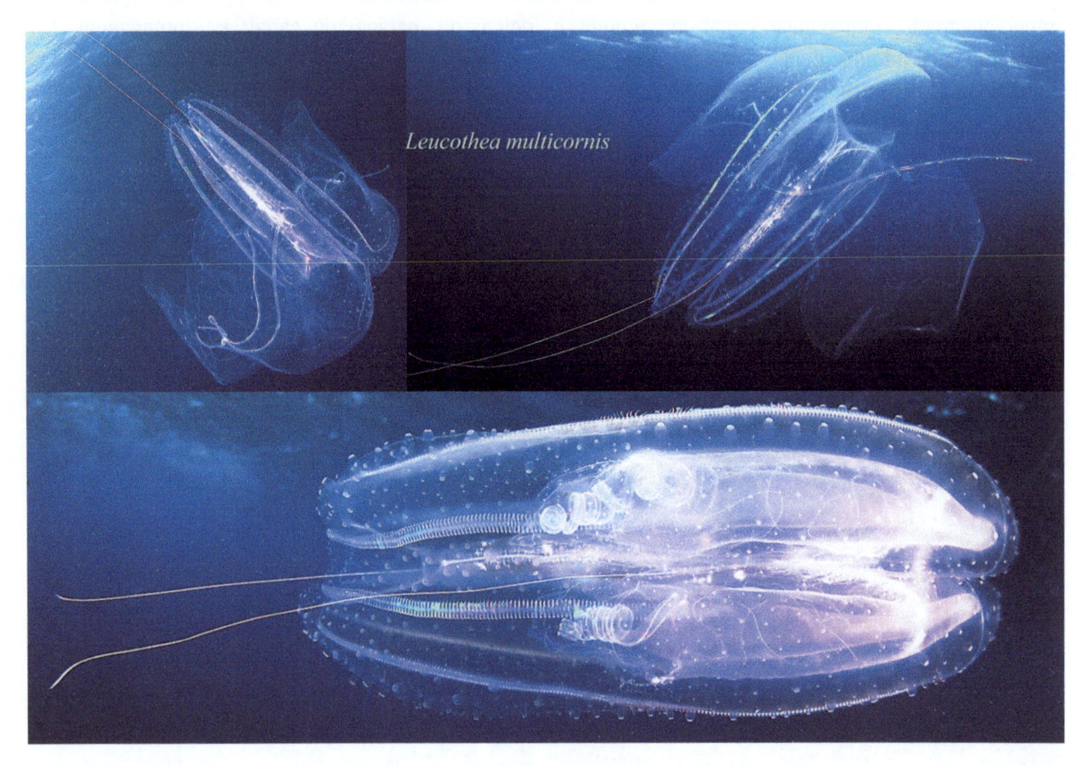

Fig. 79 *Leucothea multicornis*, Red Sea. (Photo: Alexander Semenov©)

Fig. 80 *Leucothea multicornis*. (Photo: Alexander Semenov©)

Fig. 81 *Leucothea ochracea*. (Photo: Susan Mears©)

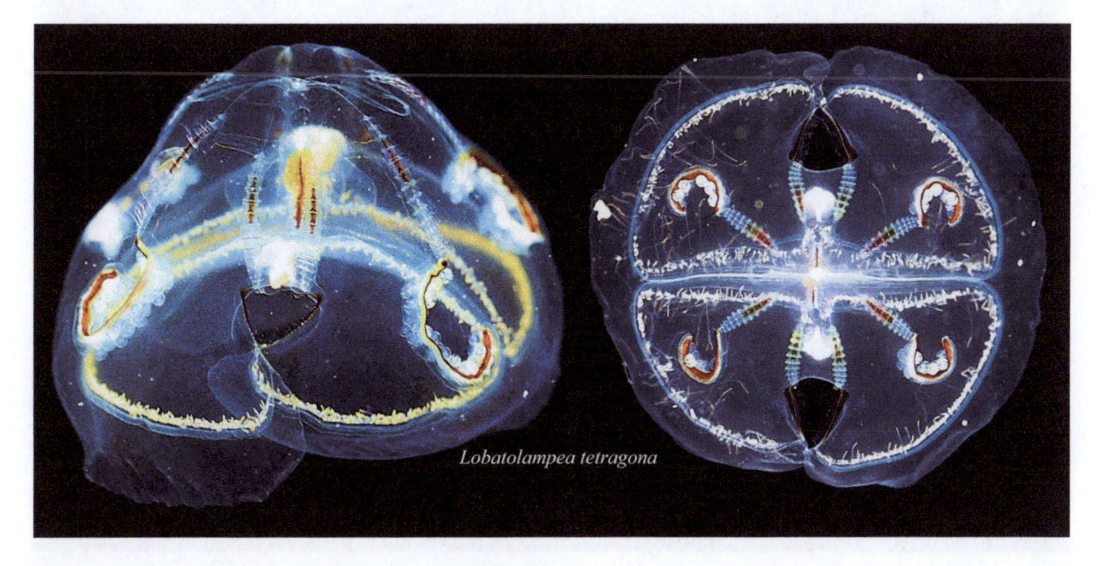

Fig. 82 Benthic *Lobatolampea tetragona*, Philippines

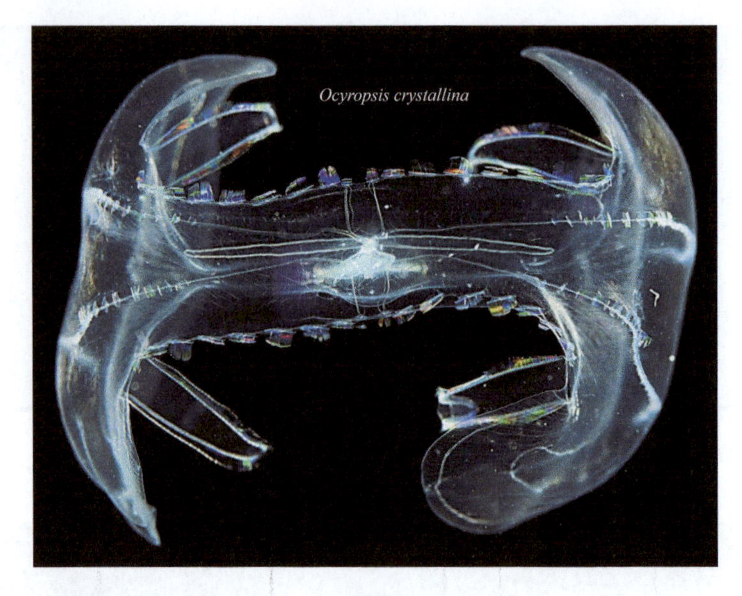

Fig. 83 *Ocyropsis crystallina*, New Caledonia

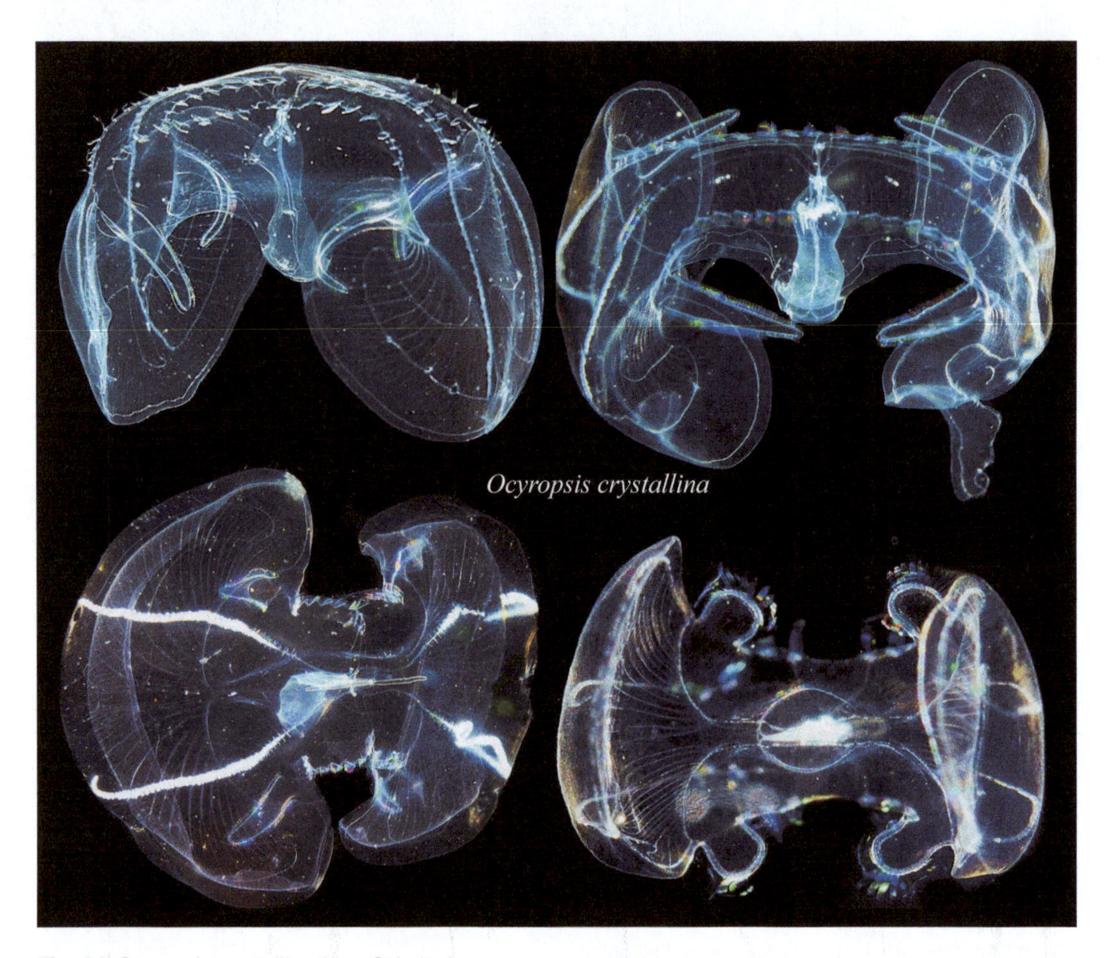

Fig. 84 *Ocyropsis crystallina*, New Caledonia

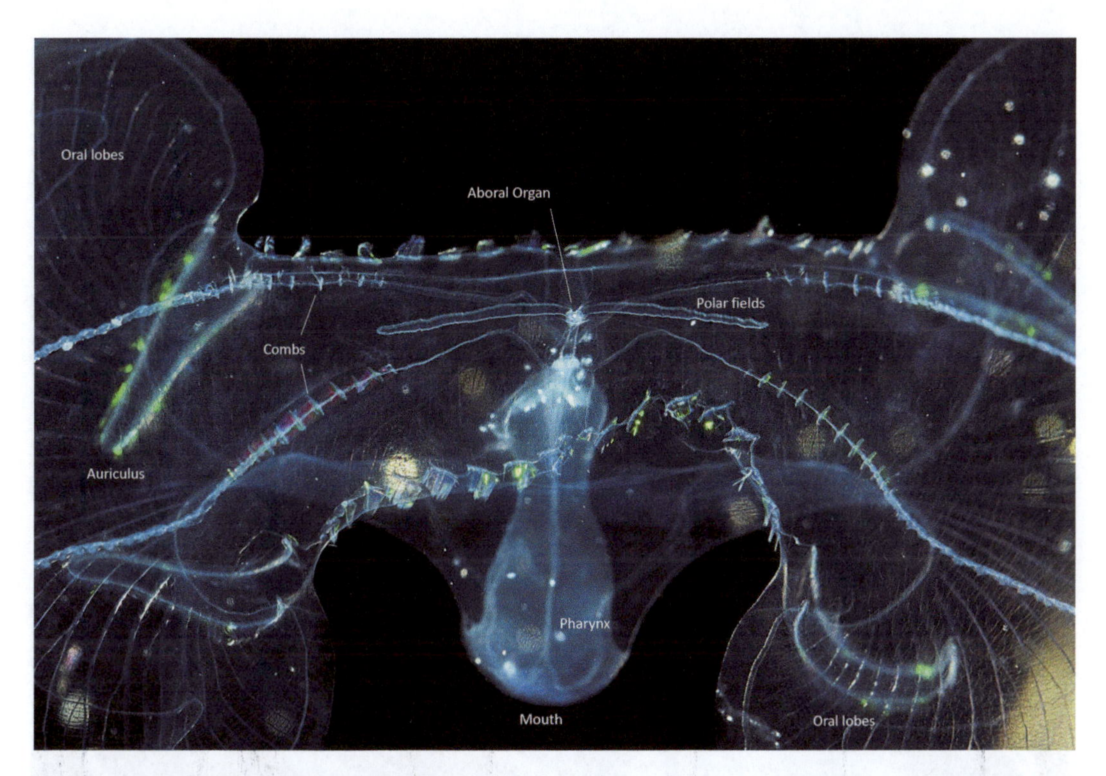

Fig. 85 *Ocyropsis crystallina*, New Caledonia

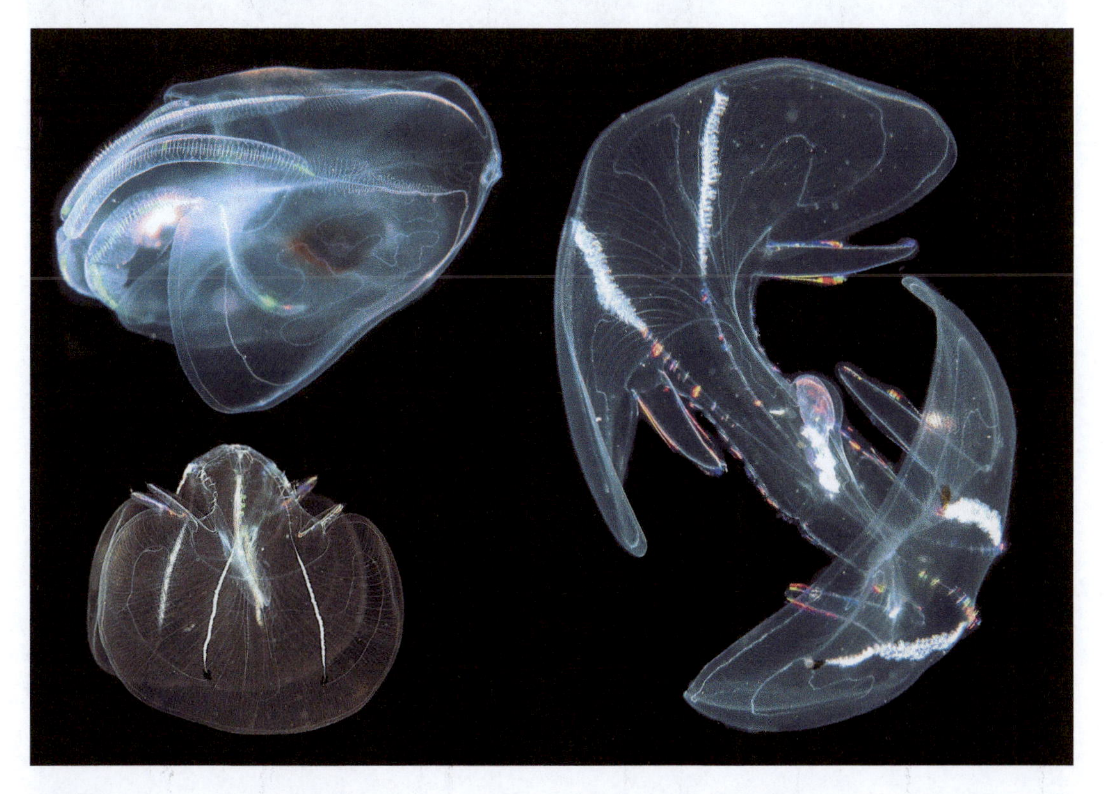

Fig. 86 The upper right photo is *Ocyropsis maculata maculata*, Florida. The images on the right and lower left are *Ocyropsis cristalina guttata*

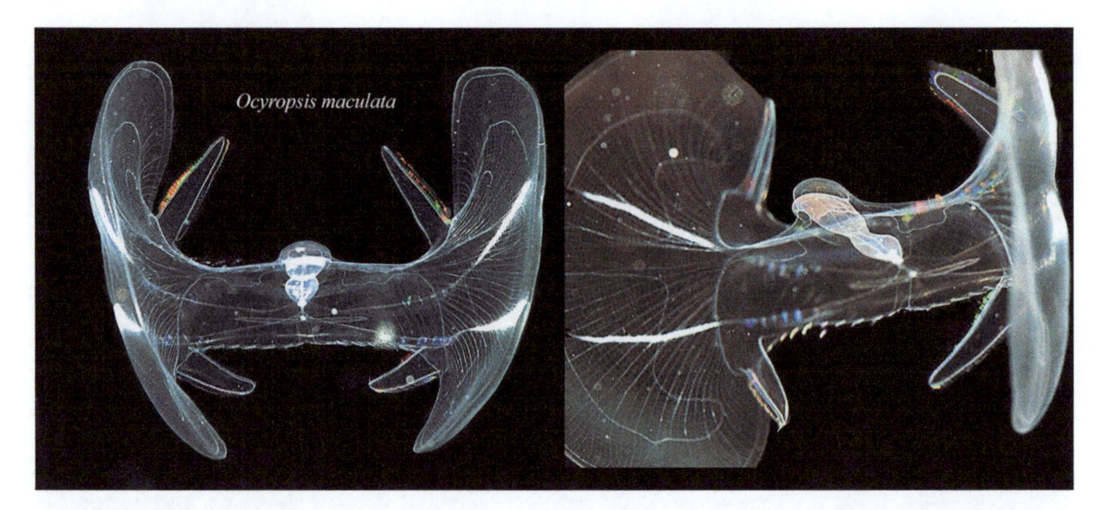

Fig. 87 *Ocyropsis maculata*, Florida

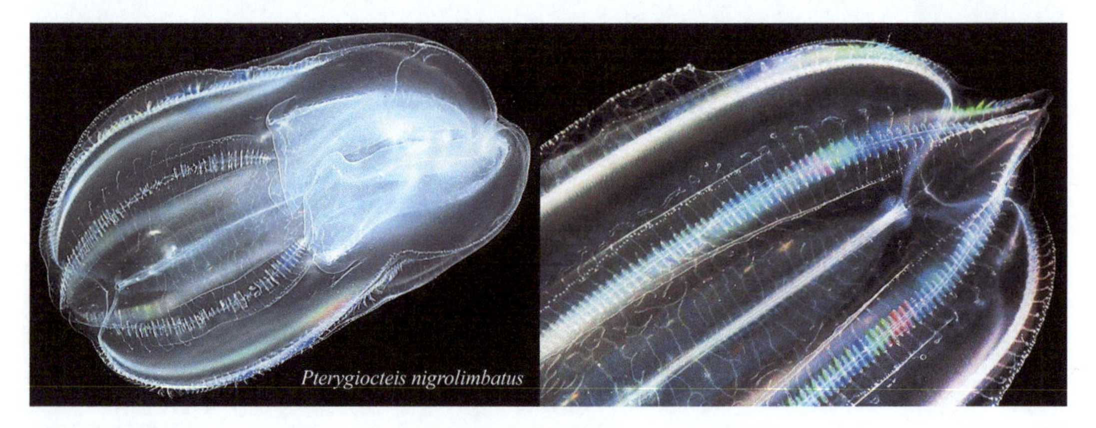

Fig. 88 Subantarctic Lobata (*Pterygiocteis nigrolimbatus*, Punta Arenas, Chile)

Fig. 89 Two species of *Beroe* from the Pacific Coasts of North (right) and South America (left). The left image shows *Beroe* from the northern part of the Humbolt Current (photo: Rigoberto Moreno Mendoza); the right photo shows *Beroe mitrata* with the orange gut from the San Juan Archipelago (Washington, USA)

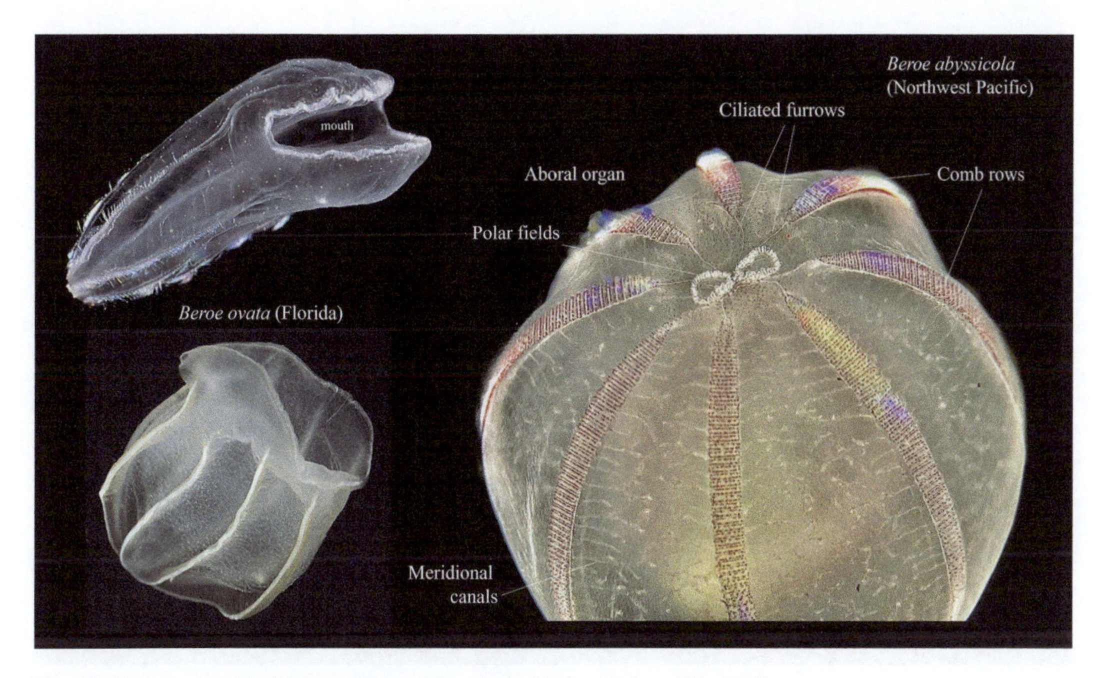

Fig. 90 *Beroe ovata* (Florida) and *Beroe abyssicola* (Friday Harbor, WA, USA)

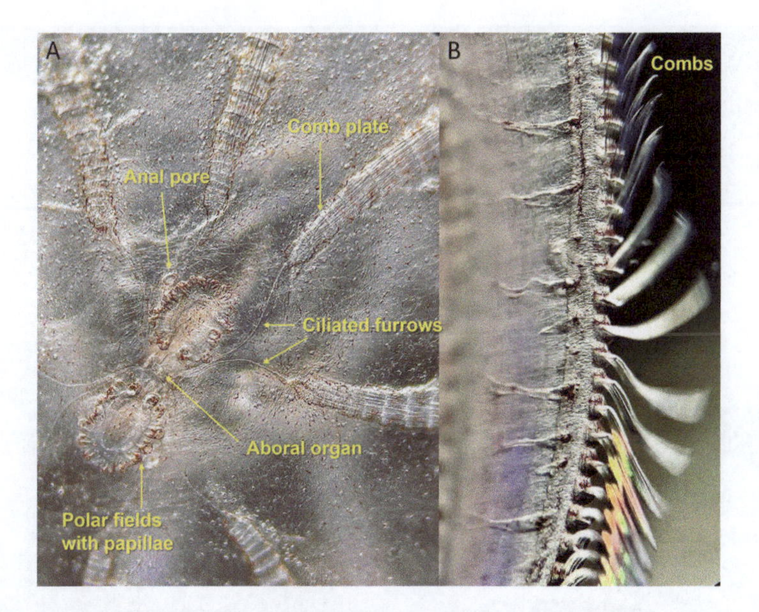

Fig. 91 (**a**) Aboral view of *Beroe abyssicola*. (**b**) Cilia of comb plates in *B. abyssicola*

Fig. 92 *Beroe abyssicola* (Friday Harbor, WA, USA) and *Beroe* sp. (Weddell Sea, Antarctica)

Fig. 93 *Beroe abyssicola*, White Sea. (Photo: Alexander Semenov©). Mortensen 1927, says among other things, that *B. abyssicola* is very similar to *Beroe cucumis.* the former is a deep sea speices and "claret" red but less colored when young, the latter is found near the surface and not intensely colored. Figure 93 shows images by A. Semenov,. I beleive he is a diver so these are near surface water images. If shot near the surface they may not be B abyssicola. The image in the lower right has very short ctene plates relative to the other specimen and is colored differently. The type photos have shorter plates

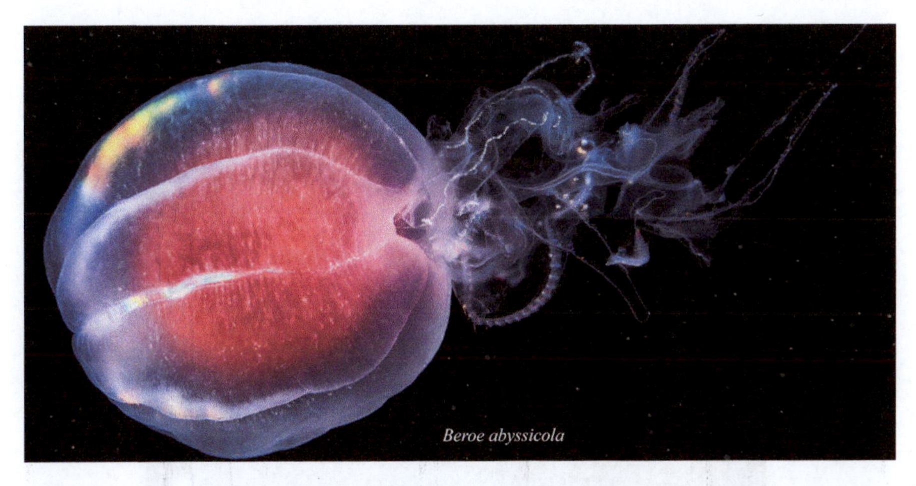

Fig. 94 *Beroe abyssicola* feeding on *Bolinopsis infundibulum*, White Sea. (Photo: Alexander Semenov©)

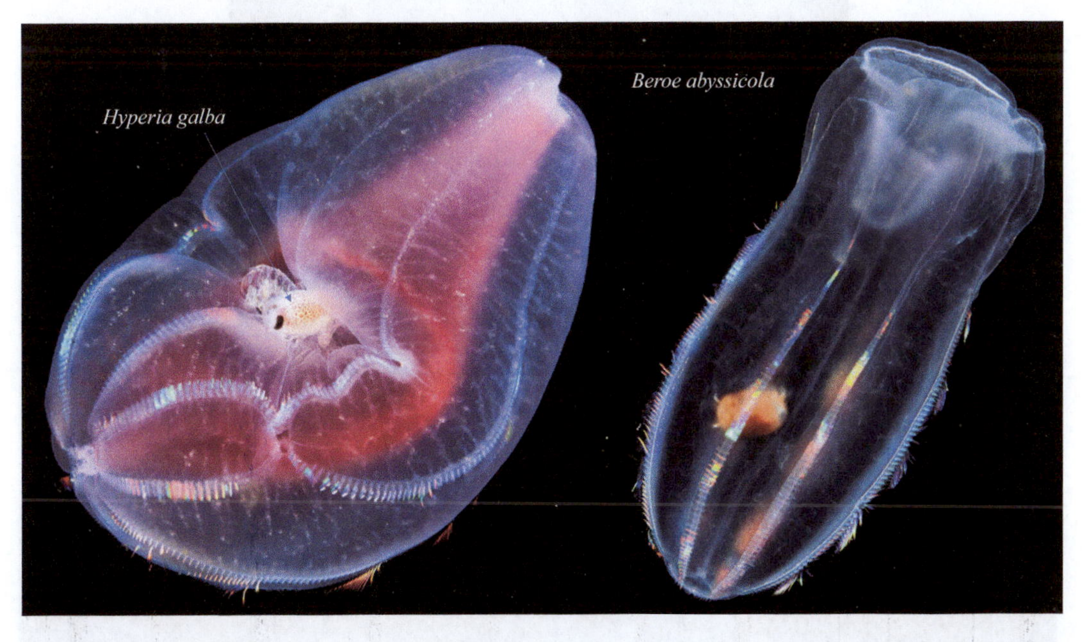

Fig. 95 *Beroe abyssicola* with the amphipod ectoparasite *Hyperia galba* (Montagu), White Sea. (Photo: Alexander Semenov©)

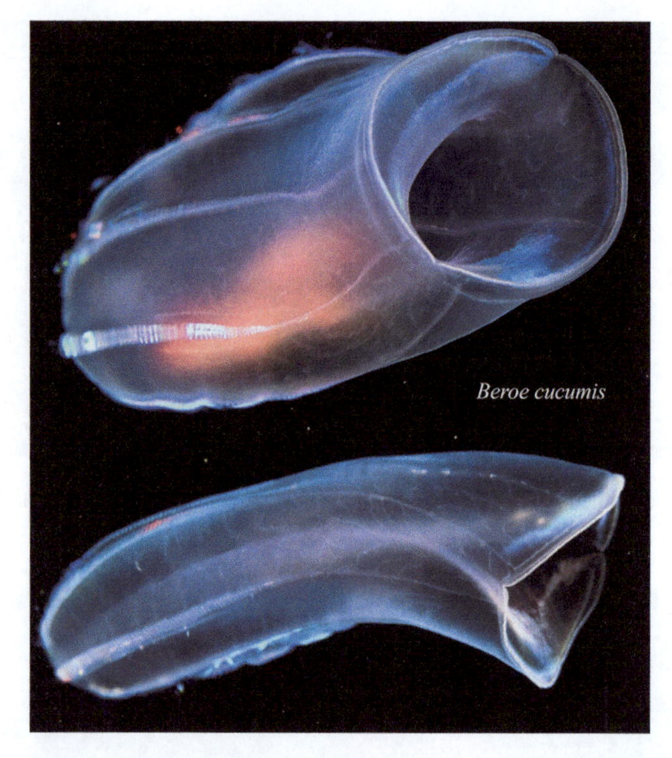

Fig. 96 *Beroe cucumis*, White Sea. (Photo: Alexander Semenov©)

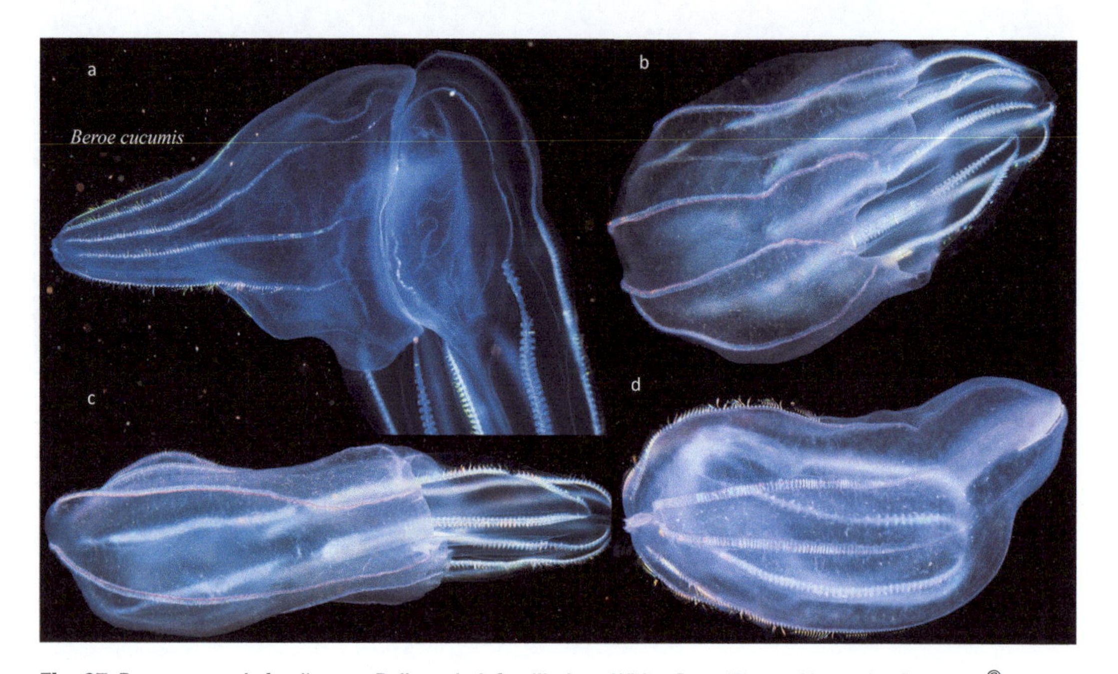

Fig. 97 *Beroe cucumis* feeding on *Bolinopsis infundibulum*, White Sea. (Photo: Alexander Semenov©)

Fig. 98 The jellyfish *Cyanea capillata* feeds on *Beroe cucumis*. (Photo: Alexander Semenov©; https:/www.flickr.com/photos/a_semenov/9147073560/in/photostream/)

Fig. 99 *Beroe foskalii*, Mediterranean Sea. (Photo: Alexander Semenov©)

Fig. 100 *Beroe ovata* – general view (**a**); *Beroe* sp. with *Bolinopsis* inside (**b**); the aboral organ of *Hormiphora beehleri* (**c**); Florida, USA

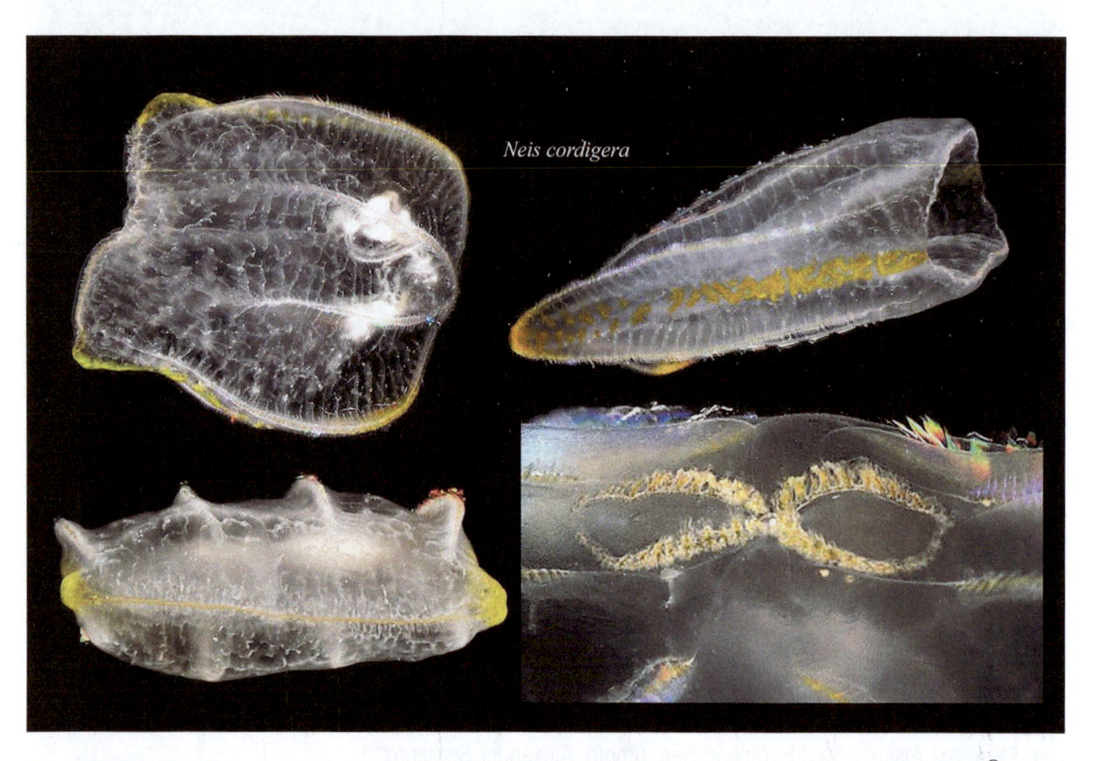

Fig. 101 Beroida: *Neis cordigera*, Brunswick River (New South Wales, Australia). (Photo: Denis Riek© (http:/www.roboastra.com/Cnidaria3/brac195.html); see also https:/bie.ala.org.au/species/https:/biodiversity.org.au/afd/taxa/be04b5e5-a9ed-49a3-b726-03e69be5188c)

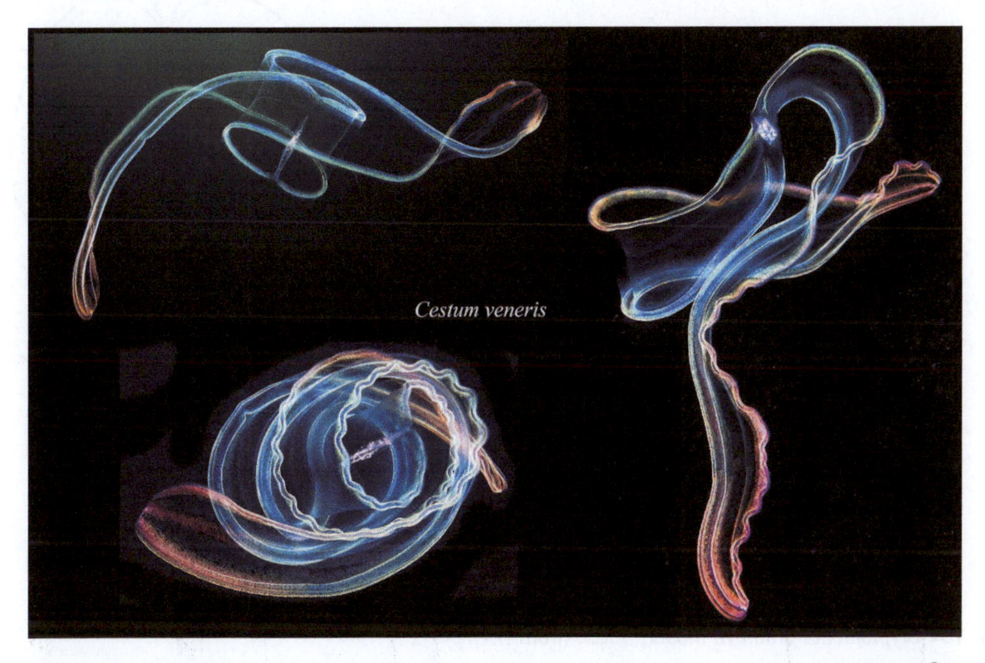

Fig. 102 Cestida: *Cestum veneris*, Ponza Island, Tyrrhenian Sea, Italy. (Photo: Alexander Semenov©. https:/www.flickr.com/photos/a_semenov/25886898121/in/photostream/, https:/www.flickr.com/photos/a_semenov/23003342862/in/photostream/, and https:/www.flickr.com/photos/a_semenov/22481800260/in/photostream/)

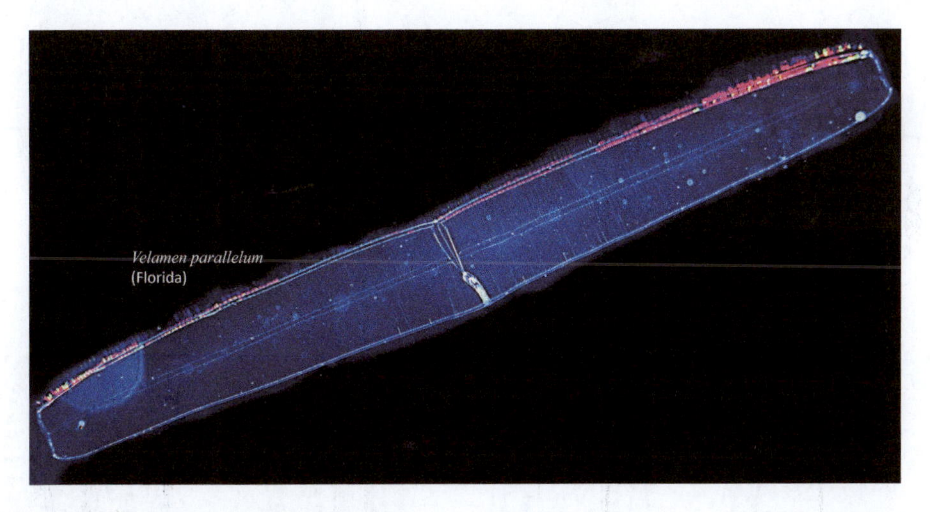

Fig. 103 Cestida: *Velamen parallelum* (Florida)

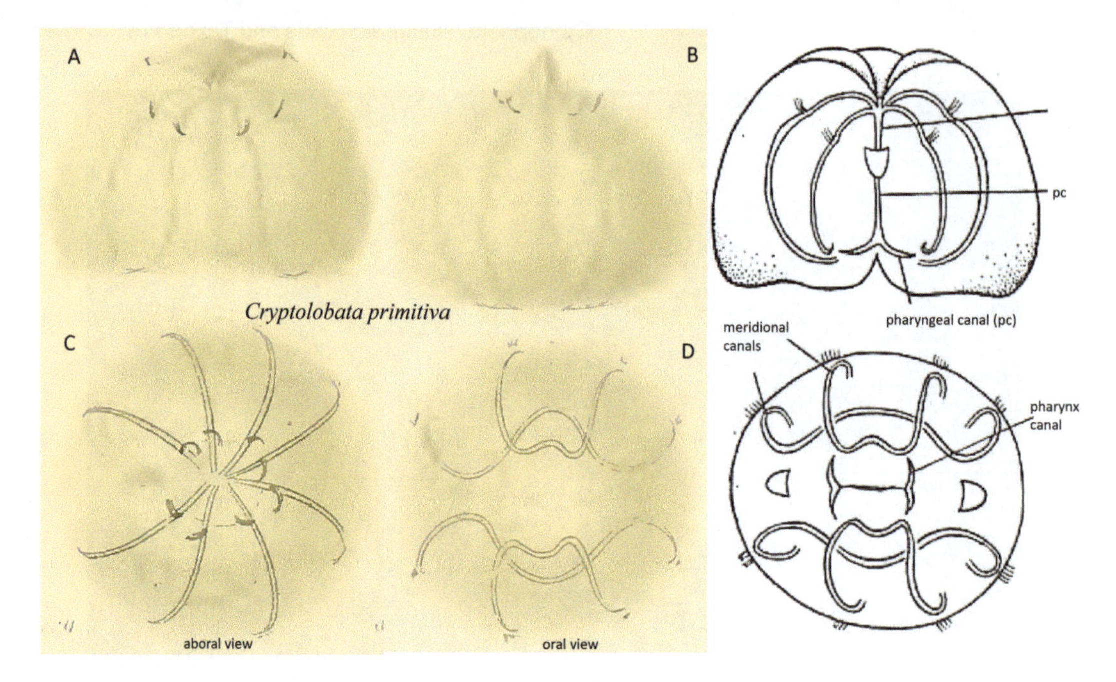

Fig. 104 *Cryptolobata primitiva* Moser, 1909. (from [71])

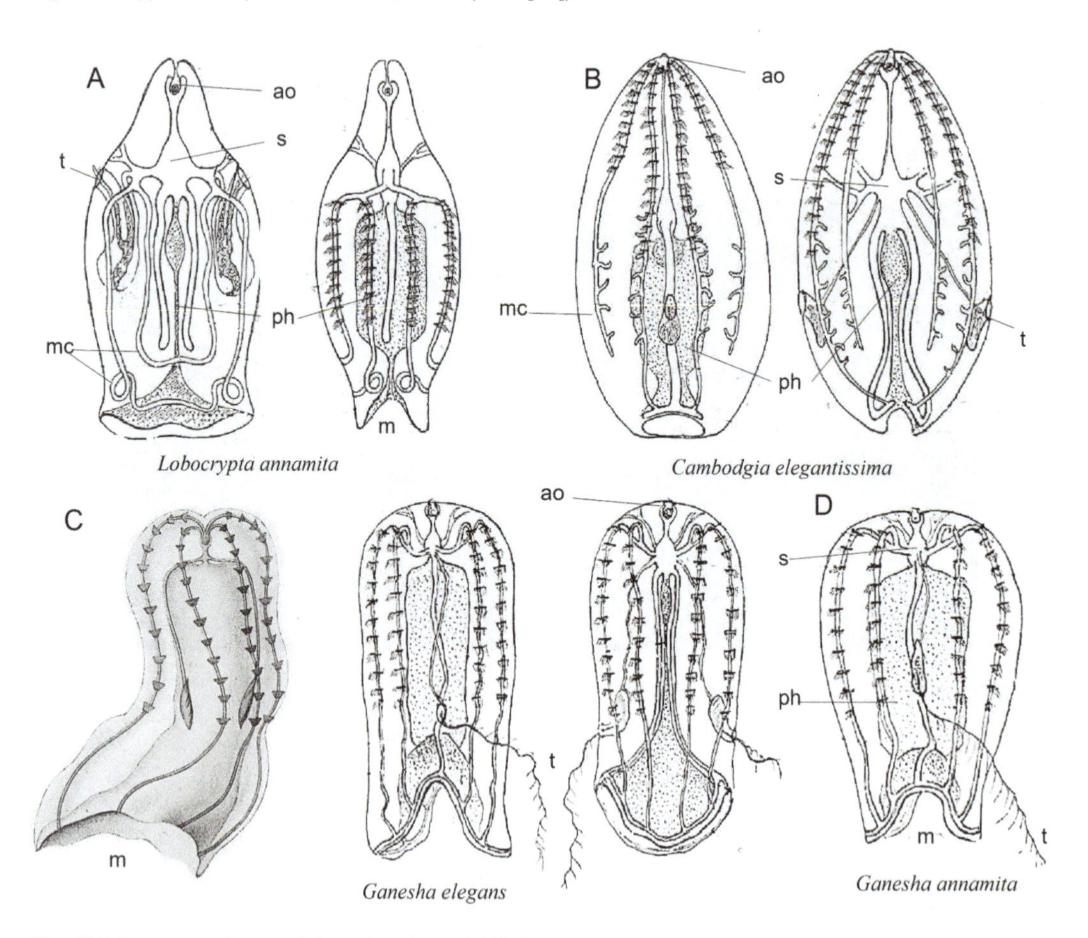

Fig. 105 Rare ctenophores of the orders Cryptolobiferida (**a** – *Lobocrypta annamita*), Cambojiida (**b** – *Cambodgia elegantissima*), Ganeshida (**c** – *Ganesha elegans*, **d** – *Ganesha annamita*). (from C. Dawydoff 1929–1946 [72], IndoPacific). Abbreviations: ao, aboral organ; m, mouth; mc, meridional canals; ph, pharynx; s, stomach; t, tentacles

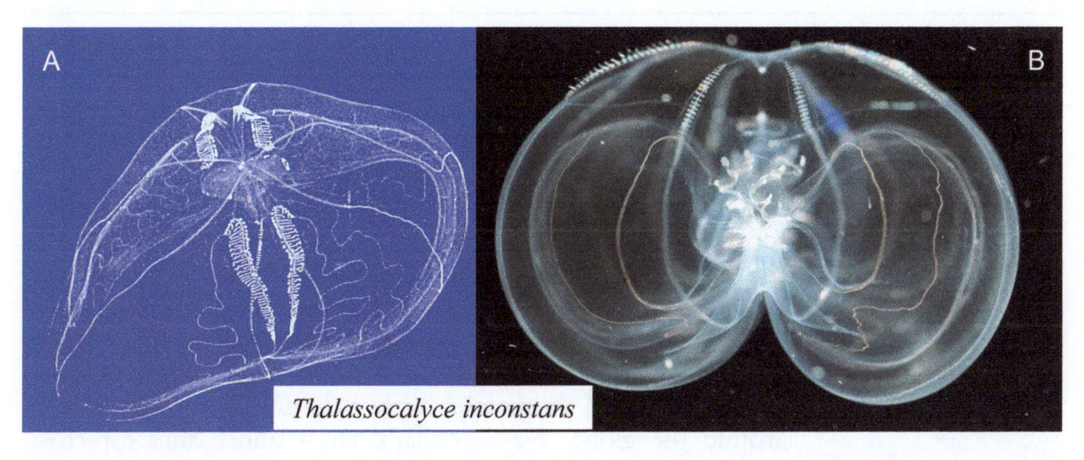

Fig. 106 Thalassocalycida: *Thalassocalyce inconstans*. ((**a**) Image courtesy of the NOAA Office of Ocean Exploration and Research, Laulima O Ka Moana. NOAA Ocean Exploration & Research from USA (https:/upload.wikimedia.org/wikipedia/commons/1/12/Ctneophore_-_Flickr_-_NOAA_Ocean_Exploration_%5E_Research.jpg); (**b**) Photo by L. Madin, Woods Hole Oceanographic Inst. (WHOI) (www.cmarz.org and https:/ocean.si.edu/ocean-life/plankton/ctenophore-feeds))

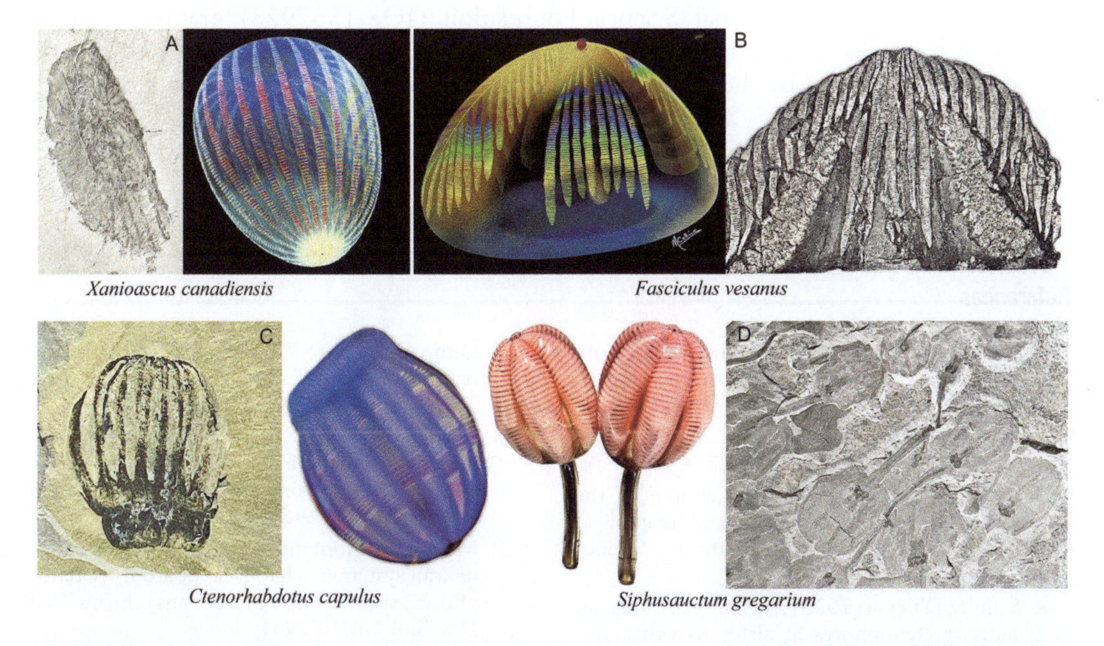

Fig. 107 Unusual fossils that have been attributed to the Ctenophora, Middle Cambrian (~506 million years ago), Burgess Shale, Canada. *Xanioascus canadiensis* (**a**, actual fossil and reconstruction), *Fasciculus vesanus* (**b**), *Ctenorhabdotus capulus* (**c**), and *Siphusauctum gregarium* (**d**). See text and [183] for details. These species and their reconstruction are at the exposition in the Royal Ontario Museum© in Toronto, Canada (see also https://burgess-shale.rom.on.ca/phylas/ctenophora/)

Acknowledgments

Many thanks to Alexander Semenov, who provided amazing images of multiple ctenophore species, and Denis Riek for the incredible photos of *Pukia* and *Neis*, Dr. Steven Haddock for *Bathyctena* and *Charistephane* photos as well as many scientists and photographers for their remarkable images of various species around the globe; we provided web links to numerous images of ctenophores. We thankful to Ocean Genome Atlas Project and the team led by Capt. Peter Molnar (vessel SAM), Tyler Meade, Matthew Stromberg, as well as Mr. James F. Jacoby (vessel Miss Phebe II) and Mr. Steven Sablonski (vessel Copacetic) with help to collect ctenophores around the globe. We also thank Dr. Claudia Mills for cross-checking some of our species identification and Nicholas Bezio for some *Beroe* identification. Of course, any errors are ours. Images of fossil ctenophores were obtained from the public exposition at the Royal Ontario Museum© in Toronto, Canada, and we deeply acknowledge all efforts in such artistic reconstructions. All other photographs are from the authors. This work was supported by the National Science Foundation (IOS-1557923) grants to LLM. Research reported in this publication was also supported in part by the National Institute of Neurological Disorders and Stroke of the National Institutes of Health under Award Number R01NS114491 (to L.L.M.). The content is solely the author's responsibility and does not necessarily represent the official views of the National Institutes of Health

References

1. Moroz LL et al (2014) The ctenophore genome and the evolutionary origins of neural systems. Nature 510(7503):109–114
2. Li Y et al (2021) Rooting the animal tree of life. Mol Biol Evol 38(10):4322–4333
3. Ryan JF et al (2013) The genome of the ctenophore *Mnemiopsis leidyi* and its implications for cell type evolution. Science 342(6164):1242592
4. Schultz DT et al (2023) Ancient gene linkages support ctenophores as sister to other animals. Nature 618(7963):110–117
5. Whelan NV et al (2015) Error, signal, and the placement of Ctenophora sister to all other animals. Proc Natl Acad Sci USA 112(18): 5773–5778
6. Whelan NV et al (2017) Ctenophore relationships and their placement as the sister group to all other animals. Nat Ecol Evol 1(11): 1737–1746
7. Dunn CW, Leys SP, Haddock SH (2015) The hidden biology of sponges and ctenophores. Trends Ecol Evol 30(5):282–291
8. Moroz LL (2018) NeuroSystematics and periodic system of neurons: model vs reference species at single-cell resolution. ACS Chem Neurosci 9(8):1884–1903
9. Moroz LL, Romanova DY (2022) Alternative neural systems: what is a neuron? (Ctenophores, sponges and placozoans). Front Cell Dev Biol 10:1071961
10. Moroz LL (2024) Brief history of Ctenophora. In: Moroz LL (ed) Ctenophores. Methods in molecular biology. Humana, New York. p 1–26
11. Hyman LH (1940) Invertebrates: Protozoa through Ctenophora, vol 1. McGraw-Hill, New York/London, p 726
12. Brusca RC, Brusca GJ (eds) (2003) Invertebrates, 2nd edn. Sinauer Associates, Inc, Sunderland, p 936

13. Brusca RC, Giribet G, Moore W (2022) Invertebrates, 4th edn. Sinauer Associates of Oxford University Press, p 1104
14. Harbison GR (1985) On the classification and evolution of the Ctenophora. In: Morris SC et al (eds) The origins and relationships of lower invertebrates. Clarendon Press, Oxford, pp 78–100
15. Podar M et al (2001) A molecular phylogenetic framework for the phylum Ctenophora using 18S rRNA genes. Mol Phylogenet Evol 21(2):218–230
16. Krumbach T (1925) Erste und einzige Klasse der Acnidaria, Vierte Klasse des Stammes der Coelenterata: Ctenophora, Rippenquallen, Kammquallen. In: Krumback WKT (ed) Handbuch der Zoologie, Protozoa, Porifera, Coelenteratea, Mesozoa. W. de Gruyter, Berlin and Leipzig, pp 902–998
17. Mills CE (1998–present) Phylum Ctenophora: list of all valid species names. Available from http://faculty.washington.edu/cemills/Ctenolist.html
18. Ospovat MF (1985) On phylogeny and classification of the phylum Ctenophora. Zool Zhurnal 64:965–974
19. Agassiz A (1874) Embryology of the ctenophorae. Mem Am Acad Arts Sci 10(3): 357–398
20. Chun C (1880) Die Ctenophoren des Golfes von Neapel, Fauna und Flora des Golfes von Neapel. W. Engelmann, Leipzig, pp 1–313
21. Hertwig R (1880) Ueber den Bau der Ctenophoren. Jenaische Z Naturwiss 14:393–457
22. Samassa P (1892) Zur histologie der ctenophoren. Arch Mikrosk Anat 40(1):157–243
23. Bethe A (1895) *Der* subepitheliale Nervenplexus der Ctenophoren. Biol Zbl 15:140–145
24. Mayer AG (1912) Ctenophores of the Atlantic coast of North America. Carnegie Inst Wash Publ 162:1–58
25. Mortensen T (1912) Ctenophora. Dan Ingolf-Exp 5A(2):1–96
26. Komai T (1922) Studies on two aberrant ctenophores: *Coeloplana* and *Gastrodes.* The Author
27. Tokioka T (1968) Ctenophora (in Japanese). Nakayama Pub. Co. Ltd, Japan, pp 205–233
28. Hernandez-Nicaise M, Franc J (1994) Embranchement des ct'enaires. In: Cie Me (ed) Trait'e de Zoologie, volume Tome III. Masson et Cie, Paris, pp 943–1075
29. Horridge GA (1974) Recent studies on the Ctenophora. In: Muscatine L, Lenhoff HM (eds) Coelenterate biology. Academic Press, New York, pp 439–468
30. Tamm SL (1982) Ctenophora. In: Electrical conduction and behavior in "simple" invertebrates. Clarendon Press, Oxford, pp 266–358
31. Tamm SL (2014) Cilia and the life of ctenophores. Invertebr Biol 133(1):1–46
32. Haddock SH (2007) Comparative feeding behavior of planktonic ctenophores. Integr Comp Biol 47(6):847–853
33. Dunlap H (1974) Ctenophora. In: Giese AC, Pearse JP (eds) Reproduction in marine invertebrates. Academic Press, New York, pp 201–265
34. Martindale MQ, Henry JQ (2015) Ctenophora. In: Wanninger A (ed) Evolutionary developmental biology of invertebrates. 1: Introduction, non-bilateria, acoelomorpha, xenoturbellida, chaetognatha. Springer Vienna, Vienna, pp 179–201
35. Hernandez-Nicaise M-L (1991) Ctenophora. In: Harrison FWFW, Westfall JA (eds) Microscopic anatomy of invertebrates: Placozoa, Porifera, Cnidaria, and Ctenophora. Wiley, New York, pp 359–418
36. Moroz LL (2015) Convergent evolution of neural systems in ctenophores. J Exp Biol 218(Pt 4):598–611
37. Moroz LL, Kohn AB (2016) Independent origins of neurons and synapses: insights from ctenophores. Philos Trans R Soc Lond Ser B Biol Sci 371(1685):20150041
38. Moroz LL, Romanova DY, Kohn AB (2021) Neural versus alternative integrative systems: molecular insights into origins of neurotransmitters. Philos Trans R Soc Lond Ser B Biol Sci 2021(376):20190762
39. Harbison G, Madin L, Swanberg N (1978) On the natural history and distribution of oceanic ctenophores. Deep-Sea Res 25(3): 233–256
40. Harbison GR, Madin LP (1982) Ctenophora. In: Parker SP (ed) Taxonomy and classification of living organisms. McGraw-Hill, New York, pp 707–715
41. Harbison GR (1996) Ctenophora. In: Gasca R, Suárez E (eds) Introducción al Estudio del Zooplancton Marino. ECOSUR, Chetumal, pp 101–147
42. Gershwin L, Zeidler W, Davie PJF (2010) Ctenophora of Australia. Mem Queensland Mus 54(3):1–45
43. Verhaegen G, Cimoli E, Lindsay D (2021) Life beneath the ice: jellyfish and ctenophores from the Ross Sea, Antarctica, with an image-based training set for machine learning. Biodivers Data J 9:e69374

44. Gibbons MJ et al (2021) Records of ctenophores from South Africa. PeerJ 9:e10697
45. Mianzan HW, Dawson EW, Mills CE (2009) Phylum Ctenophora: Comb Jellies. In: Gordon DP (ed) New Zealand inventory of biodiversity. Canterbury University Press, Christchurch, pp 49–58
46. Ralph PM, Kaberry C (1950) New Zealand coelenterates: ctenophores from Cook Strait, Zoology publications from Victoria University of Wellington, no. 3. Victoria University College, Wellington, pp 1–11
47. Benham WBS (1907) New Zealand ctenophores. Trans N Z Inst 39:138–143
48. Puente-Tapia FA et al (2021) An updated checklist of ctenophores (Ctenophora: Nuda and Tentaculata) of Mexican seas. Reg Stud Mar Sci 41:101555
49. Nogueira Junior M, Brandini FP, Codina JC (2015) Diel vertical dynamics of gelatinous zooplankton (Cnidaria, Ctenophora and Thaliacea) in a subtropical stratified ecosystem (south Brazilian bight). PLoS One 10(12): e0144161
50. Ruiz-Escobar F, Valadez-Vargas DK, Oliveira OM (2015) Ctenophores from the Oaxaca coast, including a checklist of species from the Pacific coast of Mexico. Zootaxa 3936(3):435–445
51. Gueroun SK et al (2021) Planktonic Ctenophora of the Madeira archipelago (Northeastern Atlantic). Zootaxa 5081(3):433–443
52. Kohler LG et al (2022) Gelatinous macrozooplankton diversity and distribution dataset for the North Sea and Skagerrak/Kattegat during January-February 2021. Data Brief 44: 108493
53. Oliveira OM, Feliu G, Palma S (2014) *Beroe gracilis* (Ctenophora) from the Humboldt Current System: first occurrence of this species in the southern hemisphere. Zootaxa 3827(3):397–400
54. Oliveira OM, Migotto AE (2014) First occurrence of *Beroe forskalii* (Ctenophora) in South American Atlantic coastal waters, with notes on the use of macrociliary patterns for beroid identification. Zootaxa 3779:470–476
55. Jamieson AJ, Lindsay DJ, Kitazato H (2023) Maximum depth extensions for Hydrozoa, Tunicata and Ctenophora. Mar Biol 170(3): 33
56. Ford M, Bezio N, Collins A (2020) *Duobrachium sparksae* (incertae sedis Ctenophora Tentaculata Cydippida): a new genus and species of benthopelagic ctenophore seen at 3,910 m depth off the coast of Puerto Rico. Plankton Benthos Res 15(4):296–305
57. Mills CE, Haddock SD (2007) Ctenophores. In: Carlton JT (ed) Light and Smith's manual: intertidal invertebrates of the central California coast. University of California Press, Berkeley, pp 189–199
58. Johansen E et al (2021) Assessing the value of a citizen science approach for ctenophore identification. Front Mar Sci 8:772851
59. Jaspers C et al (2018) Ocean current connectivity propelling the secondary spread of a marine invasive comb jelly across western Eurasia. Glob Ecol Biogeogr 27(7):814–827
60. Christianson LM et al (2022) Hidden diversity of Ctenophora revealed by new mitochondrial COI primers and sequences. Mol Ecol Resour 22(1):283–294
61. Fricke H (1970) Neue kriechede ctenophoren der gattung *Coeloplana* aus Madagaskar. Mar Biol 5:225–238
62. Eschscholtz F (1829) System der Acalephen. Eine ausführliche Beschreibung aller medusenartigen Strahltiere. Ferdinand Dümmler, Berlin
63. Norekian TP, Moroz LL (2019) Neural system and receptor diversity in the ctenophore *Beroe abyssicola*. J Comp Neurol 527(12): 1986–2008
64. Norekian TP, Moroz LL (2019) Neuromuscular organization of the Ctenophore *Pleurobrachia bachei*. J Comp Neurol 527(2): 406–436
65. Norekian TP, Moroz LL (2020) Comparative neuroanatomy of ctenophores: neural and muscular systems in *Euplokamis dunlapae* and related species. J Comp Neurol 528(3): 481–501
66. Norekian TP, Moroz LL (2024) Illustrated neuroanatomy of ctenophores: immunohistochemistry. In: Moroz LL (ed) Ctenophores: methods and protocols, methods in molecular biology. Humana, New York. p 147–161
67. Norekian TP, Moroz LL (2024) Scanning electron microscopy of ctenophores: illustrative atlas. In: Moroz LL (ed) Ctenophores: methods and protocols, methods in molecular biology. Humana, New York. pp 163–184
68. Lindsay DJ, Miyake H (2007) A novel benthopelagic ctenophore from 7,217m depth in the Ryukyu Trench, Japan, with notes on the taxonomy of deepsea cydippids. Plankton Benthos Res 2(2):98–102
69. Mortensen T (1932) Ctenophora from the "Michael Sars" North Atlantic deep-sea expedition 1910. Reports of the "Michael Sars" North Atlantic deep-sea expedition 1910, vol 3. Bergen Museum, pp 1–9

70. Mortensen T (1913) Ctenophora from the "Michael Sars" North Atlantic deep-sea expedition 1910. Reports of the "Michael Sars" North Atlantic deep-sea expedition 1910, vol 3. Bergen Museum, pp 1–9
71. Moser F (1909) Die Ctenophoren der deutsche Südpolar-expedition 1901–1903. Deutsche Südpolar-Expedition, XI Zoologie III:116–192, 3 pls
72. Dawydoff C (1946) Contribution a la connaissance des ctenophores pelagiques des eaux de l'Indochine. Bull Biol Fr Belg 80: 116–170
73. Lamarck J-BMd (1816) Histoire naturelle des animaux sans vertèbres. Tome second, vol 2. Verdière, Paris
74. Chun C (1879) Die im Golf von Neapel erscheinenden Rippenquallen. Mittheilungen aus der Zoologischen Station zu Neapel 1: 180–217
75. Delle Chiaje S (1841–1844) Descrizione e notomia degli animali invertebrati della Sicilia citeriore osservati vivi negli anni 1822-1830. Parts 1-8. Batteli & Co, Naples
76. Curreri G (1900) Osservazioni sui ctenofori comparenti nel Porto di Messina. Boll Della Soc Zool Ital 1(4):190–193
77. Leloup E (1938) Siphonophores et cténophores. Résultats du voyage du S.Y. Belgica en 1897-1898-1899: sous le commandement de A. de Gerlache de Gomery, Rapports scientifiques publiés aux frais du gouvernement belge, sous la direction de la Commission de la Belgica, J.-E. Buschman. J.E. Buschmann, Anvers, pp 4–12
78. Carré C, Carré D (1993) *Ctenella aurantia*, new genus and new species of mediterranean tentacled ctenophore (Ctenellidae fam. nov.) without colloblasts, but with labial suckers. Can J Zool 71:1804–1810
79. Gegenbaur C (1856) Studien über Organisation und Systematik der Ctenophoren. Archiv Naturgesch 22:163–205
80. Ceccolini F, Cianferoni F (2020) New substitute name *Attenboroughctena* nom. nov. for the genus *Ceroctena* Carre amp; Carre, 1991 (Ctenophora: Cydippida: Pleurobrachiidae). Zootaxa 4822(2):zootaxa.4822.2.12
81. Carré C, Carré D (1991) *Ceroctena bicornis* gen. sp. nov., genre et espèce nouveaux de cténophore méditerranéen (Cydippida, Pleurobrachiidae). C R Acad Sci III 313(12): 559–564
82. Agassiz L (1860) Contributions to the natural history of the United States of America, vol 3. Little, Brown and Company, Boston
83. Torrey HB (1904) Contributions from the laboratory of the Marine Biological Association of San Diego. II. The ctenophores of the San Diego region. Univ Calif Publ Zool 2(2): 45–51
84. Ghigi A (1909) Raccolte planctoniche fatte dalla R. Nave "Liguria" nel viaggio di circonnavigazione del 1903–1905. I. Ctenofori. Pubblicazioni del R. Instituto di Studi Superiori Practici e di Perfeziona-mento in Firenze Sezione di Scienze Fisiche e Naturali 2:1–24
85. Mertens KH (1833) Beobachtungen und Untersuchungen über die Beroëartigen Akalephen. In: Mémoires de l'Académie impériale des sciences de St.-Pétersbourg. 6e série, Sciences mathématiques, physiques et naturelles. L'Impr. de l'Académie Impériale des Sciences, St.-Pétersbourg, pp 479–543
86. Haeckel E (1899–1904) Kunstformen der Natur. Verlag des Bibliographiichen Initituts, Leipzig und Wien. 51, 100 illustrations
87. Gegenbaur C (1856) Studien über Organisation und Systematik der Ctenophoren. Archiv Naturgesch XXII:163–205
88. Agassiz A, Mayer AG (1902) Medusae. Report of the scientific research expedition to the tropical Pacific. U.S. Fish Commission Steamer Albatross, 1899–1900. III. Mem Mus Comp Zool Harv Coll 26:136–176
89. Chun C (1898) Die Ctenophoren der Plankton-expedition. Ergeb Plankton Exped 2:1–33
90. Moser F (1903) Die Ctenophoren der Siboga-expedition. In: Weber M (ed) Uitkomsten op zoologisch, botanisch, oceanographisch en geologisch gebied verzameld in Nederlandsch Oost-Indië, 1399–1900 aan boord H.M. Siboga onder commando van luitenant ter zee 1e kl. G. F. Tydeman. Buchhandlung und druckerei vormals E.J. Brill, Leiden, pp 1–34
91. Carré C, Carré D (1993) *Minictena luteola*, genre et espèce nouveaux de cténophore Cydippida méditerranéen a cinq types de colloblastes. Beaufortia 43(10):168–175
92. Carré D, Carré C (1993) Five types of colloblast in a cydippid ctenophore, *Minictena luteola* Carre; and Carre: an ultrastructural study and cytological interpretation. Philos Trans R Soc Lond Ser B Biol Sci 341(1298): 437–448
93. Fleming J (1822) The philosophy of zoology: or a general view of the structure, functions and classification of animals, vol 2. A. Constable, Edinburgh
94. Chun C (1889) Bericht über eine nach den Canarischen Inseln im Winter 1887–88

ausgeführte Reise. II. Beobachtungen über die pelagische Tiefen- und Oberflächenfauna des östlichen Atlantischen Oceans. Juni-Dec 1888. Sitzungsber K Preuss Akad Wiss Berl Math Phys Kl:519–553

95. Müller OF (1776) Zoologiae Danicae prodromus, seu Animalium Daniae et Norvegiae indigenarum: characteres, nomina, et synonyma imprimis popularium. Typis Hallagerii, Havni, Copenhagen
96. Moser F (1908) Cténophores de la Baie d'Amboine. Rev Suisse Zool 16:1–26
97. Mills CE (1987) Revised classification of the genus *Euplokamis* Chun, 1880 (Ctenophora: Cydippida: Euplokamidae n. fam.) with a description of the new species *Euplokamis dunlapae*. Can J Zool 65(11):2661–2668
98. Majaneva S et al (2021) Hiding in plain sight – *Euplokamis dunlapae* (Ctenophora) in Norwegian waters. J Plankton Res 43(2): 257–269
99. Krumbach T (1925) Erste und einzige Klasse der Actinaria Vierte Klasse des Stammes der Coelenterata. Ctenophora. In: Kukenthal W, Krumbach T (eds) Handbuch der zoologie. de Gruyter, Berlin, pp 905–995
100. Carus JV (1863) Coelenteraten. In: Peters WCJ, Carus JV, Gerstaecker CEA (eds) Handbuch der zoologie. W. Engelmann, Leipzig, pp 518–562
101. Carré D, Carré C (1989) *Haeckelia bimaculata* sp. nov., a new species of ctenophore (Cydippida Haeckeliidae) from the Mediterranean with cnidocysts and pseudocolloblasts. C R Acad Sci III Sci Vie 308:321–327
102. Mills CE, Miller RL (1984) Ingestion of a medusa (*Aegina citrea*) by the nematocyst-containing ctenophore *Haeckelia rubra* (formerly *Euchlora rubra*): phylogenetic implications. Mar Biol 78(2):215–221
103. Stechow E (1921) Neue Genera und Species von Hydrozoen und anderen Evertebraten. Archiv Naturgesch 87:248–265
104. Quoy JRC, Gaimard JP (1833) Zoologie IV: zoophytes. In: Zoologie. Voyage de la corvette l'Astrolabe: exécuté par ordre du roi, pendant les années 1826–1827–1828–1829/sous le commandement de J. Dumont d'Urville. J. Tastu, Paris, pp 1–390
105. Dawydoff C (1937) Les Gastrodés des eaux indochinoises et quelques observations sur leur cycle évolutif. C R Hebd Seances Acad Sci Paris 204:1088–1090
106. Lesson RP (1829) Zoophytes. In: Voyage medical autour du Monde execute par I ordre du Roi sur la Corvette de Sa Majeste la Coquille, pendant les annees 1822. 1823, 1824, et 1825…par M.L.I. Duperey, Capitaine de Fregate, Zoologie. A. Bertrand, Paris, p 151
107. Fabricius O (1780) Fauna Groenlandica, systematice sistens animalia groenlandiae occidentalis hactenus indagata, quoad nomen specificium, triviale, vernaculumque, synonyma auctorum plurimum, descriptionem, locum, victum, generationem, mores, usum capturamque singuli, pro ut detegendi occasio fuit, maximaque parte secundum proprias observationes. Ioannis Gottlob Rothe, Hafniae [= Copenhagen] & Lipsiae [= Leipzig]
108. Majaneva S, Majaneva M (2013) Cydippid ctenophores in the coastal waters of Svalbard: is it only Mertensia ovum? Polar Biol 36: 1681–1686
109. Townsend JP et al (2020) A mesopelagic ctenophore representing a new family, with notes on family-level taxonomy in Ctenophora: Vampyroctena delmarvensis gen. nov. sp. nov. (Vampyroctenidae, fam. nov.). Mar Biodivers 50(3):34
110. Arafat H et al (2018) Extensive mitochondrial gene rearrangements in Ctenophora: insights from benthic Platyctenida. BMC Evol Biol 18(1):65
111. Alamaru A et al (2017) Molecular diversity of benthic ctenophores (Coeloplanidae). Sci Rep 7(1):6365
112. Willey A (1896) On *Ctenoplana*. J Cell Sci 2(155):323–342
113. Kowalevsky A (1880) *Coeloplana metschnikowii*. Zool Anz 3(51):140
114. Dawydoff C (1930) Une nouvelle Coeloplanide de la cote sud d'Annam (Coeloplana agniae nov. sp.). Arch Zool Exp Gén Notes Rev 70:83–86
115. Song JI, Hwang SJ (2010) A new species of genus *Coeloplana* (Ctenophora: Tentaculata: Platyctenida) from Korea. Anim Syst Evol Divers 26(3):217–221
116. Mortensen T (1927) Two new ctenophores. In: Papers from Dr. Th. Mortensen's Pacific expedition, 1914–16, XXXIX. Videnskabelige Meddelelser-Dansk Naturhistorisk Forening I Kobenhavn, Copenhagen, pp 277–288
117. Krumbach T (1933) Uber eine kriechende ctenophore aus dem Golfe von Suez und ein paar Thesen über die Architektonik der Rippenquallen. Mitt Mus Naturkd Berl 19:475–479
118. Komai T (1920) Notes on *Coeloplana bocki* n. sp. and its development. Annot Zool Jpn 9: 575–584

119. Tanaka H (1932) *Coeloplana echinicola* n. sp. Mem Coll Sci Univ Kyoto Ser B 7(5): 247–249
120. Alamaru A, Brokovich E, Loya Y (2016) Four new species and three new records of benthic ctenophores (Family: Coeloplanidae) from the Red Sea. Mar Biodivers 46(1):261–279
121. Krempf A (1921) *Coeloplana gonoctena*: biologie, organisation, développement. Bull Biol Fr Belg 54:252–312
122. Devanesen D, Varadarajan S (1942) On three new species of *Coeloplana* found at Krusadai Island, Marine Biological Station, and Gulf of Mannar. J Madras Univ 14(2):181–188
123. Utinomi H (1963) *Coeloplana komaii*, a new creeping ctenophore from Sagami Bay. Jpn J Zool 14(1):15–19
124. Dawydoff C (1938) Les coeloplanides Indochinoises. Arch Zool Exp Gén 80:125–162
125. Thiel H (1968) *Coeloplana meteoris* nov. spec. (Ctenophora: Platyctenea): Bescheibung und systematische Stellung mit einem Vergleich der Gastrovascularsysteme in deiser Ordung. Meteor Forsch-Ergebn 3:1–13
126. Abbott JF (1902) Preliminary notes on *Coeloplana*. Annot Zool Jpn 4(4):103–108
127. Matsumoto GI, Gowlett-Holmes KL (1996) *Coeloplana scaberiae* sp. nov., a new benthic ctenophore (Ctenophora: Platyctenida: Coeloplanidae) from South Australia. Rec S Aust Mus 29(1):33–40
128. Matsumoto GI (1999) *Coeloplana thomsoni* sp. nov., a new benthic ctenophore (Ctenophora: Platyctenida: Coeloplanidae) from Western Australia. In: The seagrass flora and fauna of Rottnest Island, Western Australia. Western Australia Museum, Perth, pp 385–393
129. Glynn PW, Bayer FM, Renegar DA (2014) *Coeloplana waltoni*, a new species of minute benthic ctenophore (Ctenophora: Platyctenida) from south Florida. Proc Biol Soc Wash 127(2):423–436, 14
130. Rankin JJ (1956) The structure and biology of *Vallicula multiformis*, gen. et sp. nov., a platyctenid ctenophore. J Linn Soc Lond Zool 43(289):55–71
131. Korotneff A (1886) *Ctenoplana kowalevskii*. Z Wiss Zool 43:242–251
132. Dawydoff C (1929) Sur la presence du genre *Ctenoplana* dans les eaux de l'Indochine. C R Hebd Seances Acad Sci 198:1315–1316
133. Gnanamuthu C, Nair RV (1948) *Ctenoplana bengalensis* n. sp. from the Madras plankton. Proc Indian Acad Sci 27(6):153–160
134. Dawydoff C (1936) Les Ctenoplanidae des eaux de l'Indochine Francaise Etude systématique. Bull Biol Fr Belg 70:456–486
135. Yosii N (1933) *Ctenoplana* from Japan. Proc Imp Acad 9(9):539–540
136. Fricke HW, Plante R (1971) Contribution à l'étude des Cténophores Platycténides de Madagascar: *Ctenoplana* (*Diploctena* n.s. gen.) neritica n.sp. et Coeloplana (*Benthoplana* n.s. gen.) meteoris (Thiel 1968). Cah Biol Mar 12:57–75
137. Komai T (1942) *Lyrocteis* and other resembled species (in Japanese). Anim Plant 10(1–3):15–18; 109–112; 209–216
138. Shepherd B, Yong S, Wandell M (2019) Collecting and exhibiting *Lyrocteis imperatoris* Komai 1941, a sessile ctenophore from mesophotic ecosystems. Calif Acad Sci 50:3–11
139. Yamauchi S, Fujii K, Ishii R (2017) Captive breeding and rearing of *Lyrocteis imperatoris.* J Jpn Assoc Zoo Aquarium 58(1–2):1–8
140. Shepherd B, Pinheiro HT, Rocha LA (2018) Ephemeral aggregation of the benthic ctenophore *Lyrocteis imperatoris* on a mesophotic coral ecosystem in the Philippines. Bull Mar Sci 94(1):101–102
141. Komai T (1941) A new remarkable sessile ctenophore. Proc Imp Acad Tokyo 17:216–220
142. Robilliard GA, Dayton PK (1972) A new species of platyctenean ctenophore, *Lyrocteis flavopallidus* sp. nov., from McMurdo Sound, Antarctica. Can J Zool 50(1):47–52
143. Dawydoff C (1950) La nouvelle forme de Ctenophores planarises sessiles provenant de la Mer de Chine Meridionale (*Savangia atentaculata* nov. gen. nov. spec.). C R Hebd Seances Acad Sci 231(17):814–816
144. Harbison G, Madin L (1982) Ctenophora. In: Parker SP (ed) Synopsis and classification of living organisms. McGraw-Hill, New York, pp 707–715
145. Harbison GR, Madin LP (1978) *Bathocyroe fosteri* gen.nov., sp.nov.: a mesopelagic ctenophore observed and collected from a submersible. J Mar Biol Assoc U K 58(3): 559–564
146. Horita T, Akiyama H, Kubota S (2011) *Bathocyroe longigula* spec. nov., an undescribed ctenophore (Lobata: Bathocyroidae) from the epipelagic fauna of Japanese coastal waters. Zool Med Leiden 85(15):877–886
147. Bigelow HB (1912) Reports on the scientific results of the expedition to the eastern tropical Pacific, in charge of Alexander Agassiz, by the U.S. Fish Commission Steamer Albatross, from October 1904, to March 1905,

Lieutenant Commander L.M. Garrett, U.S. N., commanding. XXVI. The ctenophores. Bull Mus Comp Zool 54:369–408

148. von Lendenfeld R (1885) The metamorphosis of *Bolina chuni*. Nov. Spec. Proc Linnean Soc NSW 9:929–931
149. Agassiz A (1865) North American Acalephae, Illustrated catalogue of the museum of comparative zoölogy at Harvard College, vol 2. Welch, Bigelow, & Co, Cambridge, MA, pp 1–234
150. Moser F (1907) Neues über Ctenophoren. Mitteilung II. Zool Anz 31:449–454
151. Bigelow HB (1904) Medusae from the Maldive Islands. Bull Mus Comp Zool 39:245–269
152. Tokioka T (1964) *Bolinopsis rubripunctata* n. sp., a new lobatean ctenophore from Seto. Publ Seto Mar Biol Lab 12(1):93–99
153. Townsend JP et al (2020) Ink release and swimming behavior in the oceanic ctenophore *Eurhamphaea vexilligera*. Biol Bull 238(3):206–213
154. Komai T, Tokioka T (1940) *Kiyohimea aurita*, n. gen., n. sp., type of a new family of lobate Ctenophora. Annot Zool Jpn 19(1): 43–46
155. Matsumoto GI, Robison BH (1992) *Kiyohimea usagi*, a new species of lobate ctenophore from the Monterey submarine canyon. Bull Mar Sci 51(1):19–29
156. Hoving HJ, Neitzel P, Robison B (2018) In situ observations lead to the discovery of the large ctenophore *Kiyohimea usagi* (Lobata: Eurhamphaeidae) in the eastern tropical Atlantic. Zootaxa 4526(2):232–238
157. Harbison G, Matsumoto GI, Robison BH (2001) *Lampocteis cruentiventer* gen. nov, sp. nov.: A new mesopelagic lobate ctenophore, representing the type of a new family (class Tentaculata, order Lobata, family Lampoctenidae, fam. nov). Bull Mar Sci 68:299–311
158. Agassiz A, Mayer AG (1899) Acalephs from the Fiji Islands. Bull Mus Comp Zoöl Harv Coll 32:157–189
159. Komai T (1918) On ctenophores of the neighbourhood of Misaki. Annot Zool Jpn 9(4):451–474
160. Quoy JRC, Gaimard JP (1824–1826) Zoologie. In: Freycinet Ld (ed) Voyage au tour du monde fait par ordre du roi, sur les corvettes de S. M: l'Uranie et la Physicienne pendant les années 1817 à 1820. Chez Pillet Aîné, Paris, pp 1–712
161. Matsumoto GI (1988) A new species of lobate ctenophore, *Leucothea pulchra* sp. nov., from the California Bight. J Plankton Res 10(2):301–311
162. Uyeno D et al (2015) New records of *Lobatolampea tetragona* (Ctenophora: Lobata: Lobatolampeidae) from the Red Sea. Mar Biodivers Rec 8:e33
163. Horita T (2000) An undescribed lobate ctenophore, *Lobatolampea tetragona* gen. nov. & spec. nov., representing a new family, from Japan. Zool Mededel 73(12–33): 457–464
164. Harbison GR, Miller RL (1986) Not all ctenophores are hermaphrodites. Studies on the systematics, distribution, sexuality and development of two species of *Ocyropsis*. Mar Biol 90(3):413–424
165. Rang S (1827) Description d'un genre nouveau de la classe des Acalèphes. Bull Hist Nat Soc Linn Bordeaux 6:316–319
166. Mills CE, Alain Dubois A (2023) The taxonominal status of the nomina *Lesueuria* Milne Edwards, 1841 and Lesueuriidae Chun, 1880, and introduction of a new genus and a new family for *Lesueuria pinnata* Ralph & Kaberry, 1950, as well as an additional new species of the new genus (Ctenophora, Lobata). Bionomina 36:1–35
167. Eschscholtz JFv (1825) Bericht über die zoologische Ausbeute während der Reise von Kronstadt bis St. Peter und Paul. Isis von Oken 6:733–747
168. Johansson ML et al (2018) Molecular insights into the ctenophore genus *Beroe* in Europe: new species, spreading invaders. J Hered 109(5):520–529
169. Linnaeus C (1758) Systema Naturae per regna tria naturae, secundum classes, ordines, genera, species, cum characteribus, differentiis, synonymis, locis. Editio decima, reformata, vol 1, 10th rev edn. Laurentius Salvius, Holmiae
170. Lesson RP (1836) Mémoire sur la famille des Beroïdes (Beroidae Less.). Ann Sci Nat 25: 235–266
171. Chamisso Ad, Eysenhardt CG (1821) De animalibus quibusdam e classe vermium Linneana, in circumnavigatione Terrae, auspicante Comite N. Romanoff, duce Ottone di Kotzebue, annis 1815–1818 peracta, observatis Fasciculus secundus, reliquos vermes continens. Nova Acta Physico-Med Acad Cesar Leop-Carol 10:343–373
172. Fabricius O (1780) Fauna groenlandica. I. G. Rothe, Hafniae
173. Milne Edwards H (1841) Sur la structure et les fonctions de quelques zoophytes,

mollusques et crustacés des côtes de la France. Ann Sci Nat 2(16):193–232

174. Künne C (1939) Die *Beroe* (Ctenophora) der südlichen Nordsee, *Beroe gracilis n. sp.* Zool Anz 127:172–174
175. Péron F, Freycinet L (1807–1816) Voyage de découvertes aux Terres Australes, exécuté par ordre de sa Majesté l'Empereur et Roi, … pendant les années 1800. 1801, 1802, 1803 et 1804. Imprimerie Royale, Paris
176. Bruguière JG (1789–1792) Encyclopédie méthodique ou par ordre de matières. Histoire naturelle des vers, vol 1. Pancoucke, Paris
177. Komai T (1921) Notes on the two Japanese Ctenophores, *Lampetia pancerina* Chun and Beroe ramosa n. sp. Annot Zool Jpn 10(1): 15–18
178. Lesson RP (1843) Histoire naturelle des zoophytes. Acalèphes. Librairie Encyclopédique de Roret, Paris
179. Lesueur CA (1813) Mémoire sur quelques nouvelles espèces d'animaux mollusques et radiaires recueillis dans la Méditerranée près de Nice. Nouv Bull Sci Soc Philom Paris (2) 3 (69):281–285
180. Fol H (1869) Ein Beitrag zur Anatomie und Entwicklungsgeschichte einiger Rippenquallen. Inaugural-Dissertation zur erlangung der doctorwürde in der medicin und chirurgie der medicinischen faultät der Friedrich-Wilhelms-Universität zu Berlin, pp 1–12
181. Madin L, Harbison G (1978) Thalassocalyce inconstans, new genus and species, an enigmatic ctenophore representing a new family and order. Bull Mar Sci 28(4):680–687
182. Swift HF et al (2009) Feeding behavior of the ctenophore *Thalassocalyce inconstans*: revision of anatomy of the order Thalassocalycida. Mar Biol 156(5):1049–1056
183. Conway Morris S, Collins DH (1996) Middle Cambrian ctenophores from the Stephen Formation, British Columbia, Canada. Philos Trans R Soc Lond B 351:279–308
184. Parry LA et al (2021) Cambrian comb jellies from Utah illuminate the early evolution of nervous and sensory systems in ctenophores. iScience 24(9):102943
185. Fu D et al (2019) The Qingjiang biota-A Burgess Shale-type fossil Lagerstatte from the early Cambrian of South China. Science 363(6433):1338–1342
186. Ou Q et al (2015) A vanished history of skeletonization in Cambrian comb jellies. Sci Adv 1(6):e1500092
187. Hou X et al (1999) The Chengjiang fauna: exceptionally well-preserved animals from 530 million years ago. Yunnan Science and Technology Press
188. Hu S et al (2007) Diverse pelagic predators from the Chengjiang Lagerstätte and the establishment of modern-style pelagic ecosystems in the early Cambrian. Palaeogeogr Palaeoclimatol Palaeoecol 254(1):307–316
189. Chen JY, Zhou MY (1997) Biology of the Chengjiang fauna. Natl Mus Nat Hist 10: 11–105
190. Chen JY et al (2007) Raman spectra of a Lower Cambrian ctenophore embryo from southwestern Shaanxi, China. Proc Natl Acad Sci USA 104(15):6289–6292
191. Luo HL et al (1999) Early Cambrian Chengjiang fauna from Kunming region, China. Yunnan Science and Technology Press, Kunming
192. Chen L et al (2002) Early Cambrian Chengjiang fauna in eastern Yunnan, China. Yunnan Science and Technology Press, Kunming. 199p., in Chinese with English abstract
193. Stanley GD, Stürmer W (1987) A new fossil ctenophore discovered by X-rays. Nature 328(6125):61–63
194. Otto M (1994) Zur Frage der "Weichteilerhaltung" im Hunsruckschiefer. Geol Paleontol 28:45–63
195. Stanley GD, Sturmer W (1983) The first fossil ctenophore from the lower devonian of West Germany. Nature 303:518–520
196. Klug C et al (2021) A late-surviving stem-ctenophore from the Late Devonian of Miguasha (Canada). Sci Rep 11(1):19039
197. O'Brien LJ, Caron J-B (2012) A new stalked filter-feeder from the middle Cambrian burgess shale, British Columbia, Canada. PLoS One 7(1):e29233
198. Zhao Y et al (2019) Cambrian sessile, suspension feeding stem-group ctenophores and evolution of the comb jelly body plan. Curr Biol 29(7):1112–1125.e2
199. Zhao Y, Hou X-G, Cong P-Y (2023) Tentacular nature of the "column" of the Cambrian diploblastic *Xianguangia sinica*. J Syst Palaeontol 21(1):2215787
200. Xian-Guang H et al (2017) The Cambrian fossils of Chengjiang, China: the flowering of early animal life, 2nd edn. Wiley-Blackwell, Chichester
201. Han J et al (2016) The earliest pelagic jellyfish with rhopalia from Cambrian Chengjiang Lagerstätte. Palaeogeogr Palaeoclimatol Palaeoecol 449:166–173

202. Shu DG et al (2006) Lower Cambrian vendobionts from China and early diploblast evolution. Science 312(5774):731–734
203. Hoyal Cuthill JF, Han J (2018) Cambrian petalonamid *Stromatoveris* phylogenetically links Ediacaran biota to later animals. Palaeontology 61(6):813–823
204. Tang F et al (2011) *Eoandromeda* and the origin of Ctenophora. Evol Dev 13(5): 408–414
205. Xingliang Z, Reitner J (2006) A fresh look at *Dickinsonia*: removing it from Vendobionta. Acta Geol Sin 80(5):636–642
206. Dzik J (2002) Possible ctenophoran affinities of the Precambrian "sea-pen" *Rangea*. J Morphol 252(3):315–334
207. Zhuravlev AY, Wood RA, Penny AM (1818) Ediacaran skeletal metazoan interpreted as a lophophorate. Proc Biol Sci 2015(282): 20151860
208. Shore AJ et al (2021) Ediacaran metazoan reveals lophotrochozoan affinity and deepens root of Cambrian Explosion. Sci Adv 7(1): eabf2933
209. Mackie GO, Mills CE, Singla CL (1992) Giant axons and escape swimming in *Euplokamis dunlapae* (Ctenophora: Cydippida). Biol Bull 182(2):248–256
210. Rudman WB (1999) Benthic ctenophores. Sea Slug Forum. Available from: http://www.seaslugforum.net/find/ctenopho
211. Samimi-Namin K et al (2023) Aggregations of a sessile ctenophore, *Coeloplana* sp., on Indo-West Pacific gorgonians. Diversity 15(10):1060
212. Francisco B (2006) Aglajid eating a benthic ctenophore. [Message in] Sea Slug Forum. Sea Slug Forum. Available from: http://www.seaslugforum.net/find/15543
213. Benham WBS (1907) New Zealand Ctenophores. Transactions of the New Zealand Institute 39:138–143, pl. VII

Chapter 3

Brief History of Placozoa

Daria Y. Romanova and Leonid L. Moroz

Abstract

Placozoans are morphologically the simplest free-living animals. They represent a unique window of opportunities to understand both the origin of the animal organization and the rules of life for the system and synthetic biology of the future. However, despite more than 100 years of their investigations, we know little about their organization, natural habitats, and life strategies. Here, we introduce this unique animal phylum and highlight some directions vital to broadening the frontiers of the biomedical sciences. In particular, understanding the genomic bases of placozoan biodiversity, cell identity, connectivity, reproduction, and cellular bases of behavior are critical hot spots for future studies.

Key words Placozoa, *Trichoplax*, *Hoilungia*, Evolution, Basal Metazoa, History, Ecology, Phylogeny, Morphology, scRNA-seq, Genomics

1 Introduction

Figure 1 summarizes more than 100 years of placozoan's investigations—a fast-growing field with many surprises and discoveries [1–7]. Placozoa is a remarkable and enigmatic group critical to understanding the origin of animal innovations [6, 8, 9] (Fig. 2). We will briefly introduce placozoans and current trends toward exploring these still cryptic organisms. The key player in this endeavor is *Trichoplax*.

2 What Is *Trichoplax*?

Trichoplax adhaerens (Fig. 3) was initially discovered in a seawater aquarium in Graz, Austria, by Franz Schulze [7, 10, 11]; animal's Latin generic name is derived from the classical Greek "τρίχοξ" meaning hair and "πλάξ," plate. Thus, *Trichoplax adhaerens* can be translated as a hairy adhesive plate. *Trichoplax* is the simplest known free-living animal [1, 6, 12]. Many early studies have focused on the morphological organization and biology of *Trichoplax adhaerens*

Leonid L. Moroz (ed.), *Ctenophores: Methods and Protocols*, Methods in Molecular Biology, vol. 2757, https://doi.org/10.1007/978-1-0716-3642-8_3, © Springer Science+Business Media, LLC, part of Springer Nature 2024

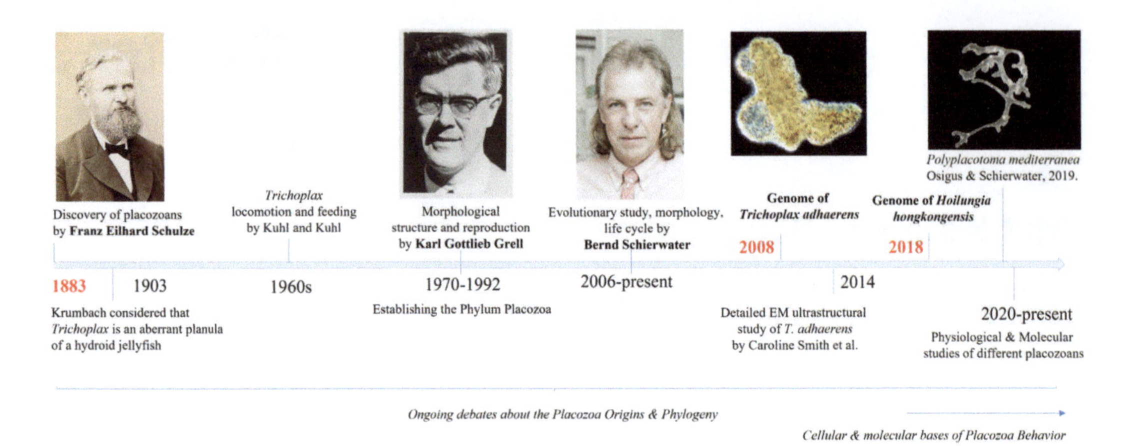

Fig. 1 The brief history of Placozoa research: from the discovery of *Trichoplax* to functional genomics; see text for details

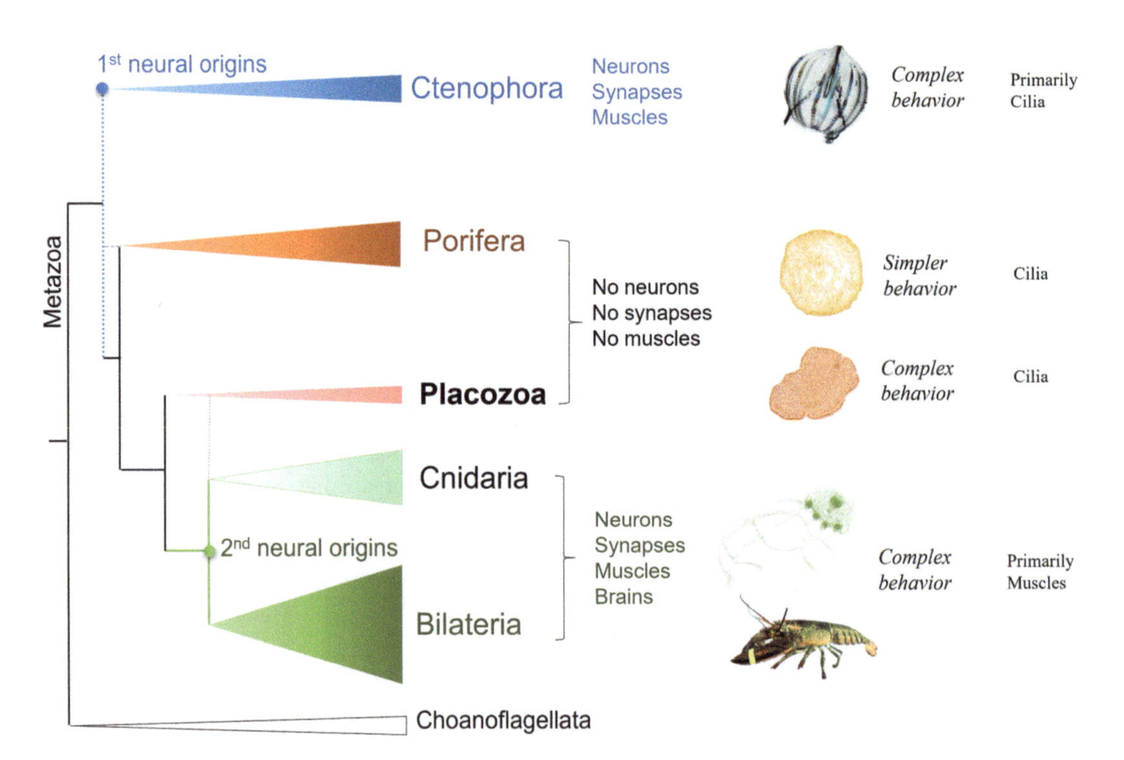

Fig. 2 Phylogeny of basal metazoan lineages, the position of Placozoa, and neural evolution. The consensus tree and schematic of the phylogenetic relationships were done according to [54–57, 130], but also *see* [54, 57, 58, 131] and text for details

[13–17]. At that time, *Trichoplax* was considered an aberrant cnidarian larva.

Krumbach studied the biology of hydroid jellyfish *Eleutheria krohni*, including their sexual reproduction [18]. And since *Trichoplax* was found in the same aquarium as *E. krohni*, Krumbach

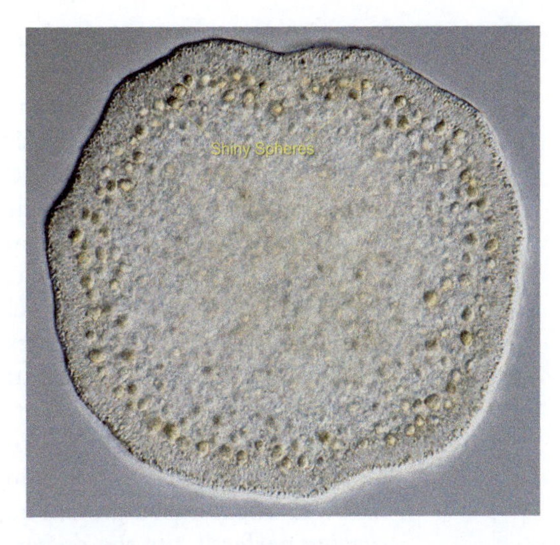

Fig. 3 The general morphology of *Trichoplax* sp. (H2 haplotype). Shiny spheres, possible toxin-releasing cells are indicated

considered that *Trichoplax* was an aberrant planula of hydroid jellyfish [18]. The most influential zoological textbook at that time stated: "Thus, *Trichoplax* and *Treptoplax*, which have the construction of planulae, were actually to be modified planulae of Hydroidea" [19]. Such perceptions reduced attention to these mysterious animals till the 1960s of the twentieth century. Monticelli described *Treptoplax reptans* from the aquarium of the Zoological Station at Naples (Napoli, Stazione Zoologica) [20, 21]. The described differences *Treptoplax* from *Trichoplax* was the apparent absence of cilia on the upper side and cellular microanatomy. However, no additional findings of this species were reported, and today, *Treptoplax* is not recognized as a valid species.

In the 1960s, Kuhl and Kuhl published papers on locomotion and feeding of *Trichoplax* [22, 23]. Karl G. Grell expanded morphological and behavioral studies of *Trichoplax* obtained from a university marine aquarium in Tübingen (originally collected from the Red Sea) and proposed the existence of a new zoological phylum.

The separate phylum Placozoa was formally established in 1971 [24], after careful microanatomical analysis of *Trichoplax adhaerens* [1, 25–27]. The name of the phylum was derived from Placula—a hypothetical two-layered organism ancestral to all animals [28]. Thus, *Trichoplax* was not cnidarian larvae anymore [29].

Several authors [24, 30–36], working independently, expanded morphological characteristics of *Trichoplax* (animals were obtained from different habitats and might be different haplotypes) with a systematic summary of classical microscopic anatomy provided by Grell and Ruthmann [1] and the description of just four cell types in the entire organism. This fact established *Trichoplax* as the simplest free-living animal (some obligate parasites might be simpler).

As a result, multiple evolutionary scenarios of early animal evolution were proposed, considering this two-layered body plan as the first transition from unicellular/colonial organisms to more complex animal organizations in other basal metazoan lineages.

In 1880s, Il'ya I. Mechnikov proposed the *functional* theory of *Parenchymella* or Phagocytella as the common ancestor of all Metazoa [14, 31, 37, 38], in parallel with Haeckel's Gastrea [39] and Butschli's Plakula [28] theories (see also [7, 40]). In two later scenarios, *Trichoplax*'s architecture can be interpreted either as a secondarily unfolded gastrula or a primarily two-layered body plan. One of the essential postulates of the Phagocytella hypothesis is the most ancestral method of animal nutrition, phagocytosis [41], followed by external digestion due to the secreted enzymes [8]. *Trichoplax* perfectly fits the scenario of preserving such an ancestral body plan [9, 27, 30, 31]. A detailed description of the morphology of fiber cells enhanced this hypothesis. Indeed, the syncytium of tetraploid fiber cells, with a gigantic mitochondrial cluster [1, 42, 43], phagocyte bacterial and algae, further illuminated the Phagocytella as a functional reconstruction of the digestion mechanisms as driving forces in earlier animal evolution. New studies provide additional facts related to feeding habits, phagocytosis, and the presence of micro-cavities as mechanisms of nutrient uptake and signaling [42, 44–47].

The presence of endosymbiotic bacteria in fiber cells was also confirmed by ultrastructural observations and genomic studies [43, 48–50]. About two consistent bacterial phylotypes, named Cand. *Grellia incantans* (Rickettsiales) and Cand. *Ruthmannia eludens* (Margulisbacteria) (see more details in [43, 49]. However, the question remains: is it a symbiotic relationship or method of nutrition? In any case, some representatives of Ediacaran Biota might have a similar feeding strategy on microbian and algal mats as in extant *Trichoplax*, and particular Ediacarans, such as *Dickinsonia*, could be placozoan-grade organisms [51, 52]. Considering a substantial diversity of life forms and strategies in placozoans [53], out-of-box re-evaluations of some late Proterozoic fossils would be interesting.

One recent phylogeny still considers placozoans as highly derived and secondarily simplified cnidarians [54]. However, most phylogenomic reconstructions view Placozoa as the sister group to the clade Cnidaria and Bilateria [29, 55–58]. Under any scenario (Fig. 2), placozoans are important model and reference organisms to understand the origins and evolution of animal tissues, cell types [6, 9, 59–68], and the nervous system in particular [60, 63, 69–72].

3 How Many Placozoans Do Exist in the World Ocean?

For more than 30 years, *Trichoplax* was the sole representative of these disk-shaped cryptic marine animals until recently [65, 73–79]. Currently, four genera of Placozoa have been confirmed [80–82], and their biogeography is presently restricted to subtropical and tropical shallow habitats (Fig. 4). Nevertheless, over 30 mitochondrial haplotypes were discovered [5, 73], and the number of cryptic species might have exceeded several dozen [3, 74, 82]. Regardless, four placozoan species have been formally described: *Trichoplax adhaerens* (mitochondrial H1 haplotype), *Hoilungia hongkongensis* or H13 [80], *Polyplacotoma mediterranea* or H0 haplotype [81], and *Cladtertia collaboinventa* or H23, nov., as the fourth newly described species of Placozoa [82].

The nuclear genomes of *Trichoplax adhaerens* [2], *Trichoplax* sp. [83], and *Hoilungia hongkongensis* [80] have been sequenced, with draft genomes from four other haplotypes (potentially novel yet undescribed species [54]), confirming distinct genomic organizations for each species. The genomes of *Polyplacotoma mediterranea* and *Cladteria collaboinventa* has not yet been sequenced, and chromosome-scale assemblies are highly desirable. The sequenced placozoan genomes provide a good starting point for comparative analyses, but a formal description of novel species is needed (even based on this digital information).

Mitochondrial genomes in placozoans are also relatively large and unusual for animals [73, 84]; they contain additional genes, ORF, spacers, and introns, not typical for the rest of Metazoa. The evolutionary and functional significance of these larger mitochondrial genomes and unknown proteins in placozoans is unknown.

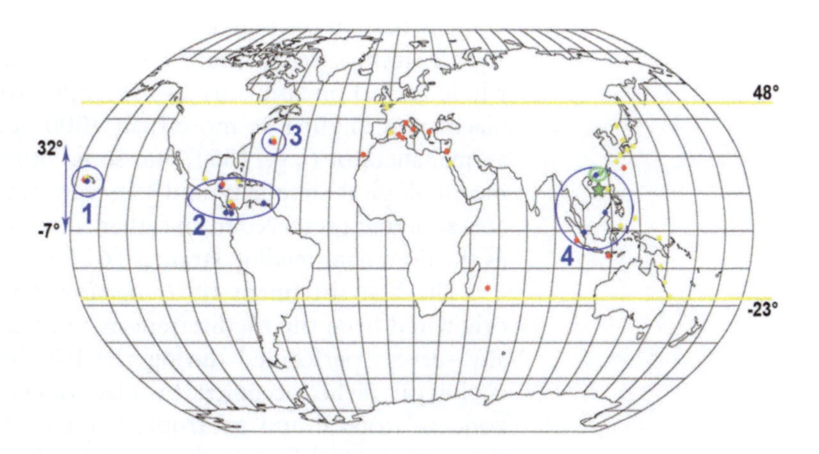

Fig. 4 Biogeography of *Trichoplax* and *Hoilungia* in the World Ocean: *Trichoplax adhaerens*, *Trichoplax* sp., *Hoilungia* sp., *Hoilungia hongkongensis* (modified from Refs. [4, 78])

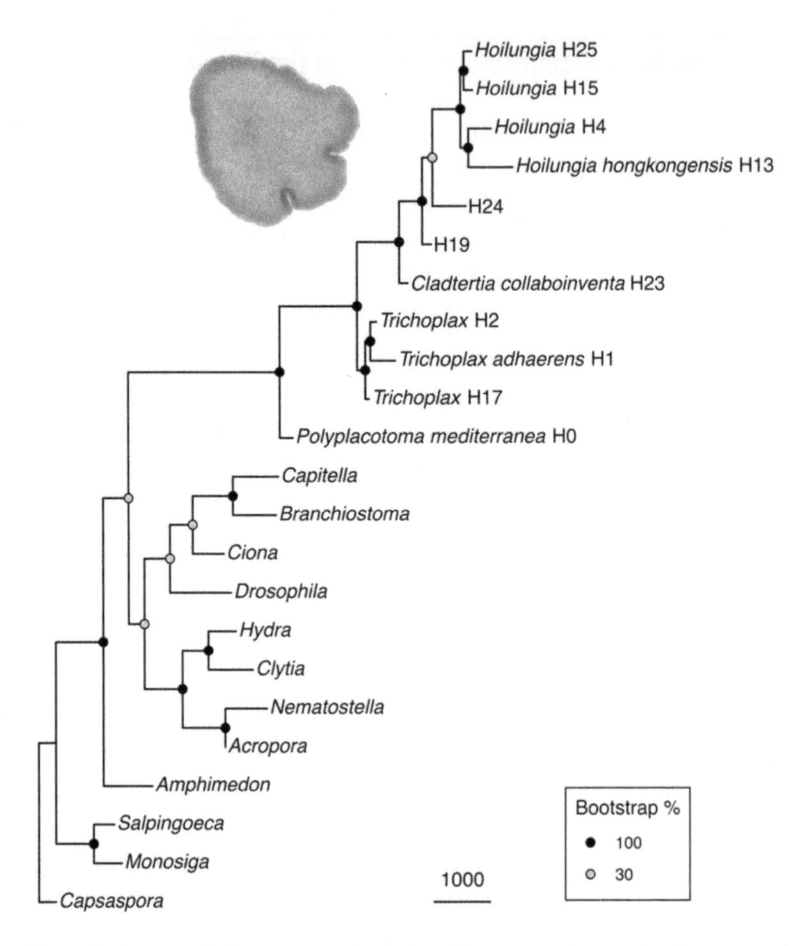

Fig. 5 Phylogeny of placozoan and relationships among known haplotypes and species (from Ref. [82]). The phylum might contain more than four genera and multiple species; see text for details. See also details in Refs. [4, 5, 80–82]

The current phylogenetic relationships among known placozoans are summarized in Fig. 5. A novel higher-level taxonomy for a whole animal phylum was also recently proposed [82]. Although placozoans might lose more than 4000 genes from the common animal ancestor (e.g., [55]) and some other features, placozoans' morphological simplicity and lifestyles can still be viewed as the ancestral trait preserved from earlier animals with external digestion as the dominant feeding strategy [61].

The first specimens of *Trichoplax* (known as H1 haplotype) originated from the Mediterranean and Red Seas. Broad biodiversity surveys performed during the last decade indicate that the ecological niche occupied by placozoans covers the sublittoral zone of tropical and subtropical waters from 0.5 to 20 meters deep and potentially contains many species [3].

Placozoans were collected from warm coastal waters in such habitats as mangroves, reefs, or shallow bays [75, 77, 78, 85–

87]. Placozoans exist under variable conditions: different salinity (20–50%), temperatures (11–27 °C), light sources, and variable pH [7, 65, 77, 78], where the growth of their populations is subject to seasonal fluctuations [85, 87, 88].

The exact microenvironments of placozoans and natural food sources are unknown. Pearse and Voigt [87] suggested that Placozoa in tropical and subtropical areas feed on absorbing organic detritus, as well as biofilms of algae and bacteria on corals and other substrates. Not surprisingly, most of our knowledge about this animal group comes from studies of these animals in culture.

Very little is known about the natural predators of placozoans. At least two species of small turbellarian-like gastropods of the genus *Rhodope* (Heterobranchia) might be specialized predators of placozoans, sharing the same habitats [89, 90]. These tiny gastropods have a highly simplified morphology with secondary loss of many molluscan traits (e.g., radula, anterior tentacles, foot, gills, mantle, and heart). Several species may prey on placozoans (e.g., *Rhodope veranii, R. marcusi, R. roskoi, R. rousei*, and *Helminthope psammobionta* [89]).

One species even has its name as the placozoan eater: ***Rhodope placozophagus*** [89]. Cuervo-González provided an interesting description of its predatory behaviors [89]. "*R. placozophagus* is a very active and fast-moving organism that, upon fortuitous contact with placozoans, immediately begins to suck the tissues of its prey using strong peristaltic movements of the buccal bulband digestive gland." The mollusc sometimes eats for up to 50 min; *R. placozophagus* "does not suck the loose interior content of its prey, which consists mainly of fibrous cells..., but instead eats the complete tissues including the upper epithelium which is covered with shiny spheres." It was also noted that "the placozoans are unaware that they are being predated and do not detect the presence of the predator or make any attempt to escape. Since placozoans are able to regenerate their structures and regrow from small fragments, predation is not necessarily detrimental to them. Occasionally, fragmented placozoans become spherical [91], which makes them unable to be caught by their predators" [89]. The last note is important since it suggests additional ecological explanations of diverse morphoforms (apart from flat placozoans) and life strategies across placozoans [53, 92].

Potentially placozoans can be attacked by various invertebrates in their ecological niches (cnidarians, nemerteans, polychaetes and acoels, platyhelminthes, molluscs, arthropods, etc.). And it might not be surprising that placozoans developed an antipredator chemical defense in the form of toxin-containing cells located in the upper layer of animals; these structures were earlier called "shiny spheres" [46, 93]. Jackson and Buss [93] provided evidence that "shiny spheres" contain potential toxins, which might repel predators. Cuervo-González also reported that "acoela, small snails or copepods abruptly turned and moved in another direction upon contact" with *Trichoplax* in laboratory experiments [89].

4 The Hidden Complexity of Placozoans

Described morphological simplicity of placozoans contrasts with the relatively complex behavioral repertoire of placozoans. Even initial observations of these disklike animals revealed highly dynamic changes in their locomotion and shapes (Fig. 6). And these dynamics are strongly affected by food sources and environmental conditions (Fig. 7). High-density food substrates induced social-type organization (Fig. 8) when many animals aggregated with collective behaviors [53, 94]. Both well-coordinated behavioral feeding and locomotory patterns have been described [85, 95–102], including chemotaxis [103] and phototaxis [53]. Amino acids [71, 104] and peptides [105] can induce and integrate certain aspects of locomotion and feeding. The signaling role of gaseous nitric oxide (NO) was also proposed [106, 107].

Complex behaviors [94, 95, 97, 99, 108] imply complex cellular organization and intercellular communications. However, little is known about the connectomes, morphological and ultrastructural differences across placozoans [109]. Virtually all microanatomical descriptions were performed on *Trichoplax adhaerens* (see

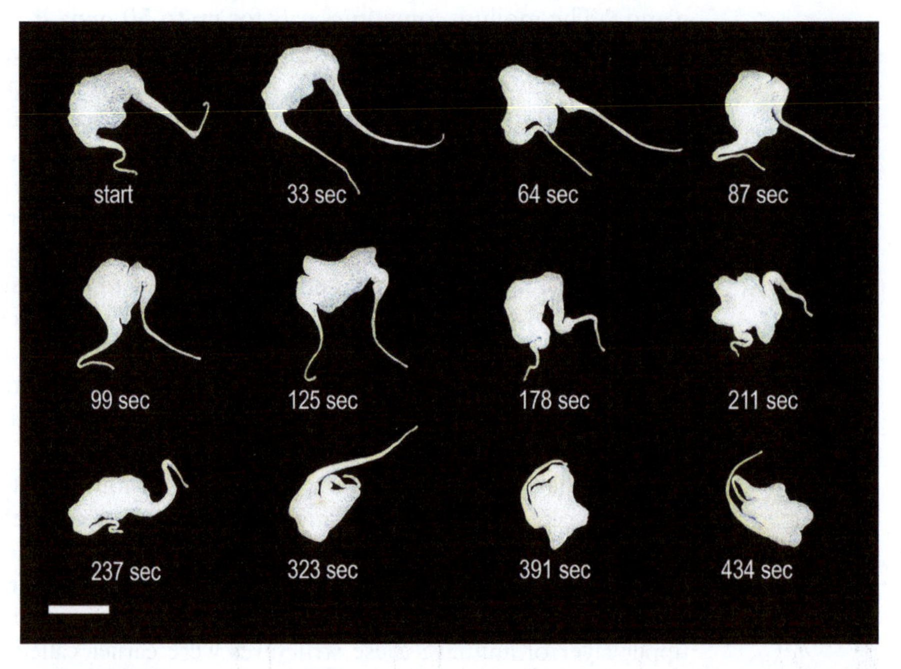

Fig. 6 Dynamic changes in the body shape of *Trichoplax* sp. (H2 haplotype). Note unusual morphological features such as "pseudopodia"-like elongations in a single individual during ~7 min of time-lapse (from Ref. [53]). Scale bar: 200 μm

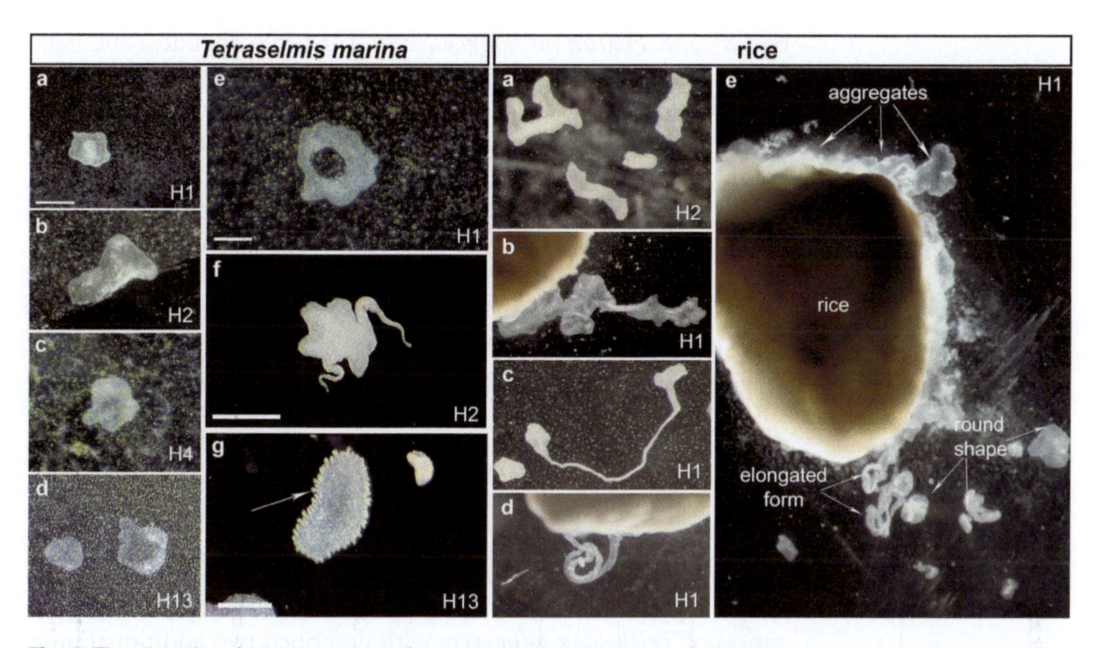

Fig. 7 The diversity of placozoan body forms. Illustrated examples of different haplotypes (indicated in each photo) cultured on two feeding substrates (the gree algae, *Tetraselmis marina* [left box], and rice grains [right boxes)). *Tetraselmis marina* box shows both canonical placozoans body shapes (**a**–**d**) and unusual morphologies (animals with a "hole" in the middle of the body (**e**), elongated "pseudopodia"-like structures (**f**, *see* also Fig. 6), and animals with numerous small ovoid formations in the rim area (**g**). The "rice box" also shows varieties of animal's morphology in cultures that use rice as a food source. The aggregations of animals can often be found around grains (rice box a–d: different shapes of placozoans (a–b), including those with highly elongated forms c–d). Scale bar: *Tetraselmis marina* box: 500 μm; (from Ref. [53]). Rice box: grain, 1 mm

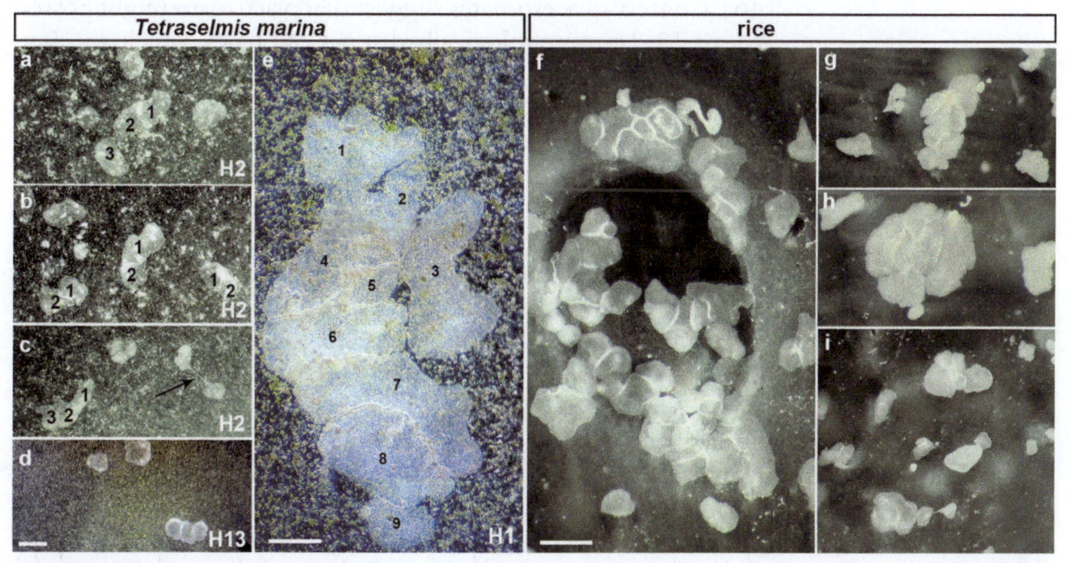

Fig. 8 Social feeding behavior during long-term culturing on a high density of algal (*Tetraselmis marina* box) and rice (for H1-f) substrates. The aggregation of animals depended upon the density of the substrates. Aggregates often included 2–15 individuals. For example, there were aggregates of 2–3 animals in H2 (a–c) on *Tetraselmis marina* and 9 H1 individuals (e) at the same substrate, but H2 often forms aggregates for 10–15 animals around rice grains (g–i). This behavioral pattern was observed for all haplotypes (H1, H2, H13). The arrow in c indicates the fission of *Trichoplax* (from Ref. [53]). Scale bar: a–d, 100 μm; e, 1 mm; f–i, 500 μm

below) and *Hoilungia hongkongensis* [46]. No chemical and electrical synapses (with either innexins or connexins) were reported from placozoans.

Trichoplax and kin have no symmetry and no organs, and their flat body is organized into three distinct cell layers with a prominent dorsoventral organization. Dorsal and ventral homologization between Placozoa and classical bilaterian axes is not evident [110] and might not be appropriate terms for placozoans. Thus, most literature distinguishes the upper vs. lower (substrate facing) sides in *Trichoplax* and kin. Even the first morphological analyses of Placozoa showed the absence of canonical neurons and muscles with four somatic cell types described: (i) upper epithelial cells with a single cilium, (ii) lower [substrate-facing] epithelial cells with a single cilium and microvilli, (iii) gland [secretory] cells with multiple vesicles, and (iv) fiber cells in the middle layer, which can contact other cell types [1].

Smith and colleagues using advanced fast freezing electron microscopy, provided a detailed study of the ultrastructural organization of *Trichoplax adhaerens* with described two additional morphological cell types [111]: (v) lipophil cells with exceptionally large vesicles (likely releasing digestive enzymes) and high-lipid content and (vi) crystal cells with aragonite form of calcium carbonate, which can be gravity sensors and contribute to geotaxis [98]. Additional types and subtypes exist [44, 98, 112], and future systematic studies are needed. Comparable investigations of *Hoilungia* also illuminated the growing diversity of cell types [46]. Conservative estimates suggest that placozoans might possess at least 12–16 morphologically distinct cell types, which are summarized in Fig. 9.

Some of these cells, such as neuroid cells and even fiber cells forming syncytial-type meshworks, might be functional analogs of neurons. However, the absence of recognized chemical and electrical synapses implies multiple levels of volume-type transmission, which is recently discussed [61, 113, 114] as the most ancestral type of signaling in animals and beyond. Peptidergic cells are likely sources of such non-synaptic signaling [61, 105, 113], although other signal molecules have also been implicated in the control of placozoan behaviors [61, 104, 107, 115].

At least in some placozoan cells, the electrical excitability, diversity of ion channels [77, 89–92], and ultrafast mechanical contractions [74] indicate additional mechanisms for cellular interactions and behavior control. Integrating morphological, single-cell genomics [59], and behavioral data is needed to understand hidden molecular complexity and cellular bases of behaviors in these seemingly simpler animals. However, we would not be surprised that about 50 cell types and states would be discovered in placozoans [60].

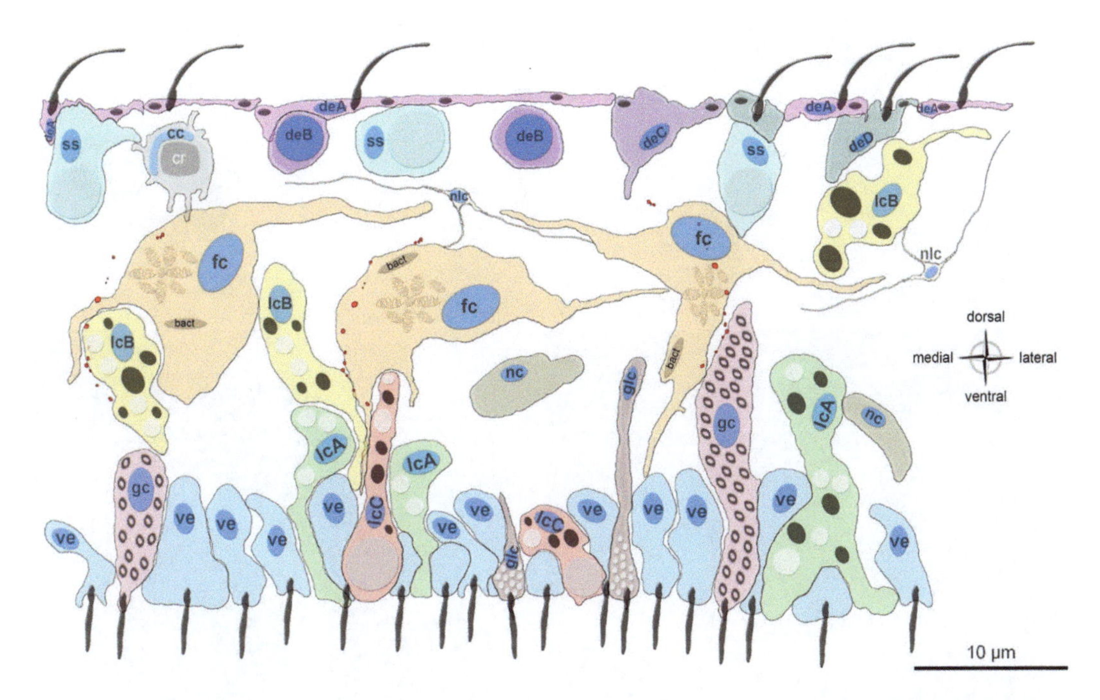

Fig. 9 The morphological diversity of cell types at *Hoilungia*. Abbreviations: ve, lower ["ventral"] epithelium; de, upper ["dorsal"] epithelium with likely subpopulations of cells (A, B, C); lc, lipophil cells with subpopulations (lcA, lcB. lcC); gc, gland cells; glc, putative mucus-released cells; fc, fiber cells, nlc, neuroid-like cells; cc, crystal cells; cr, crystal; ss, shiny spheres; p, pigment; inc, inclusion or vesicles; nc, additional cell type. The schematic diagram combines data from several ultrastructural observations [1, 44, 46, 98, 111, 132, 133]. Scale bar: 10 μm

5 Life Strategies in Placozoa

Placozoa can reproduce both sexually and vegetatively. Two types of asexual reproduction are known: by dividing the body (fragmentation or fission) and budding; the latter mode includes the formation of "swarmers"—small "larval-type" progeny [91, 116]. During fission, the process can take 1–3 days and ends with forming two often equal individuals [30, 91, 117]. Swarmer-type forms were also described for four placozoan species, including *Hoilungia* [53]. However, other more spherical forms and structures in placozoans occur (Fig. 10). The exact relationships among these not-flat-like forms and stages of placozoans and their classification are unknown. Figure 11 shows schematic relationships of different morphological and, perhaps, developmental forms (*see* details in [53]).

Sexual reproduction is possible in placozoans [83, 118–120]. All studied haplotypes (H1, H2, H4, and H16) showed likely events of early embryogenesis (up to the 64–128 cell stage; *see* also [1, 119]). Initial analysis suggests that subtypes of cells can form

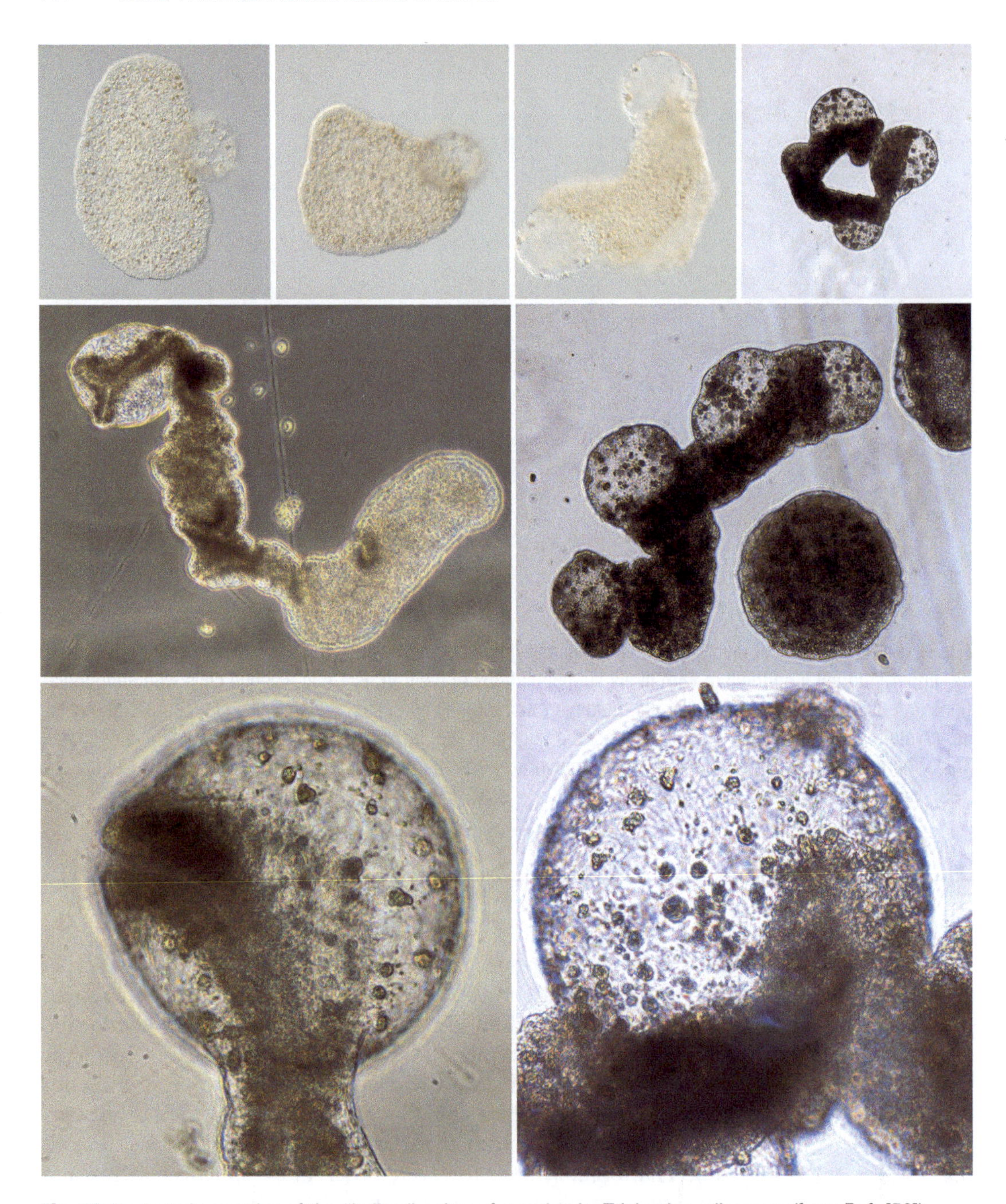

Fig. 10 Illustrated examples of the "hollow" sphere formation in *Trichoplax adhaerens* (from Ref. [53])

oocytes [1, 119]. It is a sporadic process that creates many experimental challenges. The activation of oocyte formation might be facilitated by a high population density, lack of food, stress, degenerative phases of culturing, and possible temperature changes [119, 121]. Oocytes could be formed from cells of the ventral epithelium, usually, one oocyte per individual [121], and the oocyte size can reach 70–120 μm in diameter [1, 119]. After fertilization, a

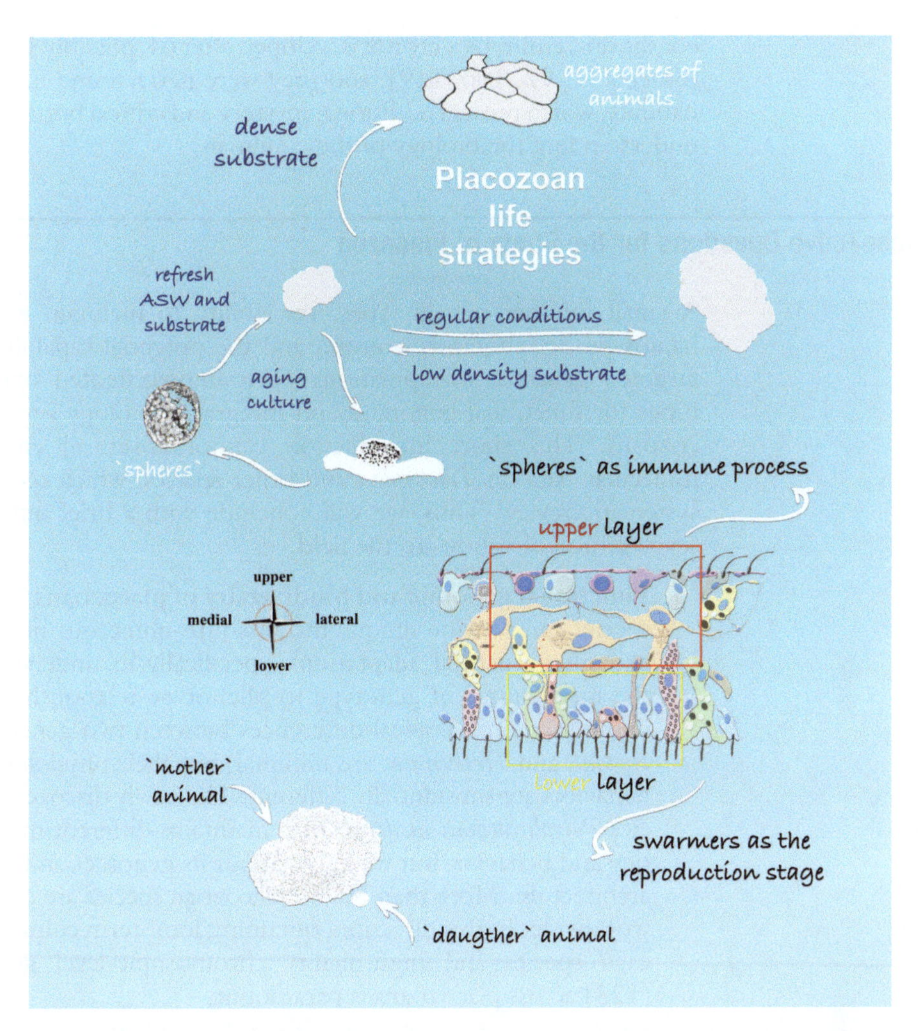

Fig. 11 Life strategies in Placozoa: Schematic representation of feeding and reproductive stages (from Ref. [53]). The density of food substrate predominantly determines formations of different morphological stages, and dense algal substrate leads to aggregates from multiple animals ("social" feeding behavior). In aging culture, the appearance of specialized spherical structures was often observed. The formation of small juvenile animals or "swarmers" might have different etiology and development from other spherical type structures, as shown in the schematic cross-section of *Trichoplax* [53]

protective membrane is formed, and first divisions might occur inside the maternal organism [1, 121, 122]. Grell and Benwitz indicated that fiber cells might assume nutritive functions, acting as trophocytes [1, 121]. Even more interesting, the oocyte itself also developed "short, pseudopodia-like projections that phagocytize parts of trophocytes' extensions," and bacteria, potentially from fiber cells, can be found in the endoplasmic reticulum of oocytes [1, 119]. The cellular origin of the sperm cell has not been established; either fiber cells or novel type of small non-flagellated cells might be involved [1]. Actual "true" sperm cells have never been convincingly identified. Under laboratory conditions, embryos were not developed after 64, possibly 128 blastomeres [1, 92, 116, 119], and they were never found in natural habitats, which present additional mystery and critical bottleneck in understanding the biology of these animals.

6 Prospective Questions for the Study of Placozoa

A small number of cell types, the ability to maintain multiple haplotypes in laboratory culture, and the potential capabilities of targeted genomic manipulations open unprecedented opportunities for functional genomics and integrative biology across placozoans. This short introduction cannot cover all emerging molecular data on *Trichoplax* and other species, which requires a systematic review. Thus, we will conclude with a brief outline of seven critical directions to the field.

1. **Functional genomic and biodiversity of placozoans**. Cryptic species can provide unique insights into numerous molecular and ultrastructural adaptations, specifically in understanding the mechanisms of genotype-to-phenotype relationships. For example, morphological differences between two genera, *Trichoplax* and *Hoilungia*, are minimal, but their physiology and behaviors are substantially different. The newly discovered species *Polyplacotoma mediterranea* maintains different morphology and behavior, but we know about its genomes and cellular architecture. More than 30–50 placozoan species are expected worldwide. In this direction, obtaining long-term culturing for each species and high-quality chromosome-level genomes [123] across placozoans is paramount.
2. **Characterization of molecular/genomic diversity of cell types across Placozoa**. Single-cell "omics" approaches are essential to understand the mechanisms underlying the diversity and function of cell types across different placozoan species. Novel approaches include single-cell transcriptomics or scRNA-seq [59] and still little-understood proteomics, metabolomic, and lipidomic profiling of individual cell types under different functional states during the development and regeneration.
3. **Identification of cellular homologs across placozoan species and beyond**. Could cross-species homologization be reliably established across placozoa? The brief answer is yes. But finding homologs between placozoan cell types and cell types in other phyla is an open question.
4. **Understanding life strategies across Placozoa** at all levels of their organization. Some aspects of these studies have been outlined recently [53]. However, we know little about life

history transitions in Placozoa. Placozoans might not be the only traditional flat animals; multiple spheroid and other forms have been discovered with unknown mechanisms and unknown relationships [91, 92, 116]. Some placozoans' life forms or developmental stages might not be benthic; both swimming swarmers and other forms might exist. We know practically nothing about sexual reproduction in placozoans [83, 119]. Early embryonic-like stages have been discovered morphologically [1], but it was hard to induce them and study them in culture conditions.

5. **Deciphering cellular bases of behavior in these nerveless animals**. Both individual and even social behavior have been described in placozoans. However, mechanisms of behavioral integration and intercellular communications in these animals, without synapses, are unknown, and different models of non-synaptic communication have been proposed [61, 100]. It is still the tip of the iceberg, and discoveries of many signaling molecules are expected, such as nitric oxide, glycine, glutamate, GABA, ATP, and others [71, 104, 107, 115], in addition to small peptides [105, 124].
6. **Mechanisms of the placozoan immunity and potential symbiogenesis with prokaryotes**. The initial survey of immune-type systems in placozoans reveals the apparent simplicity of known metazoan immune complement [125]. With various bacteria located in fiber and other cell types [43, 49, 50, 126], the underlying biology of these relationships is still unknown.
7. **Generating transgenic placozoan line**. This is a novel and open-ended frontier in the field. The morphological simplicity of placozoans provides unprecedented opportunities to study the genotype-to-phenotype relationships at all levels of organizations: from cells to behavior.

To sum up, we enthusiastically encourage the field collections and developments of long-culture conditions for all placozoan species (and haplotypes) and the "transfer" of future interdisciplinary research to the sea—to natural habitats in existing ecosystems around the globe. Some practical protocols have been published [127–129]. This book provides additional method chapters that are hopefully useful for further investigation of this prebilaterian group of animals.

Acknowledgments

This work was supported in part by the Human Frontiers Science Program (RGP0060/2017) and National Science Foundation (IOS-1557923) grants to L.L.M. Research reported in this publication was also supported in part by the National Institute of Neurological Disorders and Stroke of the National Institutes of Health under award number R01NS114491 (to L.L.M). D.R. was supported by the Russian Science Foundation grant (23-14-00050). The content is solely the authors' responsibility and does not necessarily represent the official views of the National Institutes of Health.

References

1. Grell KG, Ruthmann A (1991) Placozoa. In: Harrison FW (ed) Microscopic anatomy of invertebrates. Wiley-Liss, New York, pp 13–27
2. Srivastava M et al (2008) The *Trichoplax* genome and the nature of placozoans. Nature 454(7207):955–960
3. Schierwater B, DeSalle R (2018) Placozoa. Curr Biol 28(3):R97–R98
4. Schierwater B et al (2021) The enigmatic Placozoa Part 2: exploring evolutionary controversies and promising questions on earth and in space. BioEssays 43(10):e2100083
5. Schierwater B et al (2021) The enigmatic Placozoa Part 1: exploring evolutionary controversies and poor ecological knowledge. BioEssays 43(10):e2100080
6. Syed T, Schierwater B (2002) *Trichoplax adhaerens*: discovered as a missing link, forgotten as a hydrozoan, re-discovered as a key to metazoan evolution. Vie Milieu 52:177–187
7. Schierwater B et al (2010) *Trichoplax* and Placozoa: one of the crucial keys to understanding metazoan evolution. In: Key transitions in animal evolution. CRC Press, Boca Raton, pp 289–326
8. Nielsen C (2019) Early animal evolution: a morphologist's view. R Soc Open Sci 6(7): 190638
9. Grell K (1981) *Trichoplax adhaerens* and the origin of Metazoa. In: Lincei ADC (ed) Origine dei Grandi Phyla dei Metazoi. Accademia Nazionale dei Lincei, Convegno Intern, pp 101–127
10. Schulze FE (1883) *Trichoplax adhaerens*, nov. gen., nov. spec. Zool Anz 6:92–97
11. Schulze FE (1891) Uber *Trichoplax adhaerens*. Phys Abh Kgl Acad Wiss Berl:1–23
12. Rassat J, Ruthmann A (1979) *Trichoplax adhaerens* F.E. Schulze (Placozoa) in the scanning electron microscope. Zoomorphologie 72:59–72
13. Graff L (1891) Die Organisation der Turbellaria acoela. W. Engelmann, Leipzig
14. Metschnikoff E (1886) Embryologische Studien an Medusen. Ein Beitrag zur Genealogie der Primitiv-Organe. A. Holder, Wien
15. Noll F (1890) Über das Leben niederer Seetiere, pp 85–87
16. Stiasny G (1903) Einige histologische details über *Trichoplax adhaerens*. Zeitschr wiss Zool 75:430–436
17. Gabrowski T (1903) Morphogenetische Studien als Beitrag zur Methodologie zoologischer Forshung. Jena
18. Krumbach T (1907) *Trichoplax*, die umgewandelte Planula einer Hydramedusae. Zool Anz Suppl 31:450–454
19. Hyman LH (1940) Invertebrates: Protozoa through Ctenophora, vol 1. McGraw-Hill, New York/London, p 726
20. Monticelli FS (1893) *Treptoplax reptans* n.g., n.sp. Atti dell'Academia dei Lincei, Rendiconti 5(II):39–40
21. Monticelli FS (1896) Adelotacta zoologica. 2. *Treptoplax reptans* Montic. Mitt Zool Stat Neapel 12:444–462
22. Kuhl W, Kuhl G (1966) Untersuchungen uber das bewegungsverhalten von *Trichoplax adhaerens* F. E. Schulze (Zeittransformation: Zeitraffung), vol 56. Z. Morph. U. Okol Tiere, pp 417–435
23. Kuhl W, Kuhl G (1963) Bewegungsphysiologische Untersuchungen an *Trichoplax Adhaerens* F.E.Schulze. Zool Anz Suppl 26:460–469
24. Grell KG (1971) Trichoplax adhaerens F.E. Schulze und die Entstehung der Metazoen. Naturwiss Rundschau 24:160–161
25. Grell K (1972) Eibildung und Furchung von *Trichoplax adhaerens* FE Schulze (Placozoa). Zoomorphology 73:297–314
26. Grell KG, Benwitz G (1971) Die Ultrastruktur von *Trichoplax adhaerens* F. E. Schulze. Cytobiologie 4:216–240

27. Grell KG, Benwitz G (1981) Additional investigations on the ultrastructure of *Trichoplax adhaerens* F.E. Schulze (Placozoa). Zoomorphology 98(1):47–67
28. Butschli O (1884) Bemerkungen zur Gastraeatheorie. Morphol Jb 8:415–427
29. Ender A, Schierwater B (2003) Placozoa are not derived cnidarians: evidence from molecular morphology. Mol Biol Evol 20(1): 130–134
30. Malakhov VV (1990) Enigmatic groups of marine invertebrates: *Trichoplax*, Orthonectida, Dicyemida, Porifera. Moscow State University Press, Moscow, p 144. (in Russian)
31. Ivanov AV (1973) *Trichoplax adhaerens*, a Phagocitella-like animal. Zoologiceskij Zurnal (Zool J) 52:1117–1130. (in Russian)
32. Malakhov VV, Nezlin LP (1983) *Trichoplax* – a living model of the origin of multicellular organisms. Nauka 3:32–41. (in Russian)
33. Ivanov DL, Malakhov VV, Tsetlin AB (1980) Fine morphology and ultrastructure of the primitive multicellular organism *Trichoplax* sp. 1. Morphology of adults and vagrants according to the data of scanning electron microscopy. Zool Zhurnal 59(12):1765. (in Russian)
34. Ivanov DL, Malakhov VV, Tsetlin AB (1980) New find of a primitive multicellular organism *Trichoplax* sp. Zool Zhurnal 59:1735–1739. (in Russian)
35. Ivanov DL et al (1982) Fine morphology and ultrastructure of the primitive multicellular organism *Trichoplax* sp. 2. Ultrastructure of adults. Zool Zhurnal 61:645–652. (in Russian)
36. Okshtein IL (1987) On the biology of *Trichoplax* sp. (Placozoa). Zool Zhurnal 66(3):339
37. Ivanov AV (1968) The origin of the multicellular animals: phylogenetic essays. Nauka, Leningrad, p 288. (in Russian)
38. Metchnikoff EI (2013) In: Tauber AI, Williamson D, Gourko H (eds) The evolutionary biology papers of Elie Metchnikoff. Springer, Cham
39. Haeckel E (1874) Die Gastrea-Theorie, die phylogenetische Classification des Thierreichs und die Homologie der Keimblatter. Jena Z Naturw 8:1–55
40. Nielsen C (2012) Animal evolution: interrelationships of the living phyla. Oxford University Press, Oxford
41. Metschnikoff E (1892) La Phagocytose Musculaire Ann de L'institut Pasteur 6:1–12
42. Buchholz K, Ruthmann A (1995) The mesenchyme-like layer of the fiber cells of *Trichoplax adhaerens* (Placozoa), a syncytium. Z Naturforsch C Biosci 50c:282–285
43. Hadfield MG, McFall-Ngai MJ (2021) *Trichoplax* and its bacteria How many are there? Are they speaking? In: Bosch TCG, Hadfield MG (eds) Cellular dialogues in the Holobiont. CRC Press, Boca Raton, pp 35–48
44. Mayorova TD et al (2019) The ventral epithelium of *Trichoplax adhaerens* deploys in distinct patterns cells that secrete digestive enzymes, mucus or diverse neuropeptides. Biol Open 8(8)
45. Romanova DY (2019) Cell types diversity of H4 haplotype Placozoa sp. Mar Biol J 4(1): 81–90
46. Romanova DY et al (2021) Hidden cell diversity in Placozoa: ultrastructural insights from *Hoilungia hongkongensis*. Cell Tissue Res 385(3):623–637
47. Ruthmann A, Behrendt G, Wahl R (1986) The ventral epithelium of *Trichoplax adhaerens* (Placozoa): cytoskeletal structures, cell contacts and endocytosis. Zoomorphology 106:115–122
48. Driscoll T et al (2013) Bacterial DNA sifted from the *Trichoplax adhaerens* (Animalia: Placozoa) genome project reveals a putative rickettsial endosymbiont. Genome Biol Evol 5(4):621–645
49. Gruber-Vodicka HR et al (2019) Two intracellular and cell type-specific bacterial symbionts in the placozoan *Trichoplax* H2. Nat Microbiol 4(9):1465–1474
50. Kamm K et al (2019) Genome analyses of a placozoan rickettsial endosymbiont show a combination of mutualistic and parasitic traits. Sci Rep 9(1):17561
51. Sperling EA, Vinther J (2010) A placozoan affinity for *Dickinsonia* and the evolution of late Proterozoic metazoan feeding modes. Evol Dev 12(2):201–209
52. Hoekzema RS et al (2017) Quantitative study of developmental biology confirms *Dickinsonia* as a metazoan. Proc Biol Sci 284(1862): 20171348
53. Romanova DY et al (2022) Expanding of life strategies in Placozoa: insights from long-term culturing of *Trichoplax* and *Hoilungia*. Front Cell Dev Biol 10:10
54. Laumer CE et al (2018) Support for a clade of Placozoa and Cnidaria in genes with minimal compositional bias. elife 7:7
55. Moroz LL et al (2014) The ctenophore genome and the evolutionary origins of neural systems. Nature 510(7503):109–114
56. Whelan NV et al (2017) Ctenophore relationships and their placement as the sister group

to all other animals. Nat Ecol Evol 1(11): 1737–1746

57. Laumer CE et al (1906) Revisiting metazoan phylogeny with genomic sampling of all phyla. Proc Biol Sci 2019(286):20190831
58. Telford MJ, Moroz LL, Halanych KM (2016) Evolution: a sisterly dispute. Nature 529(7586):286–287
59. Sebe-Pedros A et al (2018) Early metazoan cell type diversity and the evolution of multicellular gene regulation. Nat Ecol Evol 2(7): 1176–1188
60. Moroz LL (2018) NeuroSystematics and periodic system of neurons: model vs reference species at single-cell resolution. ACS Chem Neurosci 9(8):1884–1903
61. Moroz LL, Romanova DY, Kohn AB (2021) Neural versus alternative integrative systems: molecular insights into origins of neurotransmitters. Philos Trans R Soc B B 20190762:1–21
62. Scientists, G.C.O et al (2014) The Global Invertebrate Genomics Alliance (GIGA): developing community resources to study diverse invertebrate genomes. J Hered 105(1):1–18
63. Schierwater B et al (2009) The diploblast-bilateria sister hypothesis: parallel revolution of a nervous systems may have been a simple step. Commun Integr Biol 2(5):403–405
64. Schierwater B, de Jong D, Desalle R (2009) Placozoa and the evolution of Metazoa and intrasomatic cell differentiation. Int J Biochem Cell Biol 41(2):370–379
65. Schierwater B (2005) My favorite animal, *Trichoplax adhaerens*. BioEssays 27(12): 1294–1302
66. Brooke NM, Holland PW (2003) The evolution of multicellularity and early animal genomes. Curr Opin Genet Dev 13(6):599–603
67. Collins AG (1998) Evaluating multiple alternative hypotheses for the origin of Bilateria: an analysis of 18S rRNA molecular evidence. Proc Natl Acad Sci U S A 95(26): 15458–15463
68. Osigus HJ et al (2022) Studying placozoa WBR in the simplest metazoan animal, *Trichoplax adhaerens*. Methods Mol Biol 2450:121–133
69. Varoqueaux F, Fasshauer D (2017) Getting nervous: an evolutionary overhaul for communication. Annu Rev Genet 51:455–476
70. Striedter GF et al (2014) NSF workshop report: discovering general principles of nervous system organization by comparing brain maps across species. J Comp Neurol 522(7): 1445–1453
71. Nikitin MA et al (2023) Amino acids integrate behaviors in nerveless placozoans. Front Neurosci 17:1125624
72. Moroz LL, Romanova DY (2022) Alternative neural systems: what is a neuron? (ctenophores, sponges and placozoans). Front Cell Dev Biol 10:1071961
73. Miyazawa H et al (2021) Mitochondrial genome evolution of placozoans: gene rearrangements and repeat expansions. Genome Biol Evol 13(1)
74. Voigt O et al (2004) Placozoa – no longer a phylum of one. Curr Biol 14(22):R944–R945
75. Signorovitch AY, Dellaporta SL, Buss LW (2006) Caribbean placozoan phylogeography. Biol Bull 211(2):149–156
76. Ball EE, Miller DJ (2010) Putting placozoans on the (phylogeographic) map. Mol Ecol 19(11):2181–2183
77. Eitel M, Schierwater B (2010) The phylogeography of the Placozoa suggests a taxon-rich phylum in tropical and subtropical waters. Mol Ecol 19(11):2315–2327
78. Eitel M et al (2013) Global diversity of the Placozoa. PLoS One 8(4):e57131
79. Kawashima T et al (2022) Observing phylum-level metazoan diversity by environmental DNA analysis at the Ushimado Area in the Seto Inland Sea. Zool Sci 39(1):157–165
80. Eitel M et al (2018) Comparative genomics and the nature of placozoan species. PLoS Biol 16(7):e2005359
81. Osigus HJ et al (2019) *Polyplacotoma mediterranea* is a new ramified placozoan species. Curr Biol 29(5):R148–R149
82. Tessler M et al (2022) Phylogenomics and the first higher taxonomy of Placozoa, an ancient and enigmatic animal phylum. Front Ecol Evol 10:10
83. Kamm K et al (2018) *Trichoplax* genomes reveal profound admixture and suggest stable wild populations without bisexual reproduction. Sci Rep 8(1):11168
84. Dellaporta SL et al (2006) Mitochondrial genome of *Trichoplax adhaerens* supports placozoa as the basal lower metazoan phylum. Proc Natl Acad Sci U S A 103(23): 8751–8756
85. Ueda T, Koya S, Maruyama YK (1999) Dynamic patterns in the locomotion and feeding behaviors by the placozoan *Trichoplax adhaerens*. Biosystems 54:65–70
86. Nakano H (2014) Survey of the Japanese coast reveals abundant placozoan populations in the Northern Pacific Ocean. Sci Rep 4: 5356

87. Pearse VB, Voigt O (2007) Field biology of placozoans (*Trichoplax*): distribution, diversity, biotic interactions. Integr Comp Biol 47(5):677–692
88. Maruyama YK (2004) Occurrence in the field of a long-term, year-round, stable population of placozoans. Biol Bull 206(1):55–60
89. Cuervo-González R (2017) *Rhodope placozophagus* (Heterobranchia) a new species of turbellarian-like Gastropoda that preys on placozoans. Zool Anz 270:43–48
90. Riedl R (1959) Beiträge zur Kenntnis der Rhodope veranii, Teil I. Geschichte und Biologie. Zoologischer Anzeiger 163:107–122
91. Thiemann M, Ruthmann A (1991) Alternative modes of asexual reproduction in *Trichoplax adhaerens* (Placozoa). Zoomorphology 110(3):165–174
92. Thiemann M, Ruthmann A (1990) Zoomorphology spherical forms of *Trichoplax adhaerens* (Placozoa). Zoomorphology 110(1): 37–45
93. Jackson AM, Buss LW (2009) Shiny spheres of placozoans (*Trichoplax*) function in antipredator defense. Invertebr Biol 128(3): 205–212
94. Fortunato A, Aktipis A (2019) Social feeding behavior of *Trichoplax adhaerens.* Front Ecol Evol 7:7
95. Smith CL, Pivovarova N, Reese TS (2015) Coordinated feeding behavior in *Trichoplax*, an animal without synapses. PLoS One 10(9): e0136098
96. Senatore A, Reese TS, Smith CL (2017) Neuropeptidergic integration of behavior in *Trichoplax adhaerens*, an animal without synapses. J Exp Biol 220(Pt 18):3381–3390
97. Armon S et al (2018) Ultrafast epithelial contractions provide insights into contraction speed limits and tissue integrity. Proc Natl Acad Sci U S A 115(44):E10333–E10341
98. Mayorova TD et al (2018) Cells containing aragonite crystals mediate responses to gravity in *Trichoplax adhaerens* (Placozoa), an animal lacking neurons and synapses. PLoS One 13(1):e0190905
99. Smith CL et al (2019) Coherent directed movement toward food modeled in *Trichoplax*, a ciliated animal lacking a nervous system. Proc Natl Acad Sci U S A 116(18): 8901–8908
100. Romanova DY et al (2020) Sodium action potentials in Placozoa: insights into behavioral integration and evolution of nerveless animals. Biochem Biophys Res Commun 532(1):120–126
101. Seravin LN (1989) Orientation of invertebrates in three-dimensional space: 4. Reaction of turning from the dorsal side to the ventral one. Zoologicheskij zhurnal (Zool J) 68:18–28. (in Russian)
102. Seravin LN, Gerasimova ZP (1998) Features of the fine structure of *Trichoplax adhaerens*, feeding on dense plant substrates. Cytologiya 30:1188–1193. (in Russian)
103. Romanova DY et al (2020) Glycine as a signaling molecule and chemoattractant in *Trichoplax* (Placozoa): insights into the early evolution of neurotransmitters. Neuroreport 31(6):490–497
104. Moroz LL et al (2021) Evolution of glutamatergic signaling and synapses. Neuropharmacology 199:108740
105. Varoqueaux F et al (2018) High cell diversity and complex peptidergic signaling underlie Placozoan behavior. Curr Biol 28(21): 3495–3501 e2
106. Moroz LL, Mukherjee K, Romanova DY (2023) Nitric oxide signaling in ctenophores. Front Neurosci 17:1125433
107. Moroz LL et al (2020) The diversification and lineage-specific expansion of nitric oxide signaling in Placozoa: insights in the evolution of gaseous transmission. Sci Rep 10(1):13020
108. Davidescu MR et al (2023) Growth produces coordination trade-offs in *Trichoplax adhaerens*, an animal lacking a central nervous system. Proc Natl Acad Sci U S A 120(11): e2206163120
109. Guidi L et al (2011) Ultrastructural analyses support different morphological lineages in the phylum Placozoa Grell, 1971. J Morphol 272(3):371–378
110. DuBuc TQ, Ryan JF, Martindale MQ (2019) "Dorsal-ventral" genes are part of an ancient axial patterning system: evidence from *Trichoplax adhaerens* (Placozoa). Mol Biol Evol 36(5):966–973
111. Smith CL et al (2014) Novel cell types, neurosecretory cells, and body plan of the early-diverging metazoan *Trichoplax adhaerens.* Curr Biol 24(14):1565–1572
112. Smith CL, Mayorova TD (2019) Insights into the evolution of digestive systems from studies of *Trichoplax adhaerens.* Cell Tissue Res 377(3):353–367
113. Moroz LL (2021) Multiple origins of neurons from secretory cells. Front Cell Dev Biol 9:669087
114. Moroz LL, Romanova DY (2021) Selective advantages of synapses in evolution. Front Cell Dev Biol 9:726563

115. Moroz LL et al (2020) Microchemical identification of enantiomers in early-branching animals: lineage-specific diversification in the usage of D-glutamate and D-aspartate. Biochem Biophys Res Commun 527(4): 947–952
116. Thiemann M, Ruthmann A (1988) *Trichoplax adhaerens* Schulze, F. E. (Placozoa) – the formation of swarmers. Z Naturforsch C Biosci 43(11–12):955–957
117. Zuccolotto-Arellano J, Cuervo-Gonzalez R (2020) Binary fission in *Trichoplax* is orthogonal to the subsequent division plane. Mech Dev 162:103608
118. Charlesworth D (2006) Population genetics: using recombination to detect sexual reproduction: the contrasting cases of Placozoa and *C. elegans*. Heredity (Edinb) 96(5):341–342
119. Eitel M et al (2011) New insights into placozoan sexual reproduction and development. PLoS One 6(5):e19639
120. Signorovitch AY, Dellaporta SL, Buss LW (2005) Molecular signatures for sex in the Placozoa. Proc Natl Acad Sci U S A 102(43):15518–15522
121. Grell KG, Benwitz G (1974) Electronenmikroskopische Beobachutungen uber das Wachstum der Eizele und die Bildung der "Befruchtungsmembran" von *Trichoplax adhaerens* F.E. Schulze (Placozoa). Z Morphol Tiere 79:295–310
122. Grell KG (1972) Eibildung und Furchung von *Trichoplax adhaerens* F.E. Schulze (Placozoa). Z Morphol Tiere 73:297–314
123. Hoencamp C et al (2021) 3D genomics across the tree of life reveals condensin II as a determinant of architecture type. Science 372(6545):984–989
124. Nikitin M (2015) Bioinformatic prediction of *Trichoplax adhaerens* regulatory peptides. Gen Comp Endocrinol 212:145–155
125. Kamm K, Schierwater B, DeSalle R (2019) Innate immunity in the simplest animals – placozoans. BMC Genomics 20(1):5
126. Klinges JG et al (2019) Phylogenetic, genomic, and biogeographic characterization of a novel and ubiquitous marine invertebrate-associated Rickettsiales parasite, Candidatus Aquarickettsia rohweri, gen. nov., sp. nov. ISME J 13(12):2938–2953
127. Smith CL et al (2021) Microscopy studies of placozoans. Methods Mol Biol 2219:99–118
128. Heyland A et al (2014) *Trichoplax adhaerens*, an enigmatic basal metazoan with potential. Methods Mol Biol 1128:45–61
129. Gauberg J, Senatore A, Heyland A (2021) Functional studies of *Trichoplax adhaerens* voltage-gated calcium channel activity. Methods Mol Biol 2219:277–288
130. Li Y et al (2021) Rooting the animal tree of life. Mol Biol Evol 38(10):4322–4333
131. Redmond AK, McLysaght A (2021) Evidence for sponges as sister to all other animals from partitioned phylogenomics with mixture models and recoding. Nat Commun 12(1): 1783
132. Smith CL, Reese TS (2016) Adherens junctions modulate diffusion between epithelial cells in *Trichoplax adhaerens*. Biol Bull 231(3):216–224
133. Mayorova TD et al (2021) Placozoan fiber cells: mediators of innate immunity and participants in wound healing. Sci Rep 11(1): 23343

Chapter 4

Stable Laboratory Culture System for the Ctenophore *Mnemiopsis leidyi*

Joan J. Soto-Angel, Eva-Lena Nordmann, Daniela Sturm, Maria Sachkova, Kevin Pang, and Pawel Burkhardt

Abstract

Ctenophores are marine organisms attracting significant attention from evolutionary biology, molecular biology, and ecological research. Here, we describe an easy and affordable setup to maintain a stable culture of the ctenophore *Mnemiopsis leidyi*. The challenging delicacy of the lobate ctenophores can be met by monitoring the water quality, providing the right nutrition, and adapting the handling and tank set-up to their fragile gelatinous body plan. Following this protocol allows stable laboratory lines, a continuous supply of embryos for molecular biological studies, and independence from population responses to environmental fluctuations.

Key words Breeding, Ctenophora, Comb jelly, Cultivation, Husbandry, Protocol, Sea walnut

1 Introduction

Ctenophores (comb jellies) are candidates for being one of the earliest extant lineage of animals and thus hold a key phylogenetic position to study the origin of animals, their nervous system and cell types [1–8]. Reconstructions of ancient gene linkages inferred from high-quality genomes of a ctenophore, two sponges, and three unicellular relatives of animals tipped the scales on ctenophores as the most probable sister group to all other metazoan lineages [9]. Moreover, the high invasive potential and predatory impact of some ctenophore species have led to increasing attention to this phylum in marine ecosystem studies [10]. Although ctenophores have been reported already in 1671 [11], the first species were described in the late 1700s [12–14], while the formal description of the taxon Ctenophora was in 1829 [15, 16]. *Mnemiopsis*

Authors Joan J. Soto-Angel and Eva-Lena Nordmann have equally contributed to this chapter.

Leonid L. Moroz (ed.), *Ctenophores: Methods and Protocols*, Methods in Molecular Biology, vol. 2757,
https://doi.org/10.1007/978-1-0716-3642-8_4, © Springer Science+Business Media, LLC, part of Springer Nature 2024

leidyi was described in 1865 [17], ca. 200 years after the first reports of ctenophore species.

Ctenophore cultivation may be challenging since most species are extremely fragile and sensitive to abrupt changes in environmental conditions. Indeed, conventional aquaria are unsatisfactory for mid- to long-term ctenophore cultivation (e.g., [18]). This might have been a major reason for most studies up until recently to resort to wild-caught animals and their direct offspring, instead of keeping a steady culture in the laboratory. Given these methodological constraints, pioneer cultivation attempts aimed to develop new culture methods provided with gentle but constant water flow and tanks without sharp edges that helped to minimize animal damage [19–21]. These systems were later modified and are currently successfully used in many jellyfish exhibitions in public aquariums worldwide [22, 23]. Subsequent contributions focused on the trophic and reproductive biology of selected ctenophore species, making available the first culture procedures [24, 25]. In the last years, detailed protocols have been made available, including spawning and embryo collection [26, 27], a new spawning and cydippid larvae rearing method [28], a laboratory aquaculture system for behavior characterization [29], and protocols to maintain and isolate *M. leidyi* in vitro cell cultures [30] and to study whole body regeneration [31]. Those contributions mostly focused on a few aspects of ctenophore cultivation. Very recently, comprehensive protocols covering all aspects of long-term ctenophore husbandry in the lobate ctenophores *M. leidyi* [32] and *Bolinopsis mikado* [33] have been published. However, those setups needed elaborated recirculating systems with the concomitant high amount of seawater (and other resources) required. The aim of this contribution is to pave the way for a successful *M. leidyi* husbandry by providing an alternative culture system that is simple, accessible, and efficient.

The lobate ctenophore *Mnemiopsis leidyi* is an emerging model system for molecular biology studies. Its genome [2], single-cell transcriptome [34], temporal developmental expression profiles [35], methylome [36], as well as molecular biological techniques such as in situ hybridization [37, 38], immunostaining [38, 39], gene knockdown by morpholinos [40, 41], and gene manipulations by CRISPR-Cas9 [42] recently became available. A stable culture system for *M. leidyi* enables opportunities for molecular biology, behavioral studies, and evolutionary research [43–46].

Following the pioneer work of Baker and Reeve [25], Pang and Martindale [26], Salinas-Saavedra and Martindale [27], and the more recent Patry et al. [28], Ramon-Mateu et al. [31], Presnell et al. [32], and Ikeda et al. [33], we here report an affordable setup and step-by-step protocol for culturing the ctenophore *M. leidyi*. The first generation was originally sampled from Kristineberg, Sweden, Baltic Sea, and corresponds to the same population as the specimen of the recently published whole-organism single-cell RNA sequencing study [34]. Our protocol covers all aspects of a long-term, multigenerational *M. leidyi* culture, including facility setup, sampling, daily routine, handling, breeding, hatchery, and nursery. Following this protocol, we have maintained a culture of *M. leidyi* with thousands of specimens belonging to twenty-one generations. In addition, the adult setup and nursery methods described here have been successfully assayed and can be equally applied to the common northern comb jelly *Bolinopsis infundibulum*, a species which laboratory culture had not been described until now.

2 Materials

2.1 Tank Setup, Transfer, and Nursery

- 6 L transparent seawater tanks (CAT# RFSCW8, CAMBRO®).
- 15 L transparent seawater tanks (CAT# RFSCW18, CAMBRO®).
- Rectangular aquarium, at least 50 cm wide.
- Aquarium water chiller.
- Water pump.
- Gear motor(s), 15 RPM.
- 25 mL serological pipettes (e.g., CAT# 734-0343, VWR® Europe).
- Flexible silicone hose, 5–6 mm diameter.
- Light tubes (neutral white to sunlight, 4500–5500 K).
- Plug-in timer.
- Transfer pipettes, 3.5 mL (e.g., CAT# 86.1171 Sarstedt®).
- 60 mL crystalizing dish without spout (e.g., CAT# 216-1862, VWR® Europe).
- 300 mL crystalizing dish with spout (e.g., CAT# 216-1815, VWR® Europe).
- 90 mm tissue culture dish (e.g., CAT# 734-2795, VWR® Europe).

The culture system consists of transparent, cylindrical, open at the top, 6–15 L, seawater tanks, which are partially submerged in a larger rectangular aquarium (Fig. 1). The latter is filled with fresh water and refrigerated through an aquarium water chiller coupled to a pump. The flow rate should be in accordance with the water chiller requirements provided by the manufacturer (*see* **Note 1**). The culture tanks are therefore refrigerated by water bath (*see* **Note 2**) and kept at constant temperature, adjustable in a range between 14 and 18 °C (*see* Subheading 2.2 on water quality for other water

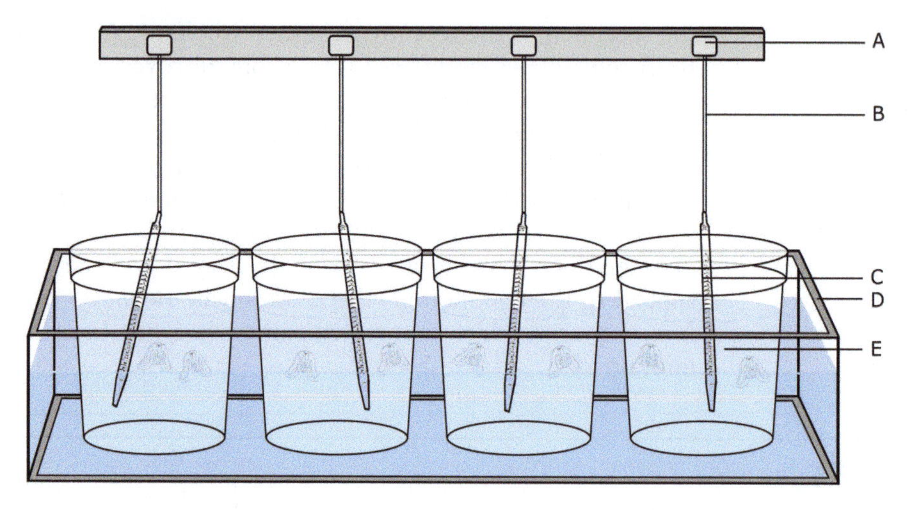

Fig. 1 Tank setup for *Mnemiopsis leidyi* culture. (**a**) gear motor, (**b**) flexible silicone hose, (**c**) 25 mL serological pipette, (**d**) rectangular aquarium, (**e**) seawater tank. Water in the rectangular aquarium acts as water bath and is thermostated through an aquarium water chiller (not shown)

parameters). Circular current flow is generated by a rotating and oscillating 25 mL serological pipettes (*see* **Note 3**), ca. 350 mm long and 13 mm thick. The motion pipettes are connected through a flexible silicone hose to a small gear motor running constantly at 15 RPM. Light tubes (4500–5500 K) are mounted beneath the transparent bottom of the rectangular aquarium and their switching on and off is controlled by a plug-in timer, with a regime of 12 h light and 12 h dark. *See* Subheading 3.1 for transfer and Subheading 3.2.3 for details on nursery.

2.2 Water Quality

- Mechanical filter cylinders: Mesh 20, 10, and 1 μm.
- Active charcoal filter, ca. 0.45 μm.
- Ultraviolet lamps and irradiation cylinder/panel.
- Portable pH and conductivity meter (e.g., WTW™ ProfiLine™ pH/Cond 3320; CAT# 15206778, FisherScientific®).
- Carboys, 20 L (e.g., CAT# 216-1700, VWR® Europe).

Purified Sea Water (PSW) used both for the main *Mnemiopsis* culture and auxiliary cultures is obtained by filtrating seawater through mechanical filters (20, 10, and 1 μm mesh-size, in that sequence), later purified through an active charcoal filter (0.45 μm mesh-size), and finally UV treated. Water outlet and all recipients used must be kept clean (*see* **Notes 4–6**). Water parameters (temperature, pH, and salinity) are measured before water usage (*see* **Note 7**). Optimum values for temperature between 14 and 18 °C, salinity between 28‰ and 30‰, and pH between 7.8 and 8.1.

2.3 Auxiliary Cultures

- Living *Brachionus* (rotifers).
- Concentrated algae (e.g., *Nannochloropsis*-based products, RG complete™ or similar).
- *Artemia* cysts.
- *Artemia* hatcher or equivalent.
- 6 L seawater transparent tanks (CAT# RFSCW8, CAMWEAR®, CAMBRO®).
- Lid for 6 L seawater tanks (CAT# RFSCWC6, CAMWEAR®, CAMBRO®).
- Air pumps or equivalents.
- Frozen copepods, 0.7–1.8 mm.
- Centrifuge tubes, 50 mL (e.g., CAT# 62.547.254, Sarstedt®).
- Microcentrifuge tubes, 1.5 mL (e.g., CAT# 211-2164, VWR® Europe).
- 10 mL serological pipettes (e.g., CAT# 7342-0352, VWR® Europe).
- Transfer pipettes, 3.5 mL (e.g., CAT# 86.1171 Sarstedt®).
- 100 mL plastic beaker (e.g., CAT# 213-1623, VWR® Europe).
- 2 L beaker with spout (e.g., CAT# 213-3404, VWR® Europe).
- Flexible silicone hose, 5–6 mm diameter.
- High-capacity filters (200–300 mL) 70 μm, 150 μm, and 500 μm.

Ctenophores are fed daily with a variety of prey items, including at least two of the following: *Brachionus* sp. (Rotifera), freshly hatched *Artemia* (brine shrimp) (instar I stage), *Artemia* 24 h post-hatching (hph) (instar II stage), and copepods. *Brachionus* permanent culture is kept in 6 L tanks with lid and constant aeration. Freshly hatched *Artemia* nauplii are used directly, with no supplements. *Artemia* nauplii ca. 24 hph are enriched with living algae in order to increase its nutritional profile. This step does not require keeping algae cultures, as concentrated algae (e.g., RG complete) that are used to feed *Brachionus* can be conveniently used to feed *Artemia* 24 hph as well.

3 Methods

3.1 Acquisition and Handling

Regardless of whether the ctenophores were sampled in the wild or obtained from a previous culture, handling procedures require extreme care, as the animals are particularly fragile and can easily be injured (*see* **Note 7**). In order to relocate a specimen from the sea to a container or from a container to another, proceed as described in Subheading 3.1.2. Note that the smaller individuals, specifically the cydippid larvae are more robust than the lobed, larger individuals (*see* **Note 8**). Laboratory gloves are generally not recommended for the culture, since even non-powdered gloves seem to leave trace residues in the water. If gloves are needed, we recommend to thoroughly rinse them with PSW before they are used.

3.1.1 Sampling

Some shallow-water ctenophore species (e.g., *Pleurobrachia pileus*, *Beroe ovata*, *Mnemiopsis leidyi*, and *Bolinopsis infundibulum*) can be easily sampled on board of a vessel that allows reaching specimens by hand. Alternatively, most species can be obtained by snorkeling or by underwater SCUBA diving when appropriate. *Mnemiopsis leidyi* is a coastal blooming species, relatively easy to spot and sampled from the shore, both in natural locations and in harbors all around its habitat. *Mnemiopsis leidyi* is widely distributed, forming native populations in western Atlantic, and introduced populations in the North Sea, the Baltic Sea, and the Mediterranean Sea [47]. Previous mentioned methods are the most adequate to sample *M. leidyi*. However, the use of nets is necessary when its abundance is low. Net sampling is generally a stressing procedure for the animals that often results in mechanically injured specimens. If nets are strictly required, the use of a relatively large mesh size (i.e., 800 μm), slow towing speed (between 0.5 and 1 knots, i.e., 0.3–0.5 m/s), and large (e.g., 3 L), closed, non-filtering cod-end can minimize severe damage. Ensure to keep the sampled ctenophores first in the original seawater (you may want to take water with you) (*see* **Note 7**), and only gradually acclimate the sampled animals to the new laboratory conditions, keeping water parameters as close as possible to the original water source.

3.1.2 Transferring Adult Ctenophores

Whether the adult ctenophores are directly obtained from the wild or another culture, proceed as follows (for larvae transfer, *see* Subheading 3.2.3).

- Ensure that the receiving tank is already available and filled with appropriate water (seawater from the sampling site in case of sampling, fresh PSW in case of culture), at a similar temperature, salinity, and pH than the original one.
- Select an appropriate transfer container (e.g., a 1 L jar/bottle for sampling or a 50–100 mL beaker for transferring), preferably rounded and without any sharp edges or spout, with an opening large enough to maneuver easily with the specimen (e.g., 60 mL crystalizing dish without spout).
- Rinse your hands/laboratory gloves thoroughly with warm fresh water and dry them with a cloth that leaves no residues.
- Rinse the transfer container with PSW at similar temperature (±3 °C) to the one where the animals are staying.

- Slowly submerge the transfer container into the water avoiding turbulences or abrupt water intake.
- Let the transfer container fill completely with water before continuing (no large bubbles should remain), and move it toward the specimen to be transferred (either laterally or vertically depending on the depth and position of the animal). Avoid touching, moving, turning, or pushing the animals neither with your hands nor with the transfer container, especially with the edges. If needed, use the transfer container to create a slow current to move the ctenophore to the desired position.
- Once the specimen is completely within the transfer container, and as long as it is in a safe position to avoid damaging the lobes, lift the transfer container as vertically as possible and take it out from the water carefully.
- If needed, remove as much water as possible from the transfer container (*see* **Note 9**). Healthy *M. leidyi* tolerate slight levels of compression between the water surface and the bottom of the container (*see* below).
- If the beaker with the animal is placed on a not-completely clean surface, make sure to clean it before proceeding.
- Introduce the transfer container with the animal into the new tank, and avoid turbulences and/or rapid water flow. This can be done by ensuring that the water level between the transfer container and the receiving tank is aligned before submerging. Simultaneously turn and lift the transfer container carefully, slowly releasing both the water and the animal into the receiving container (*see* **Note 9**). Avoid pouring the animals or making them slide down through the surface of the transfer container.
- Observe the recently transferred animal. If there are some signs of damage, proceed as described in **Note 10**.

3.2 The Culture

For a description of tank setup and water quality, *see* Subheadings 2.1 and 2.2, respectively. The system presented here has allowed us to successfully keep an affordable long-term *Mnemiopsis* culture. Routines and protocols described are specifically designed to minimize the time and costs invested, maximize the production, and allow an easy access and collection of the animals. Their implementation has allowed us to get several thousands of individuals belonging to 21 generations from an original set of five individuals and to get several thousands of viable eggs and larvae per week. Several individuals have been kept to date for more than 3.5 years. Occasional loss of some individuals is compensated by the rapid growth rate and high production of larvae. Thus, the described culture system allows producing a considerable surplus of specimens available for experimentation.

Table 1
Overview of the weekly feeding and washing routines for *Mnemiopsis leidyi* culture

	Day 1	Day 2	Day 3	Day 4	Day 5
Artemia	Feed *Artemia* ≈ 48 hph with algae. Feed ctenophores with enriched *Artemia* (2–3 doses). Set up *Artemia* (morning)	Feed ctenophores with recently hatched *Artemia* (2–3 doses)	Feed *Artemia* ≈ 24 hph with algae. Feed ctenophores with enriched *Artemia* (2–3 doses). Set up *Artemia* (morning)	Feed ctenophores with recently hatched *Artemia* (2–3 doses)	Feed *Artemia* ≈ 24 hph with algae. Feed ctenophores with enriched *Artemia* (2–3 doses). Set up *Artemia* (morning)
Brachionus	Feed *Brachionus* with algae. Feed ctenophores with *Brachionus* (1 dose)	Feed ctenophores with *Brachionus* (1 dose). Feed *Brachionus* with algae	Wash *Brachionus* culture. Feed *Brachionus* with algae	Feed ctenophores with *Brachionus* (1 dose). Feed *Brachionus* with algae	Feed ctenophores with *Brachionus* (1 dose). Feed *Brachionus* with algae
Frozen copepods	Thaw frozen copepods. Feed ctenophores with frozen copepods (2–3 small doses)	Thaw frozen copepods. Feed ctenophores with frozen copepods (2–3 small doses)			
Rearing tanks: Juveniles and adults	Remove debris	Wash ctenophore culture. Remove debris	Remove debris	Remove debris	Partial water exchange (if needed). Remove debris
Nursery (larvae)	Add additional water (if needed)		Wash beakers		Add additional water (if needed)

hph hours post-hatching

The recommended weekly plan (Table 1) requires a basic maintenance that includes feeding (3–4 times/day, 5 days/week), cleaning and washing (debris removed daily, weekly partial water exchange, and weekly full water exchange), and water quality monitoring when appropriate. Therefore, five consecutive days per week are required once the culture is established. An overview of the daily routines is illustrated in Fig. 2.

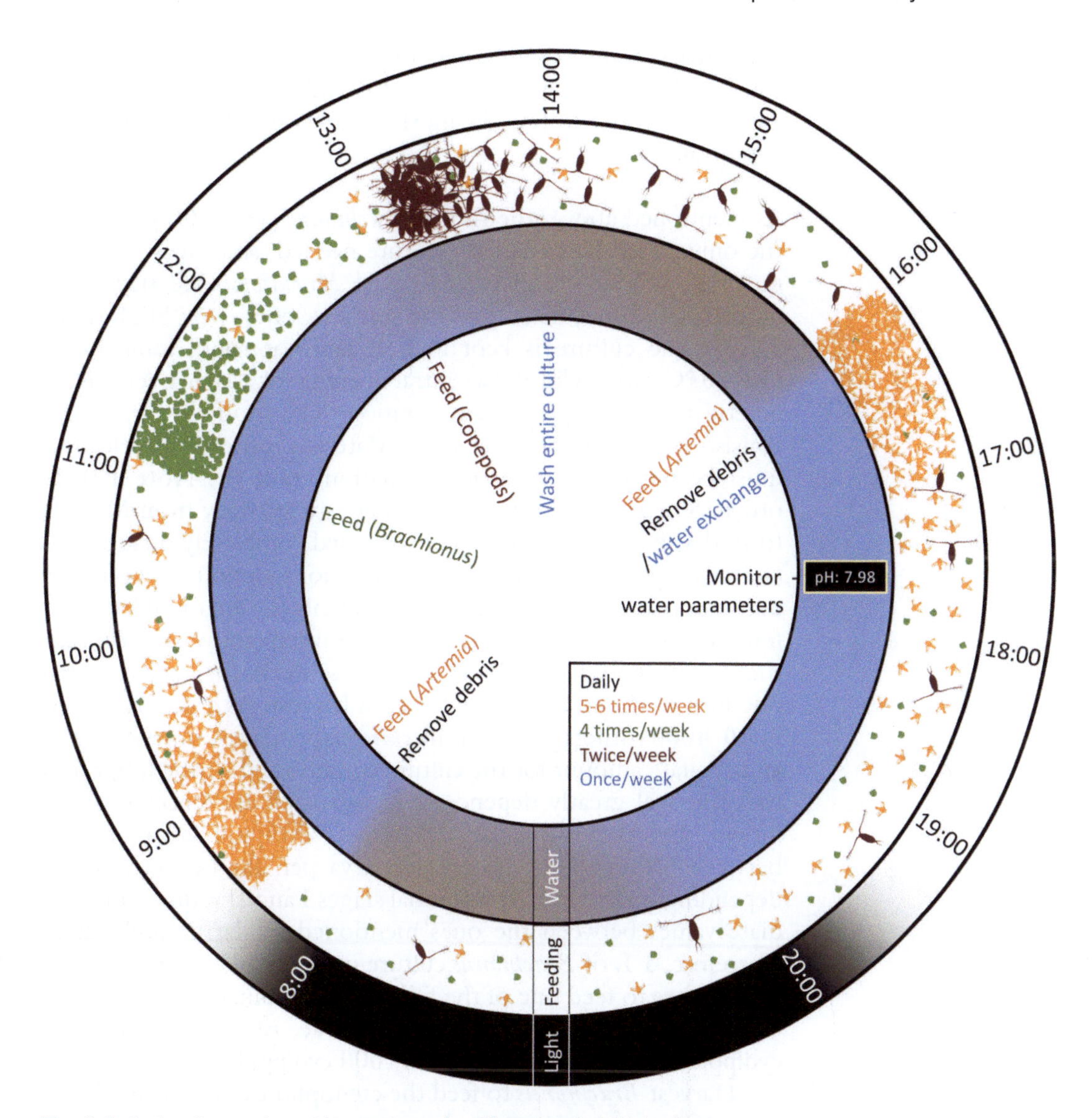

Fig. 2 Daily feeding and washing regimes for *Mnemiopsis leidyi* culture

Ctenophore density in the tanks can vary depending on the needs and the space availability, with maximum values adjusted according to their average size. Ideally, density of medium to large adult ctenophores (>12 mm aboral-oral length excluding lobes) should not exceed one specimen/L. Smaller individuals and larvae can be kept at much higher densities (*see* **Note 11**).

3.2.1 Rearing

Feeding

An overview of the feeding regime is described in Subheading 2.3 and illustrated in Fig. 2. *Mnemiopsis* specimens should always have some prey available in the tank, but overfeeding must be avoided, as the uneaten food will deteriorate water quality. An optimal strategy consists of splitting daily feeding amounts into several doses.

Indicative amounts of food per ctenophore are given for each food source. In any case, we recommend to individually assess different amounts in order to find the most suitable ones for a given culture condition.

Brachionus

As mentioned above, the rotifer *Brachionus* (150–400 μm size) is the only permanent auxiliary culture needed, since it is simple to maintain and suitable for feeding all *Mnemiopsis* life stages and particularly convenient for the earliest ones (*see* Subheading 3.2.3). The culture is kept in 6 L tanks at room temperature (18–20 °C), with a lid and a central opening for a 10 mL serological pipette providing permanent aeration. The food for *Brachionus* consists of concentrated algae (*see* Materials section). Overfeeding must be prevented to avoid population crash (*see* **Note 12**). In order to maximize production, rotifers are fed daily from Monday to Friday. Each *Brachionus* tank is washed completely once a week (*see* protocol below). Under these conditions, optimum population densities (ca. 200.000–400.000 individuals/L) are obtained just a few days after the start of the culture. Subsequently, each *Brachionus* culture tank is harvested 4–5 times/week, for approximately one fifth of the total volume for each harvest. As a general guide, 2000–4000 rotifers per adult *Mnemiopsis* per day (4 times/week) is an adequate amount for the culture conditions described, but note that this will greatly depend on factors such as ctenophore size, fitness, and water quality parameters. Cydippid larvae require between 50 and 200 rotifers per larva per day (4 times/week) depending on the size. Transitional stages I and II require intermediate values between the ones mentioned for larvae and adults. Therefore, 1 L of *Brachionus* culture in optimum densities potentially allows to feed one of the following combinations (for 1 day): (1) 150 adults, (2) 1500 cydippid larvae, (3) 100 adults +500 cydippid larvae, or (4) 50 adults +1000 cydippid larvae.

Harvest *Brachionus* to feed the ctenophores (*see* **Note 13**):

- Turn off aeration by removing the pipette and let algae aggregates sink down.
- After approximately 10 min, take the required amount of *Brachionus* culture with a 1 L beaker, with an upper limit of one fifth the total culture tank volume (*see* **Note 14**).
- Filter and rinse gently through a 70 μm mesh above the sink with 2 L of PSW in order to remove the remaining algae (*see* **Note 15**).
- In order to maximize the algae removal, leave the filter in a 1 L beaker with 400 mL of PSW, wait for 10–20 min and repeat the previous step.

- Fill up the used *Brachionus* culture tank with PSW at room temperature, with the same water volume that has been taken out.
- Place back the aeration pipette and use it to mix the culture.
- Check that the aeration pipette reaches the bottom of the tank and that air flow is adequate.
- Feed the Ctenophores with *Brachionus* using a new transfer pipette previously washed (in and out surfaces) with PSW, 1 dose/day (see above for indicative amounts).

Washing the *Brachionus* culture:

- Prepare the number of 6 L tank(s) needed with PSW at room temperature.
- Turn off aeration from the old tank by removing the pipette and let the algae sink down.
- After approximately 10 min, filter the complete *Brachionus* culture through a 70 μm mesh, avoiding the most basal section where the algae aggregates are deposited (*see* **Note 16**).
- Rinse the *Brachionus* in the filter with an additional 1 L of PSW (*see* **Note 15**).
- If bubbles (i.e., dead *Brachionus* and debris) are present in large amounts, remove them with the aid of a transfer pipette.
- Wash off the *Brachionus* from the 70 μm mesh into the new tank with PSW.
- Install a new 10 mL serological pipette and set them for intermediate levels of aeration.
- Check that the aeration pipette reaches the bottom of the tank and that air flow is adequate.

Feeding *Brachionus*:

- Add approximately 1–2 mL of concentrated living algae (refrigerated) per every 5 L of *Brachionus* culture to a 100 mL plastic beaker with 30–40 mL of PSW and mix with a transfer pipette. If the following days the *Brachionus* culture will not be fed, use twice the indicated amount of concentrated algae (but *see* **Note 12**).
- Feed *Brachionus* culture with the prepared algae using a transfer pipette. Add an equivalent amount to each tank. Should differences in color density be observed, add a larger amount to the less greenish (more transparent or more brownish) tanks.
- Mix thoroughly the water in the bucket using the aeration pipette.
- Check that the aeration pipette reaches the bottom of the tank and that air flow is adequate.

Artemia

Artemia (brine shrimp) is a widely used food source in aquaculture. A great variety of protocols for cysts hatching and nauplii larvae harvesting are easily accessible and normally provided by the manufacturer. Therefore, only some considerations regarding the use of *Artemia* for *Mnemiopsis* culture are provided:

- Special care must be taken when harvesting and rinsing *Artemia* to wash off the majority of the accompanying heterotrophic bacteria and the unhatched cysts.
- Recently hatched *Artemia* (Instar I stage, 350–550 μm, 18–22 h after cysts incubation at 24–26 °C, not feeding) is the most nutritious form when the nauplii are not fed and is therefore preferred over older nauplii when no enrichment is performed.
- Instar II stage (450–650 μm, 26–30 h after incubation at 24–26 °C, feeding) are metanauplii that have already consumed most of their reserves and are ready to ingest food. The enrichment at this stage through feeding with algae rich in highly unsaturated fatty acids (HUFA) has shown multiple benefits in aquaculture [48, 49], and *M. leidyi* spawning seems to be boosted when enriched nauplii are used as food source. For this purpose, concentrated algae such as the widely used *Nannochloropsis*-based solutions or RG complete (*Brachionus* food) can be employed. The enrichment process requires a few drops of concentrated algae and takes a minimum of 1 h (ideally 3–4 h) since the algae mixture is added.
- Most attempts of feeding *M. leidyi* with adult *Artemia* resulted unsatisfactory, probably due to the high swimming speed of the later. We therefore do not recommend using *Artemia* older than 48 hph.
- Before feeding the ctenophores with either Instar I or II stages, the nauplii should be rinsed thoroughly with PSW through a 150 μm mesh filter.
- A larger number of small doses of *Artemia* nauplii is preferred over a single large dose. This will allow ctenophores to constantly have available prey, reduce debris, and will keep the nauplii in their optimal nutritional profile just before being added to the ctenophore culture. For doses later in the day, keep Instar I refrigerated at 4–5 °C.
- As an indicative value, 200–1000 *Artemia* nauplii per adult *Mnemiopsis* and day (5 times/week) is an adequate amount for the culture conditions described, but note that this will greatly depend on factors such as ctenophore size, fitness, and water quality parameters.
- At its maximum growth rate, *M. leidyi* can start feeding *Artemia* when they are 10 days old (but *see* Subheading 3.2.3).

Copepods

Copepods are a convenient prey for *M. leidyi*, as they constitute a major component of their natural diet [50]. Specimens fed with copepods reach larger sizes and produce a larger offspring. However, copepod cultivation in a scale sufficient to feed a ctenophore culture with several hundred adult animals does not only demand a considerable amount of time (requiring an additional algae culture) but also a considerable amount of space (since a too high density may lead in a crash of the culture). Frozen copepods can be an alternative. Depending on the supplier, copepods can range from 0.5 to 1.8 mm. Some aspects must be taken into account when feeding with copepods:

- Avoid thawing and freezing multiple times. To achieve this, copepods can be split into smaller volumes (e.g., microcentrifuge tubes 1.5 mL or 50 mL centrifuge tubes) before freezing, and thaw only the corresponding amount needed at a time.
- Leave the frozen fraction to thaw slowly and then rinse them thoroughly with PSW using a large mesh (e.g., 500 μm).
- Copepods have to be gradually supplied to the ctenophores, particularly preventing overfeeding, since dead copepods will directly be a starter for heterotrophic bacterial growth.
- Avoid using frozen copepods on a daily basis with the described setup. The best results can be achieved when using them the day before and the day of washing, so all the uneaten copepods will be removed after washing (Table 1).
- Water circulation in the tank has to be strong enough to keep them floating until they are captured by the ctenophores before settling down.
- After several hours, the uneaten copepods have to be removed from the bottom of the tank (*see* Subheading 3.2.1) to avoid declining water quality.
- Small ctenophores (<5 mm) do not predate efficiently on large copepods, as they require smaller prey (*see* Subheading 3.2.3).
- For other recommendations, check the specifications provided by the manufacturer.
- Make sure frozen copepods do not contain any cryopreservant or other supplements, since accumulation in the ctenophores can lead to unwanted long-term consequences in their health and reproduction.

Washing

Clean water is fundamental to keep healthy specimens (*see* **Note 17**). Since the setup described here does not include a continuous water exchange (as occurs with some Kreisel systems with open circulation), the washing is crucial. The entire culture is washed once a week, and partial water exchange can be done additionally to ensure appropriate water quality levels (Table 1). Considerable variations of water parameters must be avoided to reduce the stress already caused by the washing procedure itself (*see* **Notes 7** and **18**).

- Prepare the tanks with PSW well in advance (but *see* **Note 6**) and adjust the temperature to the existing one in the culture (*see* **Note 5**).
- If necessary, add some reverse osmosis (or deionized) water to the receiving tank in order to decrease the salinity. The contrary (i.e., increasing the salinity in the receiving tank) is not necessary unless the variation exceeds 5‰ (*see* **Note 7**).
- Stop, carefully remove and discard the old motion pipette.
- To transfer the specimens from a previous to a new tank, proceed as described in Subheading 3.1.2.
- If evidences of damage are observed, proceed as described in **Note 10**.
- Install a new motion pipette and adjust it to achieve the right angle of rotation-oscillation (*see* **Note 3**).
- Rinse the old tank thoroughly with warm water and a brush. Avoid using soap or detergents.

3.2.2 Breeding

Under the parameters and feeding regime described above (*see* Subheadings 2.1 and 2.3), *M. leidyi* specimens reach maturity in ca. 3–4 weeks (ca. 12 mm long in oral-aboral length, *see* Fig. 3), with total lengths of 22 ± 9 mm after 1 month depending on temperature and food availability (*see* **Note 19**).

Several attempts to understand the factors inducing spawning in *M. leidyi* have been published over the last decades [26, 51]. Despite most of them coincide in the need of a timeframe of ca. 6–8 h of complete darkness, recent contributions have shown that seasonal and population variations might condition spawning in *M. leidyi* [27] and even under a 24 h light regime, spawning can occur [52]. In addition, food conditions play an important role in reproduction rates, both at spawning level and egg viability [53]. Our experience also confirms these findings but with an interesting addition: Highest spawning and egg viability rate are only achieved after a period of consecutive weeks with a varied diet (Fig. 4). In contrast, a feeding regime based exclusively on a single prey type seriously compromised reproductive success (Fig. 4). We have also observed that the spawning is not 100% synchronous, and basal spawning occurs throughout the day, at a minimum rate of 1–2 eggs per hour in healthy specimens. In general, larger ctenophores produce larger offspring [25, 52]. However, the variation is sometimes considerably wide [52] (Fig. 4), and the unpredictability requires increasing the number of replicates in order to ensure an

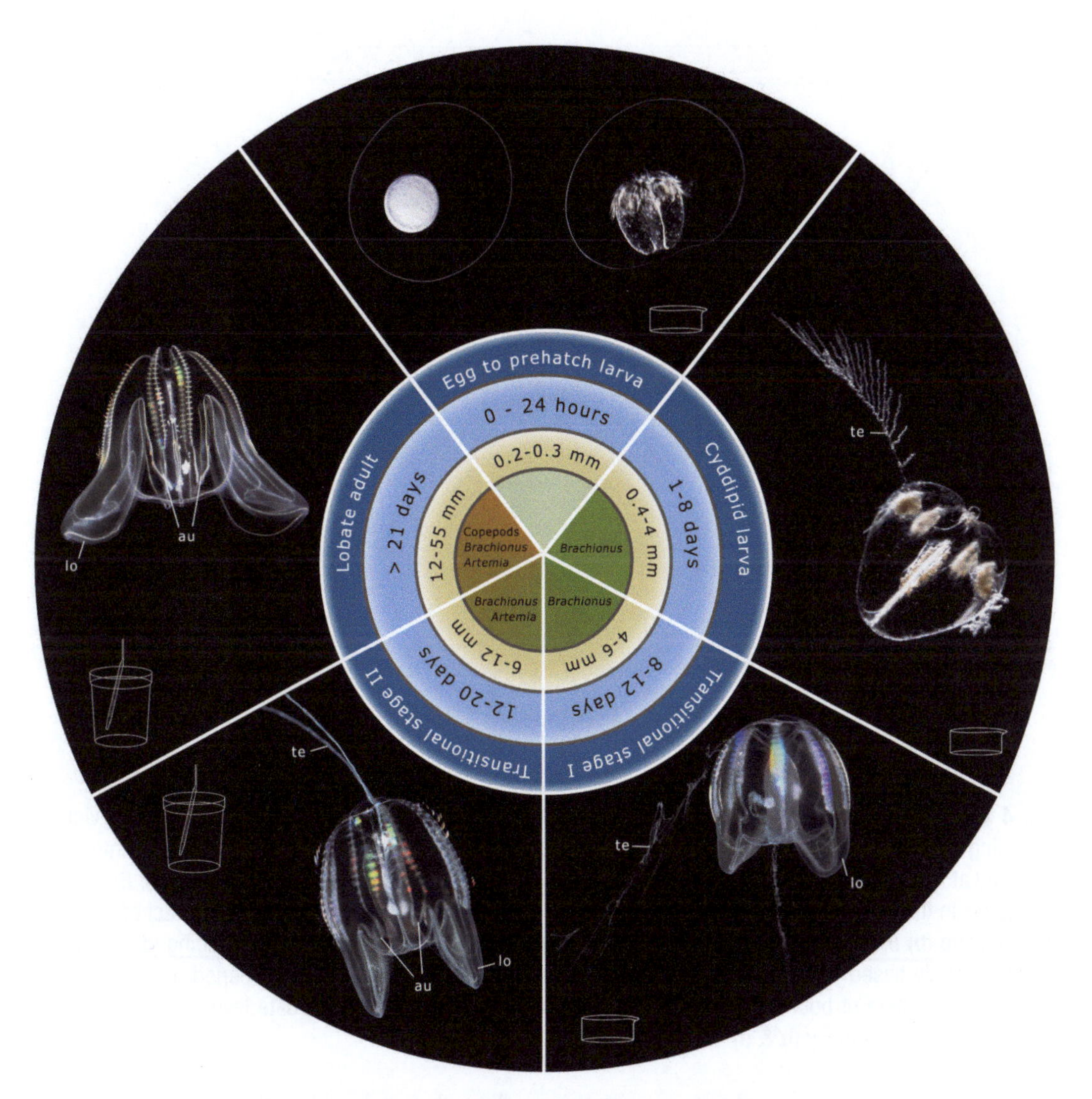

Fig. 3 Life cycle of *Mnemiopsis leidyi* from egg to maturity including major developmental stages, time, size, and main feeding prey. Transition stages are intermediate forms in which the tentacles of a cydippid larva co-occur with the lobes of a lobed adult in the same specimen. Transition stage I is characterized by the presence of early lobes until the onset of auricle development, while in transition stage II exhibits notable auricles and the lobes are clearly expanded. *te* tentacles, *lo* lobes, *au* auricles. (Figure design inspired by Ramondenc et al. 2019 [55])

elevated production of offspring. Taking these aspects into account, proceed as follows:

- Prepare as many 300 mL beakers as necessary at least 1 day in advance. Fill them with PSW and leave them at room temperature with a lid. This will allow to create a biofilm on the bottom of the beaker to prevent later egg adhesion (*see* **Note 20**).
- Select one or two healthy (*see* **Note 17**), well-fed (Fig. 4) adults (>30 mm) and place them into a 2 L beaker filled with PSW at

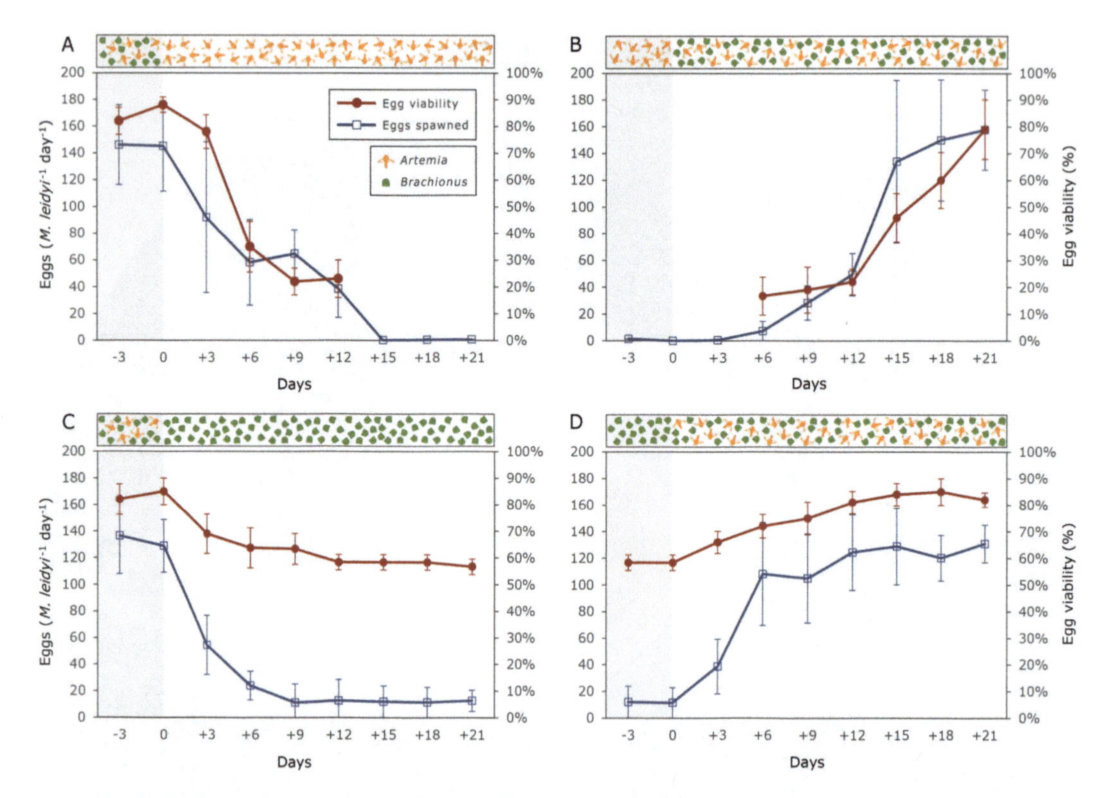

Fig. 4 Variation in spawning intensity (number of eggs spawned per individual; $n = 4$) and egg viability according to feeding regime. Egg viability refers to percentage of eggs that successfully developed into larvae observed at least 32 h after fertilization ($n =$ all spawned eggs). Note that the presence/absence of *Brachionus* or *Artemia* in the diet have dissimilar reaction time in egg production and egg viability: The effect of removing (**a**) or adding (**b**) *Brachionus* is slower than the corresponding effect of removing (**c**) or adding (**d**) *Artemia*. A diet exclusively based on *Artemia* (**a**) causes virtually no spawning after day 15. A period over 2–3 weeks using a combination of both prey items is therefore necessary to achieve adequate levels of spawning and high values of egg viability (**b** & **d**)

the same conditions than the tank where they were taken from (*see* **Note 21**).

- Leave them overnight.
- The next day, check for signs of spawning. Most ctenophores spawn around 5–8 h after programmed sunset, which can be adjusted at convenience.
- Once the spawning is complete, eggs will float during 2–4 h before sinking. Transfer the eggs into the previously prepared 300 mL beakers, while they are still floating (*see* **Note 20**). Attempts of transferring eggs laying in the bottom of the beaker for several hours can result in very low survival rate (<10%).
- At room temperature, eggs will normally hatch 18–24 h after fertilization [25, 26]. However, in some cases, this period can be longer, up to ca. 32 h according to our observations.

3.2.3 Nursery

Once the critical hatching phase has been successfully overcome (*see* **Note 20**), the survival rate of the early hatched cydippid larvae to the adult stage under the conditions described here is very high (>80%). The cydippid larvae (1–8 days old) and transitional stage I (8–12 days old) are raised at room temperature in 300 mL beakers (e.g., crystalizing dish) provided with a lid (e.g., the base of a tissue culture dish) to avoid water evaporation. The beakers used over the first week can be the same where the larvae hatched (but *see* **Note 22**). The nursery stage normally takes ca. 2 weeks (Fig. 3) but can be extended up to 2 years if necessary by regulating food availability. The only food source required during the first weeks is *Brachionus* (Fig. 3).

- As soon as the early hatched larvae are observed, feed them with *Brachionus* 4 times/week (*see* Subheading 3.2.1).
- Under food saturation conditions, larvae can grow as quick as 0.5–1 mm/day (but *see* **Note 23**).
- If necessary (in case of high evaporation rates due to low humidity environment), add water every second day by pipetting PSW water to the side of the beaker to mix the water properly and allow the *Brachionus* to be evenly distributed in the beaker.
- Wash the beakers once a week by transferring the larvae to a new beaker filled with PSW (*see* **Notes 22**, **24**, and **25**).
- Add *Brachionus* after every washing.
- After ca. 2 weeks, the specimens reach the transitional stage II, characterized by the presence of auricles, and clearly formed lobes (Fig. 3). At this point, the ctenophores can start to prey efficiently on *Artemia* nauplii and therefore are ready to be moved to a larger tank as described in Subheading 2.1. For transferring, proceed as described in Subheading 3.1.2.

4 Notes

1. If desired, tanks can be maintained at room temperature, as long as this factor is controlled, preventing considerable circadian variations. Different *M. leidyi* populations have been successfully cultured using higher temperatures [25, 27, 51].
2. Water temperature from the tanks containing animals will only be close to the outer water in the aquarium (and therefore to the indication in the water chiller) as long as both water levels are kept as close as possible. However, beware of undesired tank flotation.
3. The flow speed can be easily adjusted by modifying pipette oscillation. The more inclined the position of the motion

pipette, the stronger the water flow. Completely vertical position of the motion pipette results in only rotation and thus very reduced current speed. Stronger flow (more inclined pipette) is adequate for smaller specimens (including eggs and larvae), whereas low current (closer to vertical position) is more appropriate for larger stages, injured and/or delicate specimens.

4. Hoses, beakers, bottles, tanks, carboys, or any other device used for storage or transport water and/or move animals must be chemical free (including soap or detergent).
5. Rinse any recipient with PSW before using, especially when they are used for transfer ctenophores or remove debris from the tanks.
6. Avoid storing seawater during several consecutive days, even PSW. Optimum water quality depends on temperature and light exposure. As a general recommendation, PSW shall be used within a week since preparation.
7. Considerable variations in water density (i.e., changes of ±5‰ salinity and/or ±5 °C temperature) can injure the animals. In general, a water change from lower to higher salinity is stressing for the animals, resulting in partial or total disintegration. On the contrary, ctenophores adapt well from high to low salinity, even when sinking to the bottom of the recipient. If this occurs, the animal will compensate its buoyancy in approximately 1–2 h after the water change.
8. Once the eggs hatch and as long as the parental generation is adequately fed (*see* Fig. 4), *M. leidyi* larvae are particularly resistant and resilient to manipulation, water exchange, and variations in water parameters (pH, salinity, and temperature). However, avoid transferring them abruptly (without a previous acclimatization period) to water temperatures under 10 °C. Larvae up to a 1–2 mm long can be easily pipetted and filtered, without causing any damage.
9. Some circumstances might require transferring the least possible amount of water. These include other plankton accompanying the sample, *M. leidyi* eggs and larvae produced in the tanks, or poor water quality in the established culture due to unwanted presence of chemicals, debris, algae, or bacteria. When visible, the unwanted particles or organisms can be manually removed by pipetting. If they are either too small or too abundant to be removed, proceed diluting several times before placing the ctenophore into its final destination (*See* **Note** 7).
10. *Mnemiopsis leidyi* specimens can sometimes show evidence of damage after being transferred into a new location. This does not necessarily mean that the new water is not adequate, as they might be reacting to the conditions in the previous tank or to

the manipulation itself. Three scenarios can occur: (1) If more than 30% of the ctenophores in a bucket show signs of severe damage (i.e., disintegrating entire lobes, or apical region), stop the motion pipette and leave the bucket untouched until the next day. Any attempt of performing a new transfer at this point will result in further damage. The following day, transfer the whole group to a new tank and observe the response. Repeat if necessary. (2) If less than 10–15% of the ctenophores in a group show signs of severe damage, proceed as 1, but do not attempt a water exchange the next day, as the problem is not affecting the whole group and most probably not due to the new water used. In any case, monitor the state of the other ctenophores. (3) If a single (or few) ctenophore shows some signs of slight damage (e.g., in the apical section of the lobes or a small hole in one lateral), stop the motion pipette for some hours, observe the progression, and if it looks better, switch on the motion pipette again by the end of the day. If not, wait until the next day to activate the rotation or transfer it to a separate beaker and monitor it.

11. Hundreds of larvae (<3 mm) can be kept at 6 L tanks. When specimens reach a size between ca. 10 and 20 mm in total length, the maximum recommended density is 5 specimens/L. For animals between 20 and 30 mm, a maximum density of 2–3 specimens/L is recommended.
12. Different algae products come in different concentrations, and manufacturer's indications should be an adequate starting point. In any case, avoid overfeeding *Brachionus* culture as this might lead to crashes of the population due to high levels of ammonia and nitrite. Optimum color of the water must be slightly darker than cucumber flesh green but not as dark as cucumber peel. Daily harvesting also helps to keep appropriate water quality and a stable population over time.
13. Ideally, *Brachionus* should be harvested prior to feeding the rotifer culture, in order to reduce the amount of algae transferred from the *Brachionus* culture to the Ctenophore culture. However, harvesting starved *Brachionus* (i.e., brownish water) translates into poor-quality food. In this later case, add a small amount of algae at least 1 h before harvesting.
14. Focus on the upper section of the tank containing *Brachionus* (8–9 cm). Prevent mixing during harvesting in order to avoid collecting algae precipitates laying in the bottom of the tank.
15. If necessary, place the filter into a beaker with some PSW before proceeding to rinse the harvested sample. *Brachionus* is particularly sensitive to drying out. After a few seconds of exposure to air, they become floating, thus inconvenient to be used, and a source of debris and contamination.

16. If the algae aggregates are very abundant, and/or there is a need to recover as much *Brachionus* as possible, wash off the entire *Brachionus* culture through a combination of 250 μm and 70 μm mesh. This will allow to discard most algae aggregates (250 μm mesh) while retaining *Brachionus* (70 μm mesh).
17. There are two key features that allow to categorize a *M. leidyi* specimen as healthy: When they spend most of the time with the lobes widely opened (in contrast to closed, resembling a walnut) and the auricles are clearly long and pointy, clearly surpassing the oral end (in contrast to very reduced or nonexistent, and/or rounded).
18. A common practice consists of adding a large dose of food 2–3 h prior to washing. Properly fed animals are less sensitive to washing procedures. However, some food should also be available after the water exchange.
19. Attempts of raising cydippid larvae into the fertile adults exclusively feeding with *Brachionus* have been satisfactory. However, the larval production at this stage is lower than in larger specimens fed with other food sources.
20. According to prior culturing attempts, manipulating and/or transferring eggs and early larvae result in very limited survival rate [25]. Certainly, egg hatching is a bottleneck in ctenophore culture. If needed, the survival rate can be improved (by a 15–20%) by transferring the eggs while still floating to the aforementioned beakers with a biofilm cover. This prevents eggs being stuck to the bottom of the recipient. Alternatively, the use of agar-coated petri dish has been described as an effective measure [26, 51].
21. Some studies have pointed out that egg viability is more reduced when a single individual self-fertilized the eggs [52, 54]. Our results do not concur in this regard, showing high viability also when spawned individually (Fig. 4).
22. Optimum conditions for nursery stage depend on the size of the specimens. Early hatched larvae can be kept by hundreds in a 300 mL beaker, but this requires a larger amount of food, with the subsequent water quality impoverishment. In addition, a large amount of larvae also leads to a slower growth rate, while the smaller densities correspond to higher rates, up to 0.5–1 mm/day in conditions of food saturation. Our recommendation is to use maximum starting densities of 30–40 larvae in 300 mL for the first week and split the culture the following week (15–20 larvae of 8–12 days old in 300 mL).
23. Growth rates are very different within the same offspring, resulting in ctenophores ranging from a few mm to 15 mm after the first 2 weeks.

24. *Mnemiopsis leidyi* eggs and larvae can be transferred using a transfer pipette. Should larvae be bigger than 2 mm, cut the tip of the transfer pipette to avoid damaging while transferring. If the tentacles are extended, induce contraction by gently touching the larva or by gently pumping water on it before pipetting. Avoid pipette transferring for lobed specimens larger than 7–8 mm, use a beaker instead.
25. *Mnemiopsis leidyi* eggs and larvae smaller than 4 mm can be transferred and/or concentrated as well by carefully filtrating through a 70 μm mesh. Avoid exposure to air by submerging the mesh into a small amount of PSW.

Acknowledgments

The authors want to express their gratitude to Miguel Candelas (L'Oceanogràfic, Spain) for his valuable recommendations on feeding regimes and Anne Aasjord, Kjerstin Nilsen Nøkling and Eilen Myrvold (Michael Sars Centre) for their accurate advice on water quality aspects. We also thank Mari Bergsvåg (Michael Sars Centre) for taking the picture of the lobed adult ctenophore in Fig. 3 and Aino Hosia and Luis Martell (University Museum of Bergen) for sharing their expertise on sampling delicate gelatinous zooplankton. This work was supported by the Michael Sars Centre core budget and the European Research Council Consolidator Grant (101044989, "ORIGINEURO") awarded to P.B.

References

1. Dunn CW et al (2008) Broad phylogenomic sampling improves resolution of the animal tree of life. Nature 452(7188):745
2. Ryan JF et al (2013) The genome of the ctenophore *Mnemiopsis leidyi* and its implications for cell type evolution. Science 342(6164): 1242592
3. Moroz LL et al (2014) The ctenophore genome and the evolutionary origins of neural systems. Nature 510(7503):109
4. Jékely G, Keijzer F, Godfrey-Smith P (2015) An option space for early neural evolution. Philos Trans R Soc Lond Ser B Biol Sci 370(1684):20150181
5. King N, Rokas A (2017) Embracing uncertainty in reconstructing early animal evolution. Curr Biol 27(19):R1081–R1088
6. Burkhardt P, Jekely G (2021) Evolution of synapses and neurotransmitter systems: the divide-and-conquer model for early neural cell-type evolution. Curr Opin Neurobiol 71: 127–138
7. Burkhardt P (2022) Ctenophores and the evolutionary origin(s) of neurons. Trends Neurosci 45(12):878–880
8. Burkhardt P et al (2023) Syncytial nerve net in a ctenophore adds insights on the evolution of nervous systems. Science 380(6642):293–297
9. Schultz DT et al (2023) Ancient gene linkages support ctenophores as sister to other animals. Nature 618(7963):110–117
10. Costello J et al (2012) Transitions of *Mnemiopsis leidyi* (Ctenophora: Lobata) from a native to an exotic species: a review. Hydrobiologia 690(1):21–46
11. Mertens KH (1833) Beobachtungen und Untersuchungen über die beroeartigen Akalephen. University of California
12. Forskål P (1775) Descriptiones animalium, avium, amphibiorum, piscium, insectorum, vermium; quae in itinere orientali observavit. Ex Office Mölleri, p 164

13. Müller O (1776) Zoologiae Danicae Prodromus, seu Animalium Daniae et Norvegiae Indigernarum characteres, nomina, et synonyma imprimis popularium. Copenhagen, Hallager for the author, p 282
14. Fabricius O (1780) Fauna Groenlandica. I. G. Rothe, p 452
15. WoRMS Editorial Board, World Register of Marine Species (WoRMS). Available from http://www.marinespecies.org at VLIZ. Accessed 03 July 2020
16. Eschscholtz JF (1829) System der Acalephen. Eine ausführliche Beschreibung aller Medusenartigen Strahlthiere... Mit 16 Kupfertafeln. F. Dümmler
17. Agassiz A (1865) Illustrated catalogue of the museum of Comparative Zoology, at Harvard College. No. II, North American Acalephae, Cambridge
18. Nagabhushanam A (1959) Feeding of a ctenophore, *Bolinopsis infundibulum* (OF Müller). Nature 184(4689):829–829
19. Greve W (1968) The "planktonkreisel", a new device for culturing zooplankton. Mar Biol 1(3):201–203
20. Greve W (1970) Cultivation experiments on North Sea ctenophores. Helgoländer Meeresun 20(1):304
21. Ward WW (1974) Aquarium systems for the maintenance of ctenophores and jellyfish and for the hatching and harvesting of brine shrimp (*Artemia salina*) larvae. Chesap Sci 15(2): 116–118
22. Raskoff KA et al (2003) Collection and culture techniques for gelatinous zooplankton. Biol Bull 204(1):68–80
23. Knowles T (2016) The history of jelly husbandry at the Monterey Bay Aquarium. Der Zoologische Garten 85(1):42–51
24. Hirota J (1972) Laboratory culture and metabolism of the planktonic ctenophore, Pleurobrachia bachei A. Agassiz. In: Biological oceanography of the northern North Pacific Ocean. Idemitsu Shoten, Tokyo
25. Baker L, Reeve M (1974) Laboratory culture of the lobate ctenophore *Mnemiopsis mccradyi* with notes on feeding and fecundity. Mar Biol 26(1):57–62
26. Pang K, Martindale MQ (2008) Mnemiopsis leidyi spawning and embryo collection. Cold Spring Harbor Protocols 2008(11):pdb.prot5085
27. Salinas-Saavedra M, Martindale MQ (2018) Improved protocol for spawning and immunostaining embryos and juvenile stages of the ctenophore *Mnemiopsis leidyi*. Protoc Exch. https://doi.org/10.1038/protex.2018.092
28. Patry WL et al (2020) Diffusion tubes: a method for the mass culture of ctenophores and other pelagic marine invertebrates. PeerJ 8:e8938
29. Courtney A, Merces GO, Pickering M (2020) Characterising the behaviour of the ctenophore *Pleurobrachia pileus* in a laboratory aquaculture system. bioRxiv
30. Dieter AC, Vandepas LE, Browne WE (2022) Isolation and maintenance of in vitro cell cultures from the ctenophore *Mnemiopsis leidyi* (M. leidyi). In: Blanchoud S, Galliot B (eds) Whole-body regeneration: methods and protocols. Springer, New York, pp 347–358
31. Ramon-Mateu J et al (2022) Studying ctenophora WBR using *Mnemiopsis leidyi*. Methods Mol Biol 2450:95–119
32. Presnell JS et al (2022) Multigenerational laboratory culture of pelagic ctenophores and CRISPR–Cas9 genome editing in the lobate *Mnemiopsis leidyi*. Nat Protoc 17(8): 1868–1900
33. Ikeda S et al (2022) An effective method to mass culture a lobate ctenophore *Bolinopsis mikado*. Plankton Benthos Res 17(4):343–348
34. Sebé-Pedrós A et al (2018) Early metazoan cell type diversity and the evolution of multicellular gene regulation. Nat Ecol Evol 2(7):1176
35. Levin M et al (2016) The mid-developmental transition and the evolution of animal body plans. Nature 531(7596):637–641
36. de Mendoza A et al (2019) Convergent evolution of a vertebrate-like methylome in a marine sponge. Nat Ecol Evol 3(10):1464–1473
37. Pang K, Martindale MQ (2008) Ctenophore whole-mount in situ hybridization. Cold Spring Harb Protoc 2008(11):pdb.prot5087
38. Sachkova MY et al (2021) Neuropeptide repertoire and 3D anatomy of the ctenophore nervous system. Curr Biol 31(23):5274–5285 e6
39. Pang K, Martindale MQ (2008) Ctenophore whole-mount antibody staining. Cold Spring Harb Protoc 2008(11):pdb.prot5086
40. Yamada A et al (2010) Highly conserved functions of the Brachyury gene on morphogenetic movements: insight from the early-diverging phylum Ctenophora. Dev Biol 339(1): 212–222
41. Jokura K et al (2019) CTENO64 is required for coordinated paddling of ciliary comb plate in ctenophores. Curr Biol 29(20):3510-+
42. Presnell JS, Browne WE (2021) Kruppel-like factor gene function in the ctenophore *Mnemiopsis leidyi* assessed by CRISPR/Cas9-mediated genome editing. Development 148(17):dev199771

43. Sutherland KR et al (2014) Ambient fluid motions influence swimming and feeding by the ctenophore *Mnemiopsis leidyi*. J Plankton Res 36(5):1310–1322
44. Presnell JS et al (2016) The presence of a functionally tripartite through-gut in Ctenophora has implications for metazoan character trait evolution. Curr Biol 26(20):2814–2820
45. Vandepas LE et al (2017) Establishing and maintaining primary cell cultures derived from the ctenophore *Mnemiopsis leidyi*. J Exp Biol 220(7):1197–1201
46. Tamm SL (2019) Defecation by the ctenophore *Mnemiopsis leidyi* occurs with an ultradian rhythm through a single transient anal pore. Invertebr Biol 138(1):3–16
47. Shiganova T et al (2001) Population development of the invader ctenophore *Mnemiopsis leidyi*, in the Black Sea and in other seas of the Mediterranean basin. Mar Biol 139(3): 431–445
48. Rees J-F et al (1994) Highly unsaturated fatty acid requirements of *Penaeus monodon* postlarvae: an experimental approach based on Artemia enrichment. Aquaculture 122(2–3): 193–207
49. Brett M, Müller-Navarra D (1997) The role of highly unsaturated fatty acids in aquatic foodweb processes. Freshw Biol 38(3):483–499
50. Granhag L, Møller LF, Hansson LJ (2011) Size-specific clearance rates of the ctenophore *Mnemiopsis leidyi* based on in situ gut content analyses. J Plankton Res 33(7):1043–1052
51. Freeman G, Reynolds GT (1973) The development of bioluminescence in the ctenophore *Mnemiopsis leidyi*. Dev Biol 31(1):61–100
52. Sasson DA, Ryan JF (2016) The sex lives of ctenophores: the influence of light, body size, and self-fertilization on the reproductive output of the sea walnut, *Mnemiopsis leidyi*. PeerJ 4:e1846
53. Jaspers C, Møller LF, Kiørboe T (2015) Reproduction rates under variable food conditions and starvation in *Mnemiopsis leidyi*: significance for the invasion success of a ctenophore. J Plankton Res 37(5):1011–1018
54. Sasson DA, Jacquez AA, Ryan JF (2018) The ctenophore *Mnemiopsis leidyi* regulates egg production via conspecific communication. BMC Ecol 18(1):12
55. Ramondenc S et al (2019) From egg to maturity: a closed system for complete life cycle studies of the holopelagic jellyfish *Pelagia noctiluca*. J Plankton Res 41(3):207–217

Chapter 5

Illustrated Neuroanatomy of Ctenophores: Immunohistochemistry

Tigran P. Norekian and Leonid L. Moroz

Abstract

Ctenophores or comb jellies are representatives of an enigmatic lineage of early branching metazoans with complex tissue and organ organization. Their biology and even microanatomy are not well known for most of these fragile pelagic and deep-water species. Here, we present immunohistochemical protocols successfully tested on more than a dozen ctenophores. This chapter also illustrates neural organization in several reference species of the phylum (*Pleurobrachia bachei*, *P. pileus*, *Mnemiopsis leidyi*, *Bolinopsis microptera*, *Beroe ovata*, and *B. abyssicola*) as well as numerous ciliated structures in different functional systems. The applications of these protocols illuminate a very complex diversification of cell types comparable to many bilaterian lineages.

Key words Ctenophora, Electrophysiology, Behavior, Neurotransmitters, Neuropeptides, Nitric oxide, *Pleurobrachia*, *Bolinopsis*, *Mnemiopsis*, *Beroe*, Neurons, Gap junctions

1 Introduction

For over a century, ctenophores were often considered the sister group of cnidarians and received less attention than other basal metazoans. The recent genomic revolution and sequencing of ctenophore genomes [1–4] and multiple transcriptomes [1, 5] started hot debates about their identity, relationships with other animals, and origins of ctenophore innovations. The consensus has not been reached [6–9]. However, the emerging conclusion is that ctenophores are the earliest surviving animal lineages, independently developing enormous complexity in their tissue and organ organization and behavior [10–12]. There is a reasonable case that ctenophores independently evolved neurons, synapses, muscles, and mesoderm [1, 12, 13].

Despite the recent rise in interest, we know very little about the microanatomical organization in ctenophores, with the latest

Authors Tigran P. Norekian and Leonid L. Moroz have equally contributed to this chapter.

Leonid L. Moroz (ed.), *Ctenophores: Methods and Protocols*, Methods in Molecular Biology, vol. 2757,
https://doi.org/10.1007/978-1-0716-3642-8_5, © Springer Science+Business Media, LLC, part of Springer Nature 2024

systematic review of their ultrastructure published in 1991 [10]. Even basic microanatomy is unknown for most of these species. Plus, many of these pelagic organisms are highly fragile, which by itself presents a significant challenge for histochemical and molecular characterization in situ [14].

Immunohistochemistry is a relatively simple and powerful tool for any comparative biological study. This approach is still very needed to characterize ctenophore cells and tissues. Protocols and recommendations were reported for larval stages of *Mnemiopsis* [15, 16]. Here, we summarize practical immunohistochemistry protocols for ctenophores. We successfully tested this methodological approach on 14 adult ctenophore species and their larvae, focusing on their neuromuscular organization [17–21].

This chapter also provides illustrative examples of neural nets and receptor types in several reference species of the phylum (Figs. 2, 3, 4, 5, 6, 7 and 8). These are *Pleurobrachia bachei*, *P. pileus*, *Mnemiopsis leidyi*, *Bolinopsis microptera*, *Beroe ovata*, and *B. abyssicola* collected in the Northwestern Pacific and Atlantic (Fig. 1). In addition, we show ciliated structures in the aboral, swimming, digestive systems, and meridional canals of ctenophores (Figs. 3 and 9). The applications of these protocols illuminate a very complex diversification of cell types in ctenophores. The level of anatomical complexity within this phylum is comparable to many bilaterian lineages.

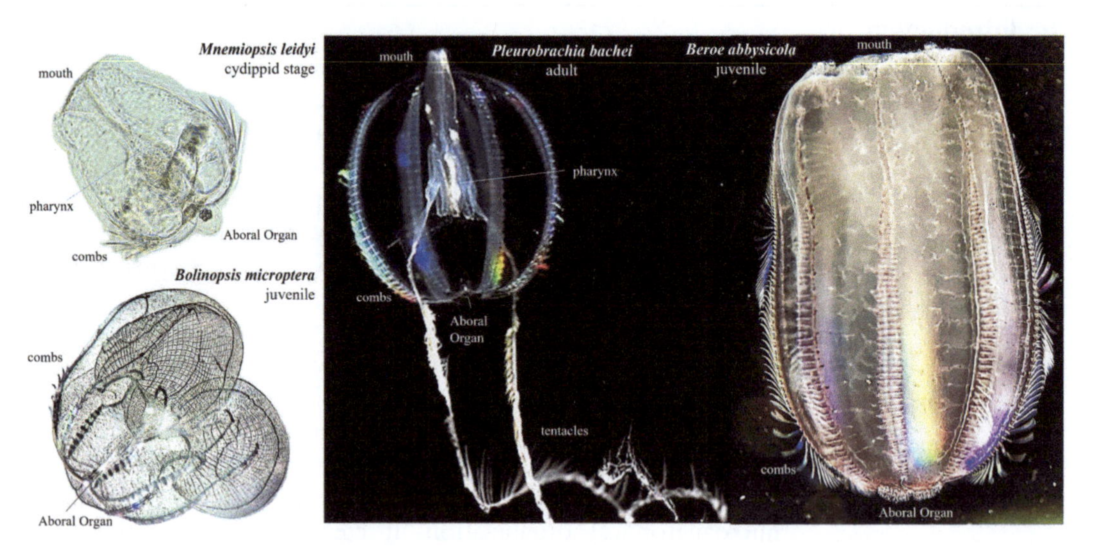

Fig. 1 *Mnemiopsis leydi* cydippid stage, juvenile and adult stages of *Bolinopsis microptera, Pleurobrachia bachei, and Beroe abbysicola* images with main organs

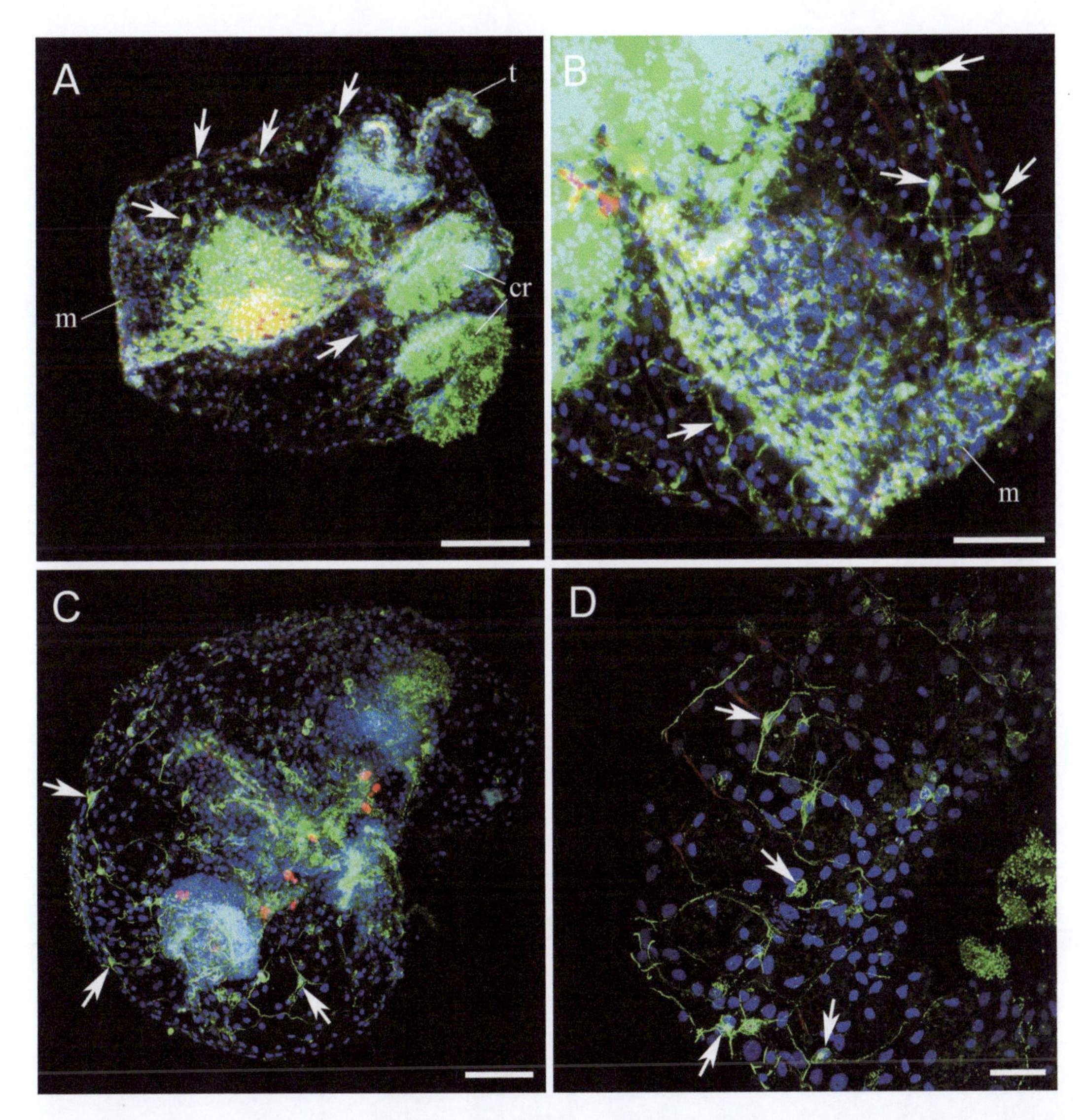

Fig. 2 The nervous system in early ctenophore development (revealed by anti-tubulin immunoreactivity in green; see text and [17, 21]). (**a**, **b**) Hatched 4-day-old *Mnemiopsis* cydippid stage. Note individual neurons (arrows) that started forming the subepithelial neural network. (**c**, **d**) *Bolinopsis* 3-day cydippid after hatching. Arrows point to the subepithelial neurons that begin to form the neural network. Abbreviations: *cr* comb row, *m* mouth, *t* tentacle. Scale bars: (**a**) 50 μm; (**b**) 30 μm; (**c**) 50 μm; (**d**) 20 μm

2 Materials

1. Phosphate-buffered saline (PBS)—0.2 M stock solution, pH = 7.6.
2. Stock 16% paraformaldehyde aqueous solution in 10 mL ampules (Ted Pella, Catalog # 18505, EM grade). Working concentration—4% paraformaldehyde in 0.1 M PBS.

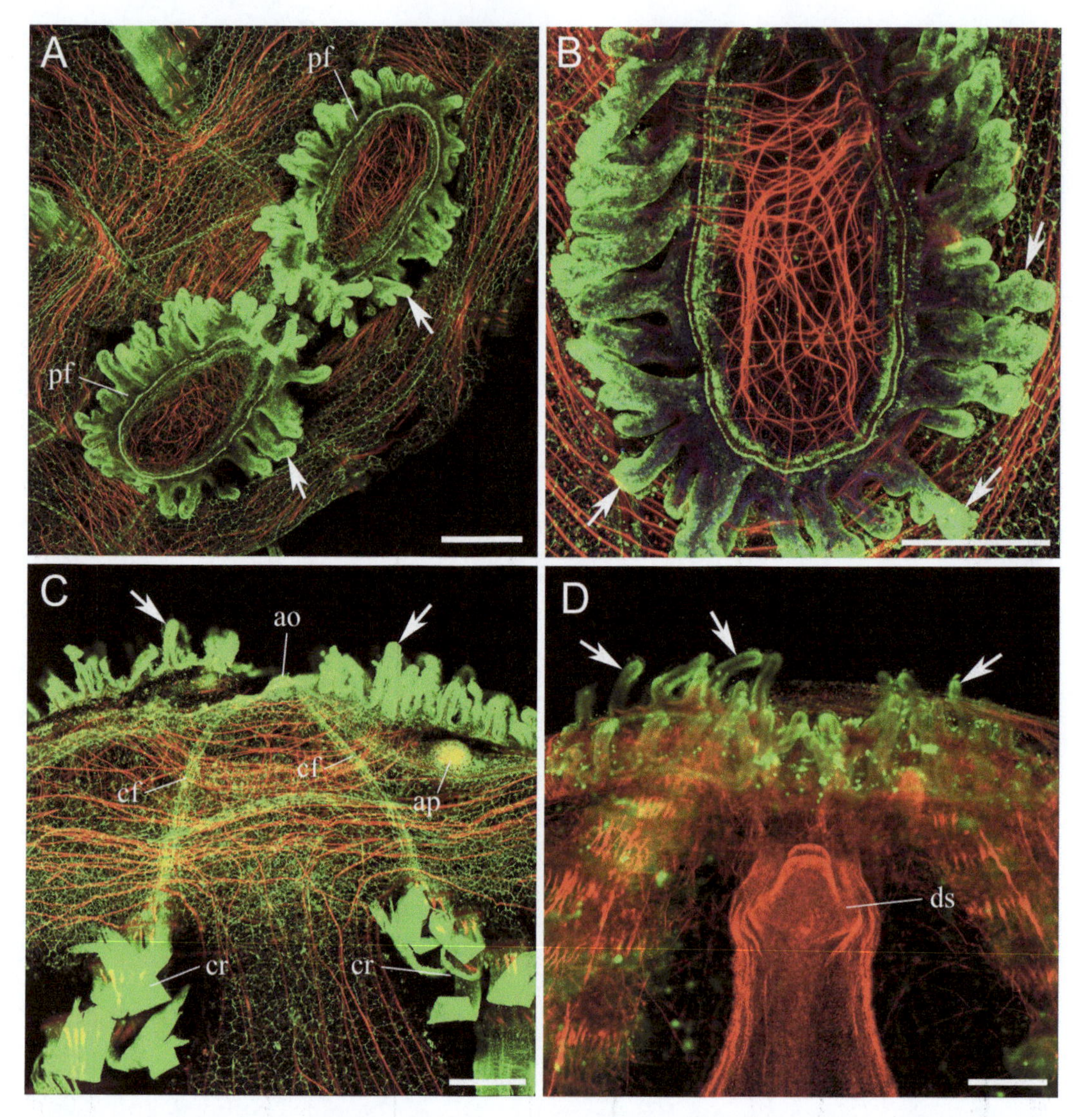

Fig. 3 Aboral organ and polar fields in *Beroe ovata*. (**a**, **b**) Polar fields with tall lobes (arrows) forming a crown are brightly labeled by tubulin AB (green), horizontal view. Note a subepithelial neural network covering the entire surface area, which is also stained by AB in green. Phalloidin-labeled muscle fibers are red. (**c**, **d**) Optical cross-section of the aboral area. Arrows point to the polar field lobes. See more details in [18, 20]. Abbreviations: *ao* aboral organ, *ap* anal pore, *cf* ciliated furrow, *cr* comb row, *ds* digestive system, *pf* polar field. Scale bars: 200 μm

3. Goat serum as a blocking solution (Sigma, Catalog#: G9023). Working concentration—6% goat serum in 0.1 M PBS.
4. Primary antibodies. Rat monoclonal anti-tubulin antibody (AbD Serotec, Cat# MCA77G).
5. Secondary antibodies. Goat anti-rat IgG antibodies: Alexa Fluor 488 conjugated (Molecular Probes, Invitrogen, Cat#

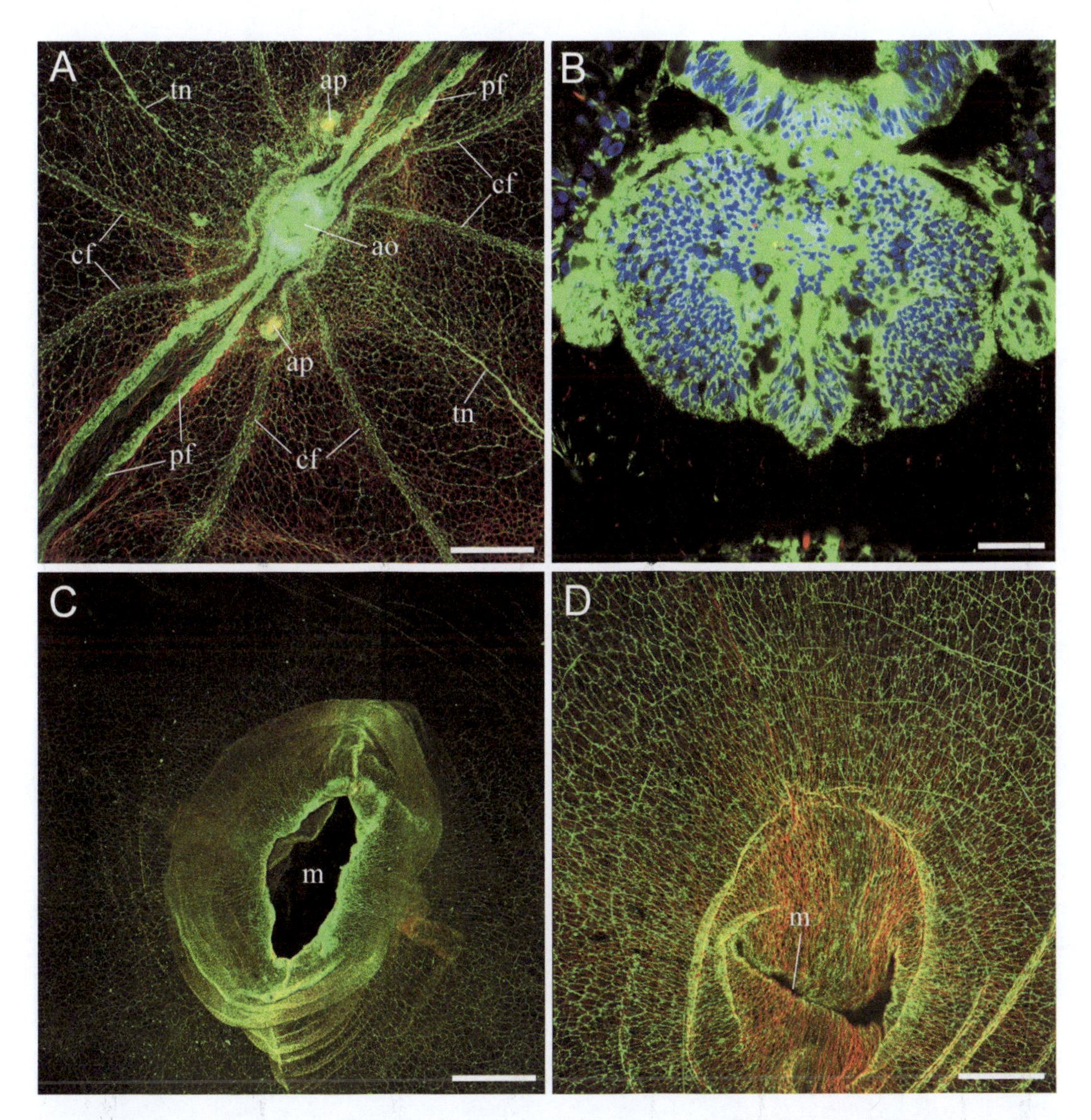

Fig. 4 The aboral organ and mouth in *Pleurobrachia*. (**a**) Aboral organ (*ao*) and polar fields (*pf*) in *Pleurobrachia pileus* (horizontal view). (**b**) Side view of the aboral organ in *Pleurobrachia pileus*. The aboral organ consists of many tightly packed and immunoreactive to tubulin AB cells, whose nuclei are stained blue by DAPI. (**c**, **d**) Mouth (*m*) area of the *Pleurobrachia bachei*. Note a subepithelial neural network covering the entire surface area, which is labeled by tubulin AB in green; phalloidin—red. See additional details in [19, 20]. Abbreviations: *ao* aboral organ, *ap* anal pore, *cf* ciliated furrow, *m* mouth, *pf* polar field, *tn* tentacular nerve. Scale bars: (**a**) 200 μm; (**b**) 20 μm; (**c**) 500 μm; (**d**) 300 μm

A11006) and Alexa Fluor 568 conjugated (Molecular Probes, Invitrogen, Cat# A11077).

6. Phalloidin (Alexa Fluor 488 and Alexa Fluor 568 phalloidins from Molecular Probes, Catalog#: A-12379 and A-12380) for labeling muscle fibers.

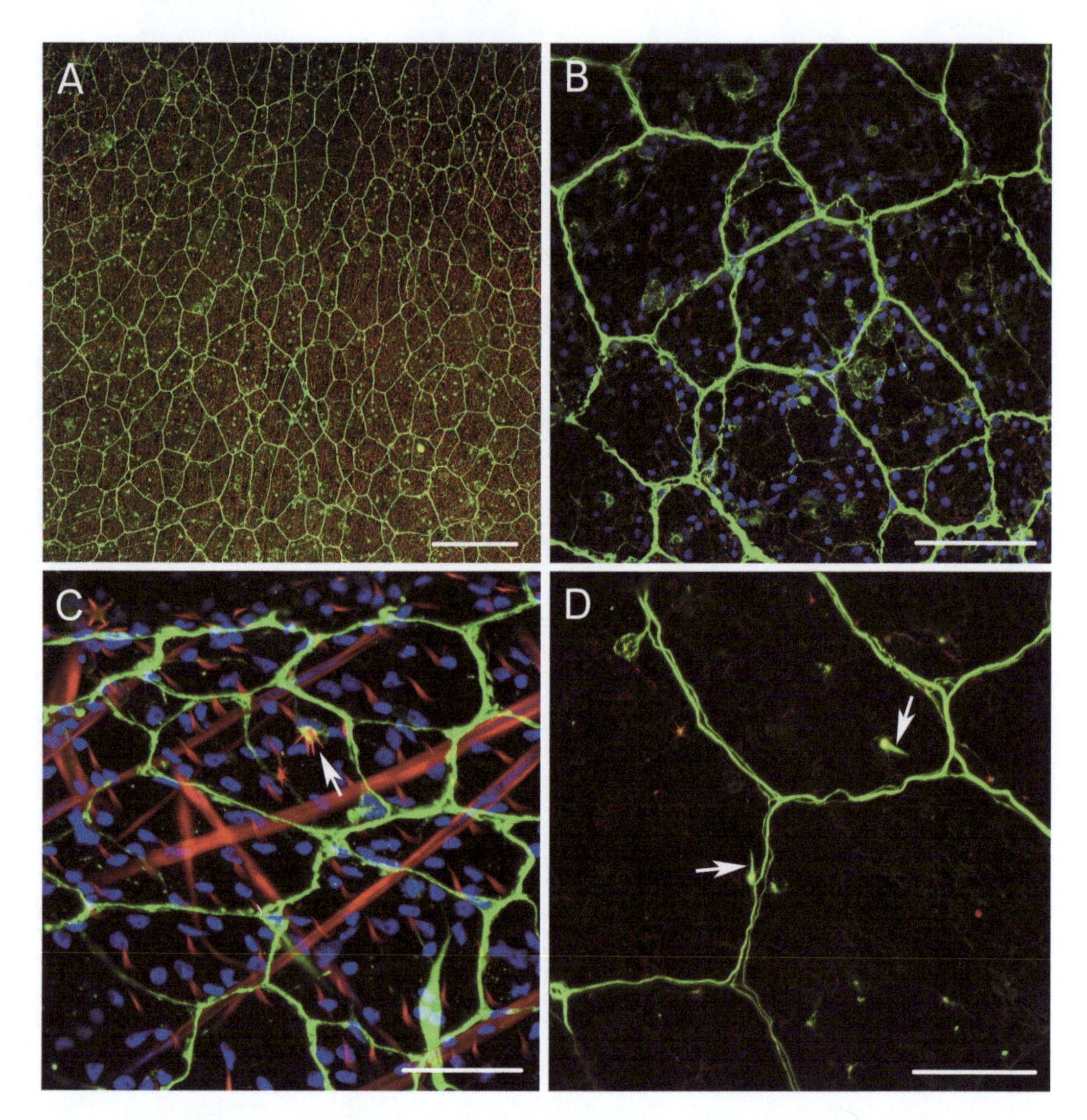

Fig. 5 Subepithelial neural network in different ctenophore species labeled by tubulin AB (green). The phalloidin labeling is red, while DAPI is blue. (**a**, **b**) Neural network in the body wall of adult *Pleurobrachia bachei*. (**c**) Subepithelial neural network in *Beroe abyssicola* pharynx wall (green) with parietal muscle fibers shown in red. The arrow points to the receptor with multiple stereocilia. (**d**) Subepithelial neural network in *Euplokamis* with surface receptors identified by arrows. Scale bars: (**a**) 200 μm; (**b**) 35 μm; (**c**) 50 μm; (**d**) 20 μm

7. Mounting medium VECTASHIELD Hard-Set Mounting Medium with DAPI (Vector Labs, Cat# H-1500).
8. Glass microscope slides and coverslips.

2.1 Equipment

1. Nikon Research Microscope Eclipse E800 with Epi-fluorescence using standard TRITC and FITC filters.
2. Nikon C2 laser scanning confocal microscope.

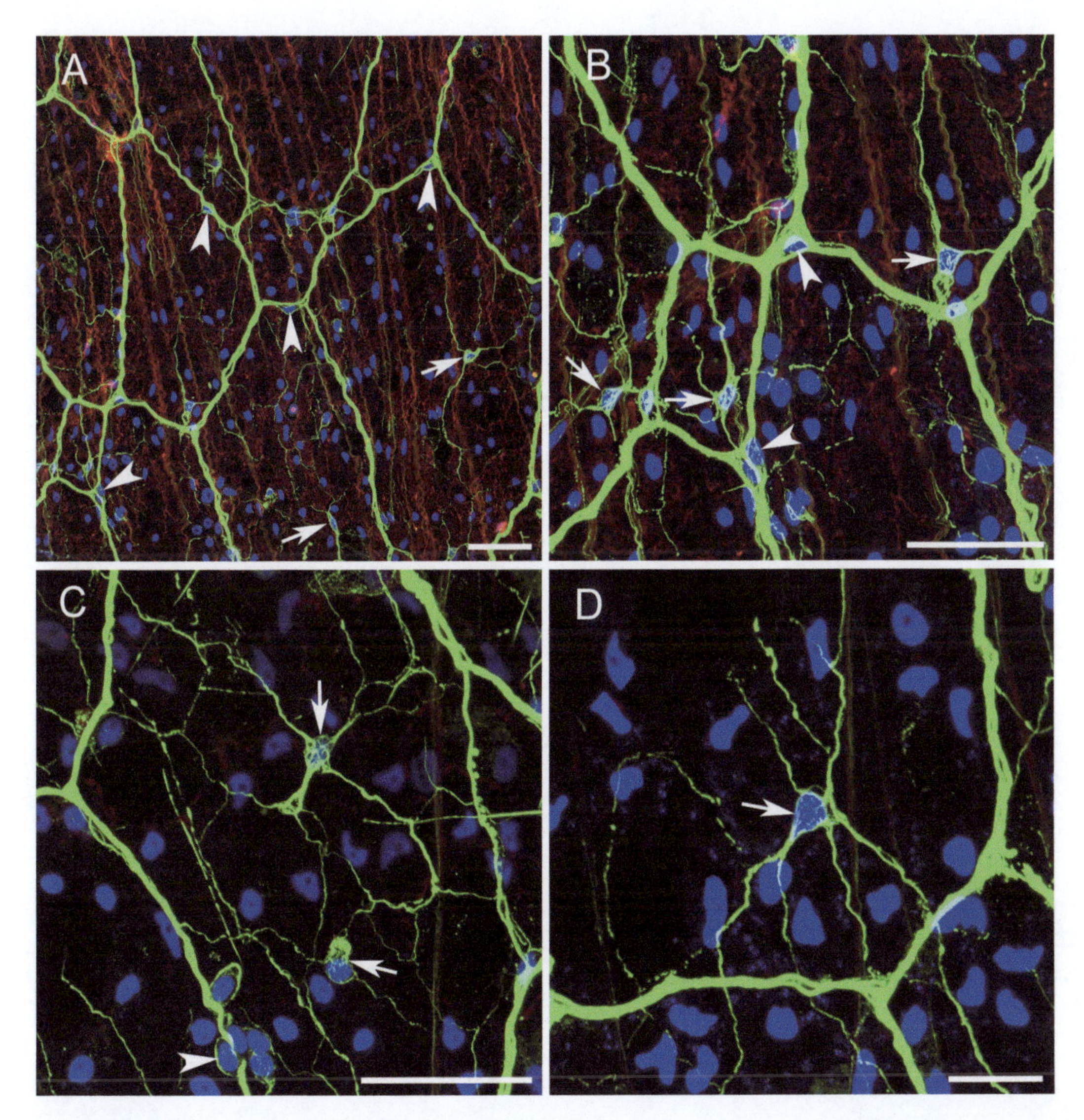

Fig. 6 Subepithelial neural network in *Pleurobrachia pileus* (tubulin AB, green; phalloidin labeling is red, while DAPI is blue). Note the individual neurons with long processes between major strands of the network (arrows). Arrowheads point to the nuclei inside the strands of the network. Scale bars: (**a**, **b**, and **c**) 25 μm; (**d**) 10 μm

3 Methods

Fixation—12 h

1. Before fixation and dissection, animals were incubated in high Mg^{2+} seawater (0.3 M $MgCl_2$ in filtered seawater in 1:1 ratio) for 30–40 min. It was essential for such muscular animals like *Beroe* with strong withdrawal reactions (*see* **Note 1**).

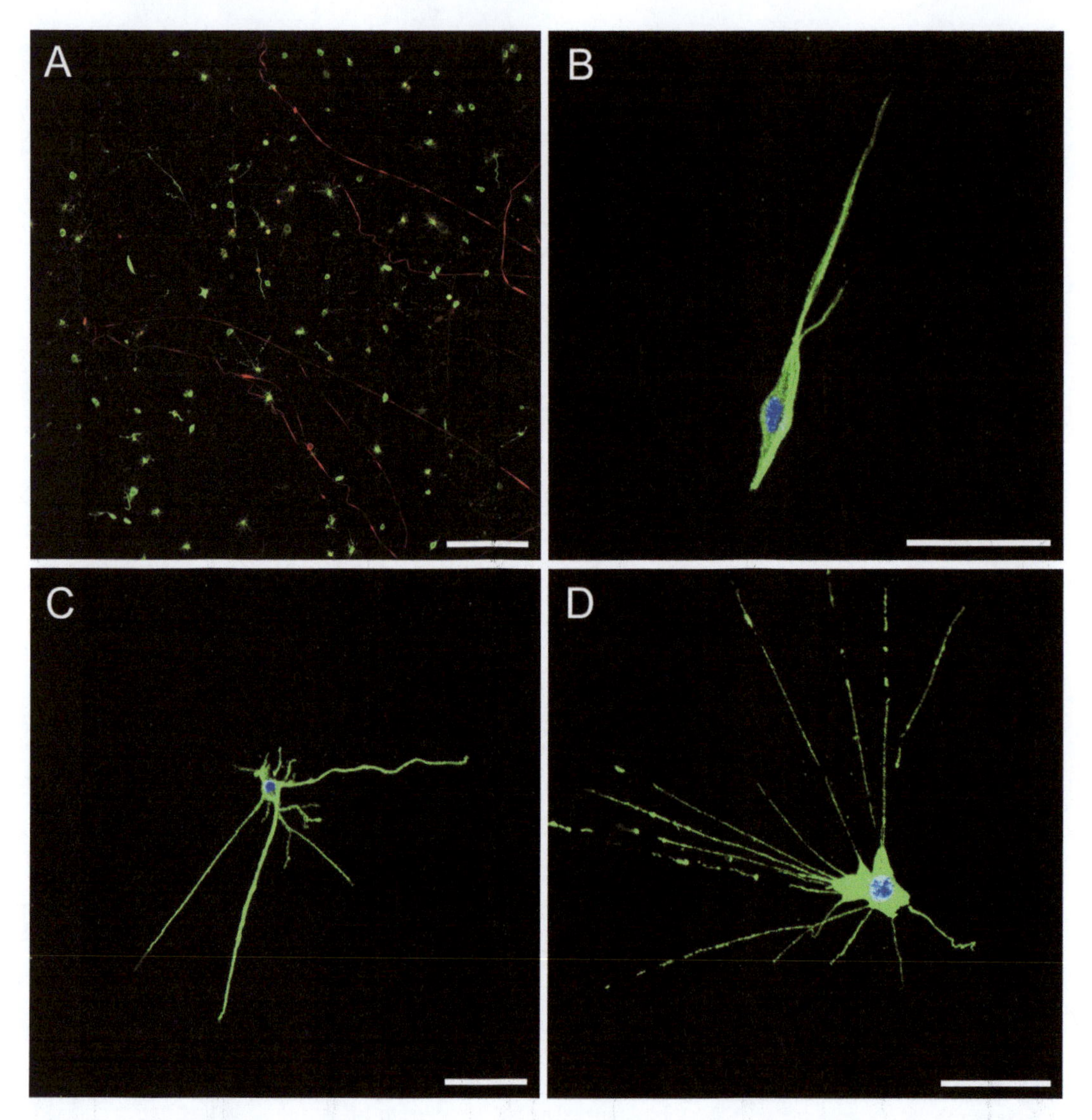

Fig. 7 Mesogleal neurons in *Pleurobrachia pileus* (tubulin AB, green; phalloidin labeling is red, while DAPI is blue). (**a**) Numerous tubulin-immunoreactive neural type cells in the mesogleal region. (**b**) Bipolar mesogleal neuron-like cell. (**c**, **d**) Multipolar mesogleal neural type cells. Scale bars: (**a**) 100 μm; (**b**, **c**, and **d**) 20 μm

2. Large ctenophores were dissected in smaller pieces, while small animals were fixed whole (all *Pleurobrachia* and smaller *Beroe*). The focus during dissecting was on specific organs and areas—like the aboral organ with polar fields, comb rows, mouth, etc.
3. The tissue was placed in 20–30 mL vials with closed lids.
4. Seawater or high Mg^{2+} seawater was removed from the vials with a suction pipette, and 4% paraformaldehyde in PBS was immediately added into vials (*see* **Note 1**).

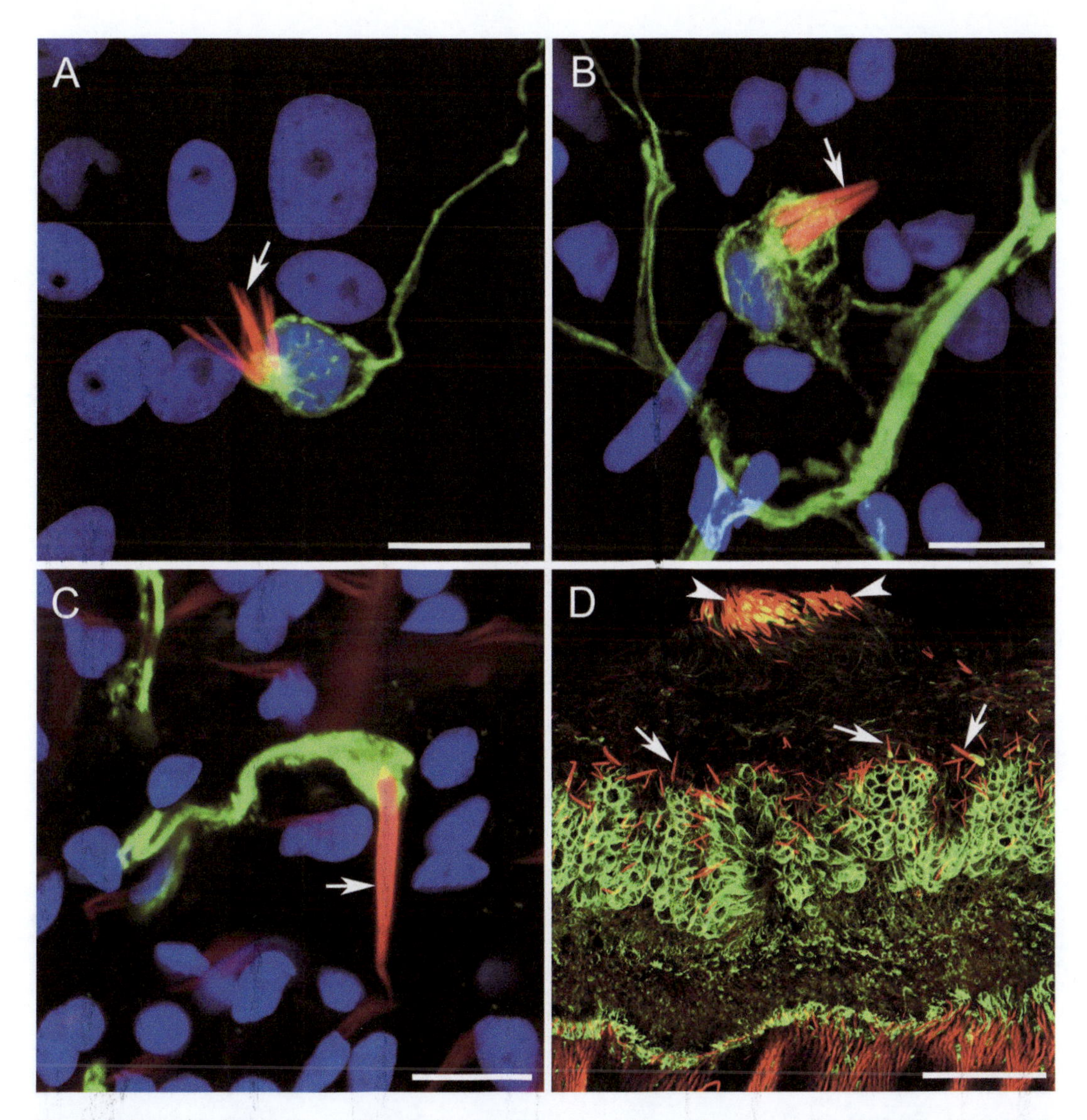

Fig. 8 Surface receptors in *Beroe abyssicola*. (**a**, **b**) Receptors with multiple stereocilia labeled by phalloidin in red (arrows), DAPI is blue. The cell body of the receptor is labeled by tubulin AB (green). (**c**) Receptor with a single large stereocilium labeled by phalloidin (arrow) on the top of tubulin-ir cell. (**d**) Numerous receptors with a single large and labeled by phalloidin cilium (arrows) are found on the lips and in large congregates (arrowheads) on the outside body surface. Scale bars: (**a**, **b**, and **c**) 10 μm; (**d**) 100 μm

5. The vials with tissue in fixative solution were placed in the refrigerator and left overnight or for 12 h at 5 °C. The tissue can stay in a fixative for a day. However, long fixations for 2 days or more are not advisable.
6. After fixation, the tissue was rinsed in 0.1 M PBS. The fixative was removed from vials with suction pipettes and properly disposed of—under the hood and using gloves. Then clean PBS was added for 40 min.

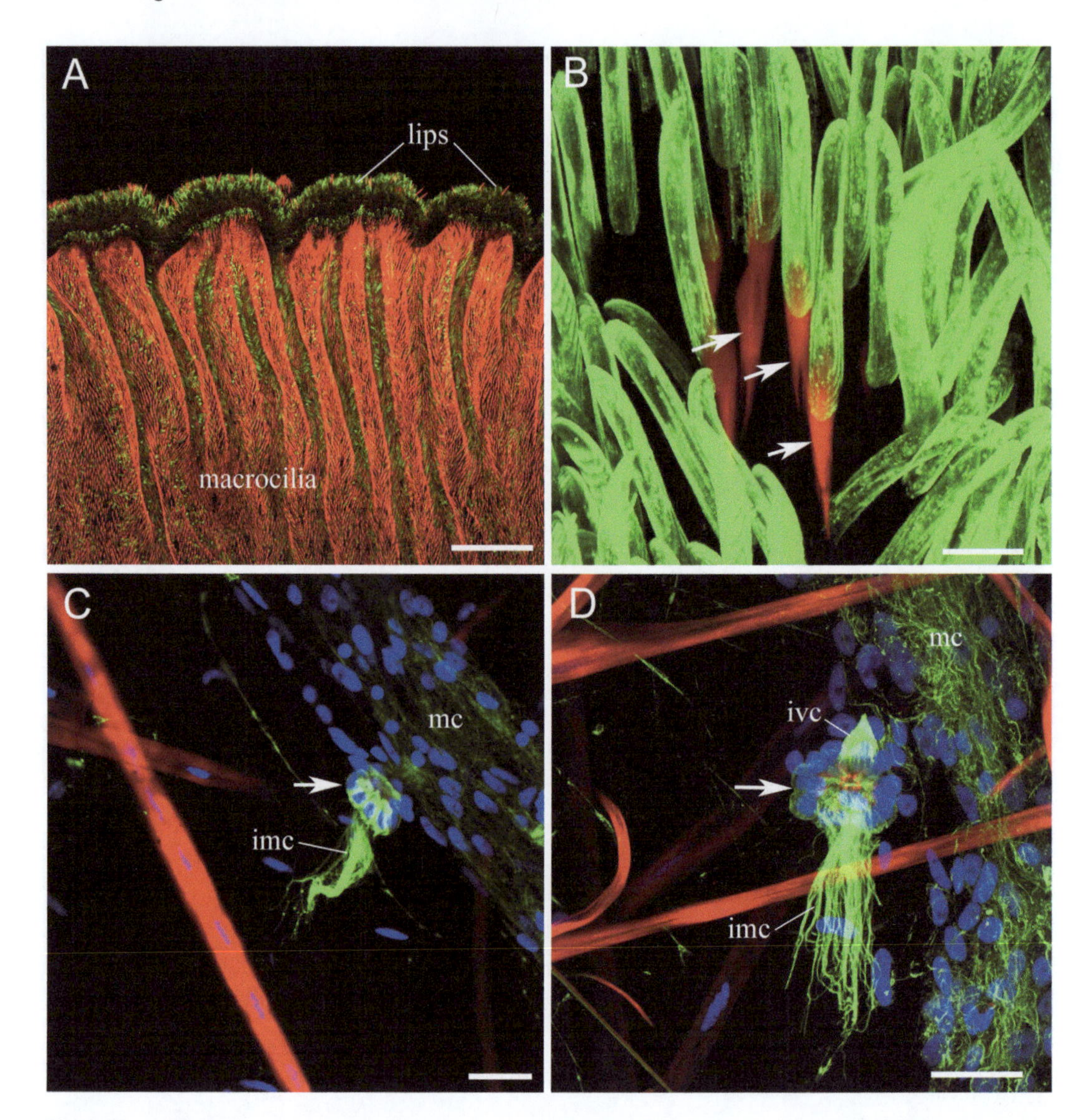

Fig. 9 Non-neuronal structures in ctenophores: Macrocilia and ciliated pores in meridional canals. Tubulin AB is green, phalloidin is red, while DAPI is blue. (**a**, **b**) Microcilia inside the mouth of *Beroe abyssicola*. The long cylindrical body of macrocilia is stained by tubulin AB and attached to the phalloidin-labeled bundles of actin (arrows) inside the macrociliary cells. (**c**, **d**) Meridional canal pores (ciliated rosettes) in *Beroe abyssicola.* Each pore consists of two superimposed rings (arrows) with eight ciliated cells. Two groups of cilia are labeled by tubulin antibody: shorter intravascular cilia, which project into the meridional canal, and much longer intramesogleal cilia that protrude into the mesoglea. Muscle fibers are stained in red. Abbreviations: *mc* meridional canal, *imc* intramesogleal cilia, *ivc* intravascular cilia. Scale bars: (**a**) 200 μm; (**b**) 10 μm; (**c**, **d**) 20 μm

7. There should be at least three rinses with a total minimum time of 2 h. At this stage, the tissue can be stored in a refrigerator for a few days.

   Blocking solution—12 h to 1 day

8. After fixation and rinses, the large pieces of tissue and the whole fixed small animals were dissected and trimmed as necessary for the final viewing. The tissue was removed from the vials, placed in the small Petri dishes, and looked at under a dissecting microscope. The final size of tissues should be relatively flat and small to be viewed effectively under the confocal microscope later. At the same time, one should keep in mind the necessity to preserve the functional continuity of the data (e.g., not to cut polar fields in half).
9. The cleaned and trimmed pieces of tissue were then placed again in the 20–30 mL vials with the clean 0.1 M PBS containing 0.2% Triton-X 100. Incubation of the tissue in Triton-X 100 solution for 6–12 h was necessary for improving the subsequent antibody penetration (*see* **Note 2**).
10. PBS in vials was then replaced with 6% goat serum in PBS as a blocking solution. The specimens were kept in a refrigerator at 5 °C for 12 h (up to 1 day).

    Primary antibodies—2 days
11. The old blocking solution was removed from the vials.
12. The primary antibodies were then diluted in 6% goat serum in PBS (blocking solution) and added into vials with specimens. The final dilution depended on the type of primary antibodies. For example, we used the final dilution of 1:40 for the rat monoclonal anti-tubulin antibody. The tissue should stay in primary antibodies for about 2 days in a refrigerator at 5 °C. Depending on the tissue samples' size, the primary body incubation could be reduced to 1 day or increased to 3 days. From time to time, some shaking of the vials is advisable during incubation.
13. At the end of incubation, the primary antibody solution was removed from the vials and replaced with fresh 0.1 M PBS. The rinsing of specimens should last minimum 12 h, better 1 day, and include a minimum 3–4 rinses. All long-term steps are carried in a refrigerator.

    Secondary antibodies—1 day
14. Following the multiple rinses in PBS after primary antibody incubation, the specimens were placed in the secondary antibodies—one of two types of goat anti-rat IgG antibodies: Alexa Fluor 488 conjugated or Alexa Fluor 568 conjugated, at a final dilution 1:20 in 0.1 M PBS (*see* **Note 3**). Incubation in secondary antibody was carried in the refrigerator for minimum 12 h, up to 1 day.
15. The secondary antibodies were then removed from vials and replaced with 0.1 M PBS. There should be at least three rinses in PBS for 12 h.

    Phalloidin labeling—8 h to a day

16. In order to obtain additional information and visualize muscle fibers in ctenophores along with immunolabeling, we used the well-known marker phalloidin, which binds to F-actin. Depending on the type of secondary antibodies, we used Alexa Fluor 488 or Alexa Fluor 568 conjugated phalloidins. The tissue was incubated in phalloidin solution in PBS for 8 h at a final dilution 1:80 in a refrigerator at 5 °C.
17. The specimens were then washed in 0.1 M PBS several times for 6 h.

    Mounting and Viewing

18. Following the PBS rinses, the pieces of tissue were transferred onto the microscope glass slides. Each piece of tissue was oriented for the best viewing position. The PBS was then carefully removed and a mounting medium was added to the slides. We preferred to use a hard-set mounting medium with DAPI to visualize the cell nuclei. The tissue was then covered with a cover glass.
19. For the mounting medium to harden sufficiently, it was necessary to wait for about 2 h. However, it was better to wait at least 6 h before viewing for DAPI staining to penetrate the entire thickness of the tissue.
20. The slides with mounted tissue were viewed first at the microscopes with epi-fluorescence to identify and select the best specimens and areas inside each specimen for the following detailed scanning on the confocal microscope.
21. The slides could be stored in closed booklets in a refrigerator for several weeks for later viewing. However, a signal loss would inevitably occur after prolonged exposure to the laser—especially fast the loss of signal developed for the phalloidin labeling.

The illustrated examples of immunolabeling in different ctenophore species are summarized in Figs. 2, 3, 4, 5, 6, 7, 8 and 9. For more detailed descriptions of microanatomy, we recommend several earlier and recent publications [10, 11, 17–36].

4 Notes

1. Paraformaldehyde fixation worked very well for many ctenophore groups such as *Pleurobrachia*, *Beroe*, *Hormifora*, and *Euplokamis*. However, ctenophores from two genera, *Mnemiopsis* and *Bolinopsis*, which are the most fragile, full of water, and least dense, were very difficult to fix. Their tissue was disintegrating during fixation in paraformaldehyde. Nevertheless, the larva stages of these species were much denser and tougher and could be easily fixed and processed for

immunolabeling (Figs. 1a and 2—*Mnemiopsis* and *Bolinopsis* larvae). For adult *Mnemiopsis* and *Bolinopsis,* we have developed a different fixation protocol, which showed promising results. The fixative solution contained 1 part of 6% paraformaldehyde in filtered seawater and 1 part of 100% ethanol with added 0.5% acetic acid. The prepared solution was then placed in the freezer to reach −20 °C. The specimens were cooled in a refrigerator to 5 °C; then the seawater was removed with a suction pipette, and a fixative solution at −20 °C was added into the vials with ctenophore tissue. The vials were kept at −20 °C in the freezer for 1 h and then transferred into the refrigerator at 5 °C for overnight (12 h) fixation. The following steps were usual, including rinses and incubations in primary and secondary antibodies.

2. Ctenophore tissue is not very dense and easily penetrable for primary and secondary antibodies. Therefore, we used Triton-X 100, widely accepted in immunochemistry protocols for improved antibody penetration, for only 6–12 h during incubation in the blocking solution. The only exception was the early larva stages, which had much tougher tissue. We used 0.1 M PBS with 0.2% Triton-X 100 during blocking solution and primary antibody incubations. We also used Triton-X 100 in our novel fixation protocol, which was used for adult *Mnemiopsis* and *Bolinopsis* and contained 1:1 paraformaldehyde and ethanol with 1% acetic acid at −20 °C.
3. Some obvious reminders. A blocking solution should be made from the serum of the animal, which was used to produce secondary antibodies. Secondary antibodies are made against the animal in which the primary antibody was produced. If the secondary antibody were conjugated with Alexa Fluor 488, then phalloidin should be labeled with a fluorescent label with a different wavelength—Alexa Fluor 568 (and vice versa).

Acknowledgments

This work was supported in part by the Human Frontiers Science Program (RGP0060/2017) and the National Science Foundation (IOS-1557923) grants to L.L.M. Research reported in this publication was also supported in part by the National Institute of Neurological Disorders and Stroke of the National Institutes of Health under award number R01NS114491 (to LLM). The content is solely the authors' responsibility and does not necessarily represent the official views of the National Institutes of Health.

References

1. Moroz LL et al (2014) The ctenophore genome and the evolutionary origins of neural systems. Nature 510(7503):109–114
2. Ryan JF et al (2013) The genome of the ctenophore *Mnemiopsis leidyi* and its implications for cell type evolution. Science 342(6164): 1242592
3. Hoencamp C et al (2021) 3D genomics across the tree of life reveals condensin II as a determinant of architecture type. Science 372(6545):984–989
4. Schultz DT et al (2021) A chromosome-scale genome assembly and karyotype of the ctenophore *Hormiphora californensis.* G3 (Bethesda) 11(11):jkab302
5. Whelan NV et al (2017) Ctenophore relationships and their placement as the sister group to all other animals. Nat Ecol Evol 1(11): 1737–1746
6. Telford MJ, Moroz LL, Halanych KM (2016) Evolution: a sisterly dispute. Nature 529(7586):286–287
7. Halanych KM et al (2016) Miscues misplace sponges. Proc Natl Acad Sci U S A 113(8): E946–E947
8. Redmond AK, McLysaght A (2021) Evidence for sponges as sister to all other animals from partitioned phylogenomics with mixture models and recoding. Nat Commun 12(1):1783
9. Li Y et al (2021) Rooting the animal tree of life. Mol Biol Evol 38(10):4322–4333
10. Hernandez-Nicaise M-L (1991) Ctenophora. In: Harrison FWFW, Westfall JA (eds) Microscopic anatomy of invertebrates: Placozoa, Porifera, Cnidaria, and Ctenophora. Wiley, New York, pp 359–418
11. Tamm SL (1982) Ctenophora. In: Electrical conduction and behavior in "simple" invertebrates. Clarendon Press, Oxford, pp 266–358
12. Nielsen C (2019) Early animal evolution: a morphologist's view. R Soc Open Sci 6(7): 190638
13. Moroz LL, Kohn AB (2016) Independent origins of neurons and synapses: insights from ctenophores. Philos Trans R Soc Lond Ser B Biol Sci 371(1685):20150041
14. Mitchell DG, Edgar A, Martindale MQ (2021) Improved histological fixation of gelatinous marine invertebrates. Front Zool 18(1):29
15. Pang K, Martindale MQ (2008) Ctenophore whole-mount antibody staining. CSH Protoc 2008:pdbprot5086
16. Sachkova MY et al (2021) Neuropeptide repertoire and 3D anatomy of the ctenophore nervous system. Curr Biol 31(23):5274–5285 e6
17. Norekian TP, Moroz LL (2016) Development of neuromuscular organization in the ctenophore *Pleurobrachia bachei.* J Comp Neurol 524(1):136–151
18. Norekian TP, Moroz LL (2019) Neural system and receptor diversity in the ctenophore *Beroe abyssicola.* J Comp Neurol 527(12): 1986–2008
19. Norekian TP, Moroz LL (2019) Neuromuscular organization of the Ctenophore *Pleurobrachia bachei.* J Comp Neurol 527(2):406–436
20. Norekian TP, Moroz LL (2020) Comparative neuroanatomy of ctenophores: neural and muscular systems in *Euplokamis dunlapae* and related species. J Comp Neurol 528(3): 481–501
21. Norekian TP, Moroz LL (2021) Development of the nervous system in the early hatching larvae of the ctenophore *Mnemiopsis leidyi.* J Morphol 282(10):1466–1477
22. Tamm SL (2014) Cilia and the life of ctenophores. Invertebr Biol 133(1):1–46
23. Jager M et al (2011) New insights on ctenophore neural anatomy: immunofluorescence study in *Pleurobrachia pileus* (Muller, 1776). J Exp Zool B Mol Dev Evol 316B(3):171–187
24. Jager M et al (2013) Evidence for involvement of Wnt signalling in body polarities, cell proliferation, and the neuro-sensory system in an adult ctenophore. PLoS One 8(12):e84363
25. Bullock TH, Horridge GA (1965) Structure and function in the nervous systems of invertebrates, vol 1719. Freeman, San Francisco
26. Chun C (1880) Die Ctenophoren des Golfes von Neapel. Fauna Flora Neapel, Monogr, pp 1–313
27. Hernandez-Nicaise ML (1973) The nervous system of ctenophores. II. The nervous elements of the mesoglea of beroids and cydippids (author's transl). Z Zellforsch Mikrosk Anat 143(1):117–133
28. Hernandez-Nicaise ML (1973) The nervous system of ctenophores. I. Structure and ultrastructure of the epithelial nerve-nets. Z Zellforsch Mikrosk Anat 137(2):223–250
29. Hernandez-Nicaise ML (1973) The nervous system of ctenophores. III. Ultrastructure of synapses. J Neurocytol 2(3):249–263
30. Hernandez-Nicaise ML, Amsellem J (1980) Ultrastructure of the giant smooth muscle

fiber of the ctenophore *Beroe ovata*. J Ultrastruct Res 72(2):151–168
31. Horridge GA (1964) The giant mitochondria of ctenophore comb plates. Q J Microsc Sci 105:301–310
32. Horridge GA (1964) Presumed photoreceptive cilia in a ctenophore. Q J Microsc Sci 105:311–317
33. Horridge GA (1965) Non-motile sensory cilia and neuromuscular junctions in a ctenophore independent effector organ. Proc R Soc Lond Biol 162:333–350
34. Horridge GA (1974) Recent studies on the Ctenophora. In: Muscatine L, Lenhoff HM (eds) Coelenterate biology. Academic, New York, pp 439–468
35. Horridge GA, Mackay B (1964) Neurociliary synapses in *Pleurobrachia* (Ctenophora). Q J Microsc Sci 105:163–174
36. Hyman LH (1940) Invertebrates: protozoa through Ctenophora, vol 1. McGraw-Hill, New York/London, p 726

Chapter 6

Scanning Electron Microscopy of Ctenophores: Illustrative Atlas

Tigran P. Norekian and Leonid L. Moroz

Abstract

Scanning electron microscopy (SEM) is a powerful tool for ultrastructural analyses of biological specimens at their surface. With comb jellies being very soft and full of water, many methodological difficulties limit their microanatomical studies via SEM. Here, we describe SEM protocols and approaches successfully tested on ctenophores *Pleurobrachia bachei* and *Beroe abyssicola*. Our SEM investigation revealed the astonishing diversity of ciliated structures in all major functional systems, different receptor types, and complex muscular architecture. These protocols can also be practical for various basal bilaterian lineages such as cnidarians.

Key words Ctenophora, Receptors, Neurons, Muscles, Digestive system, Cilia, *Pleurobrachia*, *Bolinopsis*, *Mnemiopsis*, Development, Evolution

1 Introduction

G. Adrian Horridge was the pioneer of electron microscopic investigation of ctenophores [1–6], with the first insights into the unique organization of these enigmatic animals and their nervous system. Mari-Luz Hernandez-Nicaise provided a further detailed ultrastructural analysis of ctenophore cells, neurons, and synapses [7–11]. These morphological and subsequent functional studies [12, 13] significantly contributed to the view that Ctenophora is a distinct lineage of early branching metazoans, sister to all other Metazoa [14–18]. Ctenophores might have developed muscular, neural, and other integrative systems independently from the rest of the animals [19–23]. Nevertheless, we know little about the identity and functions of many cell types and innovations within the ctenophore lineage. The ongoing single-cell transcriptomics in ctenophores [24] and future cell-type atlases should be inherently

Authors Tigran P. Norekian and Leonid L. Moroz have equally contributed to this chapter.

Leonid L. Moroz (ed.), *Ctenophores: Methods and Protocols*, Methods in Molecular Biology, vol. 2757,
https://doi.org/10.1007/978-1-0716-3642-8_6,

© Springer Science+Business Media, LLC, part of Springer Nature 2024 linked to the ultrastructural organization of particular cell populations.

Toward this goal, scanning electron microscopy (SEM) is a relatively fast and essential tool for higher-resolution imaging of virtually any structure. The resolution of electron microscopy is thousands of times higher than light microscopy for obvious reasons—the wavelength of electrons used for probing the tissue is much smaller than that of visible light (photons). There are two main challenges in SEM research, especially for fragile marine animals with high saltwater content: (1) vacuum and the need for dehydration and (2) creating the proper contrast. The beam of electrons in SEM must function in a relatively high vacuum, and therefore, any traces of evaporating water should be eliminated via complete dehydration of biological samples. This process of dehydration can easily destroy the surface structure of the tissue. Also, soft biological tissues are usually not sufficiently electron-dense and require additional techniques for creating a good contrast for visualization.

This chapter describes the SEM protocols for processing one of the most challenging animals for this technique—comb jellies. Ctenophores are composed of >95% water; therefore, fixation and dehydration usually cause significant tissue deformation. Nevertheless, we successfully adopted SEM protocols for both juvenile and adult stages of two ctenophore species: *Pleurobrachia bachei* and *Beroe abyssicola* (Fig. 1), which we describe here. These approaches revealed the astonishing diversity of ciliated structures, complex muscular architecture, rare mesogleal neurons, a variety of different receptor types, and other unique structures (*see* Figs. 2, 3, 4, 5, 6, 7, 8, 9, 10, 11, 12, 13, 14, 15, 16, 17, 18 and 19). Additional details can be found elsewhere [25–27]. These protocols can also be practical for various basal bilaterian lineages such as cnidarians [28].

2 Materials

1. High Mg^{2+} seawater—333 mM magnesium chloride was added to filtered seawater at 1:1 ratio.
2. Phosphate-buffered saline (PBS)—0.2 M stock solution, pH = 7.6.
3. Sodium bicarbonate 2.5% solution.
4. Cacodylate buffer—0.2 M stock solution, pH = 7.4, prepared from sodium cacodylate in distilled water with added HCl for pH balance.

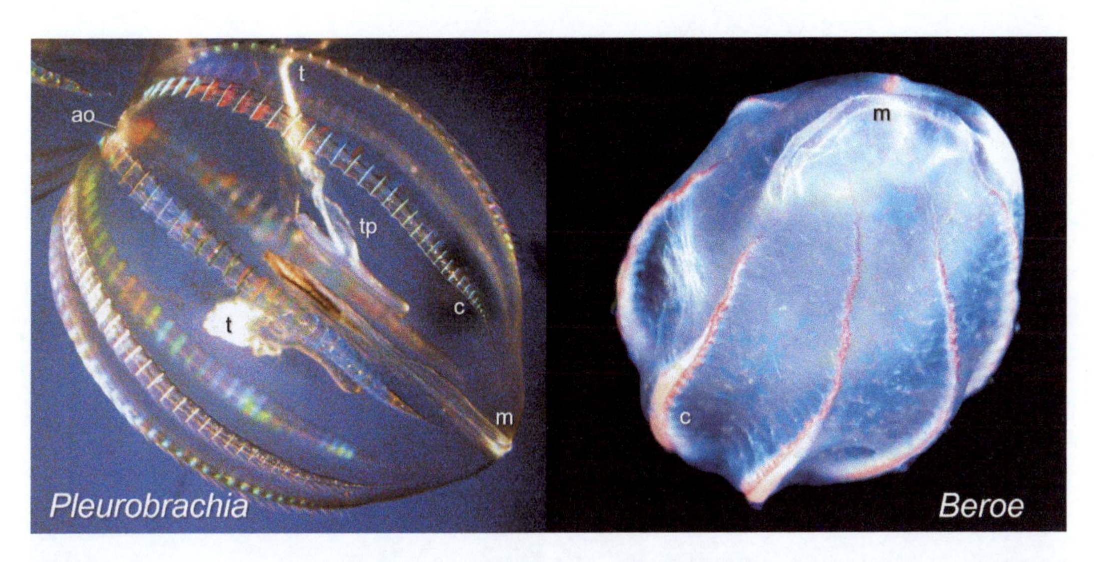

Fig. 1 *Pleurobrachia bachei* and *Beroe abyssicola* images with main organs. Abbreviations: *ao* the aboral organ, *m* mouth, *c* comb plates, *t* tentacles, *tp* tentacle pocket

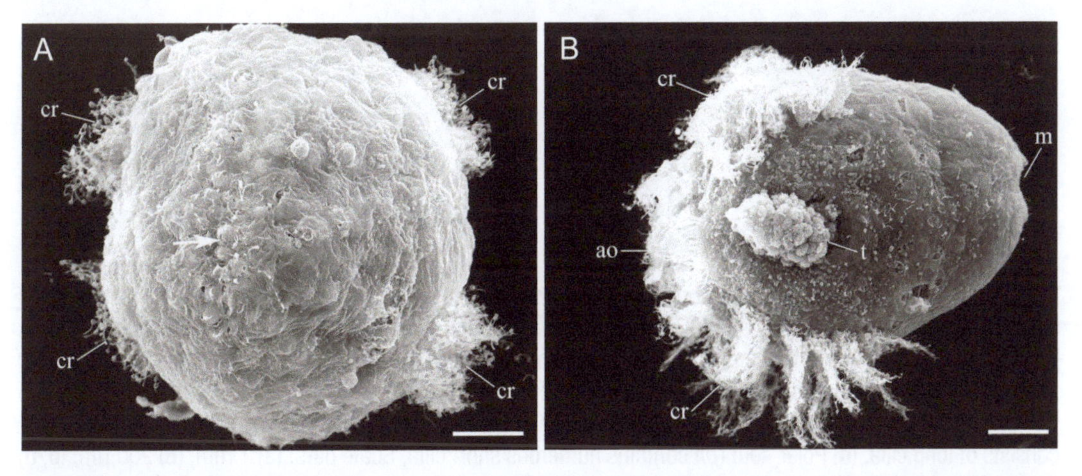

Fig. 2 *Pleurobrachia* embryos (SEM). (**a**) One-day-old embryo extracted from the egg capsule. Note that there are no tentacles outside the body, and the mouth is closed (arrow). (**b**) Three-day-old hatched larva with tentacles (*t*) and opened the mouth (*m*). Note also that there are only four ciliated rows (*cr*) in embryos (eight in adults, Fig. 1a). Scale bars: 20 μm

5. Glutaraldehyde—8% aqueous stock solution in 10 mL ampules, EM grade (Ted Pella, Catalog # 18421). The working concentration—2.5% glutaraldehyde in 0.1 M PBS.
6. Osmium tetroxide—4% aqueous stock solution in 5 mL ampules, EM grade (Ted Pella, Catalog # 18463). The working concentration—2% osmium tetroxide in 1.25% sodium bicarbonate.
7. Ethanol—100% solution.

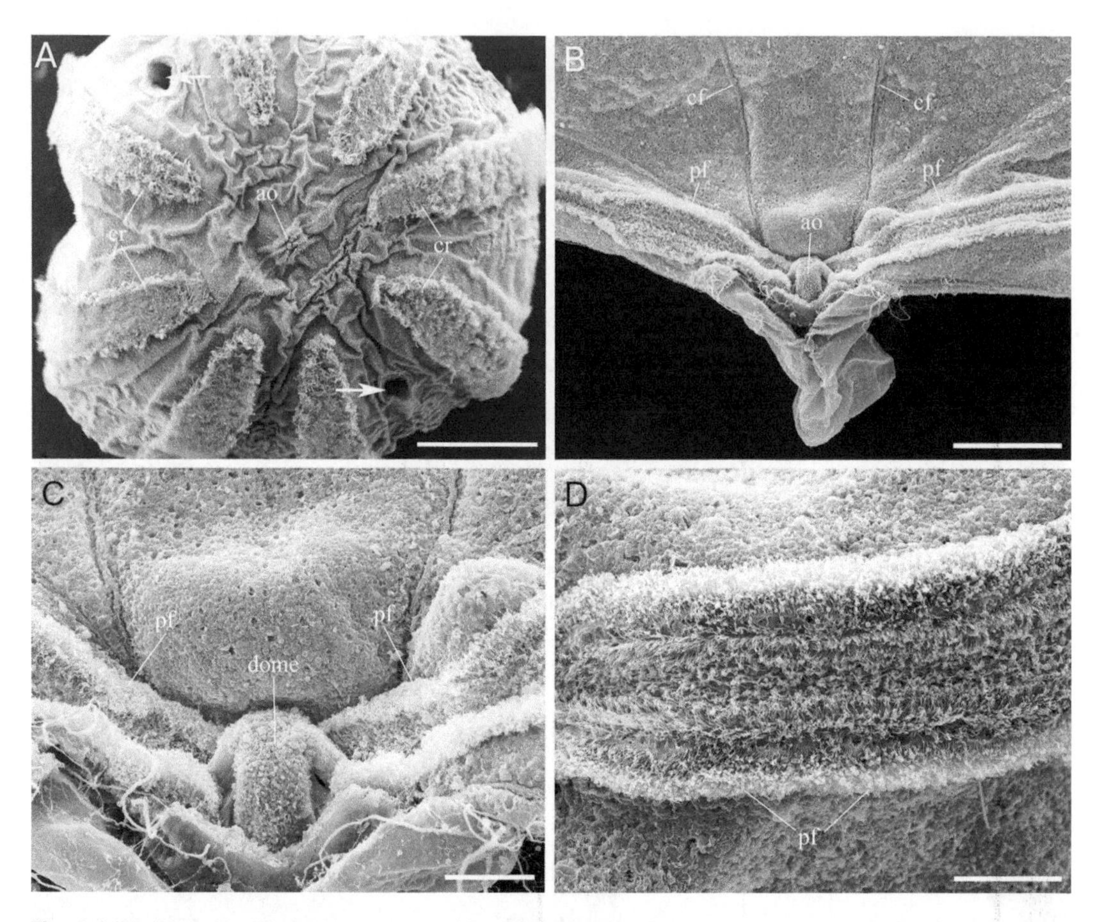

Fig. 3 Adult *Pleurobrachia* from the aboral side. (**a**) The whole *Pleurobrachia* preserved after fixation shows some shrinkage but remains mostly intact. Comb rows (*cr*) and aboral organ (*ao*) are clearly visible. Arrows point to the openings of the tentacle pockets. (**b**, **c**) Aboral organ (*ao*) area with polar fields (*pf*). The cut across *Pleurobrachia* was made next to the aboral organ along the polar fields. Note also the ciliated furrows (*cf*) running from the aboral organ to the comb rows. The aboral organ is covered by a protective dome that consists of long cilia. (**d**) Polar field (*pf*) contains numerous short cilia. Scale bars: (**a**) 1 mm; (**b**) 200 μm; (**c**, **d**) 50 μm

8. Liquid CO_2 for critical point drying.
9. Platforms for SEM viewing and containers for them.

2.1 Equipment

1. Critical point dryer—Samdri-790 (Tousimis Research Corporation).
2. Metal coating—Sputter Coater (SPI Sputter).
3. SEM microscope NeoScope JCM-5000 (JEOL Ltd., Tokyo, Japan).

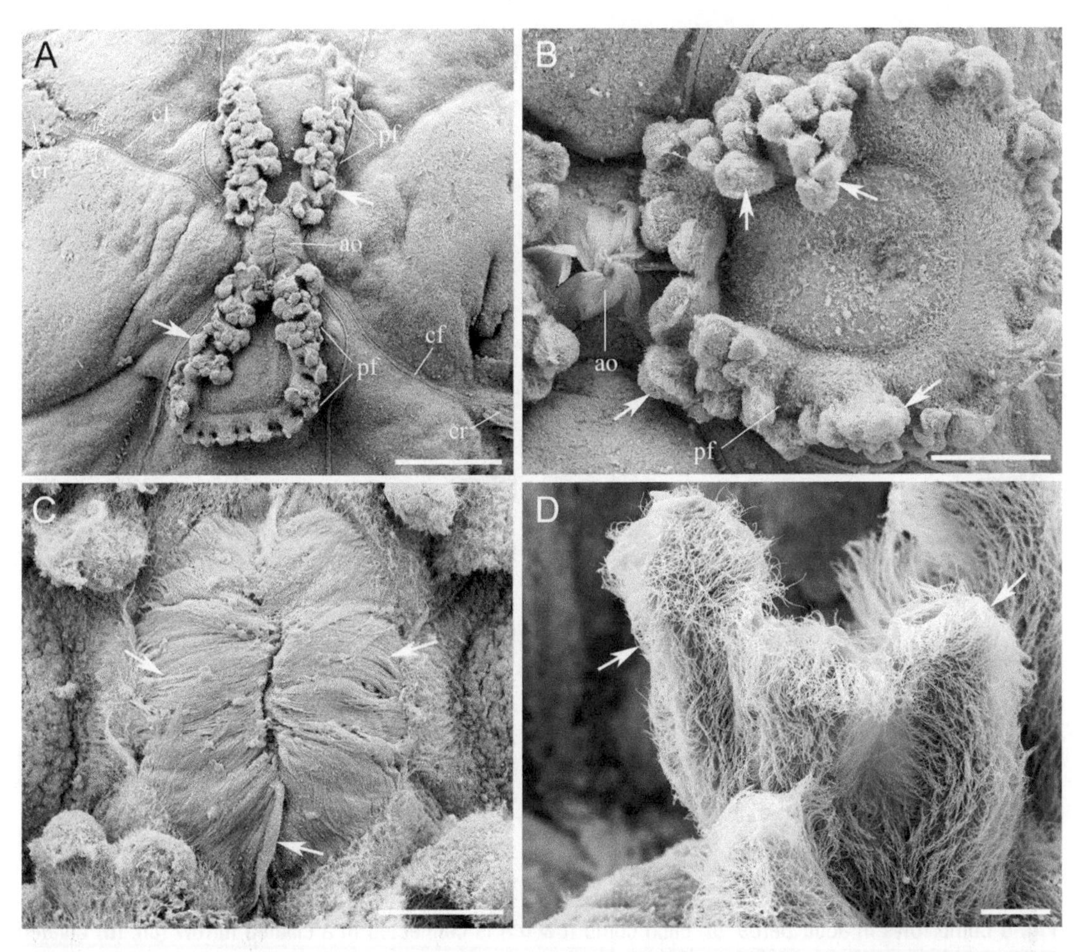

Fig. 4 Aboral organ and polar fields in adult *Beroe*. (**a**, **b**) Aboral organ (*ao*) and polar fields (*pf*) show proper, not withdrawn, shape after 1 h in high Mg^{2+} solution before fixation. Eight ciliated furrows (*cf*) are clearly visible, connecting the aboral organ with the comb rows (*cr*). Note that polar fields in *Beroe* have a crown of tall lobes (arrows) around their periphery. (**c**) The aboral organ is covered by a protective dome consisting of numerous long cilia (arrows). (**d**) The lobes (arrows) of polar fields are covered by long cilia. Scale bars: (**a**) 500 μm; (**b**) 200 μm; (**c**) 100 μm; (**d**) 20 μm

3 Methods

Primary fixation in glutaraldehyde—4 to 12 h

1. Before fixation, animals were incubated in high Mg^{2+} seawater (333 mM magnesium chloride was added to filtered seawater at 1:1 ratio) for about 1 h to completely relax the tissue and block any possible synaptic excitatory inputs to the muscle fibers. Otherwise, fixative caused strong muscle contraction in live ctenophores, withdrawal of some organs, and loss of natural and relaxed anatomical state. It was self-evident in more

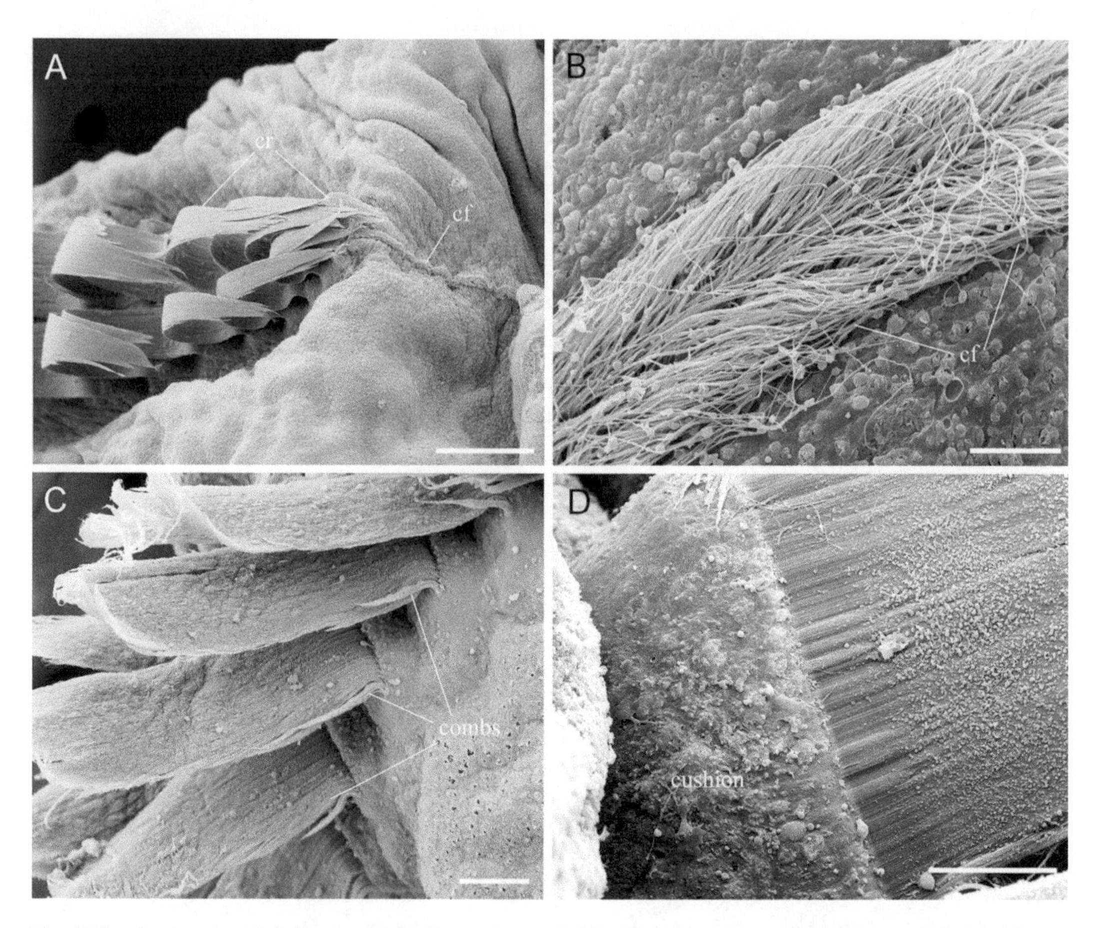

Fig. 5 Comb rows in adult *Beroe*. (**a**) Each comb row (*cr*) is connected to a ciliated furrow (*cf*). (**b**) Ciliated furrow (*cf*) represents a narrow band of tightly packed thin cilia. (**c**) Swim cilia arranged in rows of combs or ctenes. (**d**) At the base of each comb, there is a cushion consisting of polster cells that carry swim cilia. Scale bars: (**a**) 200 μm; (**b**) 10 μm; (**c**) 100 μm; (**d**) 50 μm

muscular *Beroe* compared to *Pleurobrachia*. For example, an aboral organ with polar fields in *Beroe* would be withdrawn entirely inside the body during fixation. Only prolonged high Mg^{2+} incubation beforehand would allow us to fix its proper anatomical state and view it on SEM.

2. After bathing in the high Mg^{2+} seawater, adult *Pleurobrachia* or small adult *Beroe* (0.5–5 cm long) were placed in 20–30 mL vials with closed lids.
3. The seawater was then removed from vials with a suction pipette, and 2.5% glutaraldehyde in 0.1 M phosphate-buffered saline (pH = 7.6) was immediately added (*see* **Notes 1** and **2**). The vials were then left for 4 h at room temperature or 12 h in a refrigerator at 5 °C.
4. Following fixation, the specimens were washed in 2.5% sodium bicarbonate solution. The fixative was removed from vials with

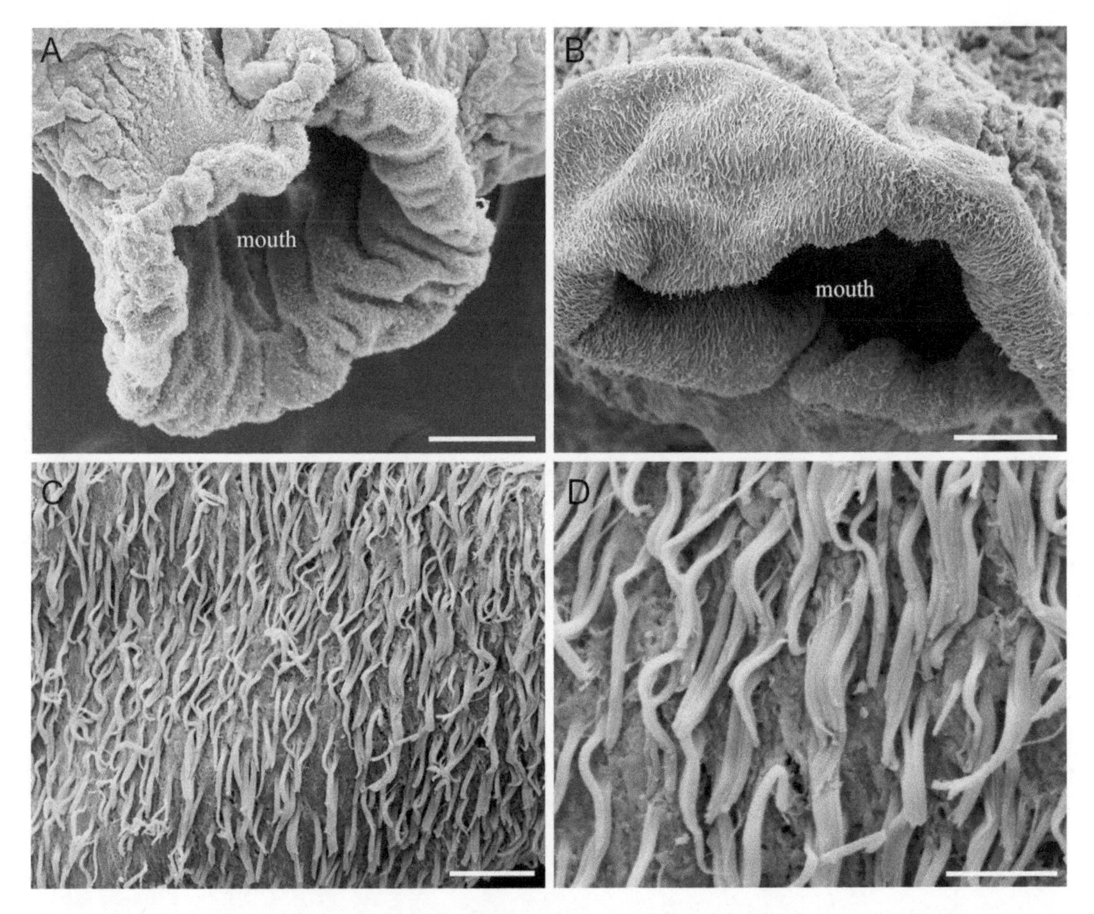

Fig. 6 Mouth area of adult *Pleurobrachia*. (**a**, **b**) Mouth shows significant shrinkage. (**c**, **d**) The inside surface of the lips is covered by cilia. Scale bars: (**a**) 200 μm; (**b**) 100 μm; (**c**) 20 μm; (**d**) 10 μm

suction pipettes and adequately disposed of—under the hood and with gloves. Then clean 2.5% sodium bicarbonate solution was added.

5. There should be at least three rinses in 2.5% sodium bicarbonate with a total wash time of 2 h. At this stage, the tissue can be stored in a refrigerator for a few days.

Secondary fixation in osmium tetroxide—3 to 4 h

6. Before the secondary fixation, the larger fixed animals were dissected and trimmed as necessary. The tissue was removed from the vials, placed in the small Petri dishes, and looked at under a dissecting microscope. The final size of the tissue should reflect the maximum field of view for the SEM—in our case, not more than 1 cm, and the anatomical functionality of dissected pieces. The smallest animals under 1 cm could be processed whole without dissection.

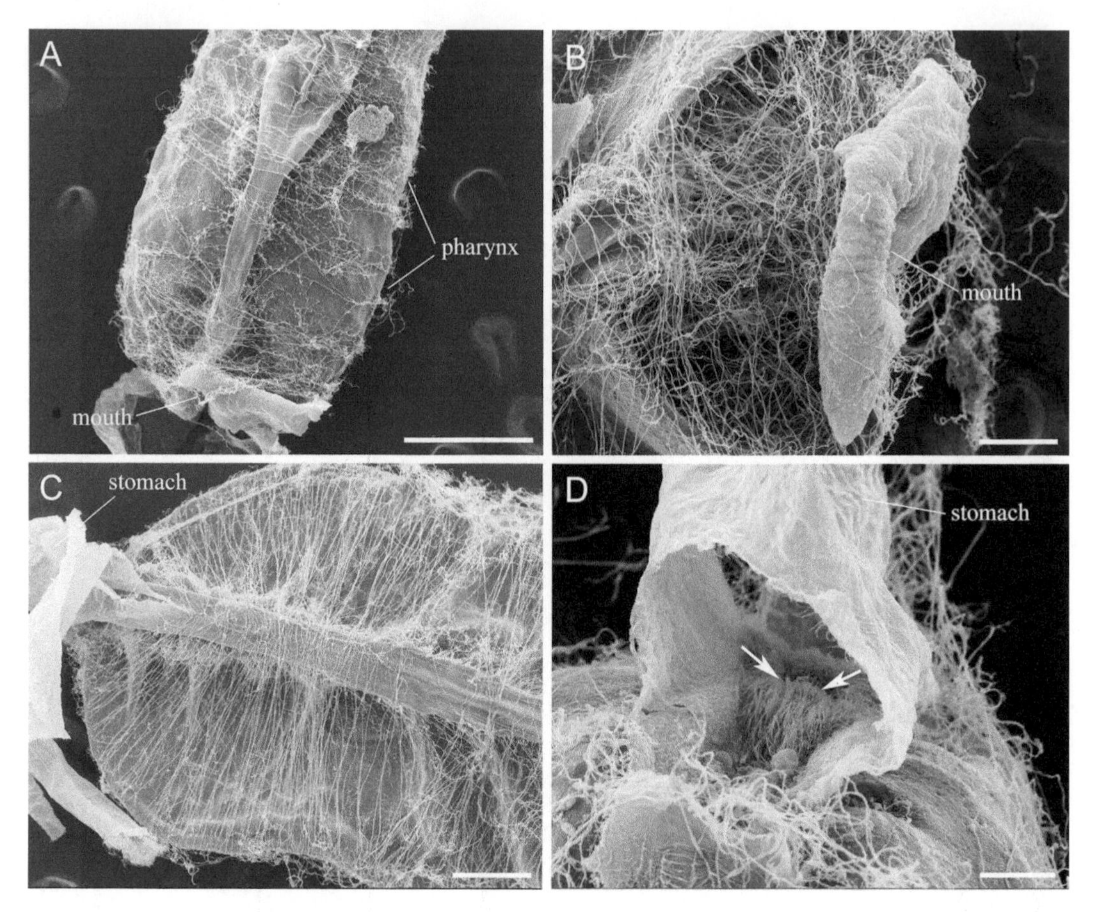

Fig. 7 Pharynx in *Pleurobrachia*. (**a**) Pharynx is frequently separated from the rest of the tissue during processing for SEM. (**b**) Mouth side of the pharynx with the lips still attached. (**c**) Stomach side of the pharynx. Note numerous thin, long muscle fibers that run through the mesogleal area and connect to the pharynx. (**d**) Pharynx and stomach are separated by a sphincter, which is densely covered with long cilia (arrows). Scale bars: (**a**) 500 μm; (**b**) 100 μm; (**c**) 200 μm; (**d**) 50 μm

7. The dissected pieces of tissue were then placed again in the 20–30 mL vials with the clean 2.5% sodium bicarbonate solution.
8. For secondary fixation, 2% osmium tetroxide in 1.25% sodium bicarbonate solution was added into the vials with ctenophore tissue. The vials were then left for 3 h at room temperature. The working solution of 2% osmium tetroxide should be prepared just before use—it does not last long, for a maximum of 2–3 days in a refrigerator. Osmium tetroxide is very toxic and should be handled carefully under the hood with gloves. The vials with tissue should be covered with a light-blocking container—osmium tetroxide is sensitive to light.

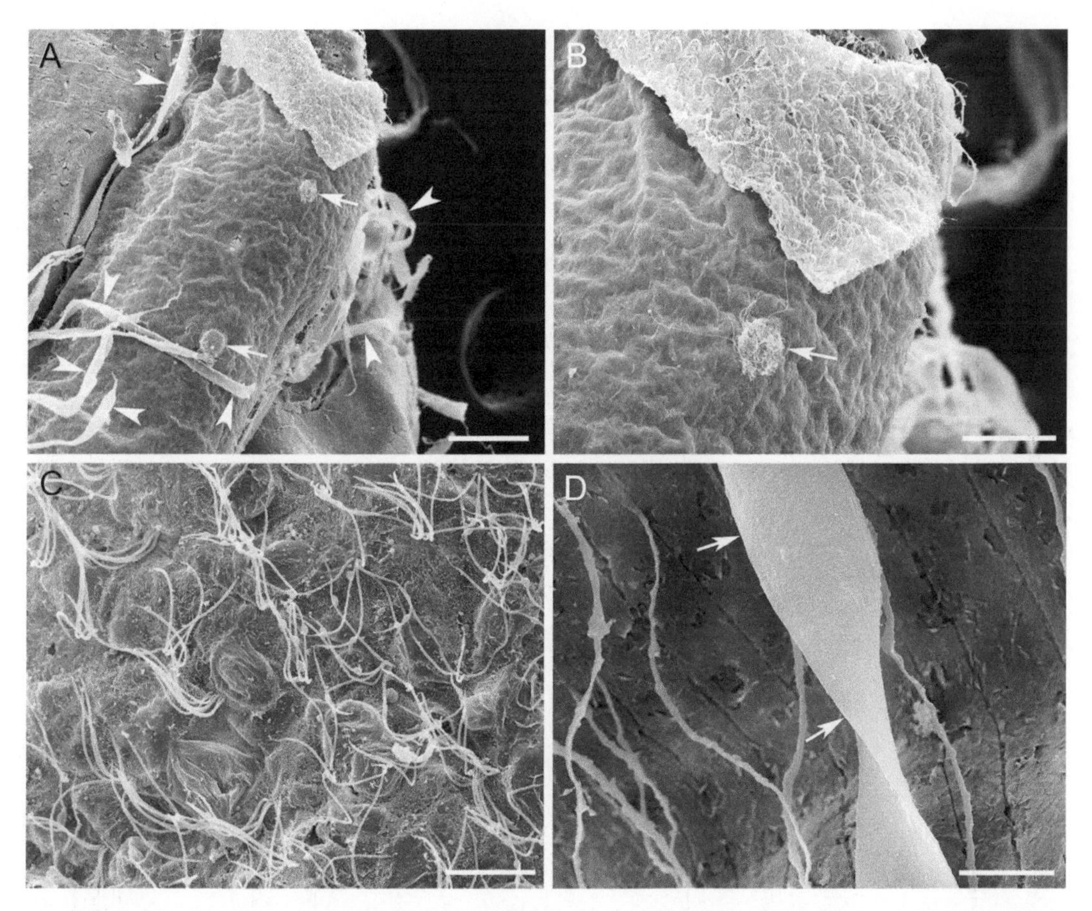

Fig. 8 Meridional canals in *Pleurobrachia*. (**a**, **b**) Meridional canals have pore complexes (ciliated rosettes) along their length, indicated by arrows. Arrowheads point to the long and flat muscle fibers connected to the body wall above the meridional canals along their entire length. Note that a strip of the meridional canal wall on the top of the image was flipped over to reveal its internal surface. (**c**) The internal surface of the meridional canal wall is covered with groups of long and flexible cilia. (**d**) Higher resolution image of a wide and flat long muscle fiber next to the meridional canal. Scale bars: (**a**) 100 μm; (**b**) 50 μm; (**c**) 10 μm; (**d**) 20 μm

9. After postfixation, osmium tetroxide should be appropriately disposed of, and the tissue should be rinsed several times with distilled water over 1 h.

Dehydration in ethanol—3 h

10. Next step was dehydration in ethanol. The dehydration series takes the tissue through the following ethanol concentrations: 30%, 50%, 70%, 90%, 100%, and second 100%. First, distilled water was removed with a suction pipette, and 30% ethanol was added to vials. Then 30% ethanol was removed, 50% ethanol was added, etc. Each step of incubation at a specific concentration should last about 30 min. Some shrinkage of the ctenophore tissue inevitably occurs during dehydration in ethanol

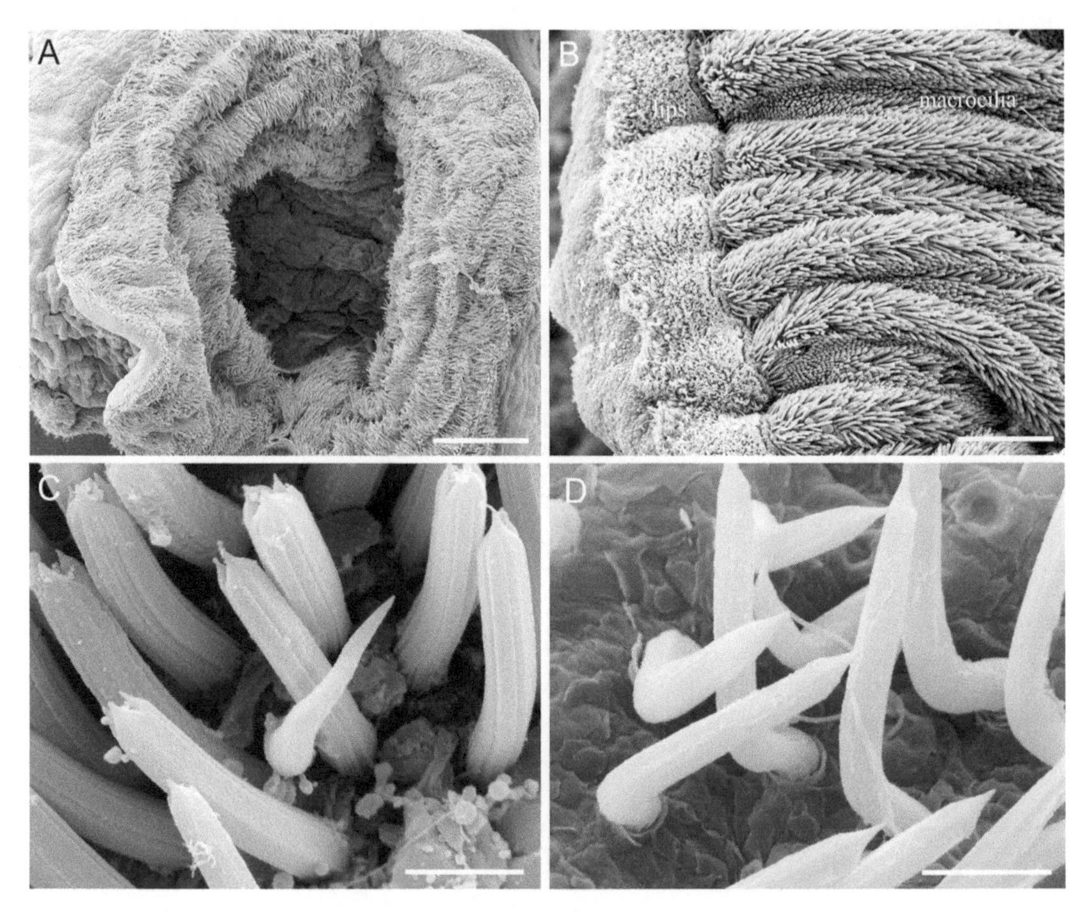

Fig. 9 Mouth of adult *Beroe* with macrocilia. (**a**) Opened mouth. (**b**) Macrocilia covers the entire area inside the mouth, from the lips to the entrance of the pharynx. (**c**) Closer to the lips, the macrocilia are always straight and display three sharp teeth at the end. (**d**) Further into the pharynx, the macrocilia density is reduced; they have a single sharp end and are usually bent like a hook pointing toward the inside of the pharynx. Scale bars: (**a**) 200 μm; (**b**) 100 μm; (**c**, **d**) 5 μm

(Figs. 6a, b and 9a). Also, the entire ctenophore or a large piece of tissue frequently breaks into smaller pieces at this stage. It happens because dehydration causes some loss of tissue flexibility, and tissue becomes more fragile, especially with areas that do not have dense and tough organs, such as ctene rows, pharynx, and tentacle pockets. This fact should not discourage us from proceeding further.

11. If there is a need to pause the protocol for a while, the best way to store specimens for a few days is in ethanol concentration of 70% in a refrigerator at 5 °C.

Critical point drying—2 h

12. The next step after dehydration was drying the tissue. The ctenophore specimens still in 100% ethanol were placed in a

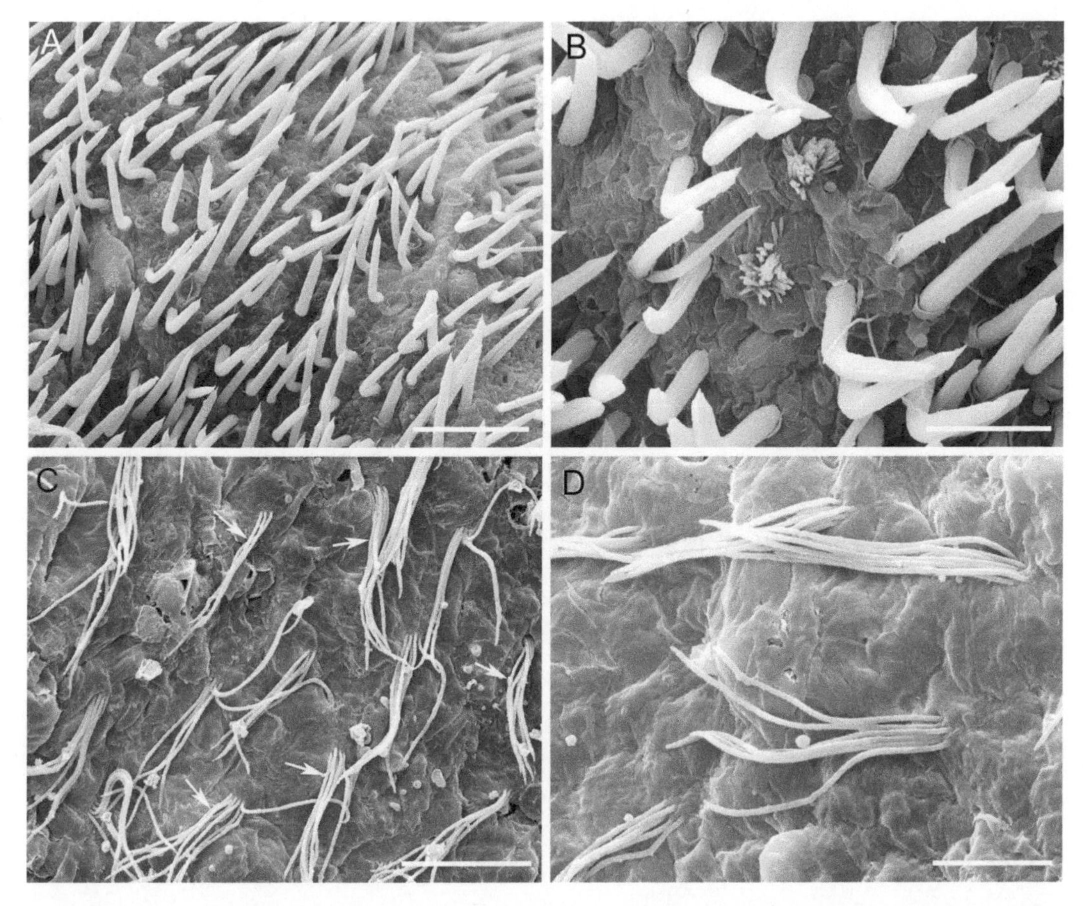

Fig. 10 Surface of the pharynx in adult *Beroe*. (**a**, **b**) Passing the mouth area and at the entrance of the pharynx, the macrocilia density is significantly reduced, and they become much shorter. (**c**, **d**) Further into the pharynx, macrocilia disappear entirely, and only groups of thin flexible cilia outline its surface. Scale bars: (**a**) 20 μm; (**b**, **c**) 10 μm; (**d**) 5 μm

critical point drying equipment chamber (*see* **Note 3**). We used mostly Samdri-790 dryer from Tousimis Research Corporation, but many different models are available on the market.

13. As a first step, ethanol should be replaced by liquid CO_2. It is important to ensure that all of the ethanol was replaced by CO_2. Depending on the thickness of the tissue and the number of specimens in the chamber, the duration of incubation in liquid CO_2 and the number of rinses should be increased. The samples will be ruined for good SEM imaging if some ethanol is left in the tissue. Usually, 30–40 min of incubation and 4–5 rinses were sufficient.
14. After passing the critical point, the liquid CO_2 evaporated during the drying procedure. For CO_2, the critical point is achieved at 35 °C and 1200 psi. Then the pressure was slowly

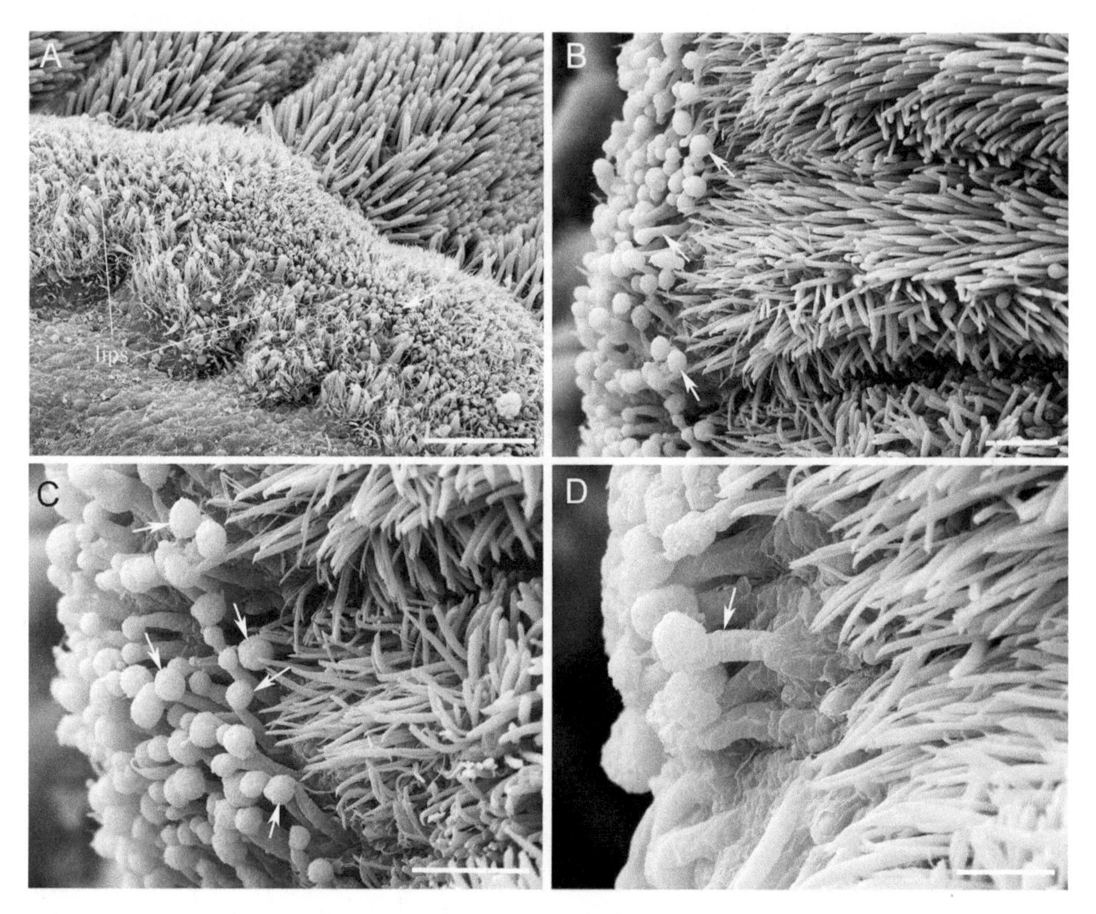

Fig. 11 The interlocking mechanism that keeps the lips closed in *Beroe*. In addition to adhesive epithelial cells, the lips contain numerous papillae-like structures (arrows), which mechanically fasten the lips together like a jigsaw puzzle. Some of the "locking" papillae are shorter than others and are spread over the entire lips area. Scale bars: (**a**) 50 μm; (**b**, **c**) 20 μm; (**d**) 10 μm

reduced to the atmospheric value. It was imperative to reduce it very slowly over at least 30 min.

15. The completely dry specimens were collected from the equipment chamber and placed under the dissecting scope for further refining. Additional breaking of original tissue could occur during critical point drying and some additional shrinkage. However, in some experiments, we managed to preserve a whole round-shaped body of *Pleurobrachia* for SEM imaging (Fig. 3a). In general, in highly muscular *Beroe*, the tissues did not go through as much breaking and shrinkage during the drying process as in other ctenophores, like *Pleurobrachia*.

Mounting tissue on viewing platforms—2 h

16. The tissue was placed on the holding platforms for SEM viewing during the next step. That was when the initial overview of

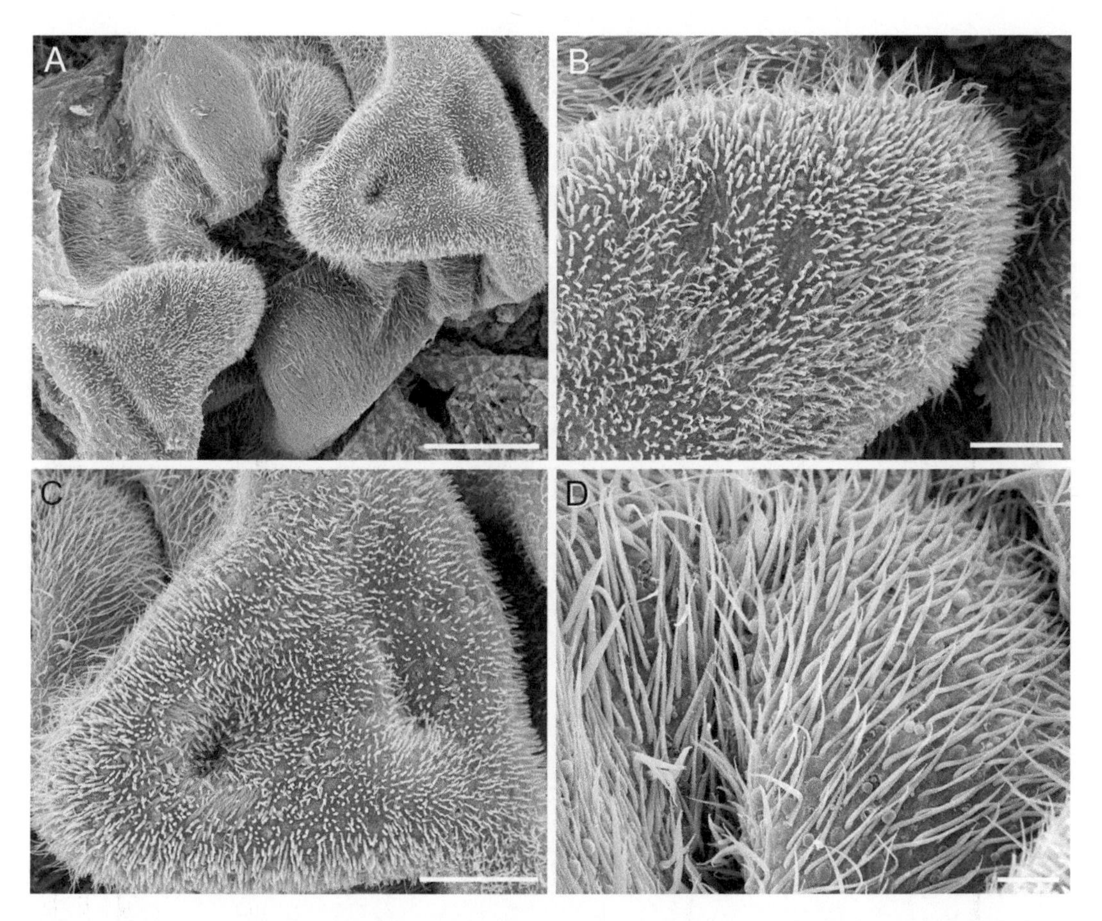

Fig. 12 The aboral end of the digestive system in *Beroe*. Note two symmetrical lobes and a highly ciliated surface. Scale bars: (**a**) 200 μm; (**b**) 50 μm; (**c**) 100 μm; (**d**) 20 μm

all dry samples occurred; the best pieces were selected for SEM viewing, carefully cleaned and dissected if necessary, and placed on the platform in a proper orientation. It should be done under a stereomicroscope using fine dissecting tools. At this stage, some dissecting can be done—for example, cutting open the tentacle pocket and revealing the withdrawn inside tentacles (Fig. 19d) or removing a piece of the integument and looking inside the mesogleal region at numerous muscle fibers attached to the pharynx of *Pleurobrachia* (Figs. 7 and 16), or mesogleal fibers attached to the outside wall of *Beroe* (Fig. 18).

Metal coating—1 h

17. Before inserting the platforms into SEM, the specimens should be processed for a metal coating to create a proper electron density on the surface and, therefore, good contrast for visualization. We used Sputter Coater from SPI Sputter. One crucial detail is that better metal coating leads to better imaging. So,

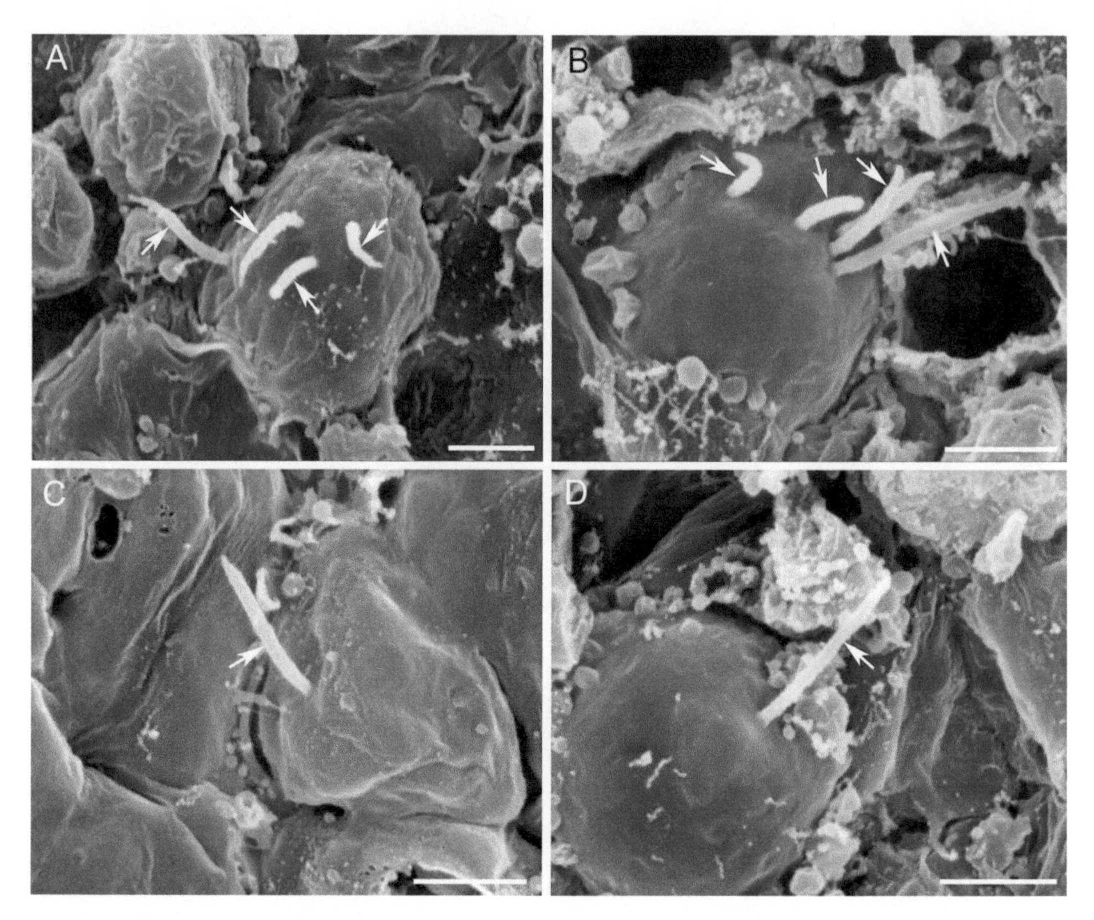

Fig. 13 Surface receptors in *Pleurobrachia*. (**a**, **b**) Receptors with multiple cilia. (**c**, **d**) Receptors with a single cilium. Arrows point to individual cilia. Scale bars: 2 μm

we ran the samples through metal coating twice, rotating the platforms 180° in between to ensure that tissue was covered at all angles.

SEM imaging

18. SEM observations and recordings were done on a portable, desktop SEM microscope NeoScope JCM-5000 (JEOL Ltd., Tokyo, Japan). It was straightforward to use, without additional cooling systems, and powerful enough for most biological purposes. Except for some tissue shrinkage, the preservation of the surface structure in *Pleurobrachia* and *Beroe* was very good and provided a lot of important information [25–27]. Excellent images were obtained from ciliated structures such as swim cilia (Fig. 5), sensory cells on the surface of the body (Figs. 13 and 14), macrocilia inside the mouth of *Beroe* (Figs. 9 and 10). We also had an interesting view of internal areas such as mesoglea with different internal

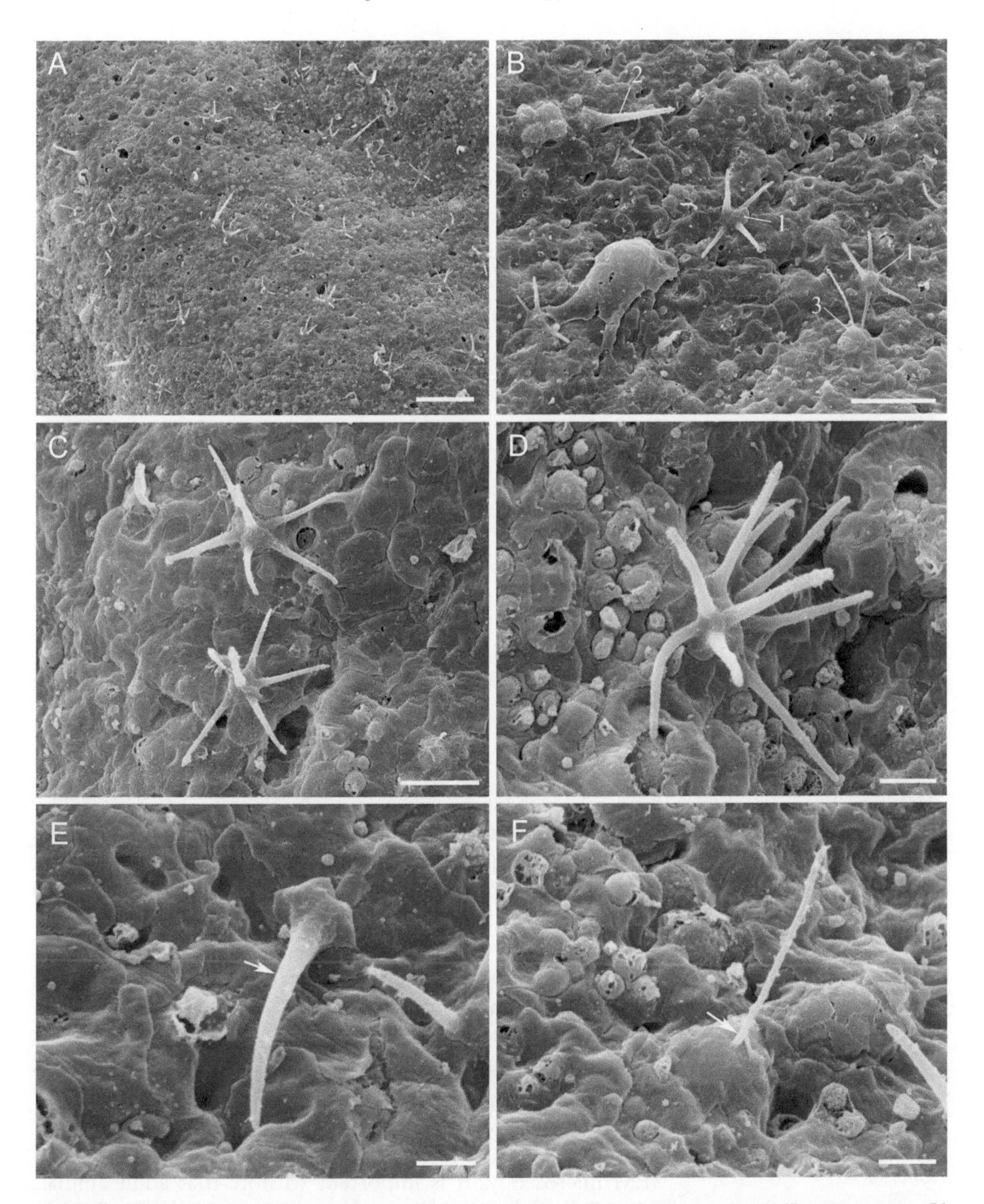

Fig. 14 Surface receptors in *Beroe.* (**a**) The surface of the body is covered with numerous ciliated receptors. (**b**) There are three types of receptors: receptors with multiple cilia (type *1*), receptors with a single large and thick cilium (type *2*), and receptors with a single thin cilium on the apical part of a cell (type *3*). (**c**, **d**) Type 3 receptor with a group of 3–9 cilia. (**e**) Type 2 receptor with a thick and long single cilium (arrow). (**f**) Type 3 receptor with a single thin and long cilium (arrow). Scale bars: (**a**) 20 μm; (**b**) 10 μm; (**c**) 5 μm; (**d**, **e**, **f**) 2 μm

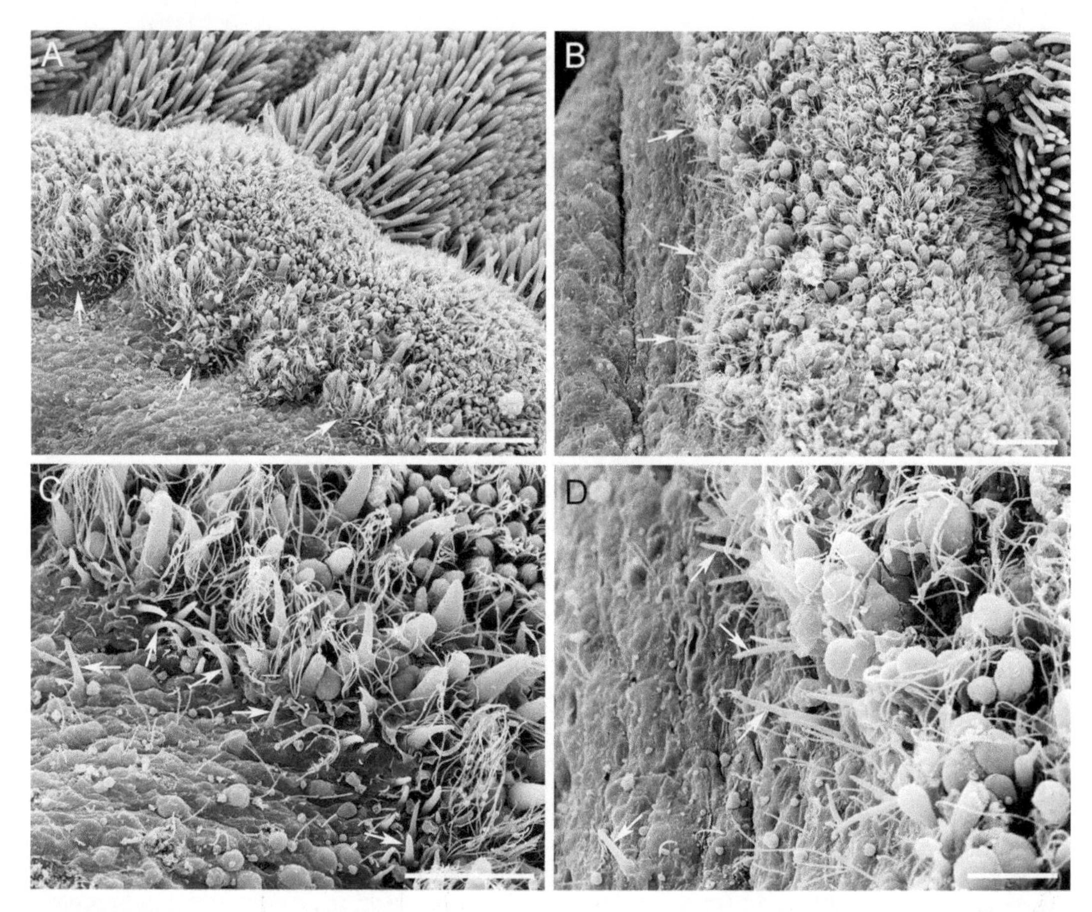

Fig. 15 Receptors on the edge of the mouth in *Beroe*. (**a**, **b**) The lips area contains numerous receptors with the highest concentration on their outside edge (arrows). (**c**, **d**) There are receptors with a single large and thick cilium (arrows)—presumably mechanoreceptors of type 2. Next to them, there are also numerous receptors with a thin long cilium (type 3). Scale bars: (**a**) 50 μm; (**b**, **c**) 20 μm; (**d**) 10 μm

organs, numerous muscle fibers (Figs. 7, 8, 16, 17 and 18), and even individual neurons (Fig. 17). *See also* **Note 4**.

4 Notes

1. We attempted to fix six different species of ctenophores for SEM microscopy. It is important to note that adult animals from *Bolinopsis* and *Mnemiopsis* genera could not be fixed in glutaraldehyde. They dissolved in the fixative, with cells losing connection to each other and their tissue losing its integrity. These two species have some of the most fragile and full of water bodies among ctenophores, and we found them inappropriate for SEM investigation. Animals from four other genera—*Pleurobrachia*, *Beroe*, *Euplokamis*, and *Hormiphora*—have much tougher bodies and demonstrated good fixation

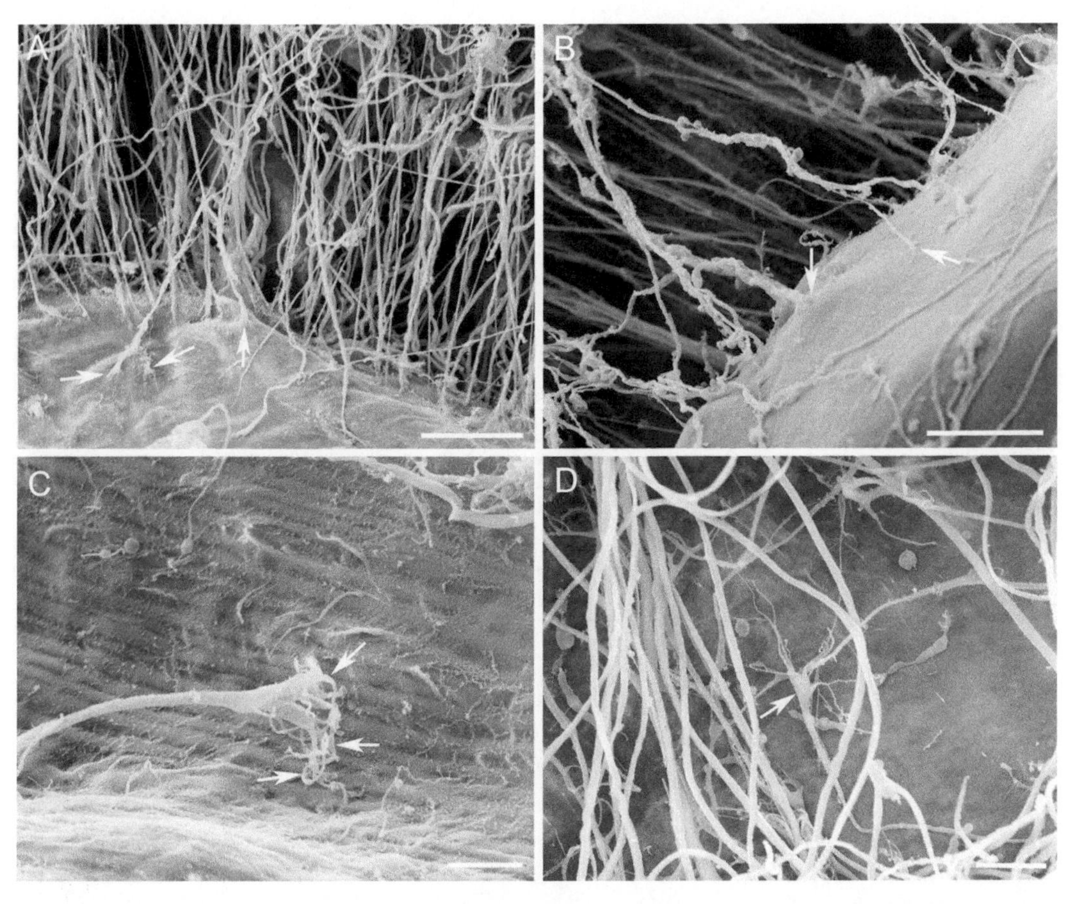

Fig. 16 Muscle fibers in the mesoglea of *Pleurobrachia*. (**a**, **b**) Many mesogleal muscle fibers are firmly attached to the surface of the pharynx (arrows indicate the attachment point) and well preserved after the drying process. (**c**) There is always some branching and widening of the muscle fiber at the point of contact to increase the binding surface (arrows). (**d**) Mesogleal neuronal-like processes (arrow) with extensive branching are among muscle fibers on the surface of the pharynx. Scale bars: (**a**, **b**) 50 μm; (**c**) 10 μm; (**d**) 20 μm

results. Most SEM research was done on *Pleurobrachia bachei* and *Beroe abyssicola* [25–27].

2. The primary fixative, 2.5% glutaraldehyde, could be prepared in 0.1 M cacodylate buffer (prepared from sodium cacodylate in distilled water). In this case, the subsequent washing steps also should be carried in 0.1 M cacodylate buffer, and 4% osmium tetroxide should be diluted in 0.2 M cacodylate buffer at 1:1 ratio. Both PBS and cacodylate buffer gave very good fixation results for SEM. Cacodylate buffer has both positive and negative sides. The negative side is its toxicity and carcinogenic effects. One must be cautious during pH adjustment by adding HCl—it causes the release of arsenic and should be processed under a hood and while wearing gloves. The positive side is resistance to bacterial contamination and the ability to store samples for long periods of time.

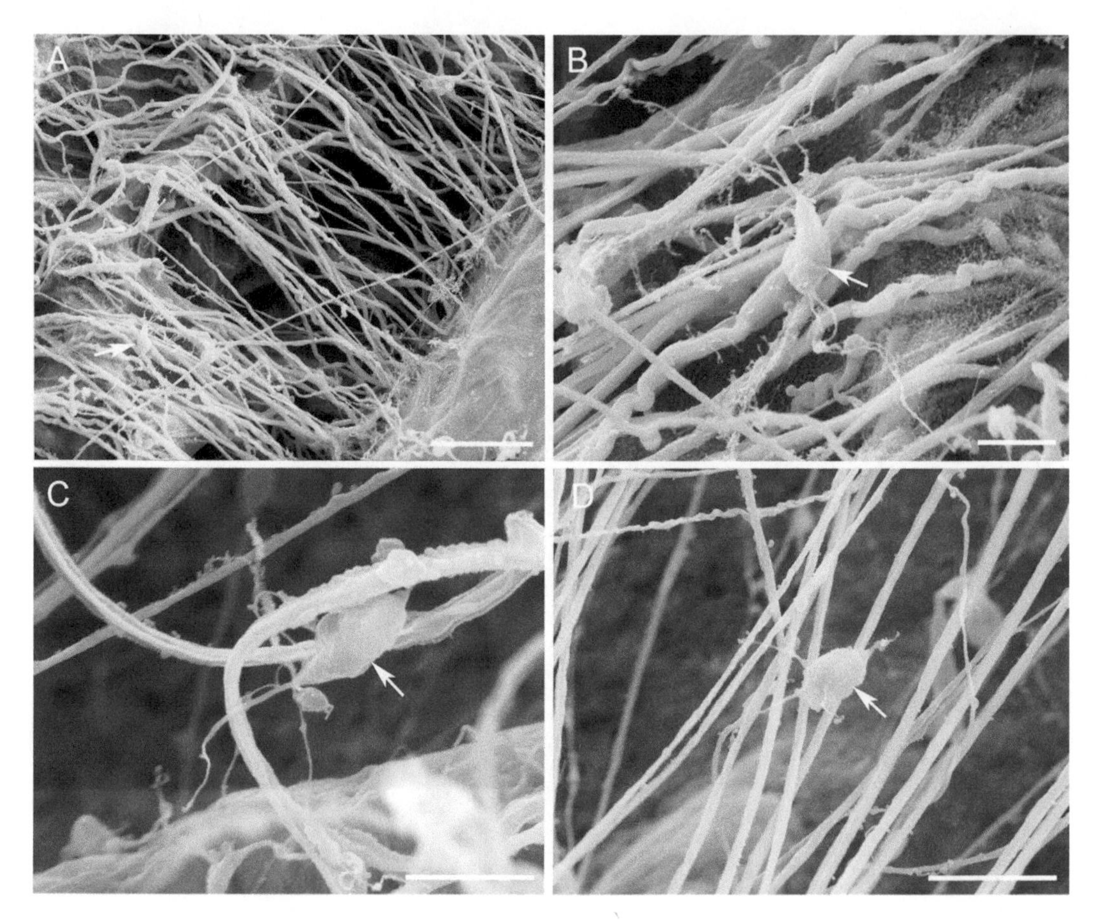

Fig. 17 Mesogleal neural-like cells (arrows) among muscle fibers in the pharynx area of *Pleurobrachia*. Scale bars: (**a**) 50 μm; (**b**, **c**) 10 μm; (**d**) 20 μm

3. There are several possible ways of drying tissue for SEM. Some insects with their hard exoskeleton, fish scales, or teeth of mammals and other vertebrates, etc., do not require any special drying technique or even dehydration—just a simple air drying would suffice. However, for most soft-bodied animals, tissue drying is crucial for SEM. Critical point drying is the most common method of drying biological specimens for SEM. The regular air drying of biological specimens can cause severe deformation of the surface structure due to the considerable surface tension present at the phase boundary as the liquid evaporates. Water, for example, has a very high surface tension to air. During critical point drying, this surface tension is reduced to zero. It is achieved by choosing a specific temperature and pressure for suitable inert fluid (liquid CO_2 is universally used today) along the boundary between the liquid and gaseous phases, where liquid and vapor have the same density— the critical point. For CO_2, the critical point is achieved at 35 °C and 1200 psi. When the temperature is raised

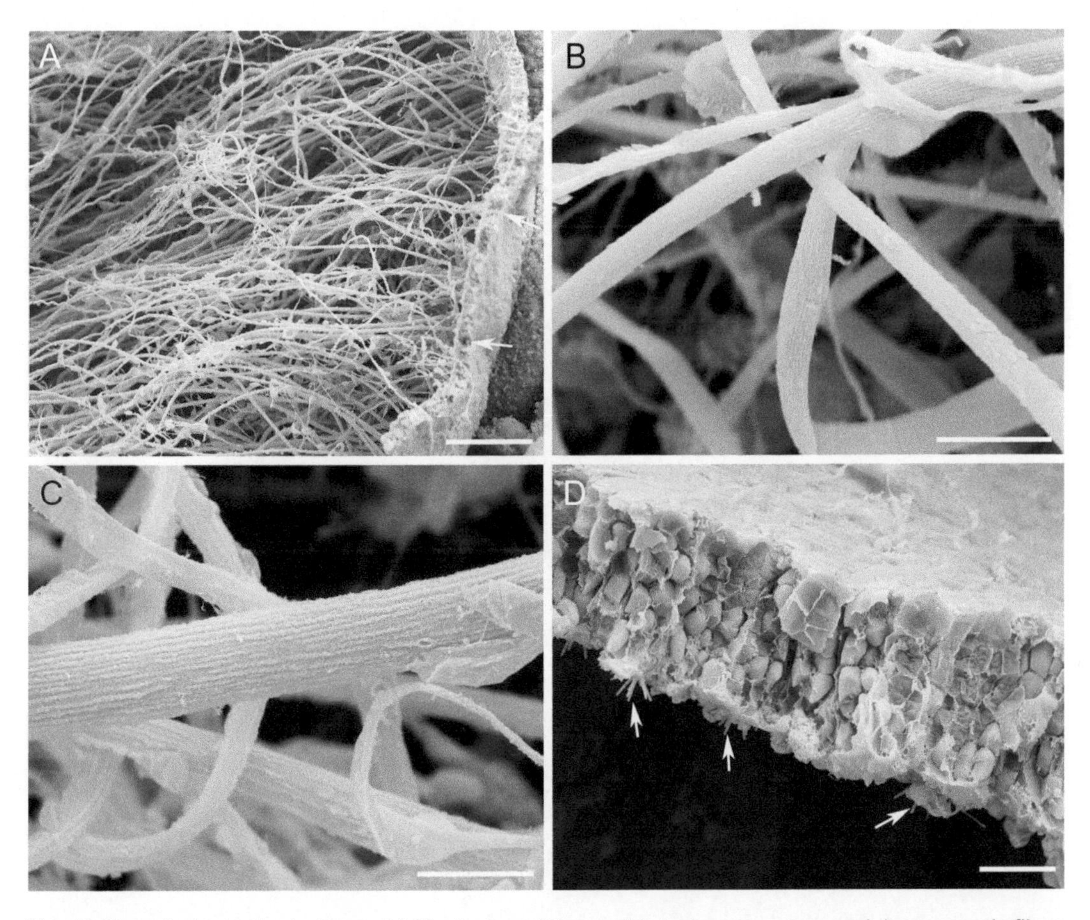

Fig. 18 Mesogleal muscles in *Beroe*. (**a**) The body wall (arrows) is broken open to reveal dense muscle fibers inside the mesogleal area. (**b**, **c**) Long muscle fibers have a variable thickness. (**d**) A glimpse at the cross-section of the body wall. Note the receptors with multiple cilia (arrows) located on the outer surface of the body. Scale bars: (**a**) 100 μm; (**b**) 20 μm; (**c**, **d**) 10 μm

above the critical temperature, liquid CO_2 changes to vapor without a change of density and no surface tension and, therefore, completely preserves the surface structure of the tissue.

In addition to the critical point drying technique, the chemical drying procedure has been successfully used for some biological specimens. One example is the use of hexamethyldisilazane (HMDS), which has been described as a good alternative. After dehydration, ethanol is replaced with HMDS, and the tissue is then left under the hood for complete air drying. The surface tension for HMDS during evaporation is much smaller than for water, and therefore, preservation of the surface structure is much better. We tried to use HMDS with our ctenophore samples as a much simpler alternative but found that the preparation quality was significantly lower than after the critical point drying. Therefore, the decision

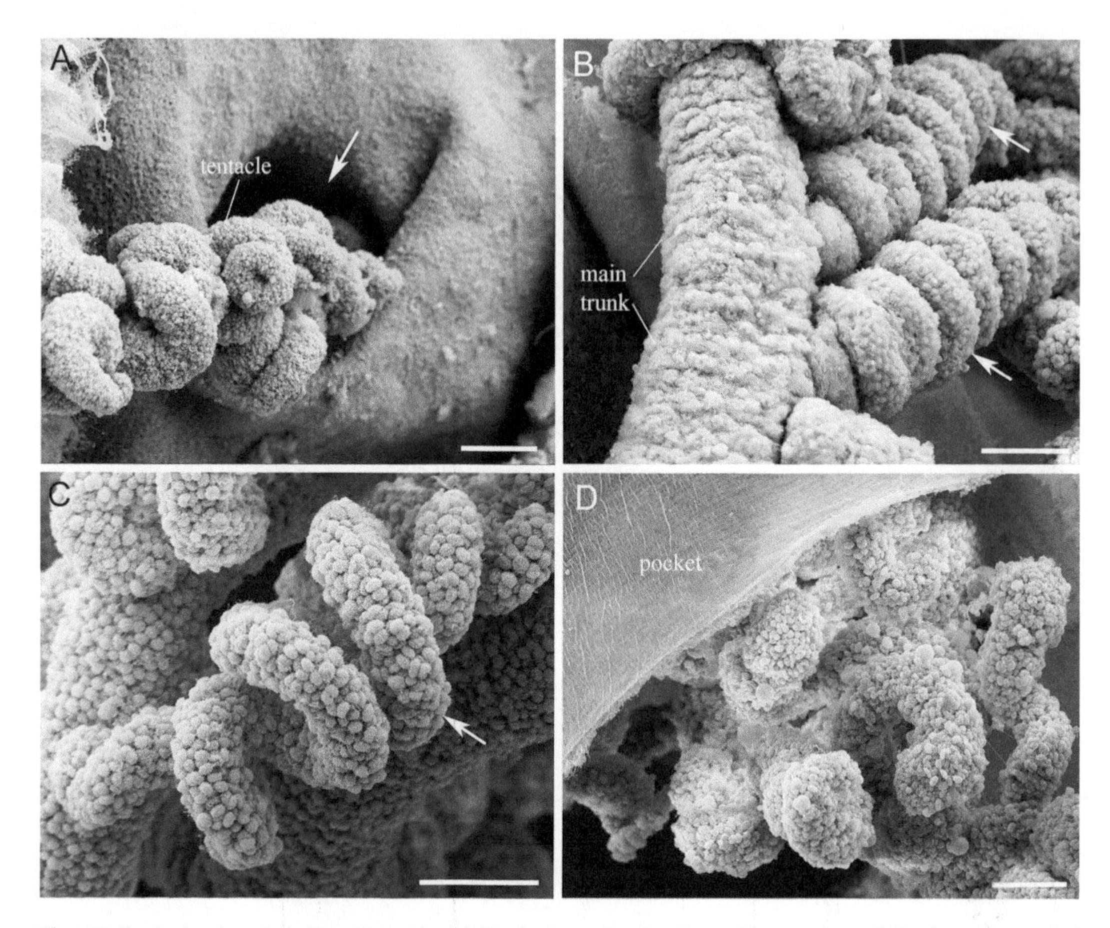

Fig. 19 Tentacles in adult *Pleurobrachia*. (**a**) Tentacle protruding from the opening of the tentacle pocket (arrow). (**b**, **c**) Each tentacle consists of the main trunk and secondary tentacles packed in tight spirals (arrows). Small round-shaped colloblasts are visible along the main trunk and secondary tentacles. (**d**) The tentacle pocket is cut open and shows a withdrawn tentacle inside. Scale bars: (**a**) 100 μm; (**b**, **c**, **d**) 50 μm

has been made to use only critical point drying with ctenophores.

4. We used SEM to analyze not only the tissue of adult ctenophores but also looked at the *Pleurobrachia* embryos during different stages of their development [25]. The basic protocol for processing embryos is very similar to that of adult animals. There are only two details that should be stressed here. Embryos are very small and, most of the time, are suspended in the water column, making it very difficult to rinse them and change solutions. So, we used for that purpose the 2 mL Eppendorf tubes and centrifuge to separate them from a liquid (500 rotations per minute for live embryos and 1000 rotations per minute for fixed embryos for 2–3 min) and then change solutions multiple times. Also, it is necessary to have a large number of embryos fixed initially because many of them will be lost during processing (up to 30–40% of the original number).

Another challenging detail—scanning embryos of the very early stages. *Pleurobrachia* embryos hatch on approximately the third day. To look at one-day or two-day-old embryos, it is necessary to remove them from the egg by fine forceps under the dissecting scope (Fig. 2a). Otherwise, the eggshell is not penetrable for SEM electrons. It is a very time-consuming work. However, after hatching on day 3, this problem does not exist (Fig. 2b).

Acknowledgments

This work was supported in part by the Human Frontiers Science Program (RGP0060/2017) and the National Science Foundation (IOS-1557923) grants to LLM. Research reported in this publication was also supported in part by the National Institute of Neurological Disorders and Stroke of the National Institutes of Health under award number R01NS114491 (to LLM). The content is solely the authors' responsibility and does not necessarily represent the official views of the National Institutes of Health.

References

1. Horridge GA (1964) The giant mitochondria of ctenophore comb plates. Q J Microsc Sci 105:301–310
2. Horridge GA (1964) Presumed photoreceptive cilia in a ctenophore. Q J Microsc Sci 105:311–317
3. Horridge GA (1965) Non-motile sensory cilia and neuromuscular junctions in a ctenophore independent effector organ. Proc R Soc Lond Biol 162:333–350
4. Horridge GA (1974) Recent studies on the Ctenophora. In: Muscatine L, Lenhoff HM (eds) Coelenterate biology. Academic, New York, pp 439–468
5. Horridge GA, Mackay B (1964) Neurociliary synapses in *Pleurobrachia* (Ctenophora). Q J Microsc Sci 105:163–174
6. Horridge GA, Tamm SL (1969) Critical point drying for scanning electron microscopic sthdy of ciliary motion. Science 163(3869):817–818
7. Hernandez-Nicaise M-L (1991) Ctenophora. In: Harrison FWFW, Westfall JA (eds) Microscopic anatomy of invertebrates: Placozoa, Porifera, Cnidaria, and Ctenophora. Wiley, New York, pp 359–418
8. Hernandez-Nicaise ML (1973) The nervous system of ctenophores. II. The nervous elements of the mesoglea of beroids and cydippids (author's transl). Z Zellforsch Mikrosk Anat 143(1):117–133
9. Hernandez-Nicaise ML (1973) The nervous system of ctenophores. I. Structure and ultrastructure of the epithelial nerve-nets. Z Zellforsch Mikrosk Anat 137(2):223–250
10. Hernandez-Nicaise ML (1973) The nervous system of ctenophores. III. Ultrastructure of synapses. J Neurocytol 2(3):249–263
11. Hernandez-Nicaise ML, Amsellem J (1980) Ultrastructure of the giant smooth muscle fiber of the ctenophore *Beroe ovata*. J Ultrastruct Res 72(2):151–168
12. Tamm SL (1982) Ctenophora. In: Electrical conduction and behavior in "simple" invertebrates. Clarendon Press, Oxford, pp 266–358
13. Tamm SL (2014) Cilia and the life of ctenophores. Invertebr Biol 133(1):1–46
14. Halanych KM et al (2016) Miscues misplace sponges. Proc Natl Acad Sci U S A 113(8): E946–E947
15. Li Y et al (2021) Rooting the animal tree of life. Mol Biol Evol 38(10):4322–4333
16. Redmond AK, McLysaght A (2021) Evidence for sponges as sister to all other animals from partitioned phylogenomics with mixture models and recoding. Nat Commun 12(1):1783

17. Telford MJ, Moroz LL, Halanych KM (2016) Evolution: a sisterly dispute. Nature 529(7586):286–287
18. Whelan NV et al (2017) Ctenophore relationships and their placement as the sister group to all other animals. Nat Ecol Evol 1(11): 1737–1746
19. Moroz LL (2015) Convergent evolution of neural systems in ctenophores. J Exp Biol 218 (Pt 4):598–611
20. Moroz LL et al (2014) The ctenophore genome and the evolutionary origins of neural systems. Nature 510(7503):109–114
21. Moroz LL, Kohn AB (2016) Independent origins of neurons and synapses: insights from ctenophores. Philos Trans R Soc Lond Ser B Biol Sci 371(1685):20150041
22. Moroz LL (2014) The genealogy of genealogy of neurons. Commun Integr Biol 7(6): e993269
23. Moroz LL, Romanova DY, Kohn AB (1821) Neural versus alternative integrative systems: molecular insights into origins of neurotransmitters. Philos Trans R Soc Lond Ser B Biol Sci 2021(376):20190762
24. Sebe-Pedros A et al (2018) Early metazoan cell type diversity and the evolution of multicellular gene regulation. Nat Ecol Evol 2(7): 1176–1188
25. Norekian TP, Moroz LL (2016) Development of neuromuscular organization in the ctenophore *Pleurobrachia bachei.* J Comp Neurol 524(1):136–151
26. Norekian TP, Moroz LL (2019) Neural system and receptor diversity in the ctenophore *Beroe abyssicola.* J Comp Neurol 527(12): 1986–2008
27. Norekian TP, Moroz LL (2019) Neuromuscular organization of the Ctenophore *Pleurobrachia bachei.* J Comp Neurol 527(2):406–436
28. Norekian TP, Moroz LL (2020) Atlas of the neuromuscular system in the Trachymedusa *Aglantha digitale*: insights from the advanced hydrozoan. J Comp Neurol 528(7): 1231–1254

Chapter 7

DNA Isolation Long-Read Genomic Sequencing in Ctenophores

David Moraga Amador, Andrea B. Kohn, Yelena Bobkova, Nedka G. Panayotova, and Leonid L. Moroz

Abstract

Long-read sequencing has proven the necessity for high-quality genomic assemblies of reference species, including enigmatic ctenophores. Obtaining high-molecular-weight genomic DNA is pivotal to this process and has proven highly problematic for many species. Here, we discuss different methodologies for gDNA isolation and present a protocol for isolating gDNA for several members of the phylum Ctenophora. Specifically, we describe a Pacific Biosciences library construction method used in conjunction with gDNA isolation methods that have proven successful in obtaining high-quality genomic assemblies in ctenophores.

Key words Genome, Ctenophora, *Pleurobrachia*, *Mnemiopsis*, *Beroe*, *Bolinopsis*, Evolution, Basal metazoans, PacBio sequencing, Long-read sequencing

1 Introduction

Long-read DNA sequencing technologies for long-range mapping [1–5] were introduced about a decade ago. Initially, these technologies were vastly more expensive than their short-read counterparts, and their use was mainly restricted to small genomes. However, recent advances in throughput and data quality have expanded their use, thus enabling the analysis of large, complex genomes at a resolution that had not been possible before by other sequencing methods. These long-range technologies are rapidly advancing the field with improved reference genomes, more comprehensive variant identification, and more complete views of transcriptomes and epigenomes [1, 3, 6]. Long-read sequencing improves mappability for resequencing and simplifies de novo assembly at the chromosome-scale level, including ctenophore

Authors David Moraga Amador and Andrea B. Kohn have equally contributed to this chapter.

Leonid L. Moroz (ed.), *Ctenophores: Methods and Protocols*, Methods in Molecular Biology, vol. 2757,
https://doi.org/10.1007/978-1-0716-3642-8_7, © Springer Science+Business Media, LLC, part of Springer Nature 2024

species [7–9], which convincingly confirms the phylogenetic position of comb jellies as sister to the rest of animals [7].

The most developed of the long-read technologies, Pacific Biosciences (PacBio) sequencing, is also referred to as single-molecule, real-time (SMRT) sequencing [10]. It does not rely on an amplification step for either library construction or sequencing. The product of the library construction process is adapter-ligated DNA in its native form, leading to more uniform coverage across the target genome. Palindromes and low-diversity regions of the genome can be effectively resolved. However, these advantages come at the cost of stringent sample requirements. To take full advantage of the benefits of SMRT sequencing, samples must be free of impurities that may potentially inhibit library construction or sequencing reactions, retain high integrity throughout the isolation process (i.e., undamaged high-molecular-weight DNA), and be supplied in sufficient amounts (typically 10–20 micrograms, but now these requirements are reduced to about ng range) to ensure appropriate library construction yield. The most current PacBio platform (https://www.pacb.com/products-and-services/sequel-system/) in its current state of development (beginning of 2019, with version 3 chemistry and SMRT Link 6.1 software) produces some of the longest average polymerase read lengths available in the industry (average > 30,000 bp at that time and signigicantly longer now), with more than half the data in reads >45 kilobases (Kb).

This chapter discusses and describes DNA isolation and library construction protocols that are suitable for genomic analysis of ctenophores through PacBio long-read sequencing.

2 Materials

2.1 Overview of Genomic DNA Isolation Methods

The numerous genomic DNA (gDNA) isolation methods described [6, 11–14] reflect that no universal protocol will work for every possible sample and all probable experiments. One's choice depends largely on two main factors: the source (i.e., nature) of the sample and the requirements of the downstream application. Relatively straightforward procedures can attain efficient isolation of high-quality and pure gDNA from mammalian cell cultures. The same is true for plasmid DNA extraction from recombinant DNA constructs. However, other sample sources (e.g., plant, soil, and marine organisms) can be more challenging. Steps involving chemical, enzymatic, or physical treatments often require much optimization to obtain gDNA of the necessary quality, purity, and yield that subsequent procedures may use. Many marine organisms have mucous membranes and gelatinous bodies, making obtaining pure, high-quality gDNA difficult. Besides, many marine organisms are not easily separated from contaminants or do not have cell-dense tissues readily available.

There are several critical steps in the process of gDNA isolation: effective disruption of cells or tissue, separation of cellular debris from the gDNA, denaturation of nucleoprotein complexes, inactivation of nucleases including DNases, purification of gDNA, and quality control (purity, quantity, integrity assessment) of the resultant product.

Early procedures for DNA extraction were developed from CsCl gradient centrifugation in which DNA is separated and isolated on a density gradient [15, 16]. Today, most protocols have been developed into commercial kits that perform the DNA or RNA extraction processes faster, cheaper, and easier while generating adequate quality products.

Many specialized gDNA (and/or RNA) isolation methods have been developed with attention to the specific nature of the sample and the desire for automation. Generally, they are divided into solution-based, column-based, and magnetic-particle-based protocols. Due to the fragile nature of ctenophores, the common, solution-based acid guanidinium thiocyanate-phenol-chloroform extraction protocol such as the Invitrogen™ TRIzol™ (Cat # 15596026) reagent did not generate high-quality gDNA. The TRIzol™ reagent can simultaneously isolate RNA, DNA, and protein from diverse biological sources, but it requires a lot of starting material, uses harsh and toxic reagents, and is more time-consuming than other kits.

The column-based protocols of gDNA isolation may be the most popular and convenient today. The column-based methods are divided into two categories, anion-exchange chromatography and silica-membrane technology. One of the most popular kits for reasons of brevity, low price, and convenience is the DNeasy family of products from QIAGEN (Cat # 69504). The DNeasy membrane is a silica-based membrane column that can undergo centrifugation in a plastic microcentrifuge tube. DNA is selectively bound to the membrane while contaminants pass through. The DNA-containing column is then washed with an ethanolic solution (i.e., 70–80% ethanol, 20–30% water), and the DNA is eluted off the column with 100% water or low-salt (10 mM Tris, pH 8.0) solution. The major disadvantage of this method is the centrifugation step inevitably compromises the gDNA size that can be isolated. Another disadvantage is the sample size; the column is placed in a 1.5 mL microcentrifuge tube, and whole ctenophores sometimes are too large to fit into the tube. However, DNeasy is ideal for isolating <20 Kb gDNA.

More recently, a variety of magnetic-particle-based kits have become available. One example of this type of kits is the QIAGEN MagAttract® H.M.W. kit (Cat # 67563). This product is as convenient to use as the DNeasy kit, with the added advantage of being suitable for automation. Also, the QIAGEN MagAttract® protocol seems better suited for generating high-molecular-weight gDNA.

However, the yield was ten times lower than the QIAGEN Genomic-tip kits. Of note, we tested other kits such as the Jet-Flex™ genomic DNA purification kit (Cat # A30701, Thermo-Fisher Scientific) and OMNIPREP™ (Cat # 786-136, G-BIOSCIENCES), but none of these protocols produced high-quality gDNA from ctenophores.

3 Methods

3.1 High-Molecular-Weight gDNA Isolation Protocol for Ctenophores

Here we present a protocol for DNA isolation from ctenophores that resulted in high-molecular-weight gDNA. This document uses the terms extraction, isolation, and purification interchangeably. However, isolation encompasses the process of extraction and purification of gDNA.

In our experience, the most successful protocol to generate high-molecular-weight ctenophore gDNA was with the QIAGEN Genomic-tip kit (Cat # 10262) [17]. Gel electrophoresis and Agilent TapeStation analysis showed gDNA fragments in the >50 Kb range, an integrity level sufficient for demanding downstream applications, including constructing large-insert sized gDNA libraries for PacBio sequencing. The QIAGEN Genomic-tip uses a solid-phase anion-exchange resin, which yields high-quality high-molecular-weight gDNA through a gravity-fed column. The protocol was followed according to the manufacturer's recommendations with a few modifications (where noted).

1. Evaluating necessary steps to ensure the isolation of high-quality gDNA is critical; *see* **Notes 1–3**.
2. Before DNA extractions for genomic analysis, all animals were microscopically examined for potential ectoparasites such as copepods and then washed three times for at least 30 min in filtered seawater (FSW) that the animals are native to. Use fresh animals when possible to proceed further (*see* **Note 4**).
3. The QIAGEN Genomic-tip comes in various sizes, but the 500/G is preferred because of the large volume capacity. Animals were placed in a lysis buffer containing 200 μg/mL of RNAse A as well as both protease and proteinase K at concentrations of 1 mg/mL. Samples are incubated at 50 °C with gentle shaking as recommended by the manufacturer's protocol. However, the incubation time is shortened to the point when the animal has just dissolved. Longer incubations (e.g., 2 h as recommended by the manufacturer) resulted in degraded gDNA.
4. The lysate is loaded onto an equilibrated gravity-fed column. The column is washed with a medium-salt buffer to remove all contaminants like traces of RNA and protein.

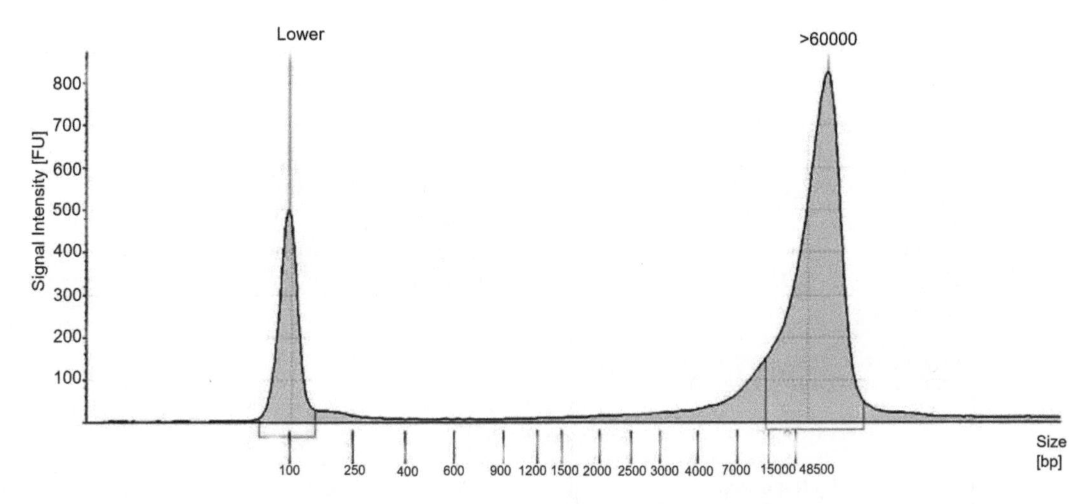

Fig. 1 *Beroe abyssicola* gDNA, as seen on the Agilent 2200 TapeStation (genomic tape) system. The peak at 100 bp, labeled "lower" is a size marker. The average gDNA size is estimated as >60,000 Kb

5. The gDNA is then eluted from the QIAGEN Genomic-tip with a high-salt buffer.
6. Finally, the eluted gDNA is desalted and concentrated by isopropanol precipitation.
7. The purified gDNA is then brought up in Dnase-/RNase-free water (Cat # AM9938, ThermoFisher Scientific). Interestingly, the use of T.E. or Tris buffers resulted in degradation of ctenophore gDNA.
8. Quality and quantity of genomic DNA are analyzed on a Qubit® 2.0 Fluorometer (ThermoFisher Scientific) and genomic DNA analysis ScreenTape (Cat # 5067-5365) on an Agilent 2200 TapeStation system (Cat # G2964AA). See section below on "Evaluation of Genomic DNA." Using this protocol, we routinely isolated DNA from ctenophores that showed a >60 Kb peak in the Agilent TapeStation; *see* Fig. 1.
9. Proceed to the construction of the sequencing library as soon as DNA isolation is completed and quality control is acceptable. For additional considerations, *see* **Notes 5–8**.

3.2 Evaluation of Genomic DNA

3.2.1 Quantitative Assessment

Absorbance-based methods (e.g., NanoDrop or equivalent) have traditionally been a favorite in most molecular biology labs. However, these methods are not adequate as they lack specificity and almost always overestimate the gDNA concentration. U.V. absorbance measurements are not selective and cannot distinguish DNA, RNA, or protein. Absorbance values are easily affected by other contaminants (e.g., free nucleotides, salts, and organic compounds) and variations in base composition; *see* Fig. 2. In addition, the sensitivity of spectrophotometric methods is often inadequate, prohibiting the quantitation of DNA and RNA at low

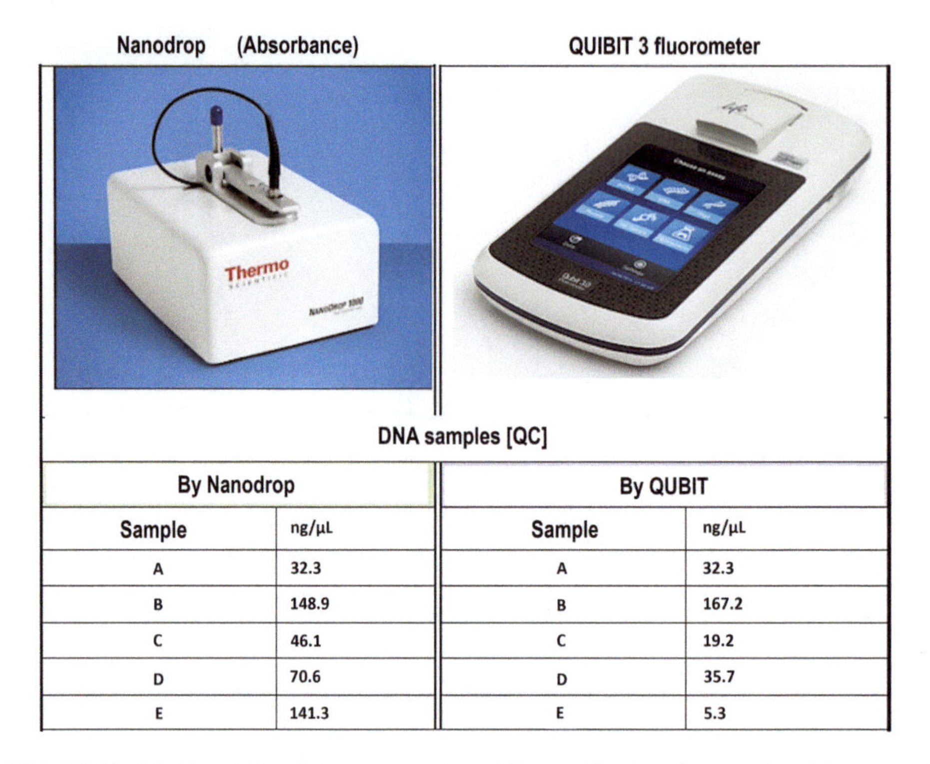

By Nanodrop		By QUBIT	
Sample	ng/μL	Sample	ng/μL
A	32.3	A	32.3
B	148.9	B	167.2
C	46.1	C	19.2
D	70.6	D	35.7
E	141.3	E	5.3

Fig. 2 Comparison of Qubit and NanoDrop methods for gDNA quantification. Samples A and B appeared to be very pure preps as the NanoDrop and Qubit values were in good agreement. These samples behaved well during library construction and sequencing. Samples D and E were very problematic and required additional purification. Initial attempts for library construction failed. Yields after extensive purification were low, but samples were eventually sequenced successfully

concentrations. The most sensitive and reliable gDNA concentration measurements are obtained using double-stranded gDNA (dsDNA)-specific reagents, such as those in the PicoGreen or Qubit assay [18–21]. Although the PicoGreen assay is more accurate, the much faster Qubit dsDNA assay is both cheaper and adequate for most purposes. It employs a fluorescent, DNA-binding dye that enables reliable, sensitive (down to 0.1 ng/μL) and specific quantitation of small amounts of dsDNA. The dye shows a minimal binding to single-stranded gDNA (ssDNA) and RNA For more recommendations when performing Qubit assays; *see* **Notes 9–12**.

3.2.2 Purity Assessment

A full assessment of the purity of a gDNA isolation prep is not a trivial matter. However, there are a few practical ways for doing this using the NanoDrop (or equivalent) by looking at three main parameters: OD ratios at 260/280 and 260/230 and the scanning pattern over the 220–350 nm range. These are a few things to consider:

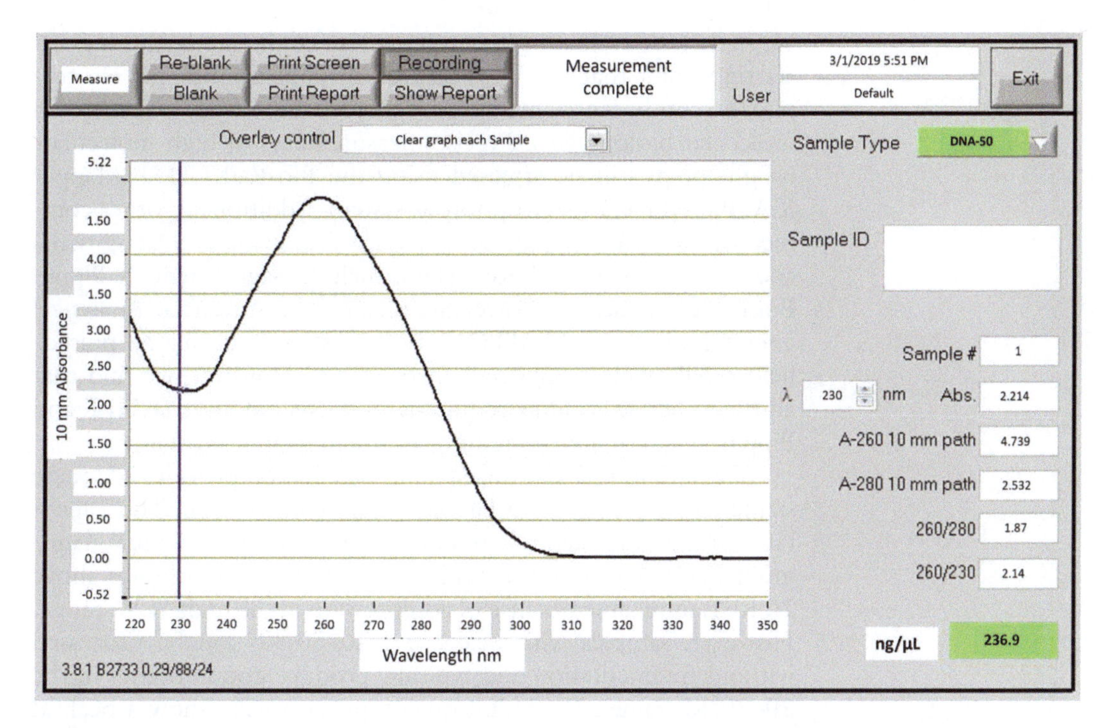

Fig. 3 Typical scan pattern and metrics for a pure gDNA preparation when seen on the NanoDrop

1. OD 260/A280 ratio of ~1.8 is generally accepted as "pure" for DNA. Some preps may have OD 260/A280 that may be as high as 2.2 (this should not be of concern).
2. OD 260/A230 ratio of 2.0–2.5 is generally accepted as "pure" for nucleic acid.
3. A low OD 260/A280 ratio may be the result of a contaminant such as protein or a reagent such as phenol. Although peptide bonds have an absorbance maximum at 280 nm, the presence of proteins in the sample can contribute significantly to the absorbance value at 260 nm.
4. A low A260/A230 ratio may be the result of a contaminant absorbing at 230 nm or less. Such contaminants include carbohydrates, residual phenol, residual guanidine, and/or glycogen. On the other hand, a high A260/A230 ratio may be the result of a dirty pedestal or using a blank solution that is not of similar ionic strength as the sample solution.
5. The 220–350 nm scan of a pure DNA prep will typically have a valley at 230 nm and a peak at 260 nm that extends down to baseline absorption at ~300 nm; *see* Fig. 3.

3.2.3 Qualitative (Size) Assessment

gDNA quality is of crucial importance for long-read sequencing. An isolation protocol should ideally avoid causing any DNA handling that can result in depurination, the formation of interstrand cross-links, nicks, etc. There are no quick and straightforward ways of thoroughly assessing all these parameters.

Nevertheless, evaluating the average size of the material is a good starting point that can be measured through routine gel electrophoresis and densitometry procedures, commonly available in molecular biology labs. Precise measurements of high-molecular-weight preps can be attained using the Bio-Rad CHEF Mapper X.A. Pulsed Field Electrophoresis system. Additionally, other commercially available systems are capable of resolving DNA fragments and smears up to ~50 Kb. These include Sage Science's Pippin Pulse Electrophoresis Power Supply and the Advanced Analytical Technologies, Inc. FEMTO Pulse (a fast and automated pulsed-field capillary electrophoresis instrument). However, we find that all these methods either require large DNA amounts (CHEF and Pippin Pulse) or are too tedious to use in routine workflows.

We have found that the Agilent TapeStation provides adequate sizing data for 20–30 Kb PacBio library construction. The TapeStation (genomic "tapes") requires very little material (few nanograms) and takes but a few minutes to run. The resolution above 20 Kb is unreliable because of compression in the high M.W. range. However, samples with peaks at >40–50 Kb can be processed without fragmentation and typically produce good libraries in the 20–30 Kb range. Very high-quality preps usually show a peak at >60 Kb (*see* Fig. 1).

3.3 Pacific Biosciences® Library Construction

3.3.1 PacBio Sample Requirements for Long-Insert Library Construction

Since long-insert (>20 Kb) library construction protocols for PacBio sequencing do not utilize any amplification, the quality of the input DNA will be directly reflected in the extent of sequencing success or failure. Any DNA damage (e.g., abasic sites, inter-strand cross-links, nicks, etc.) or contaminants in the DNA preparation (e.g., single-stranded DNA, RNA, proteins, polysaccharides, dyes, salts, etc.) will negatively affect the library construction process. Pure, high-quality gDNA is imperative for obtaining long read lengths and overall optimal sequencing performance. For additional information on the handling of gDNA samples for library construction, *see* **Notes 13–23**.

3.3.2 Pacific Biosciences® Library Construction Background

Once high-quality DNA has been isolated; long-insert sequencing libraries must be appropriately constructed with attention to the PacBio sequencing chemistry requirements. Here, we describe a protocol to generate 20–30 Kb SMRTbell sequencing libraries for four different ctenophores: *Pleurobrachia bachei*, *Beroe abyssicola*, *Bolinopsis microptera*, and *Mnemiopsis leidyi.*

The gDNA isolation protocol Subheadings 3.1 and 3.2 resulted in relatively low concentration solutions. However, because DNA was eluted in water, samples could be concentrated (if desired) by simple volume reduction on a SpeedVac with no significant change in the salt concentration. Typically, DNA preparations of adequate quality with a concentration >10 ng/μL were directly submitted to the MoBio PowerClean step without volume reduction.

3.3.3 PacBio Library Construction Protocol

Large-insert, SMRTbell (PacBio) libraries were constructed according to the recommended protocol by PacBio (*see* Fig. 4 for workflow), with a few important modifications as follows:

1. gDNA preparations from ctenophores were evaluated as described in Subheadings 3.2.1, 3.2.2 and 3.2.3. Samples with a concentration of at least 10 ng/μL, size >30 Kb, OD 260/280 = 1.8–2.0, and OD 260/230 > 2.0 were submitted to a MoBio PowerClean purification step (QIAGEN, Cat # 12877-50). The final elution step was substituted by 0.6X AMPure (Beckman Coulter, Cat # A63880); *see* **Note 23**.
2. The MoBio cleaned gDNA was quantified (Qubit) and sized on the TapeStation (genomic tape). This QC step was necessary because the MoBio procedure often resulted in significant fragmentation and loss of material. The extent of gDNA loss was sample dependent, from 30% to 95% in extreme cases (e.g., *Mnemiopsis* gDNA). The MoBio procedure was performed on as many DNA preps as necessary to obtain ~5 micrograms of cleaned material.
3. G-tubes (Covaris Inc. Cat # 520079) were used to fragment DNA preps that were still >30 Kb after the MoBio step. If necessary, this step required dilution of DNA in Tris-EDTA buffer (up to 160 uL). The samples were processed without G-tube fragmentation if the average DNA size was <30 Kb.
4. Whenever possible, five micrograms of sheared and concentrated DNA (140 ng/μL) were used for the subsequent SMRT bell library construction steps. However, in some cases, libraries were constructed with as little as 2 micrograms of input DNA. The protocol details for the various types of libraries are described in PacBio documents (P/N 100-286-000 Version 10 January 2018), except that size selection was performed on the SageELF, rather than the BluePippin™ (*see* **step 5**, in Fig. 4). The library construction steps included: ExoVII treatment, DNA damage repair, end repair, blunt-end ligation of SMRT bell adaptors, and ExoIII/ExoVII treatment.
5. As outlined in Fig. 4, the library construction procedure typically resulted in 1.2–1.5 micrograms of SMRT bell library (i.e., 25–30% yield, except when tested on *Mnemiopsis*). The final library was size-selected in the SageELF™ instrument (Cat # ELD 7510), using 0.75% agarose gel cassettes and the 1–18 Kb v2 cassette definition program. The desired SageELF fractions were cleaned using AMPure magnetic beads (0.6X AMPure sample ratio) and eluted in 15 uL of 10 nM Tris HCl, pH 8.0. Library fragment size was estimated by the Agilent TapeStation (genomic DNA tapes), and these data were used for calculating molar concentrations. Typically, fractions in

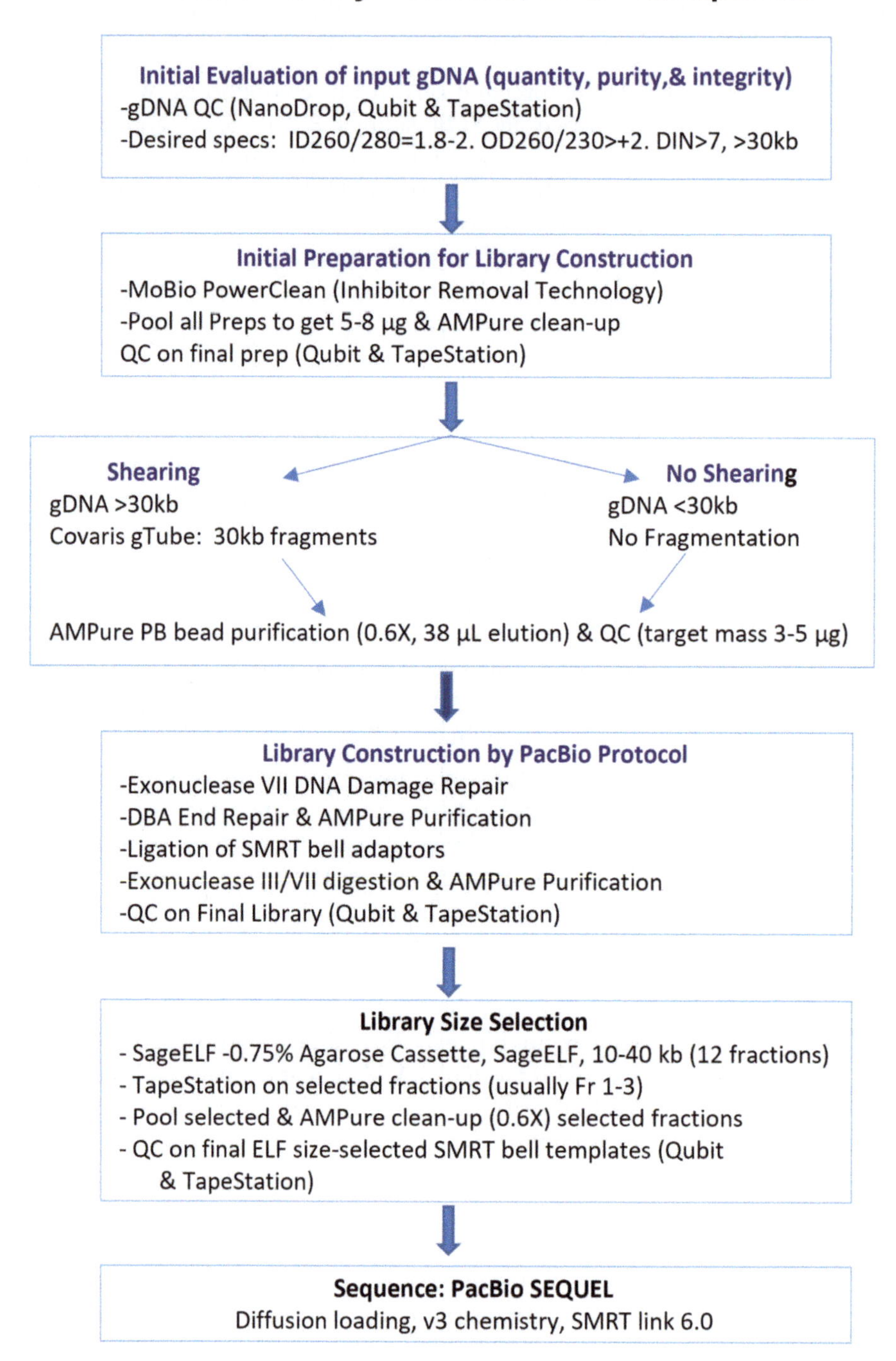

Fig. 4 Flow diagram of long-insert library construction steps for PacBio sequencing of ctenophores

wells 1, 2, and 3 contained library fragments in the 20–35 Kb range; *see* Fig. 5. Between 4 and 10 pM of the library was loaded onto the PacBio SEQUEL sample plate for sequencing.

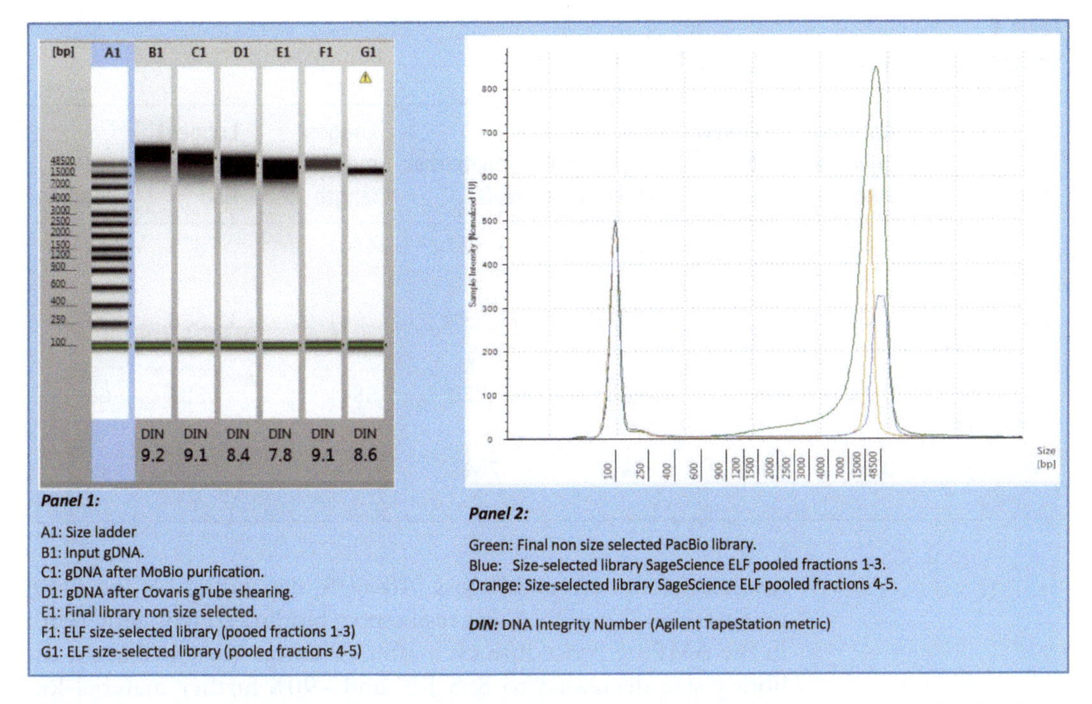

Fig. 5 Electrophoretic bands and peaks profiles for various PacBio library construction steps. Panel 1 shows the pseudogel graph of the indicated steps. Panel 2 displays the electrophoretic peak for the final non-size-selected library (largest, green peak) and two different size-selected library peaks that were created by pooling the indicated E.L.F. fractions

All other steps for sequencing were done according to the recommended protocol by the PacBio sequencing calculator.

3.4 Conclusions

Long-read sequencing promises to enable rapid advances in the study of ctenophore genomes. However, the fragile body plan of these organisms, combined with their unique mesoglea, offers many technical challenges when attempting to meet the stringent sample quality requirements for long-read sequencing technologies (e.g., PacBio). Among other difficulties to contend with, these organisms typically have a high mucilage content and lack cell-dense tissues. These factors make it hard to obtain high-quality DNA PacBio long-insert libraries. In particular, they require relatively large quantities of DNA, which must also be of high purity and integrity.

We have described a protocol that can be broadly used for high-quality gDNA isolation and PacBio library construction in ctenophores. This procedure was used with comparable success on several species (*Pleurobrachia bachei*, *Beroe abyssicola*, *Bolinopsis microptera*). The protocol was also tested on *Mnemiopsis leidyi*. However, this species yielded marginal library and sequencing results (i.e., lower library yields, shorter polymerase reads, and lower sequencing yields). MoBio cleaning of *Mnemiopsis* DNA

Table 1
Sequencing metrics for ctenophore: PacBio Sequel

Species	Approx. genome size	Gb per SMRT cell	Polymerase read length	Polymerase N50	Longest subread length	Longest subread N50	%P1 loading
Pleurobrachia bachei	0.2 Gb	3.4	6563	14,500	4361	7375	45
Beroe abyssicola	1 Gb	9.6	13,013	25,250	7981	12,536	72
Bolinopsis microptera	0.2 Gb	9.5	14,385	52,750	5841	9583	66
Mnemiopsis leidyi	0.16 Gb	1.6	3453	7648	3205	6839	46

resulted in a considerable loss (40–60%, depending on the DNA preparation). The DNA was resilient to binding to magnetic beads in the AMPure procedure even after cleaning. Most strikingly, the library size decreased to 3–5 Kb, and ~90% further material loss occurred after the Exonucleases III and VII step at the end of the library construction procedure. This behavior was consistent with damaged DNA upon isolation and/or during the library construction process. *Mnemiopsis* DNA library construction was attempted several times on freshly isolated DNA (never frozen) and on DNA preps that had been preserved at −80 °C. The results were the same for reasons that need further investigation. Interestingly, besides the library construction challenges with *Mnemiopsis*, the "surviving" library material produced average polymerase reads that were about one-fifth of the length of those generated by sequencing other ctenophores (e.g., *Beroe* and *Bolinopsis*).

Most of the gDNA isolation and library construction protocols steps were adapted from the manufacturer's manuals and procedures. However, several key steps in the workflow (Fig. 4) were modified, which resulted in relatively streamlined, robust, and efficient results. For DNA isolation, the QIAGEN Genomic-tip kit--based method was used with a few modifications to ensure sufficient purity and integrity of the final preparation. The most critical factors were gentle handling, optimized RNase/Proteinase K digestion conditions, proper washing of the lysate on the gravity-fed column, and prompt desalting-concentrating of the final DNA solution. The PacBio, long-insert library construction procedure was modified to include a stringent gDNA cleanup step (MoBio), and a library size selection by the SageELF rather than by the BluePippin™. Three out of four ctenophores species tested, yielded outstanding sequencing results that have enabled novel genomic analyses. Table 1 shows the sequencing run metrics obtained in the PacBio SEQUEL for all four species tested, using v2.5 chemistry and SMRT link 5.1 software.

4 Notes

1. Only use wide-bore pipette tips when handling DNA and pipette very slowly to reduce shearing.
2. Minimize or eliminate any high-heating steps during isolation or preparation of DNA.
3. Minimize or eliminate high-speed vortexing; use gentle mixing techniques such as slow inversion.
4. Preferably, perform DNA extraction/isolation as soon as the tissue has been obtained. Otherwise, use tissue that has been flash-frozen with liquid nitrogen and stored at −80 °C. Alternatively, tissues may be adequately preserved when placed in a number of commercially available products [22]. However, these options should be carefully tested.
5. Include an extra cleanup step before library construction (e.g., MoBio PowerClean, same as QIAGEN DNeasy PowerClean Cleanup kit Cat # 12877-50).
6. Minimize or eliminate the number of freeze/thaw cycles with your sample to reduce DNA damage.
7. Allow sufficient thawing time for aliquots of DNA, as partially frozen DNA is prone to shearing.
8. Shipping DNA: Overnight shipping at 4 °C is preferred. However, if shipping overnight is not an option, flash-freeze the DNA sample with liquid nitrogen and ship frozen. Alternatively, some commercial products (e.g., DNAstable Plus) allow for the shipping of DNA at room temperature.
9. Assure that bubbles are not introduced into the sample at the reading as this can affect the results. Slight tapping on the tube wall or brief centrifugation will often help dissipate bubbles.
10. Samples should be diluted or concentrated as needed to remain within the quantitative range of the assay. If you get a concentration value as "too high" or "too low," it means that your sample is out of range.
11. The assay should be performed at room temperature, and the assay tubes must be at room temperature when the reading is taken. Do not hold assay tubes in your hand for too long while trying to read the samples.
12. The same Qubit protocol was used throughout the entire library construction process. The Qubit assay kits come in

quantification ranges: broad range and high sensitivity. The quantification results differ somewhat between the two kits.

13. The gDNA sample should be dissolved in 10 mM Tris, pH 7.5–8.0 at a minimum concentration of 30–50 ng/μL. However, for ctenophores, we used DNase/RNase-free water.
14. The gDNA sample needs to be double-stranded. Single-stranded DNA will not be ligated to the adaptor in the template preparation process and can interfere with quantitation and polymerase binding. For this reason, DNA must be quantified by fluorescence-based reagents such as PicoGreen or Qubit, which only detect double-stranded DNA.
15. A minimum of freeze-thaw cycles for your gDNA sample.
16. The gDNA samples do not need to be exposed to high temperatures (i.e., >65 °C for 1 h can cause a detectable decrease in sequence quality) or pH extremes (<6 or >9).
17. The gDNA sample should not be exposed to intercalating fluorescent dyes or ultraviolet radiation. If purified from a gel fragment, ethidium bromide, and UV must be avoided for staining and visualization. We recommend using SYBR safe with visualization on a blue lightbox (long wavelength).
18. An OD 260/280 ratio of approximately 1.8–2.0. OD 260/230 ratio higher than 2.0 is recommended for the gDNA sample.
19. The gDNA sample should be above 30 Kb in size if possible.
20. The gDNA sample should not contain insoluble material.
21. The gDNA sample should not contain RNA contamination.
22. The gDNA sample should not contain chelation agents (i.e., EDTA), divalent metal cations (i.e., Mg^{2+}), denaturants (guanidinium salts, phenol), or detergents (S.D.S., Triton-X100, CTAB).
23. The gDNA sample should not contain carryover contamination from the starting organism/tissue (heme, humic acid, polysaccharides, polyphenols, etc.) that may affect library construction and sequencing performance. For preparations containing a low level of contaminants, it may be sufficient to perform a 1X AMPure bead cleanup or a high-salt phenol-chloroform wash following gDNA extraction (http://www.pacb.com/wp-content/uploads/2015/09/Shared-Protocol-Guidelines-for-Using-a-Salt-Chloroform-Wash-to-Clean-Up-gDNA.pdf). Unfortunately, DNA preps from ctenophores contain significant levels of impurities. Preparations with OD 260/80 or 260/230 < 1.8, and with significantly different Qubit versus NanoDrop values, are good candidates for more stringent cleanup procedures with reagents such as MoBio

PowerClean (QIAGEN DNeasy PowerClean Cleanup kit Cat# 12877-50). In our experience, the use of the MoBio PowerClean DNA Clean Up Kit before library construction resulted in significantly improved sequencing results in the PacBio. In some cases, samples that had failed to sequence were "rescued" by a MoBio cleanup step. Other cleanup methods, namely, Zymo Research (Cat # D41010) and Clontech (Cat #740230.10) were tested. However, these methods resulted in significantly greater DNA fragmentation and loss than the MoBio kit.

Acknowledgments

This work was supported in part by the Human Frontiers Science Program (RGP0060/2017) and the National Science Foundation (IOS-1557923) grants to L.L.M. Research reported in this publication was also supported in part by the National Institute of Neurological Disorders and Stroke of the National Institutes of Health under award number R01NS114491 (to L.L.M). The content is solely the authors' responsibility and does not necessarily represent the official views of the National Institutes of Health.

References

1. Beaulaurier J, Schadt EE, Fang G (2019) Deciphering bacterial epigenomes using modern sequencing technologies. Nat Rev Genet 20(3):157–172
2. Goodwin S, McPherson JD, McCombie WR (2016) Coming of age: ten years of next-generation sequencing technologies. Nat Rev Genet 17(6):333–351
3. Jing Y et al (2019) Hybrid sequencing-based personal full-length transcriptomic analysis implicates proteostatic stress in metastatic ovarian cancer. Oncogene 38(16):3047–3060
4. Taylor TL et al (2019) Rapid, multiplexed, whole genome and plasmid sequencing of foodborne pathogens using long-read nanopore technology. bioRxiv:558718
5. Wenger AM et al (2019) Highly-accurate long-read sequencing improves variant detection and assembly of a human genome. bioRxiv:519025
6. Wagner Mackenzie B, Waite DW, Taylor MW (2015) Evaluating variation in human gut microbiota profiles due to DNA extraction method and inter-subject differences. Front Microbiol 6:130
7. Schultz DT et al (2023) Ancient gene linkages support ctenophores as sister to other animals. Nature 618(7963):110–117
8. Schultz DT et al (2021) A chromosome-scale genome assembly and karyotype of the ctenophore *Hormiphora californensis*. G3 (Bethesda) 11(11):jkab302
9. Hoencamp C et al (2021) 3D genomics across the tree of life reveals condensin II as a determinant of architecture type. Science 372(6545):984–989
10. Vermeesch JR et al (2018) Single molecule real-time (SMRT) sequencing comes of age: applications and utilities for medical diagnostics. Nucleic Acids Res 46(5):2159–2168
11. de Kok JB et al (1998) Use of real-time quantitative PCR to compare DNA isolation methods. Clin Chem 44(10):2201–2204
12. Nacheva E et al (2017) DNA isolation protocol effects on nuclear DNA analysis by microarrays, droplet digital PCR, and whole genome sequencing, and on mitochondrial DNA copy number estimation. PLoS One 12(7): e0180467
13. Psifidi A et al (2015) Comparison of eleven methods for genomic DNA extraction suitable for large-scale whole-genome genotyping and

long-term DNA banking using blood samples. PLoS One 10(1):e0115960

14. Varma A, Padh H, Shrivastava N (2007) Plant genomic DNA isolation: an art or a science. Biotechnol J 2(3):386–392
15. Garger SJ, Griffith OM, Grill LK (1983) Rapid purification of plasmid DNA by a single centrifugation in a two-step cesium chloride-ethidium bromide gradient. Biochem Biophys Res Commun 117(3):835–842
16. Maniatis T, Fritsch EF, Sambrook J (1982) Molecular cloning, 1st edn. Cold Spring Harbor Laboratory
17. Moroz LL et al (2014) The ctenophore genome and the evolutionary origins of neural systems. Nature 510(7503):109–114
18. Ahn SJ, Costa J, Emanuel JR (1996) PicoGreen quantitation of DNA: effective evaluation of samples pre- or post-PCR. Nucleic Acids Res 24(13):2623–2625
19. Mardis E, McCombie WR (2017) Library quantification: fluorometric quantitation of double-stranded or single-stranded DNA samples using the qubit system. Cold Spring Harb Protoc 2017(6):pdb.prot094730
20. Nakayama Y et al (2016) Pitfalls of DNA quantification using DNA-binding fluorescent dyes and suggested solutions. PLoS One 11(3):e0150528
21. Rengarajan K et al (2002) Quantifying DNA concentrations using fluorometry: a comparison of fluorophores. Mol Vis 8:416–421
22. Gray MA, Pratte ZA, Kellogg CA (2013) Comparison of DNA preservation methods for environmental bacterial community samples. FEMS Microbiol Ecol 83(2):468–477

Chapter 8

RNA Isolation from Ctenophores

Andrea B. Kohn, Yelena Bobkova, and Leonid L. Moroz

Abstract

RNA-seq or transcriptome analysis of individual cells and small cell populations is essential for virtually any biomedical field. Here, we examine and discuss the different methods of RNA isolation specific to ctenophores. We present a convenient, inexpensive, and reproducible protocol for RNA-seq libraries that are designed for low quantities of samples. We demonstrated these methods on early (one, two, four, eight cells) embryonic and developmental stages, tissues, and even a single aboral organ from the ctenophore *Pleurobrachia bachei* and other ctenophore species (e.g., *Mnemiopsis*, *Bolinopsis*, and *Beroe*).

Key words Ctenophora, Single cells, Transcriptome, RNA-seq, *Pleurobrachia*, Development

1 Introduction

RNA sequencing (RNA-seq) has and will continue to profoundly impact medicine and clinical and basic research. It has become indispensable for virtually any biomedical field, providing an unbiased view of the entire molecular machinery within a biological system of interest. The rapid technological progress has catapulted RNA-seq to be the most sensitive tool for gene expression analysis today. The number of different RNA-seq protocols has grown exponentially from its inception over a decade ago [1–3]. To perform an RNA-seq experiment, a robust sequencing library has to be constructed. Illumina, Inc. provides a diversity of protocols: >50 RNA-seq and >30 single-cell RNA-seq (scRNA-seq) library construction techniques (https://www.illumina.com/techniques/sequencing/rna-sequencing.html). Depending on the amount of starting material, total RNA, message RNA (mRNA), small RNA, targeted RNA, RNA exome-capture, and single-cell, the scope of possible RNA-seq methods is overwhelming.

Here we present a workflow convenient for simpler workplaces and expedition/fieldwork conditions, starting with animal collections, RNA isolation, and high-quality RNA-seq library generation (Fig. 1). We also offer options in the workflow for cloning genes of

Leonid L. Moroz (ed.), *Ctenophores: Methods and Protocols*, Methods in Molecular Biology, vol. 2757,
https://doi.org/10.1007/978-1-0716-3642-8_8, © Springer Science+Business Media, LLC, part of Springer Nature 2024

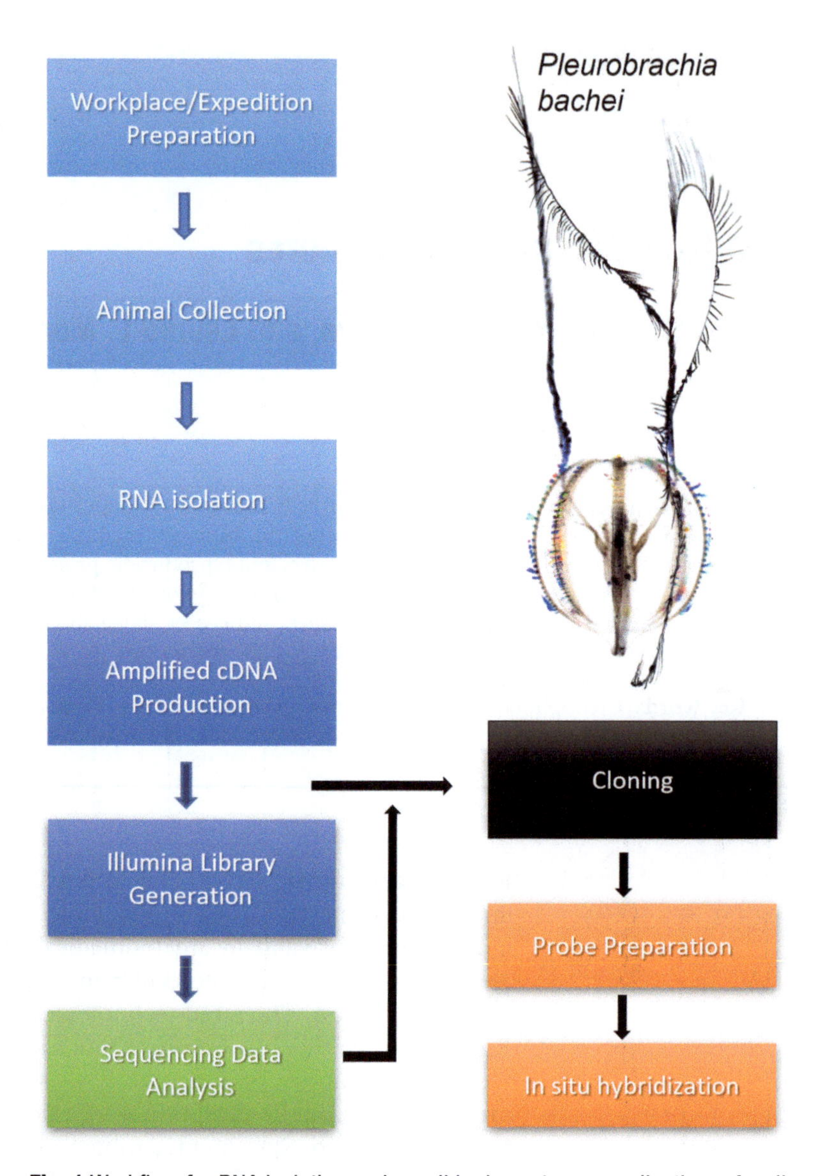

Fig. 1 Workflow for RNA isolation and possible downstream applications. Amplified cDNA can be used as a template for cloning (*see* **Note 3**). Gene sequences are obtained from the analyzed sequencing data

interest and probe production for validation with in situ hybridization (Fig. 1); see also the respective chapter in this book. Libraries can be constructed from very low input <40 pg of total RNA, and our protocol produces high-quality sequencing data. As illustrative examples, the starting material included early (one, two, four, eight cells) embryonic and developmental stages, tissues, and even a single aboral organ from the ctenophore *Pleurobrachia bachei* [4] as well as other ctenophores and mollusks [5–9]. We validated these results with hundreds of in situ hybridizations using probes designed from many of these sequencing projects.

2 Materials

2.1 RNA Isolation

RNeasy Micro Kit (Cat # 74104, QIAGEN).

RNeasy Mini Kit (Cat # 74004, QIAGEN).

QIAshredder (Cat # 79654, QIAGEN).

gDNA Eliminator Spin Columns (Cat # 74134, QIAGEN).

RNase-free DNase Set (Cat # 79254, QIAGEN).

Nuclease-free Water (not DEPC-treated) (Cat # AM9937, ThermoFisher Scientific).

RNaseZap™ RNase Decontamination Solution (Cat # AM9780, ThermoFisher Scientific).

MagaZorb® total RNA mini-prep kit (Cat # MB2004, Promega).

Maxwell® RSC simply RNA tissue kit (Cat # AS1340, Promega).

NUCLEOSPIN RNA II from Macherey-Nagel (Fisher Cat # NC9581114).

E-Gel® SizeSelect™ 2% Agarose (Cat # G6610-02, ThermoFisher Scientific).

Qubit™ RNA BR Assay Kit (Cat # Q10210, ThermoFisher Scientific).

Qubit™ RNA HS Assay Kit (Cat # Q32852, ThermoFisher Scientific).

Qubit™ dsDNA BR Assay Kit (Cat # Q32850, ThermoFisher Scientific).

RNA ScreenTape (Cat # 5067-5576, Agilent Technologies).

RNA ScreenTape Sample Buffer (Cat # 5067-5577, Agilent Technologies).

RNA ScreenTape Ladder (Cat # 5067-5578, Agilent Technologies).

High Sensitivity RNA ScreenTape (Cat # 5067-5579, Agilent Technologies) total.

High Sensitivity RNA ScreenTape Sample Buffer (Cat # 5067-5580, Agilent Technologies).

High Sensitivity RNA ScreenTape Ladder (Cat # 5067-5581, Agilent Technologies).

2.2 Library Construction

Advantage® UltraPure PCR dNTP Mix (Cat # NC9287432, ThermoFisher Scientific).

SMARTScribe Reverse Transcriptase (Cat # 639536, Takara Bio).

Advantage 2 PCR Kit (Cat # 639207, Takara Bio).

Rnase Inhibitor (Cat # N8080119, Takara Bio).

Table 1
Adaptors and primers for library construction

Primer name	Primer sequence
CapTS oligonucleotide	5′-AGCAGTGGTATCAACGCAGAGTACrGrGrG-3′[a]
CapT30 primer	5′-AAGCAGTGGTATCAACGCAGAGTACT(30)-3′
Cap PCR primer	5′-AAGCAGTGGTATCAACGCAGAGT-3′

[a]The r is designated as a ribonucleotide

SMART® cDNA Library Construction Kit (Cat # 63490, Takara Bio).

SPRIselect Reagent (Cat # B23318, Beckman Coulter).

NEBNext® Ultra II DNA Library Prep Kit for Illumina® (Cat # E7645S, New England Biolabs).

NEBNext® Ultra II multiplex oligos (Cat # E7335S, New England Biolabs).

NEBNext® Ultra II multiplex oligos Dual Index Primers (Cat # E7600S, New England Biolabs).

High Sensitivity D1000 Screen tape (Cat # 5067-5584, Agilent Technologies).

High Sensitivity D1000 Reagents (Cat # 5067-5585, Agilent Technologies).

High Sensitivity D1000 Ladder (Cat # 5067-5587, Agilent Technologies).

High Sensitivity D1000 Sample Buffer (Cat # 5067-5603, Agilent Technologies).

TOPO-TA cloning kit (Cat # K203001, ThermoFisher Scientific).

1 kb DNA ladder (Cat # N3232L, New England Biolabs)

Primers 0.2 uM scale HPLC purified, IDT, Integrated DNA Technologies, Inc. (*see* Table 1 for all primer and adaptor sequences).

2.3 Equipment

LoBind tubes 0.5 uL, 1.5 uL (Cat # 80077-236, 80077-230 Eppendorf, VWR International).

DynaMag™-2 magnet (Cat # 123-21D, ThermoFisher Scientific).

MicroTUBE AFA fiber Screw-Cap 6 × 16 mm (Cat # NC0380760/520096, Covaris).

Agilent TapeStation™ 4200 or Bioanalyzer (Cat # G2991AA, Agilent Technologies).

Galaxy Ministar Centrifuge (120 V, 50/60 Hz) (Cat # 93000-196, V.W.R. International).

Sonicating bath (2- to 3-L tank; 80 W, 40 kHz transducer) (Cat # 15-335-20, Fisher).

Qubit® 2.0-4.0 Fluorometer (Cat # Q33238, ThermoFisher Scientific).

PCR machine, MJ Research Thermo Cycler (Cat # PTC-100, MJ Research).

Covaris M220 Focused-ultrasonicator (Cat # 4482277, Covaris).

E-Gel® iBase™ and E-Gel® Safe Imager™ Kit (Cat # G8152ST, ThermoFisher Scientific).

3 Methods

3.1 Sample Preparation

3.1.1 Workplace Preparation

Being aware of the risk of Ribonuclease (RNase) contamination is critical before any animal is collected and RNA isolated. These enzymes degrade RNA and can have detrimental effects on any downstream processes. RNases are found in practically every cell type for both prokaryotes and eukaryotes. They are the toughest enzymes to inactivate and can retain activity after freeze-thaw cycles and even autoclaving. For more details on working with RNA, *see* **Note 1**.

Fieldwork is often required for collecting many novel and rare species of ctenophores and is a challenging environment for every demand of molecular biology laboratories. Decontaminating and managing RNases in these environments is of utmost importance. All surfaces, including benchtops, equipment, and even pipettes, should be assumed to be contaminated with RNases. Decontaminate your work area with 70% ethanol, 4% bleach or RNAaseZap to control RNase contamination, but make sure none of these reagents come in contact with your isolated RNA sample. For more details on setting up an appropriate workstation for fieldwork, (*see* **Note 1**).

3.1.2 Specimen Preparation

Before RNA isolation, all ctenophores should be microscopically examined for potential ectoparasites or food (such as copepods) and then washed three times for at least 30 min in filtered seawater (FSW) obtained from the same location the animals are collected. Animals are anesthetized in 60% (volume/body weight) isotonic $MgCl_2$ (337 mM). Specific tissues are surgically removed with sterile fine forceps and scissors and processed for RNA isolations (*see* **Note 1** for more details on dealing with RNase contamination).

3.2 RNA Isolation

High-quality RNA is the first and often the most critical step in performing many molecular techniques. Here, we will focus on total RNA isolations because it is the most reliable method when working with ultralow amounts of starting material and obtaining dedicated RNA-seq libraries sufficient for deep transcriptome sequencing (e.g., [4, 5, 8, 10, 11]). When considering the total RNA isolation process, several technologies are available including organic extraction methods, centrifuge column-based, magnetic particle based, and direct lysis. Given the fragile nature of ctenophores, column-based technology has worked best.

QIAGEN's RNeasy kits, which are column-based, have produced high-quality total RNA from ctenophores for all downstream processes. The RNeasy kits come in two varieties depending on the amount of RNA input: the mini-kit which is designed to isolate up to 100 μg of total RNA and the micro kit designed to isolate up to 45 μg total RNA. The key to both kits is not to overload the columns and clog them. An option with the RNeasy kits is the QIAshredder that can be used for simple and rapid homogenization of cell and tissue lysates but is usually not needed for ctenophores given their fragile bodies. The gEliminator spin column is another option for removing genomic DNA (gDNA) contamination and is also an excellent alternative. See QIAGEN's webpage for a list of all RNease kits (https://www.qiagen.com).

Before beginning any protocol, read all the manufacturer's manuals. All procedures presented follow the manufacturer's recommendation except where noted. All solution preparations must be adhered including adding β-mercaptoethanol (β-ME) to Buffer RLT (*see* **Note 2**), adding four volumes of ethanol to the RPE, and preparing DNase I stock solution if the gEliminator spin column is not used. Here we present the general scheme for isolating total RNA from ctenophores with the RNeasy kits.

3.2.1 RNA Isolation Protocol (QIAGEN's RNeasy Kits)

1. Sample lysis is performed in prepared RLT buffer and loaded onto a RNeasy column membrane.
2. Genomic DNA contamination is removed either by a DNase 1 step on the column or a gEliminator spin column.
3. Ethanol is added to the lysate, producing conditions that promote selective binding of RNA to the RNeasy column membrane.
4. The total RNA binds to the column, contaminants are efficiently washed away, and high-quality RNA is eluted in RNase-free water.
5. This can also be a pause point, and samples can be stored at − 80 °C without degradation.

It should be noted that the Promega MagaZorb® total RNA mini-prep kit also produced high-quality RNA from ctenophores. And for automation, the Maxwell® RSC simplyRNA tissue kit also gave very good quality RNA. Nevertheless, the kits described above cannot be used effectively with pigmented tissue like retina or chromatophores. For these purposes, we employed Nucleospin RNA II from Macherey-Nagel. Many RNA isolation/extraction kits are available, and all need to be tested for your species of interest. Here, we present the kits that worked well to isolate total RNA from ctenophores.

3.2.2 Quality Control for Isolated RNA

All molecular biology reactions require high-quality pure (free of all contaminants including gDNA) and precise amounts of RNA for optimal downstream application performance. RNA samples are run on an Agilent® Technologies TapeStation™ or Bioanalyzer™ according to the manufacturer's recommendation. The Agilent® TapeStation™ instrument automatically analyzes RNA concentration and integrity through a combination of microfluidics, capillary electrophoresis, and fluorescent dye that binds to nucleic acid (see https://www.agilent.com for more details). The TapeStation™ has two different tapes depending on the amount of input RNA: the RNA ScreenTape for the sensitivity of 5 ng/μL and a high sensitivity RNA ScreenTape for measured concentrations down to 100 pg/μL. The analysis algorithm for both Agilent® Technologies TapeStation™ and Bioanalyzer™ can provide information about RNA integrity (the RNA Integrity Number, or RIN) with a maximum value of 10 when analyzing total RNA. Significant decreases in the RIN are indicative of degraded total RNA. High-quality RNA will have a minimal amount of degradation. A typical ctenophore sample has both 18 s and 28 s ribosomal peaks, compared to the hidden break in the *Aplysia* ribosomal RNA. A high sensitivity RNA ScreenTape was used to analyze isolated total RNA from the combs rows in *Mnemiopsis leidyi*, (*see* Fig. 2).

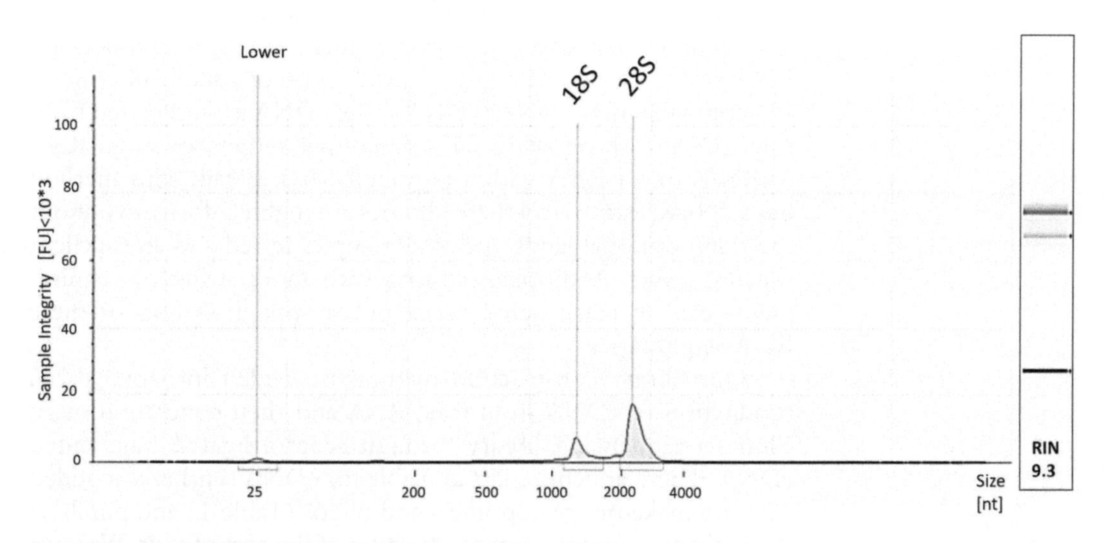

Fig. 2 *Mnemiopsis leidyi* total RNA isolated from a small piece of a comb row run on a high sensitivity RNA ScreenTape. This is high-quality total RNA with a RIN of 9.3

The concentration of RNA is determined on a Qubit® 2.0 fluorometer. Concentration can be obtained for the Agilent® RNA. ScreenTape is not as accurate as the Qubit® measurement. The Qubit assays use dyes that bind selectively to DNA, RNA, or protein, thus making the assay more sensitive than any UV absorbance methodology; see https://www.thermofisher.com for more information. There are two types of assays depending on the concentration of input RNA, a broad range kit for measuring a quantification range of 20–1000 ng and a high sensitivity kit for the 5–100 ng range. The sample in Fig. 2 had a Qubit RNA concentration of 12.9 ng/μL.

Once the quality and quantity are satisfactory, this material is used to construct a sequencing library. The ultimate test of high-quality RNA is the quality of the sequencing data produced. Lower indicates the marker provided in the kit.

3.3 Library Construction for RNA-seq

RNA-seq is a precise tool to measure the gene expression profile of the transcriptome (all expressed RNAs) in a single cell, a population of cells, a tissue, or a whole organism. There are numerous commercially and noncommercially available kits or methods for performing RNA-seq available. However, many kits are designed for sequencing cores. These kits are usually for high-throughput production in 96- or 384-well formats. Most importantly, they are costly. We present a convenient, inexpensive, and reproducible protocol for RNA-seq that is designed for a small number of samples, even one sample. Our protocol is applicable for single cells and ultralow amounts of starting material.

Today direct sequencing of RNA is possible with technologies like Pacific Biosystems and Oxford Nanopores Technologies, Inc. However, these methods are not applicable yet to single cells or small amounts of starting material. Since we are not sequencing RNA directly, we go through the production of a stable intermediate, complementary DNA (cDNA). The cDNA is synthesized from total RNA with an oligo dT primer that targets message RNA (mRNA) or all RNA with a polyadenylated 3′ tail. This method has a 3′ bias, but we feel these libraries are highly informative about the transcriptional landscape of all samples tested and an excellent starting point. Additional libraries with more stringent requirements can be constructed based on the initial analysis of these RNA-seq libraries.

Our library construction method is divided into two parts, production of cDNA from total RNA and then generation of an Illumina sequencing library through adaptor-ligated fragmented cDNA. There are entire kits available for cDNA synthesis at added cost; we make our own primers and oligos (Table 1) and purchase the individual components at a fraction of the cost of a kit. We use a commercial kit to make the sequencing library NEBNext® Ultra II DNA Library Prep Kit for Illumina® for convenience.

Library construction starts with total RNA isolation (*see* Subheading 3.2) that is reverse transcribed by a specific reverse transcriptase enzyme to first-strand cDNA with an oligo dT primer. The second strand of the cDNA is produced through a template-switch method as described [8, 12, 13]. The cDNA is amplified through a polymerase chain reaction (PCR) and then purified. At this point, a small amount of the cDNA is reserved for downstream applications like cloning (*see* **Note 3**). The resultant amplified cDNA is then converted to an Illumina sequencing library with NEBNext® Ultra II DNA Library Prep Kit for Illumina® and the NEBNext® Multiplex Oligos for Illumina according to manufacturer's recommendations. We present this protocol for one sample, but if more than one library is being made, it is best to make master mixes; *see* **Note 4** for master mix setup.

3.3.1 Library Construction Protocol

cDNA Synthesis

1. For isolated RNA, combine the following in a 0.5- or 0.2-mL LoBind tube:

   3.5 μL RNA (2 ng–1 μg total RNA)

   1 μL 5X first-strand buffer

   1 μL of 3′ CapT30 Primer (10 μM) (Table 1)

   5.5 μL total volume

2. For single cells, obtain single cell and add to 4.5 uL of 1X first-strand buffer.

   Sonicate for 20 s and add the following:

   1 μL of 3′ CapT30 primer (10 μM)

   5.5 μL total volume

   Mix contents and spin the tubes briefly in a microcentrifuge.

3. Incubate the tubes at 72 °C in a hot-lid thermal cycler for 3 min.

4. Add the following reagents in the order shown:

   1.0 μL 5X first-strand buffer

   0.25 μL DTT (20 mM)

   1.0 μL dNTP Mix (10 mM)

   1.0 μL CapTS oligonucleotide (12 μM) (Table 1)

   0.25 μL RNase inhibitor

   1.0 μL SMARTScribe reverse transcriptase (100 U)

   10.0 μL total volume

   Add the reverse transcriptase just prior to use. Mix well by pipetting, and spin the tube briefly in a microcentrifuge.

5. Incubate the tubes at 42 °C for 1.5 h (not to exceed 2 h).

6. Prepare the PCR reagents in the order shown and add them to the above cDNA sample:

   74.0 μL RNase- and DNase-free H_2O

   10.0 μL 10X Advantage 2 PCR buffer

   2.0 μL dNTP Mix (10 mM)

   2.0 μL Cap PCR primer (10 μM) (Table 1)

   2.0 μL 50X Advantage 2 polymerase mix

   90 μL total volume

7. Amplify in thermal cycling using the following program:

   95 °C 1 min, one cycle

   The following steps perform 12–18 cycles, depending on the amount of starting material (usually 12–16 cycles for tissue, 16–18 cycles for single cells)

   95 °C 15 s

   65 °C 30 s

   68 °C 3 min

   Hold 4 °C ∞*

8. Run an E-Gel® SizeSelect™ 2% agarose with 5 uL of the amplified cDNA.
9. Save 10 uL of this cDNA for further use; *see* **Note 3**.

Purification of Amplified cDNA

SPRIselect bead purification is a proprietary paramagnetic bead-based chemistry for purifying DNA fragments to the desired size for library preparation.

1. Make fresh 80% ethanol with RNase- and DNase-free H_2O.
2. Place amplified cDNA sample (85 mL) in 1.5 mL LoBind tube.
3. Add 51 μL of SPRIselect® reagent (0.6 × sample volume).
4. Pipet up and down five times to thoroughly mix the bead suspension with the DNA.
5. Incubate the mixture at room temperature for 5 min.
6. Place the tube in a magnetic rack such as the DynaMag™-2 magnet for 3 min or until the solution is clear of brown tint when viewed at an angle.
7. Carefully remove and discard the supernatant without disturbing the bead pellet.
8. Without removing the tube from the magnet, add 500 μL of freshly prepared 80% ethanol.
9. Incubate for 30 s on the magnet. After the solution clears, remove and discard the supernatant without disturbing the pellet.

10. Repeat **steps** 7–**9** for a second wash.
11. To remove residual ethanol, pulse-spin the tube, place it back in the magnetic rack, and carefully remove any remaining supernatant with a 20-μL pipettor without disturbing the pellet.
12. Keeping the tube on the magnet, air-dry the beads at room temperature for 2 min or until the pellet appears dry.
13. Remove the tube from the magnetic rack and add 50 μL of RNase- and DNase-free H_2O directly to the pellet to disperse the beads. Pipet the suspension up and down five times, then vortex the sample for 10 s, mix thoroughly, and incubate at room temperature for 5 min.
14. Pulse-spin and place the tube in the magnetic rack for at least 1 min until the solution clears. Transfer the supernatant containing the eluted DNA to a tube (50 μL) microTUBE AFA Fiber Screw-Cap.

Fragmentation of cDNA

Shear purified cDNA with a Covaris M220 Focused-ultrasonicator.

1. Place tube in M220 Focused-ultrasonicator, and sonicate to 400 bp size.
2. Fragmented cDNA is ready for library preparation for Illumina sequencing.

Library Preparation for Illumina Sequencing

The next steps of this protocol use the NEBNext® Ultra DNA Library Prep Kit for Illumina® because it can be scaled down to produce one library, and the high sensitivity is sufficient for even a single cell. A new kit has been introduced but not tested by us with a fragmentation buffer, so the Covaris sonication step would not be necessary; see https://www.neb.com/products/ for a list of all products available. Ultralow RNA input is an idea with this kit because the end-repair, dA-tailing, and linker-ligation are performed sequentially without purification between the steps, thus minimizing material loss. To avoid index hopping or index switching, a specific type of misassignment of the indexed library, we recommend the dual index primer also set by NEB.

We run an E-Gel® SizeSelect™ 2% agarose with 2 uL of the amplified library before the final purification step to see if more cycles of PCR are needed (*see* Fig. 3a). If nothing is seen on the gel, a couple of cycles are added.

Quality Control of Sequencing Libraries

1. The competed libraries are run on a high sensitivity D1000 ScreenTape (*see* Fig. 3b).
2. Qubit assay is also performed to determine the concentration of the libraries.
3. Further quality control is performed by our sequencing core before the libraries are run on the sequencer.

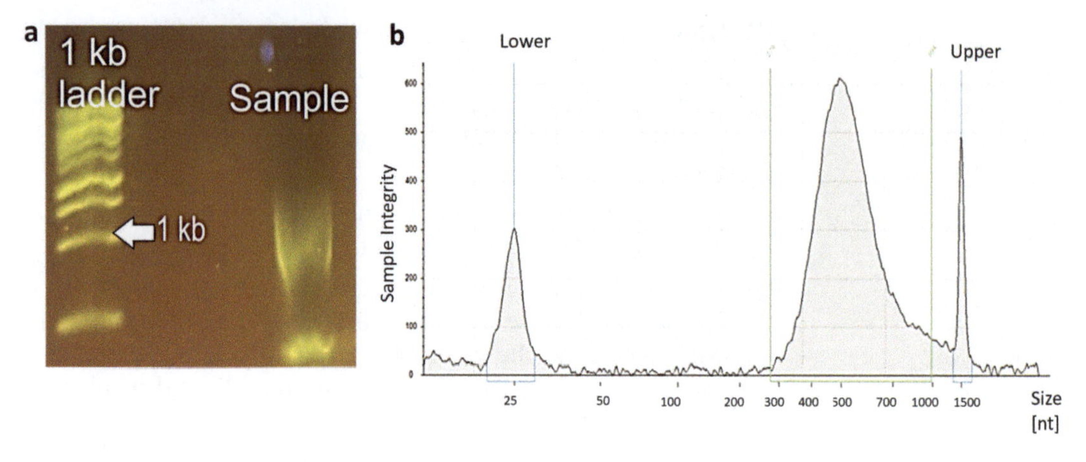

Fig. 3 (**a**) *Bolinopsis infundibulum* skin RNA-seq library run on an E-gel before SPRIselect bead purification after 16 cycles of amplification and before bead purification. The size of smear is between 500 bp and 1 kb. (**b**) Same RNA-seq library after bead purification run with a D1000 ScreenTape and Agilent 2200 TapeStation system. The average size is 542 bp with a concentration of 85 nM. Lower and upper refer to the markers provided in the kit

3.4 Conclusions

High-quality RNA can be isolated from ctenophores despite their fragile body plan and unique mesoglea challenges. Described RNA-seq libraries generate high-quality data, and we validated our libraries with hundreds of in situ hybridizations based on analysis of the RNA-seq data.

RNA-seq data analysis is challenging and not the scope of discussion here. One of the examples is https://neurobase.rc.ufl.edu/Pleurobrachia which has been described elsewhere [4]. This webpage has multiple RNA-seq datasets that have been assembled and annotated. All projects are searchable by keywords, PFAM domains, KEGG pathways, GO ontology, and BLAST. All data is also downloadable. This is the publicly available database for ctenophore transcriptomic data.

4 Notes

1. Proper technique when working with RNA is one of the most critical and challenging tasks in this process because the ramifications will considerably impact the quality of your data. First, always wear gloves and change them often because of skin RNases coined “fingerases.” Besides ensuring the workplace is contamination-free of RNases, care needs to be given to instruments and all equipment used. Metal and glass are extremely difficult to decontaminate. For metal instruments, wipe with ethanol, and then flame if possible. New plastic containers should be used instead of glass. It is best to use unopened,

petri dish sleeves, prepackages, plastic pestles for dissolving tissue if needed, unopened transfer pipettes, unopened pipette tips, and unopened bagged RNase-free LoBind Eppendorf tubes of 0.5 uL and 1.5 uL. Use reagents and nuclease-free supplies, including water and reserved chemicals for RNA work only. Designate a "low-traffic" area of the lab away from air vents or open windows as an "RNase-free zone," and place all your instruments, equipment, and supplies in this area.

2. During field trips and expeditions, we prepare QIAGEN RNA isolation kits (RNeasy) lysis buffer RLT with β-ME in RNase-free LoBind 1.5 mL Eppendorf tubes, and freeze in bags that get shipped on dry ice to locations. Both of QIAGEN's RNeasy kits, mini and micro, use the same buffer and volume, so the type of isolation can be determined in the lab. We will isolate tissue or animals for future RNA isolation on-site. Most critical in this process is the tissue for the RNA isolation must be completely dissolved in the RLT buffer before freezing and shipping. Small disposable on-time use petals can be used to help dissolve tissue. Or sonication in short bursts of 5–10 s alternated with placement on ice until the tissue is dissolved is also an alternative. Once the tissue is dissolved, freeze, and then ship samples on dry ice. If time and conditions are appropriate, RNA isolation may be performed on-site.
3. Validation of sequencing data is critical. We reserved some of the amplified cDNA for cloning genes of interest. The amplified cDNA is highly concentrated and diluted at least 1:50 and used as a template in a PCR reaction to cloning. Primers are designed from the RNA-seq data. After bands are amplified, they are purified and cloned into a cloning vector such as TOPO-TA (see https://www.thermofisher.com/us/en/home/life-science/cloning/ta-cloning-kits.html for details). The TOPO vector is ideal for probes made for in situ hybridization.
4. Master mixes save time and money. There is always a 0.5 μL overage in the volumes to accommodate pipette error. Also, it is often inaccurate to pipette very small amounts of reagent, so a master mix circumvents this issue. An example of a setup of a master mix with two, four, and six samples for the cDNA synthesis is shown here in Table 2.

Acknowledgments

We would like to thank Dr. David A. Amador Moraga and the UF/ICBR core staff for their dedication, support, and hard work to accommodate challenging sequencing demands. This work was supported in part by the Human Frontiers Science Program

Table 2
Master mix setup

1×	Reagent	2.5×	4.5×	6.5×
1.0 μL	5X first-strand buffer	5.0 μL	9.0 μL	13 μL
0.25 μL	DTT (20 mM)	0.625 μL	1.0 μL	1.625 μL
1.0 μL	dNTP mix (10 mM)	2.5 μL	4.5 μL	6.5 μL
1.0 μL	CapTS oligonucleotide (12 μM)	2.5 μL	4.5 μL	6.5 μL
0.25 μL	RNase inhibitor	0.625 μL	1.0 μL	1.625 μL
1.0 μL	SMARTScribe reverse transcriptase (100 U)	2.5 μL	4.5 μL	6.5 μL
	Add 4.5 μL of master mix per reaction			

(RGP0060/2017) and National Science Foundation (IOS-1557923) grants to L.L.M. Research reported in this publication was also supported in part by the National Institute of Neurological Disorders and Stroke of the National Institutes of Health under Award Number R01NS114491 (to L.L.M). The content is solely the authors' responsibility and does not necessarily represent the official views of the National Institutes of Health.

References

1. Nagalakshmi U et al (2008) The transcriptional landscape of the yeast genome defined by RNA sequencing. Science 320(5881): 1344–1349
2. Mortazavi A et al (2008) Mapping and quantifying mammalian transcriptomes by RNA-Seq. Nat Methods 5(7):621–628
3. Lister R et al (2008) Highly integrated single-base resolution maps of the epigenome in Arabidopsis. Cell 133(3):523–536
4. Moroz LL et al (2014) The ctenophore genome and the evolutionary origins of neural systems. Nature 510(7503):109–114
5. Whelan NV et al (2017) Ctenophore relationships and their placement as the sister group to all other animals. Nat Ecol Evol 1(11): 1737–1746
6. Moroz LL, Kohn AB (2010) Do different neurons age differently? Direct genome-wide analysis of aging in single identified cholinergic neurons. Front Aging Neurosci 2:6
7. Moroz LL, Kohn AB (2015) Unbiased view of synaptic and neuronal gene complement in ctenophores: are there pan-neuronal and pan-synaptic genes across Metazoa? Integr Comp Biol 55(6):1028–1049
8. Kohn AB et al (2013) Single-cell semiconductor sequencing. Methods Mol Biol 1048:247–284
9. Moroz LL, Kohn AB (2013) Single-neuron transcriptome and methylome sequencing for epigenomic analysis of aging. Methods Mol Biol 1048:323–352
10. Whelan NV, Kocot KM, Halanych KM (2015) Employing phylogenomics to resolve the relationships among Cnidarians, Ctenophores, Sponges, Placozoans, and Bilaterians. Integr Comp Biol 55(6):1084–1095
11. Whelan NV et al (2015) Error, signal, and the placement of Ctenophora sister to all other animals. Proc Natl Acad Sci U S A 112(18): 5773–5778
12. Matz MV (2002) Amplification of representative cDNA samples from microscopic amounts of invertebrate tissue to search for new genes. Methods Mol Biol 183:3–18
13. Zhu YY et al (2001) Reverse transcriptase template switching: a SMART approach for full-length cDNA library construction. BioTechniques 30(4):892–897

Chapter 9

Gene Expression Patterns in the Ctenophore *Pleurobrachia bachei*: In Situ Hybridization

Andrea B. Kohn, Yelena Bobkova, and Leonid L. Moroz

Abstract

In situ hybridization is a powerful and precise tool for revealing cell- and tissue-specific gene expression and a critical approach to validating single-cell RNA-seq (scRNA-seq). However, applying it to highly fragile animals such as ctenophores is challenging. Here, we present an in situ hybridization protocol for adult *Pleurobrachia bachei* (Cydippida)—a notable reference species representing the earliest-branching metazoan lineage, Ctenophora, sister to the rest of Metazoa. We provided expression patterns for several markers of cell phenotypes, as illustrated examples. The list includes predicted small secretory molecules/neuropeptides, WntX, genes encoding RNA-binding proteins (Musashi, Elav, Dicer, Argonaut), Neuroglobin, and selected transcription factors such as BarX. Both cell- and organ-specific expression of these genes further support the convergent evolution of many ctenophore innovations, which are remarkably distinct from tissue and organ specification in other basal metazoan lineages.

Key words In situ hybridization, Gene expression, Ctenophores, Genome, Wnt signaling, Dicer, Argonaut, Neuroglobin, Musashi, ELAV, Neuropeptides, *Pleurobrachia*, *Mnemiopsis*, Neurons, Evolution

1 Introduction

Despite the recent progress in comparative biology, comb jellies or ctenophores are still one of the most elusive clades among the basal Metazoa [1–3]. The phylum Ctenophora is critically important to understand the origins of animal innovations and signaling as well as to resolve the relationships among early-branching metazoans [1, 4–16].

Following the genome sequencing and assembly, we broadly used in situ hybridization (ISH) to validate genomic predictions and characterize tissue-specific expression profiling in the ctenophore *Pleurobrachia bachei* [17, 18]. However, *Pleurobrachia*, in particular, and ctenophores, in general, are incredibly challenging preparations to perform ISH and even to fix and process [19]. Their tissues are very fragile, and ISH in adults is virtually

Leonid L. Moroz (ed.), *Ctenophores: Methods and Protocols*, Methods in Molecular Biology, vol. 2757,
https://doi.org/10.1007/978-1-0716-3642-8_9, © Springer Science+Business Media, LLC, part of Springer Nature 2024

impossible for most species. Unsurprisingly, most efforts focused on larval stages using *Mnemiopsis leidyi* as a model [20].

ISH was first described in 1969 by Joseph G. Gall [21, 22]. ISH is based on a simple principle where a nucleic acid probe with a reporter localizes the complementary RNA or DNA sequence within a preserved biological tissue. The nucleic acid probe can be a double-stranded DNA (dsDNA), single-stranded DNA (ssDNA), RNA (riboprobes), or synthetic oligonucleotides such as peptide nucleic acid (PNA), morpholino, and locked nucleic acid (LNA). The molecular reporter for the probes can be radioactive isotopes such as ^{32}P, ^{35}S, or, ^{3}H or nonradioactive labels including biotin, digoxigenin (DIG), and fluorescent dyes (for fluorescent in situ hybridization—FISH). Detection of the reporter attached to the in situ probes allows visualization of a specific DNA or RNA sequence in a cell, tissue section, or whole animal. As a result, cell/tissue gene expression can be estimated using highly specific riboprobes, and numerous approaches have been successfully used both in developmental biology and neuroscience.

We present an ISH protocol successfully tested using several fragile and gelatinous ctenophores like *Pleurobrachia bachei* as an illustrative example (*see* also [23, 24]). This protocol is derived from work on other invertebrates [25], updated and revised [26], and validated [17, 18]. Our ISH protocol can be easily adapted to other fragile animals as well.

2 Materials

2.1 Reagents

2.1.1 Probe Generation Reagents

- Not I-HF with Buffer 4, BSA (Cat # R0189S, New England BioLabs).
- PmeI with Buffer 4, BSA (Cat # R0560S, New England BioLabs).
- pCR®4-TOPO Vector (Cat # K4575-J10, Invitrogen/Life Technologies).
- DIG (digoxigenin) RNA Labeling Mix (Cat # 11277073910, Sigma Aldrich).
- Fluorescein RNA Labeling Mix (Cat # 11685619910, Sigma Aldrich).
- T3 RNA polymerase (Cat # 11031163001, Sigma Aldrich).
- T7 RNA polymerase (Cat # 10881767001, Sigma Aldrich).
- Turbo DNaseI (Cat # AM1907, ThermoFisher Scientific).
- 7.5 M Lithium Chloride (Cat # 9480, ThermoFisher Scientific).
- MiniElute PCR Purification Kit (Cat # 28004, Qiagen).
- Qubit® RNA Assay Kit (Cat # Q32852, ThermoFisher Scientific).

- Qubit® DNA Assay Kit (Cat # Q32850, ThermoFisher Scientific).
- RNAse Inhibitor (Roche # 03-335-399-001).

2.1.2 *In Situ* Hybridization Reagents

- 10× phosphate buffered saline (PBS) (Cat # BP399-1, ThermoFisher Scientific).
- Formaldehyde 37% (Cat # BP531-500, ThermoFisher Scientific).
- Methanol (MeOH) (Cat # BP1105-1, ThermoFisher Scientific).
- Ethanol, 100% (EtOH) (Cat # NC9789925, ThermoFisher Scientific).
- Triton X 100 (Cat # NC9903183, ThermoFisher Scientific).
- Tween-20 (Cat # BP337-100, ThermoFisher Scientific).
- 1 M $MgCl_2$, RNase-free (Cat # 9530G, ThermoFisher Scientific).
- 1 M Tris pH 8.0 (Cat # AM9856, ThermoFisher Scientific).
- 0.5 M EDTA pH 8.0 (Cat # 9260G, ThermoFisher Scientific).
- 5 M NaCl, RNase-free (Cat # 9759, ThermoFisher Scientific).
- Yeast tRNA (Cat # 15401-029, ThermoFisher Scientific).
- SSC 20× (saline-sodium-citrate buffer) (Cat # 15557-036, ThermoFisher Scientific).
- Denhardt Solution 50× (Cat # D2532, Sigma Aldrich).
- Goat Serum (Cat # G9023-10ML, Sigma Aldrich).
- Levamisole (Cat # 31742, Sigma Aldrich).
- Albumin from bovine serum, BSA (Cat # A9647-50G, Sigma Aldrich).
- Sodium dodecyl sulfate (SDS), 20% solution (Cat # 75832, Affymetrix).
- NBT (4-nitro-blue-tetrazolium-chloride)/BCIP (5-bromo-4-chloro-3-indolyl-phosphate) stock solution (Cat # 11681451001, Sigma Aldrich).
- Anti-digoxigenin-alkaline phosphatase (AP), Fab fragments (Cat # 11093274910, Sigma Aldrich).
- Anti-fluorescein-AP, Fab fragments (Cat # 11426338910, Sigma Aldrich).
- BM Purple AP substrate (Cat # 11 442 074 001, Sigma Aldrich).
- Fast Red TR/Naphthol AS-MX AP substrate tablet set (Cat # F-4523, Sigma Aldrich).
- Vector Red AP Substrate kit (Cat # SK-5100, Vector Laboratories).

- TSA Plus Fluorescein kit (Cat # NEL741E001KT, PerkinElmer).
- Methyl salicylate (Cat # M6752-250ML, Sigma Aldrich).
- Permount (Cat # SP15-100, ThermoFisher Scientific).
- VECTASHIELD Mounting Medium (Cat # H-1000, Vector Laboratories).
- RNase-free and DNAase-free water (Cat # 10977-015, ThermoFisher Scientific).

2.2 Equipment

- Qubit® 2.0 Fluorometer (Cat # Q32866, Life Technologies).
- Belly Dancer orbital shaker (Cat # Z377554, Sigma Aldrich).
- Shaking incubator (Cat # H-7700-5, BioExpress).
- 24-well plate (Cat # 720084, ThermoFisher Scientific).
- Nalgene sterile disposable filter units 0.2 μM (Cat # 097403A, ThermoFisher Scientific).
- Nalgene syringe filter, 0.2 μM (Cat # 192-252-0, ThermoFisher Scientific).
- 60 mL syringes (Cat # 13-689-8, ThermoFisher Scientific).
- Corning 50 mL centrifuge tubes (Cat # 05-538-68, ThermoFisher Scientific).
- Dissecting tools.
- Microscope of choice.

2.3 Stock Solution Preparation

Since this protocol involves working with RNA, all solutions need to be RNase-free (*see* **Note 1** for tips on working with RNA). For dilutions, Milli-Q H_2O can be used because it is RNase- and DNase-free. We prepare most working solutions in Corning 50 mL centrifuge tubes and store them at the appropriate temperature, 4 °C or −20 °C.

2.3.1 Stock Solutions

- Prepare a liter of filter sea water (FSW) (*see* **Note 2** for preparation of FSW).
- Make three separate concentrations of methanol (MeOH). However, we prefer to use PTW, not water, to dilute the MeOH.
- 30%, 50%, 70% MeOH
- Add 10 mL, 20 mL, and 30 mL MeOH to different Corning 50 mL tubes.
- Fill each to 40 mL Milli-Q H_20.

- 1× PBS
- 5 mL 10× PBS
- Fill to 50 mL with Milli-Q H_2O.

- Four percent formaldehyde/paraformaldehyde in FSW or 1× PBS (*see* **Note 3** for a comparison of formaldehyde and paraformaldehyde)
- 5.4 mL of 37% formaldehyde/paraformaldehyde
- Fill to 50 mL with FSW or 1× PBS.

- PTW
- 50 μL Tween-20
- Fill to 50 mL with 1× PBS.
- Hybridization buffer
- 50% formamide: 25 mL formamide
- 5 mM EDTA: 0.5 mL 0.5 M EDTA
- 5× SSC: 12.5 mL 20× SSC
- 0.1% Tween-20: 50 μL Tween-20
- 0.5 mg/mL tRNA: 25 mg tRNA (*see* **Note 4** for preparation of tRNA)
- 1× Denhardt: 1 mL of 50× Denhardt (0.02% Ficoll, 0.02% polyvinylpyrrolidone, 0.02% BSA).
- Fill to 50 mL with Milli-Q H_2O.

- PBT (250 mL)
- 1× PBS: 25 mL (10× PBS)
- 0.1% Triton X 100: 0.25 mL
- BSA (albumin) 2 mg/mL: 500 mg
- Fill to 250 mL with Milli-Q H_2O.

- 10% goat serum (*see* **Note 5** for preparation and storage of goat serum)
- 0.4 mL of GS
- Fill to 4 mL with PBT.

- 1% goat serum
- 0.04 mL GS
- Fill to 4 mL with PBT.

- Detection Buffer (**Critical; Always Make Fresh**)
- 100 mM NaCl: 1 mL of 5 M NaCl.
- 0.1% Tween-20: 50 μL of Tween-20
- 50 mM $MgCl_2$: 2.5 mL of 1 M $MgCl_2$
- 100 mM Tris: 5 mL of 1 M Tris pH = 8

- 1 mM levamisole (*see* **Note 6** for preparation of levamisole)
- Fill to 50 mL with Milli-Q H_2O and then filter.
- NBT (4-nitro-blue-tetrazolium-chloride)/BCIP (5-bromo-4-chloro-3-indolyl-phosphate) Stock solution (*see* **Note 7** for alternative preparations of NBT/BCIP)
- *Fluorescent Label*
- Developing solution
- 100 mM NaCl: 1 mL of 5 M NaCl
- 0.2% Tween-20: 100 μL of Tween-20
- Fill to 50 mL with Milli-Q H_2O.
- Adjust pH to 8.2–8.5 with 10 M NaOH; filter.
- Detection buffer—Vector Red Alkaline Phosphatase kit
- To 5 mL of developing solution, add the following:
- Two drops of reagent 1; mix.
- Two drops of reagent 2; mix.
- Two drops of reagent 3; mix.
- Inactivation Buffer
- 10 mM HCl: 100 μL 5 N HCl
- 0.2% Tween-20: 100 μL Tween-20
- Fill to 50 mL Milli-Q H_2O.
- Prepare hazardous waste bottles for all reagents according to regulations.

3 Methods

3.1 Riboprobe Generation for ISH

The generation of an ISH riboprobe depends on the vector used to clone the transcript and eventually make a probe. The vector determines the restriction sites used for linearizing the plasmid and the choice of polymerase to synthesize the RNA. One can use a PCR product to generate a probe too, but PCR amplification would be required every time to synthesize a probe. This is impractical if many repetitive experiments are designed at different intervals. We recommend making bacterial stocks of all clones, so bacterial culture can be grown and plasmid isolated to make the probes as needed. All riboprobes are constructed in the antisense direction.

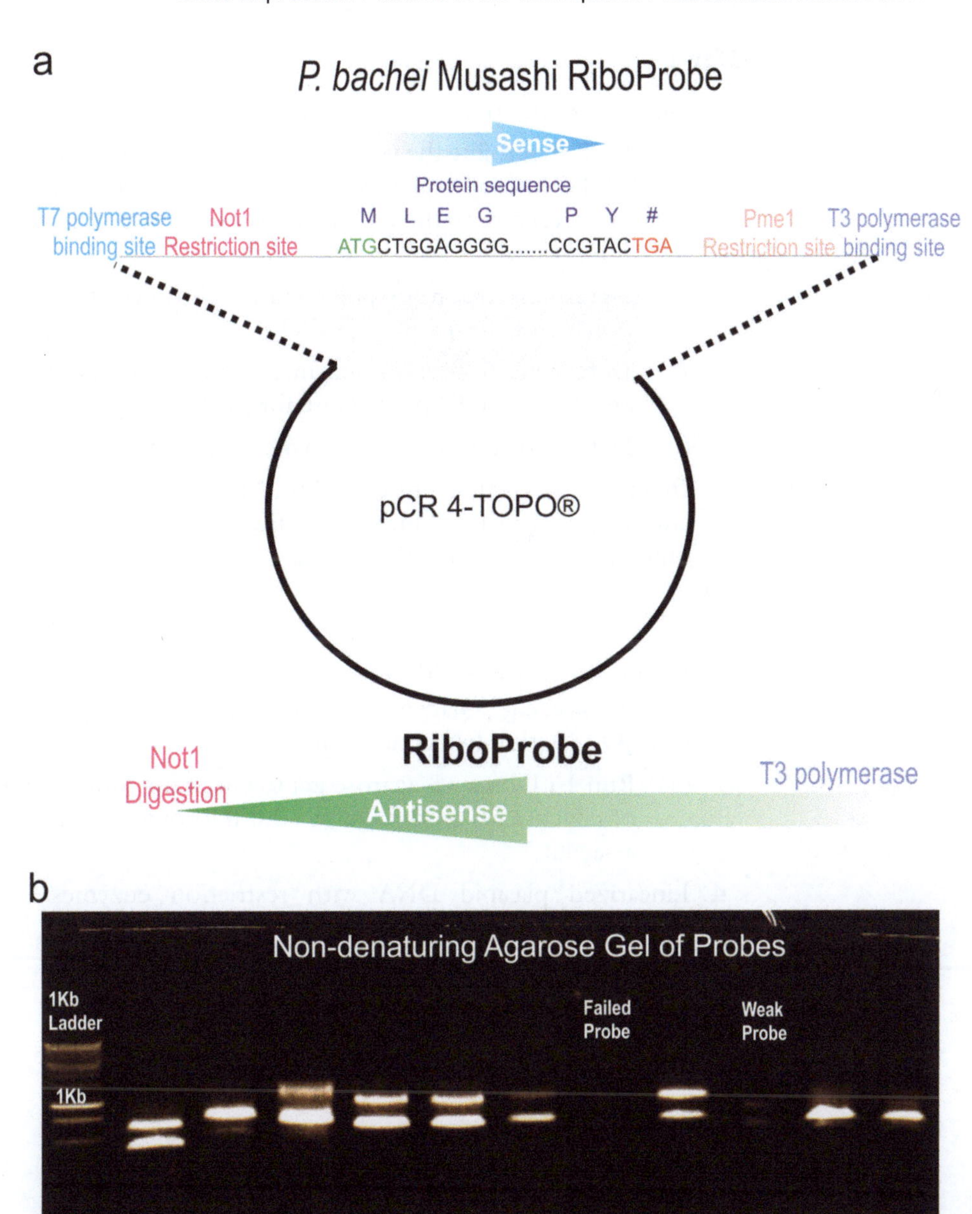

Fig. 1 Strategy for construction of a probe and quality control. (**a**) Probe construction. There are three key steps to making a probe. (**1**) The selection of a target sequence and the vector for cloning: Here, we use pCR®4-TOPO. (**2**) Sequencing with either the T7 or T3 polymerase primer (these sites are specific to the vector). Then verify the direction of the insert. Is it in the sense or antisense direction relative to the primer used to sequence? To determine the sense or antisense direction, the predicted amino acid sequence is obtained from the cloned sequence. In this case, the sequence primer was T7, and the direction of the insert is in the sense direction relative to the predicted protein sequence. (**3**) Making the probe. An antisense riboprobe is generated with the opposite side polymerase, T3. The plasmid is first linearized with the restriction site at the 3′ end of the coding sequence. In this example, Not1 is used. Now, if the sequence would had been in the antisense direction relative to the sequencing primer, the riboprobe would had been made with the T7 polymerase and the plasmid and would had been linearized with Pme1. (**b**) Quality control of a probe. The integrity of the probe is evaluated on a non-denaturing agarose gel. Some probes will have two bands because

Day 1

1. Determine the strategy for making the probe as in Fig. 1a
   (a) First, determine the vector the clone is in. We use pCR®4-TOPO most frequently. Determine the direction of the insert relative to the vector. Antisense must be selected for riboprobe.
   (b) Determine the RNA polymerase: T3 polymerase or T7 polymerase for pCR®4-TOPO.
   (c) Determine restriction enzyme: Notl for T3 polymerase and PmeI for T7 polymerase for pCR®4-TOPO.
   (d) Determine if restriction enzyme cuts insert.
2. Inoculate a 3 mL culture of Luria broth with appropriate antibiotic and clone of interest for the probe, then grow overnight at 37 °C with gentle shaking.

Day 2

3. Isolate the plasmid with miniprep from culture (use miniprep kit of choice, e.g., Qiagens, QIAprep) elute with 50 μL of Milli-Q H_2O (30 μL if low concentration).
   (a) Run 1 μL on a 1% agarose gel to check for concentration, or check 2 μL on the Qubit® 2.0 fluorometer with DNA assay kit.
4. Linearized plasmid DNA with restriction enzymes using Notlor PmeI depending on direction.

   × μL for 1 μg plasmid DNA

   5.0 μL 10× buffer #4

   1.0 μL Not-HF (200 units) or PmeI (200 units) enzyme both using buffer 4

   0.5 μL of 100× BSA

   × μL Milli-Q H_2O (add to a total of 50 μL)

   50 μL total
5. Incubate at 37 °C for 1 h.
   (a) Run a 2 μL on a 1% agarose gel to see if it is linearized; also run 2 μL of uncut plasmid to compare.
   (b) Purify the linearized plasmid to remove enzyme and salts using MiniElute PCR Purification Kit.

Fig. 1 (continued) the RNA is not denatured, and the true size cannot be determined. One should look to see degradation and weak or the absence of the signal (if the probe failed). An alternative and convenient way is to use the Qubit® 2.0 fluorometer to determine the concentration and the Agilent TapeStation to determine the quality of the probe

(c) Add five volumes (250 μL) of PB; mix well and place in the column provided.

(d) Centrifuge for 1 min at 1000 g, and then turn to maximum speed for 1 min discarding flow-through (*see* **Note 8** pertaining to centrifuge timing).

(e) Place column back in the tube and add 750 μL of PE.

(f) Centrifuge at maximum speed for 1 min; discard flow-through.

(g) Then centrifuge at maximum speed for an additional 1 min to dry column.

(h) Place column in a new 1.5 mL tube and add 10–14 μL of Milli-Q H_2O, depending on concentration.

(i) Incubate for 1–3 min.

(j) Centrifuge for 1 min at 1000 g, and then turn to maximum speed for 1 min (*see* **Note 8** pertaining to centrifuge timing).

(k) The tube now contains linearized plasmid and template for transcription; store at −20 °C.

6. Run 1 μL on a 1% agarose gel to check for concentration or check 2 μL on Qubit® 2.0 fluorometer with DNA assay kit (can stop here).

Day 3 (can also be done on Day 2 if time permitting)

7. In vitro transcription reaction (for digoxigenin or fluorescein).

   × μL for 1 μg linearized plasmid DNA

   2.0 μL 10× transcription buffer (in polymerase kit)

   2.0 μL 10× DIG RNA labeling mix

   Or 2.0 μL 10× fluorescein RNA labeling mix

   2.0 μL RNA polymerase, 20 units/μL (T3 or T7)

   1.0 μL RNAse inhibitor

   × μL Milli-Q H_2O (add to total of 20 μL)

   20 μL total

8. Incubate at 37 °C for 2 h.

   (a) Check 1 μL on a 1 % non-denaturing agarose gel for concentration.

9. Add 2 μL TURBO 10× buffer and 1 μL of TURBO DNase (two units/μL); incubate for 20–30 min at 37°C (*see* **Note 9** for reducing background).

10. Add 2.3 μL of resuspended DNase inactivation reagent, and incubate for 5 min at room temperature (RT). Centrifuge at 10,000 g for 1.5 min and transfer to clean 1.5 mL tube.

11. Run 1 μL on a 1% non-denaturing agarose gel for concentration (*see* Fig. 1b for examples of good and bad probes).
12. Check 1 μL on Qubit® 2.0 fluorometer with RNA assay kit; use 0.1–1.0 μg/mL in next step.
13. The timing of the probe construction can be performed concurrently with the first day of the ISH protocol. However, caution needs to be taken that a high-quality probe is prepared and ready when the prehybridization is started because the protocol cannot be delayed after this point.

3.2 Whole-Mount ISH of the Ctenophore, Pleurobrachia bachei

We used adult *P. bachei* animals, which can be small and sturdy to withstand fixation. These animals have very little connective tissue; therefore, challenges with permeability issues are minimal. Our protocol is broken up into 5 days which can be shortened if the time dedicated to the protocol is flexible. We include colorimetric and fluorescent development. This protocol was validated using more than 200 in situ hybridizations.

Make all stock solution (Subheading 2.3) in advance of starting this protocol except where noted. All washes are at room temperature (RT) unless otherwise noted. A flow diagram of the whole protocol with a timeline is in Fig. 2.

Day 1 (fixation of the specimen)

1. Fix the whole specimen in 4% formaldehyde in filtered sea water (FSW) overnight at 4 °C (*see* **Note 3**). The FSW should be, if possible, the same water animals are found in the wild.
2. Place no more than ten animals in a 50 mL conical tube. Critical is all the animals are covered with liquid. To mix, hold on the side and rotate gently (*see* **Note 10** for mixing strategy).

Day 2 (dehydration of specimen)

3. Rinse 3× for 10 min in PTW (PBST) at RT.
4. To mix, hold on side and rotate gently. Dispose of all solutions in hazardous waste properly.
5. Wash in 1:1 methanol (MeOH)/PTW (to equilibrate to MeOH) 10 min at RT.
4. Place specimen tube on its side in 100% MeOH and store at −20 °C for at least 2 h, but no more than a week (*see* **Note 11** for reducing background)

Day 3 (rehydration of specimen and hybridization)

6. Rehydrate specimens for 10 min each in MeOH/PTW 3:1, 1:1, 1:3, and 0:1 at RT.

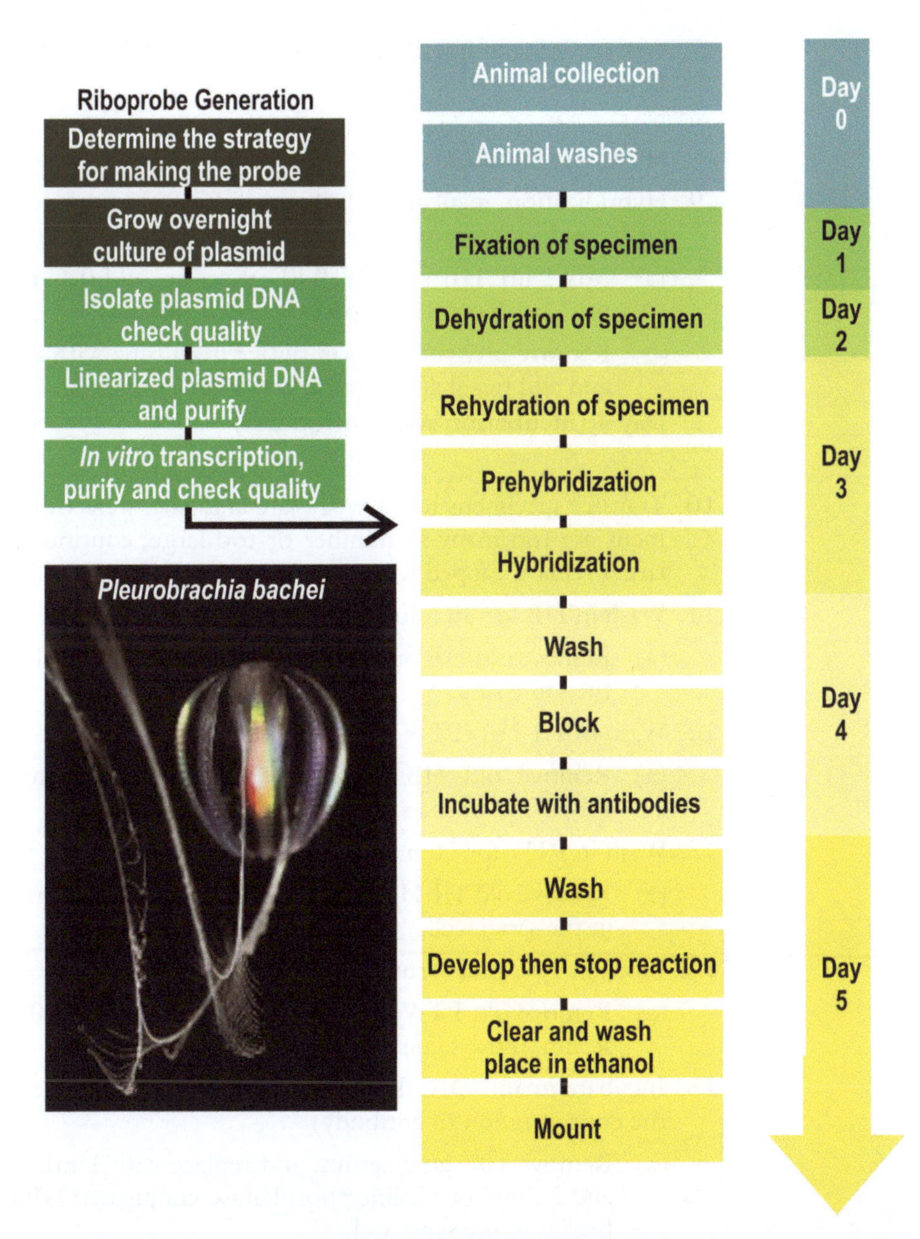

Fig. 2 Flow diagram for in situ hybridization with a timeline. Different steps to the ISH are indicated in the boxes. Colors correlate to timing and the appropriate days. The timing of this protocol can be variable. For first-time users, it is best if the probe is made and ready to be used before the in situ hybridization protocol starts because if your probe is not of acceptable quality, nothing can proceed. Once one is experienced, probe generation can be done in parallel with the Day 1 of the in situ hybridization protocol. Day 1 can be a combination of collecting and washing animals followed by fixation of the specimens overnight. The next day can be a combination of Day 2 and 3, so the dehydration, hydration, and hybridization can be performed on the same day as long as the appropriate times are followed as suggested in the protocol. The next day is a short day with washing and incubation of antibodies. We do not suggest trying to shorten this step. The development is going to be the most variable for timing. Development can be very rapid for something like tubulin or a neuropeptide or very slow for a receptor or channel and depends on the expression level of your transcript of interest. Be prepared to have a long day of development for a successful result

7. Wash in 1:1 solution of hybridization buffer (HB) and PTW for 15 min at RT.
8. Prehybridization, incubate in HB (no probe) buffer for 1 h at 60 °C (*see* **Note 12**).
9. Hybridization, incubate in HB with DIG-RNA probe O/N at 60 °C and rock gently if possible.
   (a) Mix 1 mL HB with 2–10 μL of probe (add 0.1–1.0 μg/mL) (*see* **Note 13** on probe concentration).
   (b) Remove prehybridization buffer from tube with animals, and add freshly prepared HB to cover specimen.

Day 4 (incubation with antibodies)

10. Transfer specimens to 24-well plate and label wells or if specimens are too many in number or too large, continue in the larger conical capped tubes (*see* **Note 14** on 24-well plates).
11. Wash in HB for 30 min at 60 °C.
    (a) Remove old HB buffer and replace with 1 mL fresh HB in the same well.
12. Wash in 1:1 HB/PTW for 30 min at 60 °C.
    (a) Remove old HB buffer and replace with 1 mL 1:1 HB/PTW in the same well.
13. Wash in PTW for 30 min at RT.
    (a) Remove old 1:1 HB/PTW and replace it with 1 mL PTW in the same well.
14. Block in 10% goat serum (GS) for 60 min at RT.
    (a) Remove old PTW and replace it with 1 mL of 10% goat serum in the same well.
15. Incubate in anti-DIG 1/2000 at 4 °C O/N (*see* **Note 15** on the concentration of antibody).
    (a) Remove 10% goat serum, and replace with 1 mL 1% GS and 1:2000 of alkaline phosphatase-conjugated DIG antibodies in the same well.

Day 5 (development)

16. Wash 4× 30 min in PBS at RT.
    (a) Make detection buffer and aliquot 1 mL into a clean well for each sample. When ready to develop, add 20 μL of NBT/BICP mix until dissolved. It should be yellow in color! Be sure the NBT/BCIP is fully dissolved before adding samples. *Now* add samples. **Put on ice and cover with tin foil** (*see* **Note 16** about detection buffer).

17. Watch for appropriate color development (*see* **Note 17** for light considerations).
18. Prepare wells with PBS transfer animals to fresh wells with PBS to stop (*see* **Note 18** about crystal formation).
19. Wash in 4% formaldehyde in MeOH 30 min at RT (*see* **Note 19** for clearing times).
    (a) The timing of this step depends on the strength of the probe signal and the background. This time is for a high probe signal with low background.
20. Wash 3× 10 min in 100% ethanol (EtOH) at RT.
21. Store in 100% EtOH at 4 °C.
22. Mount.
    (a) Add animals to methyl salicylate until they sink in a vial (*see* **Note 20** for working with methyl salicylate).
    (b) Put animal onto microscope glass, clean, absorb methyl salicylate leftovers, add a drop of Permount, and put on the coverslip.

Figures 3, 4, 5, and 6 illustrate in situ hybridization experiments with *permanent* markers for selected genes.

3.2.1 Fluorescent Detection

An alternative to the colorimetric single labeled DIG system is fluorescent detection using Vector Red Alkaline Phosphatase kit. Fluorescent detection can be with direct or indirect detection. This protocol is quick and gives excellent results. The TSA Plus Fluorescein kit also gave excellent results.

Day 4 (a direct method of detection)
Follow the above protocol for Day 4 up to **step 14** and proceed with the following:

1. Wash 5× 15 min with PBT at RT.
2. Make detection buffer and aliquot 1 mL into a clean well for each sample at RT.
3. Watch for appropriate color development. The vector red product is visible by the eye (it can be used for nonfluorescent ISH) and in the rhodamine, Cy3 channel (*see* **Note 21**).
4. Stop the reaction by placing PBT.
5. Wash 6× 20 min in PBT (~2 h of washes).
6. Mount in VECTASHIELD mounting medium.
7. For viewing counterstains with DAPI to visualize nuclei.

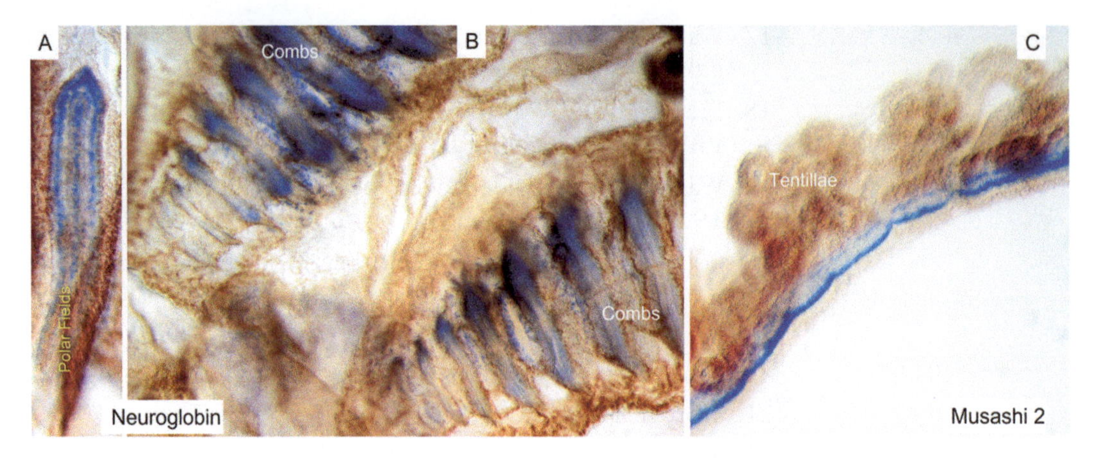

Fig. 3 *Pleurobrachia* Neuroglobin (**a**, **b**) and Musashi-2 (**c**) gene expression patterns. In situ hybridization was performed on whole-mount preparations using DIG-labeled probes (blue staining) for two specific genes (*see* details in [18]). High expression levels are observed in nonneuronal cells within comb plates and tentacles. Neuroglobin is also expressed in the cells of the polar fields, known as putative chemosensory structures in ctenophores

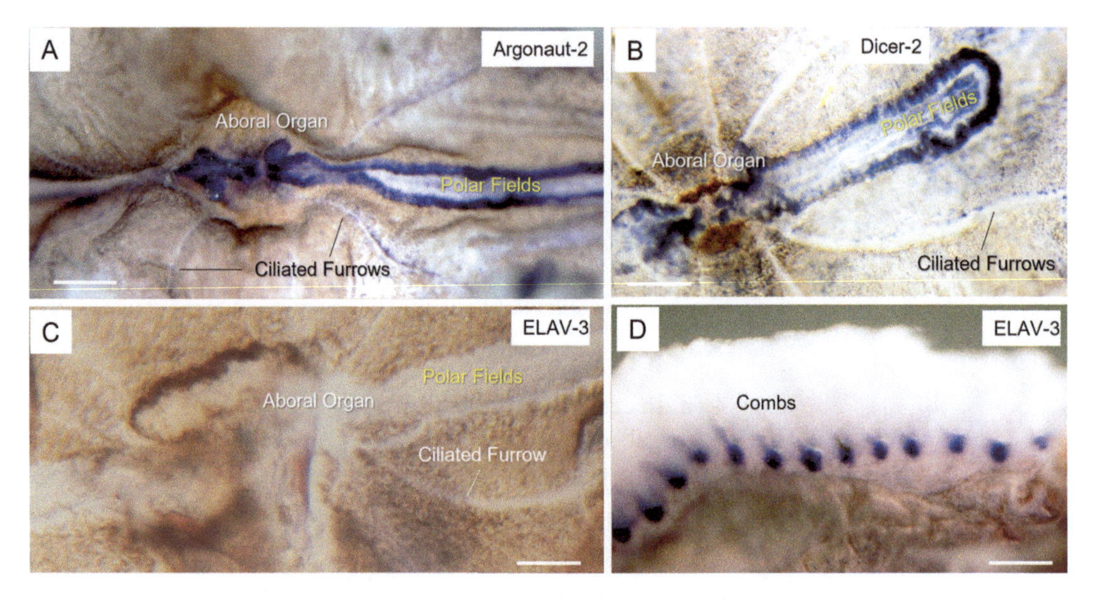

Fig. 4 Argonaute (**a**, JN202326) and Dicer-2 (**b**, JN202325) genes are selectively expressed in the aboral organ, polar fields, and cells close to the ciliated furrows and skin (weak expression) of *Pleurobrachia bachei*. All these structures are associated with sensory and integrative functions. *Pleurobrachia* ELAV-3 (JN202319, RNA-binding protein recognized as pan-neuronal markers in many bilaterians) is not expressed in neurons or organs enriched with neurons such as the aboral organ and polar fields or cells with a neuronal-like appearance (**c**). However, the highest levels of ELAV-3 expression were detected in the adult comb plates (**d**); modified from [17]. In situ hybridization was performed on whole-mount preparations using DIG-labeled probes (blue staining). Scale bars: 500 μM

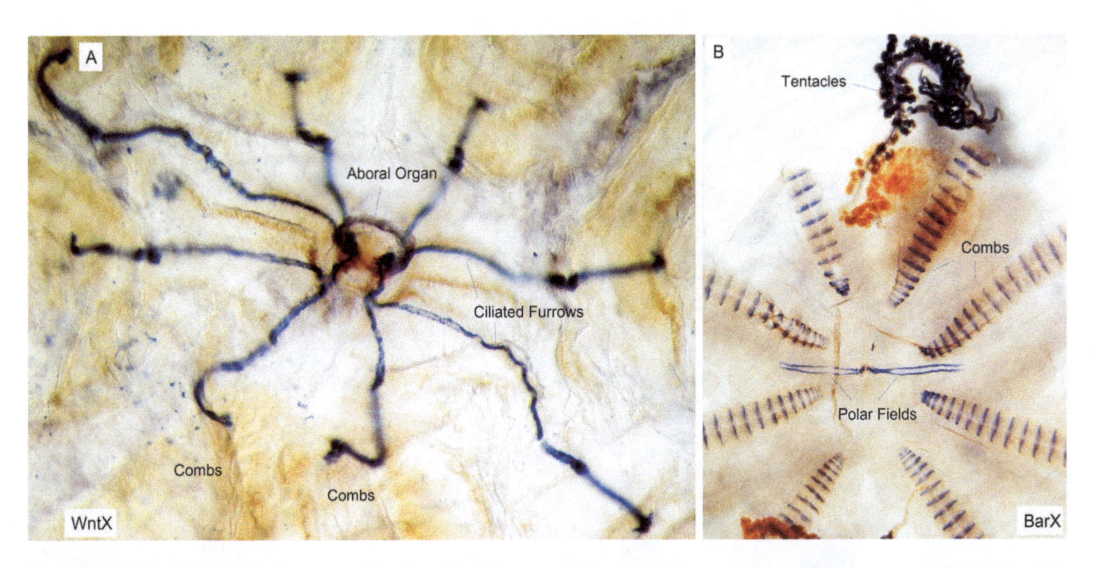

Fig. 5 WntX is selectively expressed in the aboral organ (AO) and major conductive pathways of *Pleurobrachia* (**a**), indicating its involvement in integrative and neural-like functions (in situ hybridization on a whole-mount preparation using DIG-labeled probes (blue staining)). One of the highest WntX expressions is found in AO and ciliated furrows, whereas the polar fields showed a moderate expression level associated with their central regions. (**b**) The homeobox transcription factor BarX is differentially expressed in tentacles, comb plates, and polar fields. (Modified from [17])

3.2.2 Double/Fluorescent Label

Multiple simultaneous hybridizations can be performed with this same protocol by using combinations of digoxigenin-, biotin-, and fluorochrome-labeled (fluorescein) probes to different transcripts of interest. Such multiprobe experiments are possible because of the different fluorescent dye coupled antibodies, including fluorescein or FITC (fluorescein isothiocyanate; yellow), rhodamine or TRITC (tetramethylrhodamine isothiocyanate; red), and AMCA (amino-methylcoumarin acetic acid; blue).

Figures 7 and 8 illustrate in situ hybridization experiments with fluorescent markers for selected genes.

4 Notes

1. Obtaining high-quality, intact RNA is the first and often the most critical step in performing successful ISH experiments. Labeled RNA is produced during the in vitro transcription reaction. It is crucial to be aware of the risk of RNase contamination. RNases are found in practically every cell type and can be tough enzymes to inactivate. The primary sources of RNases within most laboratory environments are microorganisms such as bacteria, fungi, their spores, and human contamination. Some basic precautions include wearing gloves throughout the experiment and changing them if they have come in contact with skin. Designate an RNase-free area of work to wash the

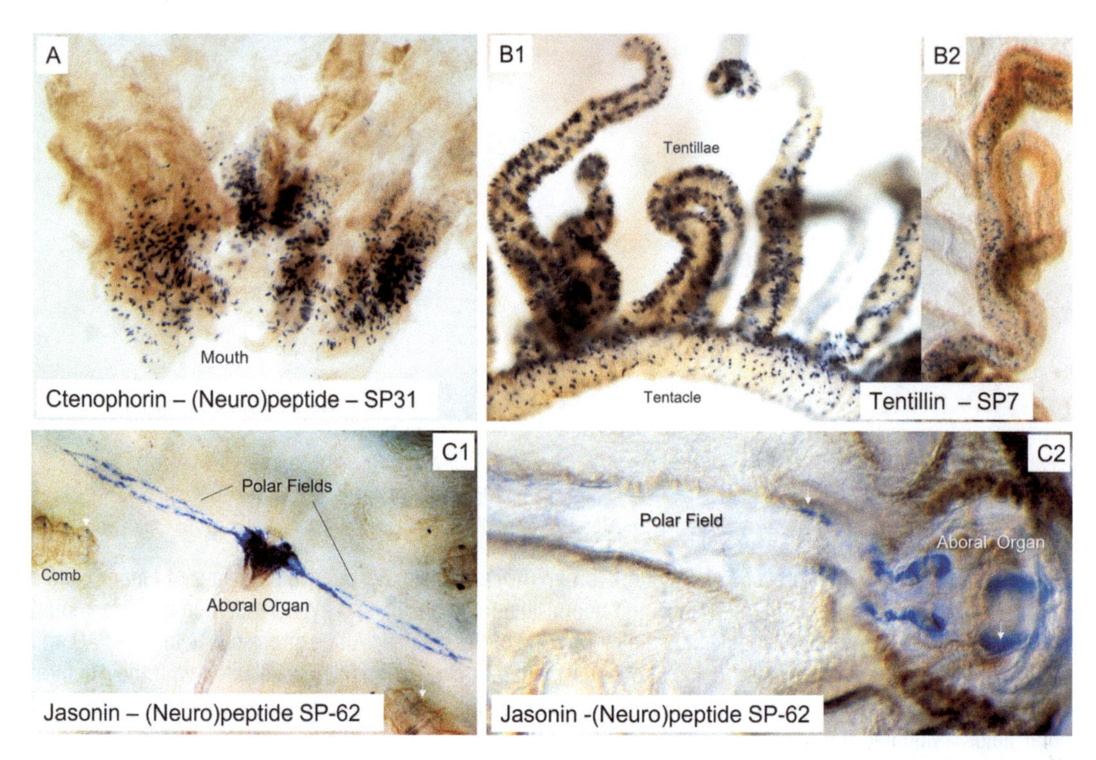

Fig. 6 Expression of genes encoding putative secretory molecules in *Pleurobrachia bachei* (DIG-labeled in situ hybridization (ISH)). We selected these transcripts using a computational pipeline and RNA-seq profiling from different adult tissues. (Modified from [17]). Ctenophorin (**a**), or secretory (neuro)peptide 31 (SP-31, JQ700341), is uniquely expressed in polarized cells around the mouth of *Pleurobrachia*, and we found its homologs in all ctenophore species we sequenced [17]. Tentillin, (**b1**, **b2**) (SP-7, JQ700317) is a *Pleurobrachia*-specific gene, uniquely expressed in polarized secretory-like cells in tentillae and tentacles. Jansonin (**c1**, **c2**) (SP-62, JQ700372) is also a species-specific gene, primarily expressed in the aboral organ and polar fields

benchtop after each experiment with RNaseZap or similar products. It is good to have a dedicated set of pipettors and equipment used solely for RNA work. Use fresh packaged filtered pipette tips and tubes guaranteed to be RNase-free. It is also possible to order and use RNase-free chemicals and reagents. Liquid solutions can be treated with diethylpyrocarbonate (DEPC); however, DEPC degradation products can inhibit in vitro transcription; therefore, obtaining RNase-free (non-DEPC treated) reagents is best. By definition, Milli-Q water is RNase-free if the equipment is maintained correctly.

2. Prepare at least a liter of filter sea water (FSW) with a Nalgene sterile disposable filter unit and 0.2 μM filter. If possible, use the same water the animals were obtained or use the water the animals were collected and shipped. Store capped containers at 4 °C for a week, and check each time for any bacterial growth.

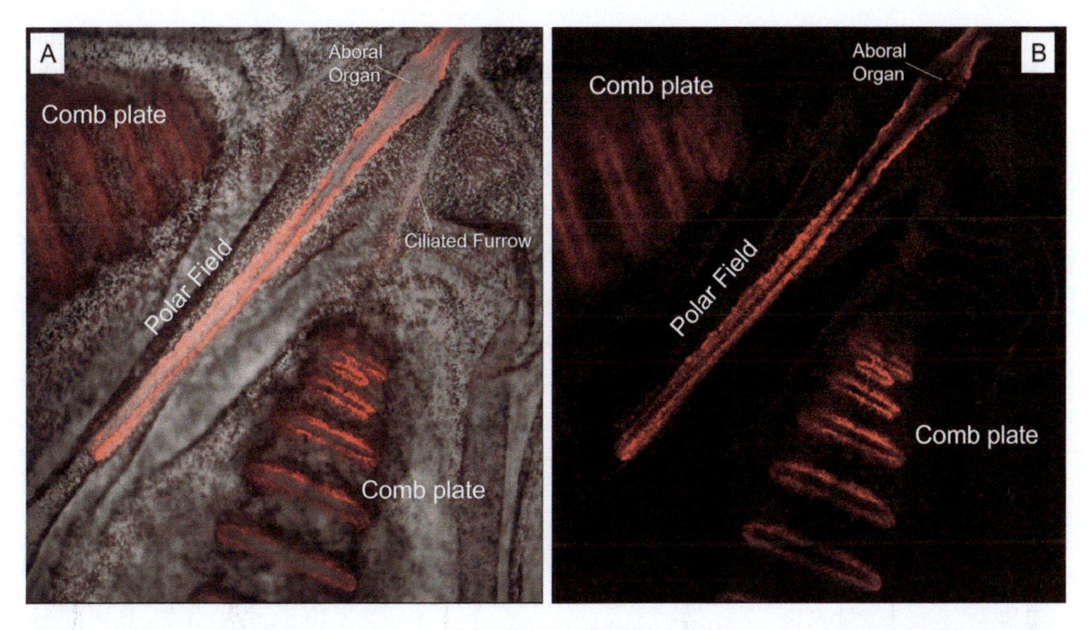

Fig. 7 β-Tubulin mRNA is differentially expressed in the polar fields (PFs) and comb rows of *Pleurobrachia bachei*. We recommend that the fluorescent in situ hybridization (ISH) (FISH) protocol should be first tested with highly abundant genes (e.g., β-tubulin) as a positive control of the procedure. The bright staining of the same structures can be observed even with background illumination using a convenient transmission light source (**a**) to visualize all structures in this whole-mount preparation from the aboral side. The protocol was performed with a probe labeled with Dig. However, the Vector Red Alkaline Phosphatase kit was used for development and then visualized on the confocal microscope (**b**)

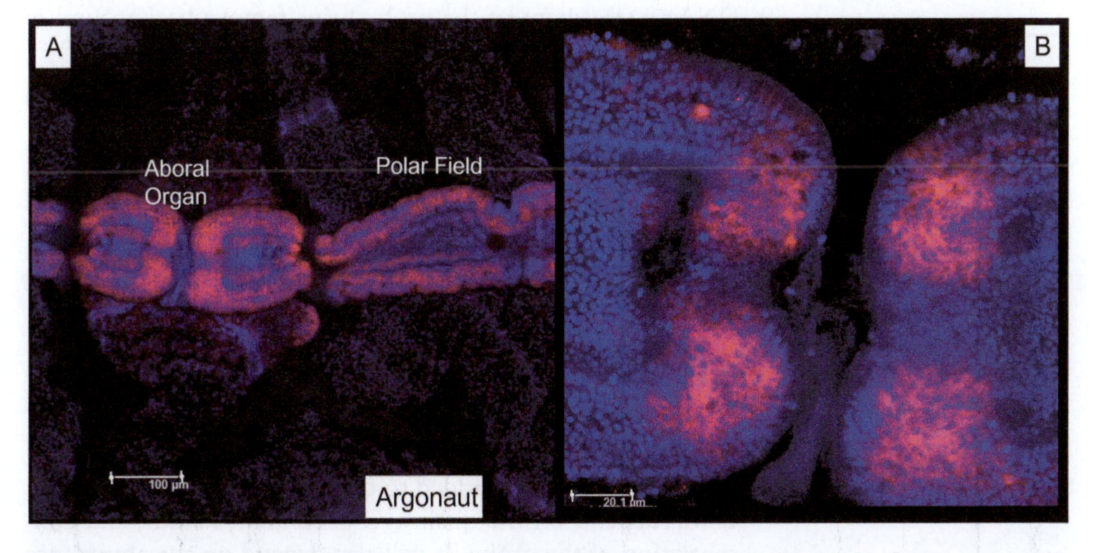

Fig. 8 Differentially expressed Argonaute in *Pleurobrachia bachei* (red labeling with fluorescent in situ hybridization (FISH) protocol). Dark blue fluorescence is DAPI nuclear staining. Low magnifcation of the aboral organ and polar field (**a**) and high magnification of the aboral organ (**b**)

3. The distinction between paraformaldehyde (PFA) and formaldehyde is that PFA is the polymerization product of formaldehyde. PFA comes in small glass ampoules and is more expensive than formaldehyde, but both have similar results for ISH. Working with PFA or formaldehyde requires a well-ventilated area or hood.
4. ThermoFisher Scientific tRNA is convenient because 25 mg of tRNA comes in individual vials. We add 1 mL of RNase-free H_2O and dissolve it according to manufacture recommendations.
5. Goat serum arrives frozen in large quantities. Make 1.0 mL aliquots in 1.5 mL centrifuge tubes, and keep frozen at −20 °C until ready to use. This eliminates the time of freeze-thaw and possible contamination. We do the same thing for 50× Denhardt.
6. Prepare 200 mM levamisole in the receiving bottle, and then aliquot 0.25 mL in 1.5 mL centrifuge tubes; keep frozen at −20 °C until ready to use.
7. We prefer the NBT (4-nitro-blue-tetrazolium-chloride)/BCIP (5-bromo-4-chloro-3-indolyl-phosphate) stock solution because everything is premixed to the appropriate concentration, and it is in ready to use liquid form. Keep reagent in the dark.
8. We centrifuge first at a slower rate to allow the liquid to absorb into the filter. Then we increase the speed to elude the product.
9. DNAase treatment of the probe helps remove background in ISH.
10. Gently rotating the animals in the Corning 50 mL tubes will provide better results.
11. The longer the animals are stored (usually after 2 weeks) in 100% MeOH at −20 °C, the greater chance of background developing.
12. We store about ten animals (depending on size) per Corning 50 mL tube in 100% MeOH at −20 °C until prehybridization. We do not use more than three animals per probe. We place one to three animals in autoclaved test tubes 16 × 125 mM or new plastic tubes similar in size. After the prehybridization, the animals/tissue samples might shrink.
13. The amount of probe added to the detection buffer is variable. It will depend on the quality, G-C content, length, and concentration of the probe. It will also depend on the abundance of the transcript of interest. Sigma Aldrich recommends between 0.1 and 1.0 μg/mL. We usually start with 0.2–0.4 μg/mL.

14. 24-well plates are very convenient for ISH experiments. Typically, we use a row or column per sample and then transfer the sample in its dedicated row or column. Transfer the sample to a clean well when possible. Fragile specimens may be damaged with many transfers.
15. The amount of antibody added to the incubation can be quite variable and depends on many factors. With a probe concentration of 0.2–0.4 μg/mL, a 1 to 2000 dilution of alkaline phosphatase-conjugated DIG antibodies works well for us.
16. Detection buffer is one of the most important buffers in the ISH protocol. Not only is it essential to make at the correct concentrations, but pH is also critical. The pH should be as close to 9.5 ± 0.1 pH units. Plus, this solution needs to be filtered. We use a 60 mL syringe with a Nalgene syringe filter, 0.2 μM. Detection buffer can be prepared fresh on the day of detection.

    Do not keep detection buffer longer than 2 weeks as it will become cloudy, but we prefer freshly prepared buffer.
17. This is a light-sensitive reaction so try and keep in the dark until you need to monitor the development. We often wait for at least 30 min before visually examining the development. In general, we briefly examine the preparations under a stereo microscope and cover them for additional development either at +4 °C or at room temperature—the selection of developing conditions depends upon the abundance of a target transcript and the intensity of a signal.
18. In some ctenophores, we had seen crystal formation at times when animals were kept for any length of time in PBS at 4 °C. When this has been an issue, we have stopped the development reaction in 4% formaldehyde in MeOH. This is not the most efficient way to stop the reaction because this solution does not wash the samples and PBS, but it helps reduce the crystal formation.
19. Development times are highly variable. The 4% formaldehyde in MeOH and 100% EtOH are both clearing steps. A considerable amount of background/tissue coloration might be lost. For the first time using this protocol, try several development times to evaluate what works best for your transcript of interest and specimen.
20. Methyl salicylate is an organic ester found in natural products. However, caution should be applied when handling this chemical, and the use of a well-ventilated room and a hood is required.
21. We were successful with the direct method of detection and Vector Red AP Substrate kit. This provides both white light and fluorescence options to visualize expression.

The timing of this protocol can be variable. For first-time users, it is best if the probe is made and ready to be used before the start of the in situ hybridization protocol because if your probe is not of acceptable quality, nothing can proceed. Once one is experienced, probe generation can be done in parallel with the Day 1 of the in situ hybridization protocol. Day 1 can be a combination of collecting and washing animals followed by fixation of the specimens overnight. The next day can be a combination of Day 2 and 3, so the dehydration, hydration, and hybridization can be performed on the same day as long as the appropriate times are followed as suggested in the protocol. The next day is a short day with washing and incubation of antibodies. We do not suggest trying to shorten this step. The development is going to be the most variable for timing. Development can be very rapid for something like tubulin or a neuropeptide or very slow for a receptor or channel and depends on the level of expression of your transcript of interest. Be prepared to have a long day of development for a successful result.

5 Anticipated Results

In situ hybridization techniques can be long and arduous experiments, especially when optimizing conditions for perfect results. However, the information they provide is invaluable with cell-specific spatial resolution. We present the ISH protocol for fragile small gelatinous animals such as ctenophores. We defined six basic components to successful ISH experiments: the experimental design, fixation of cell/tissue/animal, pretreatment and permeabilization of cells, binding of specific RNAs to a labeled riboprobe, amplification of the reporter through immunological detection (for indirect detection only), and visualization. Although we successfully tested these protocols for dozens of different preparations, if poor initial results for new species are obtained, it is relatively straightforward to return to the six key stages of ISH to systematically evaluate and adjust required parameters starting with the detection of relatively abundant transcripts such as cytoskeleton proteins or secretory molecules (Figs. 6 and 7).

6 Illustrated Examples: Quest for Neuronal Genes in Ctenophores

As illustrated examples, we selected several genes as markers of distinct cellular populations focusing on the still ongoing quest to identify and characterize unique neural populations in ctenophores. The systematic genomic survey in *Pleurobrachia* and related ctenophore species strongly suggested independent origins of neural systems in this lineage [7, 10, 17, 27, 28] and an apparent lack of pan-neuronal genes across Metazoa [9, 18]. Indeed, such ctenophore gene homologs as Neuroglobin, ELAV, and Musashi, known to be expressed in bilaterian nervous tissue, do not specifically label neurons in *Pleurobrachia* but many other cell types in combs and tentacles (Figs. 3b, c and 4c, d). Although Neuroglobin marked small cells in the polar field (putative chemosensory organ with neurons) and within the comp plates (Fig. 3a, b), their identity remains to be determined. The overall labeling is quite different from mapping neural cells in *Pleurobrachia* and other studied ctenophores, with most neurons localized in the skin and mesoglea [29–32]. Surprisingly, genes encoding different aspects of RNA processing (including small RNAs) such as Argonaute and Dicer (Figs. 4a, b and 8) are associated with the polar fields and the aboral organ (contains statocyst and might be an analog of the elementary brain)—structures with dense neuronal populations.

WntX is expressed specifically in ciliated furrows (Fig. 5a), which are nonneuronal elements (*see* also [23] for a closely related species—*P. pileus*). The homeobox transcriptional factor BarX known to be involved in vertebrate mesenchymal transitions (e.g., [33]) is differentially expressed in combs, polar fields, and tentacles (Fig. 5b).

It was indicated that ctenophore neural systems might be peptidergic in their nature with a diversity of neuron-specific small secretory peptides—neuropeptides [10, 17, 34–36]. Indeed, some predicted *Pleurobrachia* neuropeptides expressed in neural-type structures (Fig. 6), as reported earlier [17], are similar to recent data on *Mnemiopsis* [35]. Nevertheless, two predicted neuropeptides are *Pleurobrachia* innovations (SP-7, SP-62), and their homologs have not been identified in the *Mnemiopsis*. In contrast, SP-31 has a homolog in *Mnemiopsis* with similar patterns of expression in the mouth-gut areas in both species—this was the reason for its original name as Ctenophorin [17].

In conclusion, analysis of cell-specific gene expression patterns, especially in adult ctenophores, is critical in deciphering neuronal organizations in basal metazoan lineages and animal innovations in general. In situ hybridization was and will remain a helpful tool in this endeavor.

Acknowledgments

This work was supported in part by the Human Frontiers Science Program (RGP0060/2017) and National Science Foundation (IOS-1557923) grants to L.L.M. Research reported in this publication was also supported in part by the National Institute of

Neurological Disorders and Stroke of the National Institutes of Health under Award Number R01NS114491 (to L.L.M). The content is solely the authors' responsibility and does not necessarily represent the official views of the National Institutes of Health.

References

1. Hernandez-Nicaise M-L (1991) Ctenophora. In: Harrison FWFW, Westfall JA (eds) Microscopic anatomy of invertebrates: Placozoa, Porifera, Cnidaria, and Ctenophora. Wiley, New York, pp 359–418
2. Nielsen C (2012) Animal evolution: interrelationships of the living phyla. Oxford University Press, Oxford
3. Tamm SL (1982) Ctenophora. In: Electrical conduction and behavior in "simple" invertebrates. Clarendon Press, Oxford, pp 266–358
4. Li Y et al (2021) Rooting the animal tree of life. Mol Biol Evol 38(10):4322–4333
5. Redmond AK, McLysaght A (2021) Evidence for sponges as sister to all other animals from partitioned phylogenomics with mixture models and recoding. Nat Commun 12(1):1783
6. Moroz LL et al (2021) Evolution of glutamatergic signaling and synapses. Neuropharmacology 199:108740
7. Moroz LL (2018) NeuroSystematics and periodic system of neurons: model vs reference species at single-cell resolution. ACS Chem Neurosci 9(8):1884–1903
8. Telford MJ, Moroz LL, Halanych KM (2016) Evolution: a sisterly dispute. Nature 529(7586):286–287
9. Moroz LL, Kohn AB (2016) Independent origins of neurons and synapses: insights from ctenophores. Philos Trans R Soc Lond Ser B Biol Sci 371(1685):20150041
10. Moroz LL, Romanova DY, Kohn AB (1821) Neural versus alternative integrative systems: molecular insights into origins of neurotransmitters. Philos Trans R Soc Lond Ser B Biol Sci 2021(376):20190762
11. Nielsen C (2019) Early animal evolution: a morphologist's view. R Soc Open Sci 6(7): 190638
12. Jekely G, Budd GE (2021) Animal phylogeny: resolving the slugfest of ctenophores, sponges and acoels? Curr Biol 31(4):R202–R204
13. Daley AC, Antcliffe JB (2019) Evolution: the battle of the first animals. Curr Biol 29(7): R257–R259
14. Littlewood DTJ (2017) Animal evolution: last word on sponges-first? Curr Biol 27(7):R259–R261
15. Dunn CW (2017) Ctenophore trees. Nat Ecol Evol 1(11):1600–1601
16. Halanych KM et al (2016) Miscues misplace sponges. Proc Natl Acad Sci U S A 113(8): E946–E947
17. Moroz LL et al (2014) The ctenophore genome and the evolutionary origins of neural systems. Nature 510(7503):109–114
18. Moroz LL, Kohn AB (2015) Unbiased view of synaptic and neuronal gene complement in ctenophores: are there pan-neuronal and pan-synaptic genes across metazoa? Integr Comp Biol 55(6):1028–1049
19. Mitchell DG, Edgar A, Martindale MQ (2021) Improved histological fixation of gelatinous marine invertebrates. Front Zool 18(1):29
20. Pang K, Martindale MQ (2008) Ctenophore whole-mount in situ hybridization. CSH Protoc 2008:pdb.prot5087
21. Gall JG, Pardue ML (1969) Formation and detection of RNA-DNA hybrid molecules in cytological preparations. Proc Natl Acad Sci U S A 63(2):378–383
22. Pardue ML, Gall JG (1969) Molecular hybridization of radioactive DNA to the DNA of cytological preparations. Proc Natl Acad Sci U S A 64(2):600–604
23. Jager M et al (2013) Evidence for involvement of Wnt signalling in body polarities, cell proliferation, and the neuro-sensory system in an adult ctenophore. PLoS One 8(12):e84363
24. Alie A et al (2011) Somatic stem cells express Piwi and Vasa genes in an adult ctenophore: ancient association of "germline genes" with stemness. Dev Biol 350(1):183–197
25. Jezzini SH, Bodnarova M, Moroz LL (2005) Two-color in situ hybridization in the CNS of *Aplysia californica*. J Neurosci Methods 149(1):15–25
26. Moroz LL, Kohn AB (2015) Analysis of gene expression in neurons and synapses by multicolor in situ hybridization. In: Hauptmann G (ed) In situ hybridization methods. Humana Press, New York, pp 293–317
27. Moroz LL (2014) The genealogy of genealogy of neurons. Commun Integr Biol 7(6): e993269

28. Moroz LL, Romanova DY (2022) Alternative neural systems: what is a neuron? (Ctenophores, sponges and placozoans). Front Cell Dev Biol 10:1071961
29. Norekian TP, Moroz LL (2016) Development of neuromuscular organization in the ctenophore *Pleurobrachia bachei*. J Comp Neurol 524(1):136–151
30. Norekian TP, Moroz LL (2019) Neural system and receptor diversity in the ctenophore *Beroe abyssicola*. J Comp Neurol 527(12):1986–2008
31. Norekian TP, Moroz LL (2019) Neuromuscular organization of the Ctenophore *Pleurobrachia bachei*. J Comp Neurol 527(2):406–436
32. Norekian TP, Moroz LL (2020) Comparative neuroanatomy of ctenophores: neural and muscular systems in *Euplokamis dunlapae* and related species. J Comp Neurol 528(3):481–501
33. Makarenkova HP, Meech R (2012) Barx homeobox family in muscle development and regeneration. Int Rev Cell Mol Biol 297:117–173
34. Moroz LL (2021) Multiple origins of neurons from secretory cells. Front Cell Dev Biol 9:669087
35. Sachkova MY et al (2021) Neuropeptide repertoire and 3D anatomy of the ctenophore nervous system. Curr Biol 31(23):5274–5285 e6
36. Burkhardt P, Jekely G (2021) Evolution of synapses and neurotransmitter systems: the divide-and-conquer model for early neural cell-type evolution. Curr Opin Neurobiol 71:127–138

Chapter 10

Characterization of the Mitochondrial Proteome in the Ctenophore *Mnemiopsis leidyi* Using MitoPredictor

Viraj Muthye and Dennis V. Lavrov

Abstract

Mitochondrial proteomes have been experimentally characterized for only a handful of animal species. However, the increasing availability of genomic and transcriptomic data allows one to infer mitochondrial proteins using computational tools. MitoPredictor is a novel random forest classifier, which utilizes orthology search, mitochondrial targeting signal (MTS) identification, and protein domain content to infer mitochondrial proteins in animals. MitoPredictor's output also includes an easy-to-use R Shiny applet for the visualization and analysis of the results. In this article, we provide a guide for predicting and analyzing the mitochondrial proteome of the ctenophore *Mnemiopsis leidyi* using MitoPredictor.

Key words Mitochondria, Proteome, MitoPredictor, Ctenophora, *Mnemiopsis*, Random Forest, Machine learning

1 Introduction

Best known for their role in energy metabolism [1], mitochondria are involved in a multitude of cellular processes, including Fe/S cluster biosynthesis [2], amino acid metabolism [3], lipid metabolism [4], apoptosis [5, 6], and cellular signaling [7]. The diverse mitochondrial functions require more than a thousand proteins, the vast majority of which are encoded in the nuclear genome and transported into the mitochondria via several import pathways [8, 9]. Collectively, these nuclear-encoded mitochondrial proteins constitute the bulk of the "mitochondrial proteome," an inclusive collection of proteins with at least some mitochondrial functions. A few additional proteins are encoded in the mitochondrial genome—mtDNA, which is replicated, repaired, and expressed mostly independently from the nuclear genome [8]. Information on the mitochondrial proteome should be useful for understanding the functional repertoire of mitochondria. However, our ability to distinguish mitochondrial proteins remains limited.

Leonid L. Moroz (ed.), *Ctenophores: Methods and Protocols*, Methods in Molecular Biology, vol. 2757, https://doi.org/10.1007/978-1-0716-3642-8_10, © Springer Science+Business Media, LLC, part of Springer Nature 2024

Mitochondrial proteomes have been experimentally characterized for a few animals [10–13], plants [14–17], fungi [18], and protists [19]. In animals, experimental characterization of the mitochondrial proteome is limited to model animal species, including *Homo sapiens* [10, 11], *Mus musculus* [10, 11], *Caenorhabditis elegans* [12], and *Drosophila melanogaster* [13]. Most animal phyla—including the four non-bilaterian phyla (Ctenophora, Porifera, Placozoa, and Cnidaria)—lack experimental mitochondrial proteomic data. This is unfortunate, because non-bilaterian taxa represent most of the oldest lineages in the animal phylogenetic tree and thus are essential for understanding the full extent of animal diversity. They also display some of the most unusual mitochondrial genome organization among animals [20].

Computational techniques provide a complementary approach to experimental characterization of mitochondrial proteomes. They attempt to infer mitochondrial proteins from genomic and transcriptomic data, which are becoming increasingly available even for non-bilaterian taxa. Several subcellular localization predictors have been developed and can be divided into four broad categories:

- **Homology-based predictors** (e.g., OrthoFinder [21], InParanoid [22], Proteinortho [23]): One way to predict mitochondrial proteins is by finding orthologs in experimentally characterized mitochondrial proteomes. The advantage of this approach is that it can detect mitochondrial proteins, which lack identifiable mitochondrial targeting signal (MTS), as is the case for most mitochondrial inner-membrane proteins and all outer-membrane proteins. Its disadvantage is that it would fail to identify novel proteins without known orthologs and might be misled by instances of protein subcellular re-localization.
- **N-terminus mitochondrial targeting signal (MTS) predictors** (e.g., TargetP [24], MitoFates [25]): Another common method for inferring mitochondrial proteins is by searching for the mitochondria targeting signal (MTS) at the protein's N-terminus. MTS is a short 10–90 amino acid sequence, enriched in positively charged amino acids that directs proteins toward the mitochondria. Once inside the mitochondrial matrix, the MTS is cleaved, and the mature protein is formed [25]. Because MTS predictors do not rely on homology, they can identify novel mitochondrial proteins. However, these predictors tend to suffer from a high false-positive rate and would not identify mitochondrial proteins possessing a noncanonical MTS.
- **Full protein sequence feature-based predictors** (e.g., CELLO2.5 [26], ngLOC [27], SubMitoPred [28]): Some predictors utilize features extracted from the whole protein sequences, like amino acid composition, protein domain

composition, hydrophobicity, etc., to infer mitochondrial localization. CELLO uses n-peptide composition, partitioned amino acid composition, g-gap dipeptide composition, and local amino acid composition; ngLOC uses the density distribution of peptide sequences of a fixed length (*n-grams*); while SubMitoPred—protein domain composition and several additional features extracted from the input protein sequences.

- **Ensemble predictors** (e.g., MitoPredictor [29], SubCons [30]): Ensemble methods use machine learning algorithms to integrate multiple sources of information for predicting mitochondrial proteomes. MitoPredictor is a random forest-based predictor, which utilizes (1) orthology, (2) MTS prediction, and (3) protein domain information. SubCons uses a random forest classifier to integrate results from four independent programs: CELLO2.5, SherLoc2 [31], LocTree2 [32], and MultiLoc2 [33].

While each of the four categories of predictors has its advantages and disadvantages, ensemble methods tend to outperform other methods because they integrate information from multiple sources. In this article, we provide a guide for predicting and analyzing the mitochondrial proteome in the ctenophore *Mnemiopsis leidyi* using our recently developed ensemble predictor MitoPredictor. MitoPredictor provides several advantages over other ensemble predictors and has been shown to outperform SubCons, a widely used ensemble protein predictor of animal protein subcellular localization. In addition to running the analysis, MitoPredictor provides an easy-to-use R Shiny applet for visualization and exploration of the results. The applet also includes information on existing experimentally characterized mitochondrial proteomes from human, mouse, *C. elegans*, and *D. melanogaster*.

Ctenophores are the prime candidate for the characterization of their mitochondrial proteome. They constitute an ancient lineage in animal evolution and, possibly, the sister group to the rest of the animals [34]. They also have some of the most unusual mitochondrial genomes among animals [35]. While nuclear and mitochondrial genomes from multiple ctenophores have been sequenced (Tables 1 and 2), no mitochondrial proteomes have been experimentally characterized from this group.

Here, we used MitoPredictor to predict and analyze the mitochondrial proteome of *M. leidyi*. This ctenophore has one of the smallest mitochondrial genomes in animals (~10 kb) that lacks all tRNA genes as well as *atp6* and *atp8* [35]. Additionally, mtDNA-encoded ribosomal RNA molecules exhibit extremely reduced primary and secondary structures. *M. leidyi* mtDNA also exhibits a very high rate of mitochondrial sequence evolution. It is likely that some of these changes would necessitate changes in its mitochondrial proteome. For example, we have shown that mt-ribosomal

Table 1
Publicly available nuclear genomes from Ctenophora

Species	GenBank ID
Pleurobrachia bachei	GCA_000695325.1 [36]
Bolinopsis microptera	GCA_026151205.1 [34, 37]
Hormiphora californensis	GCA_020137815.1 [38]
Beroe forskalii	GCA_011033025.1 [39]
Beroe ovata	GCA_946803715.1
Mnemiopsis leidyi	GCA_000226015.1 [40]

Table 2
Publicly available mitochondrial genomes from Ctenophora

Species	GenBank ID
Beroe forskalii isolate Bf201311	MG655624.1 [39]
Beroe forskalii isolate Bf201606	MG655623.1 [39]
Beroe forskalii isolate Bf201706	NC_038065.1 [39]
Hormiphora californensis isolate Hc2	MN544301.1 [38]
Hormiphora californensis isolate Hc1	NC_045864.1 [38]
Hormiphora californensis isolate 20161213-T1	CM035363.1 [38]
Pleurobrachia bachei	JN392469.1 [41]
Beroe cucumis	NC_045305.1 [42]
Bolinopsis microptera isolate Bmic1	CM047416.1 [37]
Mnemiopsis leidyi	NC_016117.1 [35]

proteins encoded in the nucleus have undergone expansions, possibly to compensate for the reduction in the rRNA structures [35]. While *M. leidyi* lacks an experimentally characterized mitochondrial proteome, there is a high-quality nuclear genome, which is deposited at the *Mnemiopsis* Genome Project Portal (https://research.nhgri.nih.gov/mnemiopsis/) [40, 43].

2 Materials and Methods

2.1 Software Required

1. MitoPredictor: https://github.com/virajmuthye/mitopredictor.
2. Proteinortho v5.16b: https://www.bioinf.uni-leipzig.de/Software/proteinortho/.

3. TargetP v1.1: http://www.cbs.dtu.dk/services/TargetP-1.1/index.php.
4. MitoFates: http://mitf.cbrc.jp/MitoFates/cgi-bin/top.cgi.
5. CD-HIT [44]: http://weizhongli-lab.org/cd-hit/.
6. R (≥3.0.2): https://www.r-project.org/.
7. R studio: https://rstudio.com/.
8. Perl: https://www.perl.org/get.html.

2.2 Databases Required

1. Pfam-A.hmm: ftp://ftp.ebi.ac.uk/pub/databases/Pfam/releases/ (select latest release).
2. *Mnemiopsis leidyi* protein models from the *Mnemiopsis* Genome Portal: https://research.nhgri.nih.gov/mnemiopsis/download/download.cgi?dl=proteome. There are two categories of protein models in the portal: *filtered* and *unfiltered*. Filtered protein models were used for the analysis.

2.3 Identifying Mitochondrial Proteins Using MitoPredictor

MitoPredictor consists of seven BASH scripts and an R script that process input data and invoke the computational tools listed above (Subheading 2.1) to predict mitochondrial proteins. Each step of MitoPredictor requires only a single command: e.g., *step1_prep.bash*, which would run this script using default settings. The information on changing the default settings are specified within each individual script. Please see **Note 4** prior to running MitoPredictor.

The input file for MitoPredictor is a set of amino acid sequences from the query species in a FASTA format. The FASTA format consists of at least two lines for each entry: (1) the first line beginning with the ">" symbol and containing the name of and some additional (optional) information about the entry and (2) the amino acid sequence. An example of a protein entry in the FASTA format is given below:

```
>sp|Q8ITI5|FOXG1_MNELE Forkhead box protein G1
OS=Mnemiopsis leidyi OX=27923 GN=FOXG1 PE=2 SV=1
MVVTTATKPHPFSIENILKSASPKPQKPLFSYNALIA-
MAISQSPLKKLTLSEIYDFIIET
FPYYRDNKKGWQNSIRHNLSLNKCFVKVPRHYNDPGKGNYWMLNPNS-
DEVFIGGKLRRRP
GQNGGSLESYMHLKTRTSPYQRGDTVCKRDRVVYLSNSGAGNCQ-
FYQPVPCSSPTAMLSR
SSLIVQTSPTTLTIPHHPPQNYIQTLSPNVSPVRVGNQTSPRQ-
SALPSSSLPLLPSPPSL
PSKLSSPPLPSLSTNLPSPPLDTGDILLSPFLRQTVDSTSQNFLEH-
MIRIREQVQRSGHL
ALAQSRAVVYQPIPRKSL
```

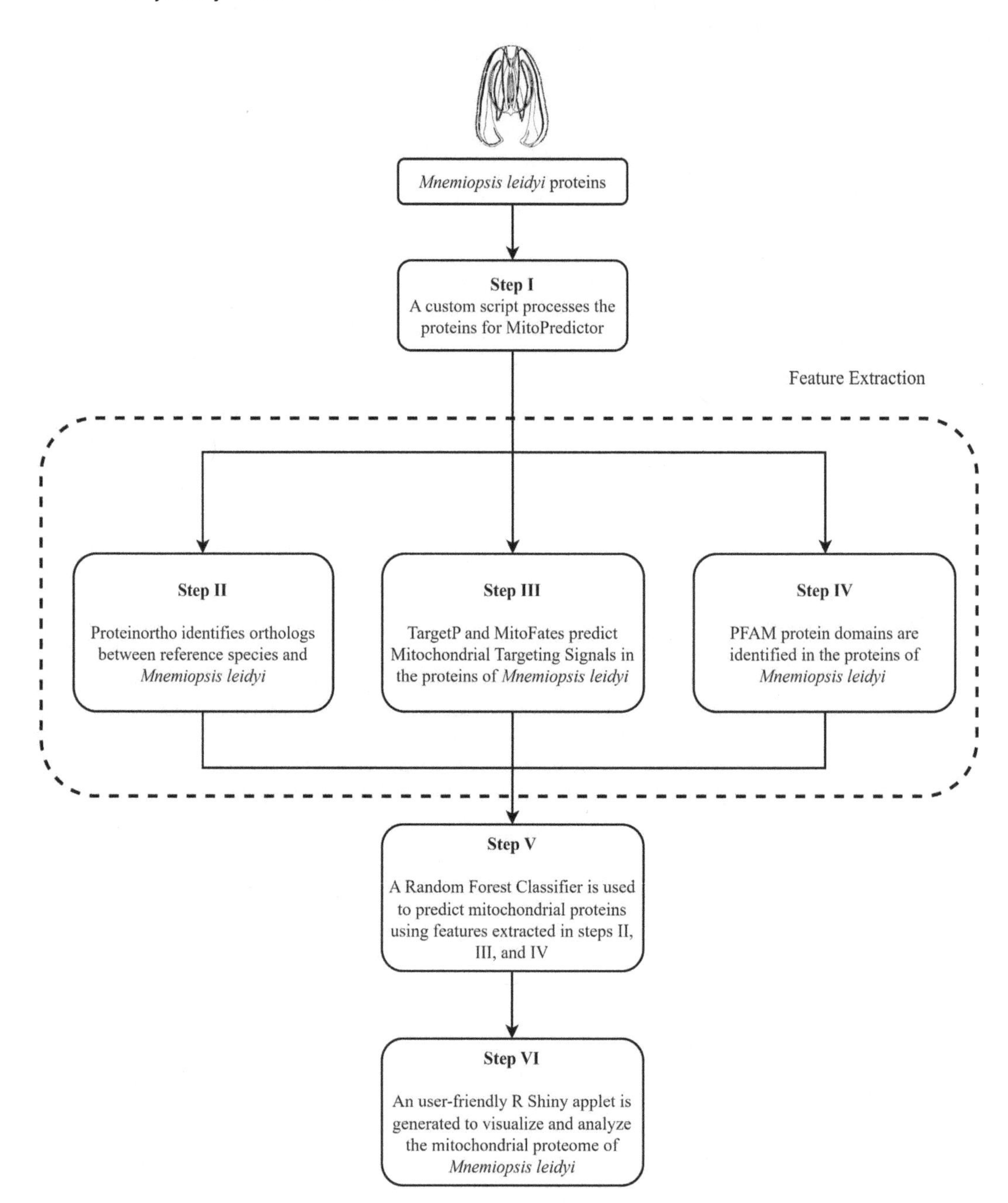

Fig. 1 Flowchart outlining the basic steps used by MitoPredictor for predicting mitochondrial proteins in the ctenophore *Mnemiopsis leidyi*. The silhouette of *M. leidyi* was taken from PhyloPic V2 (www.phylopic.org)

The software and databases listed above need to be downloaded and installed prior to running MitoPredictor. The input proteome, i.e., protein models from the *Mnemiopsis* Genome Project Portal, should be placed in the MitoPredictor directory. MitoPredictor has been developed and tested on a Linux cluster but should be transferable to other systems. *See* Fig. 1 for a flowchart.

2.3.1 Step I: Preprocessing Mnemiopsis Proteins

This step is run as *step1_prep.bash*. (*see* **Notes 1**, **2**, and 7 prior to running step I.)

- MitoPredictor will identify the input file which should have the ".fasta" file extension.
- Next, all potential fragments, i.e., proteins below 100 amino acids in length, are removed.
- CD-HIT clusters the remaining proteins at 98% similarity. This cutoff can be changed in the script "prep.bash" inside the prep folder. These proteins are used as input for step II (orthology) and step IV (protein domain analysis).
- All proteins without methionine at position 1 are removed. The remaining proteins are used as an input for step III (MTS prediction). Removal of potential fragments that are missing a complete N-terminus is aimed at reducing the false-positive rate in step II.

2.3.2 Step II: Orthology Analysis

This step is run as *step2_ortho.bash*.

- The script invokes Proteinortho, which identifies groups of orthologous proteins (OGs) in the reference species (human, mouse, *C. elegans*, *D. melanogaster*, and *S. cerevisiae*) and the query species (*M. leidyi*) (*see* **Note 3**). Default settings are used for Proteinortho: e-value: $1e^{-05}$; identity: 25%; minimum coverage of best BLAST hit alignments: 50%; algebraic connectivity: 0.1; and purity: 0.1 (these values can be changed in the "*run_-proteinortho.bash*" script in the "orthology" folder). A four-letter abbreviation is used to denote each of the five reference species (Table 3).
- One feature is extracted in this step: OrthoScore (OS). For each *M. leidyi* protein, the OS is 1 if an ortholog is found in at least one reference mitochondrial proteome, and 0 if not.

2.3.3 Step III: MTS Analysis

This step is run as *step3_mts.bash*.

- TargetP and MitoFates predict MTS in *M. leidyi* proteins. Default settings are used for TargetP and MitoFates but can be changed in the "*runTargetP.bash*" and "*runMitoFates.bash*" scripts in the "mts" folder.
- For each *M. leidyi* protein, five features are extracted from the results generated by the two programs:
  - (i) From TargetP: mTP (probability that the protein possesses an MTS), sTP (probability that the protein possesses a signal peptide), other (probability that the protein localized to any other location)

Table 3
The four-letter abbreviations that were used to denote each reference species in MitoPredictor

Species	Abbreviation
Homo sapiens	hsap
Mus musculus	mmus
Caenorhabditis elegans	cele
Drosophila melanogaster	dmel
Saccharomyces cerevisiae	scer

(ii) From MitoFates: mfprob (probability that the protein possesses an MTS) and net charge

2.3.4 Step IV: Protein Domain Analysis

This step is run as *step4_domain.bash.*

- Protein domains are identified in *M. leidyi* proteins using the Perl script *pfam_scan.pl* (ftp://ftp.ebi.ac.uk/pub/databases/Pfam/Tools/) and the Pfam-A.hmm libraries.
- Based on the protein domain content of each *M. leidyi* protein, a domain score (DS) is assigned to that protein. The DS ranges from 0 to 1, where a score of 0 indicates that the protein domain exists only in non-mitochondrial proteins in all reference animal species, while a score of 1 indicates that the protein domain has been found only in mitochondrial proteins in the majority of the reference species.

2.3.5 Step V: Prediction of Mitochondrial Proteins Using a Random Forest Classifier

This step is run as *step5_identify_and_concat.bash.*

- In this step, a random forest classifier is used with the features extracted in steps II, III, and IV to predict mitochondrial proteins. The classifier is implemented in R ("*rf.R*") and uses the R package "randomForest" (https://www.stat.berkeley.edu/~breiman/RandomForests/).

2.3.6 Step VI: Make Basic Statistics of the Predicted Mitochondrial Proteome

This step is run as *step6_make_stats.bash.*

- MitoPredictor generates basic statistics of the predicted *M. leidyi* mitochondrial proteome and data for the R Shiny applet (Fig. 2).
- Step VI generates two primary outputs:

  (i) The "FINAL.MATRIX" file: This file serves as an input for the R Shiny applet "mito_app.R." The R Shiny applet can be

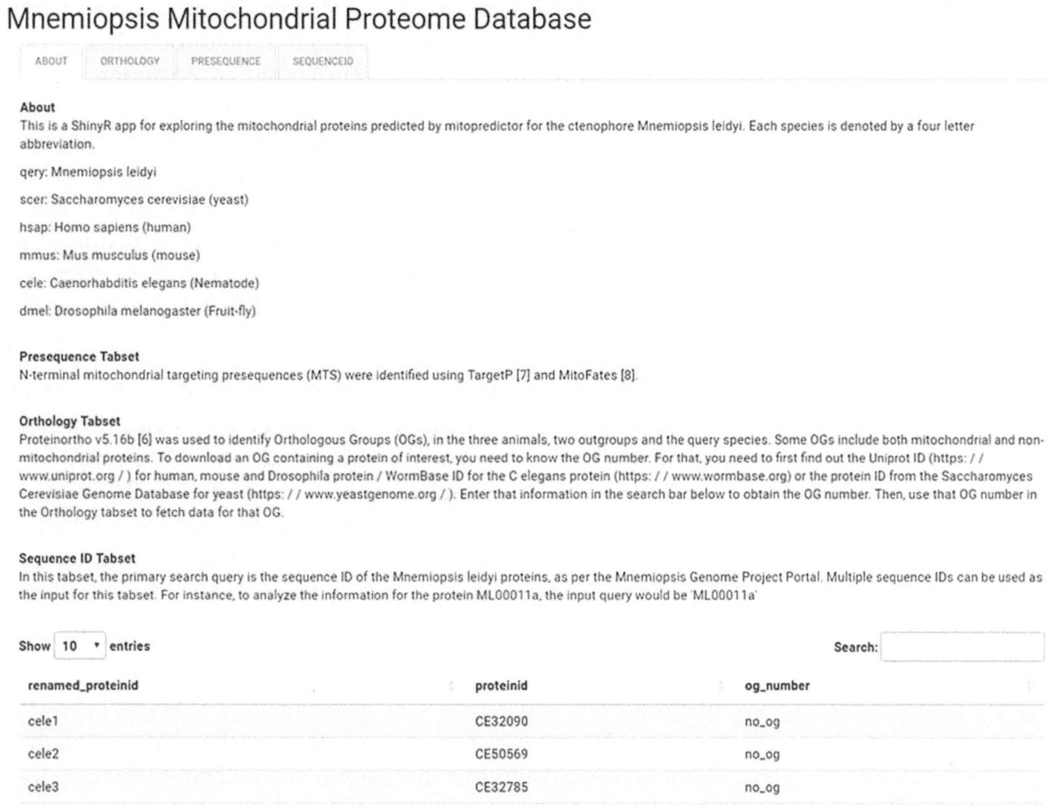

Mnemiopsis Mitochondrial Proteome Database

ABOUT | ORTHOLOGY | PRESEQUENCE | SEQUENCEID

About
This is a ShinyR app for exploring the mitochondrial proteins predicted by mitopredictor for the ctenophore Mnemiopsis leidyi. Each species is denoted by a four letter abbreviation.

qery: Mnemiopsis leidyi

scer: Saccharomyces cerevisiae (yeast)

hsap: Homo sapiens (human)

mmus: Mus musculus (mouse)

cele: Caenorhabditis elegans (Nematode)

dmel: Drosophila melanogaster (Fruit-fly)

Presequence Tabset
N-terminal mitochondrial targeting presequences (MTS) were identified using TargetP [7] and MitoFates [8].

Orthology Tabset
Proteinortho v5.16b [6] was used to identify Orthologous Groups (OGs), in the three animals, two outgroups and the query species. Some OGs include both mitochondrial and non-mitochondrial proteins. To download an OG containing a protein of interest, you need to know the OG number. For that, you need to first find out the Uniprot ID (https: / / www.uniprot.org /) for human, mouse and Drosophila protein / WormBase ID for the C elegans protein (https: / / www.wormbase.org) or the protein ID from the Saccharomyces Cerevisiae Genome Database for yeast (https: / / www.yeastgenome.org /). Enter that information in the search bar below to obtain the OG number. Then, use that OG number in the Orthology tabset to fetch data for that OG.

Sequence ID Tabset
In this tabset, the primary search query is the sequence ID of the Mnemiopsis leidyi proteins, as per the Mnemiopsis Genome Project Portal. Multiple sequence IDs can be used as the input for this tabset. For instance, to analyze the information for the protein ML00011a, the input query would be 'ML00011a'

Show 10 entries Search:

renamed_proteinid	proteinid	og_number
cele1	CE32090	no_og
cele2	CE50569	no_og
cele3	CE32785	no_og

Fig. 2 The About tabset of the R Shiny applet generated by MitoPredictor. The applet consists of four tabsets: About, Orthology, Presequence, and Sequence ID

accessed in R Studio via the "Run App" button or by giving the R command runApp ("[directory where the app files are stored]"). The applet can be used to visualize, analyze, and download the mitochondrial proteins from *M. leidyi* and the reference species.

(ii) The "stats" folder: This folder contains information on the predicted *M. leidyi* mitochondrial proteome. More information is provided in Subheading 2.5.

2.3.7 Step VII: Clean Intermediate Files and Prepare MitoPredictor for the Next Analysis

This step is run as *step7_clean.bash*.

This step removes intermediate and result files generated during the MitoPredictor run in preparation for the next analysis. It is important to save the results of the run (the "FINAL.MATRIX" file, "mito_app.R" file, and the "stats" folder) before running this step.

2.4 R Shiny Applet

The R Shiny applet allows for the visualization and analysis of the predicted mitochondrial proteome. It is designed to facilitate comparative analysis of the predicted mitochondrial proteome (in the present case, the *M. leidyi* mitochondrial proteome) with the reference mitochondrial proteomes. The "FINAL.MATRIX" file generated in MitoPredictor step VI above serves as the input for the applet. The R Shiny applet is organized into four tabsets: (1) About, (2) Orthology, (3) Presequence, and (4) SequenceID (Fig. 2). The "About" tabset provides general information and description of other tabsets. Here, we describe the remaining three tabsets:

- **Orthology**: The orthology tabset allows users to analyze results of orthology analysis from Proteinortho. Users can download information on a particular orthology group (OG) of interest. The primary search query in this tabset is the OG number assigned during analysis. The OG number can be fetched by using either the Uniprot ID for human, mouse, and *D. melanogaster* or the *C. elegans* protein ID from WormBase.
- **Presequence**: This tabset includes MTS prediction results from TargetP and MitoFates for all the proteins from reference and query species.
- **SequenceID**: This tabset is meant for visualization and analysis of query proteins. In this case, users can extract information for a list of *M. leidyi* proteins. The input for this tabset is the protein ID assigned to the protein on the *Mnemiopsis* Genome Project Portal (e.g., "ML000111a").

2.5 Using the Information in the "Stats" Folder to Analyze Mitochondrial Proteomes

The "stats" folder includes information on the predicted mitochondrial proteome. Here, we describe how files in the "stats" folder can be used for the functional annotation of mitochondrial proteins and for the further analysis of the predicted mitochondrial proteome.

- **Stats.txt**: The file "Stats.txt" provides a brief overview of the predicted mitochondrial proteome (Table 4). Based on this file, we can see that MitoPredictor predicted 1161 mitochondrial proteins in *M. leidyi*. Interestingly, the number of *M. leidyi* proteins possessing a recognizable MTS was low (320), with most of the mitochondrial proteins predicted primarily by the results of the orthology search (982). Among *M. leidyi* proteins with detected MTS, nearly half (144/320) had no ortholog in any of the reference species. One thousand five protein domains were identified in the predicted mitochondrial proteome of *M. leidyi*. The mitochondrial carrier domain ("Mito_carr") was the most abundant protein domain, present in 23 mitochondrial proteins.

Table 4
Statistics for the predicted mitochondrial proteomes that are listed in the file "Stats.txt"

Total number of inferred mitochondrial proteins in *M. leidyi*	1161
Number of *M. leidyi* proteins orthologous to known mitochondrial proteins	982
Number of *M. leidyi* proteins without MTS but with orthologs in reference mitochondrial proteomes	806
Number of *M. leidyi* proteins with MTS	320
Number of *M. leidyi* proteins with MTS but no orthologs in reference mitochondrial proteomes	144
Number of unique protein domains in mitochondrial proteins	1005
Number of mitochondrial proteins with at least one domain	1052
Top ten most abundant protein domains in *M. leidyi* mitochondrial proteins (only the most abundant protein domain shown here)	Mito_carr,23

Table 5
List of resources for functional annotation of mitochondrial proteins

Species	Resources
Human	PantherDB, ConsensusPathDB, StringDB, UniProt
Mouse	PantherDB, ConsensusPathDB, StringDB, UniProt
C. elegans	PantherDB, StringDB, WormBase [51]
D. melanogaster	PantherDB, StringDB, FlyMine [52]

The gene lists generated by the MitoPredictor output can be used directly as input for these resources

- **species_specific.matrix**: The file "species_specific.matrix" provides information on *M. leidyi* proteins predicted to possess an MTS with no orthologs in reference mt-proteomes. These represent potential *M. leidyi*-specific or ctenophore-specific mitochondrial proteins (but *see* **Note 5**). The protein domain composition of these proteins, provided in the file, can be used to predict their potential function. The FASTA sequences of these proteins can be found in the file "query_species_specific. fasta." This file can be uploaded in tools like PANNZER [45] (*see* **Note 6**) or dcGO [46] for functional annotation.

The *Mnemiopsis* Genome Project Portal also has several tools for functional analysis of *M. leidyi* proteins. For example, to identify the function of the species-specific *M. leidyi* protein "ML189344a," a user can enter this name in the "View Gene Page" at the *Mnemiopsis* Genome Project Portal (https://research.nhgri.nih.gov/mnemiopsis/genes/genewiki.cgi). The "View Gene Page" includes information such as functional annotation by Argot2 and Blast2GO and BLAST searches against various databases.

- **domain_count.matrix**: Protein domain composition is known to be useful in functional annotation of proteins and predicting subcellular localization. This file contains a list of protein domains predicted to be present in at least one *M. leidyi* protein and the number of mitochondrial proteins containing each protein domain. The file "Stats.txt" contains a list of the top 5 most abundant protein domains in the predicted mitochondrial proteome. Here, we see that "Mito_carr" was the most abundant protein domain, found in 23 predicted mitochondrial proteins.
- **<Species>_query_mitoorthologs.txt:** For each reference species, the "stats" folder contains a list of mitochondrial proteins with and without orthologs in *M. leidyi*. For instance, the file "hsap_query_mitoorthologs.txt" contains a list of human mitochondrial proteins which possess orthologs in *M. leidyi*. This file can be used as input for several functional annotation tools, like ConsensusPathDB [47], PantherDB [48, 49], and StringDB [50]. Below, we provide a list of resources which can be used to analyze these lists provided by MitoPredictor as output (Table 5).

3 Notes

1. The input file for step I requires a ".fasta" file extension. Step I (processing) is *not* a mandatory step. The file "prep.bash" in the "prep" folder provides information on how to skip step I. However, the presence of a large number of C-terminus fragments will result in errors in MTS prediction by TargetP and MitoFates, which require complete N-terminus for accurate prediction of MTS. Therefore, we highly recommend not skipping step I.
2. The CD-HIT cutoff used in this analysis is 98%. This cutoff can be changed in the file "prep.bash." While changing the cutoff used in CD-HIT in this step, it is important to make the appropriate corresponding change to the choice of word size (Table 6).
3. While Proteinortho was selected as the tool to infer orthology, there have been several orthology predictors that have been developed since then, like OrthoFinder. However, MitoPredictor has not been tested using OrthoFinder or any other software.
4. Steps II, III, and IV can be run in parallel to reduce the runtime of MitoPredictor. In step V, the random forest classifier

Table 6
The word size for threshold used, as recommended by the CD-HIT manual

Word size (−*n*)	Threshold (cutoff)
5	0.7–1
4	0.6–0.7
3	0.5–0.6
2	0.4–0.5

requires the features extracted from steps II–IV and should be run only after all the previous steps are finished.

5. In the output of MitoPredictor, a "species-specific" mitochondrial protein is a protein that lacks orthologs in the five reference proteomes (human, mouse, *C. elegans*, *D. melanogaster*, and yeast).

   However, some of these "species-specific" mitochondrial proteins can be orthologous to mitochondrial or non-mitochondrial proteins in reference species, which are missed by the orthology search applied by Proteinortho. Some "species-specific" mitochondrial proteins could also be orthologs of mitochondrial or non-mitochondrial proteins from species not included in MitoPredictor. Finally, it is also possible that some of these "species-specific" proteins are false positives of MTS prediction.

   We recommend running Protein BLAST (BLASTP) on "query_species_specific.fasta" after running MitoPredictor. BLASTP can be downloaded and run locally or accessed at (https://blast.ncbi.nlm.nih.gov/Blast.cgi). In *M. leidyi*, 23 of 144 species-specific proteins produced no hits in the NR database (nonredundant protein sequences), while 121 showed similarity to some known proteins in other species. It is important to note that a BLASTP hit does not necessarily imply the presence of a homolog.

6. Pannzer can be accessed via the Pannzer2 web server (http://ekhidna2.biocenter.helsinki.fi/sanspanz/). The webserver accepts a file of protein sequences in FASTA format as an input. GO predictions from four predictors (ARGOT, RM3, JAC, and HYGE) are provided by Pannzer2. ARGOT has been shown to be the top performing predictor.

7. In case users have a transcriptome for the species of interest, TransDecoder can be used to predict protein models prior to running MitoPredictor (https://github.com/TransDecoder/TransDecoder/wiki). TransDecoder, along with its instructions for usage, can be downloaded from https://github.com/TransDecoder/TransDecoder/wiki. TransDecoder

generates several output files. The output file with the ".trans-decoder.pep" file extension should be used as the input file for MitoPredictor. We also recommend using CD-HIT to cluster the predicted proteins at 100% similarity.

4 Analysis of the Predicted *M. leidyi* Mitochondrial Proteome

4.1 Conservation and Loss of Common Animal Mitochondrial Proteins

Any investigation into a novel mitochondrial proteome should infer potential losses and gains of function. To study the loss of mitochondrial function, one can consider the orthologous groups (OGs) that contain proteins from the reference species, but not the species of interest (*M. leidyi*). In our study, we identified 255 OGs containing at least 1 mitochondrial protein from at least 3 out of the 4 reference animal species and yeast. One hundred seventy-nine of 255 of them included at least 1 predicted mitochondrial protein from *M. leidyi*. However, 76 did not include any *M. leidyi* ortholog, suggesting a potential loss in this species (Fig. 3a). Next, we analyzed the function of the human mitochondrial proteins contained in these 76 OGs using PantherDB. The top 5 enriched GO biological process terms were "tRNA aminoacylation for mitochondrial protein translation (GO:0070127)," "mitochondrial translational termination (GO:0070126)," "asparaginyl-tRNA aminoacylation (GO:0006421)," and "[4Fe-4S] cluster assembly (GO:0044572)" (Fig. 3c). This observation is consistent with our previous study that found that the loss mitochondrial tRNA genes in *M. leidyi* co-occurred with the loss of most mitochondrial aminoacyl tRNA synthetases [53].

4.2 Analysis of Mitochondrial Carrier Proteins in M. leidyi

Mitochondria can be subdivided into four main compartments: mitochondrial outer membrane, mitochondrial inner membrane, mitochondrial intermembrane space, and mitochondrial matrix. The mitochondrial outer membrane is permeable to certain solutes (molecular mass less than 5 kDa). However, the mitochondrial inner membrane is impermeable to most molecules. Mitochondrial carrier proteins are hydrophobic inner-membrane proteins involved in the transport of specific substrates, like ADP, ATP, NAD+, succinate, fumarase, malate, and phosphate, across the mitochondrial inner membrane that are present in all major eukaryotic groups [54, 55].

In the four reference animal species, the number of mitochondrial carrier proteins ranges from 44 in *C. elegans* to 54 in human. We found that the number of mitochondrial carrier proteins in *M. leidyi* (35) was lower than those in the bilaterian animals, but similar to that in yeast (Fig. 3b). Mitochondrial carrier proteins not identified in *M. leidyi* included the tricarboxylate carrier protein, mitochondrial uncoupling protein 4, mitochondrial brown fat

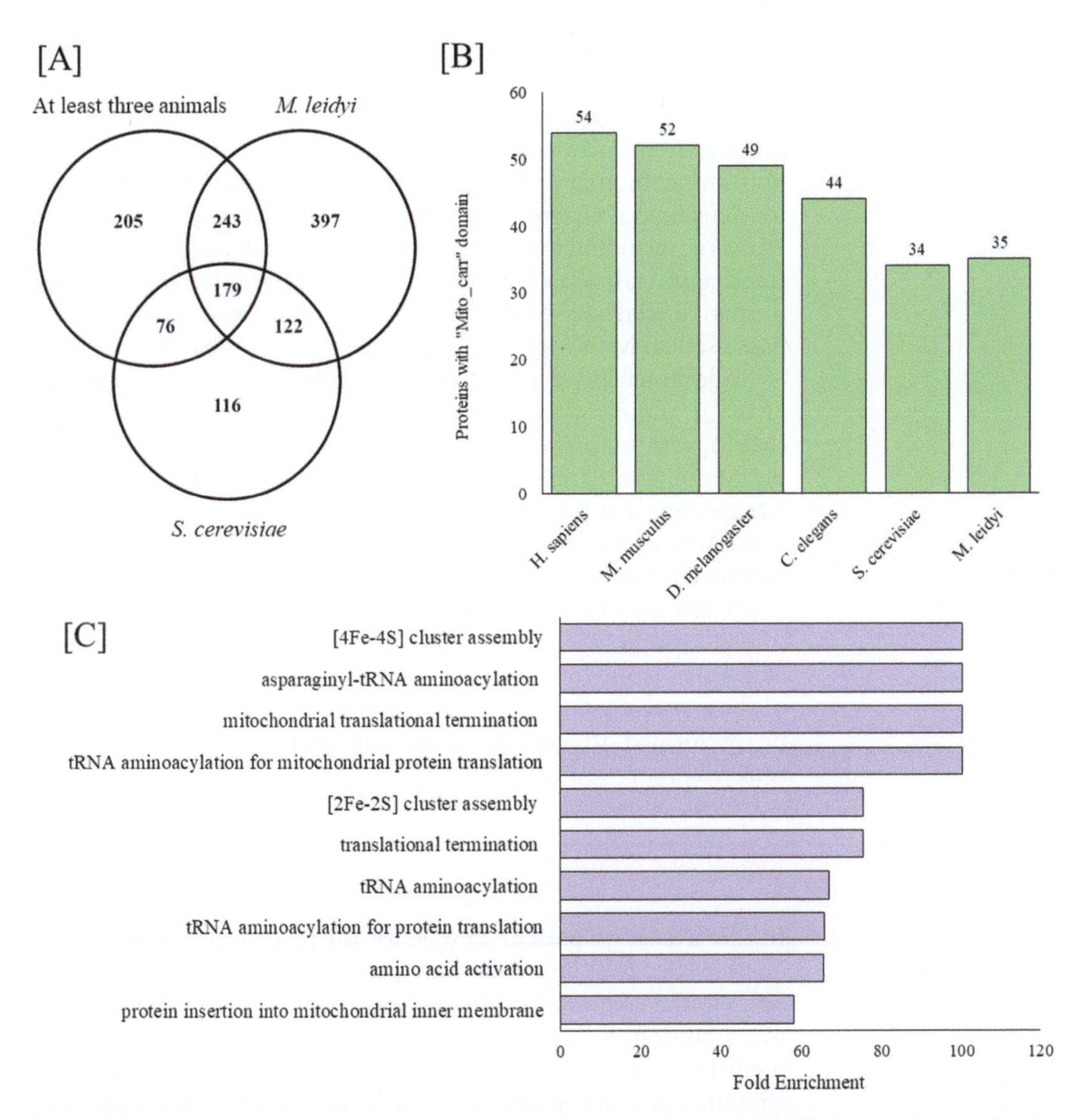

Fig. 3 (**a**) Venn diagram of OGs (groups of orthologous proteins predicted by Proteinortho) containing mitochondrial proteins from at least three out of the four reference animal species, *M. leidyi*, and yeast. (**b**) The number of proteins with "Mito_carr" protein domains. (**c**) Functional annotation of proteins from 76 OGs that contain mitochondrial proteins from the majority of reference animals and yeast but not from *M leidyi*. Functional annotation was performed using the human proteins from these 76 OGs in PantherDB

uncoupling protein 1, and mitochondrial ornithine transporter 2. Further analysis would be required to identify whether and how these substrates are imported into the mitochondria in *M. leidyi*.

4.3 Species-Specific Mitochondrial Proteins in M. leidyi

Out of 320 *M. leidyi* proteins predicted to possess an MTS, 144 lacked an ortholog in any reference mitochondrial proteomes. Sixty-three of these 144 proteins possessed at least 1 predicted protein domain. These "species-specific" proteins can represent instances of neolocalization of cytosolic proteins, known mitochondrial proteins with high rates of sequence evolution, or novel mitochondrial proteins. Mitochondrial neolocalized proteins, i.e., predicted mitochondrial proteins in *M. leidyi* with only non-mitochondrial orthologs in reference species, are the easiest to detect and analyze. We identified several such proteins in *M. leidyi*, including five cytosolic ribosomal proteins, DNA methyltransferase 1-associated protein 1 (DMAP1), and NAD-dependent protein deacetylase sirtuin-2 (SIR2). All these proteins in *M. leidyi* possessed an MTS sequence predicted at high probability.

Additionally, we detected a *M. leidyi* protein containing an "MFS_1" domain, belonging to the Major Facilitator Superfamily (MFS), with no orthologs in any reference proteomes. The Major Facilitator Superfamily is the largest group of secondary membrane transporters that is responsible for the transport of a wide range of substrates across various membranes [56, 57]. BLASTP identified potential homologs of the *M. leidyi* MFS protein in several prokaryotes, but only three eukaryote species. Interestingly, the majority of the eukaryotic sequences possessed an MTS as predicted by MitoFates (four of six sequences). Experimental characterization of these predicted species-specific mitochondrial proteins needs to be performed to confirm their mitochondrial localization and function.

5 Conclusion

In this article, we present a guide for the prediction and analysis of nuclear-encoded mitochondrial proteins (mitochondrial proteome) in the ctenophore *M. leidyi* with MitoPredictor. MitoPredictor, a random forest-based classifier, uses orthology search, MTS prediction, and protein domain information for inferring mitochondrial proteins in animals. In addition, MitoPredictor provides an easy-to-use R Shiny applet for the visualization and analysis of the results. The output of MitoPredictor is also well-suited for further functional analysis using several existing tools, like PantherDB and ConsensusPathDB. MitoPredictor identified 1162 mitochondrial proteins in *M. leidyi*. One hundred fifty-six of these predicted mitochondrial proteins did not possess an ortholog in the five reference species included in MitoPredictor. Several well-conserved mitochondrial proteins were not identified in *M. leidyi*, including multiple mitochondrial aminoacyl tRNA synthetases, multiple proteins involved in the assembly of electron transport chain complexes, and several mitochondrial carrier proteins. While this article focuses on the ctenophore *M. leidyi*, MitoPredictor can be useful for predicting and analyzing mitochondrial proteomes of other animal species.

References

1. Hatefi Y (1985) The mitochondrial electron transport and oxidative phosphorylation system. Annu Rev Biochem 54:1015–1069. https://doi.org/10.1146/annurev.bi.54.070185.005055
2. Stehling O, Wilbrecht C, Lill R (2014) Mitochondrial iron–sulfur protein biogenesis and human disease. Biochimie 100:61–77. https://doi.org/10.1016/j.biochi.2014.01.010
3. Guda P, Guda C, Subramaniam S (2007) Reconstruction of pathways associated with amino acid metabolism in human mitochondria. Genomics Proteomics Bioinformatics 5(3):166–176. https://doi.org/10.1016/S1672-0229(08)60004-2
4. Mayr JA (2015) Lipid metabolism in mitochondrial membranes. J Inherit Metab Dis 38(1):137–144. https://doi.org/10.1007/s10545-014-9748-x
5. Oberst A, Bender C, Green DR (2008) Living with death: the evolution of the mitochondrial pathway of apoptosis in animals. Cell Death Differ 15(7):1139–1146. https://doi.org/10.1038/cdd.2008.65
6. Wang C, Youle RJ (2009) The role of mitochondria in apoptosis. Annu Rev Genet 43(1):95–118. https://doi.org/10.1146/annurev-genet-102108-134850
7. Chandel NS (2014) Mitochondria as signaling organelles. BMC Biol 12(1):34. https://doi.org/10.1186/1741-7007-12-34
8. Meisinger C, Sickmann A, Pfanner N (2008) The mitochondrial proteome: from inventory to function. Cell 134(1):22–24
9. Wiedemann N, Pfanner N (2017) Mitochondrial machineries for protein import and assembly. Annu Rev Biochem 86(1):685–714. https://doi.org/10.1146/annurev-biochem-060815-014352
10. Calvo SE, Clauser KR, Mootha VK (2016) MitoCarta2.0: an updated inventory of mammalian mitochondrial proteins. Nucleic Acids Res 44(D1):D1251–D1257. https://doi.org/10.1093/nar/gkv1003
11. Smith AC, Robinson AJ (2016) MitoMiner v3.1, an update on the mitochondrial proteomics database. Nucleic Acids Res 44(D1):D1258–D1261. https://doi.org/10.1093/nar/gkv1001
12. Li J, Cai T, Wu P et al (2009) Proteomic analysis of mitochondria from Caenorhabditis elegans. Proteomics 9(19):4539–4553
13. Hu Y, Comjean A, Perkins LA, Perrimon N, Mohr SE (2015) GLAD: an online database of gene list annotation for drosophila. J Genomics 3:75–81. https://doi.org/10.7150/jgen.12863
14. Heazlewood JL, Howell KA, Whelan J, Millar AH (2003) Towards an analysis of the rice mitochondrial proteome. Plant Physiol 132(1):230–242. https://doi.org/10.1104/pp.102.018986
15. Salvato F, Havelund JF, Chen M et al (2014) The potato tuber mitochondrial proteome. Plant Physiol 164(2):637–653. https://doi.org/10.1104/pp.113.229054
16. Millar AH, Sweetlove LJ, Giegé P, Leaver CJ (2001) Analysis of the Arabidopsis mitochondrial proteome. Plant Physiol 127(4):1711–1727. https://doi.org/10.1104/pp.010387
17. Rao RSP, Salvato F, Thal B, Eubel H, Thelen JJ, Møller IM (2017) The proteome of higher plant mitochondria. Mitochondrion 33:22–37. https://doi.org/10.1016/j.mito.2016.07.002
18. Cherry JM, Hong EL, Amundsen C et al (2012) Saccharomyces Genome Database: the genomics resource of budding yeast. Nucleic Acids Res 40(D1):D700–D705. https://doi.org/10.1093/nar/gkr1029
19. Gawryluk RMR, Chisholm KA, Pinto DM, Gray MW (2014) Compositional complexity of the mitochondrial proteome of a unicellular eukaryote (Acanthamoeba castellanii, supergroup Amoebozoa) rivals that of animals, fungi, and plants. J Proteome 109:400–416. https://doi.org/10.1016/j.jprot.2014.07.005
20. Lavrov DV, Pett W (2016) Animal mitochondrial DNA as we do not know it: mt-genome organization and evolution in nonbilaterian lineages. Genome Biol Evol 8(9):2896–2913. https://doi.org/10.1093/gbe/evw195
21. Emms DM, Kelly S (2019) OrthoFinder: phylogenetic orthology inference for comparative genomics. Genome Biol 20(1):238. https://doi.org/10.1186/s13059-019-1832-y
22. Persson E, Sonnhammer ELL (2022) InParanoid-DIAMOND: faster orthology analysis with the InParanoid algorithm. Bioinformatics 38(10):2918–2919. https://doi.org/10.1093/bioinformatics/btac194
23. Lechner M, Findeiß S, Steiner L, Marz M, Stadler PF, Prohaska SJ (2011) Proteinortho: detection of (co-)orthologs in large-scale analysis. BMC Bioinformatics 12(1):124. https://doi.org/10.1186/1471-2105-12-124

24. Emanuelsson O, Nielsen H, Brunak S, von Heijne G (2000) Predicting subcellular localization of proteins based on their N-terminal amino acid sequence. J Mol Biol 300(4): 1005–1016. https://doi.org/10.1006/jmbi.2000.3903
25. Fukasawa Y, Tsuji J, Fu SC, Tomii K, Horton P, Imai K (2015) MitoFates: improved prediction of mitochondrial targeting sequences and their cleavage sites *[S]. Mol Cell Proteomics 14(4): 1113–1126. https://doi.org/10.1074/mcp.M114.043083
26. Yu CS, Chen YC, Lu CH, Hwang JK (2006) Prediction of protein subcellular localization. Proteins: Struct Funct Bioinform 64(3): 643–651. https://doi.org/10.1002/prot.21018
27. King BR, Guda C (2007) ngLOC: an n-gram-based Bayesian method for estimating the subcellular proteomes of eukaryotes. Genome Biol 8(5):R68. https://doi.org/10.1186/gb-2007-8-5-r68
28. Kumar R, Kumari B, Kumar M (2018) Proteome-wide prediction and annotation of mitochondrial and sub-mitochondrial proteins by incorporating domain information. Mitochondrion 42:11–22. https://doi.org/10.1016/j.mito.2017.10.004
29. Muthye V, Kandoi G, Lavrov DV (2020) MMPdb and MitoPredictor: tools for facilitating comparative analysis of animal mitochondrial proteomes. Mitochondrion 51:118–125. https://doi.org/10.1016/j.mito.2020.01.001
30. Salvatore M, Warholm P, Shu N, Basile W, Elofsson A (2017) SubCons: a new ensemble method for improved human subcellular localization predictions. Bioinformatics 33(16): 2464–2470. https://doi.org/10.1093/bioinformatics/btx219
31. Briesemeister S, Blum T, Brady S, Lam Y, Kohlbacher O, Shatkay H (2009) SherLoc2: a high-accuracy hybrid method for predicting subcellular localization of proteins. J Proteome Res 8(11):5363–5366. https://doi.org/10.1021/pr900665y
32. Goldberg T, Hamp T, Rost B (2012) LocTree2 predicts localization for all domains of life. Bioinformatics 28(18):i458–i465. https://doi.org/10.1093/bioinformatics/bts390
33. Blum T, Briesemeister S, Kohlbacher O (2009) MultiLoc2: integrating phylogeny and Gene Ontology terms improves subcellular protein localization prediction. BMC Bioinformatics 10(1):274. https://doi.org/10.1186/1471-2105-10-274
34. Schultz DT, Haddock SHD, Bredeson JV, Green RE, Simakov O, Rokhsar DS (2023) Ancient gene linkages support ctenophores as sister to other animals. Nature 618(7963): 110–117. https://doi.org/10.1038/s41586-023-05936-6
35. Pett W, Ryan JF, Pang K et al (2011) Extreme mitochondrial evolution in the ctenophore Mnemiopsis leidyi: insight from mtDNA and the nuclear genome. Mitochondrial DNA 22(4):130–142. https://doi.org/10.3109/19401736.2011.624611
36. Moroz LL, Kocot KM, Citarella MR et al (2014) The ctenophore genome and the evolutionary origins of neural systems. Nature 510(7503):109–114. https://doi.org/10.1038/nature13400
37. Johnson SB, Winnikoff JR, Schultz DT et al (2022) Speciation of pelagic zooplankton: invisible boundaries can drive isolation of oceanic ctenophores. Front Genet 13:970314. https://doi.org/10.3389/fgene.2022.970314
38. Schultz DT, Francis WR, McBroome JD, Christianson LM, Haddock SHD, Green RE (2021) A chromosome-scale genome assembly and karyotype of the ctenophore Hormiphora californensis. G3 (Bethesda) 11(11). https://doi.org/10.1093/g3journal/jkab302
39. Schultz DT, Eizenga JM, Corbett-Detig RB, Francis WR, Christianson LM, Haddock SHD (2020) Conserved novel ORFs in the mitochondrial genome of the ctenophore *Beroe forskalii*. PeerJ 8:e8356. https://doi.org/10.7717/peerj.8356
40. Ryan JF, Pang K, Schnitzler CE et al (2013) The genome of the ctenophore Mnemiopsis leidyi and its implications for cell type evolution. Science 342(6164):1242592. https://doi.org/10.1126/science.1242592
41. Kohn AB, Citarella MR, Kocot KM, Bobkova YV, Halanych KM, Moroz LL (2012) Rapid evolution of the compact and unusual mitochondrial genome in the ctenophore, Pleurobrachia bachei. Mol Phylogenet Evol 63(1): 203–207. https://doi.org/10.1016/j.ympev.2011.12.009
42. Wang M, Cheng F (2019) The complete mitochondrial genome of the Ctenophore *Beroe cucumis*, a mitochondrial genome showing rapid evolutionary rates. Mitochondrial DNA B Resour 4(2):3774–3775. https://doi.org/10.1080/23802359.2019.1580165
43. Moreland RT, Nguyen AD, Ryan JF et al (2014) A customized Web portal for the genome of the ctenophore Mnemiopsis leidyi. BMC Genomics 15(1):316. https://doi.org/10.1186/1471-2164-15-316

44. Fu L, Niu B, Zhu Z, Wu S, Li W (2012) CD-HIT: accelerated for clustering the next-generation sequencing data. Bioinformatics 28(23):3150–3152. https://doi.org/10.1093/bioinformatics/bts565
45. Koskinen P, Törönen P, Nokso-Koivisto J, Holm L (2015) PANNZER: high-throughput functional annotation of uncharacterized proteins in an error-prone environment. Bioinformatics 31(10):1544–1552. https://doi.org/10.1093/bioinformatics/btu851
46. Fang H, Gough J (2013) dcGO: database of domain-centric ontologies on functions, phenotypes, diseases and more. Nucleic Acids Res 41(D1):D536–D544. https://doi.org/10.1093/nar/gks1080
47. Herwig R, Hardt C, Lienhard M, Kamburov A (2016) Analyzing and interpreting genome data at the network level with ConsensusPathDB. Nat Protoc 11(10):1889–1907. https://doi.org/10.1038/nprot.2016.117
48. Mi H, Muruganujan A, Ebert D, Huang X, Thomas PD (2019) PANTHER version 14: more genomes, a new PANTHER GO-slim and improvements in enrichment analysis tools. Nucleic Acids Res 47(D1):D419–D426. https://doi.org/10.1093/nar/gky1038
49. Mi H, Muruganujan A, Casagrande JT, Thomas PD (2013) Large-scale gene function analysis with the PANTHER classification system. Nat Protoc 8(8):1551–1566. https://doi.org/10.1038/nprot.2013.092
50. Szklarczyk D, Gable AL, Lyon D et al (2019) STRING v11: protein–protein association networks with increased coverage, supporting functional discovery in genome-wide experimental datasets. Nucleic Acids Res 47(D1):D607–D613. https://doi.org/10.1093/nar/gky1131
51. Harris TW, Arnaboldi V, Cain S et al (2020) WormBase: a modern Model Organism Information Resource. Nucleic Acids Res 48(D1):D762–D767. https://doi.org/10.1093/nar/gkz920
52. Lyne R, Smith R, Rutherford K et al (2007) FlyMine: an integrated database for Drosophila and Anopheles genomics. Genome Biol 8(7):R129. https://doi.org/10.1186/gb-2007-8-7-r129
53. Pett W, Lavrov DV (2015) Cytonuclear interactions in the evolution of animal mitochondrial tRNA metabolism. Genome Biol Evol 7(8):2089–2101. https://doi.org/10.1093/gbe/evv124
54. Palmieri F (1994) Mitochondrial carrier proteins. FEBS Lett 346(1):48–54. https://doi.org/10.1016/0014-5793(94)00329-7
55. Palmieri F, Pierri CL, De Grassi A, Nunes-Nesi A, Fernie AR (2011) Evolution, structure and function of mitochondrial carriers: a review with new insights. Plant J 66(1):161–181. https://doi.org/10.1111/j.1365-313X.2011.04516.x
56. Pao Stephanie S, Paulsen Ian T, Saier Milton H (1998) Major facilitator superfamily. Microbiol Mol Biol Rev 62(1):1–34. https://doi.org/10.1128/mmbr.62.1.1-34.1998
57. Yan N (2015) Structural biology of the major facilitator superfamily transporters. Annu Rev Biophys 44(1):257–283. https://doi.org/10.1146/annurev-biophys-060414-033901

Chapter 11

Expression and Functional Analysis of Ctenophore Glutamate Receptor Genes

Mark L. Mayer

Abstract

The functional analysis of ctenophore neurotransmitter receptors, transporters, and ion channels can be greatly simplified by use of heterologous expression systems. Heterologous expression allows the characterization of individual membrane proteins, expressed at high levels in cells, where background activity by endogenous ion channels and transporters is with few exceptions minimal. The goal of such experiments is to gain an in-depth understanding of the behavior and regulation of individual molecular species, which is challenging in native tissue, but especially so in the case of ctenophores and other marine organisms. Coupled with transcriptome analysis, and immunohistochemical studies of receptor expression in vivo, experiments with heterologous expression systems can provide valuable insight into cellular activity, prior to more challenging functional studies on native tissues.

Key words Xenopus oocytes, Voltage clamp, Ion channels, Glutamate receptors

1 Introduction

Cation permeable glutamate receptor ion channels (iGluRs) are the major mediators of excitatory synaptic transmission in the brain of all animals examined to date; in addition, they mediate synaptic transmission at the neuromuscular junction of insects and crustaceans and probably play similar roles in ctenophores [1]. Assembled as tetramers, iGluR subunits have a unique domain organization that is distinct from other neurotransmitter receptors and easily recognized when searching genome databases [2]. In detail, iGluRs have two extracellular domains, commonly named the amino terminal (ATD) and ligand binding (LBD) domain, which are connected by short unstructured linkers and fused with an ion channel domain formed by three membrane spanning α-helices, with a characteristic reentrant pore loop and pore helix found also in many voltage-gated ion channels (Fig. 1). The ATD and LBD have a characteristic clam-shell fold and likely evolved from precursors of soluble bacterial periplasmic binding proteins. The ion

Leonid L. Moroz (ed.), *Ctenophores: Methods and Protocols*, Methods in Molecular Biology, vol. 2757,
https://doi.org/10.1007/978-1-0716-3642-8_11, © Springer Science+Business Media, LLC, part of Springer Nature 2024

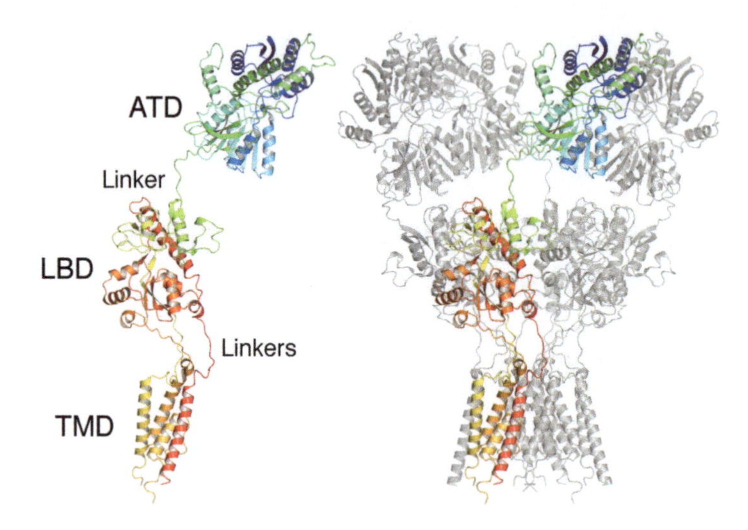

Fig. 1 Domain organization in an iGluR subunit (left) and the assembly of an iGluR tetramer with the structures of three subunits drawn in transparent gray shading (right). The ATD and LBD are globular domains that can be individually isolated and expressed as soluble proteins. In the intact protein, these domains are connected by linkers to each other and to the ion channel formed by the TMD. Crystal structures for the LBD of two ctenophore iGluRs, ML032222a and PbiGluR3, have been solved and revealed the presence of endogenous glycine in the agonist binding site [3]

channel domain is inserted into the lower lobe of the LBD with the 1st two transmembrane α-helices and pore helix interrupting the sequence of the LBD [4]. Simpler bacterial iGluRs, which lack the ATD and which have only two membrane spanning α-helices and the pore loop, have also been characterized [5], but to date, the addition of the ATD and third transmembrane domain has been reported only in eukaryotes and likely evolved before the last common ancestor of ctenophores and probably before the separation of animals and plants, because multiple full-length iGluRs have also been discovered in diverse plant species [6].

2 Heterologous Expression Systems

Two systems, each with unique advantages and limitations, are currently widely used for functional analysis of iGluRs and other membrane proteins: *Xenopus laevis* oocytes injected with mRNA, or less commonly cDNA injected into the nucleus, and mammalian cell lines, most commonly HEK 293 cells, transiently transfected with cDNA. The oocyte system is robust, requires very simple equipment for both expression and functional analysis using two-electrode voltage clamp, and is relatively low cost. Because high quality oocytes are available from commercial sources, it is no longer necessary to maintain an aquarium for housing a colony of animals, and only a simple chilled incubator is required for oocyte storage and heterologous expression. The major disadvantages of the oocyte system are the requirement of an additional step after isolation of cDNAs to prepare mRNA for microinjection, and the large size of oocytes, which precludes rapid solution exchange, limiting kinetic studies on ligand-gated ion channels. The use of HEK 293 cells has the advantage that all forms of patch clamp recording are possible, including whole-cell recording with its ability to control the ionic composition of the cytoplasm, and due to the small size of these cells, solution exchange on the millisecond timescale is possible. The disadvantages are the need for more complex and expensive equipment, including high-power microscopes, preferably with fluorescence optics, an air table for vibration isolation, better quality micromanipulators, and the associated need for a cell culture facility with laminar flow hoods, incubators with temperature and CO_2 control, and the cost of sterile plasticware and expensive cell culture media. Here, I focus on use of *Xenopus* oocytes for analysis of ctenophore glutamate receptor ion channels (iGluRs), but the approaches described could serve as a guide for studying other ctenophore receptors, ion channels, and transporters.

3 Construct Design and Expression

A prerequisite for heterologous expression is the availability of high-quality gene sequence information for the protein(s) of interest. For our analysis of glutamate receptor ion channels, we initially selected seven *M. leidyi* iGluR subunits genes [7], excluding nine others, which either lacked predicted signal peptides or which had large insertions or deletions compared to consensus iGluR subunit sequences [3]. We used native cDNAs transcribed from *M. leidyi* mRNA, for which the 5′ and 3-untranslated regions were truncated or removed, and also synthetic genes with vertebrate (*H. sapiens*) codon optimization when native cDNAs were not available; this has the advantage of potentially avoiding low receptor density in cases where differences in codon usage between species impact heterologous expression. All constructs were cloned into the polylinker of pGEMHE [8], a plasmid engineered with strong promoters and polyadenylation signals that is widely used for efficient translation in *Xenopus* oocytes, using standard techniques to prepare DNA templates for mRNA synthesis (Fig. 2).

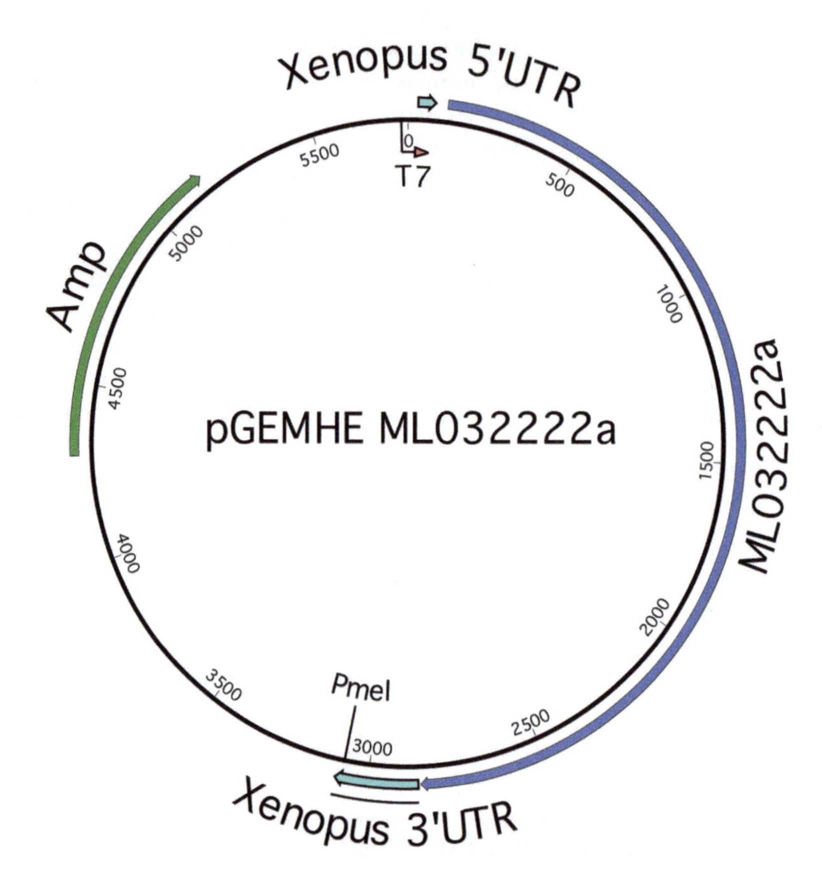

Fig. 2 The pGEMHE construct for the ctenophore iGluR ML032222a, the cDNA for which was inserted between vector encoded 5′ and 3′ untranslated regions of the *Xenopus* β-globin gene which flank the polylinker. The plasmid is grown under ampicillin selection, linearized with the blunt cutter PmeI, and used as a template for mRNA synthesis using T7 polymerase and the Ambion mMessage mMachine transcription kit

4 Preparation of mRNA and Microinjection

Prior to mRNA synthesis, pGEMHE constructs containing ctenophore iGluRs should be linearized by digestion with a single cutter restriction enzyme targeting a site in the 3′ polylinker; we used PmeI, but if other restriction enzymes are used to linearize the plasmid, either blunt cutters or enzymes leaving a 5′ overhang should be chosen; 3′ overhangs should be avoided. The restriction enzyme plasmid reaction mix should then be digested with proteinase K and SDS, followed by purification with phenol:chloroform extraction using standard protocols for molecular biology. Capped mRNA synthesis under control of the T7 promoter is performed using the Ambion mMessage mMachine transcription kit, following the manufacturer's instructions; typical yields are >100 μg per reaction. Following synthesis, the mRNA is precipitated using isopropyl alcohol and then stock solutions prepared using DEPC-treated water; 5 μL aliquots at a concentration of 1 μg/μL are stored at −80 °C using RNAse-free plastic ware.

Defolliculated stage 5–6 oocytes, 1.2–1.3 mM diameter, can be purchased from Ecocyte Bioscience, Austin, TX, and upon delivery should be immediately stored in an incubator set at 16–18 °C. For oocyte injection and storage, we use a solution containing in mM: 96 NaCl, 2 KCl, 1.8 $CaCl_2$, 1 $MgCl_2$, 2.5 Na-pyruvate, and 5 HEPES pH 7.6, with 5 μg/L Gentamicin; fresh batches should be made weekly. On the day of delivery, individual oocytes should be selected prior to injection by visual inspection under a stereomicroscope with variable magnification, checking for complete removal of the follicular cell layer, and avoiding oocytes which have low turgor pressure. Healthy stage 5 oocytes have a dark animal pole, a pale vegetal pole, with for mature stage 6 oocytes a pigment free equatorial band separating the two poles. A chamber for microinjection can be fabricated using a Petri dish onto which a small section of plastic mesh grid, cross section around 1 mM, has been glued at its corners, such that individual oocytes sit in each well of the grid. It is important to use a "cold" light source with fiber optic light guides to illuminate the microinjection chamber to avoid heating the oocyte. To transfer oocytes to and from the injection chamber, it is convenient to use a Drummond 200 μL disposable calibrated glass micropipette connected to a 1 mL syringe with flexible silicone tubing, with the fire-polished end facing the oocyte.

Before selecting oocytes for injection, individual mRNA aliquots should be thawed and stored on ice prior to dilution to the desired concentration using DEPC-treated water. Immediately before use, glass microinjection needles should be prepared using a micropipette puller; because the glass is melted during micropipette fabrication, the tips are sterile and free of RNAse. The microinjection needles are then broken back to a tip diameter of around 15 μM under a high-power microscope. If the tip is too small, it is easily clogged, and it can be difficult to aspirate viscous RNA solutions; if the tip is too large, the oocyte can be damaged and will subsequently undergo lysis.

A Drummond Nanoject programmable injector set at a volume of 50 nL is commonly used for microinjection. The injection needle is first filled with sterile mineral oil, ensuring that no bubbles are present, and then inserted onto the shaft of the Nanoject motor assembly, which is mounted on a micromanipulator set at an angle of about 60°. A 1–2 μL aliquot of RNA is pipetted onto the clean side of a freshly cut piece of parafilm, and under visual control using a stereomicroscope, the shaft of the injector slowly retracted to suck up the RNA into the glass injection needle. This should be monitored by following the meniscus at the oil/RNA interface as it moves up the pipette. It is essential that no air enter the pipette, either from the top of the assembly or from the bottom.

When the needle is filled, press the inject button to check that RNA is expelled from the tip, and then quickly position it below the liquid surface in the injection chamber to avoid drying and clogging. Oocytes are manipulated in the injection chamber, such that their dark animal poles are topside, and the injection needle brought down, so that it presses against the egg, creating a dimple near the interface between the animal and vegetal poles. A further gentle push of the needle is used to puncture the oocyte, after which it should be slightly withdrawn to relieve some of the tension in the membrane, and then the inject button pushed. The oocyte will noticeably swell during the injection, after which the needle can be gently retracted and repositioned above the next oocyte.

Typically, the injection of 25–30 oocytes per construct provides sufficient material for 1–2 days of analysis. Following injection, oocytes are stored at 16–18 °C in 24-well tissue culture dishes, typically 5 oocytes per well, in an incubator, in which a beaker of sterile water is placed to increase humidity and checked daily while changing the storage solution, removing damaged oocytes as necessary. Injected oocytes are typically incubated for around 2–4 days to allow protein expression before being used for recording. Although 50 nL is routinely used as an injection volume in most labs, the quantity of mRNA injected must be determined for individual receptor species in pilot experiments. The maximum should never exceed 50 ng, and at this high level, it is occasionally possible to trigger nonspecific expression of endogenous *Xenopus* RNAs; a good starting amount for a previously uncharacterized RNA would be 10 ng. For RNAs that express much better than average, it is possible to use 100 pg or less. Ideally, the amount of RNA injected should be adjusted to yield current amplitudes of around 1 μA. Because it is not uncommon for ion channels and receptors from distant species such as ctenophores to express at levels too low for functional analysis following heterologous expression, it is also useful to test the coexpression of multiple receptor subunits from a gene family, even if these are predicted to function as homomers; this, of course, is essential for ion channels, which assemble as obligate heteromers, which is often not known during the initial stages of analysis.

5 Two-Electrode Voltage Clamp Analysis

Standard electrophysiological techniques used for functional analysis of ion channel activity using *Xenopus* oocytes require a two-electrode voltage clamp amplifier, a stereomicroscope with variable magnification, a "cold" light source with fiber optic light guides for illumination, micromanipulators to hold amplifier head stages and microelectrodes, a computer with an analogue to digital converter and software for data acquisition and analysis, a chamber for holding the oocyte, a means for changing the extracellular solution to apply ligands, and a table on which the equipment is mounted. The equipment used varies widely in different laboratories, but the Axon™ Guide available from Molecular Devices provides a comprehensive introduction to the principles, theory, and practice of electrophysiological experiments, with information on equipment manufacturers for necessary components and supplies. An excellent source of detailed protocols specifically for the use of *Xenopus* oocytes in electrophysiological experiments is also available [9]. In our experiments, we used a custom designed chamber with a volume of 5 μL to allow relatively rapid solution exchange but with economical volumes of solutions, which were applied using a homemade assembly of polyimide-coated quartz tubing positioned adjacent to the oocyte, with solution switching achieved using the Lee Company solenoid valves under computer control.

We use a standard extracellular solution of composition (in mM): 100 NaCl, 1 KCl, 2 $CaCl_2$, 1 $MgCl_2$, 5 HEPES, pH 7.6. To voltage clamp an oocyte, it must be impaled with two microelectrodes, using a procedure similar to that for microinjection. The electrodes are pressed against the egg under visual control, creating a dimple, and then a further gentle push is used to puncture the oocyte. After this, the electrodes should be slightly withdrawn to relieve some of the tension in the dimple, before turning on the voltage clamp. The holding current should typically be less than 50 nA at −60 mV.

Xenopus oocytes have unique properties, which require a few considerations compared to other preparations used for electrophysiological experiments. To maintain voltage clamp control, it is necessary to use low-resistance microelectrodes filled with 3 M KCl, the leakage of which can damage oocytes, increase the holding current, and lead to unstable recordings; to avoid this, we use agarose plugs in the tips to prevent electrolyte leakage into the cell after impalement by two microelectrodes [10]; this has the additional advantage that the same electrodes can be used to record from multiple oocytes. Because constructs that express well can generate membrane currents of amplitudes exceeding 10–20 μA, we used a fixed gain virtual-ground circuit to voltage clamp the recording chamber at ground, with an extracellular voltage-sensing microelectrode placed near the oocyte surface; this is beneficial even with lower amplitude currents, but as noted above, it is preferable to decrease the amount of RNA injected to avoid such large currents. The gain of the main voltage clamp circuit used to control the oocyte membrane potential should be sufficiently high that, at the maximum current recorded, deviations of less than 1 mV occur.

The expression of endogenous ion channels by *Xenopus* oocytes, with a few exceptions, has little impact on the analysis of the activity of heterologous ion channels. The exceptions are a robust endogenous calcium-activated chloride current, gap junction hemi channels, and stretch-activated ion channels. The latter are activated only at extreme membrane potentials, or with very low extracellular divalent cation concentrations, or obviously with membrane deformation. The calcium-activated chloride current is more of a problem, because its activation distorts responses for heterologous expression of Ca-permeable ion channels, making accurate analysis of ion channel activity impossible (Fig. 3). On occasion, this can be useful for a yes/no decision about the functional expression of heterologous constructs and their calcium permeability, because the calcium-activated chloride current can substantially increase the amplitude of currents from heterologous ion channels, but this should otherwise be avoided. This can be achieved by either lowering the extracellular Ca^{2+} concentration to 50 μM or by substituting $CaCl_2$ with $BaCl_2$, which does not activate the endogenous calcium-activated chloride current (Fig. 3). Although not unique to oocytes, it is important to be aware that the cytoplasm contains high micromolar concentrations of the polyamines spermine and spermidine and that these act as voltage-dependent permeable blockers of iGluRs and some other cation permeable ion channels [11, 12]. As a result, the conductance-voltage plot for ctenophore iGluRs shows a characteristic biphasic rectification as polyamines enter and block and then pass through the ion channel pore as the membrane potential is depolarized.

The large majority of iGluRs from numerous species characterized to date undergo rapid and profound desensitization. The only way to accurately record such responses is to use outside out patch recording with rapid solution exchange on the sub-ms timescale. In the case of the ctenophore iGluR ML032222a, the onset of desensitization was sufficiently slow that by exchanging the recording solution at >150 bath volumes per minute, desensitizing responses to glycine could be recorded using *Xenopus* oocytes. For other iGluR subtypes, application of the lectin concanavalin A (Sigma Type IV), 0.6 mg/mL dissolved in recording solution and 0.2 μM filtered, applied for 4 min after oocytes are impaled, frequently attenuates desensitization and is always worth trying, allowing analysis of ligand sensitivity, ion selectivity, and voltage dependence [13, 14]; this was successful for the *M. leidyi* glutamate-activated iGluR ML05909a [3]. In the case of ML032222a, concanavalin A had no effect on desensitization, and to obtain steady-state currents for analysis of current-voltage plots, we introduced Cys mutants in the ligand binding domain dimer interface to block conformational changes that initiate desensitization. This approach has been used to study iGluRs from other species [13, 15] and was facilitated by prior structural knowledge obtained by solving the crystal of the

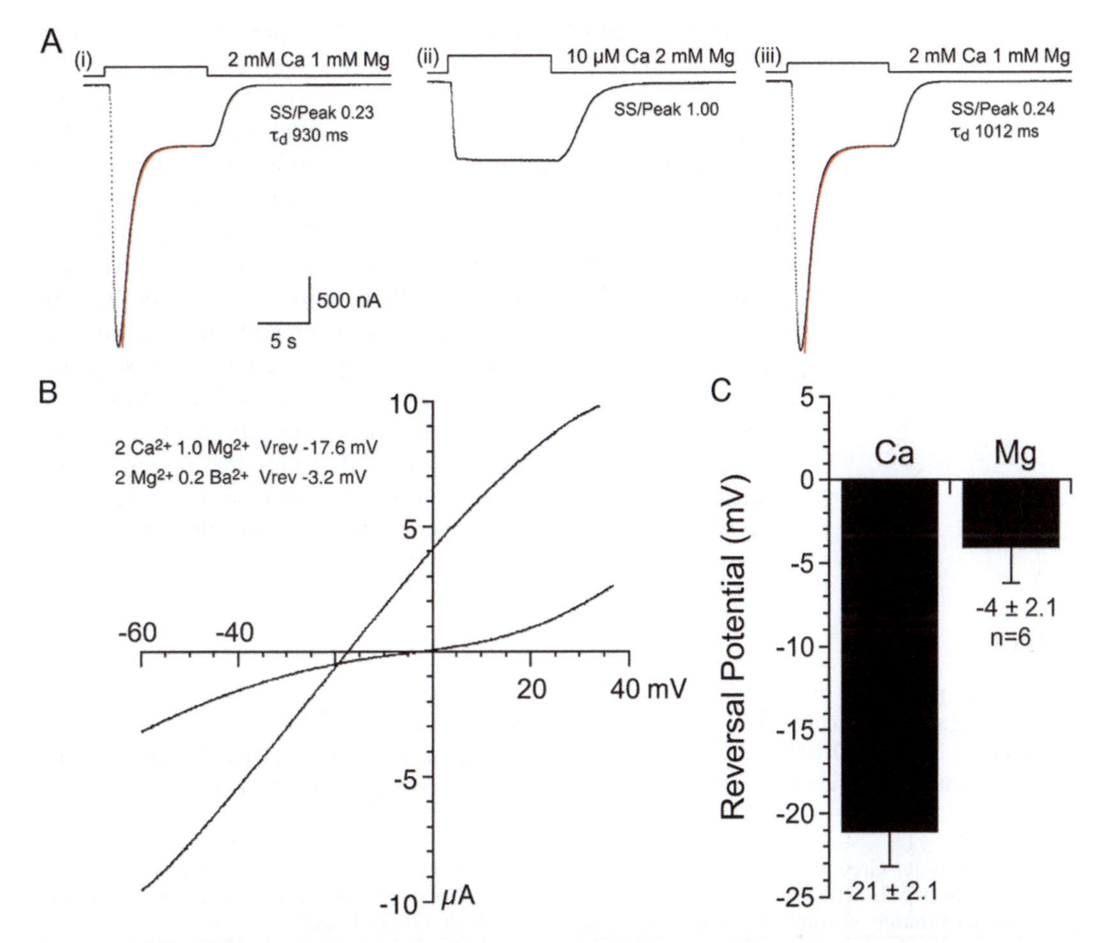

Fig. 3 iGluR responses recorded from *Xenopus* oocytes showing artifacts due to activation of calcium-dependent chloride channels. (**a**) Shows apparently desensitizing responses for which the transient inward current is instead due to activation of endogenous calcium-dependent chloride channels for iGluR responses recorded with 2 mM extracellular Ca^{2+}; switching to 50 μM Ca^{2+} to reduce calcium influx through iGluRs current reversibly eliminates this artifact. (**b**) Current voltage plots of iGluR responses recorded with either 2 mM Ca^{2+} and 1 mM Mg^{2+} or 0.2 mM Ba^{2+} and 2 mM Mg^{2+} in the extracellular solution show different reversal potentials. (**c**) On average, the reversal potential shifts from −21 mV to −4 mV, close to the value expected for a nonselective cation permeable iGluR, when activation of endogenous calcium-dependent chloride channel is blocked by replacement of Ca^{2+} by Ba^{2+}, due to differences between the equilibrium potential for Cl^- ions and cations

ML032222a ligand binding domain, but with sequence alignments used to identify equivalent residues, it should be applicable to other ctenophore iGluRs.

6 Screening for Ligand Binding

In our initial screens to assay ligand binding profiles for *M. leidyi* iGluR subunits, we attempted to use proteolysis protection assays for purified ligand binding domains before performing functional analysis with full-length receptors; this was not successful for ML032222a because the LBD was extraordinarily stable in the presence of trypsin and chymotrypsin. We discovered that this unexpected result was due to binding of endogenous glycine only after the ML032222a LBD was purified and crystallized in the presence of glutamate and its structure solved, revealing glycine in the binding site. After solving the structure, we were able to predict and experimentally confirm that other ctenophore iGluRs also bind glycine with very high affinity and to characterize their responses by expression in *Xenopus* oocytes. Subsequent amino acid sequence alignments, combined with structural information, indicate that some ctenophore iGluRs do not bind either glutamate or glycine but may bind novel ligands, indicating that a broader screen beyond traditional neurotransmitter candidates might be warranted. This note of caution might apply also to the study of other ctenophore neurotransmitter receptors.

References

1. Moroz LL et al (2014) The ctenophore genome and the evolutionary origins of neural systems. Nature 510:109–114
2. Hollmann M, Heinemann S (1994) Cloned glutamate receptors. Annu Rev Neurosci 17: 31–108
3. Alberstein R, Grey R, Zimmet A, Simmons DK, Mayer ML (2015) Glycine activated ion channel subunits encoded by ctenophore glutamate receptor genes. Proc Natl Acad Sci U S A 112:E6048–E6057
4. Sobolevsky AI, Rosconi MP, Gouaux E (2009) X-ray structure, symmetry and mechanism of an AMPA-subtype glutamate receptor. Nature 462:745–756
5. Chen GQ, Cui C, Mayer ML, Gouaux E (1999) Functional characterization of a potassium-selective prokaryotic glutamate receptor. Nature 402:817–821
6. Chen J et al (2016) Evolutionary and expression analysis provides evidence for the plant glutamate-like receptors family is involved in woody growth-related function. Sci Rep 6: 32013
7. Ryan JF et al (2013) The genome of the ctenophore Mnemiopsis leidyi and its implications for cell type evolution. Science 342:1242592
8. Liman ER, Tytgat J, Hess P (1992) Subunit stoichiometry of a mammalian K+ channel determined by construction of multimeric cDNAs. Neuron 9:861–871
9. Dascal N (2001) Voltage clamp recordings from Xenopus oocytes. Curr Protoc Neurosci Chapter 6:Unit 6 12
10. Schreibmayer W, Lester HA, Dascal N (1994) Voltage clamping of Xenopus laevis oocytes utilizing agarose-cushion electrodes. Pflugers Arch 426:453–458
11. Lopatin AN, Makhina EN, Nichols CG (1994) Potassium channel block by cytoplasmic polyamines as the mechanism of intrinsic rectification. Nature 372:366–369
12. Bowie D, Mayer ML (1995) Inward rectification of both AMPA and kainate subtype glutamate receptors generated by polyamine-mediated ion channel block. Neuron 15:453–462
13. Lomash S, Chittori S, Brown P, Mayer ML (2013) Anions mediate ligand binding in Adineta vaga glutamate receptor ion channels. Structure 21:414–425
14. Li Y et al (2016) Novel functional properties of Drosophila CNS glutamate receptors. Neuron 92:1036–1048
15. Weston MC, Schuck P, Ghosal A, Rosenmund C, Mayer ML (2006) Conformational restriction blocks glutamate receptor desensitization. Nat Struct Mol Biol 13: 1120–1127

Chapter 12

Expression, Purification, and Determination of Sensitivity to Calcium Ions of Ctenophore Photoproteins

Lyudmila P. Burakova, Svetlana V. Markova, Natalia P. Malikova, and Eugene S. Vysotski

Abstract

Light-sensitive Ca^{2+}-regulated photoproteins of ctenophores are single-chain polypeptide proteins of 206–208 amino acids in length comprising three canonical EF-hand Ca^{2+}-binding sites, each of 12 contiguous residues. These photoproteins are a stable complex of apoprotein and 2-hydroperoxy adduct of coelenterazine. Addition of calcium ions to photoprotein is only required to trigger bright bioluminescence. However, in contrast to the related Ca^{2+}-regulated photoproteins of jellyfish their capacity to bioluminescence disappears on exposure to light over the entire absorption spectral range of ctenophore photoproteins. Here, we describe protocols for expression of gene encoding ctenophore photoprotein in *Escherichia coli* cells, obtaining of the recombinant apoprotein of high purity and its conversion into active photoprotein with synthetic coelenterazine as well as determination of its sensitivity to calcium ions using light-sensitive Ca^{2+}-regulated photoprotein berovin from ctenophore *Beroe abyssicola* as an illustrative case.

Key words Ctenophores, Light-sensitive Ca^{2+}-regulated photoproteins, Coelenterazine, Bioluminescence, Sensitivity to Ca^{2+}

1 Introduction

Ctenophores (comb jellies) are found in oceans worldwide, and practically all the species are bioluminescent [1]. Their bright bioluminescence is determined by Ca^{2+}-regulated photoproteins, which share many properties with those inherent to hydromedusan photoproteins. However, in contrast to hydromedusan photoproteins, they appeared to be sensitive to light, i.e., the exposure of ctenophore photoproteins to light over the entire absorption spectral range leads to the loss of their capability to bioluminescence [2, 3]. These light-sensitive Ca^{2+}-regulated photoproteins were first isolated and partially characterized in 1974 [4, 5]. For the past two decades, the genes encoding Ca^{2+}-regulated photoproteins from four species of ctenophores have been cloned [6]. These are bolinopsin from *Bolinopsis infundibulum* [7], berovin from

Leonid L. Moroz (ed.), *Ctenophores: Methods and Protocols*, Methods in Molecular Biology, vol. 2757, https://doi.org/10.1007/978-1-0716-3642-8_12, © Springer Science+Business Media, LLC, part of Springer Nature 2024

Beroe abyssicola [8], bathocyrovin from *Bathocyroe fosteri* [9], and mnemiopsin from *Mnemiopsis leidyi* [10]. The analysis of amino acid sequences deduced from nucleotide sequences showed that ctenophore photoproteins are single-chain polypeptides of 206–208 amino acids in length, depending on protein with a degree of sequence identity being 84–88%, and, similar to hydromedusan photoproteins, contain three canonical EF-hand Ca^{2+}-binding sites, each consisting of 12 residues. Although the basic properties of ctenophore and hydromedusan photoproteins are identical to a great extent, the sequence identity between these proteins turned out to be very low [6]. To date, the spatial structures of ctenophore photoproteins have been determined only for apoproteins loaded by Ca^{2+}, Mg^{2+}, or Cd^{2+} [11–13]. These structures clearly showed ctenophore photoproteins to retain the same two-domain scaffold characteristic of the hydromedusan photoproteins like obelin or aequorin despite great differences of their amino acid sequences. Although the overall tertiary structures of ctenophore and hydromedusan photoproteins revealed a close similarity, the residues composing the substrate-binding cavity and consequently involved in stabilization of 2-hydroperoxycoelenterazine molecule, catalytic reaction, and emitter formation turned out to be drastically different. It is highly probable that these very distinctions account for the differences in properties of ctenophore and hydromedusan photoproteins, such as sensitivity to light, absorption spectrum, and requirement of alkaline pH for conversion of apoprotein into active photoprotein with synthetic coelenterazine [6].

The Ca^{2+}-regulated photoproteins draw interest due to their broad in vivo and in vitro analytical applications [14]. Nonetheless, the main use of them bases on the ability to emit light on Ca^{2+} binding, which is greatly suitable for measurements of calcium ions within living cells [15]. Now, among the known and characterized photoproteins [16] only aequorin from jellyfish *Aequorea victoria* is widely used to monitor intracellular Ca^{2+}, though there are a number of shortcomings that limit its utility. The most significant one is that Mg^{2+} under physiological concentrations slows down the rate of light response of aequorin on a sudden change of $[Ca^{2+}]$ and decreases its sensitivity to calcium [16], this most likely being due to competition of Mg^{2+} with calcium for Ca^{2+}-binding sites. Bioluminescence of other photoproteins is less sensitive to magnesium ions [16]. Taking into account a great analytical potential of photoproteins, the determination of sensitivity to Ca^{2+} and effect of Mg^{2+} on this property is a very important characteristic of any Ca^{2+}-regulated photoprotein.

Here, we present the protocols to express gene encoding ctenophore photoprotein in *Escherichia coli* cells, obtaining of the recombinant apoprotein of high purity and its conversion into active photoprotein with synthetic coelenterazine as well as determination of its sensitivity to calcium ions both without and in the presence of physiological concentrations of magnesium ions using light-sensitive Ca^{2+}-regulated photoprotein berovin from ctenophore *Beroe abyssicola* as an illustrative case.

2 Materials

2.1 E. coli *Strain*

1. XL1-Blue competent cells (Cat# 200249, Agilent).
2. BL21-CodonPlus (DE3)-RIPL competent cells (Cat# 230280, Agilent).

2.2 Reagents

1. PfuSE DNA polymerase 5000 U/mL (Cat# E363, SibEnzyme).
2. PfuSE-buffer DNA polymerase (Cat# B310, SibEnzyme).
3. dNTP 10 mM (Cat# N025, SibEnzyme).
4. pET22b(+) plasmid (Cat# 70765, Sigma-Aldrich).
5. FauND I (Nde I isoschizomer) 10,000 U/mL (Cat# E009, SibEnzyme).
6. Xho I 20,000 U/mL (Cat# R0146S, New England Biolabs).
7. SE-buffer Y 10× (Cat# B005, SibEnzyme).
8. BSA 10 mg/mL 100× (Cat# B9001S, New England Biolabs).
9. Alkaline phosphatase 20,000 U/mL (Cat# E328, SibEnzyme).
10. DNA ligase T4 200,000 U/mL (Cat# E319, SibEnzyme).
11. SE-buffer DNA ligase T4 10× (Cat# B302, SibEnzyme).
12. Agarose, low gelling temperature (Cat# A9414, Sigma-Aldrich).
13. Agarose (Cat# A9539, Sigma-Aldrich).
14. Acetic acid, glacial (Cat# AX0077, Sigma-Aldrich).
15. Ethidium bromide (Cat# E7637, Sigma-Aldrich).
16. QIAquick PCR Purification Kit (Cat# 28106, Qiagen).
17. QIAquick Gel Extraction Kit (Cat# 28706, Qiagen).
18. QIAprep Spin Miniprep Kit (Cat# 27106, Qiagen).
19. Coelenterazine (Cat# 303, NanoLight Technology).
20. Tris(hydroxymethyl)aminomethane (Cat# 252859, Sigma-Aldrich).
21. BIS–Tris propane (Cat# B4679, Sigma-Aldrich).
22. Sodium chloride (Cat# S5886, Sigma-Aldrich).
23. Potassium chloride (Cat# P9541, Sigma-Aldrich).
24. Calcium chloride (Cat# C4901, Sigma-Aldrich).
25. Magnesium chloride (Cat# C4880, Sigma-Aldrich).

26. Potassium hydroxide (Cat# P5958, Sigma-Aldrich).
27. Ethylenediaminetetraacetic acid (Cat# 431788, Sigma-Aldrich).
28. Ethylene-bis(oxyethylenenitrilo)tetraacetic acid (Cat# E3889, Sigma-Aldrich).
29. Triton X-100 (Cat# 1610407, Bio-Rad).
30. Tryptone (Cat# 403682, PanReac).
31. Bacto Yeast Extract (Cat# 212750, BD Bioscience).
32. Agar (Cat# 212303, PanReac).
33. IPTG (Cat# S012, SibEnzyme).
34. Urea (Cat# A-1049, AppliChem).
35. Ampicillin-sodium salt (Cat# A0166, Sigma-Aldrich).
36. Piperazine-N,N′-bis(2-ethanesulfonic acid) (Cat# 80635, Sigma-Aldrich).
37. Chelex-100 chelating resin (Cat# C7901, Sigma-Aldrich).
38. DC Protein Assay Kit II (Cat# 5000112, Bio-Rad).

2.3 Equipment

1. Thermostat (37 °C with an accuracy of ±0.2 °C) of choice.
2. Shaking incubator (37 °C ± 0.5 °C, 50–400 rpm) of choice.
3. MJ Mini Thermal Cycler (Cat# 345-0412, Bio-Rad) or similar.
4. Stirred water bath (42 °C with an accuracy of ±0.1 °C) of choice.
5. Microcentrifuge (16,000 g with fixed angle rotor) of choice.
6. Centrifuge (21,000 rpm, 1–4 L, 0–40 °C) with rotor supplied by bottle adapters (4 × 250, 6 × 50, or 24 × 15 mL) of choice.
7. Ultrasonic disintegrator (44 kHz, 1000 W) of choice.
8. ÄKTA pure 25 M chromatographic system (Cat# 29-0182-26, GE Healthcare Life Science) or similar.
9. HiTrap DEAE Fast Flow column (Cat# 17-5055-01, GE Healthcare Life Science) or similar.
10. HiTrap Q HP column (Cat# 17-1154-01, GE Healthcare Life Science) or similar.
11. HiTrap Desalting column (Cat# 17-1408-01, GE Healthcare Life Science) or similar.
12. Superloop 10 mL, M6 (Cat# 19-7585-01, GE Healthcare Life Science) or similar.
13. Superloop 50 mL, M6 (Cat# 19-7850-01, GE Healthcare Life Science) or similar.
14. Cary Eclipse fluorescence spectrophotometer (Agilent Technologies) or similar.

15. Double beam UV-VIS spectrophotometer (spectral range 185–900 nM) of choice.
16. Luminometer equipped with a temperature-stabilized cuvette block supplied by neutral-density filters with different transmission coefficients of choice.
17. Amicon Ultra 15 mL Centrifugal Filters (Cat# UFC901024, Merck Millipore Ltd) or similar.
18. Corning 50 mL centrifuge tubes (Cat# CLS430829, Sigma-Aldrich) or similar.
19. Corning 15 mL centrifuge tubes (Cat# CLS430791, Sigma-Aldrich) or similar.
20. Erlenmeyer baffled cell culture flask 1 L (Cat# **CLS431403**, Sigma-Aldrich).
21. Hamilton syringe, 1–20 μL (Cat# 84301, Hamilton).
22. Petri Dish, 90 × 15 mM of choice.
23. Inoculating loop, 1 μL × 70 mM of choice.
24. Chromatography column, 1 × 30 cm of choice.
25. Glass luminometer cuvettes with flat bottom 12 × 35 mM (Cat# FS60931 12, Fisher Scientific).
26. Quartz absorption cuvettes, semi Micro (Cat# Z600288, Sigma-Aldrich) or similar.
27. Quartz fluorescence cuvettes, semi Micro (Cat# Z600253, Sigma-Aldrich) or similar.

2.4 Experimental Overview

Although many protocols can be applied for expression of genes encoding ctenophore photoproteins in *E. coli* cells as well as for their purification [9, 10, 14], we suggest the ones that use, e.g., the pET system (pET22b(+) vector) (Fig. 1) and the appropriate *E. coli* strain such as BL21-CodonPlus (DE3)-RIPL to express any ctenophore apoprotein gene and the protocol we developed to produce light-sensitive photoproteins of high purity. Application of the suggested protocols has several advantages over the others: (1) the yield of high purity active photoprotein amounts to approximately 20 mg per liter of *E. coli* cell culture; (2) nearly the entire apophotoprotein is accumulated in inclusion bodies and after washing procedure can be easily extracted as an apoprotein of high purity by 6 M urea (Fig. 2c); (3) the minor impurities are removed from apophotoprotein sample extracted from inclusion bodies with ion-exchange chromatography in the presence of 6 M urea (Fig. 2a); and (4) the last ion-exchange chromatography step allows separation of the active photoprotein from apophotoprotein and coelenterazine excess (Fig. 2b), which are always present in active photoprotein samples after apoprotein conversion into active form with synthetic coelenterazine.

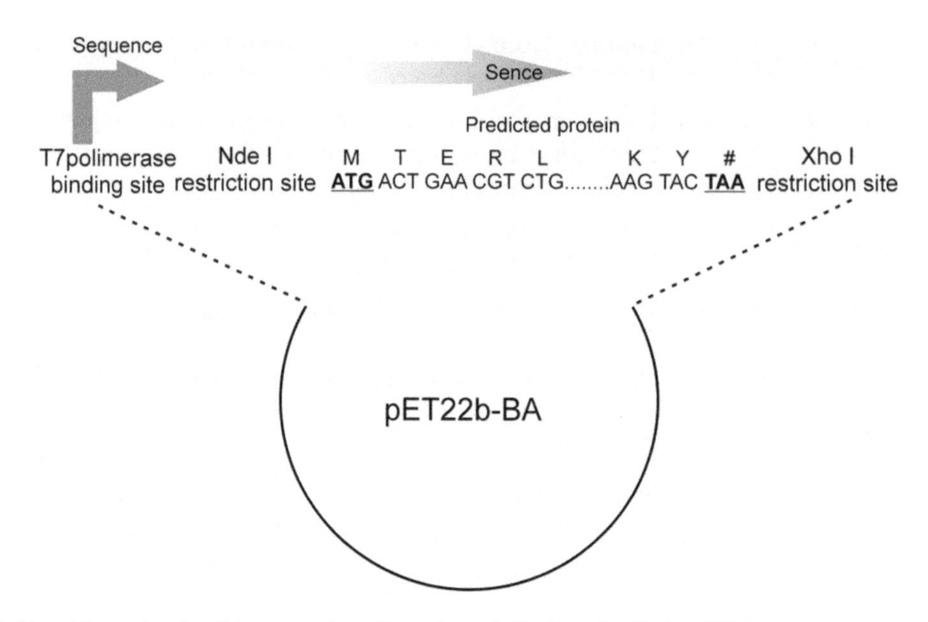

Fig. 1 Plasmid construction for expression of apo-berovin in *E. coli* cells in pET22b(+) vector. Apo-berovin gene coding sequence was inserted under promoter control of T7 RNA polymerase using FauND I (Nde I isoschizomer) and Xho I vector sites

The high purity photoproteins obtained by using these protocols can be used in biochemical and crystallographic studies as well as for development of analytical assays.

2.5 Preparation of Solutions

2.5.1 Stock Solutions

1. TAE buffer 50× pH 8.0.
2. 2 M glucose.
3. 10 mM coelenterazine in ultrapure 100% dimethyl sulfoxide (DMSO). Store at −20 °C in aliquots.
4. Ampicillin 200 g/L.
5. 1 M IPTG.
6. 1 M Tris–HCl pH 7.2.
7. 1 M Tris–HCl pH 8.5.
8. 1 M Tris–HCl pH 9.0.
9. 5 M NaCl.
10. 0.5 M EDTA pH 8.0.
11. 20% Triton X-100.
12. 1 M $CaCl_2$.
13. 1 M KCl.
14. 1 M $MgCl_2$.
15. 1 M KOH.
16. 1 M PIPES pH 7.0.

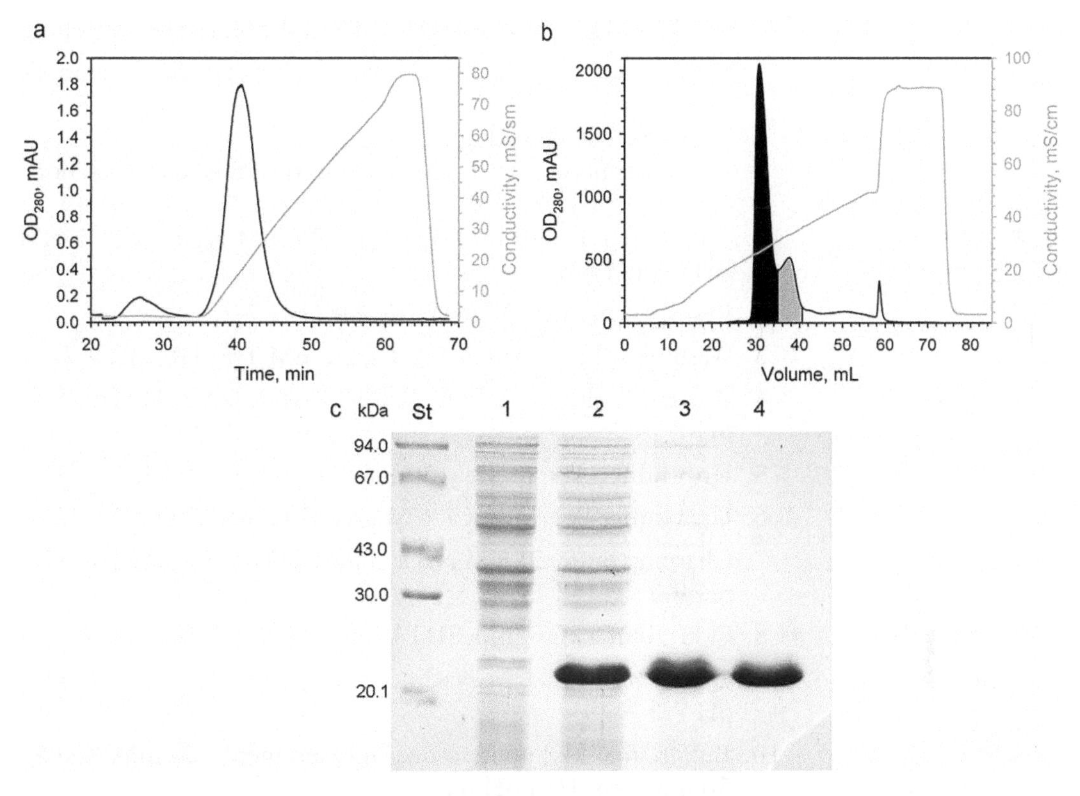

Fig. 2 Example to illustrate obtaining of high purity berovin. (**a**, **b**) Ion-exchange chromatography of apo-berovin (**a**) and active berovin (**b**). The black lines show elution profiles for corresponding proteins; gray lines—gradient of NaCl. There are two fractions at purification of active berovin (b): active berovin (black) and apo-berovin (gray). (**c**) SDS-PAGE analysis of pET22-berovin expression in *E. coli* BL21-CodonPlus (DE3)-RIPL cells and apo-berovin purification. Lanes: 1 and 2, whole-cell lysates before and after IPTG induction, respectively; 3, 6 M urea extract of inclusion bodies; 5, apo-berovin after purification on DEAE Sepharose Fast Flow column (12.5% polyacrylamide gel stained with Coomassie Brilliant Blue)

2.5.2 Culture Mediums

1. LB broth (Lysogeny broth).

   10 g tryptone, 5 g Bacto Yeast Extract, 7.5 g NaCl.

   Fill up to 1 L with distilled H_2O.

   Sterilize 250 mL by autoclaving for 60 min at 1 atm overpressure (121 °C) in 500 mL heat-resistant glass flask.

2. LB agar.

   7.5 g Bactoagar.

   Fill up to 0.5 L with LB broth.

   Sterilize 250 mL by autoclaving for 60 min at 1 atm overpressure (121 °C) in 500 mL heat-resistant glass flask.

3. SOB medium (Super Optimal Broth).

   20 g tryptone, 5 g Bacto Yeast Extract, 0.5 g NaCl, 0.186 g KCl.

   Fill up to 1 L with distilled H_2O.

   Sterilize 250 mL by autoclaving for 60 min at 1 atm overpressure (121 °C) in 500 mL heat-resistant glass flask.

Add $MgCl_2$ and $MgSO_4$ to 10 mM of final concentration.

4. SOC medium (Super Optimal Broth with catabolic repression).

   0.05 L SOB medium.

   Add glucose (sterile) up to 20 mM of final concentration.

2.5.3 Buffers

1. TAE buffer 1×: 40 mM Tris base, 20 mM acetic acid, 1 mM EDTA, pH 8.0.
2. Resuspend buffer: 20 mM Tris–HCl pH 7.2.
3. Washing solution 1: 0.9% NaCl, 20 mM Tris–HCl pH 7.2.
4. Washing solution 2: 1% Triton X-100, 20 mM Tris–HCl pH 7.2.
5. Urea buffer A: 6 M urea, 20 mM Tris–HCl pH 7.2.
6. Urea buffer B: 1 M NaCl, 6 M urea, 20 mM Tris–HCl pH 7.2.
7. Charging buffer: 0.5 M NaCl, 5 mM EDTA, 50 mM Tris–HCl pH 9.0.
8. Q HP buffer A: 5 mM EDTA, 20 mM Tris–HCl pH 7.2.
9. Q HP buffer B: 1 M NaCl, 5 mM EDTA, 20 mM Tris–HCl pH 7.2.
10. Buffer for bioluminescence measurement: 2 mM $CaCl_2$, 50 mM Tris–HCl pH 8.5.
11. Buffer for sample dilution: 50 mM BIS–Tris propane-HCl pH 8.5.
12. Buffer for spectral measurements: 100 mM $CaCl_2$, 50 mM BIS–Tris propane-HCl pH 8.5.
13. KCl-buffer: 150 KCl, 5 mM PIPES, pH 7.0.
14. Pass KCl-buffer twice through freshly washed beds of Chelex-100 chelating resin packed in a 1 × 30 cm column to remove the trace amounts of Ca^{2+}.
15. 20 mM EGTA 10× stock solution: 3.842 g EGTA, 7.345 g PIPES, 1.428 g KCl, 60 mL 1 M KOH (previously passed through Chelex-100), some volume of Milli-Q water, mix, adjust carefully pH of the solution with KOH to 7.0 at 20 ° C, and then fill up to 500 mL with Milli-Q H_2O (*see* **Note 1**).
16. 20 mM Ca^{2+}-EGTA 10× stock solution: 3.842 g EGTA (99% purity), 7.345 g PIPES, 10 mL 1 M $CaCl_2$, 80 mL 1 M KOH (previously passed through Chelex-100), some volume of Milli-Q water, mix, adjust carefully pH of the solution to 7.0 at 20 °C, and then fill up to 500 mL with Milli-Q H_2O.
17. 2 mM EGTA buffer.

    50 mL of 20 mM EGTA 10× stock solution.

    Fill up to 500 mL with KCl-buffer.

Table 1
Volumes of buffers specified in 17 and 18 (or 19 and 20) to prepare test solutions with Ca^{2+} concentrations below 10^{-5} M

$[Ca^{2+}]_{free}$, M[a]	2 mM Ca^{2+}-EGTA (mL) 2 mM Ca^{2+}-EGTA with 1 mM Mg^{2+} (mL)	2 mM EGTA (mL) 2 mM EGTA with 1 mM Mg^{2+} (mL)
1.28×10^{-8} (1.31×10^{-8})	2	67
2.57×10^{-8} (2.63×10^{-8})	4	67
4.31×10^{-8} (4.41×10^{-8})	6	60
8.62×10^{-8} (8.82×10^{-8})	10	50
1.29×10^{-7} (1.32×10^{-7})	15	50
2.59×10^{-7} (2.65×10^{-7})	24	40
4.31×10^{-7} (4.41×10^{-7})	30	30
8.62×10^{-7} (8.82×10^{-7})	40	20
1.29×10^{-6} (1.32×10^{-6})	45	15
2.57×10^{-6} (2.63×10^{-6})	60	10
4.22×10^{-6} (4.32×10^{-6})	60	6
8.00×10^{-6} (8.17×10^{-6})	60	3
1.11×10^{-5} (1.13×10^{-5})	60	2

[a]Free Ca^{2+} concentrations in the presence of 1 mM Mg^{2+} are shown in parentheses

18. 2 mM Ca^{2+}-EGTA buffer.
    50 mL of 20 mM Ca^{2+}-EGTA 10× stock solution.
    Fill up to 500 mL with KCl-buffer.
19. 2 mM EGTA buffer with 1 mM $MgCl_2$.
    50 mL of 20 mM EGTA 10× stock solution, 0.5 mL of 1 M $MgCl_2$.
    Fill up to 500 mL with KCl-buffer.
20. 2 mM Ca^{2+}-EGTA buffer with 1 mM $MgCl_2$.

50 mL of 20 mM Ca^{2+}-EGTA 10× stock solution, 0.5 mL of 1 M $MgCl_2$.

Fill up to 500 mL with KCl-buffer.

21. Preparation of test solutions with $[Ca^{2+}]$ below 10^{-5} M

    Mix different volumes of 2 mM EGTA and 2 mM Ca^{2+}-EGTA buffers as indicated in Table 1 (*see* **Note 2**).

22. Dilute 1 M $CaCl_2$ in a Chelex scrubbed solution of KCl-buffer getting about 13 test solutions of different Ca^{2+} concentrations from 10^{-6} to 10^{-2} M

3 Methods

We advise to prepare a museum of expression strains with ctenophore photoproteins genes with a view of saving time in case of necessity to obtain new batch of protein. *E. coli* strains are stored in 20% glycerol (final concentration) at −70–80 °C.

3.1 *E. coli* Strain for Apophotoprotein Gene Expression

3.1.1 Cloning Vector and Gene Template Preparation

1. PCR amplification of apo-berovin coding DNA.

   1.0 μL (10 ng) pTriplEx2-BA plasmid [8] containing full-size apo-berovin gene as a template.

   5.0 μL 10× PfuSE-buffer (in polymerase kit).

   1.0 μL 10 mM dNTP.

   1.0 μL PfuSE DNA polymerase, 5 units/μL.

   1.5 μL 10 μM forward-specific primer containing FauND I site.

   1.5 μL 10 μM reverse-specific primer containing Xho I site.

   39.0 μL Milli-Q H_2O (total volume is 50 μL).

2. Proceed amplification in a thermal cycler (e.g., MJ Mini Thermal Cycler) at 95 °C for 1 min, followed by 20 cycles of 95 °C for 20 s, 56 °C for 30 s, and 72 °C for 1min.

   Load 1 μL on a 1% agarose gel to verify synthesis.

3. Purify PCR product using PCR Purification Kit of choice (e.g., QIAquick PCR Purification Kit); elute with 30 μL of kit elution buffer.

4. Digest separately 1 μg of the pET22b(+) vector and 30 μL of amplified apo-berovin coding sequence with restriction enzymes FauND I and Xho I.

   1.0 μL FauND I (20 units), 1.0 μL of Xho I (20 units), 5 μL 10× SE-buffer Y, and 0.5 μL 100× BSA.

   Add Milli-Q H_2O to bring each volume to 50 μL.

5. Incubate 1 h at 37 °C.

6. Add 1 μL alkaline phosphatase to the mixture for pET22b(+) vector only, and incubate 1 h at 37 °C.

7. Run both mixtures on 1% low gelling temperature agarose gels in 1× TAE buffer to purify.
8. Cut gel strips containing the desired DNA fragments.
   (a) Extract and clean-up both DNAs from gel with purification kit of choice (e.g., QIAquick Gel Extraction Kit); elute with 30 μL of kit elution buffer.
   (b) Now tubes contain linearized and dephosphorylated pET22b(+) and restricted DNA apo-berovin coding sequence for cloning; store samples at −20 °C.
9. Run 1 μL of each sample on a 1% agarose gel alongside DNA markers to check for concentration.

3.1.2 Cloning Apo-berovin Coding Sequence to the Vector

1. Place purified restricted pET22b(+) vector and DNA fragment of apo-berovin coding sequence into 0.2 μL tube at the molar ratio 1:2 on ice.
   2.0 μL 10× SE-buffer DNA ligase T4 (in ligase kit).
   1.0 μL PfuSE DNA ligase T4, 200 Units/μL.
   × μL Milli-Q H_2O (add to total of 20 μL).
2. Incubate overnight at 16 °C.
3. Heat the ligation mixture for 10 min at 65 °C.
4. Place a tube with 50 μL XL1-Blue competent cells on ice.
5. Add 5 μL of the ligation mixture into the tube, and flick it a few times gently to mix.
6. Incubate 30 min on ice.
7. Heat-shock the DNA-cell suspension at 42 °C for 45 s.
8. Return the tube on ice for 2 min.
9. Add 250 μL SOC medium.
10. Incubate 1 h at 37 °C.
11. Plate cells on LB plates containing the 50 μg/mL ampicillin to select for the transformed cells containing the DNA of interest.

3.1.3 Expression Plasmid Purification

1. Pick up one colony from the plate to 5 mL LB broth containing the 50 μg/mL ampicillin, and grow overnight in a shaking incubator at 37 °C with vigorous shaking at 220 rpm.
2. Centrifuge for 5 min at 3000 g.
   (a) Extract and clean-up plasmid DNA from cells using miniprep kit of choice (e.g., QIAprep Spin Miniprep Kit); elute with 50 μL of kit elution buffer (use 30 μL if concentration is low).
   (b) Now tube contains recombinant pET22-BA plasmid (Fig. 1); store at −20 °C.

3. Run 1 μL on a 1% agarose gel to check for concentration.
4. Confirm a validity of the obtained expression construction for apo-berovin by sequencing.

3.1.4 *E. coli* Expression Strain Preparation

1. Place a tube with 20 μL BL21-CodonPlus (DE3)-RIPL competent cells on ice.
2. Add 1 μL of plasmid DNA pET22b-BA into the tube, and flick it a few times gently to mix.
3. Incubate 30 min on ice.
4. Heat-shock the DNA-cell suspension at 42 °C for 25 s.
5. Return the tube on ice for 2 min.
6. Add 100 μL SOC medium.
7. Incubate 1 h at 37 °C.
8. Plate cells on LB plates containing the 200 μg/mL ampicillin to select for the transformed cells containing the DNA of interest.
9. Pick up one colony from the plate to 5 mL LB broth containing the 200 μg/mL ampicillin and grow to OD_{600} = 0.6 in a shaking incubator at 37 °C with vigorous shaking at 220 rpm.
10. Add sterile glycerol to cell culture (1:4, v/v) and aliquot to 1.5 mL tubes.
11. Tubes containing strain for berovin gene expression store at –80 °C.

3.2 Obtaining of High Purity Active Photoprotein

3.2.1 Growing of Cells

1. Scrape an aliquot of the frozen bacteria strain suspension from the tube with museum cells by the inoculating loop to the Petri dish with solidified LB-agar containing 200 μg/mL ampicillin and incubate overnight at 37 °C.
2. Place 1 fresh colony into 2 mL of LB broth containing 200 μg/mL ampicillin, and grow overnight in a shaking incubator at 37 °C with vigorous shaking at 220 rpm.
3. Load night culture of cells into 1 L Erlenmeyer flask containing 200 mL LB broth containing 200 μg/mL ampicillin.
4. Grow cells in incubator at 37 °C with vigorous shaking at 220 rpm until the culture reaches an OD_{600} = 0.6–0.7.
5. Add 0.2 mL 1 M IPTG.
6. Grow cells for 3 h at the same conditions.
7. Transfer cell culture into proper centrifuge tube and cool.
8. Centrifuge at 4 °C for 15 min at 4000 g and discard supernatant (*see* **Note 3**).

3.2.2 Washing of Inclusion Bodies

1. Add 5 mL Resuspend buffer to the cell pellet, resuspend pellet, and disrupt cells with ultrasound at 44 kHz for 20 s × 6 on ice.
2. Centrifuge at 4 °C for 10 min at 10,000 g, and discard supernatant.
3. Add 5 mL Washing solution 1 to the precipitate, resuspend, centrifuge at 4 °C for 10 min at 10,000 g, and discard supernatant.
4. Add 5 mL washing solution 2 to the precipitate, resuspend, centrifuge at 4 °C for 10 min at 10,000 g, and discard supernatant.
5. Add 5 mL resuspend buffer to the precipitate, resuspend, centrifuge at 4 °C for 10 min at 10,000 g, and discard supernatant (*see* **Note 4**).

3.2.3 Extraction of Apophotoprotein from Inclusion Bodies

1. Add 10 mL urea buffer A to washed inclusion bodies; incubate for 30 min at 4 °C.
2. Centrifuge solution at 4 °C for 20 min at 10,000 g; transfer carefully a supernatant to clear tube.

3.2.4 Purification of Apophotoprotein

1. Wash HiTrap DEAE Fast Flow column sequentially with 15 mL urea buffer A, 10 mL urea buffer B, and 25 mL urea buffer A.
2. Load urea extract on column using 10 mL Superloop.
3. Run ion-exchange chromatography using a program for HiTrap DEAE Fast Flow column (5 mL): 15 mL urea buffer A, 35 mL 0–80% urea buffer B, 5 mL 100% urea buffer B, 15 mL urea buffer A, flow rate 2 mL/min.
4. Collect fraction containing apophotoprotein in Corning 15 mL centrifuge tube (Fig. 2a) (*see* **Note 5**).
5. Desalt 0.5-mL aliquot of eluted protein on HiTrap Desalting column equilibrated with resuspend buffer.
6. Measure OD_{280} of 1 mL of desalted protein using UV-VIS spectrophotometer, and determine concentration of apophotoprotein using an extinction coefficient calculated with ProtParam Tools ($\varepsilon_{280} = 42{,}530\ M^{-1}\ cm^{-1}$ for apo-berovin) [17].
7. Concentrate the protein sample collected after ion-exchange chromatography up to approximately 1 mL by centrifugation at 5000 g and 4 °C using Amicon Ultra 15 mL Centrifugal Filters.
8. Calculate molar concentration of protein in the concentrated solution.

3.2.5 Conversion of Apoprotein into Active Photoprotein

1. Add aliquot of 10 mM coelenterazine solution into charging buffer (apophotoprotein:coelenterazine molar ratio 1:1.1).
2. Add immediately 1 volume of the concentrated apophotoprotein solution to 9 volumes of charging buffer with coelenterazine, and mix quickly.
3. Incubate mixture at 4 °C overnight in the dark (*see* **Note 6**).
4. Add 9 volumes of Q HP buffer A to 1 volume of the mixture (*see* **Note 7**).

3.2.6 Purification of Active Photoprotein

1. Wash HiTrap Q HP column sequentially with 15 mL Q HP buffer A, 15 mL Q HP buffer B, and 15 mL Q HP buffer A.
2. Load mixture diluted in Q HP buffer A to the column using 50 mL Superloop.
3. Run ion-exchange chromatography using a program for Q HP 5 column: 10 mL Q HP buffer A, 50 mL 0–50% Q HP buffer B, 5 mL 50% Q HP buffer B, 10 mL 100% Q HP buffer B, 20 mL 5 mL 50% Q HP buffer A, flow rate 2 mL/min, temperature 4–12 °C (*see* **Note 8**).
4. Collect fractions separately in Corning 15 mL centrifuge tubes (Fig. 2b), and store them at 4 °C in the dark.
5. Determine photoprotein concentration using DC Protein Assay Kit II.

3.3 Bioluminescent Measurements

3.3.1 Specific Bioluminescent Activity

1. Prepare luminometer cuvettes with 490 μL buffer for bioluminescence measurements.
2. Place cuvette into cell of the luminometer.
3. Load 10 μL solution of photoprotein into Hamilton syringe and inject sample into cuvette (*see* **Note 8**).
4. Repeat measurements 3 times (Fig. 3) (*see* **Note 9**).
5. Calculate total emitting light from a sample for each measurement.
6. Calculate specific bioluminescent activity as an average of 3 total emitting light values divided to concentration of photoprotein (RLU/mg).

3.3.2 Bioluminescence Spectrum

1. Setup proper parameters of computer program (*see* **Note 10**).
2. Load 900 μL of buffer for spectral measurements into cuvette of fluorescence spectrophotometer.
3. Place the cuvette in the cell of fluorescence spectrophotometer.
4. Add 100 μL of photoprotein solution (protein concentration 0.1 mg/mL or higher) into cuvette.
5. Rapidly inject 20 μL of Ca^{2+}-buffer into cuvette and start measurement.

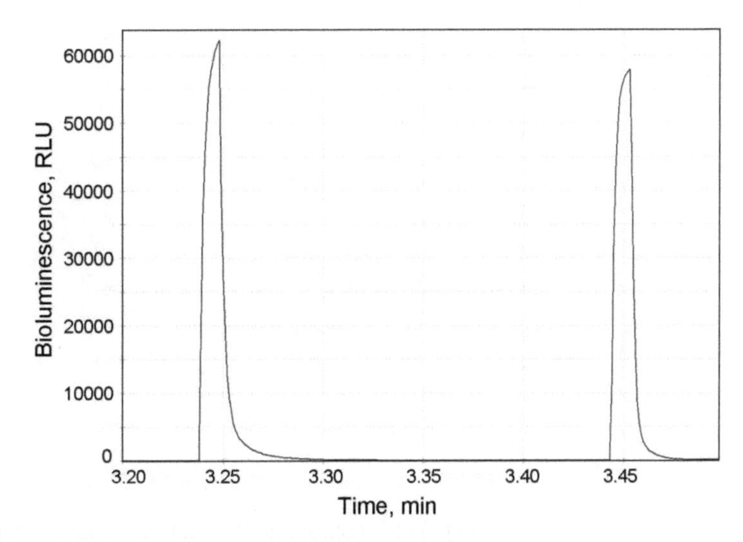

Fig. 3 Berovin light signals obtained with luminometer as an example of measurements of bioluminescent activity

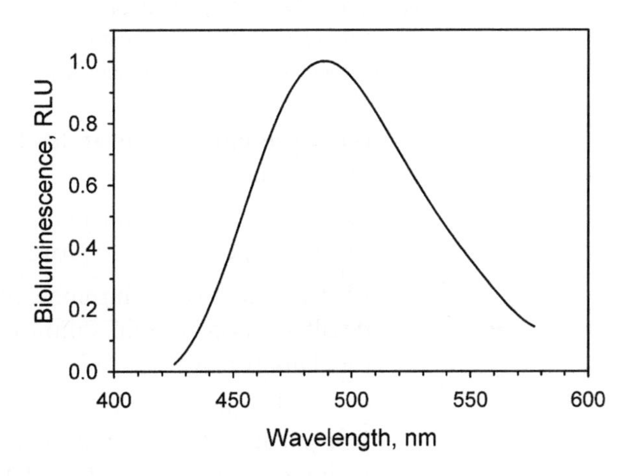

Fig. 4 Bioluminescence spectrum of recombinant berovin

6. Normalize light intensities at different wavelengths for maximal light intensity.
7. Make a plot to *display* a dependence of photoprotein bioluminescence on wavelengths (Fig. 4).

3.4 Calcium Concentration-Effect Relation

3.4.1 Bioluminescence Activity at Different Ca^{2+} Concentrations

1. Prepare 0.5–1.0 mL of purified sample of photoprotein with concentration not less than 5×10^{-6} M, and remove EDTA by gel filtration using a plastic column equilibrated with the KCl-buffer previously passed twice through freshly washed beds of Chelex-100 chelating resin (*see* **Note 8**).
2. Identify photoprotein-containing fractions by the bioluminescence assay (*see* **Note 11**).
3. If experiment is carried out in the presence of Mg^{2+}, add $MgCl_2$ up to the final concentration of 1 mM to the desalted

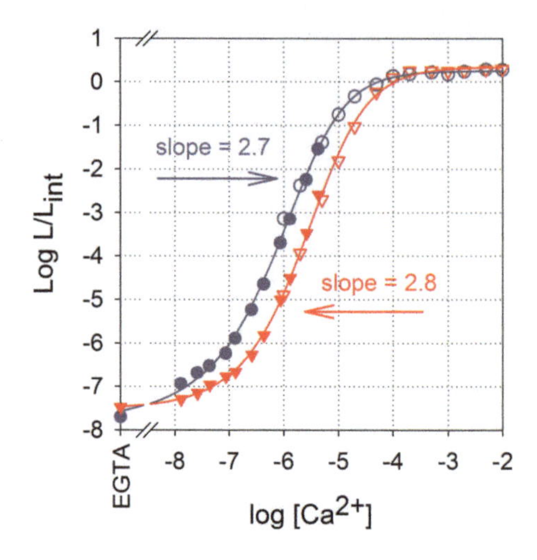

Fig. 5 Ca^{2+} concentration-effect curves for recombinant berovin without (circles) and with 1 mM Mg^{2+} (triangles). Filled symbols, Ca-EGTA buffers; open symbols, dilutions of $CaCl_2$. L, light intensity at the particular Ca^{2+} concentration, L_{int}, total light intensity at saturating Ca^{2+} concentration. A maximum slope of Ca^{2+} concentration-effect curves more than two indicates that the binding of three calcium ions is required to trigger berovin bioluminescence

photoprotein sample at least 1 h before the beginning of measurements.

4. Keep prepared photoprotein sample at 4 °C or on ice throughout the experiment (*see* **Note 12**).
5. Measurement of bioluminescence activity. Inject 10 μL of the desalted photoprotein solution into 1 mL of the test solution by Hamilton syringe 1–20 μL (*see* **Notes 13–15**).
6. Convert light intensity (L) measurements to units of L/L_{int}: first calculate L/L_{max}, and then multiply by the maximum peak-to-integral ratio (L_{max}/L_{int}) determined from kinetic measurements performed under the same conditions and with the same sample of photoprotein (*see* **Note 16**).
7. Make a log-log plot to *display* a dependence of photoprotein bioluminescence on [Ca^{2+}] (Fig. 5).

4 Concluding Remarks

Here, we present the protocols for plasmid construction to express in *E. coli* cells the gene encoding light-sensitive photoprotein berovin from ctenophore *Beroe abyssicola*, for effective procedures of its extraction from inclusion bodies and obtaining of high purity protein as well as determination of berovin sensitivity to calcium with and without magnesium ions. Although it is obvious that other protocols can be applied for obtaining ctenophore photoproteins, we believe that the procedures described here have a number of advantages. These are high levels of expression in *E. coli* of recombinant protein using pET22b(+) vector, an accumulation of apophotoprotein in inclusion bodies because a high purity apoprotein can already be obtained after several simple washing procedures, and an ion-exchange chromatography step of active photoprotein allowing the obtaining of the charged photoprotein only, since at this chromatography step apoprotein and not bound coelenterazine as well as the products of its degradation are effectively separated from active photoprotein. Thus, after final purification step, photoprotein of high purity with no any additional amino acid sequences, which could modify its biochemical properties, can be obtained. We also need to underline the simplicity of apoprotein refolding from denatured state into active conformation by simple dilution of 6 M urea extract of inclusion bodies with a buffer. Moreover, the refolding procedure can be combined with conversion of apophotoprotein into active photoprotein. Effective refolding is most likely conditioned by the properties of the photoprotein itself, among which the small size of photoprotein molecule and the lack of intramolecular S-S bonds are most important. In conclusion, we would like to note that although the described protocols were successfully used to obtain ctenophore photoproteins, berovin and bolinopsin, only, we are certainly sure that they can be as well applied for production of other ctenophore photoproteins owing to high sequence and structure identities of these photoproteins [6].

5 Notes

1. Be especially careful with samples weighing and exact pH value adjusting (at 20 °C in described protocol), because, in the upshot, even a small error can lead to noticeable change in concentration of free Ca^{2+}.
2. Calculate concentration of free Ca^{2+} for each test solution using Maxchelator (e.g., Ca-EGTA Calculator NIST v.1.2) or similar program, which takes into account the chelator dissociation equilibrium constant at given ionic strength, temperature, and pH.
3. Cell precipitate may be stored at −20 °C for up to 1 month.
4. Washed inclusion bodies may be stored at −20 °C for up to 1 year.
5. Protein purity may be verified with SDS-PAGE on any purification step (Fig. 2c).
6. At these folding conditions, the yield of active berovin amounts to approximately 70%. It is possible to increase the active

photoprotein yield slightly by using lower apoprotein concentration (<0.1 g/L) or higher coelenterazine concentration in the charging mixture.

7. Stability of active photoprotein at neutral and weak-acid pH is higher than at alkali pH.
8. Perform all procedures with active photoprotein only at dim red light or in the dark to prevent photoprotein inactivation by light.
9. Use neutral filters or dilution of sample in case of extremely high bioluminescence activity.
10. Instrument parameters for bioluminescence spectrum measurement: emission slit width, 5 nM; scan rate, 12,000 nM/min; averaging time, 0.005 s; photoelectronic multiplier voltage, 1000 V; wavelength range, 370–600 nM; and number of repetitions, 10. All spectra are corrected for the detector spectral sensitivity using a program supplied with the instrument.
11. Use only the first few protein fractions to come off the column to avoid possible contamination with the trace amount of EDTA.
12. We recommend obtaining the desalted protein sample in the amount that would be sufficient for measuring calcium concentration-effect relations during 1–2 days, since photoprotein may be not very stable in the absence of EDTA.
13. It is important to pre-equilibrate temperature of test solution in cuvette.
14. Use 2 mM EGTA solution to measure Ca^{2+}-independent bioluminescence.
15. At low [Ca^{2+}], light signal is very slow and looks like a rise of intensity; plateau is observed during tens of minutes without decay of intensity. Thus, at low [Ca^{2+}] in order to trigger light emission, we advise to add calcium to cuvette outside of luminometer and only then to place cuvette into the luminometer. This will avoid unnecessary instrumental effects that can occur at measuring low-intensity light signals. At higher [Ca^{2+}] when light signal obviously decays, initiate luminescence directly in luminometer.
16. We find that expression of the results in terms of L/L_{int} is more useful and meaningful as compared to the commonly used L/L_{max}. Both units have the effect of normalizing the measurements for photoprotein concentration and for optical efficiency of the measuring system, but L/L_{max} tends to obscure certain potentially significant characteristics of the light signal, such as the maximum peak-to-integral ratio, which can substantially differ from one photoprotein to the next, and is useful for characterization of the photoproteins [16, 18].

Acknowledgments

This work was supported by grant No. 22-14-00125 of the Russian Science Foundation.

References

1. Haddock SHD, Case JF (1999) Bioluminescence spectra of shallow and deep-sea gelatinous zooplankton: ctenophores, medusae and siphonophores. Mar Biol 133:571–582
2. Ward WW, Seliger HH (1976) Action spectrum and quantum yield for the photoinactivation of mnemiopsin, a bioluminescent photoprotein from the ctenophores *Mnemiopsis* sp. Photochem Photobiol 23:351–363
3. Anctil M, Shimomura O (1984) Mechanism of photoinactivation and reactivation in the bioluminescence system of the ctenophore *Mnemiopsis*. Biochemistry 221:269–272
4. Ward WW, Seliger HH (1974) Extraction and purification of calcium-activated photoproteins from the ctenophores *Mnemiopsis* sp. and *Beroe ovata*. Biochemistry 13:1491–1499
5. Ward WW, Seliger HH (1974) Properties of mnemiopsin and berovin, calcium-activated photoproteins from the ctenophores *Mnemiopsis* sp. and *Beroe ovata*. Biochemistry 13:1500–1509
6. Burakova LP, Vysotski ES (2019) Recombinant Ca^{2+}-regulated photoproteins of ctenophores: current knowledge and application prospects. Appl Microbiol Biotechnol 103: 5929–5946
7. Golz S, Markova S, Burakova L, Frank L, Vysotski E (2005) Isolated photoprotein bolinopsin, and the use thereof, WO/2005/000885 (Patent)
8. Markova SV, Burakova LP, Golz S, Malikova NP, Frank LA, Vysotski ES (2012) The light-sensitive photoprotein berovin from the bioluminescent ctenophore *Beroe abyssicola*: a novel type of Ca^{2+}-regulated photoprotein. FEBS J 279:856–870
9. Powers ML, McDermott AG, Shaner N, Haddock SH (2013) Expression and characterization of the calcium-activated photoprotein from the ctenophore *Bathocyroe fosteri*: insights into light-sensitive photoproteins. Biochem Biophys Res Commun 431:360–366
10. Aghamaali MR, Jafarian V, Sariri R, Molakarimi M, Rasti B, Taghdir M, Sajedi RH, Hosseinkhani S (2011) Cloning, sequencing, expression and structural investigation of mnemiopsin from *Mnemiopsis leidyi*: an attempt toward understanding Ca^{2+}-regulated photoproteins. Protein J 30:566–574
11. Stepanyuk GA, Liu ZJ, Burakova LP, Lee J, Rose J, Vysotski ES, Wang BC (2013) Spatial structure of the novel light-sensitive photoprotein berovin from the ctenophore *Beroe abyssicola* in the Ca^{2+}-loaded apoprotein conformation state. Biochim Biophys Acta 1834:2139–2146
12. Burakova LP, Natashin PV, Malikova NP, Niu F, Pu M, Vysotski ES, Liu ZJ (2016) All Ca^{2+}-binding loops of light-sensitive ctenophore photoprotein berovin bind magnesium ions: the spatial structure of Mg^{2+}-loaded apo-berovin. J Photochem Photobiol B 154:57–66
13. Molakarimi M, Gorman MA, Mohseni A, Pashandi Z, Taghdir M, Naderi-Manesh H, Sajedi RH, Parker MW (2019) Reaction mechanism of the bioluminescent protein mnemiopsin1 revealed by X-ray crystallography and QM/MM simulations. J Biol Chem 294:20–27
14. Vysotski ES, Markova SV, Frank LA (2006) Calcium-regulated photoproteins of marine coelenterates. Mol Biol 40:355–367
15. Ottolini D, Calì T, Brini M (2013) Measurements of Ca^{2+} concentration with recombinant targeted luminescent probes. Methods Mol Biol 937:273–291
16. Malikova NP, Burakova LP, Markova SV, Vysotski ES (2014) Characterization of hydromedusan Ca^{2+}-regulated photoproteins as a tool for measurement of Ca^{2+} concentration. Anal Bioanal Chem 406:5715–5726
17. Gasteiger E, Hoogland C, Gattiker A, Duvaud S, Wilkins MR, Appel RD, Bairoch A (2005) Protein identification and analysis tools on the ExPASy Server. In: Walker JM (ed) The proteomics protocols handbook. Humana Press, pp 571–607. https://web.expasy.org/protparam/. Accessed 25 Oct 2005
18. Markova SV, Vysotski ES, Blinks JR, Burakova LP, Wang BC, Lee J (2002) Obelin from the bioluminescent marine hydroid *Obelia geniculata*: cloning, expression, and comparison of some properties with those of other Ca^{2+}-regulated photoproteins. Biochemistry 41:2227–2236

Chapter 13

Functional Screening of cDNA Expression Library for Novel Ctenophore Photoproteins

Svetlana V. Markova and Eugene S. Vysotski

Abstract

The functional screening of cDNA libraries (or functional cloning) enables isolation of cDNA genes encoding novel proteins with unknown amino acid sequences. This approach is the only way to identify a protein sequence in the event of shortage of biological material for obtaining pure target protein in amounts sufficient to determine its primary structure, since sensitive functional test for a target protein is only required to successfully perform functional cloning. Commonly, bioluminescent proteins from representatives belonging to different taxa significantly differ in sequences due to independent origin of bioluminescent systems during evolution. Nonetheless, these proteins are frequently similar in functions and can use even the same substrate of bioluminescence reaction, allowing the use of the same functional test for screening. The cDNA genes encoding unknown light-emitting proteins can be identified during functional screening with high sensitivity, which is provided by modern light recording equipment making possible the detection of a very small amount of a target protein. Here, we present the protocols for isolation of full-size cDNA genes for the novel bioluminescent protein family of light-sensitive Ca^{2+}-regulated photoproteins in the absence of any sequence information by functional screening of plasmid cDNA expression library. The protocols describe all the steps from gathering animals to isolation of individual *E. coli* colonies carrying full-size cDNA genes using photoprotein berovin from ctenophore *Beroe abyssicola* as an illustrative example.

Key words Ctenophores, Light-sensitive Ca^{2+}-regulated photoproteins, Coelenterazine, Bioluminescence, Functional cloning, Functional screening, cDNA expression library

1 Introduction

Functional cloning method to identify gene is based on the prior knowledge of function of target protein encoded by this gene. The procedure involves creation of the cDNA expression library in an appropriate host and screening of the obtained clones for enzymatic activity or any other function. Application of function-based screening approach enables discovery of truly novel proteins, which amino acid sequences would not be easily determined, for example, in the situation of shortage of biological material for obtaining sufficient amounts of pure target protein. Functional

Leonid L. Moroz (ed.), *Ctenophores: Methods and Protocols*, Methods in Molecular Biology, vol. 2757,
https://doi.org/10.1007/978-1-0716-3642-8_13, © Springer Science+Business Media, LLC, part of Springer Nature 2024

screening has been proven to be an excellent method to identify genes encoding bioluminescent proteins of the novel families [1], the diversity of which is great due to the independent origin of bioluminescent systems in different taxa [2]. High sensitivity of luminescent test provided by modern light recording equipment allows the detection of a very small amount of target protein at functional screening. The functional screening of cDNA libraries has been employed to identify genes encoding several novel bioluminescent protein families. These are photoproteins from jellyfish of class Hydrozoa, such as obelins from *Obelia longissima* and *Obelia geniculata* [3], clytin and cgreGFP from *Clytia gregaria* [4], mitrocomin from *Mitrocoma cellularia* [5], as well photoproteins from comb jelly, berovin and bolinopsin from ctenophores *Beroe abyssicola* [6] and *Bolinopsis infundibulum* [7]; luciferases from marine copepods, GpLuc from *Gaussia princeps* [8]; and several MLuc isoforms from *Metridia longa* [9–12]. Recently, identification of genes encoding luciferases from fungi was also made using the function-based screening approach [13]. Even though many isolated full-size cDNA genes [4, 6, 10–12] contain an inframe stop codon in their 5′-UTR and no consensus Shine–Dalgarno sequence upstream of the start codon, these genes were nevertheless isolated using functional bioluminescent screening. This means that the presence of a stop codon in 5′-UTR and the lack of a Shine–Dalgarno sequence near start codon do not completely eliminate the translation of these eukaryotic genes in *E. coli* cells. In addition, it also demonstrates the high sensitivity and specificity of the functional screening, because this method allows the detection of even a trace quantity of bioluminescent molecules.

Functional cloning approach, like any other method, has advantages and shortcomings. In contrast to the sequence-based screening strategy to isolate genes, the function-based approach is usually more time-consuming and expensive. In addition, its application is hardly possible if there are some hindrances for expression of genes and for correct processing of enzymes in the selected host [1, 11, 13] as well as if no sensitive functional assay for target protein is available. However, only functional screening strategy allows isolation of full-size cDNA genes encoding truly novel enzymes in the absence of any sequence data. Moreover, the activities and physicochemical properties of target proteins can be initially evaluated even during functional screening.

With photoprotein berovin from ctenophore *Beroe abyssicola* used as an illustrative example, here we present the protocols to isolate full-size cDNA genes for novel bioluminescent protein family of light-sensitive Ca^{2+}-regulated photoproteins in the absence of any sequence information by functional screening of the plasmid cDNA expression library. The protocols include description of all steps from gathering animals to isolation of individual *E. coli* colonies carrying full-size cDNA genes.

2 Materials

2.1 *E. coli* Strain

1. XL1-Blue Electroporation-Competent Cells (Cat# 200228, Agilent).

2.2 Materials

1. Whatman nitrocellulose membrane filters, circles, diam. 85 mM, pore size 0.45 μM (Cat# Z750212, Sigma-Aldrich).
2. MF-Millipore™ Membrane Filter, 0.025 μM pore size, (Cat# VSWP02500, Merck Millipore).

2.3 Reagents

1. RNAlater stabilization solution (Cat# AM7020, Thermo Fisher Scientific).
2. The Dynabeads® mRNA DIRECT™ Micro Kit (Cat# 61021, Thermo Fisher Scientific).
3. SMART® cDNA Library Construction Kit (Cat# 634901, Takara Bio).
4. RNasin ribonuclease inhibitors, 10,000 U/mL (Cat# 2511, Promega).
5. Advantage 2 PCR Kit (Cat# 639207, Takara Bio).
6. 1 kb DNA Ladder size markers (Cat# N3232S, NEB).
7. 100 bp DNA Ladder size markers (Cat# N3231, NEB).
8. Agarose (Cat# A9539, Sigma-Aldrich).
9. Ethidium bromide (Cat# E7637, Sigma-Aldrich).
10. Phenol:Chloroform:Isoamyl Alcohol 25:24:1 Saturated with 10 mM Tris, pH 8.0, 1 mM EDTA (Cat# P2069, Sigma-Aldrich).
11. SfiI 20,000 U/mL (Cat# R0123S, New England Biolabs).
12. Alkaline phosphatase 20,000 U/mL (Cat# E328, SibEnzyme).
13. QIAquick Gel Extraction Kit (Cat# 28706, Qiagen).
14. Coelenterazine (Cat# 303, NanoLight Technology).
15. Dimethyl sulfoxide (DMSO) cell culture grade (Cat#. D2650, Sigma-Aldrich).
16. Dimethylformamide (DMF) (Cat#. 227056, Sigma-Aldrich).
17. Tris(hydroxymethyl)aminomethane (Cat# 252859, Sigma-Aldrich).
18. BIS-Tris propane (Cat# B4679, Sigma-Aldrich).
19. Sodium chloride (Cat# S5886, Sigma-Aldrich).
20. Calcium chloride (Cat# C4901, Sigma-Aldrich).

21. Magnesium chloride (Cat# C4880, Sigma-Aldrich).
22. Ethylenediaminetetraacetic acid (Cat# 431788, Sigma-Aldrich).
23. Ethanol for molecular biology (Cat# 51976, Sigma-Aldrich).
24. Tryptone (Cat# 403682, PanReac).
25. Bacto Yeast Extract (Cat# 212750, BD Bioscience).
26. Agar (Cat# 212303, PanReac).
27. IPTG (Cat# S012, SibEnzyme).
28. X-Gal (Cat# S004, SibEnzyme).
29. Ampicillin-sodium salt (Cat# A0166, Sigma-Aldrich).
30. HYDRANAL®-Methanol dry (Cat# 34741, Fluka).
31. Glycerol (Cat# G5516, Sigma-Aldrich).

2.4 Equipment

1. Pellet Pestles cordless motor (Cat# Z359971, Sigma-Aldrich) with replacement motor adapter (Cat# Z359998, Sigma-Aldrich).
2. Microcentrifuge (16,000 g with fixed angle rotor) of choice.
3. Orbital shaker PSU-10i (Cat# BS-010144-AAN, BioSan).
4. Eppendorf Thermomixer C with ThermoTop (Cat# EP5382000023, Cat# EP5308000003, Sigma).
5. Thermostat (37 °C with an accuracy of ±0.2 °C) of choice.
6. Shaking Incubator (37 °C ± 0.5 °C, 50–400 rpm) of choice.
7. C1000 Touch Thermal Cycler (Cat# 185-1196, Bio-Rad) or similar.
8. Stirred water bath (42 °C with an accuracy of ±0.1 °C) of choice.
9. Centrifuge (21,000 rpm, 1–4 L, 0–40 °C) with rotor supplied by bottle adapters (4 × 250, 6 × 50, or 24 × 15 mL) of choice.
10. MicroPulser Electroporator (Cat # 1652100, Bio-Rad).
11. Electroporation cuvettes, 0.1 cm (Cat # 1652089, Bio-Rad).
12. Magnetic microtube rack, DynaMag™-2 magnet (Cat# 12321D, Thermo Fisher Scientific).
13. Double beam UV-VIS spectrophotometer (spectral range 185–900 nM) of choice.
14. Luminometer equipped with injector (or allowing manual injection of solution into cuvette) of choice.
15. Hamilton syringe, 1–20 μL (Cat# 84301, Hamilton).
16. Ultrasonic Disintegrator (44 kHz, 1000 W) of choice.
17. Disposable Polypropylene Pellet Pestles & Tube, 1.5 mL (Cat# 3411D64, Thomas Scientific).

18. PYREX® 1 mL Glass Pestle Tissue Grinder (Cat# 7724, Corning).
19. Corning 50 mL centrifuge tubes (Cat# CLS430829, Sigma-Aldrich) or similar.
20. Corning 15 mL centrifuge tubes (Cat# CLS430791, Sigma-Aldrich) or similar.
21. Erlenmeyer baffled cell culture flask 1 L (Cat# **CLS431403**, Sigma-Aldrich).
22. Petri dish, 90 × 15 mM of choice.
23. Inoculating loop, 1 μL × 70 mM of choice.
24. Glass luminometer cuvettes with flat bottom 12 × 35 mM (Cat# FS60931 12, Fisher Scientific).
25. Quartz absorption cuvettes, semi Micro (Cat# Z600288**,** Sigma-Aldrich) or similar.

2.5 Experimental Overview

Functional cloning method developed for isolation of cDNA genes encoding light-sensitive photoproteins of ctenophores (Fig. 1) includes several steps: (1) animal catching and isolation of small piece of tissue containing photocytes in the place of photoprotein localization, (2) extraction of messenger RNA (poly(A) + RNA), (3) synthesis of oligo-(dT) primed first single-stranded cDNA, (4) amplification of total cDNA with specific primers for directional cloning, (4) preparation of vector for cDNA library, (5) construction of cDNA expression library, and (6) functional screening of cDNA library (Fig. 1).

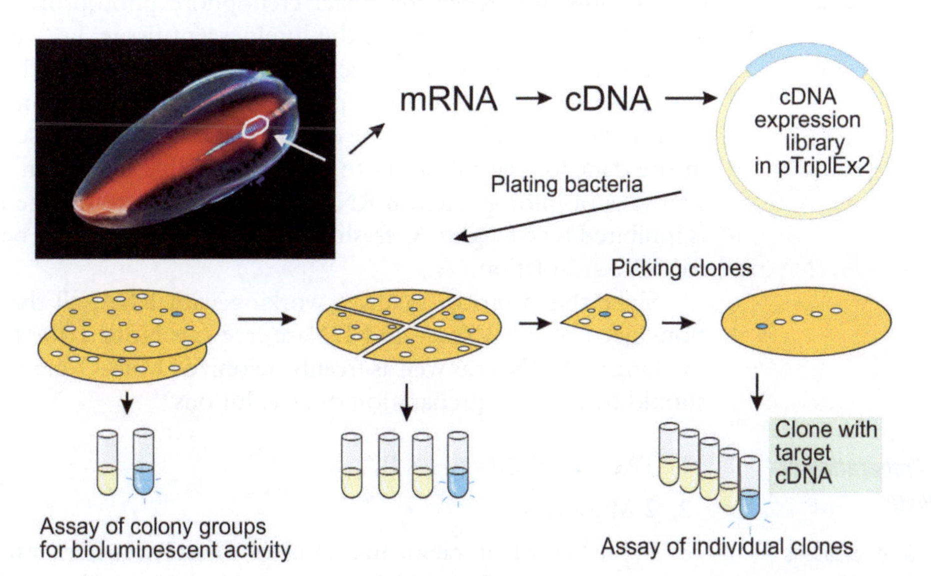

Fig. 1 Scheme for functional screening of cDNA expression library. White arrow indicates where the tissue sample was taken. Photograph of *Beroe abyssicola* (insert) by the marine biologist Alexander Semenov (Aquatilis)

The proposed protocol for creating a plasmid for full-length cDNA expression library is based on the SMART® cDNA Library Construction Kit. This kit is designed for construction of lambda phage library using λTriplex2 vector. However, for construction of cDNA library, we advise to use a plasmid vector pTriplex2, which can be easily obtained from the λTriplex2 vector supplied with kit via Cre-loxP-mediated conversion of phage to a plasmid vector after subcloning in supplied special strain *E. coli* 25.8. Application of lambda phage vector is really needed if a very small representation of the target gene in the cDNA pool is expected, because the target gene will be definitely present among a large number of original clones provided by this approach. Although the plasmid cDNA library does not provide as many clones as the lambda phage library, generation of cDNA library in a plasmid vector and its following screening are more convenient and easier. The plasmid expression vector pTriplEx2 incorporates a triple reading-frame translation cassette consisting of translation initiation signals from the *E. coli* ompA and lacZ genes, in two different reading frames, followed by a poly(T)13 transcription/translation slip site (Fig. 2). Thus, all three reading frames can be translated equally.

The success of screening of the cDNA library for the target protein gene significantly depends on the quality of the starting material for cDNA preparation. Since the SMART® cDNA Library Construction Kit provides a PCR-based method to produce a high-quality full-size cDNA, a small piece of tissue (~10 mg) is sufficient. In this view, it is desirable to use tissues with the highest expression of the target protein and therefore with target mRNA. In the case of searching for genes for novel ctenophore photoproteins, we advise to take a small piece of the luminescent tissue, i.e., the part of meridional canal (with adjacent area) running under each comb row, the walls of which contain photocytes [14]. Prior to sampling, we also recommend to keep the captured animal in aerated seawater in the dark for several hours to prevent possible inhibition of the synthesis of photoprotein mRNA, since ctenophore luminescence is inhibited by daylight. As fresh material as possible has to be used for mRNA isolation.

Since this protocol involves working with RNA, all the solutions need to be RNase- and DNase-free (*see* **Note 1** for tips on working with RNA) as well as freshly deionized Milli-Q-grade H_2O should be used for preparation of all solutions.

2.6 Preparation of Solutions

2.6.1 Stock Solutions

1. TAE buffer 50× pH 8.0.
2. 2 M glucose.
3. 0.2 mM coelenterazine in methanol. Dissolve coelenterazine in ultrapure 100% DMSO to 10 mM. Then, dilute a stock 50 times with methanol to prepare working solution. Store stock and working solutions at −20 °C in aliquots.

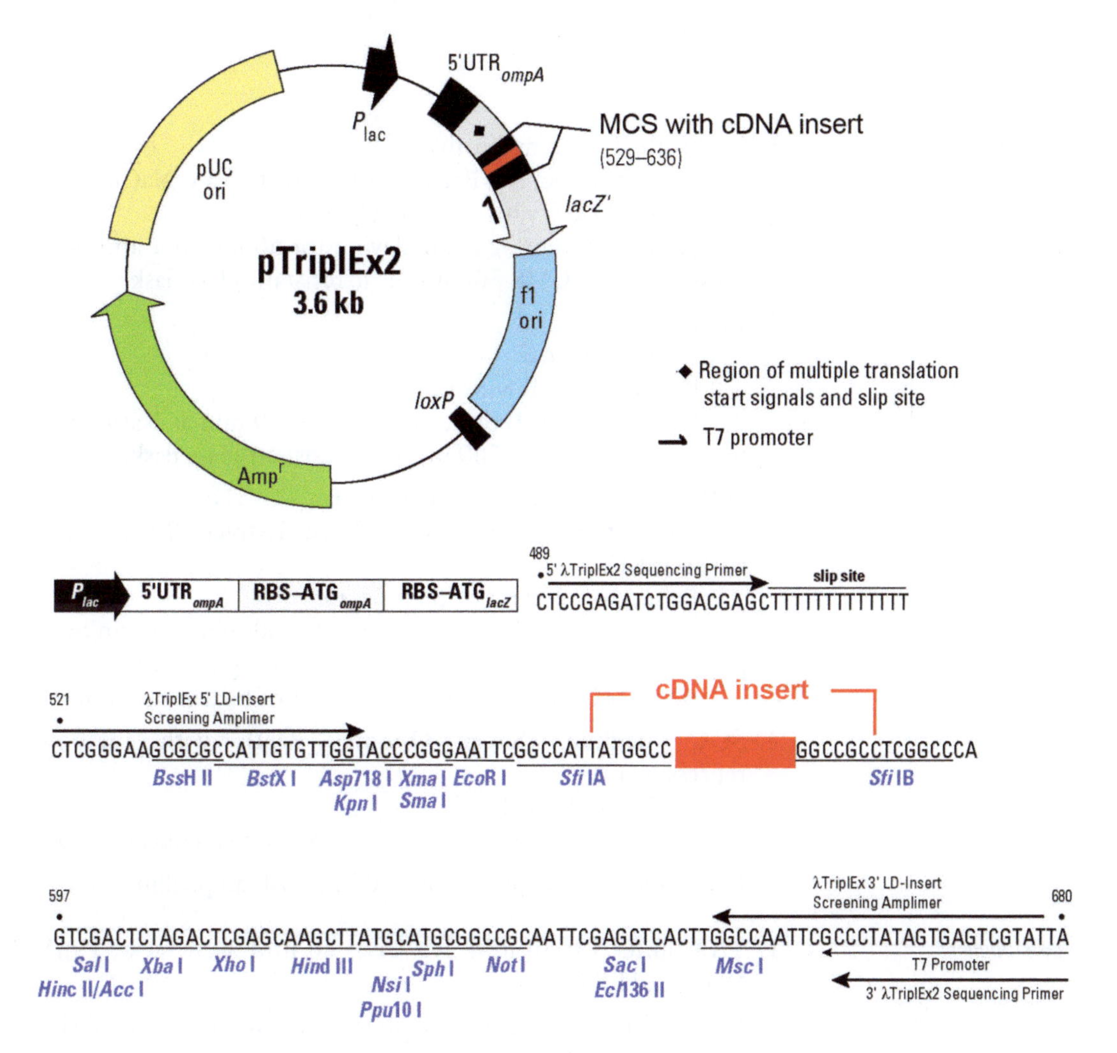

Fig. 2 Restriction map and multiple cloning site of pTriplEx2 vector

4. 80% ethanol.
5. Ampicillin 200 g/L. Dissolve in 50% ethanol. Store at −20 °C.
6. 1 M IPTG. Filter sterilized solution through 0.22-micron filter. Store at −20 °C.
7. X-Gal 20 mg/mL. Dissolve in DMF. Store at −20 °C.
8. 75% sterile glycerol.
9. 1 M Tris-HCl pH 8.5.
10. 0.5 M BIS-Tris propane pH 9.0.
11. 5 M NaCl.
12. 0.5 M EDTA pH 8.0.
13. 1 M $CaCl_2$.

14. 1 M $MgCl_2$.
15. 1 M $MgSO_4$.

2.6.2 Culture Mediums

1. LB broth (Lysogeny broth).
   10 g tryptone, 5 g Bacto Yeast Extract, 7.5 g NaCl.
   Fill up to 1 L with distilled H_2O.
   Sterilize 250 mL by autoclaving for 60 min at 1 atm overpressure (121 °C) in 500 mL heat-resistant glass flask.
2. LB agar.
   7.5 g Bactoagar
   Fill up to 0.5 L with LB broth.
   Sterilize 250 mL by autoclaving for 60 min at 1 atm overpressure (121 °C) in 500 mL heat-resistant glass flask.
3. SOB medium (Super Optimal Broth).
   20 g tryptone, 5 g Bacto Yeast Extract, 0.5 g NaCl, 0.186 g KCl.
   Fill up to 1 L with distilled H_2O.
   Sterilize 250 mL by autoclaving for 60 min at 1 atm overpressure (121 °C) in 500 mL heat-resistant glass flask.
   Add $MgCl_2$ and $MgSO_4$ to 10 mM of final concentration.
4. SOC medium (Super Optimal Broth with catabolic repression).
   0.05 L SOB medium.
   Add glucose (sterile) up to 20 mM of final concentration.
5. LB-ampicillin agar plates with 100 μg/mL ampicillin.

2.6.3 Buffers

1. TAE buffer 1×: 40 mM Tris base, 20 mM acetic acid, 1 mM EDTA, pH 8.0.
2. Buffer for bioluminescence measurement: 10 mM EDTA, 50 mM Tris-HCl pH 8.5.
3. Ca^{2+}-buffer for initiation of bioluminescence reaction: 100 mM $CaCl_2$, 100 mM Tris-HCl pH 8.5.
4. Buffer for cell sonication and conversion of apoprotein into active photoprotein (charging buffer): 0.5 M NaCl, 5 mM EDTA, 50 mM BIS-Tris propane-HCl, pH 9.0.

3 Methods

3.1 Collecting Ctenophore Specimens

1. Catch live specimens of ctenophores from a boat with a long handle bucket or plankton net, depending on comb jelly fragility, and keep in a tank containing aerated seawater at habitual temperature of these organisms.
2. Identify the gathered animals to species level using the appropriate dichotomous key to phylum *Ctenophora*.

3. In the case of failing to do it, make it later using sequencing of genes of 18S ribosomal RNA, which can be isolated from cDNA sample (*see* **Note 2**).

3.2 Direct mRNA Isolation

Isolation of high-quality mRNA is the first and most critical step in creation of the high-quality cDNA library. Therefore, as fresh material as possible should be used for mRNA isolation. An alternative approach could be storage of samples in 5–10 volumes of RNAlater stabilization solution for about a week at 4 °C before RNA isolation or quick freezing of sample in liquid nitrogen and storage at −70 °C. We advise to use a direct isolation of messenger poly(A)RNA from fresh luminescent tissue of a dark-adapted ctenophore with Dynabeads™ mRNA DIRECT™ Purification Kit (or similar) according to the manufacturer's instructions. (*see* **Note 1** for tips on working with RNA).

1. Wash magnetic Dynabeads Oligo$(dT)_{25}$ before use: pipet 200 μL (1 mg) of suspension in 500 μL of Lysis/Binding Buffer from kit, mix and place the tube on a magnetic microtube rack for 1 min, and remove supernatant with a pipet. Then, resuspend the beads in 200 μL of Lysis/Binding Buffer.
2. Homogenize ~10 mg of tissue sample in 500 μL of Lysis/Binding Buffer from kit using 1 mL glass tissue grinder and Pellet Pestles cordless motor connected with replacement motor adapter (usually 1–2 min).
3. Centrifuge the lysate at 12,000 g for 2 min in a microcentrifuge at room temperature to remove debris.
4. Transfer 500 μL of supernatant to the tube with the 200 μL of prewashed beads, and rotate on an orbital shaker for 5 min.
5. Place the sample tube on the magnet for 1 min, and then discard the supernatant with a pipet.
6. Resuspend the Dynabeads-mRNA complex in 500 μL of Washing Buffer A by careful pipetting.
7. Discard the supernatant using the magnet.
8. Repeat **steps 6** and 7.
9. Resuspend the beads with mRNA in 500 μL of Washing Buffer B, and transfer the suspension to a new tube.
10. Discard the supernatant using the magnet.
11. Resuspend the beads with mRNA in 500 μL of Washing Buffer B.
12. Discard the supernatant using the magnet.
13. For elution of mRNA, add 30 μL of 10 mM Tris-HCl, and incubate at 65–80 °C for 2 min.

14. Place the tube on the magnet and immediately transfer the supernatant to a new microcentrifuge tube, and then immediately place the tube on ice.
15. Determine the RNA concentration by absorbance at 260 nM. Estimation of mRNA concentration should take some residual rRNA into account. To assess mRNA quality, *see* **Note 3**. The eluate may be used immediately for reverse transcription or stored at −80 °C for up to 2 weeks.

3.3 cDNA Library Construction

For synthesis of cDNA and construction of cDNA library, we recommend using SMART cDNA Library Construction Kit according to the manufacturer's instructions [15] with some modifications.

3.3.1 First-Strand cDNA Synthesis

We advise performing the reverse transcription reaction immediately after the mRNA isolation. Perform the first-strand cDNA synthesis as recommended by the manufacturer, but with addition of RNasin ribonuclease inhibitors.

1. Combine the following kit reagents in sterile 0.5 mL tube (5.5 μL total):
   3 μL mRNA sample (~1 μg).
   1 μL SMART IV Oligonucleotide.
   1 μL CDS III/3′ PCR Primer.
   0.5 μL RNasin, 10 U/μL.
2. Mix by pipetting and incubate the tube at 42 °C for 2 min in the preheated thermomixer with shaking (or in an air incubator).
3. Cool the tube on ice for 2 min.
4. Add the following kit reagents to reaction tube (10.5 μL total):
   2 μL 5× First-Strand Buffer.
   1 μL 20 mM DTT.
   1 μL 10 mM dNTP mix.
   1 μL SMARTScribe MMLV Reverse Transcriptase.
5. Mix by pipetting and incubate the tube at 42 °C for 1 h in the thermomixer with shaking.
6. Place the tube on ice to terminate first-strand synthesis. First-strand cDNA can be stored at −20 °C for up to 3 months

3.3.2 cDNA Amplification

We advise to proceed directly to the PCR step of cDNA synthesis. Perform long-distance PCR (LD-PCR) according to the SMART cDNA Library Construction Kit user manual, additionally using the Advantage 2 PCR Kit and an aliquot of the resulting full-length single-stranded cDNA as a template. The optimal number of thermal cycles depends on the amount of RNA starting material (about

18–20 cycles for ~1 μg of RNA (*see* **Note 4**)). Fewer cycles generally mean fewer nonspecific PCR products.

1. Combine the kit components in the reaction tube (100 μL totally):
   2 μL First-Strand cDNA (from Subheading 3.3.1, **step 6**).
   80 μL deionized H_2O.
   10 μL 10× Advantage 2 PCR Buffer.
   2 μL 50× dNTP Mix.
   2 μL 5′ PCR Primer.
   2 μL CDS III/3′ PCR Primer.
   2 μL 50× Advantage 2 Polymerase Mix.
2. Mix and place it in a preheated (95 °C) hot-lid thermal cycler.
3. Perform LD-PCR using the cycling parameters : 95 °C for 20 s, followed by 20 cycles of 95 °C for 5 s and 68 °C for 6 min.
4. Load 5 μL on a 1.1% agarose/EtBr gel alongside a 1 kb ladder DNA size marker (0.1 μg) to verify synthesis. The ds cDNA should appear as a 0.1–4 kb smear on the gel with some distinct bands corresponding to the abundant mRNA (depends on source).
5. Proceed to the next step (Proteinase K Digestion), or store ds cDNA at −20 °C until use.

3.3.3 Preparation of cDNA to Ligation with Vector

Perform Proteinase K treatment of a half of synthesized cDNA (50 μL) to inactivate the DNA polymerase activity. Then, fulfill SfiI digestion, cDNA size fractionation on CHROMA SPIN-400 column supplied with the kit, and electrophoresis according to the SMART cDNA Library Construction Kit user manual. After electrophoresis, select the cDNA fraction with desired size distribution, precipitate cDNA, and dissolve in 7 μL of deionized H_2O, as recommended by the manufacturer's instructions [15]. After these steps, the SfiI-digested cDNA is ready to be ligated to the SfiI-digested, dephosphorylated pTriplEx2 vector.

3.3.4 Preparation of pTripLex2 Vector

Expression plasmid vector pTripLex2 can be purchased separately or generated from λTriplEx2 vector provided by SMART cDNA Library Construction Kit as SfiI-digested, dephosphorylated λTriplEx2. In the last case, the SfiI-digested cDNA ligated with the prepared λTriplEx2 should be transfected into *E. coli* BM25.8 strain provided by the kit by electroporation for Cre-lox-mediated excision of pTriplEx2 from λTriplEx2. As an option, these obtained BM25.8 clones, containing converted pTripLex2 vector with cDNA insert can also be used for screening. We advise to use the recombinant pTriplEx2 with insert about ~1.5 kb to prepare the library SfiI-digested, dephosphorylated vector. The insert increases the plasmid length and allows an efficient separation of the SfiI-digested vector from incomplete digested products by electrophoresis.

1. Digest 2 μg of the pTriplEx2 with insert with restriction enzymes SfiI: 1.0 μL SfiI (20 units), 5 μL 10× NEB-buffer 2, and 0.5 μL 100× BSA. Add Milli-Q H_2O to bring each volume to 50 μL.
2. Incubate 1 h at 37 °C.
3. Add 1 μL alkaline phosphatase and incubate 1 h at 50 °C.
4. Run the whole mixture on 1% preparative agarose gel in 1× TAE buffer to purify.
5. Cut gel strip containing the vector DNA fragment.
6. Extract and clean up DNA from gel with purification kit of choice (e.g., QIAquick Gel Extraction Kit); elute with 30 μL of kit elution buffer.
7. Now, tubes contain SfiI-digested, dephosphorylated pTriplEx2 vector for a cDNA library creation; store sample at −20 °C.
8. Check the concentration by running 1 μL on a 1% agarose/EtBr gel alongside 1 kb Ladder DNA size marker (0.5 μg).

3.3.5 Ligation of cDNA to pTriplEx2 Vector

The ratio of cDNA to vector in the ligation reaction is a critical factor for obtaining of a good quality library with a large number of independent clones. A large amount of cDNA shifts the balance toward shorter cDNA inserts, since they are more efficiently ligated. The optimal ratio of cDNA to vector in ligation reactions must be determined empirically. We recommend setting up two or three parallel ligations using three different ratios of cDNA to vector and compare the quality of libraries after bacterial clones obtaining.

1. Combine the obtained cDNA, prepared pTriplEx2 vector, and some SMART kit reagents in 5 μL volume:
   1.0 μL cDNA (or 0.5 μL, or 1.5 μL) from Subheading 3.3.3.
   1.0 μL pTriplEx2 vector (100 ng).
   0.5 μL 10× Ligation Buffer.
   0.5 μL 10 mM ATP.
   0.5 μL T4 DNA ligase.
   1.5 μL deionized H_2O to bring total volume to 5 μL
2. Incubate tubes at 16 °C overnight.
3. Heat the ligation reaction for 10 min at 65 °C.
4. To prepare the cDNA ligase mixture for electroporation of *E. coli* XL1-Blue cells, perform a drop dialysis of it against deionized H_2O using 0.025 μM disk membrane filter. Place carefully to float the filter (shine side up) onto the surface of ~30 mL of water in Petri dish. Do not touch membrane, and

pipet ligase mixture onto its center. After 1–2 h, carefully remove dialyzed DNA (the post-dialysis volume is generally larger than the original one). Store sample at −20 °C.

3.3.6 Transformation of E. coli and Plating

Use of electroporation to transform *E. coli* cells results in the highest transformation efficiencies as compared to the other methods.

1. Place sterile 0.1 cm gap electroporation cuvettes and white chamber slide of MicroPulser Electroporator on ice. Place 2 mL of SOC media into a 15 mL tube. Do this under sterile conditions.
2. Thaw 80 μL electrocompetent cells for each sample on ice, and gently mix well with a pipet with 1 μL of dialyzed ligase mixture from Subheading 3.3.5.
3. Set the electroporator for 1.8 kV (Ec1 program of MicroPulser) if using 0.1 cm cuvettes.
4. Transfer the mixture of cells and ligation to the cold cuvette (*see* **Note 5**) with a pipet. Do not create any bubbles.
5. Quickly place the cuvette in a chilled safety chamber slide, push the slide into the chamber, and pulse according to the manufacturer's guide.
6. Immediately (!) add ~1 mL of SOC media to the cuvette. Mix it quickly, but gently, with the cells. This step is very important!
7. Transfer the cell suspension to a 15 mL tube with 1 mL of SOC, and incubate at 37 °C with shaking at 150 rpm for 2 h.
8. Plate out a certain volume, e.g., 2 or 20 μL, to check a recombinant cell titer (e.g., colonies/μL) of the obtained cDNA library.
9. Incubate plates overnight at 30 °C (*see* **Note 6**) to grow recombinant bacterial colonies.

3.3.7 cDNA Library Quality Assessment

The quality of the resulting plasmid cDNA library can be estimated by the percentage of recombinant clones and the average cDNA insert size in pTriplEx2 vector. The average size of genes in cDNA library can be set by fast PCR analysis of the randomly selected recombinant colonies (~20) using vector primers. Percentage of recombinant clones of the obtained cDNA library can also be easily determined using blue-white screening for recombinants. In pTriplEx2, the cloning site is embedded in the coding sequence for the α-polypeptide of β-galactosidase lacZ' (Fig. 2). This makes it possible to use lacZ α-complementation for the blue-white color selection in appropriate host strain such as *E. coli* XL1-Blue. Plating bacteria on medium containing IPTG and chromogenic substrate X-Gal produces blue color in nonrecombinant colonies containing vector only. The ratio of white (recombinant) to blue (nonrecombinant) colonies will provide a quick estimation of efficiency of the ligase reaction.

1. Prepare Petri dishes with ampicillin for blue-white screening: add 40 μL 100 mM IPTG and 120 μL X-Gal (20 mg/mL) to the LB-agar surface of each plate, and spread over the entire surface. Dry X-Gal/IPTG-coated Petri dishes under UV-light.
2. Plate transformed competent cells and incubate at 37 °C overnight, or until blue color of colonies appears (to ~24 h).
3. Calculate the ratio of white (recombinant) to blue (nonrecombinant) colonies. (With these protocols applied, the yield of recombinant clones usually exceeds 90%).

3.4 Screening of cDNA Expression Library

Positive clones containing cDNAs for ctenophore photoproteins can be selected from the obtained cDNA expression library using a functional test for enzymatic activity, i.e., bioluminescent response of cell lysates preactivated with coelenterazine to Ca^{+2} addition. Try to complete all screening stages quickly, since XL1-Blue cells quickly lose viability on the original plate. It is recommended not to store plates with colonies for more than 2 weeks.

1. Next day, when the cell titer of the library is known, plate out the obtained unamplified cDNA library pool of *E. coli* cells on LB-ampicillin agar plates with 100 μg/mL ampicillin at a low density of ~1500 recombinant colonies per 9 cm Petri dish.
2. Incubate plates for ~24 h at 30 °C until the colonies reach a 1–1.5 mM diameter, and then refrigerate at 4 °C for 1–2 h (*see* **Note 6**).
3. Obtain an imprint of the colony patterns by applying a 0.45 μm dry nitrocellulose membrane filter (sterile) to the primary plate.
4. Then, turn over the replica filter, place it to a fresh LB-ampicillin agar plate, and leave on the plate until the clearly visible bacterial colonies appear.
5. Add 5 mL of SOC medium at replica filter plate, and scrape off/mix bacterial cells with a sterile cell spreader. To save a copy of this part of the amplified cDNA library, pipet 400 μL of cell suspension and 100 μL of 75% sterile glycerol in a sterile 1.5 mL tube. Store in 15% glycerol at −70–80 °C.
6. For gene expression, pour the 100–200 μL (depending on the initial density) cell suspension into a culture tube with 3 mL of LB medium supplemented with ampicillin (50 μg/mL), cultivate for 1 h at 37 °C until OD_{600} = ~1.0 (*see* **Note 7**).
7. Then, induce a protein synthesis with 1 mM IPTG for 1 h at 37 °C (*see* **Note 8**).

8. Prepare control sample with a similar cell density of induced *E. coli* cells containing vector only to check for background bioluminescence signal.
9. Sonicate the cell pellet from 2 mL of culture in 0.2 mL of 0.5 M NaCl, 5 mM EDTA, 50 mM BIS-Tris propane-HCl, pH 9.0 at 0 °C (*see* **Note 9**). Now crude cell lysate is ready for a functional test.
10. Add 1 μL of coelenterazine (final concentration 10^{-6} M) to the lysate, and incubate mixture at 4 °C for 4–24 h in the dark to activate ctenophore apophotoprotein.
11. Check the bioluminescence activity of primary plate samples (Subheading 3.5).
12. Make one more replica from positive primary plate. Cut the next replica into sectors for more detailed location of positive colonies and analyze as above.
13. Then, pick the colonies from positive areas to a new plate and assay small groups and individual colonies for the bioluminescence activity as above (Fig. 1).
14. Single, well-isolated, positive colonies should be picked from the last screen and used for sequencing of the cDNA inserts.

3.5 Specific Bioluminescent Assay

The charged samples of lysates must be kept in the dark to prevent inactivation due to photosensitivity of ctenophore photoproteins. It is necessary to take measurements quickly, in the dark or at dim red light. All measurements are carried out at room temperature. To get started, set the luminometer to maximum sensitivity and measure control sample to check background light signal.

1. Prepare luminometer cuvettes with 450–499 μL (volume depends on bioluminescence activity of samples) buffer for bioluminescence measurements: 10 mM EDTA, 100 mM M Tris-HCl pH 8.5.
2. Add 1–50 μL (volume depends on bioluminescence activity of samples) of sample in luminometer cuvettes with measuring buffer (500 μL total), and place cuvette into a luminometer cell.
3. Initiate the photoprotein reaction with fast injection of 0.2 mL of 100 mM $CaCl_2$, 100 mM Tris-HCl, pH 8.5 into a luminometer cell.
4. Repeat measurements 3 times.
5. Calculate specific bioluminescence activity as an average of 3 maximal values (RLU/μL).

4 Concluding Remarks

Here, we present the protocols for isolation of full-size cDNA genes encoding light-sensitive Ca^{2+}-regulated photoproteins from ctenophore *Beroe abyssicola* using functional screening of plasmid cDNA expression library. All the procedures for isolation of individual *E. coli* colonies carrying targeted genes have been described step by step. Although it may seem that they are rather laborious, time-consuming, and expensive, the result of application of functional screening strategy is very valuable and frequently cannot be obtained by other techniques, since only functional screening method allows isolation of full-size cDNA genes encoding truly novel enzymes in the absence of possibility to obtain any sequence data.

5 Notes

1. To minimize the presence of ubiquitous RNases and avoid degradation of mRNA, some basic precautions include wearing gloves throughout the experiment, usage of only clean pipettes or single-use plastic pipettes, sterilized filter tips, and new RNase-free microtubes while working with RNA. All materials coming in contact with samples must be sterile and RNasa-free. It is good to have a special working place, dedicated set of pipettors, and the equipment to be used for RNA work only. We don't recommend treating the solutions with DEPC (diethyl pyrocarbonate) to inactivate RNases, traces of which can inhibit subsequent enzymatic reactions. Avoid autoclaving as recycled steam in some autoclaves can introduce contaminants.
2. The obtained cDNA is almost always contaminated with sequences of ribosomal RNA, which can be amplified with specific primers and sequenced. The obtained sequence of 18S ribosomal RNA can be compared with the available databases. This will be a more accurate identification of the animal caught [6].
3. The integrity of mRNA can be visually approximately assessed on a denaturing formaldehyde agarose gel. Typically, mRNA appears as a smear of 0.5–6 kb, with an area of higher intensity around 1.5–2 kb, but this size distribution may be tissue specific. The mRNA content varies, depending on the source and RNA expression levels, but we can assume that this will be approximately 1–2 μg of mRNA from 1 mg of tissue in the proposed protocol.

4. If the bands are not seen on agarose gel, perhaps, the number of PCR cycles is not enough. In this case, additional cycles with the same reaction mixture can be performed. Try to minimize the number of cDNA amplification cycles, as the proportion of long full-size cDNA may decrease due to more efficient amplification of shorter molecules.
5. Cuvettes for one and the same DNA sample may be reused after thorough washing in deionized water and storing in 70% ethanol.
6. According to our observation, more recombinant clones grow after electroporation, if cells are incubated at 30 °C than at 37 °C.
7. The cultivation time before induction depends on the initial cell density and is hard to pinpoint. OD_{600} can be determined approximately by comparing with the standard tube.
8. The duration and temperature conditions of induction are selected empirically.
9. Buffer for lysate preparation is selected empirically and depends on the properties of the target protein.

Acknowledgments

This work was partially funded by RFBR, Krasnoyarsk Territory and Krasnoyarsk Regional Fund of Science, project No. 20-44-242003 and the Ministry of Science and Higher Education of the Russian Federation (project No. 0287-2021-0020).

References

1. Markova SV, Vysotski ES (2015) Coelenterazine-dependent luciferases. Biochemistry (Mosc) 80:714–732
2. Haddock SHD, Moline MA, Case JF (2010) Bioluminescence in the sea. Annu Rev Mar Sci 2:443–493
3. Markova SV, Vysotski ES, Blinks JR, Burakova LP, Wang B-C, Lee J (2002) Obelin from the bioluminescent marine hydroid *Obelia geniculata*: cloning, expression, and comparison of some properties with those of other Ca^{2+}-regulated photoproteins. Biochemistry 41:2227–2236
4. Markova SV, Burakova LP, Frank LA, Golz S, Korostileva KA, Vysotski ES (2010) Green-fluorescent protein from the bioluminescent jellyfish *Clytia gregaria*: cDNA cloning, expression, and characterization of novel recombinant protein. Photochem Photobiol Sci 9:757–765
5. Burakova LP, Natashin PV, Markova SV, Eremeeva EV, Malikova NP, Cheng C, Liu ZJ, Vysotski ES (2016) Mitrocomin from the jellyfish *Mitrocoma cellularia* with deleted C-terminal tyrosine reveals a higher bioluminescence activity compared to wild type photoprotein. J Photochem Photobiol B 162:286–297
6. Markova SV, Burakova LP, Golz S, Malikova NP, Frank LA, Vysotski ES (2012) The light-sensitive photoprotein berovin from the bioluminescent ctenophore *Beroe abyssicola*: a novel type of Ca^{2+}-regulated photoprotein. FEBS J 279:856–870
7. Golz S, Markova S, Burakova L, Frank L, Vysotski E (2005) Isolated photoprotein

bolinopsin, and the use thereof, WO/2005/000885 (Patent)

8. Szent-Gyorgyi CB, Ballou T, Dagmal E, Bryan B (1999) Cloning and characterization of new bioluminescent proteins. Part of the SPIE Conference on Molecular imaging: reporters, dyes, markers, and instrumentation. San Jose, CA, United States. Proc SPIE 3600:4–11
9. Markova SV, Golz S, Frank LA, Kalthof B, Vysotski ES (2004) Cloning and expression of cDNA for a luciferase from the marine copepod *Metridia longa*. A novel secreted bioluminescent reporter enzyme. J Biol Chem 279:3212–3217
10. Borisova VV, Frank LA, Markova SV, Burakova LP, Vysotski ES (2008) Recombinant Metridia luciferase isoforms: expression, refolding and applicability for *in vitro* assay. Photochem Photobiol Sci 7:1025–1031
11. Markova SV, Larionova MD, Burakova LP, Vysotski ES (2015) The smallest natural high-active luciferase: cloning and characterization of novel 16.5-kDa luciferase from copepod *Metridia longa*. Biochem Biophys Res Commun 457:77–82
12. Larionova MD, Markova SV, Vysotski ES (2017) The novel extremely psychrophilic luciferase from *Metridia longa*: properties of a high-purity protein produced in insect cells. Biochem Biophys Res Commun 483:772–778
13. Kotlobay AA, Sarkisyan KS, Mokrushina YA, Marcet-Houben M, Serebrovskaya EO, Markina NM, Gonzalez Somermeyer L, Gorokhovatsky AY, Vvedensky A et al (2018) Genetically encodable bioluminescent system from fungi. Proc Natl Acad Sci U S A 115:12728–12732
14. Anctil M (1985) Ultrastructure of the luminescent system of the ctenophore *Mnemiopsis leidyi*. Cell Tissue Res 242:333–340
15. Takara Bio USA (2016) SMART cDNA library construction kit user manual. Cat. No. 634901, PT3000-1

Chapter 14

Recording Cilia Activity in Ctenophores

Tigran P. Norekian and Leonid L. Moroz

Abstract

Pelagic ctenophores swim in the water with the help of eight rows of long fused cilia. Their entire behavioral repertoire is dependent to a large degree on coordinated cilia activity. Therefore, recording cilia beating is paramount to understanding and registering the behavioral responses and investigating its neural and hormonal control. Here, we present a simple protocol to monitor and quantify cilia activity in semi-intact ctenophore preparations (using *Pleurobrachia* and *Bolinopsis* as models), which includes a standard electrophysiological setup for intracellular recording.

Key words Ctenophora, Electrophysiology, Behavior, *Pleurobrachia*, *Bolinopsis*, Cilia, Locomotion

1 Introduction

Ctenophores evolved unique locomotory and integrative systems [1–6], reaching a tissue and organ complexity comparable to bilaterian animals [7]. The evolutionary scenario includes independent origins of neurons [8, 9], synapses [10, 11], and alternative integrative systems [12], where ciliated structures (not muscles) diversified as the primary effectors.

The role of cilia in ctenophores is overwhelmingly rich and essential for virtually all functions and behaviors [6, 13, 14]. One primary example is the use of cilia for complex locomotion—ctenophores move in the water column with the help of eight rows of ctenes, which consist of the large mechanically fused swim cilia [14]. The entire coordination of multiple behaviors in ctenophores is also primarily controlled by variations in the activity of swim cilia, and these mechanisms were under intensive investigation [6, 13, 15–22].

Although cilia are the major effectors in ctenophores, with presumed neuronal control, little is known about synaptic regulation and neurotransmitters controlling cilia movement. Therefore,

Authors Tigran P. Norekian and Leonid L. Moroz have equally contributed to this chapter.

Leonid L. Moroz (ed.), *Ctenophores: Methods and Protocols*, Methods in Molecular Biology, vol. 2757, https://doi.org/10.1007/978-1-0716-3642-8_14, © Springer Science+Business Media, LLC, part of Springer Nature 2024

recording cilia activity in ctenophores is essential for their behavioral and functional analyses. Here, we describe a simple protocol that was successfully used to record and quantify cilia beating in ctenophores during our investigation of the physiological roles of different transmitters [23] and their evolution [24–26].

2 Materials

1. High Mg^{2+} seawater—333 mM magnesium chloride ($MgCl_2$) is added to filtered seawater at 1:1 ratio.
2. Sylgard-coated Petri dishes (World Precision Instruments, Sylgard Silicone Elastomer, SYLG184).
3. Small steel insect pins.
4. Glass microelectrodes—borosilicate glass micropipettes for intracellular recording (World Precision Instruments, standard glass capillaries, 2 mm OD with a thin filament, 1B200F-4).
5. Potassium acetate—3 M solution.

2.1 Equipment

1. Nikon stereoscopic microscope SMZ-10A.
2. Micromanipulators (Warner Instruments, Standard Manual Control Micromanipulators, MM-33).
3. Micropipette puller (Sutter Instruments, Flaming/Brown Micropipette Puller P-97).
4. Intracellular amplifiers (A-M Systems Neuroprobe 1600 Amplifiers).
5. Chart recorder (Gould WindoGraf 940 4 Channel Recorder).

3 Methods

1. Freshly collected animals were incubated in high Mg^{2+} seawater for about 1 h to prevent muscle contractions. We used large, 1–2 cm, *Pleurobrachia bachei* and medium-sized, 3–4 cm, *Bolinopsis microptera* for these experiments.
2. The animals were then tightly pinned to a Sylgard-coated Petri dish with small steel needles to prevent body movements other than cilia beating (*see* **Note 1**). Relatively small animals (1 cm *Pleurobrachia*) were used whole without dissection; larger animals (2 cm and above) were dissected, and parts of a body wall with 2–3 cilia rows were pinned the same way to the Petri dish.
3. The high Mg^{2+} solution was washed out by fresh seawater several times during 30-min intervals to restore the normal chemical balance.

4. The Petri dish was placed in a standard electrophysiological rig on a recording platform and connected to the Ag/AgCl reference electrode.
5. We used glass microelectrodes filled with 3 M potassium acetate to record cilia beating. The sharp microelectrodes were pulled using Sutter Electrode Puller.
6. The original resistance of sharp microelectrodes (made for intracellular recordings) was 20–40 MΩ. A narrow strip of thin paper was used to carefully touch the tip of the electrode to break off the fragile sharp end. The resulting electrode was more stable to further mechanical contact and had a resistance of about 10 MΩ. Electrodes with very low resistances below 2 MΩ were unsuitable.
7. The electrodes were then connected to the micromanipulators and the intracellular amplifiers (Neuroprobe 1600, A-M Systems).
8. With the help of micromanipulators and under visual control via a dissecting microscope, the tip of the electrode was carefully placed next to the cilia combs so that during cilia beating, cilia were touching the tip of the electrode (Fig. 1). This physical contact created a brief electrical signal picked up by amplifiers and recorded on paper and in digital form using Gould Recorder (WindoGraf 980). Thus, each cilia beat was translated into a fast electrical spike, which allowed a digital recording of cilia beat frequency, but not the amplitude and power of beating (*see* **Note 2**).
9. To understand whether the cilia activity pattern was modulated by polysynaptic inputs or inherent to cilia, chemical isolation was used by bathing the preparation in high Mg^{2+} saline for 5–15 min (333 mM $MgCl_2$ was added to filtered seawater at 1: 1 ratio). High Mg^{2+} solution was applied into the recording dish using a graduated pipette attached to a long small-diameter tube. Elevated magnesium chloride solution suppresses synaptic chemical transmission and is widely used in comparative neurobiology [19, 20].

3.1 Illustrated Examples

The proposed protocol allows reliable registering and quantifying locomotory cilia activity in ctenophores. We experimented primarily with *Pleurobrachia bachei* and *Bolinopsis microptera*, although other ctenophore species, including larval and juvenile, can be used for that purpose.

Patterns of cilia beating in *Pleurobrachia* were always very variable, frequently with powerful bursts and inhibitory episodes (Fig. 2a; see also [23]). Such activity might represent intact behavior in a free-moving *Pleurobrachia* as an ambush predator [6]. Sometimes, however, the cilia beating in *Pleurobrachia* showed

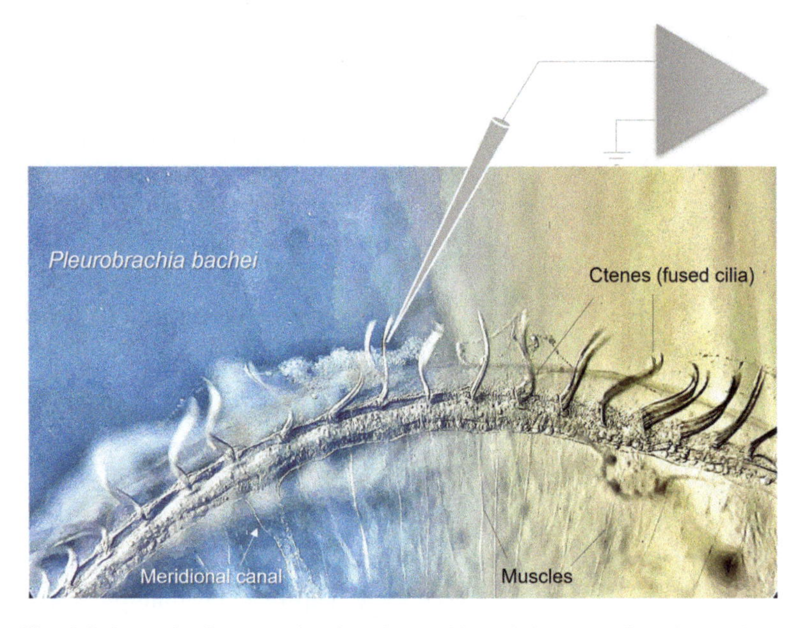

Fig. 1 Schematic diagram showing the position of the recording glass microelectrode next to the ctenes of swim cilia in *Pleurobrachia*

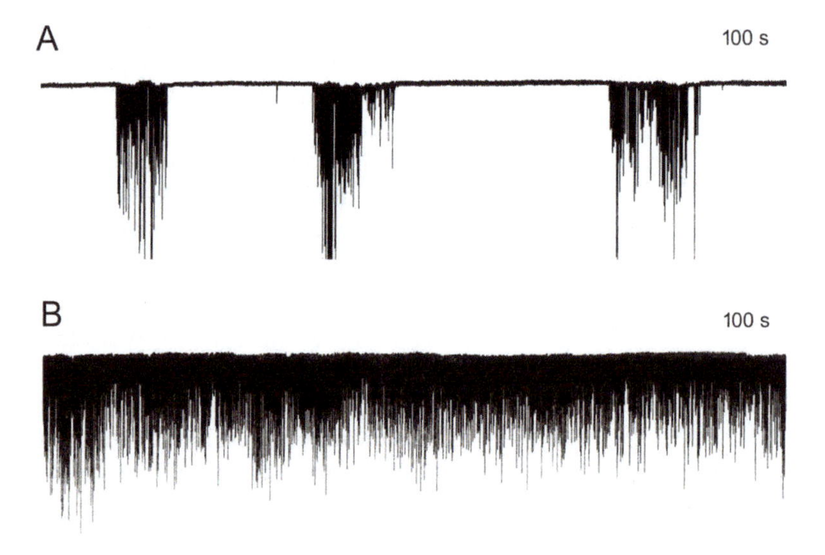

Fig. 2 Cilia beating in *Pleurobrachia bachei* is very variable and complex, similar to intact behaviors in free-moving animals. (**a**) Episodes of high-frequency bursting with periods of inhibition between them. (**b**) Tonic continuous cilia beating with possible brief episodes of acceleration. The numbers above the traces show the time of recordings

a constant active beating for a prolonged period (Fig. 2b). In contrast, *Bolinopsis* had more regular cilia beating with fewer activity patterns and lower frequency (Fig. 3a, b). Notably, the complex patterns of cilia activity were eliminated in the presence of a high

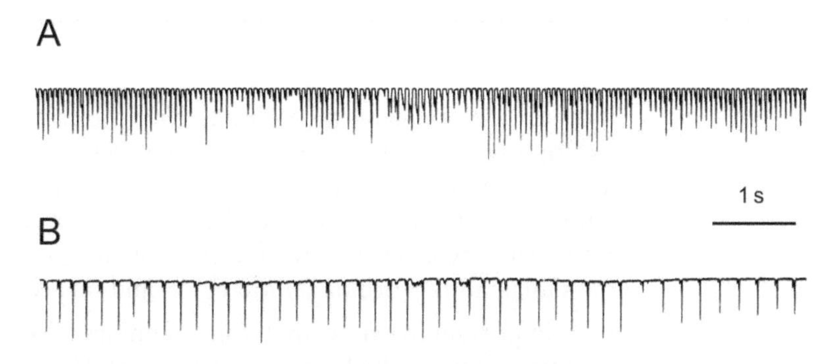

Fig. 3 Cilia beating in *Bolinopsis microptera* is less patterned and more regular than in *Pleurobrachia*. However, it demonstrates periods of higher-frequency beating (**a**) and slower activity (**b**)

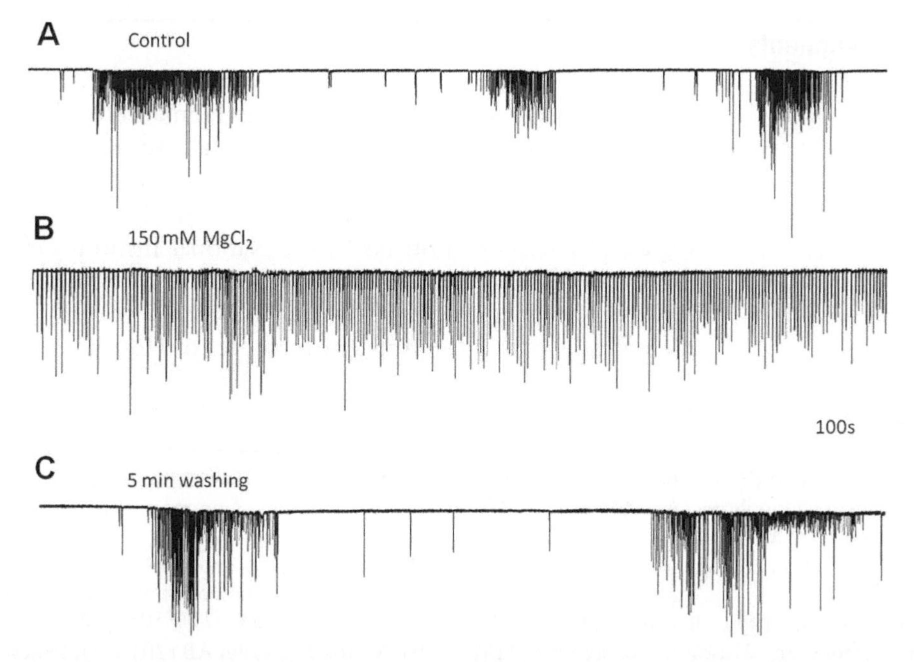

Fig. 4 High Mg^{2+} seawater blocked complex patterns of cilia beating (**a**) in *Pleurobrachia*, suggesting that synaptic inputs initiated episodes of high-frequency bursting and inhibition. (From [23]). The regular unvarying cilia beating was sustained during high Mg^{2+} solution exposure (**b**). The complete recovery was achieved within several minutes following washing in normal seawater, which fully restored episodes of bursting and inhibition (**c**). Numbers under the traces show the time of recordings

concentration of Mg^{2+} (Fig. 4; see [23]), known to suppress synaptic inputs [27, 28]. These findings indicate the presence of multifaceted chemical inputs (e.g., peptides [9, 29, 30] or nitric oxide [31]) and likely direct synaptic control of cilia, which is also confirmed by ultrastructural data with a remarkable diversity of synapses [5, 14, 32].

4 Notes

1. It was crucial for stable recording to have the ctenophore body wall tightly pinned to the Sylgard-coated Petri dish, with absolutely no movements in it, except cilia beating. If body wall muscles contracted and moved ctene row even a couple of mm away from the electrode, the electrode would stop picking up the signal, and recording continuity would be prevented.
2. It is important to note that this technique does not allow quantification of cilia beating amplitude and forces—only the frequency. The electrodes register mechanical contact with the cilia.

Acknowledgments

This work was supported in part by the Human Frontiers Science Program (RGP0060/2017) and National Science Foundation (IOS-1557923) grants to LLM. Research reported in this publication was also supported in part by the National Institute of Neurological Disorders and Stroke of the National Institutes of Health under Award Number R01NS114491 (to LLM). The content is solely the authors' responsibility and does not necessarily represent the official views of the National Institutes of Health.

References

1. Horridge GA (1965) Relations between nerves and cilia in Ctenophores. Am Zool 5:357–375
2. Horridge GA (1966) Pathways of co-ordination in ctenophores. Symp Zool Soc 16:247–266
3. Horridge GA (1974) Recent studies on the Ctenophora. In: Muscatine L, Lenhoff HM (eds) Coelenterate biology. Academic, New York, pp 439–468
4. Horridge GA, Chapman DM, Mackay B (1962) Naked axons and symmetrical synapses in an elementary nervous system. Nature 193: 899–900
5. Horridge GA, Mackay B (1964) Neurociliary synapses in *Pleurobrachia* (Ctenophora). Q J Microsc Sci 105:163–174
6. Tamm SL (1982) Ctenophora. In: Electrical conduction and behavior in "simple" invertebrates. Clarendon Press, Oxford, pp 266–358
7. Nielsen C (2012) Animal evolution: interrelationships of the living phyla. Oxford University Press, Oxford
8. Moroz LL (2014) The genealogy of genealogy of neurons. Commun Integr Biol 7(6): e993269
9. Moroz LL et al (2014) The ctenophore genome and the evolutionary origins of neural systems. Nature 510(7503):109–114
10. Moroz LL, Kohn AB (2015) Unbiased view of synaptic and neuronal gene complement in ctenophores: are there pan-neuronal and pan-synaptic genes across Metazoa? Integr Comp Biol 55(6):1028–1049
11. Moroz LL, Kohn AB (2016) Independent origins of neurons and synapses: insights from ctenophores. Philos Trans R Soc Lond Ser B Biol Sci 371(1685):20150041
12. Moroz LL, Romanova DY (2022) Alternative neural systems: what is a neuron? (Ctenophores, sponges and placozoans). Front Cell Dev Biol 10:1071961
13. Tamm SL (2014) Cilia and the life of ctenophores. Invertebr Biol 133(1):1–46
14. Hernandez-Nicaise M-L (1991) Ctenophora. In: Harrison FWFW, Westfall JA

(eds) Microscopic anatomy of invertebrates: Placozoa, Porifera, Cnidaria, and Ctenophora. Wiley, New York, pp 359–418

15. Tamm S (1973) Mechanisms of cilliary co-ordinations in ctenophores. J Exp Biol 59: 231–245
16. Tamm SL (1984) Mechanical synchronization of ciliary beating within comb plates of ctenophores. J Exp Biol 113:401–408
17. Tamm SL (1988) Calcium activation of macrocilia in the ctenophore *Beroe*. J Comp Physiol A 163(1):23–31
18. Tamm SL (1988) Iontophoretic localization of Ca-sensitive sites controlling activation of ciliary beating in macrocilia of *Beroe*: the ciliary rete. Cell Motil Cytoskeleton 11(2):126–138
19. Tamm SL, Tamm S (1981) Ciliary reversal without rotation of axonemal structures in ctenophore comb plates. J Cell Biol 89(3): 495–509
20. Tamm SL, Terasaki M (1994) Visualization of calcium transients controlling orientation of ciliary beat. J Cell Biol 125(5):1127–1135
21. Jokura K et al (2022) Two distinct compartments of a ctenophore comb plate provide structural and functional integrity for the motility of giant multicilia. Curr Biol 32(23): 5144–5152 e6
22. Jokura K et al (2019) CTENO64 is required for coordinated paddling of ciliary comb plate in ctenophores. Curr Biol 29(20): 3510–3516 e4
23. Norekian TP, Moroz LL (2023) Recording cilia activity in ctenophores: effects of nitric oxide and low molecular weight transmitters. Front Neurosci 17:1125476
24. Moroz LL et al (2021) Evolution of glutamatergic signaling and synapses. Neuropharmacology 199:108740
25. Moroz LL, Romanova DY, Kohn AB (1821) Neural versus alternative integrative systems: molecular insights into origins of neurotransmitters. Philos Trans R Soc Lond Ser B Biol Sci 2021(376):20190762
26. Moroz LL et al (2020) Microchemical identification of enantiomers in early-branching animals: lineage-specific diversification in the usage of D-glutamate and D-aspartate. Biochem Biophys Res Commun 527(4):947–952
27. Del Castillo J, Engbaek L (1954) The nature of the neuromuscular block produced by magnesium. J Physiol 124:370–384
28. Hutter OF, Kostial K (1954) Effect of magnesium and calcium ions on the release of acetylcholine. J Physiol I24:234–241
29. Sachkova MY et al (2021) Neuropeptide repertoire and 3D anatomy of the ctenophore nervous system. Curr Biol 31(23):5274–5285 e6
30. Hayakawa E et al (2022) Mass spectrometry of short peptides reveals common features of metazoan peptidergic neurons. Nat Ecol Evol 6(10):1438–1448
31. Moroz LL, Mukherjee K, Romanova DY (2023) Nitric oxide signaling in ctenophores. Front Neurosci 17:1125433
32. Burkhardt P et al (2023) Syncytial nerve net in a ctenophore adds insights on the evolution of nervous systems. Science 380(6642):293–297

Chapter 15

Electrophysiology of Ctenophore Smooth Muscle

Robert W. Meech, André Bilbaut (Deceased), and Mari-Luz Hernandez-Nicaise

Abstract

Unlike in the Cnidaria, where muscle cells are coupled together into an epithelium, ctenophore muscles are single, elongated, intramesogleal structures resembling vertebrate smooth muscle. Under voltage-clamp, these fibers can be separated into different classes with different sets of membrane ion channels. The ion channel makeup is related to the muscle's anatomical position and specific function. For example, *Beroe ovata* radial fibers, which are responsible for maintaining the rigidity of the body wall, generate sequences of brief action potentials whereas longitudinal fibers, which are concerned with mouth opening and body flexions, often produce single longer duration action potentials.

Beroe muscle contractions depend on the influx of Ca^{2+}. During an action potential the inward current is carried by Ca^{2+}, and the increase in intracellular Ca^{2+} concentration generated can be monitored in FLUO-3-loaded cells. Confocal microscopy in line scan mode shows that the Ca^{2+} spreads from the outer membrane into the core of the fiber and is cleared from there relatively slowly. The rise in intracellular Ca^{2+} is linked to an increase in a Ca^{2+}-activated K^+ conductance (K_{Ca}), which can also be elicited by iontophoretic Ca^{2+} injection. Near the cell membrane, Ca^{2+} clearance monitored using FLUO3, matches the decline in the K_{Ca} conductance. For light loads, Ca^{2+} is cleared rapidly, but this fast system is insufficient when Ca^{2+} influx is maintained. Action potential frequency may be regulated by the slowly developing K_{Ca} conductance.

Key words Ctenophore, Smooth muscle, Voltage-clamp, Patch-clamp, FLUO-3, Calcium, Ca^{2+} injection, Sodium channel, Potassium channel, Calcium-activated potassium channel

1 Introduction

The purpose of this article is to bring together the different electrophysiological studies carried out on the "giant" smooth muscle cells in the body wall of the ctenophores *Mnemiopsis* (Class, Tentaculata; Order, Lobata) and *Beroe* (Class, Nuda; Order, Beroida). Ctenophores are the earliest extant organisms to have developed true mesenchymal muscle cells. In these species the muscles do not form multicellular units, as they do for example in the Cnidaria, but remain as single multinucleate fibers running independently through the transparent gelatinous mesenchyme or mesogloea of

Leonid L. Moroz (ed.), *Ctenophores: Methods and Protocols*, Methods in Molecular Biology, vol. 2757, https://doi.org/10.1007/978-1-0716-3642-8_15, © Springer Science+Business Media, LLC, part of Springer Nature 2024

the body wall (see Note 1). Previous ultrastructural [1] and electrophysiological [2–4] studies on *Beroe ovata* have compared the longitudinal and radial muscle systems. Studies on *Mnemiopsis leidyi* have focused on two groups of longitudinal fibers that lie alongside the pharynx [5, 6].

Beroe muscle fibers fall into four main classes [2]: (a) thin pharyngeal circular fibers, (b) internal longitudinal fibers that run next to the pharynx, (c) external longitudinal fibers that run below the outer epithelium, and (d) short radial fibers that cross the body wall and make contact with neural processes at either surface. Attachments of the radial fibers to the external integument are visible with Nomarski interference contrast, EM [2] and SEM [7] techniques. However, the mesogloea is quite transparent and individual fibers (5–50 μm diameter, depending on the size of the animal) can be easily traced in whole-mount preparations using a light microscope.

In addition to these four main classes, phalloidin staining [7] has revealed other less widely distributed fiber types: (a) longitudinal fibers that spread along each comb row and penetrate the mesogloea at either end, (b) short noncontractile elements that connect each comb plate (cf. *Pleurobrachia* [8]), and (c) circular muscles that spread around the aboral organ.

In EM sections, individual longitudinal fibers are seen to have a central core (containing the nuclei, mitochondria, sacs of smooth and rough endoplasmic reticulum, Golgi bodies, and ribosomes) surrounded by a thick contractile cylinder of actin and myosin myofilaments. The myofilaments are not organized into sarcomeres, and there is no system of sarcolemmal invaginations equivalent to a T-system. In fact, the fibers closely resemble vertebrate smooth muscle [2].

As in other muscle cells, Ca^{2+} is accumulated in the tubules and cisternae of the sarcoplasmic reticulum via an ATP-dependent process [9], and release from this store may contribute to the caffeine-evoked Ca^{2+} transients previously reported [10]. The endoplasmic reticulum is made of two structurally distinct subsets: (1) the tubules and cisternae located in the axial core and (2) a diffuse network of smooth longitudinal tubules (the sarcoplasmic reticulum, SR). The SR accounts for from 0.8% to 1% of the volume of the contractile cytoplasm, but it is distributed differently in the different fiber types [11]. In longitudinal fibers, there are relatively few SR tubules at the periphery, (0.7 tubules/μm^2) compared with the center (2 tubules/μm^2), whereas in radial fibers, SR tubules have a larger diameter and are distributed evenly. They are sometimes located close to the sarcolemma, but typical junctional cisternae are found only at neuromuscular junctions and at intermuscular gap-junctions [1].

1.1 Innervation

Silver-impregnated preparations of *Beroe* show the muscles to be innervated by a loose network of fine subepithelial processes [2]. When stained with alpha-tubulin antibody, this nerve net is seen to consist of irregular polygonal units on either side of the body wall and a diffuse meshwork of neurons deep within the mesogloea [7]. An SEM study shows many fibers with thin neuronal-like processes attached to them [7]. Under EM, synapses are sufficiently numerous as to suggest that muscle cells are individually innervated [2], with most junctions occurring near the epithelia. At the presynaptic side of each junction, there is a "triad" of structures (synaptic vesicles-endoplasmic reticulum-mitochondrion), like that found in interneuronal synapses [12].

By drawing a small amount of tissue into a fine polyethylene suction pipette, it is possible to record spontaneous activity externally from the surface of the body wall [2]. The electrical impulses recorded are either small (<30 μV) biphasic events of regular form or larger (<200 μV) irregular multiphasic events associated with ripples of muscular contraction. The latter are assumed to be compound action potentials arising from muscle fibers near the recording site, while the former resemble potentials of nervous origin recorded with similar techniques from cnidarian tissues [13]. The smaller events can be evoked repeatedly by an electrical stimulus (1–2 ms duration) applied to the body wall. Such events occur singly or in short bursts and appear with a consistent latency and waveform. At 22 °C, their conduction velocity is in the range of 21–29 cm/s (mean 25 cm/s). They are observed in both inner (pharyngeal) and outer epithelia but do not cross the mesogloea. Whole-mount fluorescent antibody staining by Jager et al [14] shows a nerve net that spreads throughout the mesogloea in *Pleurobrachia pileus*, (in addition to the more compact polygonal "nerve net" in the ectodermal epithelium) and so presumably each nerve net operates independently.

What appear to be excitatory postsynaptic potentials may be recorded intracellularly from enzymically isolated muscle fragments some of which are observed, in EM sections, to have retained their synapses (M-L. Hernandez-Nicaise, unpublished). Flurries of these excitatory depolarizations summate and give rise to overshooting action potentials (*see* [15] for records). They appear similar to those recorded from elongated epithelial (polster) cells [16] associated with the comb rows of *Pleurobrachia pileus* [17]. Polster cells make contact with many synapses [12, 18, 19] and produce what appear to be depolarizing postsynaptic potentials that frequently lead to regenerative impulses that overshoot 0 mV [17].

Norekian and Moroz [8] have expanded upon the surprising complexity of the neuromuscular organization responsible for feeding, swimming, escape, and prey capture in *Pleurobrachia bachei.* There are at least nine types of neurons, five families of receptors, and multiple different groups of muscle cells. When compared to

Beroe, both species have a similar neuronal architecture [7, 8] despite their very different foraging strategies. *Pleurobrachia* "ambushes" zooplankton in a cloud of tentacles while *Beroe* is an active hunter searching out and engulfing prey with its huge mouth. These "more complex swimming patterns and hunting behaviors" are generated without any "substantial concentration of neuronal elements" apart from the aboral organ and polar fields [7]. On the other hand, there are a large number of specialized sensory structures on the oral side of the mouth, which according to Horridge [20], are necessary for the ingestion of prey. There is also a large tentacular nerve (a condensed region of the epithelial nerve net) which runs between the tentacles and the apical organ [21, 22]. The increased flexibility of the body wall of *Beroe* as compared with *Pleurobrachia* is matched by the size and density of its muscle fibers. This flexibility allows *Beroe* to consume prey that is almost as large as itself (*see* [23, 24] for descriptions of feeding).

1.2 Electrical Properties of Muscle

Ctenophore smooth muscles generate propagated action potentials with a conduction velocity that varies with fiber diameter (*Beroe*: 29-68 cm/s for diameters of 22-34 μm) [2]. Action potentials in *Beroe* longitudinal fibers peak at +20 mV and repolarize in two stages [4]. Between the two repolarizing steps, there is a plateau of variable duration (up to 15 ms at 0 mV), during which the potential declines from about +5 to −5 mV. Although action potentials in *Beroe* radial fibers also peak at +20 mV, they repolarize more quickly (duration 3 ms at 0 mV). In the *Mnemiopsis* longitudinal fibers studied by Anderson [6], action potentials had a similar amplitude to those in *Beroe* and had a similar threshold (−32 mV); they repolarize with a duration at half amplitude of 4.7 ± 0.2 ms; SEM).

Electrophysiological characteristics have been examined using current and voltage-clamp techniques in both *Mnemiopsis* [26] and *Beroe* [3, 4]. In each case, resting potentials are determined predominantly by the K^+ gradient across the muscle membrane, but their mean values are different (*Beroe* longitudinal muscles, −60 ± 1.35 mV, SEM, $n = 25$; radial muscles, −66 ± 1.37 mV, SEM, $n = 32$; *Mnemiopsis*, −51 ± 4 mV, SEM, $n = 22$; all measurements in 10 mM external K^+). In *Mnemiopsis*, Anderson [6] finds that the membrane potential follows the Nernst relationship for K^+ more closely when propionate (rather than Cl^-) is the extracellular anion and he suggests that Cl^- contributes to the resting membrane conductance. The action potential duration depends on the resting potential and increases when stimulated at 2 Hz [6]. Action potential repolarization is delayed in the presence of tetraethylammonium ions (TEA^+) but 4-aminopyridine (4-AP) has little effect.

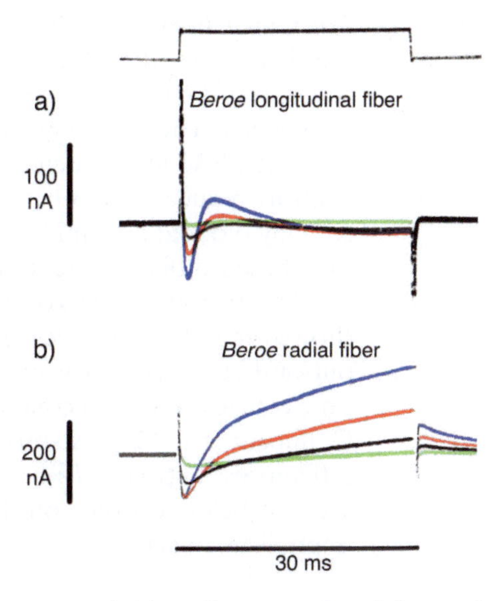

Fig. 1 Ion currents recorded from *Beroe* muscle cell fragments under voltage-clamp. Superimposed currents from longitudinal (**a**) and radial (**b**) cells during depolarizing commands to −30 mV (green), −25 mV (black), −20 mV (red), and −15 mV (blue). The voltage command protocol is shown at the top of the figure. Holding potential: −60 mV; longitudinal fiber: 400 μm long, diameter not recorded; radial fiber: 250 μm long, 80 μm diam. Cells bathed in ASW containing 2 mM Ca^{2+} at room temperature. Fibers 23 (F9–87) and 55 (F24–87)

In this respect *Mnemiopsis* fibers resemble *Beroe* radial fibers [4]. Action potentials in both species are tetrodotoxin resistant.

In *Beroe*, both longitudinal and radial fiber types generate trains of action potentials upon maintained depolarization, but the sequence is more prolonged in radial fibers. While both Na^+ and Ca^{2+} may contribute to the action potential depolarizing phase in each case, the K^+ channel blocker 4-aminopyridine (4-AP) delays repolarization significantly only in longitudinal fibers. In voltage-clamp experiments (Fig. 1), the inward current develops at about −30 mV, and in longitudinal fibers, it is followed by a large transient outward current which is responsible for the initial repolarizing phase of the action potential. Full repolarization probably depends on more slowly developing currents. Both slow and fast outward currents are carried mainly by K^+ [3]. The slower outward currents are particularly sensitive to TEA^+, and this accounts for its prolonging effect on the later phase of the action potential. TEA^+ is more effective in a Ca^{2+} free bathing medium, suggesting that a TEA^+-sensitive Ca^{2+}-activated current is one contributor [3]. Ca^{2+}-activated currents develop more quickly in *Beroe* radial muscles and it may be that, under normal conditions, the small transient outward current which is present plays only a minor role in repolarisation.

Dubas et al. [26] have used the whole-cell version of the patch-clamp technique to identify six different voltage-gated currents in

Mnemiopsis fibers. They include two transient inward currents, two transient outward currents, a delayed steady outward current, and a late outward current, evident only during prolonged depolarizing commands. The inward current components are isolated by using Cs^+ and TEA^+-filled patch pipettes to block the outward currents. Both are reduced by low external Ca^{2+} or Na^+ or by 5 mM Co^{2+} and are fully blocked by 5 mM Cd^{2+}. They can be separated by making use of their different rates of inactivation.

The two-electrode voltage-clamp study by Bilbaut et al. [3] showed that *Beroe* muscles exhibit qualitatively similar inward and outward currents to *Mnemiopsis* albeit with marked differences in time course even when conducted at similar temperatures, 18–25 °C. Bilbaut et al. [27] also provided a brief account of the qualitative differences between *Beroe* radial and longitudinal fibers. The account below expands on these findings and contains previously unpublished data.

2 Materials and Methods

2.1 Materials

1. Trypsin (type III, Sigma-Aldrich).
2. Hyaluronidase (type II, from sheep testes; most recent product number H2126 Sigma-Aldrich).
3. Bovine serum albumin (BSA, Sigma-Aldrich, fraction V).
4. FLUO-3, pentapotassium salt (originally obtained from Molecular Probes, Eugene, Oregon but now available from https://www.aatbio.com/dlist.html). ATT Bioquest (catalog number 21017)
5. Silicone resin-coated glassware (RTV 141).

2.2 Animals

Specimens of *Beroe ovata* were collected (March–May) from surface waters in the bay of Villefranche-sur-Mer, France. They were kept at 10–15 °C in large containers filled with natural seawater. The containers were cleaned daily when the water was changed.

2.3 Extracellular Recordings

Extracellular recordings were made from the surface of the body wall using a fine polyethylene suction electrode connected to a high-gain, differential, AC amplifier Tektronix (Model 5A22N)—*see* Subheading 8. Similar electrodes were used to deliver stimuli.

2.4 Intracellular Recordings

Muscle fibers in situ are continuously in motion, and so it is necessary to stabilize isolated preparations. This can be done with a syringe attached to a short length of translucent plastic tubing (6 mm diameter). At its mid-point the tube should have a 2-3 mm hole cut into its wall. The hole can be closed off with a finger so that applied suction (using the syringe) can be used to draw tissue into the plastic tubing. Once bathed in seawater and illuminated from below, muscle fibers embedded in mesogloea were viewed from above through the hole in the tube. Individual fibers can be traced for distances of up to 6 mm. Fibers severed during this process appeared to reseal, and even quite short lengths (1–2 mm) were excitable. Glass micropipettes filled with 3 M KCl and with resistances of 30–100 MΩ (10–15 MΩ if thin-walled glass tubing is used) will penetrate superficial fibers and, when connected to a unity-gain preamplifier, can be used to record resting and action potentials. Action potentials were elicited by using the micropipette to inject current into the cell via a balanced Wheatstone Bridge circuit. In these experiments, the current passed was measured as the potential across a 10 KΩ resistance to ground. Recording was restricted to only the most superficial fibers, because the mesogloea resists penetration by the micropipette and the muscle fibers are displaced as the pipette is lowered.

2.5 Isolated Muscle Cell Fragments [5]

For many experiments, it was necessary to digest the mesogloea completely and use isolated fragments of muscle. Strips of body wall, 3–4 cm long and 3 mm wide, were dissected from whole animals bathed in seawater. The long axis of the strip was made parallel to the longitudinal axis of the body (i.e., to the longitudinal fibers). The covering integument (often containing cells with a pink pigment), together with the external layer of mesogloea and some longitudinal muscles, was then removed, but the pharyngeal integument was retained. To improve the yield of radial fibers (i.e., those that cross the body wall), the longitudinal strips were recut in a transverse direction, and pieces of body wall were transferred to artificial seawater (ASW) containing (in mM): NaCl, 500; KCl, 10; $MgCl_2$, 58; $CaCl_2$, 10; buffered to pH 7.8 with Tris–HCl.

After two washes in nominally Ca^{2+}-free ASW (i.e., ASW in which Ca^{2+} had been omitted), the tissues were incubated at 30 °C for 20–30 min in 0.02% trypsin (type III, Sigma-Aldrich—*see* Subheading 8) and 0.12% hyaluronidase (type II, from sheep testes; Sigma-Aldrich) in Ca^{2+}-free ASW. When a few muscle cells were released from the main bulk of the tissue and could be seen floating in the medium, digestion was slowed by transferring pieces of tissue to cold Ca^{2+}-free ASW (10 °C) and gently agitating so as to release more fragments. The released muscle cells were transferred using a fire polished Pasteur pipette to a "recovery" medium, i.e., an ASW at 15 °C containing 2 mM Ca^{2+} and 0.2% bovine serum albumin (BSA, Sigma-Aldrich, fraction V). From now on, it was necessary to use silicone resin-coated glassware (RTV 141).

The cell fragments collected were of variable length, the fibers having been cut during either dissection or dissociation. The longest cells in a batch were more likely to be longitudinal fibers, while those with branched endings were more likely to be radial. (Final identification of the cell type depended on electrophysiological criteria. Radial fibers tended to exhibit spontaneous trains of fast action potentials; longitudinal fibers exhibit action potentials with a characteristic plateau on their falling phase. However, the ion currents seen under voltage-clamp were used as the main indicator of fiber type). The cells were retained in the cooled recovery medium for at least 2 h before being used for experimentation. This is because freshly prepared cells were highly labile. Longitudinal fibers could often be used during the 24 h following isolation; radial fibers rarely survived for more than 4 h.

For experimental purposes, isolated muscle fibers were transferred to a glass dish coated with a thin layer of silicone resin (RTV 141) and containing 2–5 mL ASW. The isolated fibers settle to the bottom of the dish and may be held in position by a recording micropipette inserted midway along their length. The flow of the bathing solution may slightly displace the free ends of the impaled fiber, but this generally did not interfere with the recording. Experiments were carried out at room temperature (18–23 °C).

For action potential recordings on isolated cells, intracellular current injection and transmembrane potential recording were performed with thin-walled double-barreled glass micropipettes as an alternative to the "bridge" circuit described above. Each barrel was filled with 3 M KCl and had an initial resistance of 5–8 MΩ. Their initial coupling resistance was 100–200 kΩ, but this would often increase somewhat after fiber penetration, so that the stray resistive voltage changes during current injection, became more obvious. The voltage recording side of the double-barreled microelectrode was connected to a high impedance amplifier and the current injected through the other barrel was monitored from a constant-current source.

2.6 Two-Electrode Voltage-Clamp to Measure Membrane Current

For voltage-clamp experiments cell fragments 400–800 μm in length were selected and penetrated with separate current-injecting and voltage-recording micropipettes. (Note: many fibers shortened somewhat during the course of the experiment with some becoming almost spherical). The micropipettes used were made from thin-walled glass tubing filled with 3 M KCl; they had resistances of 5–8 MΩ. The circuit used consisted of two high input impedance preamplifiers, one of which could be switched from recording the transmembrane potential to injecting current. The current injected was the output from a high-gain DC differential amplifier arranged such that the output was zero when the command input matched the recorded membrane potential, i.e., the command input on one side of the differential amplifier was equal and opposite to the resting potential input on the other. The command input at rest is called the "holding potential," and in most cases, in our experiments it was fixed at −60 mV. The command input can then be changed in a stepwise manner by applying rectangular commands from a pulse generator. The current required to maintain this new steady potential was measured and is equal and opposite to the currents flowing across the cell membrane.

The potential measuring micropipette was wrapped in aluminum foil to within 3 mm of the tip. We ensured that this foil shield never contacted the ASW bathing the cell—if necessary, it was insulated with a "Parafilm" cover. The shield was connected to a "driven" ground so that the shield potential followed the potential on the micropipette; this reduced the effective capacity and therefore increased the rise time of the recording side of the system. The current-supplying micropipette was also shielded, and this shield was connected to the circuit ground. This has the effect of slowing the rise time of the injected current and increases the stability of the system. The shielding between the micropipettes was required to reduce the capacitive crosstalk between the current and voltage sides of the circuit as this makes the feedback circuit unstable. Once the micropipettes had been inserted into the cell, the current electrode was switched to current-supplying mode, and the gain of the differential amplifier was adjusted to provide the greatest amplification with the minimum instability on the rising phase of the command.

The length constant of the fibers is about 2 mm [2]. Calculations from equations derived for a short cable [28] show that current injected midway along an 800 μm cylinder with a length constant of 2 mm would be expected to produce a voltage difference between the center and each end of 1%. In experiments to check the space-clamp, we penetrated cells with three micropipettes and confirmed that, even in the most elongated fibers, the errors were within 1% for depolarizing pulses up to 50 mV in amplitude. We recorded qualitatively similar currents from cells with widely different initial lengths. The records shown were not leak subtracted except where indicated.

2.7 In Situ Patch-Clamp Recording

Patch pipettes were prepared from soft-walled glass tubing using a two-step pull in the conventional way [29]. They were used without coating or further fire polishing. In all cases, the pipettes contained $5Ca^{2+}$ ASW (i.e. ASW buffered to pH 7.8 with 10 mM TRIS chloride and containing 5 mM $CaCl_2$). Pipette tip diameters were about 1 μm O.D. and light suction applied to the pipette caused the seal between glass and surface membrane to increase to 1–10 GΩ; the seal remained stable within this range when suction was released. All records were filtered with an eight-pole Bessel filter (corner frequency 1.5 KHz) and sampled at 20 KHz. Command pulses were imposed at 20 s intervals. Low-amplitude pre--pulses within the range −60 to −100 mV were scaled by PCLAMP software (Axon Instruments, Inc., Foster City, CA, USA) and used for online subtraction of leak currents. Capacitive transients were small and were similarly subtracted by the PCLAMP software. Because the patches were continuous with the sarcolemma, their intracellular surface was at the resting potential of the muscle cell fragment, taken as −55 mV. This value was based on independent intracellular recordings with high-resistance micropipettes and, on occasion, confirmed by estimates of the K^+ equilibrium potential derived from tail current analysis in cell-attached mode.

2.8 Fluorescent Dye Loading

Freshly isolated relaxed cell fragments (about 800 μm long) were attached to polylysine-coated coverslips and loaded with fluorescent Ca^{2+} indicator using a 6 MΩ glass micropipette tip-filled with a solution of 4 mM penta-K^+ FLUO-3 (Molecular Probes—*see* Subheading 8) in 3 M KCl and then back-filled with more 3 M KCl. Once inserted into a cell, the pipette provided sufficient FLUO-3 to give a steady signal in 5–10 min. This micropipette also served for voltage recording. Once the cell was sufficiently loaded with dye, the second micropipette (3 M KCl-filled) was inserted, and this provided the current injecting side of the voltage-clamp circuit. The gain of the voltage-clamp amplifier was adjusted, and different command voltages imposed. Changes in membrane current were monitored as changes in voltage across a standard resistor. A BioRad M500 confocal microscope in line scan mode permitted us to follow transient changes in internal Ca^{2+} with a temporal resolution of up to 2 ms. Most of the experiments were conducted in 5Ca^{2+}-ASW.

3 Voltage-Clamp Experiments

Beroe muscle fiber fragments (400 - 800 μm in length), prepared by enzymatic digestion, were studied using a conventional two-electrode voltage-clamp. Fibers were in a relaxed state initially but invariably shortened during the course of the experiment. In many, the contractile apparatus became detached from the membrane allowing the cells to take up a more spherical form. The currents recorded from such spheres were qualitatively similar to those recorded from more elongated cells.

Figure 1a shows the effect of 30 ms step depolarizations on a longitudinal fiber held under voltage-clamp. The voltage command protocol is shown at the top of the figure; between commands, the cell membrane was held at −60 mV. The superimposed traces show the membrane current response during command steps to −30 mV (green), −25 mV (black), −20 mV (red), and −15 mV (blue). Following the brief capacitive transient (initial large upward deflection) there is a rapidly developing inward current (downward), which is followed by an outwardly directed current (upward) that reaches a peak at about 5 ms and then inactivates to reveal the continuing inward current. Slower outward currents are not shown at this timescale (but *see* Fig. 4). In a radial fiber under the same conditions (Fig. 1b), the capacitive current is smaller and the initial inward current is followed by a steadily rising outward current. At the end of the depolarizing step when the membrane is returned to the holding potential, there is a large outward "tail" current, which is not seen in Fig. 1a. The amplitude of this tail depends on the number of channels opened by the depolarizing command. Its time course depends upon how quickly the channel population closes once the membrane has been returned to its holding potential.

Note: All figures follow the usual convention whereby inward currents are shown in the downward direction; outward currents are plotted upward.

3.1 Longitudinal Muscle Fibers

According to Hernandez-Nicaise et al. [3], longitudinal muscle cell fragments exhibit three major current components; a long-lasting inward current, an inactivating transient outward K^+ current (*see* Fig. 1a) and a more slowly developing Ca^{2+} activated K^+ current (K_{Ca}; *see* Fig. 4). The inward current can be studied in isolation by using depolarizing pre-pulses to inactivate the transient outward current and by adding TEA^+ (50 mM) to the bathing medium to block the Ca^{2+}-activated current.

3.1.1 Inward Current

Figure 2a shows current records from a longitudinal cell bathed in 100 mM TEA^+ with two superimposed traces recorded before (trace i) and after the application of 10 mM Co^{2+}. The voltage command protocol consisted of a 500 ms conditioning step to −40 mV, followed by a further step to 0 mV; between commands, the membrane potential was held at −60 mV (not shown). The long-lasting inward current present in trace (i) was blocked by the Co^{2+} (trace ii), leaving some residual transient outward current, unaffected by the conditioning step. Subtraction of trace (ii) from trace (i) provided an estimate of the time course of the inward current (*see* Fig. 2b), which is well fitted by an equation (after [30]) of the form:

$$I = I_{\max}(1 - \exp(-t/7))^3[\mathrm{A}.\exp(-t/\tau_1) + \mathrm{B}.\exp(-t/\tau_2) + \mathrm{C}.\exp(t/\tau_3)] \quad (1)$$

where I = inward membrane current. The inward current increases with a time constant of 7 ms, while the process of inactivation follows a multicomponent exponential with time constants (τ_1, τ_2, τ_3) equal to 6.8, 60, and 325 ms (determined by exponential stripping). Figure 2c shows the contributions of the three components to the overall process of inactivation as predicted by Eq. 1.

In another experiment, a longitudinal cell was bathed in 50 mM TEA^+ and depolarized to −20 mV. At this potential, there was little transient outward current (see activation curve in Fig. 3b), and any slow outward currents were blocked by the TEA^+. Once again, inactivation followed a complex time course and required three exponents in order to generate a satisfactory fit. The time constants in this case were 15, 40, and 333 ms.

3.1.2 Transient K^+ Current

The transient outward current shown in Fig. 2a is resistant to external TEA^+ and Co^{2+} so that an ASW containing both 10 mM Co^{2+} and 50 mM TEA^+ provides a pharmacological means to study it in isolation. Figure 3a shows results from one such experiment; at

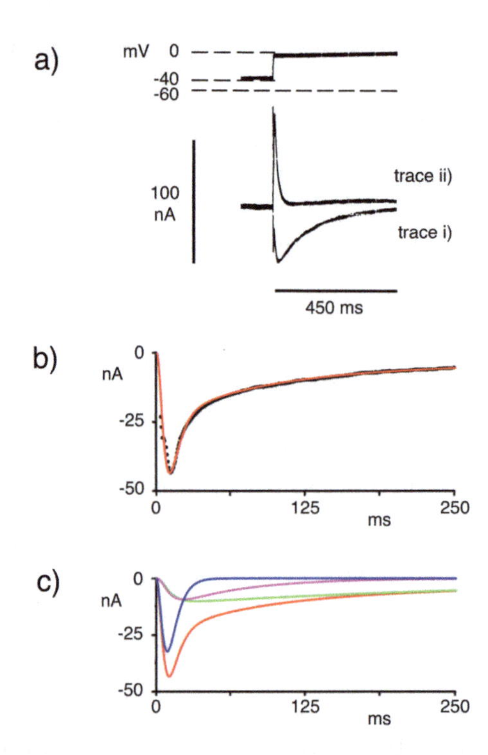

Fig. 2 Inward calcium current under voltage-clamp; longitudinal muscle. (**a**) Superimposed current records (lower traces) obtained from a cell bathed in ASW containing 10 mM Ca^{2+} and 100 mM TEA^{+} before (i) and after (ii) addition of 10 mM Co^{2+}. The voltage command protocol (upper trace) consisted of a 500 ms conditioning step to −40 mV (shown truncated), followed by a further step to 0 mV. Some residual transient current, unaffected by the conditioning step, remained. Commands imposed at 100 s intervals. Holding potential, −60 mV; temperature, 21 °C; fiber length, 500 μm. Fiber 24 (F10-82). (**b**) Analysis of the photographic records shown in (**a**). Traces were digitized and the time course of the Co^{2+}-sensitive inward current (blue data points) obtained by subtracting trace (ii) from trace (i); the continuous line was calculated from Eq. 1. The inactivation time constants, obtained by a process of exponential stripping, were 6.8 ms (τ_1), 60 ms (τ_2), and 325 ms (τ_3). (c) Time course and relative amplitudes of the different components of Ca^{2+} inactivation (blue, purple, and green lines) that contribute to the total Ca^{2+} current (orange line)

the top of the figure, the experimental protocol is shown as a variable delay between identical 60 mV depolarizing commands. In the superimposed current traces, shown below, the transient current is seen to inactivate almost completely and then fully recover following a simple exponential time course with a time constant of about 35 ms.

Despite the insensitivity of the transient outward current to TEA^{+}, examination of the "tail" current that follows a 10 ms depolarizing command (a command that produces a predominantly, but not exclusively, transient current) suggests that it is largely carried by K^{+}. "Tail" current is the term used for the current that flows at the end of a depolarizing command once the membrane has returned to its holding potential. It represents the flow of current through activated (open) channels during the short period of time

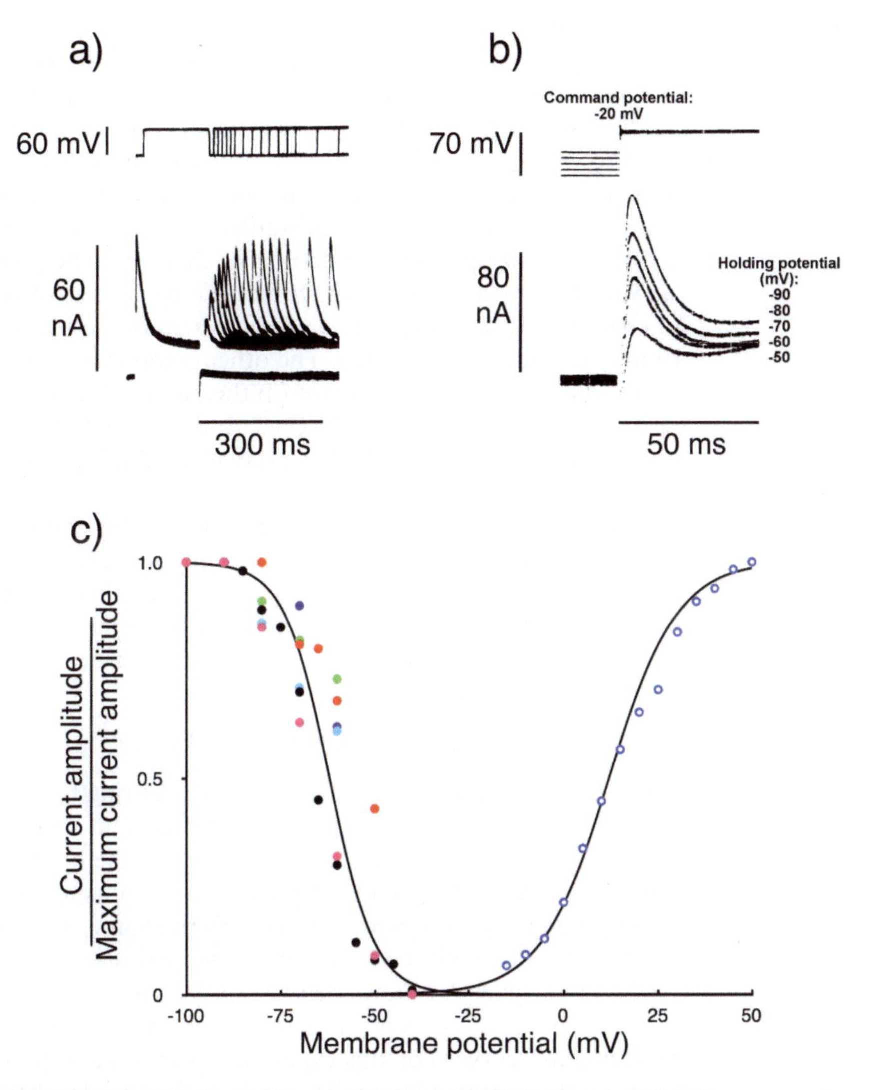

Fig. 3 Characteristics of the transient outward current; longitudinal muscle. (**a**) Time course of recovery from inactivation. Cell bathed in ASW containing both 10 mM Co^{2+} and 50 mM TEA^{+}. Membrane potential record (top) shows a series of superimposed trials made up of pairs of commands to 0 mV; trials were imposed at 30 s intervals. Time between the paired commands was varied between 1 and 300 ms. Membrane current record (bottom) shows that the peak current elicited by the first command had a constant amplitude while that associated with the second command recovered exponentially (time constant, 35 ms). Holding potential, −60 mV; temperature, 20 °C fiber diameter, 33 μm; fiber length, 800 μm. Fiber 25 (F11–82). (**b**) Effect of holding potential on the outward current. Membrane potential record (above) shows a command to −20 mV from five different holding potentials; the associated current records are superimposed and shown below. (**c**) Effect of the holding potential on transient outward current (inactivation curve; closed symbols). The proportion of transient outward current available is plotted against the holding membrane potential. Inactivation is removed by holding the cell at −100 mV; this gives the maximum transient current available (I_{max}). The current available at other holding potentials (I) is expressed as a proportion of I_{max}. The graph includes data from six muscle cell fragments identified by different coloured symbols. The continuous line through the points shows a Boltzmann curve (Equation 2) with half-inactivation at −62 mV (slope coefficient, 6 mV). The effect of the command potential on the peak transient outward current is also shown (activation curve; open symbols). I/I_{max} is shown plotted against the command potential. The Boltzmann curve (Equation 2) is drawn through the points with half-activation at +12 mV (slope coefficient, -9 mV)

that it takes for them to return to their resting (closed) state. Increasing external K^+ from 5 to 10 mM alters the equilibrium potential (measured as the tail current reversal potential) for the transient current. It shifts by 13 mV (from -78 ± 12.6 mV, SD, $n = 4$ to -65 ± 6.3 mV, SD, $n = 7$), close to that expected for a K^+ electrode (17.5 mV). Analysis of the time course of the tail current showed that it can be fitted by the sum of two exponential processes as if it flows through two populations of channels. The population carrying the most current, which we take to consist of transient K^+ channels, closes down with a time constant of 1.5 to 2 ms (at −40 mV; fibers 24, 32, and 42). The other component closes down with a much longer time constant (in the range 40–200 ms) and is partly blocked by adding 1 mM Cd^{2+} to the bathing medium. We attribute the slower current tail to a slowly developing K^+_{Ca} current (see later).

Figure 3b shows that the amplitude of the transient outward current depends on the holding membrane potential, inactivation being removed at more negative potentials (see also [3]). The relationship, which is plotted in Fig. 3c together with measurements from five other cells (filled symbols), can be fitted with a Boltzmann curve (Equation 2):

$$I = I_{max}/(1 + \exp(V_m - V_h)/z)$$

where I is the peak transient outward current during the command potential, I_{max} represents the peak outward current with channels fully available (i.e. no initial inactivation), V_m is the holding membrane potential prior to the depolarizing command, V_h is the holding potential associated with half inactivation (−62 mV), and z is the "slope coefficient" (6 mV). The activation curve (Figure 3b open symbols) shows that the peak transient outward current, which first appears at about −25 mV, also follows a "Boltzmann" curve. For this experiment the holding potential is fixed and the command potential is varied. For this data half activation occurs with commands to about +12 mV; slope coefficient -9 mV. In this case V_m is plotted as the command potential and V_h represents half activation.

3.1.3 Calcium-Activated Potassium Current

In Fig. 3a, the transient current can be seen to recover rapidly from inactivation, and this provides a way to separate the different outward current components. For Fig. 4a, the duration of the depolarizing command was set at 125 ms. On this timescale (and at this voltage, −10 mV), the inward current is concealed by the two separately developing outward currents, the slower one reaching a peak after about 125 ms. When command stimuli were presented repeatedly at 1 s intervals the amplitude of the early transient remained relatively steady, but the more slowly developing current was significantly depressed. In practice, consistent slow currents could be obtained if there was at least 100 s between command stimuli. The amplitude of the current also depended on the membrane holding potential as shown in Fig. 4b.

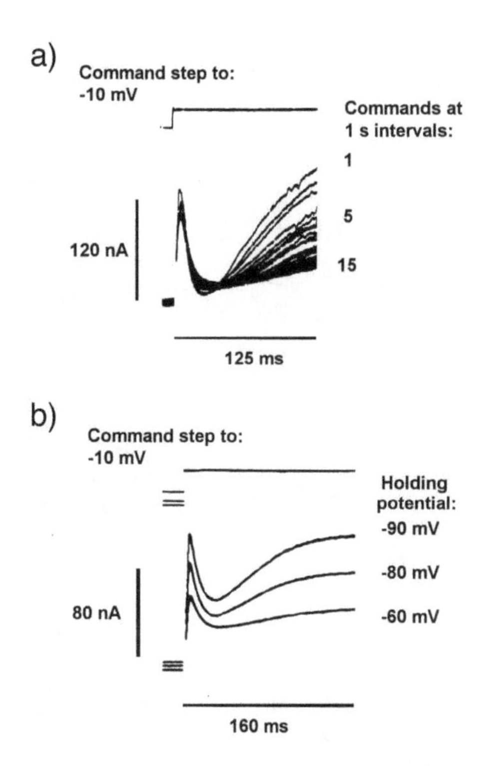

Fig. 4 Effect of stimulus repetition rate (**a**) and holding potential (**b**) on slow outward currents; longitudinal muscle. (**a**) Commands to −10 mV were applied at 1 s intervals; although the initial transient current retained a relatively constant amplitude, the later more slowly developing current was greatly reduced. Fiber bathed in 10 mM Ca^{2+}ASW. Holding potential, −60 mV; temperature, 23 °C; fiber diameter, 25 μm; fiber length, 1000 μm. Temperature, 23 °C. Fiber 17 (F7–82). (**b**) Commands to −10 mV from different holding potentials. Both outward current components were increased by increasing the holding potential from −60 to −90 mV. Fiber bathed in 2 mM Ca^{2+}ASW. Fiber 62 (F4–87)

The relationship between the inward and two outward currents is shown more clearly in the superimposed traces of Fig. 5 (top). Here the inward current remains visible as a brief deflection at the onset of the depolarizing command (the capacitative transient is not visible), and, during commands to −20 mV, it reappears once the transient outward current has inactivated. With larger depolarizations, however, it is concealed by a slow outward current, as in Fig. 4. This slow outward current is greatly reduced by a nominally Ca^{2+}-free bathing solution, although the fast component is little affected (Fig. 5, center). Returning the fiber to 2 mM Ca^{2+} ASW (Fig. 5, bottom) restores much of the slow outward current.

For the three superimposed current records shown in Fig. 6a, the holding potential was −80 mV, and the command steps were 160 ms in duration. The current generated by a command to −20 mV (shown in black) consists of a marked capacitive transient followed by the familiar fast transient and slow outward currents. At

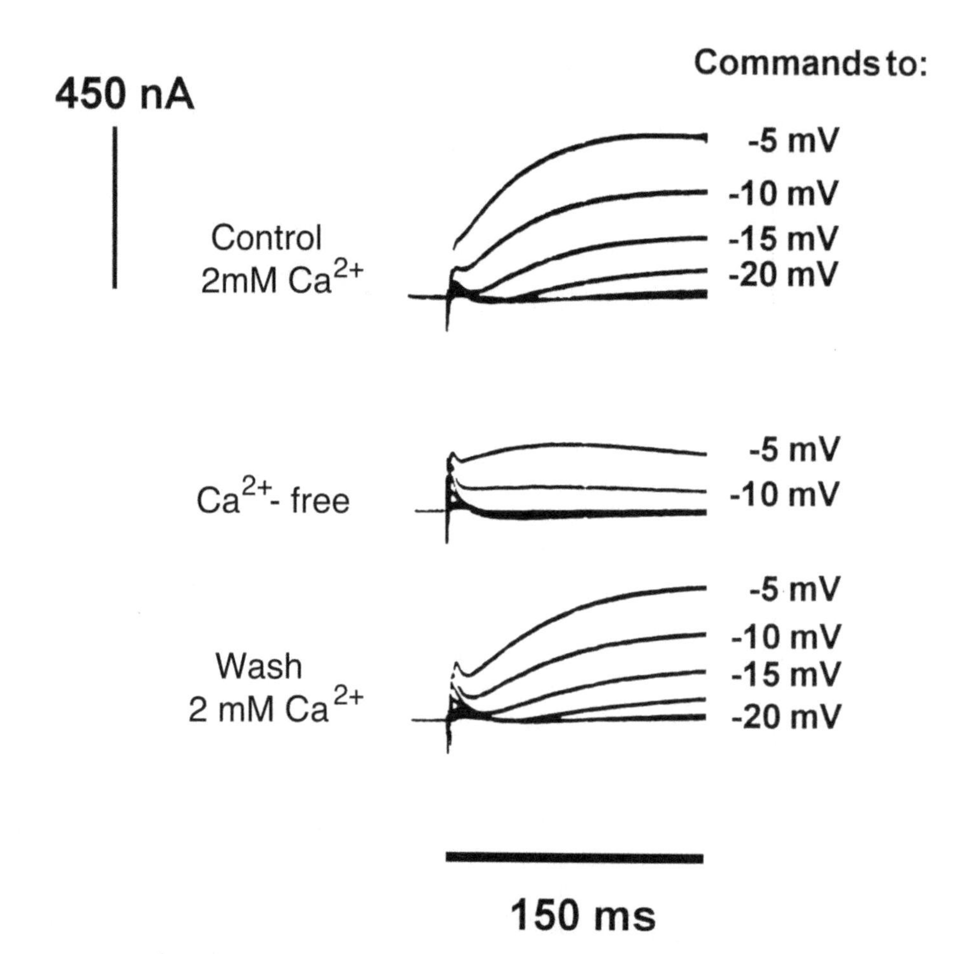

Fig. 5 Effect of Ca^{2+} free ASW on slow outward current; longitudinal muscle. Series of superimposed current records from a fiber bathed in 2 mM Ca^{2+}ASW (top), nominally 0 mM Ca^{2+}ASW (center) and returned to 2 mM Ca^{2+}ASW (bottom). Holding potential, −60 mV; command values shown at right-hand side. Fiber length 700 μm. Fiber 81 (F33–87)

0 mV (red trace), the slowly developing current reaches a peak amplitude that is twice that of the transient. With a further increase in the step depolarization (to +20 mV; blue trace), there is a remarkable change in the form of the current; the pronounced slow component is suppressed, whereas the transient component is markedly enhanced. This is the kind of behavior expected if the slow component is a K^+ current flowing through a Ca^{2+}-activated K^+ conductance, as first described for molluscan neurons [31]. The K^+ channels that provide the molecular basis for this conductance are opened by raised levels of intracellular Ca^{2+} ($[Ca^{2+}]_i$), so that if Ca^{2+} entry is reduced (see Fig. 5), the K^+ current is also reduced. Depolarizing the cell membrane to near the Ca^{2+} equilibrium potential is one way to accomplish this and consequently we take the slow current to flow through a Ca^{2+}-activated K^+ conductance (K_{Ca}).

Evidence that the slow current is carried by K^+ comes from its sensitivity to TEA^+ and the finding that the reversal potential of its "tail" is close to the K^+ equilibrium potential (mean −75 mV, $n = 3$; external potassium 10 mM); evidence for the role of Ca^{2+} is not only its dependence on the presence of external Ca^{2+} (*see* Fig. 5) but also its sensitivity to external divalent ions such as Co^{2+}.

During a maintained depolarizing command such as that shown in Fig. 6b, which is 800 ms in duration, the outward current rises to a peak at about 100 ms before slowly declining (purple data points). The time course of its decline is well fitted by a bi-exponential expression with time constants 122 and 1190 ms (continuous black line), somewhat slower than the exponential decline of the Ca^{2+} current (Fig. 2) but closer to the decline in intracellular Ca^{2+}(see Fig 12).

A current-voltage plot (open circles; Fig. 6c) measured at the peak of the slow current shows an "N" shape characteristic, similar to that seen in molluscan neurons [31]. As with molluscan neurons, the current remaining at potentials more positive than +50 mV may represent a second slowly developing outward current.

At the end of a depolarizing command, the internal Ca^{2+} returns to its resting level as it is pumped from the sarcoplasm. This is a relatively slow process (see Fig. 12), which probably accounts for the long-lasting "tail" of K^+ current (*see* Fig. 6d inset). Figure 6d (main graph) shows a plot of the (leak-corrected) tail current obtained at −35 mV after a step to −5 mV. It is well fitted by a bi-exponential line dominated by a fast component (2.7 ms time constant) but with a substantial slow component (time constant 123 ms).

3.2 Radial Muscle Fibers

When radial fibers are depolarized to −15 mV from a holding potential of −60 mV (*see* Fig. 1b), a brief inward current precedes a slowly developing current in the outward direction. The early inactivating outward current, that is such a feature of the currents recorded from longitudinal fibers, is missing, and at −15 mV, the current remains steadily outward for over 400 ms (Fig. 7a). Inactivation does become marked with more positive commands, however, and the current exhibits a maximum at about 150 ms (see also [3]). Figure 7b (black trace) shows that what is a slight "shoulder" on the rising phase when the holding potential is −60 mV (Fig. 1b) becomes a more obvious "notch" at −70 mV. A change in holding potential from −50 to −80 mV has little effect on the slow current (Fig. 7c), but with a holding potential of −40 mV, both inward and outward currents are significantly reduced (Fig. 7d). Both outward currents are reduced by 2 mM 4-aminopyridine (Fig. 7b; red trace) and reduced even further by the addition of 50 mM TEA^+ (Fig. 7b; blue trace). Both inward and outward currents are also reduced by 20 mM Co^{2+} (not shown).

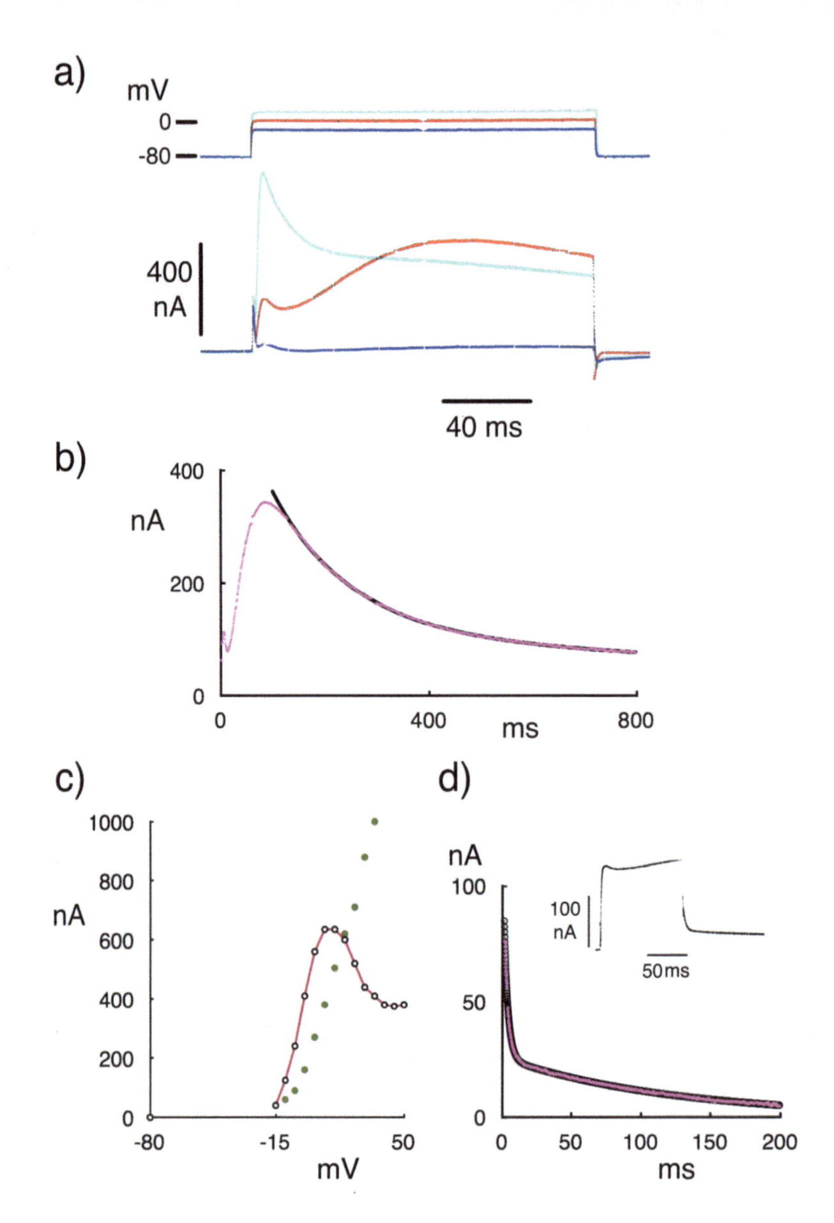

Fig. 6 Calcium-activated potassium current; longitudinal muscle. (**a**) Superimposed membrane currents (lower traces) elicited by commands to −20 (black), 0 (red), and +20 mV (blue). The slowly developing outward current is suppressed at commands to near the Ca^{2+} equilibrium potential. (**b**) Outward current recorded during a maintained depolarizing command to −5 mV. The current (purple data points) shows an initial transient rise followed by the slowly developing K_{Ca} conductance. The eventual decline in this current has two exponential components (time constants 122 ms and 1190 ms, obtained by exponential stripping), the fitted line (black) being calculated from:

$I(\text{nA}) = 580.\exp(-t/122) + 146.\exp(-t/1190)$.

The time constants were obtained by a process of exponential stripping (**c**) Peak current/voltage plot for transient (green data points) and K_{Ca} currents (blue data points connected by a red line). Fiber bathed in 10 mM Ca^{2+}-ASW; holding potential, −80 mV; temperature, 23.5 °C; fiber length, 800 μm. Fiber 12 (F5–82). (**d**) K_{Ca} tail current. Inset shows the current generated by a voltage command to −5 mV followed by repolarization to −35 mV. The slowly declining current tail was digitized, leak corrected, and plotted on the main graph (purple data points). The continuous line shows that the tail was well fitted by a bi-exponential line (time constants, 2.7 ms and 123 ms, obtained by exponential stripping) calculated from:

$I(\text{nA}) = 105.\exp(-t/2.7) + 25.5.\exp(-t/123)$,

Fiber bathed in 2 mM Ca^{2+}-ASW; holding potential, −60 mV; fiber length, 500 μm. Fiber 21 (F8–87)

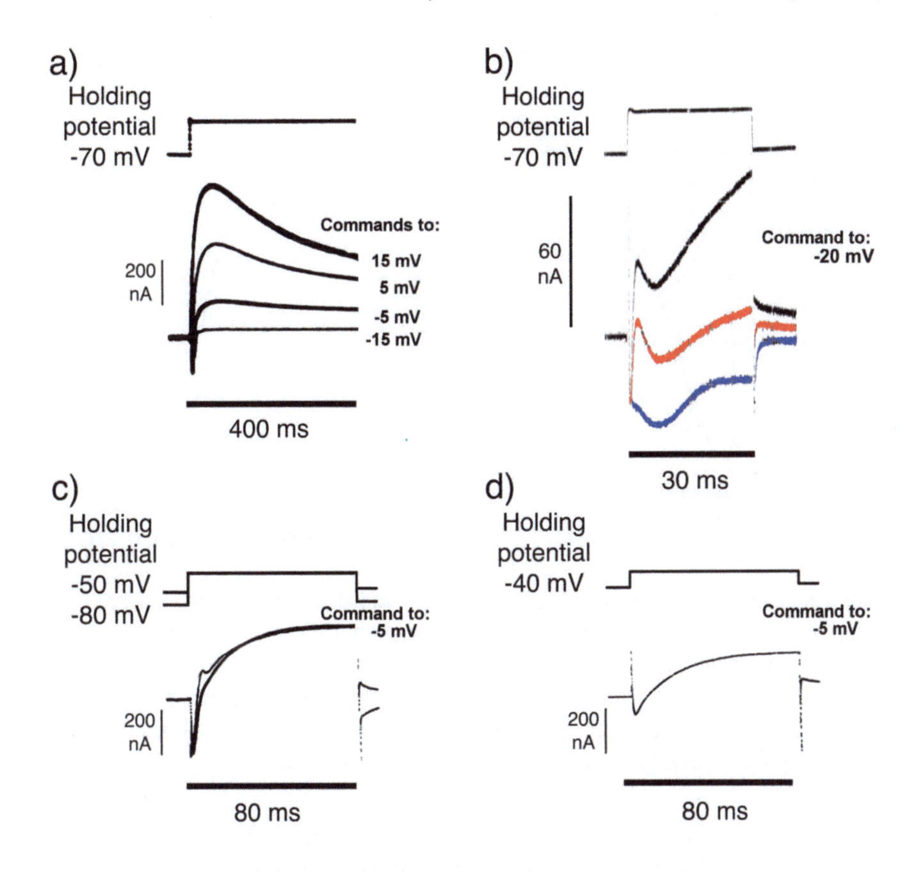

Fig. 7 Ion currents in radial muscles under voltage-clamp. (**a**) Superimposed current traces recorded during different command steps of 400 ms duration. Repolarization step omitted for clarity. Holding potential, −70 mV. Fiber 50 (F15–88). (**b**) Superimposed current traces recorded during a 30 ms command to −20 mV; cell bathed in 2 mM Ca^{2+} ASW (control, black trace); with the addition of 2 mM 4-aminopyridine (red trace); with the further addition of 50 mM TEA^{+} (blue trace). Fiber length 200 μm. Fiber 57 (F25–87). (**c**) Superimposed current traces recorded during an 80 ms command to −5 mV; effect of a change in holding potential from −50 (thicker trace) to −80 mV. (**d**) Current trace during an 80 ms command to −5 mV at a holding potential of −40 mV. (**c, d**) Fiber 50 (F15–88), bathed in 2 mM Ca^{2+} ASW + 0.1% BSA; fiber length, 500 μm. (**a**) Fiber as for (**c**) and (**d**) after repenetration with current electrode; cell contracted to a sphere 180 μm diameter

3.2.1 Fast Transient Outward Current

The notch on the rising phase of the outward current highlighted in Fig. 7b, c represents a fast, low-amplitude outward transient that reaches a peak within 7 ms of the start of the depolarizing command. The effect of the holding potential on the transient is analyzed in more detail in Fig. 8. When the holding potential is −85 mV, it is clearly visible after a step command to −5 mV but abolished when the holding potential is shifted to −35 mV (see Fig. 8a). The difference between the traces gives the notch amplitude. Measurements at different holding potentials show that the notch amplitude follows a Boltzmann relationship (Equation 2) with half inactivation at −53 mV (see Fig. 8b). In Fig. 7c, it appears to inactivate within about 20 ms, but this time is variable

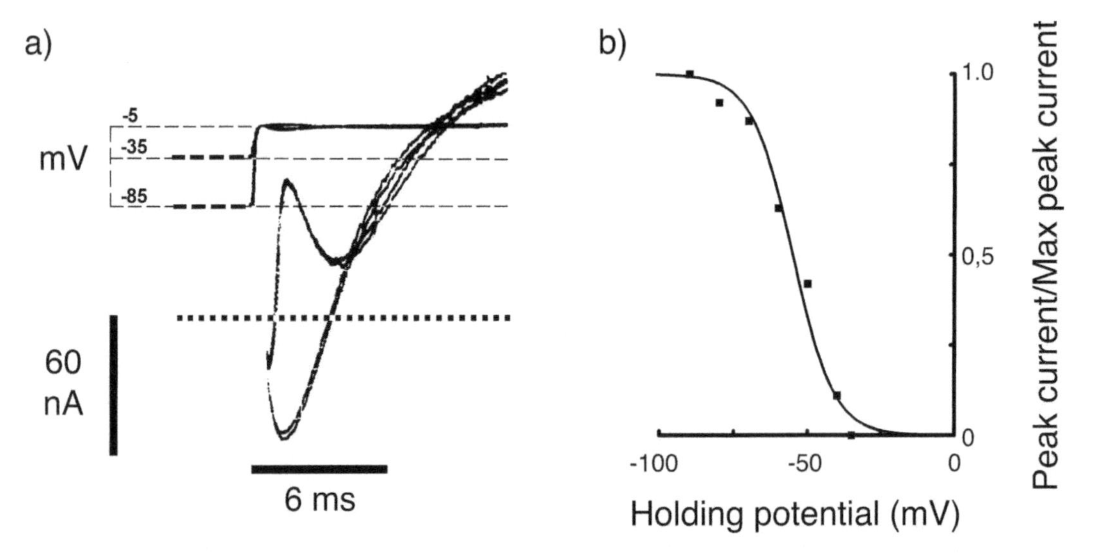

Fig. 8 Effect of holding potential on transient outward current; radial muscle. (a) The transient outward current (lower records) seen as a notch on the rising phase of the superimposed traces during a step command to −5 mV (upper record) disappears when the holding potential is shifted from −85 to −35 mV (two traces at each level). Holding potential levels are indicated by dashed lines; zero current shown by a dotted line. (b) The change in notch amplitude follows a Boltzmann relationship (Equation 2) with half inactivation at −53 mV. Fiber bathed in 2 mM Ca^{2+} ASW; fiber length, 450 μm. Fiber 76 (F13–87)

and can be as little as 5 ms (*see* Fig. 8). The fast transient current in longitudinal fibers inactivates more slowly (*see* Fig. 3b) [3].

3.2.2 Slow Outward Current

In Fig. 7, the outward current appears to consist of two major components, a fast transient and a later component that undergoes slow inactivation. There is also evidence for a maintained, i.e., non-inactivating current (Fig. 7a). The different outward current components are best separated using tail current analysis. Tail currents are fitted to exponentially declining processes by the method of exponential stripping. In most radial fibers, only two exponential components are evident with fits for the fastest being relatively consistent (mean 1.7 ± 0.3 ms, SD; $n = 6$); a satisfactory fit for the second component requires a range of values (17.5–90 ms; mean 38 ms; $n = 6$) but is consistent for individual fibers (*see* Fig. 9). Comparison of the tails after 10 and 20 ms commands suggests that the amplitude of the faster component remains relatively constant over this short length of time while that of the slower one increases steadily.

Figure 9a examines the effect of the command voltage on the two longer-lasting outward currents. During a 20 ms command that takes the membrane to +10 mV, the outward current rises slowly (blue trace), whereas during a command to +40 mV, the current rises rapidly to a steady level (black trace). Notice that the tail current after the command to +40 mV is smaller than that

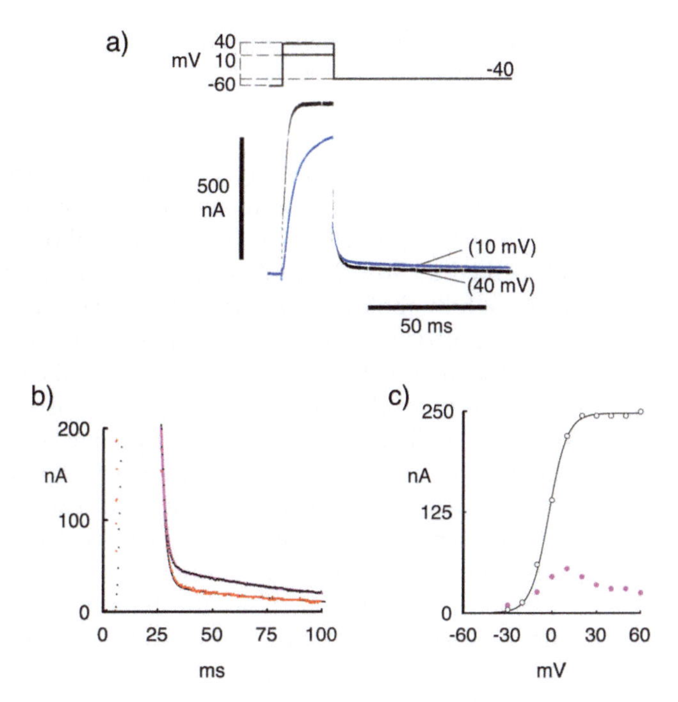

Fig. 9 Calcium-activated potassium current; radial muscle. (**a**) Superimposed current traces in response to 20 ms depolarising voltage commands (protocol shown at top). With a step command to +10 mV, the outward current rises slowly throughout the 20 ms command (blue trace); with a step command to +40 mV, the outward current rises rapidly to a steady level (black trace). Note that the tail current after the +40 mV command has a lower overall amplitude than that after the +10 mV command. (**b**) Digitized currents from a) at high gain to show that the tail currents (measured at −40 mV) are well fitted by a bi-exponential line with time constants, 2 and 121 ms (after command to +10 mV); and 2 and 67 ms (after command to +40 mV); data points shown in blue (+10 mV) and red (+40 mV); fitted lines shown in purple (+10 mV) and black (+40 mV). (**c**) Tail current amplitude at 26 ms (immediately after termination of the depolarizing command) plotted against command potential; components separated as in (**b**). The component with the longer lasting tail (filled purple data points) reaches a peak amplitude with commands to +10 mV; the faster component (open circles) increases to a steady level in a fashion described by the Boltzmann equation (Equation 2) where the membrane potential for half activation is −2 mV, and the slope coefficient is −6 mV. Fiber bathed in 2 mM Ca^{2+} ASW; fiber length, 900 μm. Fiber 20 (F9–89)

associated with the +10 mV command. The tail currents were well fitted by a bi-exponential process (Fig. 9b). When tails following different command steps were analyzed in this way, the slower component (time constant: 83 ± 11 ms SD; $n = 10$) was found to reach peak amplitude following commands to +10 mV, while the faster one (time constant: 1.9 ± 0.2 ms SD; $n = 10$) increased to a steady level (Fig. 9c) in a fashion that can be fitted with a Boltzmann curve (Equation 2), where the membrane potential for half-activation is −2 mV; and the slope coefficient is 6 mV. This analysis suggests that for larger depolarizations, most of the outward current flows through a non-inactivating voltage-gated K^+ channel, but a second class of K^+ channel is also present, whose contribution decreases the closer the command potential gets to the Ca^{2+} equilibrium potential. It seems likely that this is a Ca^{2+}-activated K^+ conductance.

3.2.3 Inward Current

When much of the K^+ current is blocked, the inward current can be seen to outlast a 30 ms command to −20 mV (blue trace; Fig. 7b). At +15 mV, it approaches the baseline after about 160 ms (Fig. 10a; cell bathed in 50 mM TEA^+). It appears that although the TEA does not fully abolish the K^+ current, there is little activation during small depolarizing commands, as judged by the low amplitude of the subsequent tail current (see change in tail currents in Fig. 7b). The duration of the revealed inward current is shorter than that in longitudinal fibers (see Fig. 2), and the time course of its inactivation is well fitted by a combination of two exponentially declining processes (time constants 14–20 and 90–120 ms). Low levels of Co^{2+} abolish the slower of the two processes and increase the time constant of the fast process to about 50 ms.

Adding 1.5 mM Cd^{2+} to the bathing medium reduces the amplitude of both inward and outward currents, with the slow outward component being most affected. The amplitude of the faster tail goes to 30% of control, while the slower one is reduced to 9% (Fig. 10b).

4 Ca^{2+} Injection Experiments

A key test for the presence of a K_{Ca} conductance is to raise Ca_i^{2+} by injection directly into the cell (*see* [32–34]). As cytoplasmic Ca^{2+} levels increase, any K_{Ca} channels present in the cell membrane will open with an increased probability, and an outward current is recorded. For the experiment shown in Fig. 10c, the muscle fiber was penetrated first with a pair of voltage-clamp electrodes (filled with 3 M KCl), and the membrane potential was held at −50 mV. The cell was then penetrated with a micropipette filled with 100 mM $CaCl_2$. As the figure shows, iontophoretic injection of current (100 nA for 3 s) into the fiber via the Ca^{2+} filled micropipette produces a slowly developing outward current that peaks about 2 s after the end of the injection. About 5 s, later the declining phase increases in rate, and thereafter, it follows an exponential course with a time constant of 4.5 s.

5 Patch-Clamp Experiments

The distribution of the different ion channel classes on the outer muscle membrane can be explored using the in situ patch-clamp technique. The voltage at the tip of the patch-clamp pipette, on the outside of the fiber, is under the control of the voltage-

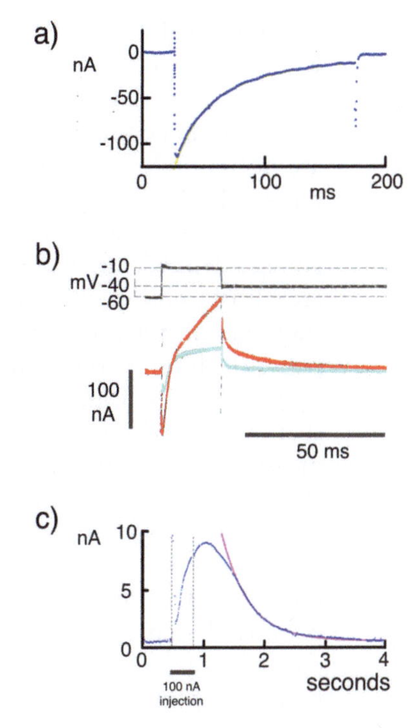

Fig. 10 Characteristics of Ca^{2+} and Ca^{2+}-activated currents; radial muscle. (**a**) Inward current from fiber bathed in 50 mM TEA^{+}. Command to +15 mV (not shown), holding potential −60 mV. Inactivation follows a bi-exponential time course with time constants of 20 and 90 ms. Fiber length, 500 μm, fiber diameter, 25 μm; temperature, 21 °C. Fiber 42 (F17–82). (**b**) Effect of 1.5 mM Cd^{2+} on radial fiber currents (blue trace). Step command to −10 mV, holding potential −60 mV (top record). Tail currents measured at −40 mV. Prior to addition of Cd^{2+} cell bathed in 2 mM Ca^{2+} ASW (red trace); fiber length 800 μm, fiber diameter 20 μm. Fiber 19 (F9–89). (**c**) Effect of Ca^{2+} injection under two electrode voltage-clamp. Holding potential −50 mV; voltage-clamp electrodes filled with 3 M KCl. Injection via 100 mM $CaCl_2$-filled micropipette. Iontophoretic injection (100 nA for 3 s) produces a slowly developing outward current that peaks about 2 s after the end of the injection. About 5 s later, the declining phase increases in rate and thereafter follows an exponential course with a time constant of 4.5 s (continuous line). Fiber 43 (FP-83)

clamp circuitry, while the intracellular surface of the patch remains at the resting potential of the muscle fiber. In the experiments reported here, the patch pipettes had a tip diameter of about 1 μm and were filled with ASW containing 5 mM Ca^{2+}. The resting potential of the fiber was taken to be −55 mV.

In one series of experiments, a total of 26 patches showed evidence of ion channel activity; in 12 patches, three current components (fast transient K^{+} currents, Ca^{2+} currents, and K^{+}_{Ca} currents) could be identified, and in four patches, fast transient K^{+} currents were seen alone. Two patches were unusual in that one had a transient K^{+} current and a Ca^{2+} current but no K^{+}_{Ca} current, while the other had a Ca^{2+} current and a K^{+}_{Ca} current but no transient K^+ current. Thus, in almost all patches, Ca^{2+} and K_{Ca} channels were found together. However, the fact that Ca^{2+} channels and K_{Ca} channels can be found separately suggests that the two populations are not entirely overlapping. We were unable to identify the currents in 8 of the 26 patches.

Figure 11a is an example of a membrane patch containing three voltage-dependent ion currents. Inward and slow outward currents are clearly visible with steps to −20 mV (blue trace), while a transient outward current is seen first as a notch on the rising phase of the slow outward current with steps to +20 mV (purple trace). The absence of an inactivating current at −20 mV suggests that this is a patch from a radial fiber. A graph of current amplitudes measured at 200 ms plotted against command potential shows the characteristic "N" shape of a Ca^{2+}-activated conductance (*see* Fig. 11b). A significant difference between these in situ patch-clamp experiments and the two-electrode voltage-clamp experiments was that the extended (100 s) period of recovery after each command step was no longer necessary, and instead, commands could be presented at 15 s intervals. The K^{+}_{Ca} tail currents had a particularly long duration (time constant: 290 ms and 10 s).

The inactivation characteristics of the inward current are shown in Fig. 11c. This shows current records from a membrane patch at different holding potentials. A plot of peak inward current at −10 mV against holding potential is shown in Fig. 11d with a best-fit Boltzmann curve drawn through the points. The maximum inward current was obtained with holding potentials greater than about −75 mV, while inactivation appeared complete around −25 mV. The amplitude of the Ca^{2+}-activated outward current was correspondingly affected by changes in the availability of this current.

6 $[Ca^{2+}]_i$ Transients Under Voltage-Clamp

We turned to conventional fluorescence techniques to measure the spread and recovery of Ca_i^{2+} after depolarising stimuli. Muscle cell fragments were loaded with fluorescent Ca^{2+} indicator and stimulated using an intracellular micropipette connected to a balanced Wheatstone Bridge. A BioRad M500 confocal microscope in line scan mode permitted us to follow transient changes in internal Ca^{2+} with a temporal resolution of up to 2 ms.

Scans of resting cells showed a higher level of fluorescence toward the center of the cell, due in part to the greater path length there, but also because the organelles in the fiber core appeared to take up the dye preferentially. To correct for this uneven distribution, fluorescence changes (ΔF) were normalized by dividing them by pre-stimulus (baseline) values (F_0) - see Equation 3.

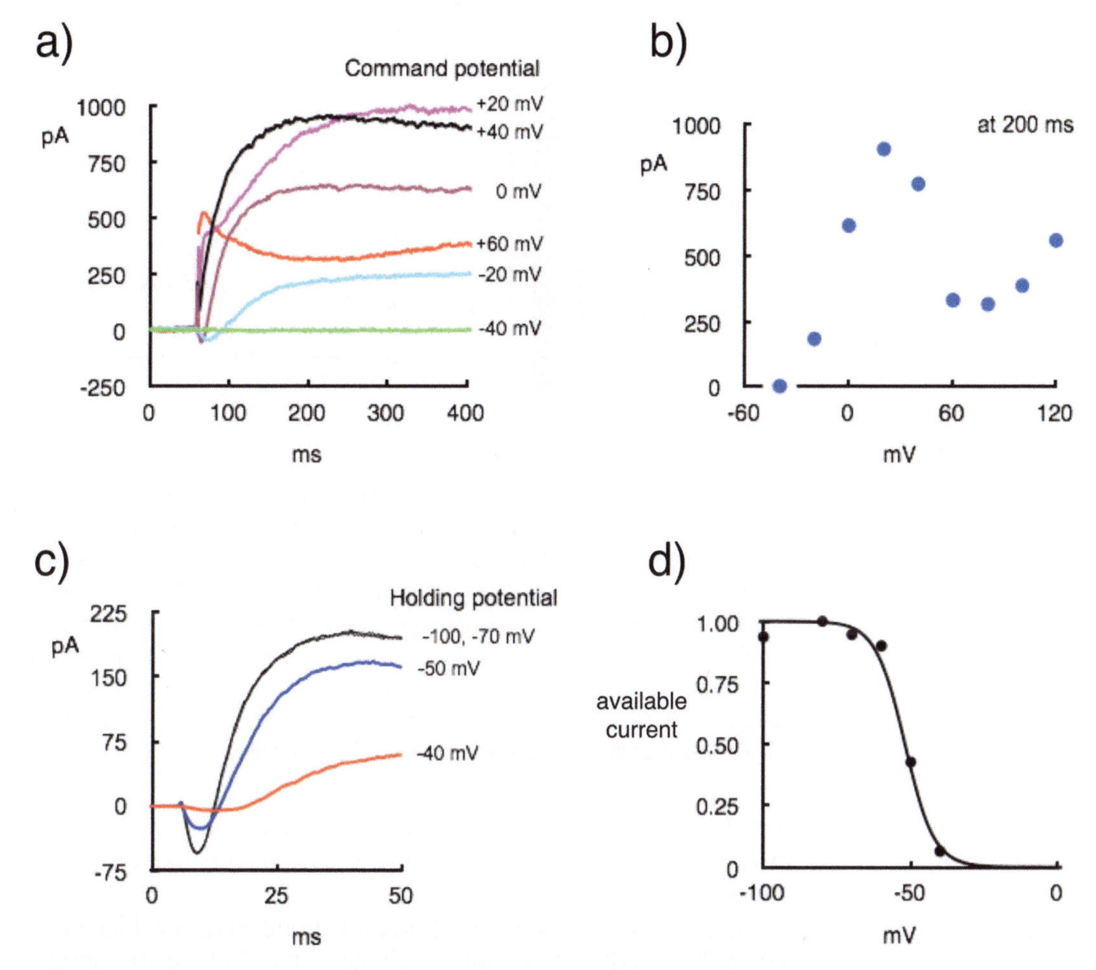

Fig. 11 Calcium and calcium-activated potassium currents recorded from in situ membrane patches. (**a**) Currents recorded in response to a sequence of depolarizing command steps (final level given at right); holding potential −80 mV; resting potential taken to be −55 mV; patch pipette filled with 5 mM Ca^{2+}-ASW; temperature, 21 °C. (**b**) Current/voltage plot of data from (**a**). Currents measured at 200 ms. (**c**) Data from another patch showing the effect of holding potential (values given at right) on the inward and slow outward current; command voltage, −10 mV; duration, 45 ms. Other recording conditions as in (**a**). (**d**) Calcium current inactivation curve with data normalized to value obtained with −80 mV holding potential. Line drawn through the data represents a best-fit Boltzmann curve (Equation 2) where the potential for half-maximal inactivation, is taken to be −52 mV; and the slope coefficient is −5 mV

$$\text{i.e.,}\ (\Delta F)_t/(F_0) = (F_t - F_0)/(F_0) \quad (3)$$

where F_t is the measured fluorescence at any time after the stimulus. The resting pre-stimulus FLUO-3 fluorescence profile was obtained by averaging the 100 or more lines that lead up to the stimulus.

Although we were able to see a change in fluorescence with a single action potential (Fig. 12a), the transients were relatively small in amplitude. Initially, we averaged the signal across the entire

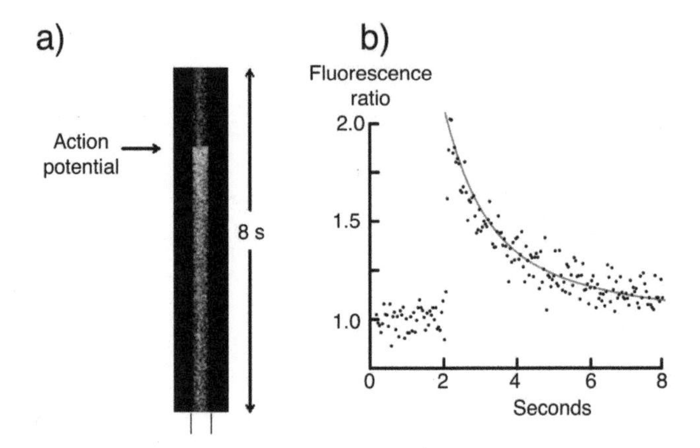

Fig. 12 Transient increase in $[Ca^{2+}]_i$ in response to a single action potential. (**a**) Fiber loaded with FLUO-3; a single action potential was elicited at the arrow using a depolarizing current step; confocal microscope in line scan mode. The alignment was such that the scan crossed the radial axis of the fiber; sequential scans (8 ms interval) are shown arranged vertically. (**b**) Fluorescence change across the entire fiber normalized by dividing by the pre-stimulus resting level (see Equation 3). Decline in fluorescence follows a bi-exponential function with time constants of 1 and 4.3 s (continuous line). Temperature, 15 °C; fiber diameter, 25 μm. Fiber number 22 (980326D)

muscle fiber and found that the recovery phase could be fitted by a bi-exponential function. In Fig. 12b the exponents had time constants of 1 s and 4.3 s.

Figure 13a shows a Ca^{2+} transient monitored in a FLUO-3-loaded longitudinal fiber under voltage-clamp. Immediately after a depolarizing command, ΔF was higher at the periphery of the fiber than at its center, and this is highlighted in a plot of the normalized fluorescence (Fig. 13b). In a 2 μm strip at the periphery of the fiber, ΔF increased to a peak during the course of the 20 ms depolarizing command, whereas in a 4 μm zone near the fiber center, the maximum appeared after a significant delay (time to peak, about 400 ms). About 750 ms after the command, the Ca^{2+} concentration appeared to be evenly spread throughout the fiber. Examination of the declining phase of the peripheral transient showed that ΔF followed a bi-exponential time course with constants of 0.13 and 3.25 s. There was some indication that the signal from the center of the fiber was briefly at a higher value than that from the periphery but eventually both periphery and center declined together. Radial fibers were found to behave in a similar manner (Fig. 14a, b).

There is evidence for Ca^{2+} induced Ca^{2+} release in *Beroe* cells (*see* Subheading 7), and it seemed possible that the brief overshoot observed at the center of the longitudinal fiber might represent some form of Ca^{2+} release mechanism. However, an attempt to

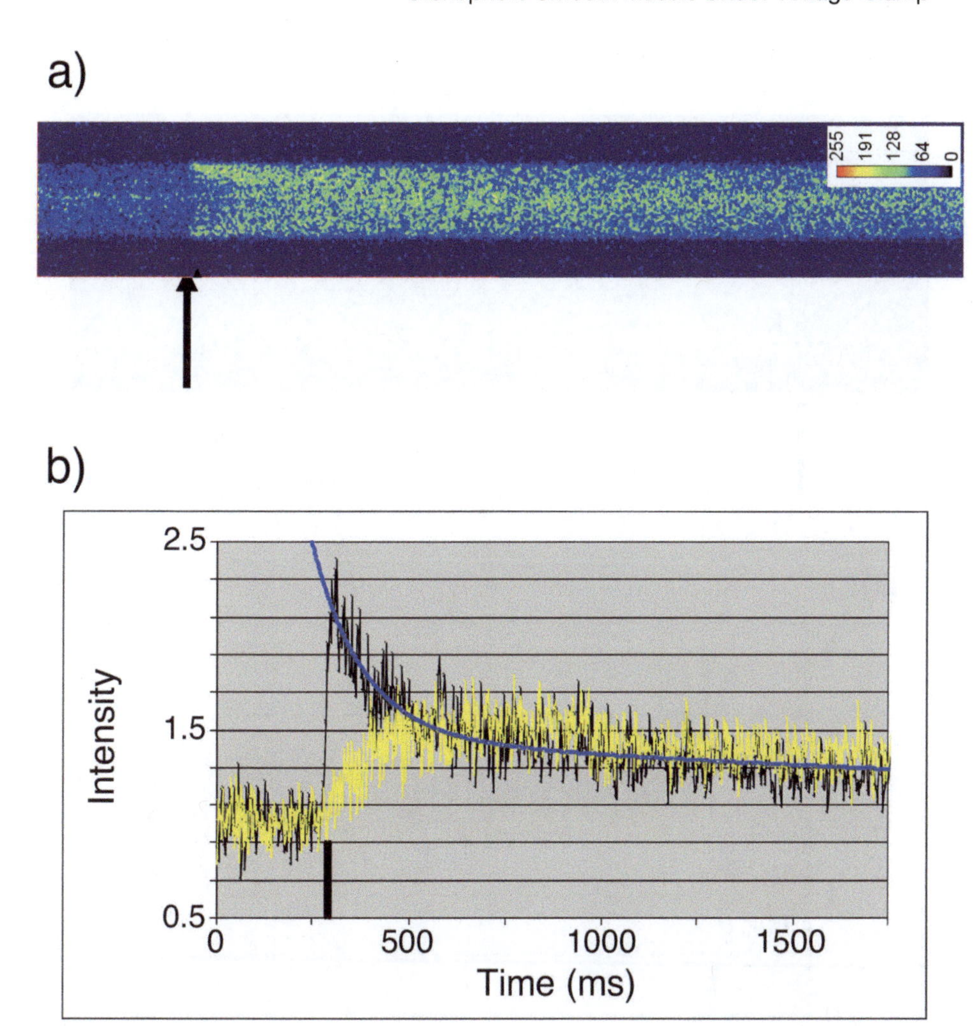

Fig. 13 Transient increase in $[Ca^{2+}]_i$ in response to voltage-clamp command; longitudinal fiber. (**a**) Fiber loaded with FLUO-3 and a single 20 ms duration voltage-clamp command imposed at the arrow; confocal microscope in line scan mode. The alignment was such that the scan crossed the radial axis of the fiber; sequential scans (2 ms interval) arranged horizontally; total scan time, 2 s. Median filter with a radius of 1 pixel used to enhance the image. (**b**) Plot of unfiltered averaged fluorescence changes; 2 μm wide region under membrane on each side of the fiber (black), 4 μm wide region at fiber center (yellow). Plot is an average of six trials and consists of three trials to 0 mV (holding potential, −60 mV), two trials to −10 mV (holding potentials, −80 and −90 mV), and a trial to −20 mV (holding potential, −80 mV); plots normalized by dividing by the pre-stimulus resting level (Equation 3). Voltage-clamp command imposed during the time of the black bar. Decline in fluorescence follows a bi-exponential time course with time constants of 130 ms and 3.25 s (continuous line). 24 μm diameter fiber. Fiber 980516; scans F, I, L, P, Q, S

augment Ca^{2+} release using the application of the plant alkaloid ryanodine produced inconclusive results.

The amplitude of ΔF depends on the holding potential before the command pulse. Figure 15a shows FLUO-3 fluorescence changes induced by 5 ms depolarizing command pulses to 0 mV with the holding potential set to −60 mV, to −80 mV, and then

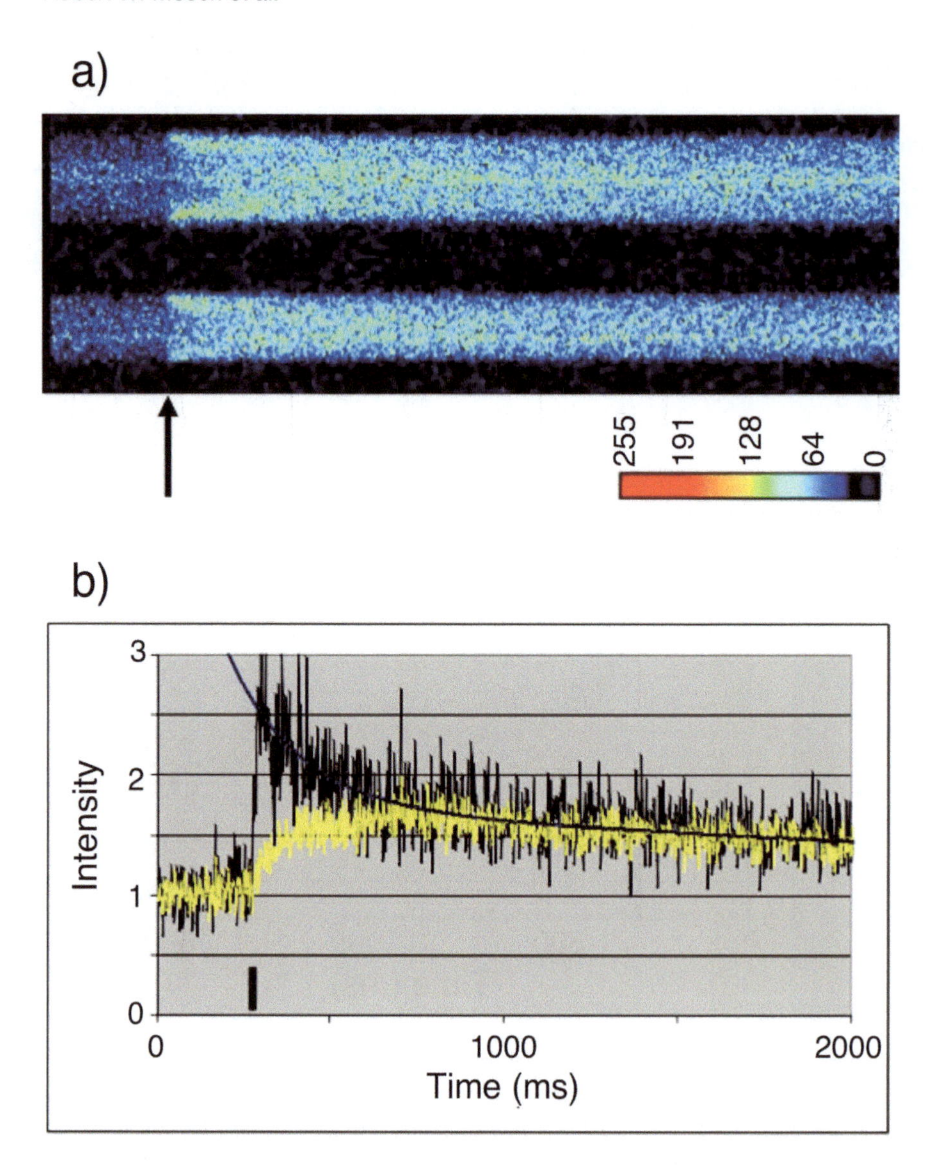

Fig. 14 Transient increase in $[Ca^{2+}]_i$ in response to voltage-clamp command; radial fiber. (**a**) Fiber loaded with FLUO-3 and a single 20 ms duration voltage-clamp command imposed at the time indicated by the arrow; command to 0 mV, holding potential, −60 mV; confocal microscope in line scan mode. Fiber formed a "U" shape and so the scan crossed the fiber in two places. The alignment was such that the scan crossed the radial axis; sequential scans (2 ms interval) arranged horizontally; total scan time, 2 s. Median filter with a radius of 1 pixel used to enhance the image. (**b**) Average plot of $[Ca^{2+}]_i$ transient monitored in both limbs of the fiber; unfiltered fluorescence changes used. The 2 μm wide sub membrane plot is an average of data from right and left sides of the two scans (black data points). The plot of the 12 μm wide central scan is also an average of data from both scans (yellow data points); plots were normalized by dividing by the pre-stimulus resting level (Equation 3). Decline in fluorescence follows a bi-exponential time course with time constants of 200 ms and 3.5 s (continuous line). Voltage-clamp command imposed during the time indicated by the black bar. Temperature, 15 °C; fiber 9805150

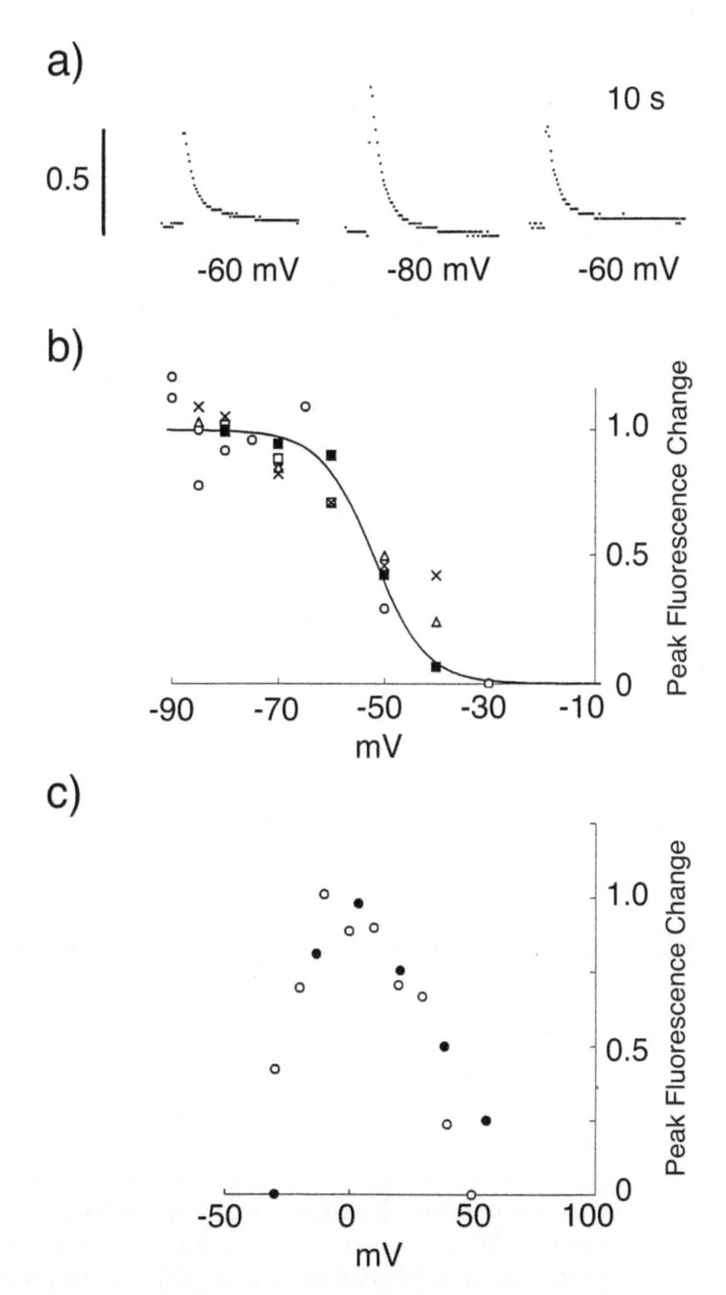

Fig. 15 Effect of membrane potential on Ca^{2+} based fluorescence changes in a FLUO-3-loaded fiber under voltage-clamp. (**a**) Change in average whole fiber fluorescence induced by a 5 ms depolarizing command to 0 mV from different holding potentials: −60, −80, and −60 mV. (**b**) Peak change in fluorescence plotted against holding potential; data from five fibers (different symbols) together with the Ca^{2+} current-inactivation curve from Fig. 11d (filled symbols) and its best-fit Boltzmann curve (Equation 2); peak change at −80 mV normalized to 1. Temperature, 18 °C. (**c**) Peak change in fluorescence plotted against command potential (filled symbols). Activation curve for the Ca^{2+} activated K^+ current derived from tail current experiment on an in situ membrane patch shown for comparison (open symbols). Tail currents were recorded at −40 mV, following depolarizing command pulses; holding potential −80 mV; resting potential taken to be −55 mV; patch pipette filled with 5 mM Ca^{2+}ASW; temperature, 21 °C. Current amplitudes were measured at the onset of repolarization and normalized so as to match the peak fluorescence change

returned to −60 mV. The ΔF shown represents the change in the average Ca_i^{2+} concentration across the whole fiber. For Fig. 15b, ΔF values from four cells were normalized (using the −80 mV data as a reference) and plotted against the holding potential (open symbols). The decrease in ΔF at more depolarized holding potentials matches the Ca^{2+} current inactivation curve from Fig. 11d (filled symbols) and its best-fit Boltzmann curve (Equation 2; solid line). ΔF also depended on the amplitude and duration of the depolarizing command. In Fig. 15c, ΔF is plotted against command potential (filled circles). Overall, the curve is bell-shaped with little or no change in ΔF at potentials above about +50 mV. It is a good match with the K^+_{Ca} current activation curve (open circles) taken from Fig. 9.

7 Discussion

The original purpose of these experiments was to explore whether ctenophore smooth muscles might make a useful "model" system. We thought that their large size could have experimental advantages. It turned out that their ion channel populations behaved in a broadly similar way to those in mammalian smooth muscle, and their handling of intracellular Ca^{2+} also appeared similar. However, following developments in genomic analysis which suggest that the Ctenophores might be the sister group to all other animals [32–36], the question arises as to whether components of ctenophore neuromuscular systems might have evolved independently. As our techniques may have encouraged us to view our observations through a "bilaterian lens" [37], it seemed necessary to present our results in more detail than heretofore. The recent findings of syncytial elements in the nervous system of *Mnemiopsis leidyi* [38] have encouraged us in this endeavor.

Perhaps it is not so surprising that the ion channels in ctenophore smooth muscles operate in such a familiar way. Propagating signals occur throughout the animal kingdom, but many of the ion channels involved have their origins in the bacteria and other early evolving life forms (*see* [15]). Examples are the Na^+-based action potentials in *Trichoplax adhaerens* [42] and the propagating Ca^{2+}-based action potentials in the hexactinellids (glass sponges) [43]. It is striking that the K^+ selectivity mechanism that evolved in the bacteria is highly conserved in metazoa of all kinds. It may also be significant that Na_{v1}-like channels, although widespread in the metazoa, are apparently absent from some early evolving species [44], and seemingly did not evolve in the ctenophores either. Ctenophores may have evolved unusual channel combinations, such as the association of Na_{v2} channels with a Ca^{2+} activated conductance, but such events recur in the metazoa. Even the independently propagating Ca^{2+}-based and Na^+-based action potentials in the giant axons of the hydromedusa *Aglantha digitale*, [45] resemble the channel combination in vertebrate thalamic neurons [46].

The main difficulty in analyzing ion channel currents in invertebrates is how to separate their sometimes-overlapping components using inadequately selective pharmacological agents developed from vertebrate research. The general TTX-insensitivity of invertebrate action potentials is one example. In the scyphomedusa *Cyanea capillata* [47], the inward current is TTX-insensitive, but it appears to be Na^+ based because it has fast kinetics and a reversal potential that varies with the extracellular Na^+ concentration. There is an inward current in Na^+-free solution (with 98 mM Ca^{2+}), but its kinetics are slow. In the hydromedusa *Polyorchis penicillatus* [48], Ca^{2+} not only carries current but also has a blocking effect on the TTX-insensitive Na^+ current responsible for the rapidly rising phase of the action potential [49].

As for outward currents, inactivating K^+ currents are widespread. In the Cnidaria they are present in the egg membranes of the sea pansy *Renilla koellikeri* [50], and in photocytes from *Obelia geniculata* [51]. Cnidarian neurons exhibit multiple classes of outward currents (*Cyanea capillata* [52], *Polyorchis penicillatus* [53]), as do myoepithelial cells (*Calliactis parasitica* [54]) and cnidocytes (*Chrysaora quinquecirrha* and *Cladonema* sp. [55]). In *Aglantha* axons, there are three kinetically distinct populations of inactivating K^+ channels, each with the same unitary conductance [56].

Evolutionary forces act "on what the ancestors have provided" [57] and in the case of ion channels, it appears that the ancestors have provided Ctenophores with a complete tool kit. Inactivating K^+ channels appear to play the same role in eggs from both ctenophores and cnidarians, for example [14] and K^+_{Ca} currents, like those in *Beroe*, play a role in the avoidance response of *Paramecium* [58]. In the next sections, we discuss ctenophore ion channel properties in more detail and suggest that they are similar to those in other invertebrate tissues.

7.1 Inward Currents in Ctenophores

Beroe longitudinal fibers in TEA^+ containing media have action potentials that can last for 400–600 ms once the fast-outward current has been inactivated. This is about the same duration as the inward current recorded under voltage-clamp (Fig. 2). The time course of inward current inactivation has three exponential components with time constants of about 10, 150, and 350 ms. In *Mnemiopsis*, the inward current has two inactivating components: one with a time constant of about 10 ms, and the other with a time constant of about 150 ms [26]. The two components have a different voltage dependence as if they operate through different pathways.

Clues to the nature of the channels that provide the membrane pathway for these currents are derived by comparing a range of different invertebrate genomes [59]. This shows that the sequences of amino acids that make up voltage-gated Ca^{2+} and Na^{+} selective channels contain four domains (domains I/II/III/IV). Each domain includes a finger-shaped loop that dips down into the interior of the cell membrane as part of the structure of the channel pore itself. At the far end of each loop is a key amino acid that is largely responsible for the selectivity of the pore. In the case of a Ca^{2+} channel, the key amino acids are all negatively charged, usually either glutamate (E) or aspartate (D), so that the four domains become E/E/E/E or E/E/D/D. The substitution of a neutral or positively charged amino acid, such as alanine (A) or lysine (K), is enough to increase the permeability of the pore to Na^{+} (*see* [60, 61]), and Na^{+} selective channels with the sequence D/E/K/A (called Na_{v1} channels) are widespread in the animal kingdom. Liebeskind et al. [59] have suggested that the ancestral pore sequence of Na^{+}-like channels is D/E/E/A. Such tetrodotoxin insensitive channels are called Na_{v2} [62], and the *Mnemiopsis leidyi* genome is known to possess two of them, called Na_{v2a} and Na_{v2b} [59, 62].

The D/E/E/A pore sequence is intermediate in structure between the Na^{+} and Ca^{2+} channel pore motifs and might be assumed to have an intermediate selectivity, excluding K^{+} but allowing both Na^{+} and Ca^{2+} to pass. In fact, the best-characterized Na_{v2} channels only pass Na^{+} when placed in Ca-free conditions. This is the situation for Na_{v2} channels from *Drosophila* (where they are known as DSC1) and the German cockroach *Blattella germanica* (where they are known as BSC1, when expressed and characterized in *Xenopus* oocytes [62, 63]). Oocytes expressing Na_{v2} from the honeybee (*Apis mellifera*; $AmCa_{v4}$ [64]) exhibit an inward current in Na-free, Ca^{2+} containing solution, but do not pass Na^{+} in Ca^{2+}-free solution even when Na^{+} is present at a high concentration (100 mM). As a consequence, Gosselin-Badaroudine et al. [64] prefer to think of Na_{v2} as a Ca^{2+} channel (and refer to it as Ca_{v4}) that is capable of carrying Na^{+} in some species. Note however that the honeybee channel differs from other Ca^{2+} channels in that it contains a 56-amino acid motif that acts as an inactivation gate more like that in many Na_{v} channels. It also has no Ca^{2+} channel β subunit interaction site [64], and it is not blocked by the Ca^{2+} channel antagonist nifedipine (10 μM)—although it is fully blocked by 1–10 mM Cd^{2+}.

Action potentials can be recorded from isolated longitudinal *Beroe* muscles in nominally Ca^{2+}-free ASW, but their amplitude is reduced, and the plateau phase is missing [4]. This also holds true for *Mnemiopsis,* even after having compensated for possible effects on the resting transmembrane potential [6]. Full-sized action potentials may be recorded from *Beroe* radial muscles in situ even after prolonged exposure to Ca^{2+}-free media [2], but this may reflect the difficulty of fully removing Ca^{2+} under these conditions; in isolated cells, there is a marked reduction in action potential overshoot [4]. Na^+-free conditions are likely to increase $[Ca^{2+}]_i$ if Na^+/Ca^{2+} exchange at the cell membrane contributes to intracellular Ca^{2+} regulation. As this could cause an increase in membrane conductance an alternative is to use a low Na^+ bathing medium and in *Mnemiopsis* muscles, action potentials were found to persist in 25% Na^+ [26]. In *Beroe* longitudinal fibers, action potentials in 25 mM Na^+ ASW were similar to controls except that their overshoot was decreased by 5–6 mV. However the fibres contracted after about 30 s of exposure and as it seems likely that $[Ca^{2+}]_i$ had been raised, an increase in the resting membrane conductance seems probable. 25 mM Na^+ ASW has the same effect in radial fibers but in a more extreme form—no action potentials could be elicited [4].

Many of these findings can be accounted for if Na_{v2}-like channels carry Ca^{2+} current during the action potential. The inward currents in *Mnemiopsis* are unlikely to be carried by Ca_{v1} channels, because Dubas et al. [26] found they were only partly blocked by high concentrations of extracellular methoxyverapamil (D-600; 100 μM) or verapamil (200 μM). These agents are relatively selective antagonists of Ca_{v1} channels if used at low concentrations. In *Beroe* radial fibers, micromolar levels of verapamil have no effect on excitability [2], suggesting that here too Na_{v2}, rather than Ca_{v1} channels, carry the inward current.

In addition to the two Na_{v2} channels already discussed, *Mnemiopsis leidyi* also has a single Ca_{v2} channel of the N/P/Q/R variety [59, 65]. Cd^{2+} and Co^{2+} are effective inward current blockers in both *Beroe* [3] and *Mnemiopsis* [26], but they do not discriminate between Na_{v2} and Ca_{v2} channels. Ca_{v2} channels are relatively insensitive to phenylalkylamines, such as verapamil and D600, or dihydropyridines, such as nitrendipine and nifedipine. Instead they may be blocked by ω-conotoxin GVIA from the cone snail *Conus geographus* or ω-agatoxin, SNX-482, from the tarantula spider *Hysterocrates gigas.* Unfortunately, neither of these two toxins has so far been tested on ctenophore muscles. Addition of 1 μM nifedipine blocks both the action potential and the Ca^{2+} transient in *Beroe* radial muscles [9], but the effect develops only slowly, total block requiring 30–40 min, and is only partly reversible (M-L Hernandez-Nicaise, personal communication).

In *Mnemiopsis*, it is possible to separate the inward current into rapidly and slowly inactivating components, because they have slightly different voltage dependences [26]. For the fast component, the inactivation time constant is 9.8 ± 1.0 ms, while for slow inactivation, it is 148 ± 9.8 ms. Both components have the same ionic selectivity, and it seems possible that they are forms of Na_{v2}. Na_{v2} channels from *Blattella germanica* inactivate to a steady level following a single exponential with a time constant of 235 ± 21 ms. Honeybee channels also have slow kinetics (time to peak about 60 ms at 0 mV) with a time constant for inactivation of 200–300 ms [64]. In *Beroe* longitudinal fibers, 90% of inactivation takes place with 6.8 and 60 ms time constants; while in radial fibers, 90% of inactivation is due to a single process with a time constant of 15 ms. So, if Na_{v2} channels are responsible for the inward current in ctenophores, they have evolved faster kinetics than those described thus far in insects.

7.2 Outward Currents in Ctenophores

Outward currents, recorded from *Mnemiopsis* fiber fragments with the whole-cell patch-clamp technique, are made up of four different components [26]. All four were blocked by replacing KCl in the recording pipette with CsCl (120 mM) and TEACl (20 mM) and by adding 4-AP (5 mM) to the bathing medium. They therefore appear to be largely carried by K^+. Three of the four currents undergo inactivation, with time scales of tens of milliseconds to several seconds. This combination of an early transient with a late outward current corresponds most closely to the situation in *Beroe* longitudinal fibers (*see* Fig. 1). Outward currents in *Beroe* fibers are also carried by K^+.

In *Mnemiopsis*, the duration of the action potential increases as the membrane potential is made less negative, indicating that repolarization is under the control of an inactivating outward current [6]. Action potential duration is also increased by external TEA^+. Once again, this resembles the situation in *Beroe* longitudinal fibers, where both TEA^+ and 4-AP lengthen the action potential even when applied separately [4]. In radial fibers, action potentials are relatively insensitive to TEA^+ or 4-AP when applied separately, and yet they produce a marked effect on the outward currents when applied together (*see* Fig. 7). Alone, 4-AP appears to reduce the slow outward component rather more than the fast transient, which suggests that it is the fast transient that is largely responsible for repolarization under these conditions. Depolarised 4-AP-treated isolated fibers fire action potentials repetitively and then "hang up" in a maintained plateau [4] as if (again, under these conditions) transient outward currents repolarize action potentials until they can no longer do so because they have become inactivated. Under normal conditions these transient currents may regulate spiking frequency. A role for Ca^{2+} activated K^+ currents is suggested by the observation that both 4-AP and TEA^+ individually can prolong action potentials in nominally Ca^{2+}-free ASW. This suggests that the reduced Ca^{2+} influx under these conditions is unable to activate sufficient outward current for repolarization (see below). The role of the slow current should be explored further, but it seems likely to be involved in longer term processes associated with the repetitive firing of action potentials.

A fast component of the early transient current, seen only in *Mnemiopsis* fibers, rises to at peak in 5 ms (at +5 mV) [26]. It may be separated from a slower component (time to peak 15 ms) by its susceptibility to a Ca^{2+}-free bathing medium containing 5 mM Cd^{2+}. Although this sensitivity might indicate Ca^{2+} activation, the voltage-dependence of inactivating K^{+} currents is sensitive to divalent ions in many preparations, including rat sensory neurons [66], rat smooth muscle [67], and neurons of the land snail *Helix aspersa* [68].

Nevertheless, there is evidence that slowly developing Ca^{2+}-activated currents are present in both *Mnemiopsis* and *Beroe.* The evidence that the slow outward current in *Beroe* muscle is a Ca^{2+}-activated K^{+} current (K^{+}_{Ca} current) may be summarized:

1. Both longitudinal and radial fibers exhibit a slow component of the outward current that is reduced when depolarizing commands approach the Ca^{2+} equilibrium potential. Such commands produce a reduced level of Ca^{2+} influx and consequently give an N-shape to the outward current-voltage relationship (*see* Figs. 6, 9, 11, and 15). This N-shape arises if the slow outward current has both calcium activated and calcium independent components. It is often considered diagnostic of the presence of a Ca^{2+}-activated conductance [31, 69].
2. The slow outward current is reversibly reduced in nominally Ca^{2+}-free saline (Fig. 5) and by divalent ions, such as Co^{2+} and Cd^{2+} (Fig. 10b).
3. Like the inward Ca^{2+} current, the Ca^{2+}-activated outward current is suppressed by a train of depolarising command stimuli at 1 s intervals (Fig. 4); in order to obtain consistent slow outward currents, depolarising commands under voltage-clamp were normally imposed at 100 s intervals. (Note that this was unnecessary in patch-clamp experiments, perhaps because the Ca^{2+} influx was limited to a small area of membrane under the patch pipette.)
4. The recovery of $[Ca^{2+}]_i$ in the muscle core after a voltage command, as monitored in FLUO-3-loaded cells, is quite slow (time constant: 3.25 s, Fig. 13; 3.5 s Fig. 14). However, near the intracellular membrane surface, recovery is initially fast with a time constant (time constant: 130 ms, Fig. 13; 200 ms, Fig. 14) that matches the decline in the K^{+}_{Ca} tail (time constant: 123 ms, after a voltage command to -5 mV, Fig. 6; 121 ms, after a voltage-command to +10 mV, Fig. 9).

We suppose that the time course of clearance of Ca^{2+} from the intracellular membrane surface, as revealed by the FLUO-3 experiments, depends on activity in two separate compartments. In the immediate vicinity of the cell membrane Ca^{2+} is rapidly cleared (time constant 130–200 ms), but towards the fiber core clearance follows a much slower time course (time constant: 3.25 s, Fig. 13; 3.5 s Fig. 14). We further suppose that during heavy or maintained Ca^{2+} loads the limited capacity of the rapid clearing mechanism, means that clearance of Ca^{2+} at the membrane surface follows that at the muscle core. The time course of this slow component of Ca^{2+} clearance is similar to the decline in outward current after iontophoretic Ca^{2+} injection (time constant: 4.5 s; Fig. 10c). However in voltage-clamp experiments the decline of the K_{Ca} tail current suggests that the Ca^{2+} at the cell membrane is cleared relatively quickly. (The exponential time constant is somewhat variable and depends on the command voltage, but generally it is no more than about 120 ms). It is possible that in our experiments the FLUO-3 (Dissociation constant for Ca^{2+} = 390 nM; Molecular Probes Handbook, 2005) is more sensitive to Ca^{2+} than are the K_{Ca} channels. It is true that in patch-clamp experiments, the K_{Ca} tail currents decline particularly slowly—time constant: 290 ms and 10 s—but this may be because a constriction of the membrane at the pipette tip provides a barrier to diffusion. Normally the intracellular space provides a large volume "sink" for Ca^{2+} entry. Presumably this contributes to the slow rise of Ca^{2+} at the surface membrane and the rapid initial stage of its decline.

The most surprising aspect of the K_{Ca} conductance is its relatively slow time course, particularly in longitudinal fibers. In some tissues, K_{Ca} channels are thought to be located next to Ca^{2+} channels so as to promote their speed of response and minimize the Ca^{2+} influx required for their activation. However, K_{Ca} channels come in a number of forms, one of which, the large conductance form (BK_{Ca}), is voltage as well as Ca^{2+} gated. In vertebrate hippocampal neurons, there is a near coincidence of BK_{Ca} channel and inward channel activation [70], and this persists despite of the presence of the Ca^{2+} chelator BAPTA it seems that the Ca^{2+} and BK_{Ca} channels are separated by less than ~30 nm which is the buffering length constant for BAPTA under these conditions [71].

In other cells, Ca^{2+} and K_{Ca} channels appear to be located further apart. In molluscan neurons, for example, K_{Ca} activation can be prevented by injecting EGTA, a somewhat slower Ca^{2+} chelator than BAPTA, into the cell cytoplasm [34]. K_{Ca} channels in molluscan neurons have a relatively low conductance (*Aplysia*, 35 ± 5 pS at $V = 0$ mV in asymmetrical K solution [72]; *Helix aspersa*, 40–60 pS in symmetrical K solution [73]) and are concerned with slower processes such as the "post-tetanic" after hyperpolarization [34]. EGTA injection only prolongs action potentials if other K channels are missing or have become inactivated (*Aplysia* [72, 74]; *Helix aspersa*, Meech unpublished). In rat sympathetic neurons, BK_{Ca} channels are directly linked to L type Ca^{2+} channels and contribute to the rapid repolarization of the action potential, whereas Ca^{2+} entry through N-type Ca channels selectively activates a low conductance K_{Ca} channel that contributes to a much slower afterhyperpolarization [75]. Intermediate conductance K_{Ca} channels are associated with Ca_{v3} channels in cerebellar Purkinje cells [76]. BK_{Ca} channels purified from rat brains form macromolecular complexes with a variety of different Ca_v channels. These include $Ca_{v1.2}$ (L-type), $Ca_{v2.1}$ (P/Q-type), and $Ca_{v2.2}$ (N-type) channels [77]. If Na_{v2} channels provide the inward pathway for Ca^{2+} in *Beroe* fibers, it would seem that that they too can link with K_{Ca} channels.

In *Beroe* smooth muscle, a substantial rise in $[Ca^{2+}]_i$ is necessary for the myofilaments to effect a generalized contraction. Ca^{2+} regulation depends on ATP-dependent exchange at the cell membrane and ATP-dependent uptake into the tubules and cisternae of the sarcoplasmic reticulum [9]. However, the SR amounts to less than about 1% of the myofilament volume [11], and so Ca^{2+} sequestration into the SR may operate at a relatively low rate. This would account for the slow overall clearing of Ca^{2+} recorded in FLUO-3-loaded cells and for the slow recovery of the K^+ conductance increase following Ca^{2+} injection. It suggests that the slow rise in K_{Ca} current in longitudinal fibers is less a consequence of high rates of Ca^{2+} sequestration and more likely because the intracellular space provides such a large "sink."

Although the relative volume of the sarcoplasmic reticulum (0.8–1% of myofilament volume) is the same for radial as for longitudinal fibers, the size, number, and distribution of the SR tubules differ considerably. In longitudinal fibers, they are thinner and more numerous than those in radial fibers. They are also less evenly distributed with most being found among the myofilaments at the core of the fiber. The increased volume of SR at the intracellular membrane surface in radial fibers might increase the rate of Ca^{2+} clearing there and account for the faster time constant of their K_{Ca} tails, but we were unable to demonstrate these higher rates in FLUO-3 experiments; clearance of Ca^{2+} in both radial and longitudinal fibers followed a similar time course.

7.3 Comparisons with Vertebrate Smooth Muscle

The giant muscle cells from the body wall of the ctenophores *Beroe* and *Mnemiopsis* resemble vertebrate smooth muscles in their arrangement of myofilaments and because they lack the striations and T-tubules characteristic of striated muscle. As with vertebrate smooth muscle, each fiber is a single cell that generates a large rapidly propagating action potential that supplies some or all of the Ca^{2+} required for contraction.

In vertebrate smooth muscle, the correlation between membrane potential, spike discharge, and tension [78] arises from the direct link between the Ca^{2+} influx during each action potential and the rise in $[Ca^{2+}]_i$ required for contraction. Smooth muscles from different vertebrate sources exhibit an enormous variability in the time course and configuration of their spike potentials, and this is reflected in the wide range of different ion currents seen under voltage-clamp. Recently, a role for Ca^{2+}-activated Cl^- channels has been explored [79]. We have not looked for this class of channel in *Beroe* fibers, but Dubas et al. [26] suggest that Cl^- might contribute to outward currents in *Mnemiopsis*. They showed that the relationship between the [K^+] in the bathing medium and the outward current reversal potential followed the Nernst equation more closely in Cl^--free media. However, there is a possibility that the effect might be caused by changes in the Donnan equilibrium between K^+ and Cl^-.

All smooth muscles share Ca^{2+} influx as a requirement for contraction, and in most cases there are at least two and possibly three components to their voltage-activated Ca^{2+} currents [80]. In cells from the stomach of the toad *Bufo marinus*, multiexponential fitting of the time course of inactivation yields three components with time constants in the order of 31, 167, and 866 ms [81]; (cf. *Beroe* longitudinal fibers Fig. 2). In *Mnemiopsis*, there is evidence for at least two components (transient and steady state), but separation is made difficult by regional variations in channel density [26]. Similar differences are found in *Beroe* radial fibers; in some fragments, the inward current inactivates within 30 ms; in other cases, a slower component is present.

Both delayed rectifier and rapidly inactivating channels have been reported in rabbit portal vein and in guinea pig bladder [82, 83]. An inward rectifier is present in guinea pig ileum [84], while in rat and rabbit aorta, there is a K^+ current activated by a drop in cytosolic ATP levels [85]. N-shaped current-voltage curves, like those described in molluscan neurons [31] and *Beroe* muscles, were reported for isolated toad stomach cells and in patch-clamp experiments on the toad *Bufo marinus* [86]. Both low and high conductance K_{Ca} channels are present in rabbit portal vein [87] and guinea pig mesenteric artery [88].

The role of the SR in excitation contraction coupling in mammalian smooth muscle is unclear (*see* [89]), but Devine et al. [90] suggest that it is sufficiently well-developed to function as a releasable Ca^{2+} store. In rabbit smooth muscle, the SR volume ranges from about 5% of cell volume in aorta and pulmonary artery to about 2% in mesenteric vein and artery and portal vein—somewhat more than in ctenophore smooth muscle, where the range is from 0.7% to 1.24% in *Beroe* and about 0.9% in *Mnemiopsis* [91]. Ca^{2+} transients in *Beroe* have a bell-shaped dependency on membrane voltage (Fig. 15), and so any Ca^{2+} release from internal stores must be proportional to Ca^{2+} influx from outside. Ca^{2+} release does occur with pharmacological treatment [10], but the confocal measurements reported here provide no clear evidence that it occurs in response to Ca^{2+} entry during a single action potential.

7.4 Summary

The association of calcium-activated conductances with calcium dependent effectors recurs throughout the animal kingdom. In the Protozoa it lies at the heart of the avoidance reaction exhibited by *Paramecium* [90] and it recurs in the contractile responses of smooth muscles in both ctenophores and vertebrates. In each case activation of K_{Ca} limits the rise of intracellular calcium by providing a negative feedback loop. In both *Paramecium* and in *Beroe* muscle the limiting action develops relatively slowly but a faster responsiveness may have evolved in some parasitic protozoans containing the homologue K_{Ca1} (see [92; 65]). Other channels, such as those responsible for the rapidly inactivating K^+ current, concerned in ctenophore longitudinal muscle fibers with action potential repolarization, are also found in the Protozoa (*Euplotes vannus* [93]; *Stylonychia* [94]) albeit in a calcium dependent form. Channels such as those carrying the Na/Ca inward current in ctenophore muscles, resemble those in other invertebrate tissues and perform similar functions. The regulation of intracellular calcium also seems broadly similar to that in vertebrate smooth muscle, although under normal conditions we saw no indication of the dynamic sub-cellular Ca^{2+} signals sometimes reported i.e. Ca^{2+} waves, Ca^{2+} sparks, Ca^{2+} puffs, etc. When it comes to their ion channel organisation the present studies suggest that there is little significant innovation in the ctenophores.

8 Notes

1. *Anatomy*: a) Mononucleate muscle units are found in some other ctenophores such as the tentacle muscle of *Pleurobrachia*, amember of the Order Cydippida [Hernandez-Nicaise M-L (1991). "Ctenophora" in Microscopic Anatomy of Invertebrates: Placozoa, Porifera, Cnidaria, and Ctenophora. eds. FW Harrison and JA Westfall (New York: Wiley), 359–418.]

   b) Stretched smooth muscle fibers of *Beroe ovata* examined by transmission electron microscopy are seen to have separate domains containing segregated actin and thick (myosin-like) filaments. These myofibril structures are probably made of two long actin filament bundles (of approximately 150 filaments) and short (2-3 μm) myosin-like bundles (of approximately 30 filaments). The myofibrils areattached at each end to the sarcolemmal membrane [Hernandez-Nicaise M-L, Nicaise G (1985) Structural evidence for contractile units in the giant smooth muscle cell of *Beroe*. Cell Motility Cytoskeleton 6: 153-158].

2. *Animals*: *Beroe ovata* were most often found in areas of the bay at Villefranche-sur-Mer, along with surface debris brought from outside by fast moving currents. Fast moving currents

could be identified by their smooth, almost oily appearance, which was in contrast to the sometimes wind-ruffled surface of the surrounding more slowly moving water. *Beroe* have been observed in the bay in January but, they were more reliably found during the months of March to May. They feed on other planktonic species and so can only be kept in good condition, without feeding, for 1 or 2 weeks.

Other species of *Beroe* (such as *B. mitrata*), which were sometimes present in the bay, had smaller diameter muscle fibers and were not used in these studies.

3. *Extracellular recording*: These experiments were carried out with a high-gain, differential AC amplifier from Tektronix (Model 5A22N). An alternative model is available from A-M Systems (Model 3000 AC/DC Differential Amplifier).

   The method for constructing suction electrodes is based on that used by Kanno [95] and by Josephson [96]. See also Mackie and Passano [97].

   The flexible tip of the electrode was made by softening a short section of polyethylene tubing (internal diameter 1.2 mm) over the end of a hot soldering iron. When sufficiently fluid, the tubing was drawn out until it tapered to a fine point suitable for electrode recording. The shaft of an 18-gauge hypodermic needle, pushed into the wide end of the electrode, provided a link to a length of flexible tubing and also a means to take the recorded signal to the amplifier by way of an attached soldered and shielded wire. The flexible tubing led to a three-way tap through which suction could be applied, via a syringe. Venting was accomplished by changing the position of the tap. Seawater was drawn into the tapered end of the suction electrode until it made contact with the metal shaft. The electrode was then lowered onto the preparation and enough suction applied to draw surface tissue into its tip. Electrical potentials were recorded between the tubing and an indifferent electrode in the bath.

   The indifferent electrode consisted of vinyl-coated silver wire (diameter 100 μm) wound around the suction electrode as close as possible to its tapered end. It was generally necessary to shield the suction pipette using grounded aluminum foil. Care must be taken to prevent the aluminum foil from shorting out the signal.

4. *Intracellular recording*: Although disturbance to the tissue is minimized by the approach described in *Methods*, two main difficulties remain. One is that the mesogloea limits accessibility to only the most superficial fibers; the other is that because the tissue as a whole is spontaneously active, it is difficult to penetrate single fibers with more than one micropipette. The slow, irregular movements appear to originate in a band of circular

muscle below the pharyngeal epithelium, and it is best to avoid including any of this tissue in the core sample. One way around the problem is to partially digest the mesogloea with hyaluronidase (hyaluronidase type II, 0.25 mg/mL, Sigma-Aldrich Chemical Co., St. Louis, Mo.) for 15–30 min. It then becomes possible to penetrate single fibers with two or more micropipettes.

5. *Enzyme digestion*: Many of the early experiments used enzyme preparations that are no longer available in the same form. In any case, different batches of enzyme required either longer or shorter periods of digestion. On some occasions, the isolated cells were fragile and collapsed upon penetration with micropipettes. If this happened, the question was whether the fragility was due to individual variety in the *Beroe* specimen used or whether it was connected with a change in the enzyme. Fresh batches of enzyme were tested using different incubation times to determine treatments that gave the most resilient muscle cell preparations. Sometimes, it was necessary to test a wide range of conditions. For example, in some experiments, we used a preliminary 10 min incubation in Ca^{2+} free-ASW containing 1.5 mg/mL hyaluronidase (type III, Sigma) at 37 °C, followed by a 10–15 min incubation in Ca^{2+} free-ASW containing 1.5 mg/mL hyaluronidase (type III, Sigma) plus 0.2 mg/mL trypsin (type III, Sigma, St Louis, MO). The procedure devised for *Beroe* muscles required modification for *Mnemiopsis* muscles [91].

6. *Two-electrode voltage-clamp*: Fibers were in a relaxed state when penetrated, but they invariably became shorter during the course of an experiment. In the longest experiments (30 min or more), the contractile apparatus appeared to become detached from the cell membrane and the cell fragment rounded up. The currents recorded from such spheres were qualitatively similar to those recorded from elongated cells, and we used them to test the effect of larger amplitude pulses. A striking property of these cells (in contrast to squid axons, for example) is the considerable time required for the K^+ channels to recover from inactivation after a depolarizing pulse. In some cases, it was necessary to space the test command pulses at 100; see intervals in order to get reproducible outward currents.

7. *Fluorescent indicators for Ca^{2+}*: Note that we used the membrane impermeant version of FLUO-3. If this is not available, it may be better to use FLUO-4.

Acknowledgments

This work was performed initially at the Station Zoologique, Villefranche-sur-Mer, France. (now l'Observatoire Océanologique de Villefranche-sur-Mer). Later experiments were carried out at the Université de Nice, Pare Valrose, Nice, France and the School of Medical Sciences, University of Bristol, Bristol, UK. It was supported by grants from the CNRS (UA 651; AB and M-LH-N) and from the Royal Society and the Wellcome Trust (RWM). We thank Chrystelle Cario, for contributing to this work, and the Director and Staff of the Station Zoologique, for providing us with laboratory space. We remember Bonnie Hurren for her support throughout. M-LH-N and RWM dedicate this chapter to the memory of André Bilbaut.

References

1. Hernandez-Nicaise M-L, Amsellem J (1980) Ultrastructure of the giant smooth muscle fiber of the ctenophore *Beroe ovata*. J Ultrastruct Res 72:151–168
2. Hernandez-Nicaise M-L, Mackie GO, Meech RW (1980) Giant smooth muscle cells of *Beroe*: ultrastructure, innervation, and electrical properties. J Gen Physiol 75:79–105
3. Bilbaut A, Hernandez-Nicaise M-L, Leech CA, Meech RW (1988) Membrane currents that govern smooth muscle contraction in a ctenophore. Nature 331:533–535
4. Bilbaut A, Meech RW, Hernandez-Nicaise M-L (1988) Isolated giant smooth muscle fibres in *Beroe ovata*: ionic dependence of action potentials reveals two distinct types of fiber. J Exp Biol 135:343–362
5. Hernandez-Nicaise M-L, Bilbaut A, Malaval L, Nicaise G (1982) Isolation of functional giant smooth muscle cells from an invertebrate: structural features of relaxed and contracted fibers. Proc Natl Acad Sci U S A 79:1884–1888
6. Anderson PAV (1984) The electrophysiology of single smooth muscle cells isolated from the ctenophore *Mnemiopsis*. J Comp Physiol B 154:257–268
7. Norekian TP, Moroz LL (2019) Neural system and receptor diversity in the ctenophore *Beroe abyssicola*. J Comp Neurol 527:1986–2008
8. Norekian TP, Moroz LL (2019) Neuromuscular organization of the ctenophore *Pleurobrachia bachei*. J Comp Neurol 527:406–436
9. Cario C, Nicaise G, Hernandez-Nicaise M-L (1996) Cytochemical localization of Ca^{2+}ATPases and demonstration of ATP-dependent calcium sequestration in giant smooth muscle fibers of *Beroe*. J Muscle Res Cell Motil 17:85–94
10. Cario C, Meech RW, Hernandez-Nicaise M-L (1995) Calcium management in giant smooth muscle cells. First evidence for a calcium-induced-calcium release. XXIII European Muscle Conference. J Muscle Res Cell Motil 16: 162
11. Cario C, Malaval L, Hernandez-Nicaise M-L (1995) Two distinct distribution patterns of sarcoplasmic reticulum in two functionally different giant smooth muscle cell of *Beroe ovata*. Cell Tissue Res 282:435–443
12. Hernandez-Nicaise M-L (1973) Les systeme nerveux des Ctenaires. I. Structure et ultrastructure des reseaux epitheliaux. Z Zellforsch 137:223–250
13. Mackie GO (1975) Neurobiology of *Stomotoca*. II. Pacemakers and conduction pathways. J Neurobiol 6:357–378
14. Jager M, Chiori R, Alié A, Dayraud C, Quéinnec E, Manuel M. (2011). New insights on ctenophore neural anatomy: immunofluorescence study in Pleurobrachia pileus (Müller, 1776). J. Exp. Zool. (Mol. Dev. Evol.) 316: 171–187
15. Meech RW (2015) Electrogenesis in the lower Metazoa and implications for neuronal integration. J Exp Biol 218:537–550. https://doi.org/10.1242/jeb.111955
16. Satterlie RA, Case JF (1978) Gap junctions suggest epithelial conduction within comb plates of ctenophore *Pleurobrachia-bachei*. Cell Tissue Res 193:87–91
17. Moss AG, Tamm SL (1987) A calcium regenerative potential controlling ciliary reversal is

propagated along the length of ctenophore comb plates. Proc Natl Acad Sci U S A 84: 6476–6480

18. Horridge GA, Mackay B (1964) Neurociliary synapses in *Pleurobrachia* (Ctenophora). Q J Microsc Sci 105:163–174
19. Horridge GA (1965) Relations between nerves and cilia in ctenophores. Am Zool 5:352–375
20. Horridge GA (1965) Macrocilia with numerous shafts from the lips of the ctenophore *Beroe*. Proc R Soc Lond B 162:351–363
21. Hernandez-Nicaise M-L (1991). "Ctenophora" in Microscopic Anatomy of Invertebrates: Placozoa, Porifera, Cnidaria, and Ctenophora. In FW Harrison and JA Westfall (eds.) (New York: Wiley), 359–418
22. Simmons, DK, Martindale, MQ. (2016). "Ctenophora" in Structure and Evolution of Invertebrate Nervous Systems. eds A Schmidt-Rhaesa, S Harzsch and G Purschke, (Oxford University Press), 48–55
23. Swanberg N (1974) The feeding behavior of *Beroe ovata*. Mar Biol 24:69–76
24. Tamm SL (1982) Ctenophores. In: Shelton GAB (ed) Electrical conduction and behavior in "simple" invertebrates. Clarendon Press, Oxford/London/New York, pp 266–358
25. Somlyo AP (1985) Excitation-contraction coupling and the ultrastructure of smooth muscle. Circ Res 57:497–507
26. Dubas F, Stein PG, Anderson PAV (1988) Ionic currents of smooth muscle cells isolated from the ctenophore *Mnemiopsis*. Proc R Soc Lond B 233:99–121
27. Bilbaut A, Hernandez-Nicaise M-L, Meech RW (1989) Ionic currents in ctenophore muscle cells. In: Anderson PAV (ed) Evolution of the first nervous systems. Plenum Press, New York, pp 299–314
28. Hodgkin AL, Nakajima S (1972) The effect of diameter on the electrical constants of frog skeletal muscle fibres. J Physiol Lond 221: 105–120
29. Hamill OP, Marty A, Neher E, Sakmann B, Sigworth FJ (1981) Improved patch-clamp techniques for high-resolution current recording from cells and cell-free membrane patches. Pflugers Arch 391:85–100
30. Hodgkin AL, Huxley AF (1952) A quantitative description of membrane current and its application to conduction and excitation in nerve. J Physiol 117:500–544
31. Meech RW, Standen NB (1975) Potassium activation in *Helix aspersa* neurones under voltage-clamp: a component mediated by calcium influx. J Physiol 249:211–239
32. Meech RW (1972) Intracellular calcium injection causes increased potassium conductance in *Aplysia* nerve cells. Comp Biochem Physiol 42A:493–499
33. Meech RW (1974) The sensitivity of *Helix aspersa* neurones to injected Ca^{2+}. J Physiol 237:259–277
34. Meech RW (1974) Calcium influx induces a post-tetanic hyperpolarisation in *Aplysia* neurones. Comp Biochem Physiol 48A:387–395
35. Dunn CW, Hejnol A, Matus DQ, Pang K, Browne WE, Smith SA, Seaver E, Rouse GW, Obst M, Edgecombe GD et al (2008) Broad phylogenomic sampling improves resolution of the animal tree of life. Nature 452:745–750. https://doi.org/10.1038/nature06614
36. Ryan JF et al (2013) The genome of the ctenophore *Mnemiopsis leidyi* and its implications for cell type evolution. Science 342:1242592
37. Moroz LL et al (2014) The ctenophore genome and the evolutionary origins of neural systems. Nature 510:109–114
38. Leys SP (2015) Elements of a "nervous system" in sponges. J Exp Biol 218:581–591. https://doi.org/10.1242/jeb.110817
39. Schultz DT, Haddock SHD, Bredeson JV et al (2023) Ancient gene linkages support ctenophores as sister to other animals. Nature 618: 110–117. https://doi.org/10.1038/s41586-023-05936-6
40. Dunn CW, Leys SP, Haddock SHD (2015) The hidden biology of sponges and ctenophores. Trends Ecol Evol 30:282–291
41. Burkhardt P et al (2023) Syncytial nerve net in a ctenophore adds insights on the evolution of nervous systems. Science 380:293–297
42. Romanova DY, Smirnov IV, Nikitin MA, Kohn AB, Borman AI, Malyshev AY, Balaban PM, Moroz LL (2020) Sodium action potentials in placozoa: insights into behavioral integration and evolution of nerveless animals. Biochem Biophys Res Commun 532:120–126., ISSN 0006-291X. https://doi.org/10.1016/j.bbrc.2020.08.020
43. Leys SP, Mackie GO, Meech RW (1999) Impulse conduction in a sponge. J Exp Biol 202:1139–1150
44. Fux JE, Mehta A, Moffat J, Spafford JD (2018) Eukaryotic voltage-gated sodium channels: on their origins, asymmetries, losses, diversification and adaptations. Front Physiol. 9: 1406. https://doi.org/10.3389/fphys.2018.01406. PMID: 30519187; PMCID: PMC6259924
45. Meech RW, Mackie GO (1993) Ionic currents in giant motor axons of the jellyfish, *Aglantha digitale*. J Neurophysiol 69:884–893

46. Llinas R, Yarom Y (1981) Properties and distribution of ionic conductances generating electroresponsiveness of mammalian inferior olivary neurons in vitro. J Physiol 315:569–584
47. Anderson PAV (1987) Properties and pharmacology of a TTX insensitive Na^+ current in neurons of the jellyfish *Cyanea capillata*. J Exp Biol 133:231–248
48. Przysiezniak J, Spencer AN (1992) Voltage-activated calcium currents in identified neurons from a hydrozoan jellyfish, *Polyorchis penicillatus*. J Neurosci 12:2065–2076
49. Spafford J, Grigoriev N, Spencer A (1996) Pharmacological properties of voltage-gated Na^+ currents in motor neurones from a hydrozoan jellyfish *Polyorchis penicillatus*. J Exp Biol 199:941–948
50. Hagiwara S, Yoshida S, Yoshii M (1981) Transient and delayed potassium currents in the egg cell membrane of the coelenterate, *Renilla koellikeri*. J Physiol 318:123–141
51. Dunlap K, Takeda K, Brehm P (1987) Activation of a calcium-dependent photoprotein by chemical signalling through gap junctions. Nature 235:60–62
52. Anderson PAV (1989) Ionic currents of the Scyphozoa. In: Anderson PAV (ed) Evolution of the first nervous systems. Plenum Press, New York, pp 267–280
53. Spencer AN (1989) Chemical and electrical synaptic transmission in the Cnidaria. In: Anderson PAV (ed) Evolution of the first nervous systems. Plenum Press, New York, pp 33–53
54. Holman MA, Anderson PAV (1991) Voltage-activated ionic currents in myoepithelial cells isolated from the sea anemone *Calliactis tricolor*. J Exp Biol 161:333–346
55. Anderson PAV, McKay MC (1987) The electrophysiology of cnidocytes. J Exp Biol 133: 215–230
56. Meech RW, Mackie GO (1993) Potassium channel family in giant motor axons of *Aglantha digitale*. J Neurophysiol 69:894–901
57. Horridge GA (1977) Mechanistic teleology and explanation in neuroethology: understanding the origins of behavior. In: Hoyle G (ed) Identified neurons and behavior of arthropods. Plenum Press, New York, pp 423–438
58. Saimi Y, Hinrichsen RD, Forte M, Kung C (1983) Mutant analysis shows that the Ca^{2+}-induced K^+ current shuts off one type of excitation in *Paramecium*. Proc Natl Acad Sci USA 80:5112–5116. 10.1073/pnas.80.16.5112
59. Liebeskind BJ, Hillis DM, Zakon HH (2011) Evolution of sodium channels predates the origin of nervous systems in animals. Proc Natl Acad Sci U S A 108:9154–9159
60. Heinemann SH, Terlau H, Stühmer W, Imoto K, Numa S (1992) Calcium channel characteristics conferred on the sodium channel by single mutations. Nature 356:441–443
61. Schlief T, Schönherr R, Imoto K, Heinemann SH (1996) Pore properties of rat brain II sodium channels mutated in the selectivity filter domain. Eur Biophys J 25:75–91
62. Gur Barzilai M, Reitzel AM, Kraus JEM, Gordon D, Technau U, Gurevitz M, Moran Y (2012) Convergent evolution of sodium ion selectivity in metazoan neuronal signaling. Cell Rep 2:242–248
63. Zhou W, Chung I, Liu Z, Goldin AL, Dong K (2004) A voltage-gated calcium-selective channel encoded by a sodium channel-like gene. Neuron 42:101–112. https://doi.org/10.1016/S0896-6273(04)00148-5
64. Gosselin-Badaroudine P, Moreau A, Simard L, Cens T, Rousset M, Collet C, Charnet P, Chahine M (2016) Biophysical characterization of the honeybee DSC1 orthologue reveals a novel voltage-dependent Ca^{2+} channel subfamily: Ca_{V4}. J Gen Physiol 148:133–145
65. Moran Y, Zakon HH (2014) The evolution of the four subunits of voltage-gated calcium channels: ancient roots, increasing complexity, and multiple losses. Genome Biol Evol 21(6): 2210–2217. https://doi.org/10.1093/gbe/evu177
66. Mayer ML, Sugiyama K (1988) A modulatory action of divalent cations on transient outward current in cultured rat sensory neurons. J Physiol 396:417–433
67. McFadzean I, England S (1992) Properties of the inactivating outward current in single smooth muscle cells isolated from the rat anococcygeus. Pflugers Arch 421:117–124
68. Gilday DME, Meech RW (1993) The effect of different divalent cations on A-currents recorded from neurones in isolated ganglia of the snail, *Helix aspersa*. J Physiol 473:175P
69. Meech RW (1976) Intracellular calcium and the control of membrane permeability. Symp Soc Exp Biol 30:161–191
70. Marrion NV, Tavalin SJ (1998) Selective activation of Ca^{2+} activated K^+ channels by co-localized Ca^{2+} channels in hippocampal neurons. Nature 395:900–905
71. Naraghi M, Neher E (1997) Linearized buffered Ca^{2+} diffusion in microdomains and its implications for calculation of $[Ca^{2+}]$ at the mouth of a calcium channel. J Neurosci 17: 6961–6973

72. Hermann A, Erxleben C (1987) Charybdotoxin selectively blocks small Ca-activated K channels in *Aplysia* neurons. J Gen Physiol 90:27–47
73. Ewald DA, Williams A, Levitan IB (1985) Modulation of single Ca^{2+}-dependent K^+-channel activity by protein-phosphorylation. Nature 315:503–506
74. Meech RW (1974) Prolonged action potentials in *Aplysia* neurons injected with EGTA. Comp Biochem Physiol 48A:397–402
75. Davies PJ, Ireland DR, McLachlan EM (1996) Sources of Ca2+ for different Ca^{2+}-activated K^+ conductances in neurones of the rat superior cervical ganglion. J Physiol 495:353–366
76. Engbers JDT, Anderson D, Asmara H et al (2012) Intermediate conductance calcium-activated potassium channels modulate summation of parallel fiber input in cerebellar Purkinje cells. Proc Natl Acad Sci U S A 109: 2601–2606
77. Berkefeld H, Sailer CA, Bildl W, Rohde V, Thumfart J-O, Eble S, Klugbauer N, Reisinger E, Bischofberger J, Oliver D, Knaus H-G, Schulte U, Fakler B (2006) BK_{Ca}-Ca_v channel complexes mediate rapid and localized Ca^{2+} activated K^+ signaling. Science 314:615–620
78. Bulbring E (1953) Measurements of oxygen consumption in smooth muscle. J Physiol 122:111–134
79. Wray S, Prendergast C, Arrowsmith S (2021) Calcium-activated chloride channels in myometrial and vascular smooth muscle. Front Physiol 12:751008. https://doi.org/10.3389/fphys.2021.751008
80. Benham CD, Hess P, Tsien RW (1987) Two types of calcium channels in single smooth muscle cells from rabbit ear artery studied with whole-cell and single-channel recordings. Circ Res 61:10–16
81. Vivaudou MB, Singer JJ, Walsh JV (1991) Multiple types of Ca^{2+} channel in visceral smooth muscle cells. Pflugers Arch 418:144–152
82. Beech DJ, Bolton TB (1989) A voltage-dependent outward current with fast kinetics in single smooth muscle cells isolated from the rabbit portal vein. J Physiol 412:397–414
83. Okabe K, Kitamura K, Kuriyama H (1987) Features of 4-aminopyridine sensitive outward current observed in single smooth muscle cells from the rabbit pulmonary artery. Pflugers Arch 409:561–568
84. Edwards FR, Hirst GDS (1988) Inward rectification in submucosal arterioles of guinea pig ileum. J Physiol 404:437–454
85. Standen NB, Quayle JM, Davies NW et al (1989) Hyperpolarizing vasodilators activate ATP-sensitive K^+ channels in arterial smooth muscle. Science 245:177–180
86. Singer JJ, Walsh JV Jr (1984) Large conductance Ca^{++} activated K^+-channels in smooth muscle cell membrane. Reduction in unitary currents due to internal Na^+ ions. Biophys J 45:68–70
87. Inoue R, Kitamura K, Kuriyama H (1985) Two Ca-dependent K-channels classified by application of tetraethylammonium distribute to smooth muscle membranes of the rabbit portal vein. Pflugers Arch 405:173–179
88. Benham CD, Bolton TB, Lang RJ, Takewaki T (1986) Calcium-activated potassium channels in single smooth muscle cells of rabbit jejunum and guinea-pig mesenteric artery. J Physiol 371:45–67
89. Wray S, Burdyga T (2010) Sarcoplasmic reticulum function in smooth muscle. Physiol Rev 90:113–178
90. Devine CE, Somlyo AV, Somlyo AP (1972) Sarcoplasmic reticulum and excitation-contraction coupling in mammalian smooth muscles. J Cell Biol 52:690–718
91. Hernandez-Nicaise M-L, Nicaise G, Malaval L (1984) Giant smooth muscle fibers of the ctenophore *Mnemiopsis leidyi*: ultrastructural study of *in situ* and isolated cells. Biol Bull 167:210–228
92. Prole, DL, Marrion NV (2012) Identification of putative potassium channel homologues in pathogenic protozoa. PLoS One. 7: e32264. doi: 10.1371/journal.pone.0032264. Epub 2012 Feb 21. PMID: 22363819; PMCID: PMC3283738
93. Krüppel T, Westermann R, Lueken W (1991) Calcium-dependent transient potassium outward current in the marine ciliate *Euplotes vannus*. Biochimica et Biophysica Acta. (BBA) – Biomembranes 1062:193–198
94. Deitmer, JW (1984). Evidence for two voltage-dependent calcium currents in the membrane of the ciliate *Stylonychia*. J.Physiol 355:137–159
95. Kanno T (1963) Electrical activity of the atrioventricular conducting tissue of the toad, studied by a minute suction electrode. Jpn J Physiol 13:97–111
96. Josephson RK (1965) Three parallel conducting systems in the stalk of a hydroid. J Exp Biol 42:139–152. https://doi.org/10.1242/jeb.42.1.139
97. Mackie GO, Passano LM (1968) Epithelial conduction in hydromedusae. J Gen Physiol 52:600–621

Chapter 16

Gap Junctions in Ctenophora

Andrea B. Kohn and Leonid L. Moroz

Abstract

Gap junction proteins form specialized intercellular communication channels, including electrical synapses, that regulate cellular metabolism and signaling. We present a molecular inventory of the gap junction proteins—innexins (INX-like) in ctenophores, focusing on two reference species, *Pleurobrachia bachei* and *Mnemiopsis leidyi.* Innexins were identified in more than 15 ctenophore species, including such genera as *Euplokamis*, *Pukia*, *Hormiphora*, *Bolinopsis*, *Cestum*, *Ocyropsis*, *Dryodora*, *Beroe*, benthic ctenophores, *Coeloplana* and *Vallicula*, and undescribed species of *Mertensiidae.* The observed diversity of innexins resulted from the independent expansion of this family from the common ancestor of ctenophores. Innexins show the conserved topology with four transmembrane domains connected by two extracellular loops, which bridge intracellular gaps. However, *INX*-like genes have highly diverse exon organization and low percentage identity for their amino acid sequences within the same species and between ctenophore species. Such a broad scope of molecular diversity differs from innexins in other phyla. We predicted posttranslational modifications in innexins: 249 and 188 for *M. leidyi* and *P. bachei*, respectively. Neither their number nor their locations were conserved within or between species. When the number of post-translational modifications is factored into the innexins' radiation, the potential for molecular and physiological diversity within gap junctions of ctenophores is almost unfathomable. RNA-seq and in situ hybridization data revealed that innexins are expressed across embryogenesis, including early cleavage stages and gastrulation. They are abundant in all adult tissues, with the highest expression level in the aboral organ (the major integrative center and the gravity sensor in ctenophores), followed by tentacles and comb plates. Nevertheless, each organ and tissue has a unique combination of innexins, suggesting their involvement in complex integrative functions and behaviors of ctenophores.

Key words Ctenophora, Innexin, Pannexin, Posttranslational modifications, *Pleurobrachia*, *Mnemiopsis*, Evolution, Phylogeny

1 Introduction

Gap junction proteins are critical components of electrical synapses, providing both fast electrical and metabolic coupling within cells across numerous animal tissues. Two distinct families of these channel proteins evolved convergently: pannexins (=innexins) and connexins [1–3]. Both families are metazoan innovations [1, 4, 5] with surprisingly similar membrane topologies [2, 4, 6].

Leonid L. Moroz (ed.), *Ctenophores: Methods and Protocols*, Methods in Molecular Biology, vol. 2757,
https://doi.org/10.1007/978-1-0716-3642-8_16, © Springer Science+Business Media, LLC, part of Springer Nature 2024

Connexins are exclusively vertebrate- and tunicate-lineage-specific genes, whereas pannexins are present both in vertebrates and invertebrates. Before their discovery in humans, invertebrate pannexins were initially named innexins—i.e., invertebrate counterparts to gap junctions. Subsequently, Panchin and colleagues found innexin homologs in vertebrates and suggested a more general name for the entire protein superfamily as pannexins [1]. Nevertheless, in the literature, the term pannexins are usually employed in research on vertebrates, whereas a more traditional innexin terminology is widely preserved in references to invertebrates.

At this moment, neither innexins nor connexins were identified outside of animals. The sequenced genomes of choanoflagellates, other holozoans, and opisthokonts, including fungi, do not contain recognizable innexins. Thus, the origin of these two families in animal evolution is elusive, and studies of basal metazoan lineages are minimal.

The phylum Ctenophora represents the earliest branch on the animal tree of life—the ctenophore-sister hypothesis [7–11], which is strongly confirmed by the recent integrative genome analyses [12]. Thus, evolution of electrical synapses would be viewed within the phylogenetic framework of the ctenophore-first scenario. Gap junctions were experimentally discovered in ctenophores using ultrastructural [13–15] and functional [16] approaches in 1970–1980s [17, 18].

The initial characterization of the *Pleurobrachia bachei* genome revealed a dozen diverse genes encoding innexins [7] in this and ten other ctenophore species (by sequencing their transcriptomes). The sponge *Amphimedon* (representing the second branching metazoan lineage) and the placozoan *Trichoplax* (the third branch)—all lack recognized gap junction proteins. Thus, pannexins/innexins were present in the common ancestor of all animals and independently diversified in ctenophores [7], some cnidarians, and most bilaterians, as evidenced by the respected phylogenetic trees [4, 19–22]. In contrast, the lineages led to sponges and placozoans lost innexins [4, 7]. Notably, there are also multiple examples of gains and losses of innexins across Metazoa (*see* Fig. 1 and [23])—reasons why placozoans, some cnidarians, echinoderms, hemichordates, and lost innexins are unclear [24]. In basal chordates, functional innexins/pannexins were "substituted" by connexins [21], which can make equally efficient electrical synapses coupling various cell populations, including neurons.

Mechanistically, both gap junction families form hemichannels of contacting cells with similar molecular architecture [4, 25–27]: four transmembrane domains, one intracellular loop, and two extracellular loops (Figs. 2, 3, and 4). The connexin hemichannel comprises six subunits (connexon), whereas eight innexins (innexon) are required for such assembly [21]. Physically opposing hemichannels from coupling cells can be stabilized by disulfide

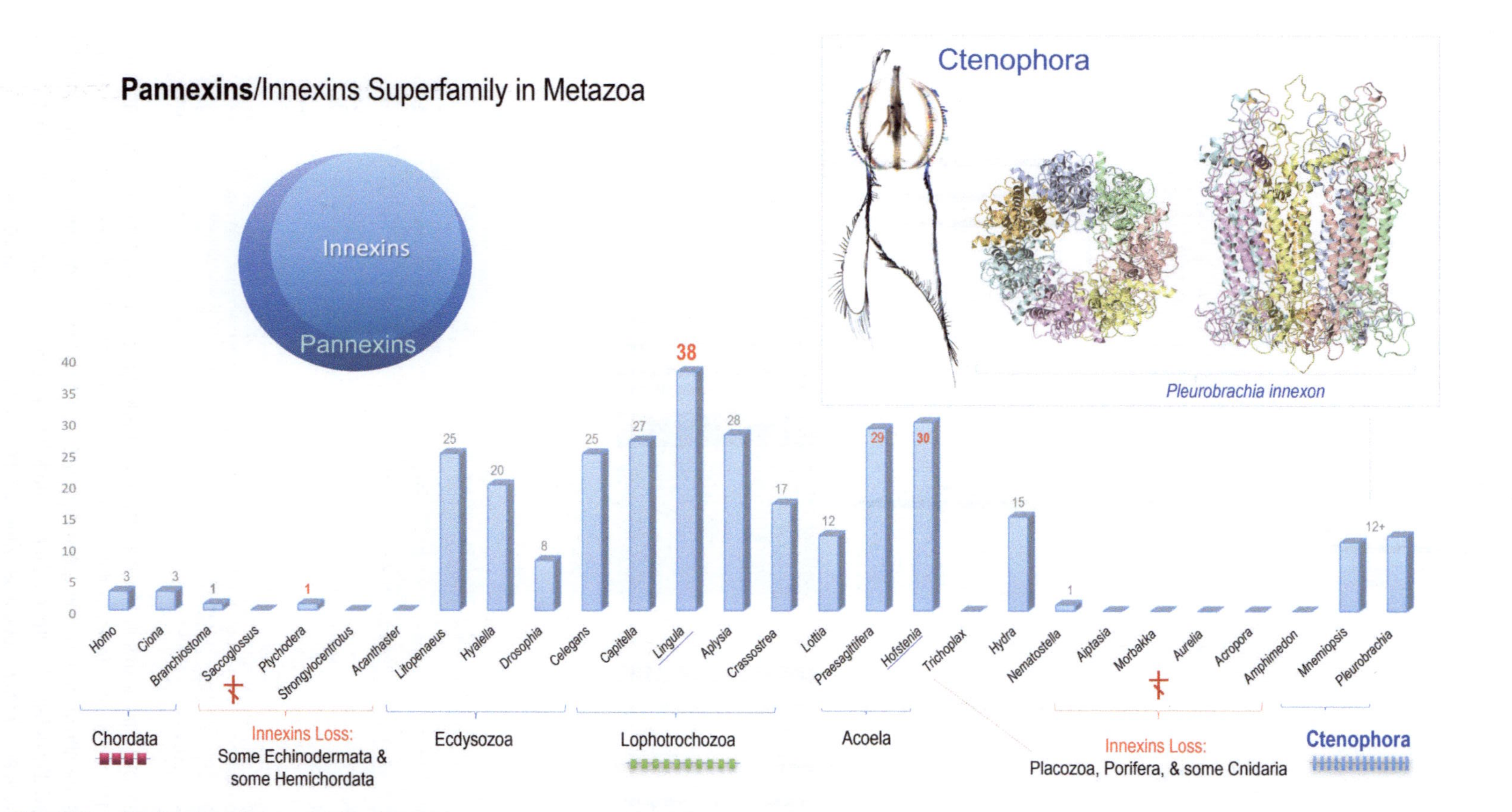

Fig. 1 The distribution of pannexin/innexin proteins across Metazoa with multiple examples of gene gains and losses. Y-axis indicates the presence and the number of genes encoding these gap junction proteins in selected reference species. Insert shows the three-dimensional structure of one innexin from the ctenophore *Pleurobrachia bachei* (photo). (Modified from Moroz et al. [23])

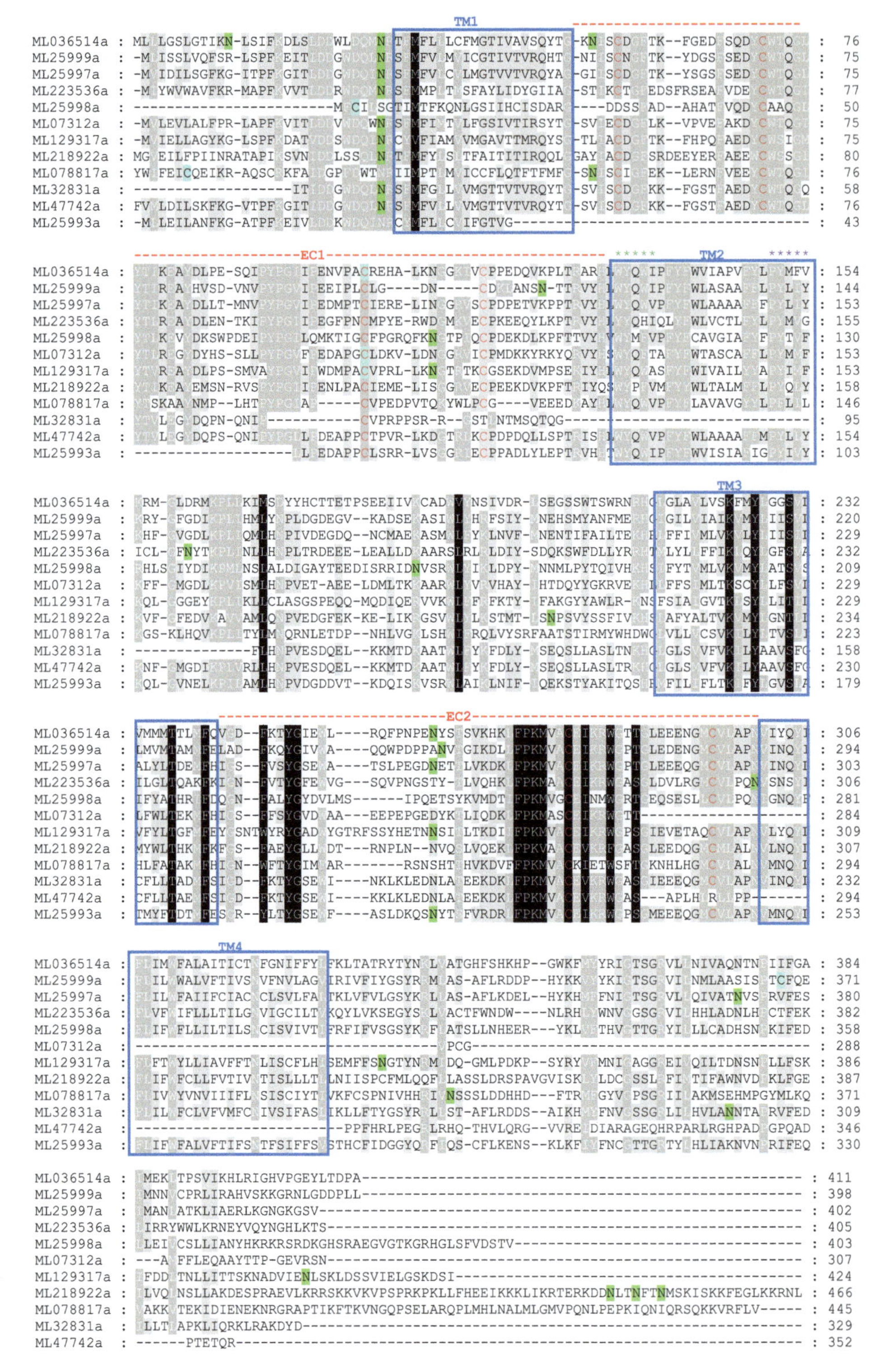

Fig. 2 Protein alignment of 12 innexins in *Mnemiopsis leidyi*. Dark blue squares outline the four transmembrane domains (TM); the red dashes indicate the extracellular loops 1 and 2 EC1 and EC2. The conserved cysteine Cys residues are shown in red. The green highlights are N-glycosylation sites, and the turquoise highlights are S-nitrosylation sites. The green asterisks mark the "innexin sequence," and the blue asterisks correspond to the canonical "PXXXW" sequence within the second transmembrane domain

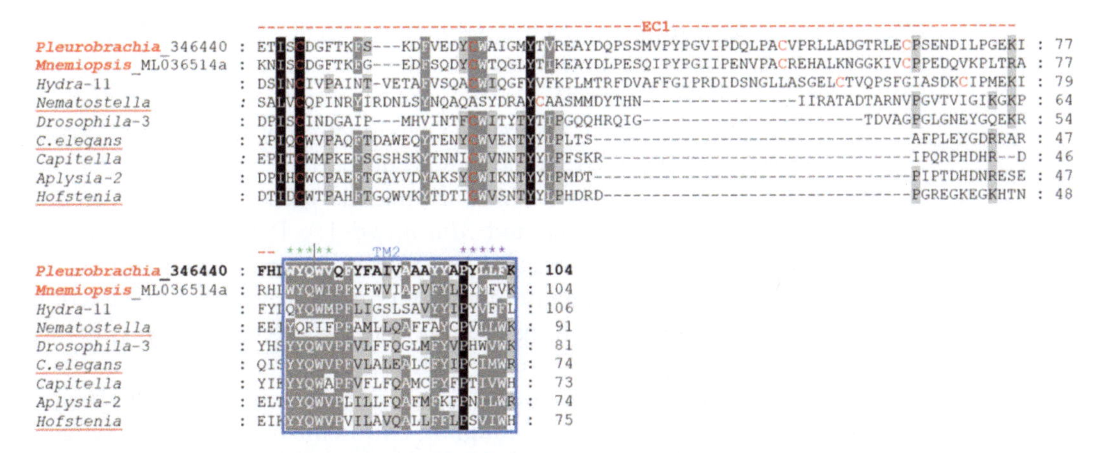

Fig. 3 Protein alignment of the extracellular loop, 1st and 2nd transmembrane domains innexins in representative species from different phyla. All sequences accession numbers are referenced in Table 2 for the phylogeny except for *C. elegans*, NP_001024407.1, *Hofstenia miamia*_g17543.t1, and *Capitella teleta*_ELT91918.1. The dark blue square outlines the TM2, and the red dashes indicate the extracellular loops 1. The conserved cysteine Cys residues are shown in red. The green asterisks indicate the "innexin sequence," and the blue asterisks indicate the "PXXXW" sequence motif within the TM2

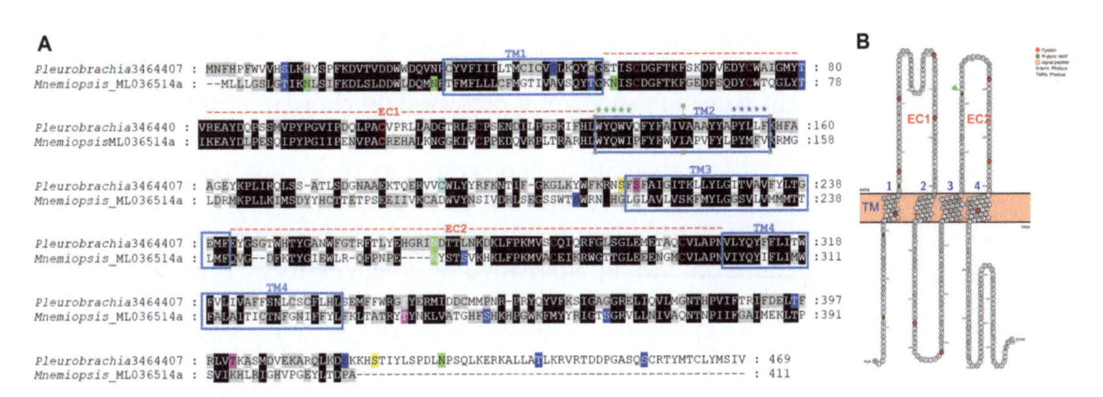

Fig. 4 (**a**) Protein alignment between representative innexins from *Pleurobrachia* (346440) and *Mnemiopsis* (ML036514a). Dark blue squares outline the four transmembrane domains (TM); the red dashes indicate the extracellular loops 1 and 2 EC1 and EC2. The conserved cysteine Cys residues are indicated in red. The green asterisks mark the "innexin sequence," and the blue asterisks correspond to the "PXXXW" sequence motif within the second transmembrane domain. The green highlights are N-glycosylation sites, the turquoise highlights are S-nitrosylation sites, the yellow highlights are S/T-cAMP- and cGMP-dependent protein kinase phosphorylation sites, and the bright blue highlights are S/T-protein kinase C phosphorylation sites and the pink highlights are S/T-protein kinase A phosphorylation sites. (**b**) Topographic representation of *Mnemiopsis* (ML036514a) with the Cys residues are indicated in red, and the PTMs are colored as described in (**a**) [47] (*See* Subheading 2)

bonds between cysteines in each extracellular loop [28]. The same cells might express multiple connexins and innexins, forming heteromeric hemichannels. Two opposing homo- and heteromeric connexons or innexons on different cells form a functional gap junction.

As a result of such combinatorial logic, N^6 and N^8 types of channels are possible for N subunits of genome-encoded connexins and innexins, respectively. It provides exceptional opportunities to generate enormous diversity of gap junction proteins and electrical synapses with different biophysical properties. For example, 14 and 12 innexin subunits, which we initially identified in two ctenophores (*Pleurobrachia* and *Mnemiopsis* [7, 19]), can theoretically form 1,475,789,056 and 429,981,696 different channels (14^8 or 12^8). It is astonishing numbers, but more likely, only a tiny fraction of possible combinations could exist in any given species yet provide a solid base for evolutionary selection.

Three pannexins identified in chordates do not form classical electrical synapses with coupling two hemichannels from contacting cells [21, 27, 29, 30]. Here, N-glycosylation prevents forming an actual conductance pore between two cells. Specifically, pannexins' first or second extracellular loop contains the Asn-X-Ser/Thr motif for asparagine glycosylation [21]. These sugars hinder intercellular gap junction formation by modifying the physical docking of the hemichannels. In other words, pannexins are membrane channels that mediate the release of low molecular weight compounds such as ATP [31, 32]. These single-channel properties might also justify the initial split in the channel terminology when the name of pannexins is preserved for chordates. At the same time, "innexins" refer to the homologous proteins in invertebrates.

More than 500 million years ago, pannexins had lost the ability to form gap junctions (due to N-glycosylation?) in ancestral chordates, which might have triggered an independent development of connexins (as an alternative or compensatory "route") and their subsequent expansion in vertebrates [21].

N-glycosylation of innexins with lots of gap junction functions is reported for some invertebrate species, such as insects (*Aedes aegypti*), nematodes (*Caenorhabditis elegans*) [33], and predicted for mollusks (*Limax valentianus*), where large fractions of innexins are glycoproteins [21, 34]. N-glycosylation sites across innexins and species are not conserved (except vertebrates with a high level of evolutionary stabilization across taxa [21]). Whether N-glycosylation in innexins can hinder or modify actual functional electrical synapses between cells remains to be determined. At least seven innexins with N-glycosylation form gap junctions in annelids, nematodes, and insects [21]. The dynamic occurrence of sugar moieties in both pannexins and innexins subunits further expands the diversity and properties of this unique class of channels associated with rapid intercellular signaling. Yet, it is well-known that "the experimental identification of N-glycosylated proteins is technically demanding, time-consuming, and expensive" [21]. Thus, computational approaches are more favorable as the first steps for many comparative studies.

This chapter summarizes the presence and distribution of innexins in representatives of the most basally branching animal phylum—Ctenophora. This ancestral lineage is of utmost importance for deciphering the origin and evolution of animals and their tissue and organ innovations, including likely independent origins and evolution of neuromuscular, sensory, and digestive systems [7, 8, 16, 18, 35, 36]. Decoding complex behaviors and reconstructing the parallel evolution of many integrative systems in this clade [15–24] are linked with a deep understanding of gap junctions and their functional plasticity across multiple ctenophore species with distinct ecological adaptations. We focus on practical implementations in analyses of innexins from two reference species of ctenophores [7, 8]: the cydippid *Pleurobrachia bachei* and *Mnemiopsis leidyi*, with brief references to other ctenophores.

2 Methods

- All sequences were identified using BLAST searches on publicly available databases, including species-specific genome browsers with cutoff values of E-value $\leq 10^{-4}$.
- All *Pleurobrachia bachei* data was assembled, annotated, and available on the University of Florida Database: http://neurobase.rc.ufl.edu/.
- All *P. bachei* gene models used in the analysis were published and described here http://neurobase.rc.ufl.edu/ [7].
- *M. leidyi* gene models were obtained at http://research.nhgri.nih.gov/mnemiopsis/jbrowse/jbrowse.cgi [8].
- Alignments of all orthologs were performed with either ClustalX2 [37] or Muscle [38] and then, if appropriate, either trimmed manually or trimmed using GBlocks.
- Further comparative analysis of individual gene families used Pfam composition [39], Gene Ontology [40], and KEGG (Kyoto Encyclopedia of Genes and Genomes) [41] to validate orthologs.
- Once alignments were obtained, basic gene trees were constructed in MEGA 6 [42]. The molecular phylogenetic analysis used the maximum likelihood (ML) method based on the Whelan and Goldman (W.A.G.) matrix-based model.
- The bootstrap consensus tree was inferred from 100 replicates. All positions containing gaps and missing data were eliminated [42]. The tree is drawn to scale.
- We screened potential serine/threonine/tyrosine phosphorylation sites from 17 different kinases using the NetPhos 3.1Server,

https://services.healthtech.dtu.dk/service.php?NetPhos-3.1 [43].

- In addition, we used the PROSITE database, https://prosite.expasy.org [44], to validate and identify N-glycosylation sites.
- SMART (Simple Modular Architecture Research Tool) was used to identify protein domain architectures, http://smart.embl-heidelberg.de/ [45].
- N-glycosylation sites were also identified using GPS-SNO1.0 http://sno.biocuckoo.org/ [46].
- The Protter web application was used to visualize the sequence topology http://wlab.ethz.ch/protter/# [47].

3 Predictions, Analyses, and Discussion

3.1 Identification of Putative Innexin-Like Proteins and Their Topology

Using reported transcriptome databases [7, 11], we identify comparable diversity of innexins in more than 15 ctenophore species, including such genera as *Euplokamis*, *Pleurobrachia*, *Pukia*, *Hormiphora*, *Bolinopsis*, *Mnemiopsis*, *Cestum*, *Ocyropsis*, *Dryodora*, *Beroe*, benthic ctenophores, *Coeloplana* and *Vallicula*, and undescribed species of *Mertensiidae* (*see* Supplement).

Twelve putative innexin-like (INX) genes were initially reported in the *P. bachei* genome, and all were found in transcriptomes [7]. Upon further examination, at least 14 *INX*-like genes might occur in *Pleurobrachia*, and 12 genes exist in the *Mnemiopsis* genome [7, 19], as summarized in Table 1.

The deduced protein sequence alignment for all *Mnemiopsis* innexins shows four transmembrane domains (T.M.) connected by two extracellular (EC) loops that bridge the intracellular gap (Figs. 2 and 4b). The signature "innexin sequence" (green asterisk) Y.Y. (X)W(Z), where X is Q, R, M, E, or S and Z is V, M, A, I, S, or T, found at the beginning of the second transmembrane domain (TM2) is not conserved in the investigated ctenophore innexins and in the hydrozoan *Hydra magnipapillata* [48] (Figs. 2 and 3). The ctenophore uniqueness is true for the PXXXW sequence (blue asterisk) at the end of the second transmembrane domain. The tryptophan (W) amino acid is "missing" in all ctenophore and *Hydra* INX-like proteins examined. Still, the proline amino acid residue is present (Figs. 2 and 3). Studies suggest that this proline residue acts like a molecular hinge to change the conformational structure in response to voltage [2].

The cysteine (Cys) amino acid residues (red letter C) in the extracellular loops are highly conserved in all innexin sequences [2, 26]. Surprisingly, two conserved pairs of Cys are in the first EC1 of both *Pleurobrachia* and *Mnemiopsis* (Figs. 2, 3, and 4). Only one other studied non-bilaterian contains two pairs of Cys

Table 1
Summary of the number of the most common predicted posttranslational modifications in *Mnemiopsis* and *Pleurobrachia* deduced INX-like proteins

Menmiopsis	# Exons	Protein kinase C (PKC)	Protein kinase A (PKA)	Casein kinase 1 (CK1)	Casein kinase 2 (CK2)	cAMP PHOSPHO	N-glycosylation	S-nitrosylation	Total
ML036514a	8	7	1	2	3	0	4	1	18
ML25999a	17	7	0	1	5	0	3	1	17
ML25997a	18	7	1	1	7	0	3	0	19
ML223536a	4	6	2	0	6	1	4	0	19
ML25998a	8	7	2	0	6	1	2	1	19
ML07312a	9	4	1	1	5	0	1	1	13
ML129317a	1	6	2	0	7	1	4	1	21
ML218922a	8	12	5	1	15	3	6	0	42
ML078817a	2	12	2	0	2	1	2	1	20
ML32831a	6	7	4	0	8	2	2	1	24
ML47742a	8	12	2	2	8	3	1	0	28
ML25993a	12	4	2	0	2	0	1	0	9
Total		91	24	8	74	12	33	7	249
Pleurobrachia									
3464407	3	7	2	1	4	2	2	1	19
3463885	14	11	1	1	10	1	3	0	27
3468487	1	2	1	0	4	1	1	0	9
3469267	9	5	1	0	9	0	2	0	17
3471871	3	2	1	0	3	0	1	0	7
3473233	1	2	1	0	0	0	1	0	4
3479642	8	2	1	0	1	0	3	0	7
3464250	11	8	3	0	7	1	4	0	23
3465454	4	8	1	0	5	0	2	0	16
3465668	8	4	2	0	5	1	0	0	12
3465979	8	3	0	0	4	1	2	1	11
3466205	9	4	1	0	8	0	0	1	14
3466313	1	2	4	0	7	0	1	1	15
3468319	5	3	0	0	4	0	0	0	7
Total		63	19	2	71	7	22	4	188

Totals are in blue
Number of exons are also noted in the corresponding *INX*-like genes

residues in the EC1 loop of their predicted innexins—*Hydra* [48, 49] (Fig. 3).

The first pair of Cys residues located in the EC1 loop is highly conserved in cnidarians and bilaterians, including the acoel *Hofstenia miamia* (Fig. 3). Interestingly, the cnidarian *Nematostella vectensis* has only one pair of Cys residues in its EC1 loop (Fig. 3). The Cys residues in the EC1 and EC2 loops are critical to forming functional channels [2].

The EC1 loop in ctenophores and *Hydra* is larger (~75 amino acid residues, Fig. 3) than other innexins described (~60 amino acid residues) [2]. Such a larger EC1 loop may accommodate the extra pair of Cys residues in the EC1. Of note, in three *Mnemiopsis* innexins, one of the extra Cys can be a site for the predicted posttranslational modification, S-nitrosylation (Fig. 4). It is consistent with the hypothesis that the gaseous messenger, nitric oxide (NO) might be a signal molecule in ctenophores [19, 50], where S-nitrosylation is one of the molecular targets for nitrergic transmission (potentially associated with NO-dependent modulation of electrical coupling across diverse cell populations in development and adult tissues).

The overall innexins' structure is similar between *P. bachei* and *M. leidyi*, including the three pairs of Cys residues (Fig. 4). But *posttranslational modifications* across innexins are not highly conserved (neither within the same species nor between these two ctenophore species).

3.2 Predicted Posttranslational Modifications (PTMs) in Innexin-Like Proteins

It has been shown that one possible PTM can alter the formation of functional gap junctions [51], so we investigated the most conserved predicted PTMs in innexins of *Pleurobrachia* and *Mnemiopsis*. To date, more than 400 different PTMs are possible [52]. Figures 2, 3, and 4 indicate potential N-glycosylation, S-nitrosylation sites, and phosphorylation (Fig. 4), and Table 1 lists the most conservative innexin modifications. Although no functional tests have been performed on any specific ctenophore innexin, PTMs suggest enormous ramifications to function [51].

The number of putative N-glycosylation sites varies considerably for *P. bachei* and *M. leidyi*, from six to zero (Table 1). The location within sequences also is inconsistent from the intracellular region N- and C-termini to the extracellular loops (Fig. 4). One N-glycosylation in the N-terminus, N28 in ML036514a, was relatively conserved across *Mnemiopsis* innexins (Fig. 2) but not in *Pleurobrachia* (Fig. 3a).

The predicted S-nitrosylation sites were far less abundant, with seven out of 12 innexins having a single S-nitrosylation site in *M. leidyi*. In contrast, only four out of 14 innexins had an S-nitrosylation site in *P. bachei*. More S-nitrosylation, potentially regulatory sites in *Mnemiopsis* might be associated with the presence of nitric oxide synthase and endogenous nitrergic transmission [50].

We also predicted the phosphorylation sites for multiple kinases (e.g., we chose a conservative set of kinases: protein kinase C (PKC), protein kinase A (PKA), casein kinase 1 (CK1), casein kinase 2 (CK2), cAMP- and cGMP-dependent protein kinase, etc.). The *Mnemiopsis* ML036514a could have 82 serine, threonine, or tyrosine phosphorylation sites from 17 different kinases (ATM, CKI, CKII, CaM-II, DNAPK, EGFR, GSK3, INSR, PKA, PKB, PKC, PKG, RSK, SRC, cdc2, cdk5, and p38MAPK), and 45 are above the threshold of probability [43]. Phosphorylation by these kinases regulates many protein functions in vertebrate connexins/pannexins, including intracellular trafficking, gap junction assembly, and stability [25, 51]. The number of phosphorylation sites from this small cohort of enzymes ranged from 3 to 27 in a single innexin. The total number of predicted PTMs in innexins in is 249 and 188 for *M. leidyi* and *P. bachei,* respectively (Table 1).

The aligned *Pleurobrachia* and *Mnemiopsis* innexins had seven predicted PKC sites; however, only two were conserved (Fig. 4). Similarly, only one N-glycosylation site was conserved between these two species (Fig. 4).

In summary, the post-translation modifications in different innexins are not conserved within the same species or between *P. bachei* and *M. leidyi.* This situation contrasts with the molluscan innexins (e.g., from *Limax valentianus* species and across the entire molluscan phylum [34]). PTMs in molluscan innexins are likely conserved, because their amino acid sequences are highly conserved too. On the other hand, the % identity between innexins for *Pleurobrachia*_346440 and *Mnemiopsis*_ML036514a (Fig. 3) is only 38%. Besides low sequence identity, the genomic organization of innexin genes in ctenophores is highly variable, with the number of exons ranging from 14 to 1 for *P. bachei* and 17 to 1 for *M. leidyi* (Table 1).

3.3 Phylogenomic Analysis of PANX/INX-Like Predicted Proteins from Metazoans

As noted, gap junctions are metazoan innovations [1, 4, 24], essential for the origin and integration of multicellular organization. Genealogical relationships among pannexin/innexin illustrate multiple events of phylum-specific radiation of these channel proteins. Besides metazoans, *INX-like* genes have been found in a few viruses, possibly resulting from lateral gene transfer between a host and intracellular parasites [53]. For example, the endoparasitic wasps *Campoletis sonorensis* transmits a polydnavirus to their caterpillars during egg-laying [53]. The sequenced ichnovirus (IV) genome reveals a significant similarity to some parts of the wasp genome, including *INX* genes [53]. Predictably, these ichnovirus INX are closely related to innexins from arthropods.

Ctenophore innexins also exhibited a remarkable independent evolutionary diversification [19, 21], forming a distinct branch on the tree topology (e.g., Fig. 5 and Table 2). There are only four close sister-type relationships between *Pleurobrachia* and *Mnemiopsis,* red lines in Fig. 5. The low % identity found between the predicted amino acid sequences may result from early diversification of ctenophore innexins. There is a higher identity (44%) between the *Drosophila* INX-2 protein and the endoparasitic wasps *Campoletis sonorensis* INX-like d1 predicted sequences than between two *Pleurobrachia*_ 3464407 and *Pleurobrachia*_ 3463885 INX-like predicted sequences.

Interestingly, there is no conservation in the exon–intron organization across *innexin/pannexin* genes [19]. The *pannexins* identified in mammals have four to five exons. However, the ctenophore *innexins* contain a far greater range of exons compared to their

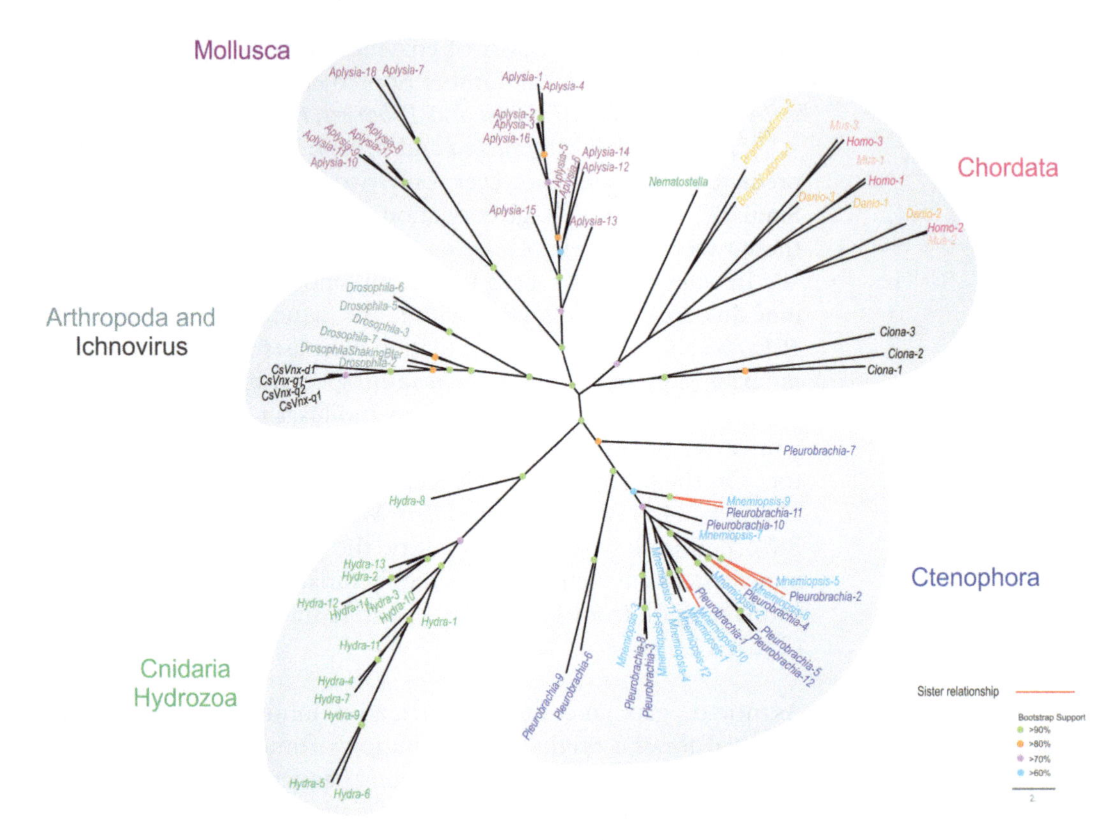

Fig. 5 Genealogical relationships among the putative protein family pannexin/innexin (PANX/INX) in metazoans (maximum likelihood method based on the Whelan and Goldman matrix-based model). The PANX/INX family shows lineage-specific diversification in major metazoan clades. Ctenophore PANX/INX proteins also share the highest identity, forming a distinct branch on the tree topology. Four close sister-type relationships exist between *Pleurobrachia* and *Mnemiopsis*, red lines. *See* Table 2 for names and accession numbers used in the analysis (*see* Subheading 2)

orthologs in Hydrozoa and bilaterians (the number of exons varies from 1 to 17, Table 1, *see* also Fig. 1) [19].

Cnidarians appear to have either dramatic losses or lineage-specific expansions of the *INX* genes (Fig. 1). The hydrozoan *Hydra* has at least 14 *INX*-like genes or even more [48]. In contrast, the anthozoan *Nematostella* has only one recognized *INX-like* gene (Fig. 5), possibly involved in the electrical coupling between blastomeres in embryos [54]. Moreover, the genomes of three other anthozoans, *Aiptasia* [55], *Acropora digitifera* [56], and *Stylophora pistillata* [57], had no *INX* genes detected. The scyphozoan *Cyanea capillata* also has no identified *INX* genes (based on our transcriptome profiling).

In summary, ctenophore innexins exhibited extraordinary phylum-specific diversification compared to other metazoans. The extreme range in the number of exons (1–17) and lack of conservation in protein identity, even between the same species, make the

Table 2
List of species accession numbers and names used on the tree in Fig. 5 to test genealogical relationships pannexin/innexin families in metazoans

Species	Database #/accession #	Name in tree
Mnemiopsis leidyi	ML036514a	*Mnemiopsis*-1
Mnemiopsis leidyi	ML07312a	*Mnemiopsis*-2
Mnemiopsis leidyi	ML078817a	*Mnemiopsis*-3
Mnemiopsis leidyi	ML129317a	*Mnemiopsis*-4
Mnemiopsis leidyi	ML218922a	*Mnemiopsis*-5
Mnemiopsis leidyi	ML223536a	*Mnemiopsis*-6
Mnemiopsis leidyi	ML25993a	*Mnemiopsis*-7
Mnemiopsis leidyi	ML25997a	*Mnemiopsis*-8
Mnemiopsis leidyi	ML25998a	*Mnemiopsis*-9
Mnemiopsis leidyi	ML25999a	*Mnemiopsis*-10
Mnemiopsis leidyi	ML32831a	*Mnemiopsis*-11
Mnemiopsis leidyi	ML47742a	*Mnemiopsis*-12
Pleurobrachia bachei	3463885	*Pleurobrachia*-1
Pleurobrachia bachei	3464250	*Pleurobrachia*-2
Pleurobrachia bachei	3464407	*Pleurobrachia*-3
Pleurobrachia bachei	3465454	*Pleurobrachia*-4
Pleurobrachia bachei	3465668	*Pleurobrachia*-5
Pleurobrachia bachei	3465979	*Pleurobrachia*-6
Pleurobrachia bachei	3466205	*Pleurobrachia*-7
Pleurobrachia bachei	3466313	*Pleurobrachia*-8
Pleurobrachia bachei	3468319	*Pleurobrachia*-9
Pleurobrachia bachei	3471871	*Pleurobrachia*-10
Pleurobrachia bachei	3468487	*Pleurobrachia*-11
Pleurobrachia bachei	3479642	*Pleurobrachia*-12
Homo sapiens	NP_056183.2	*Homo-1*
Homo sapiens	NP_443071.2	*Homo-2*
Homo sapiens	NP_443071.3	*Homo-3*
Mus musculus	NP_062355.2	*Mus-1*
Mus musculus	NP_443191.1	*Mus-2*
Mus musculus	NP_766042.2	*Mus-3*
Danio rerio	NP_957210.1	*Danio-1*

(continued)

Table 2
(continued)

Species	Database #/accession #	Name in tree
Danio rerio	NP_001243570.1	*Danio-2*
Danio rerio	XP_001919861.1	*Danio-3*
Ciona intestinalis	XP_002124490.2	*Ciona-1*
Ciona intestinalis	XP_002119287.2	*Ciona-2*
Ciona intestinalis	XP_009858431.1	*Ciona-3*
Nematostella vectensis	XP_001623899.1	*Nematostella*
Branchiostoma floridae	XP_002604233.1	*Branchiostoma-1*
Branchiostoma floridae	XP_002585873.1|	*Branchiostoma-2*
Drosophila melanogaster	NP_001162684.1	*Drosophila-2*
Drosophila melanogaster	NP_524730.1	*Drosophila-3*
Drosophila melanogaster	NP_573353.2	*Drosophila-5*
Drosophila melanogaster	NP_572374.1	*Drosophila-6*
Drosophila melanogaster	NP_788872.1	*Drosophila-7*
Drosophila melanogaster	NP_728361.1	*Drosophila-8*
Campoletis sonorensis	AAO45828.1	*CsVnx-d1*
Campoletis sonorensis	AAO45831.1	*CsVnx-q2*
Campoletis sonorensis	AAO45830.1	*CsVnx-q1*
Campoletis sonorensis	AAO45829.1	*CsVnx-g1*
Aplysia californica	NP_001191577.1	*Aplysia-1*
Aplysia californica	NP_001191579.1	*Aplysia-2*
Aplysia californica	NP_001191578.1	*Aplysia-3*
Aplysia californica	NP_001191576.1	*Aplysia-4*
Aplysia californica	NP_001191595.1	*Aplysia-5*
Aplysia californica	NP_001191594.1	*Aplysia-6*
Aplysia californica	NP_001191616.1	*Aplysia-7*
Aplysia californica	NP_001191596.1	*Aplysia-8*
Aplysia californica	NP_001191461.1	*Aplysia-9*
Aplysia californica	NP_001191462.1	*Aplysia-10*
Aplysia californica	XP_005110953.1	*Aplysia-11*
Aplysia californica	XP_005100630.1	*Aplysia-12*
Aplysia californica	XP_005109439.1	*Aplysia-13*

(continued)

Table 2
(continued)

Species	Database #/accession #	Name in tree
Aplysia californica	XP_005110669.1	*Aplysia-14*
Aplysia californica	XP_005103400.1	*Aplysia-15*
Aplysia californica	XP_005104811.1	*Aplysia-16*
Aplysia californica	XP_005110954.1	*Aplysia-17*
Aplysia californica	XP_005101166.1	*Aplysia-18*
Hydra vulgaris	NP_001274699.1	*Hydra-1*
Hydra vulgaris	XP_002160488.1	*Hydra-2*
Hydra vulgaris	XP_002166931.2	*Hydra-3*
Hydra vulgaris	XP_004213297.2	*Hydra-4*
Hydra vulgaris	XP_004212712.1	*Hydra-5*
Hydra vulgaris	XP_004212713.1	*Hydra-6*
Hydra vulgaris	XP_012561363.1	*Hydra-7*
Hydra vulgaris	XP_004208200.2	*Hydra-8*
Hydra vulgaris	XP_002170241.2	*Hydra-9*
Hydra vulgaris	XP_002164718.2	*Hydra-10*
Hydra vulgaris	XP_012566780.1	*Hydra-11*
Hydra vulgaris	XP_002165350.1	*Hydra-12*
Hydra vulgaris	XP_012554059.1	*Hydra-13*
Hydra vulgaris	XP_002170247.1	*Hydra-14*

ctenophore lineage unique. Although functional analysis of the electrical synapses in ctenophores is in its infancy, it will help to shed light on the contribution of this unusual complement of innexins associated with rapid intracellular signaling [16, 19, 58] and volume transmission (e.g., when N-nitrosylation or other post-translational modifications prevent to form electrical synapses—*see* also [23]).

3.4 Expression of Predicted INX-Like Proteins in Ctenophores

Ideally, innexins' temporal and spatial distribution is performed using both in situ hybridization and RNA-seq analyses. These two approaches are complementary; we applied them as illustrated in Figs. 6 and 7. Of note, because of more tough organization and texture, species of *Pleurobrachia* are ideal models for in situ hybridization studies, especially in adult stages. In contrast, adult *Mnemiopsis* and related lobate species are so fragile that even initial

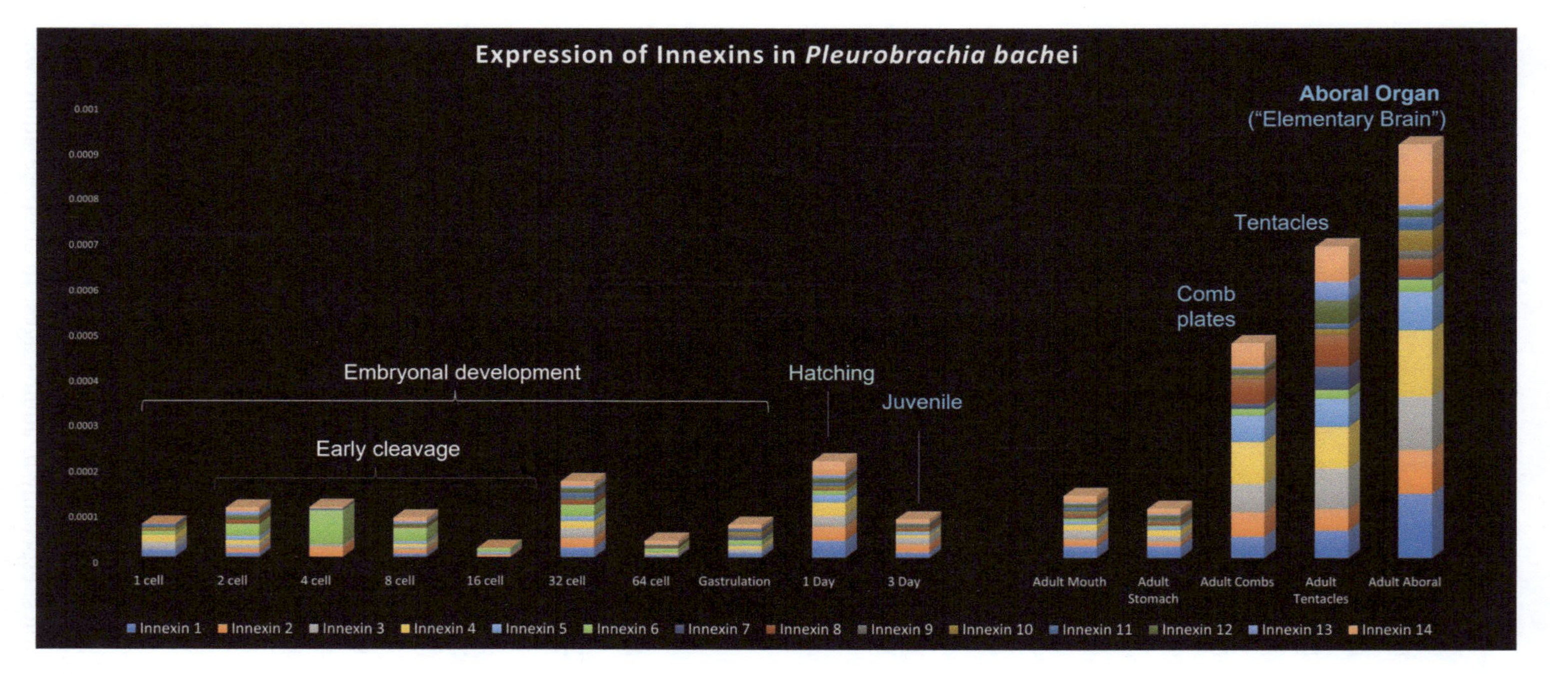

Fig. 6 RNA-seq analysis of innexin expression during development and in adult organs of *Pleurobrachia bachei*. The aboral organ has the greatest diversity and highest combined expression levels of 14 innexins, suggesting elaborated and diverse signaling pathways in this complex integrative organ—an analog of an elementary brain in ctenophores. Different innexins are co-expressed in every adult structure and developmental stage tested (it is shown as a summation of normalized frequencies of respective sequencing reads in each RNA-seq dataset; *y* axis is "expression frequency"). (Modified from Moroz et al. [7])

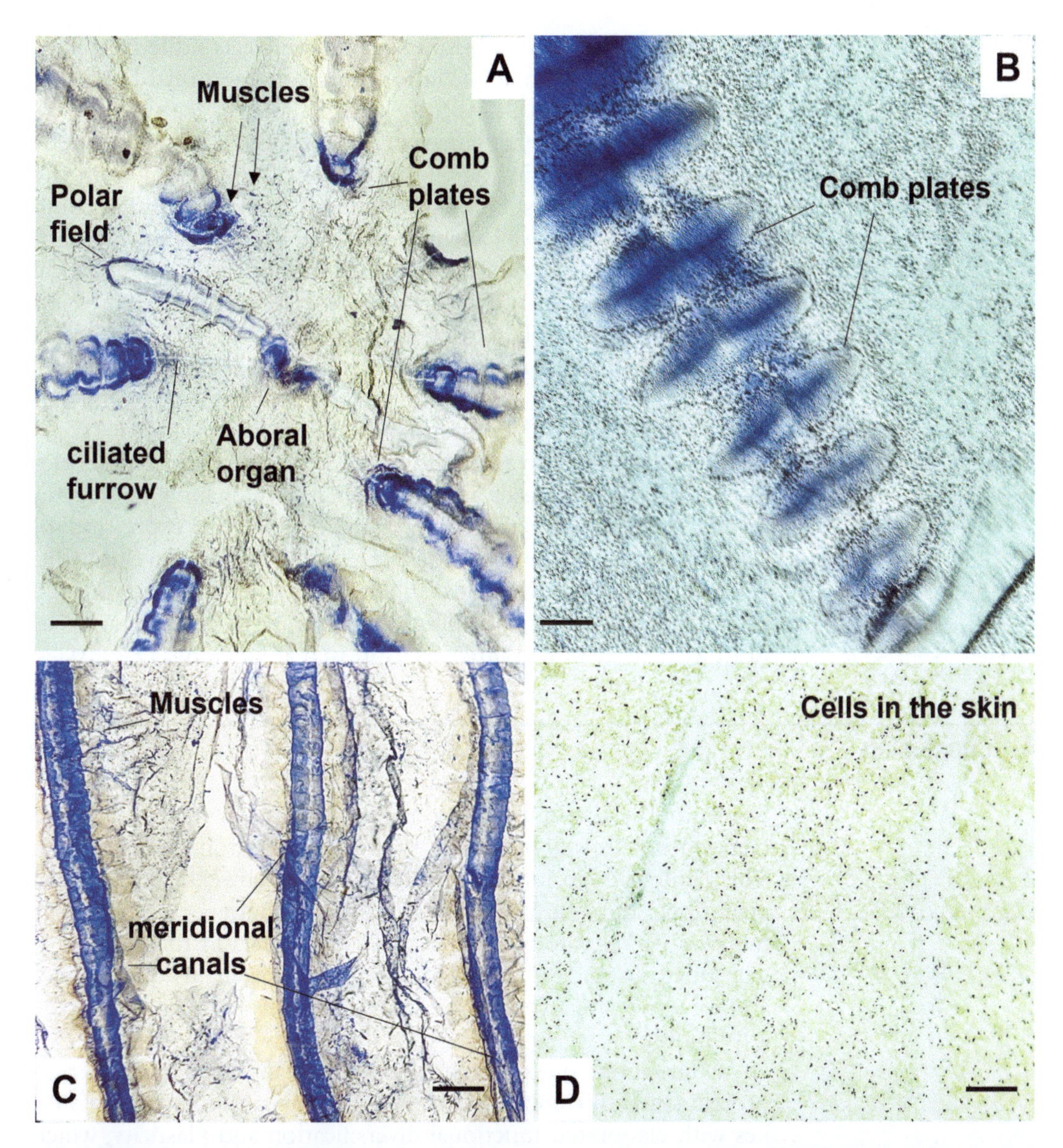

Fig. 7 Illustrative examples of spatial expression of four innexins in adult *Pleurobrachia bachei* (in situ hybridization). (**a**) The aboral view shows a specific expression of innexin 1 in the aboral organ, combs, ciliated furrows, and polar fields. Arrows indicate differential expression in selected muscle cells in sub-epithelial regions. (**b**) Particular expression of innexin 2 in comb plates. (**c**) Innexin 5 is differentiated and expressed in meridional canals underlying the comb rows (unstained). (**d**) Expression of innexin 3 in skin cells. Scale: 400 μm (**a**, **c**); 80 μm (**b**, **d**)

fixation of adult tissues is highly challenging, with rapid dissociation of virtually every structure restricting expression analysis to the earlier developmental stages only.

Multiple innexins are highly co-expressed in the adult aboral organ of *P. bachei* (Figs. 6 and 7a), also known as the major integrative center and the gravity sensor in ctenophores [18, 35,

58]. Most innexins are co-localized in combs, conductive tracts, mouth, pharynx, stomach (Fig. 6), and some neuronal-like subepithelial cells (Fig. 7d).

In ctenophores, gap junctions were previously identified by electron microscopy in the ciliated grooves, which run from the apical organ to the first comb plate of each comb row, and in comb plates themselves, as well as through the endoderm of the meridional canals [16], suggesting that the primary form of electrical coupling. Using in situ hybridization, we also detected expression of innexins in the ciliated furrows as well as in the polar fields (putative chemosensory structures) as well as selected subsets of subepithelial muscles and meridional canals (Fig. 7), consistent with earlier electron microscopy studies [13–15, 35].

Gap junctions also allow the direct exchange of small molecules among neighboring cells and sometimes form channel aggregates in the plasma membrane that might explain the expression of selected types of innexins during *Pleurobrachia* development. RNA-seq data revealed that distinct sets of innexins are expressed across embryogenesis, including early cleavage stages and gastrulation, with all identified innexins expressed at hatching. (Fig. 6).

4 Conclusion

Studies of the distribution and functions of innexins across different ctenophores are still in infancy, with more questions than answers. This family of gap junction proteins underwent extensive parallel evolution, and many preserved features might be ancestral to metazoans. Yet, the biodiversity of ctenophores in the context of gap junctions and other functional modules remains unexplored, and we anticipate an array of discoveries in this frontier of comparative biology.

Ctenophore innexins might be the most divergent gap junction protein subfamily compared to other metazoan innexins subfamilies with elaborated functional diversification and plasticity, which are predictably comparable, if not exceed, those observed in *C. elegans* [59]. When the number of PTMs is considered, the functional potential for enormous diversity within innexins of ctenophores is almost unfathomable.

Although innexins are broadly expressed across most structures, each organ and tissue has a unique combination of innexins, suggesting their involvement in complex integrative functions and behaviors of ctenophores.

Acknowledgments

This work was supported in part by the Human Frontiers Science Program (RGP0060/2017) and National Science Foundation (IOS-1557923) grants to LLM. Research reported in this publication was also supported in part by the National Institute of Neurological Disorders and Stroke of the National Institutes of Health under Award Number R01NS114491 (to LLM). The content is solely the authors' responsibility and does not necessarily represent the official views of the National Institutes of Health.

References

1. Panchin YV (2005) Evolution of gap junction proteins--the pannexin alternative. J Exp Biol 208(Pt 8):1415–1419
2. Phelan P (2005) Innexins: members of an evolutionarily conserved family of gap-junction proteins. Biochim Biophys Acta 1711(2): 225–245
3. Guiza J et al (2018) Innexins: expression, regulation, and functions. Front Physiol 9:1414
4. Abascal F, Zardoya R (2013) Evolutionary analyses of gap junction protein families. Biochim Biophys Acta 1828(1):4–14
5. Connors BW, Long MA (2004) Electrical synapses in the mammalian brain. Annu Rev Neurosci 27:393–418
6. Skerrett IM, Williams JB (2017) A structural and functional comparison of gap junction channels composed of connexins and innexins. Dev Neurobiol 77(5):522–547
7. Moroz LL et al (2014) The ctenophore genome and the evolutionary origins of neural systems. Nature 510(7503):109–114
8. Ryan JF et al (2013) The genome of the ctenophore *Mnemiopsis leidyi* and its implications for cell type evolution. Science 342(6164): 1242592
9. Li Y et al (2021) Rooting the animal tree of life. Mol Biol Evol 38(10):4322–4333
10. Whelan NV et al (2015) Error, signal, and the placement of Ctenophora sister to all other animals. Proc Natl Acad Sci U S A 112(18): 5773–5778
11. Whelan NV et al (2017) Ctenophore relationships and their placement as the sister group to all other animals. Nat Ecol Evol 1(11): 1737–1746
12. Schultz DT et al (2023) Ancient gene linkages support ctenophores as sister to other animals. Nature 618(7963):110–117
13. Hernandez-Nicaise ML, Amsellem J (1980) Ultrastructure of the giant smooth muscle fiber of the ctenophore *Beroe ovata*. J Ultrastruct Res 72(2):151–168
14. Hernandez-Nicaise ML, Nicaise G, Malaval L (1984) Giant smooth muscle fibers of the ctenophore *Mnemiopsis leidyi*: ultrastructural study of in situ and isolated cells. Biol Bull 167:210–228
15. Hernandez-Nicaise ML, Nicaise G, Reese TS (1989) Intercellular junctions in ctenophore integument. In: Anderson PAV (ed) Evolution of the first nervous systems. Plenum Press, New York, pp 21–32
16. Satterlie RA, Case JF (1978) Gap junctions suggest epithelial conduction within the comb plates of the ctenophore *Pleurobrachia bachei*. Cell Tissue Res 193(1):87–91
17. Horridge GA (1974) Recent studies on the Ctenophora. In: Muscatine L, Lenhoff HM (eds) Coelenterate biology. Academic Press, New York, pp 439–468
18. Tamm SL (1982) Ctenophora. In: Electrical conduction and behavior in "simple" invertebrates. Clarendon Press, Oxford, pp 266–358
19. Moroz LL, Kohn AB (2016) Independent origins of neurons and synapses: insights from ctenophores. Philos Trans R Soc Lond Ser B Biol Sci 371(1685):20150041
20. Moroz LL, Romanova DY (2022) Alternative neural systems: what is a neuron? (ctenophores, sponges and placozoans). Front Cell Dev Biol 10:1071961
21. Welzel G, Schuster S (2022) Connexins evolved after early chordates lost innexin diversity. elife 11:e74422
22. Yen MR, Saier MH Jr (2007) Gap junctional proteins of animals: the innexin/pannexin superfamily. Prog Biophys Mol Biol 94(1–2): 5–14
23. Moroz LL, Romanova DY, Kohn AB (2021) Neural versus alternative integrative systems: molecular insights into origins of

neurotransmitters. Philos Trans R Soc Lond Ser B Biol Sci 376(1821):20190762

24. Slivko-Koltchik GA, Kuznetsov VP, Panchin YV (2019) Are there gap junctions without connexins or pannexins? BMC Evol Biol 19(1):46
25. Pogoda K et al (2016) Regulation of gap junction channels and hemichannels by phosphorylation and redox changes: a revision. BMC Cell Biol 17(Suppl 1):11
26. Oshima A (2017) Structure of an innexin gap junction channel and cryo-EM sample preparation. Microscopy (Oxf) 66(6):371–379
27. Mikalsen SO, Kongsstovu SÍ, Tausen M (2021) Connexins during 500 million years-from cyclostomes to mammals. Int J Mol Sci 22(4):1584
28. Dahl G et al (1991) Cell/cell channel formation involves disulfide exchange. Eur J Biochem 197(1):141–144
29. Dahl G, Muller KJ (2014) Innexin and pannexin channels and their signaling. FEBS Lett 588(8):1396–1402
30. Wang J et al (2014) The membrane protein Pannexin1 forms two open-channel conformations depending on the mode of activation. Sci Signal 7(335):ra69
31. Chiu YH, Ravichandran KS, Bayliss DA (2014) Intrinsic properties and regulation of Pannexin 1 channel. Channels (Austin) 8(2):103–109
32. Dahl G (2015) ATP release through pannexon channels. Philos Trans R Soc Lond Ser B Biol Sci 370(1672):20140191
33. Sangaletti R, Dahl G, Bianchi L (2014) Mechanosensitive unpaired innexin channels in *C. elegans* touch neurons. Am J Physiol Cell Physiol 307(10):C966–C977
34. Sadamoto H et al (2021) Identification and classification of innexin gene transcripts in the central nervous system of the terrestrial slug *Limax valentianus*. PLoS One 16(4): e0244902
35. Hernandez-Nicaise M-L (1991) Ctenophora. In: Harrison FWFW, Westfall JA (eds) Microscopic anatomy of invertebrates: Placozoa, Porifera, Cnidaria, and Ctenophora. Wiley, New York, pp 359–418
36. Moroz LL (2018) NeuroSystematics and periodic system of neurons: model vs reference species at single-cell resolution. ACS Chem Neurosci 9(8):1884–1903
37. Larkin MA et al (2007) Clustal W and Clustal X version 2.0. Bioinformatics 23(21): 2947–2948
38. Edgar RC (2004) MUSCLE: a multiple sequence alignment method with reduced time and space complexity. BMC Bioinform 5: 113
39. Mistry J et al (2021) Pfam: the protein families database in 2021. Nucleic Acids Res 49(D1): D412–D419
40. Gene Ontology C (2021) The Gene Ontology resource: enriching a GOld mine. Nucleic Acids Res 49(D1):D325–D334
41. Kanehisa M et al (2012) KEGG for integration and interpretation of large-scale molecular data sets. Nucleic Acids Res 40(Database issue): D109–D114
42. Tamura K et al (2013) MEGA6: molecular evolutionary genetics analysis version 6.0. Mol Biol Evol 30(12):2725–2729
43. Blom N et al (2004) Prediction of post-translational glycosylation and phosphorylation of proteins from the amino acid sequence. Proteomics 4(6):1633–1649
44. Sigrist CJ et al (2013) New and continuing developments at PROSITE. Nucleic Acids Res 41(Database issue):D344–D347
45. Letunic I, Khedkar S, Bork P (2021) SMART: recent updates, new developments and status in 2020. Nucleic Acids Res 49(D1):D458–D460
46. Xue Y et al (2010) GPS-SNO: computational prediction of protein S-nitrosylation sites with a modified GPS algorithm. PLoS One 5(6): e11290
47. Omasits U et al (2014) Protter: interactive protein feature visualization and integration with experimental proteomic data. Bioinformatics 30(6):884–886
48. Takaku Y et al (2014) Innexin gap junctions in nerve cells coordinate spontaneous contractile behavior in *Hydra* polyps. Sci Rep 4:3573
49. Chapman JA et al (2010) The dynamic genome of *Hydra*. Nature 464(7288):592–596
50. Moroz LL, Mukherjee K, Romanova DY (2023) Nitric oxide signaling in ctenophores. Front Neurosci 17:1125433
51. D'Hondt C et al (2013) Regulation of connexin- and pannexin-based channels by post-translational modifications. Biol Cell 105(9): 373–398
52. Ramazi S, Zahiri J (2021) Post-translational modifications in proteins: resources, tools and prediction methods. Database (Oxford) 2021: baab012
53. Tanaka K et al (2007) Shared and species-specific features among ichnovirus genomes. Virology 363(1):26–35
54. Popova LB et al (2012) Gap Junctions in *Nematostella vectensis* Sea Anemone embryos. Biol Bull Rev 2(5):368–389

55. Baumgarten S et al (2015) The genome of *Aiptasia*, a sea anemone model for coral symbiosis. Proc Natl Acad Sci U S A 112:11893
56. Shinzato C et al (2011) Using the *Acropora digitifera* genome to understand coral responses to environmental change. Nature 476(7360):320–323
57. Voolstra CR et al (2017) Comparative analysis of the genomes of *Stylophora pistillata* and *Acropora digitifera* provides evidence for extensive differences between species of corals. Sci Rep 7(1):17583
58. Tamm SL (2014) Cilia and the life of ctenophores. Invertebr Biol 133(1):1–46
59. Bhattacharya A et al (2019) Plasticity of the electrical connectome of *C. elegans*. Cell 176(5):1174–1189 e16

Chapter 17

Analysis and Visualization of Single-Cell Sequencing Data with Scanpy and MetaCell: A Tutorial

Yanjun Li, Chaoyue Sun, Daria Y. Romanova, Dapeng O. Wu, Ruogu Fang, and Leonid L. Moroz

Abstract

The emergence and development of single-cell RNA sequencing (scRNA-seq) techniques enable researchers to perform large-scale analysis of the transcriptomic profiling at cell-specific resolution. Unsupervised clustering of scRNA-seq data is central for most studies, which is essential to identify novel cell types and their gene expression logics. Although an increasing number of algorithms and tools are available for scRNA-seq analysis, a practical guide for users to navigate the landscape remains underrepresented. This chapter presents an overview of the scRNA-seq data analysis pipeline, quality control, batch effect correction, data standardization, cell clustering and visualization, cluster correlation analysis, and marker gene identification. Taking the two broadly used analysis packages, i.e., Scanpy and MetaCell, as examples, we provide a hands-on guideline and comparison regarding the best practices for the above essential analysis steps and data visualization. Additionally, we compare both packages and algorithms using a scRNA-seq dataset of the ctenophore *Mnemiopsis leidyi*, which is representative of one of the earliest animal lineages, critical to understanding the origin and evolution of animal novelties. This pipeline can also be helpful for analyses of other taxa, especially prebilaterian animals, where these tools are under development (e.g., placozoan and Porifera).

Key words *Ctenophora*, Single-cell transcriptomics, Cell types, *Mnemiopsis*, Placozoa, *Trichoplax*, Porifera, Cell-type evolution, Single-cell clustering, Batch effect correction, Dimension reduction, Marker gene selection

1 Introduction

The genome of any organism operates within the context of a given cell, specific cell populations, and their microenvironments. As a result, it is imperative to integrate the complex molecular machinery of an individual cell with systemic and highly dynamic architectures of an entire organism, including its development, stereotyped, and learned behaviors. Thousands of papers and teams addressed

Yanjun Li and Chaoyue Sun contributed equally with all other contributors.

Leonid L. Moroz (ed.), *Ctenophores: Methods and Protocols*, Methods in Molecular Biology, vol. 2757,
https://doi.org/10.1007/978-1-0716-3642-8_17, © Springer Science+Business Media, LLC, part of Springer Nature 2024

these technological and conceptual challenges over decades. This endeavor eventually led to single-cell "omic" approaches, where unbiased capture of multiple molecules and molecular events should be achieved and quantified (metabolomics, lipidomics, proteomics, genomics, and transcriptomics).

The desired single-molecule quantification, classification, visualization, and integration have not yet been achieved for any single cell. On one side of the spectrum is the ongoing development of microanalytical methods, and on the other are computational methodologies.

Admittedly, each strategy initially focuses on a particular model organism or human, which is perfectly justified by itself. However, integrative approaches should use comparative data from an emerging diversity of reference species in each animal lineage (from all 33 metazoan phyla and about 100 classes) to decipher the molecular logic of living cells and to uncover principles of their communications. In this case, evolutionary selections of multiple phenotypes could be viewed as unique experiments performed for us by Mother Nature over 3.8 billion years.

Indeed, biodiversity should 'marry' single-cell multi-omics in virtually all fields of modern biomedicine [1, 2]. But the computational and information theory challenges are even more dramatic here. Apart from the quest for novel approaches incorporating machine learning and artificial intelligence in biomedicine, there is a need to implement even established computational methods into a so-called "minor" phyla framework. These still enigmatic organisms are crucial for our understanding of life as it is.

Among ~2.8 million formally described eukaryotic species, *Ctenophora* or comb jellies [3–5] have the imperative significance in deciphering the origins of virtually all animal innovations, including all organs and tissues and neuromuscular systems in particular. Ctenophores represent the earliest-branching animal lineage [6–12], which independently developed neurons, synapses, muscles, mesoderm, throughput gut, and sensory organs [8, 13–16]. However, the nature of ctenophore innovations is mainly unknown, and cell-specific approaches are highly desirable.

This chapter provides tutorial and practical guidelines for single-cell transcriptomic analysis (scRNA-seq) of non-model organisms using the sea walnut *Mnemiopsis leidyi* (Fig. 2) as an illustrated example of these approaches. *Mnemiopsis* is a powerful experimental paradigm in comparative and developmental biology [17, 18], and its genome has been recently sequenced [9].

Sebe-Pedros and colleagues provided the initial survey of cell types in *Mnemiopsis* [19] using the so-called MetaCell platform [20]. From an adult animal, 6144 cells have been analyzed, resulting in more than two dozen cell clusters, including presumed digestive, muscle, and ciliated cell types. The initial annotation was based on a subset of genes with known and, perhaps, evolutionary conserved molecular functions to predict cell identities. However, most of these cell clusters have not been annotated, which is not surprising, considering the limited information about molecular cell diversity in ctenophores. Nevertheless, the reported *Mnemiopsis* scRNA-seq dataset is an important and valuable resource for future comparative studies and reanalyses. For example, the recent mapping of *Mnemiopsis* neuropeptides [21] allowed using these scRNA-seq data in search of neuron-specific genes and factors associated with the neuronal identity in the evolutionary context [16, 22, 23].

The interest in single-cell data analysis has exploded over the last decade, resulting in over 1000 analytics software tools that have been proposed and continuously improved [24]. Several projects are specifically designed to collect and track the rapid progression of single-cell technologies, datasets, resources, and analysis software packages.

According to their statistics, R and Python are the two most popular programming languages used for implementing these software tools [25, 26], and Python is preferred in the recent trend. Some of the software packages, such as Scanpy [27], CytoScape [28], Seurat [29], Scater [30], and CellChat [31], provide general platforms for the full pipeline of single-cell analysis with the integration of rich toolboxes for various tasks; by contrast, the others are designed for some specific tasks. Although, compared to the former, the latter packages do not offer many alternative methods for different tasks along the pipeline, many of them can still perform the end-to-end analysis. In addition, several comprehensive technique reviews and tutorials are available, such as [32–34], but few of them provide a practical guide to navigating the scRNA-seq analysis landscape with the hands-on code implementations and associated result illustration.

To eliminate the gap, this chapter outlines the computational workflow of a typical single-cell transcriptomic analysis and describes in detail the primary stages using a practical case study for the available *Mnemiopsis leidyi* dataset. Here, we provide a hands-on guideline and comparison regarding the best practices for the essential analysis steps by comparing the two broadly used analysis packages, i.e., Scanpy [27] and MetaCell [20], in both theoretical and empirical manners, as complementary examples of scRNA-seq pipelines (Fig. 1) in Ctenophores and kin for future studies (Fig. 2).

Scanpy tool kit was first proposed by Wolf et al. in 2018 [27], and then it successfully became a community-driven project developed further and maintained by a broader developer community. It has become an extensive toolbox for single-cell analysis in the Python ecosystem, including methods for preprocessing, clustering, visualization, marker-genes identification, pseudotime and trajectory inference, and simulation of dynamic gene expression data.

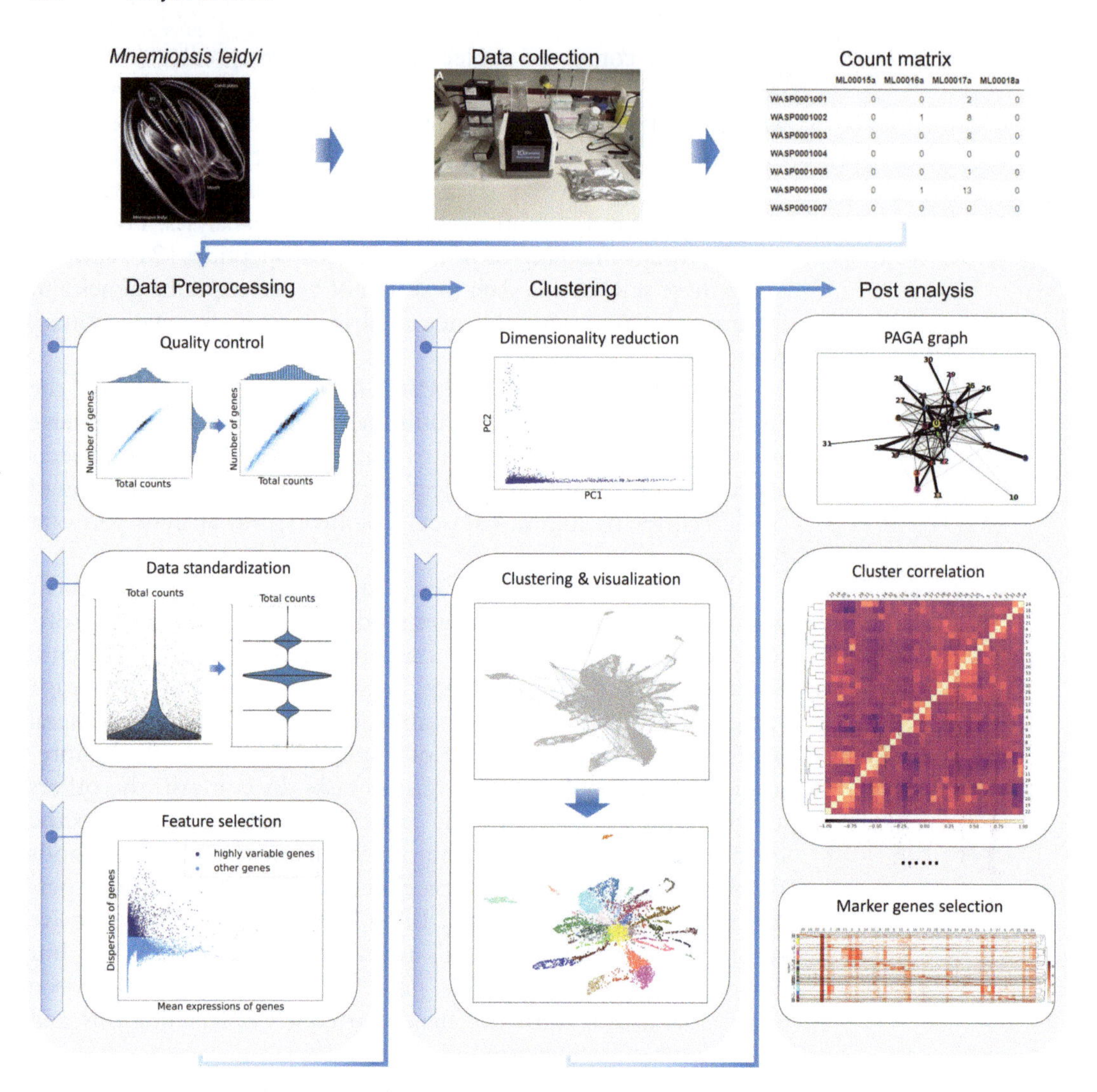

Fig. 1 Schematic of a typical scRNA-seq analysis pipeline. The raw single-cell data are acquired and processed by the single-cell sequencing technologies and constructed as count matrices, the starting point of the computational analysis pipeline. The first stage is data preprocessing, where the unreliable cells or genes are filtered out by quality control and biological or technical confounders are corrected by data standardization. Then, a feature selection step is required to select the most informatic genes. The second stage is cell clustering, where the dimensionality reduction is applied to project the cells' features to a low-dimensional space for the following cell clustering. The last stage is summarized as post-analysis, including various analyses of the generated cluster results as needed. For example, marker gene identification can annotate clusters with biological meanings, and the PAGA graph and correlation analysis can model trajectory and cluster similarity, respectively. Following our best practices, the subplots are generated using the analysis package Scanpy on the public Mnemiopsis leidyi dataset

Scanpy is also implemented in a scalable, memory-efficient, and modular form, suitable for the increasing need for large-scale single-cell data analysis. Scanpy package has already been broadly

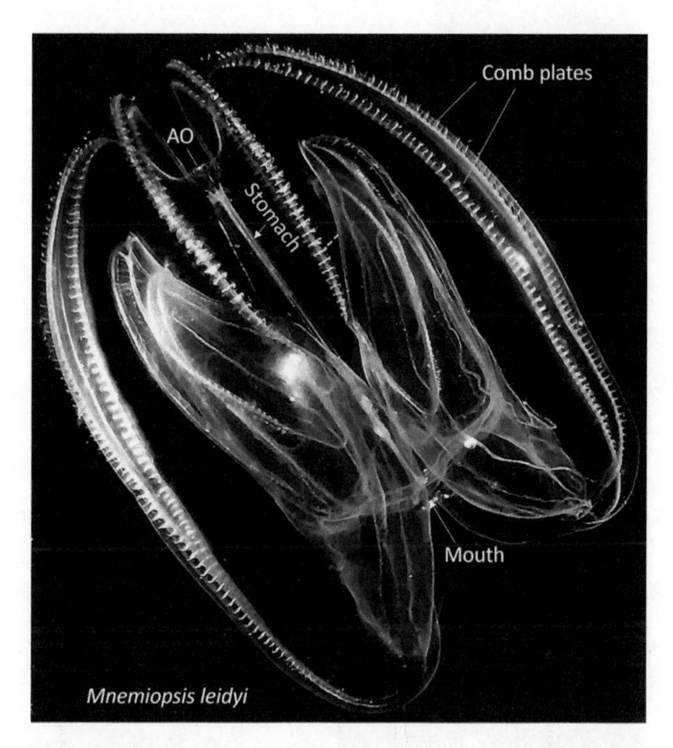

Fig. 2 *Mnemiopsis leidyi*—the experimental animal (*see* Ref. [92] 2 for details of ctenophore morphology)

utilized in many single-cell transcriptomic analyses and related research works, such as [35–37].

MetaCell is another valuable single-cell data analysis package proposed by Baran et al. in 2019 [20]. The authors defined the homogeneous groups of cells as metacells (MCs), and proposed the associated methods to compute the MCs as a graph partition using the similarities between single-cell profiles. The derived MCs could be utilized for various downstream analyses of cell types (or subtypes), transcriptional gradients, cell-cycle variation, gene and gene regulatory models, and more. Similarly, the MetaCell package has also been successfully applied in many research projects, e.g., [19, 38, 39].

This tutorial focuses on the major tasks of single-cell data analysis and utilizes the above two packages to provide a step-by-step guideline and comparison. The structure of this chapter follows the capture of single cells using 10× Genomics (as an example, *see* Figs. 3 and 4), the scRNA-seq analysis pipeline (Fig. 1). Subheading 2 introduces the data preprocessing stage for raw single-cell data, an essential stage before downstream analysis and visualization. It includes data loading, quality control, data standardization, batch effect correction, and feature selection. We also discuss the possible impacts to the scRNA analysis results caused by the different quality control and feature selection settings. Subheading 3 dissects the

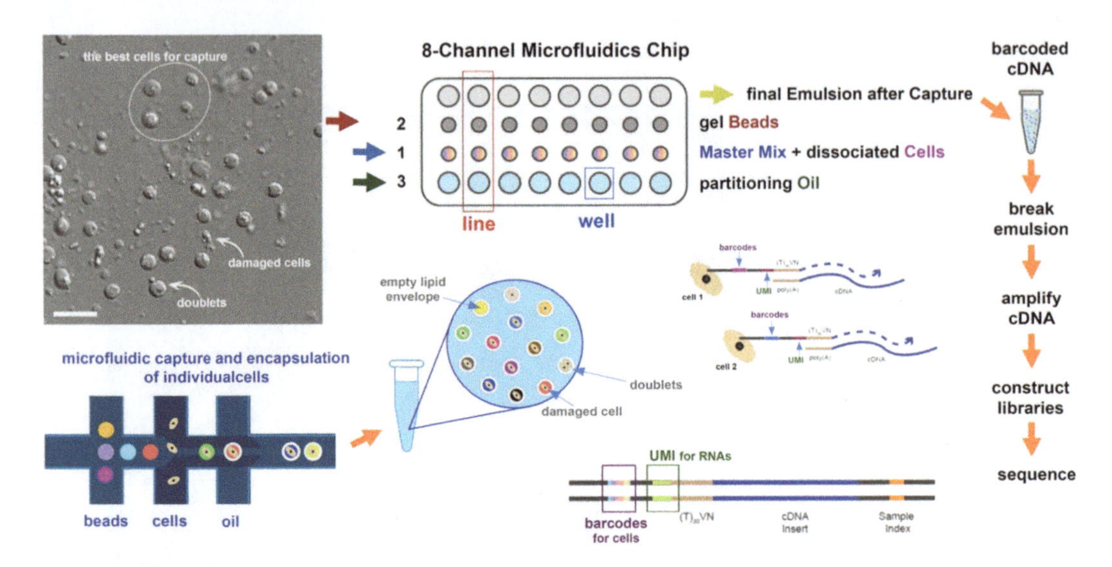

Fig. 3 Capturing of single cells using microfluidic technology of 10× Genomics

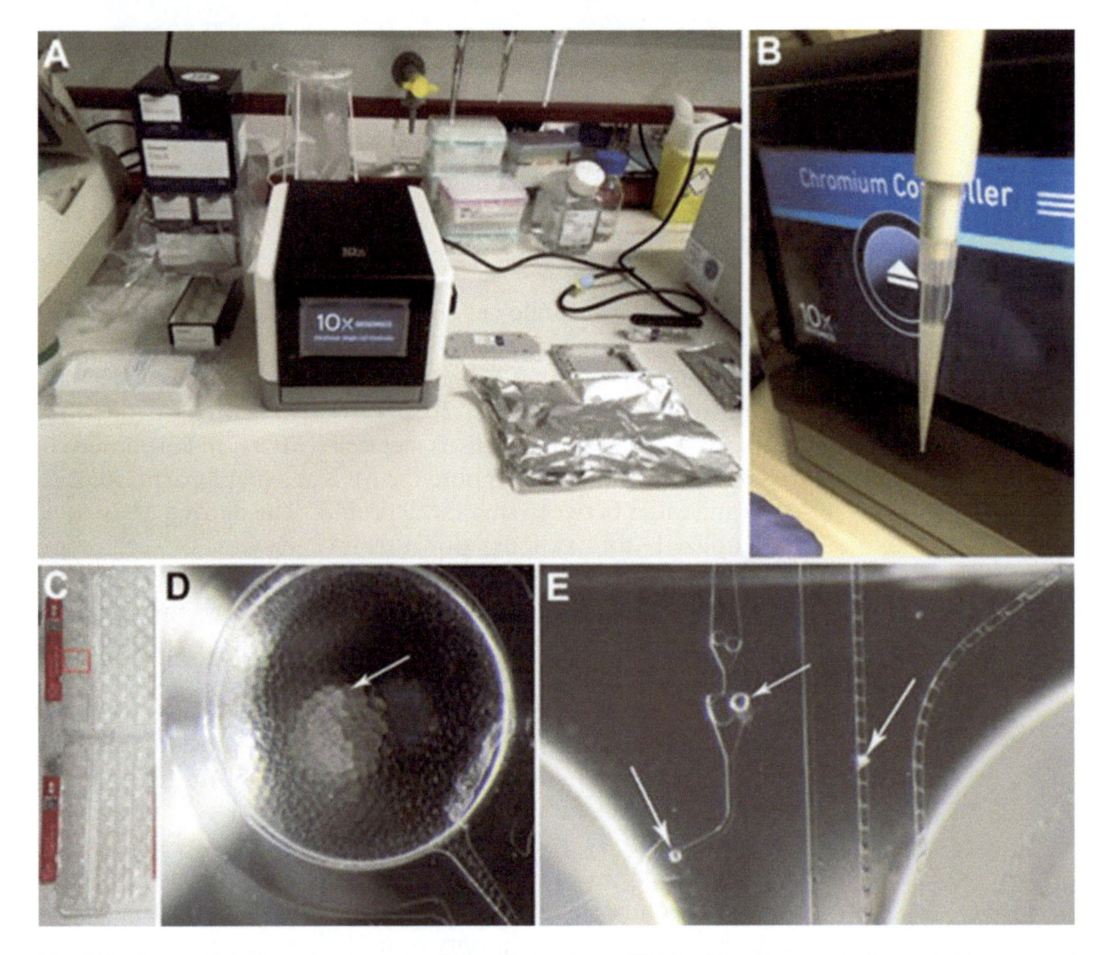

Fig. 4 Capturing of single cells using microfluidic technology. (**A**) 10× Chromium controller and chip prepared for capture *individual Mnemiopsis* cells. (**B**) After capturing, we present the final opaque emulsion with cells. (**C**) Chips for single-cell capturing. (**D**) A well with an emulsion of cells where the cattle-like structures are single-cell GEMs. (**E**) Inner organization of chip (white arrows—cells)

methods for the unsupervised cells clustering, including dimensionality reduction, clustering, and cluster visualization. Subheading 4 describes several common post-analyses for the clustering results, including cluster similarity analysis and maker gene selection. Finally, this and accompanying papers highlighted known and additional molecular markers and insight dealing with the genomic bases of cell types identified across species and phyla.

1.1 Glossary

scRNA-seq single-cell transcriptomics based on the microfluidic or related methods.

Doublets the situation when two small cells fall into one lipid droplet.

Barcodes for marking individual cells and/or transcripts. The 10× barcoded gel beads consist of a pool of barcodes that are used to separately index each cell's transcriptome. The individual gel barcodes are delivered to each cell via flow cytometry, where each cell is fed single-file along a liquid tube and tagged with a 10× gel bead.

UMI unique molecular identifier(s) mark RNA(s). Each transcript captured in the Single Cell 3' and V(D)J assay is labeled with a 10–12 bp unique molecular identifier (UMI) in addition to a 16 bp cell barcode. After sequencing, the UMI is used to distinguish sequenced reads that originate from unique mRNA molecules vs. PCR duplicates.

ERCC external RNA control consortium. ERCC standards provide a standard turn-key solution for evaluating the technical performance of any gene expression experiment.

CPM counters per million. A basic data normalization algorithm where the raw count of each gene is first divided by the total counts of all genes within its cell sample and then multiplied by 1×10^6.

***k*-NN** ***k*-nearest neighbors.** The k-NN algorithm is a non-parametric algorithm, where the label of a new point is determined by a predefined number of training samples closest in distance to it and *k* is an integer value specified by the user [40].

MNN mutual nearest neighbors. If a pair of cells is contained in each other's set of nearest neighbors, those cells are considered to be mutual nearest neighbors.

HVG highly variable genes for which a large fraction of the total expression variability is due to cell-to-cell heterogeneity.

PCA	principal component analysis. An orthogonal linear transformation projects the data onto a lower-dimensional linear space, such that the variance of the projected data is maximized [41].
t-SNE	t-distributed stochastic neighbor embedding, a non-linear dimensionality reduction technique, is commonly used for visualizing high-dimensional data in a low-dimensional space of two or three dimensions [42].
UMAP	uniform manifold approximation and projection, another nonlinear dimensionality reduction technique, similar to t-SNE and broadly used for visualization [43].
KL divergence	Kullback–Leibler divergence. A measure of statistical distance between two probability distributions.

1.2 Required Packages and Database

To run the scRNA-seq analysis pipeline presented in this tutorial locally, users need first to ensure their computational environment meets the basic system requirement and install the required packages. The version of the Scanpy utilized in this study is 1.7.2, and it was installed in a Python virtual environment with version 3.8. Scanpy does not provide any official hardware recommendation, and it could be successfully run on our desktop with a Windows 10 operating system (OS), 4 Intel Core i7-6700 CPUs, and 32G RAM.

The initial MetaCell R package with version 0.3.41 was released associated with its paper in 2019 [20], and since then, the code bases have been kept updated without adding a public new version tag. We downloaded the master branch from its GitHub repository[1] in Feb. 2022 and installed it in the R 4.1.0 environment. According to its official introduction, the MetaCell R package is tested on Linux and Mac OS but is not compatible with Windows OS. At least 16G RAM is required for a typical experiment, but for large-scale data, e.g., containing 100 K cells, the package developers suggested a dual CPU multicore workstation with 128GM RAM or more. In this tutorial, all MetaCell experiments were performed on a workstation with Ubuntu 16.04.7 LTS system (a Linux distribution system of version 16.04.7 LTS), 12 Intel Core i7-5930K CPUs, 64G RAM. The following list includes all the required packages and the database utilized in this tutorial.

[1] https://github.com/tanaylab/MetaCell/

- For sequencing data of an assembly:
  - Cell Ranger Makefastq v2.1.1 (10× Genomics).
- For analyzing RNA-seq datasets with Scanpy:
  - Miniconda3 (available at https://docs.conda.io/en/latest/miniconda) or other Python virtual environment manager tools.
  - Python 3.8 (a new virtual environment with the specific Python version can be created using the installed conda; *see* https://docs.conda.io/projects/conda/en/latest/user-guide/getting-started.html#managing-python for details).
  - Jupyter Notebook (available at https://jupyter.org, version 6.3.0).
  - Scanpy (available at https://scanpy.readthedocs.io, version 1.7.2).
  - anndata (available at https://anndata.readthedocs.io, version 0.7.6).
  - NumPy (available at https://numpy.org, version1.19.5).
  - pandas (available at https://pandas.pydata.org, version 1.2.4).
  - SciPy (available at https://scipy.org, version 1.6.2).
  - sklearn (available at https://scikit-learn.org, version 0.24.1).
  - Matplotlib (available at https://matplotlib.org, version 3.3.4).
  - Seaborn (available at https://seaborn.pydata.org, version 0.11.1).
  - Louvain (available at https://github.com/vtraag/louvain-igraph, version 0.7.0).
  - leidenalg (available at https://github.com/vtraag/leidenalg, version 0.8.4).
- For analyzing RNA-seq datasets with MetaCell:
  - R 4.1.0 (available at https://www.r-project.org).
  - RStudio (available at https://www.rstudio.com/products/rstudio, version 1.1.419).
  - metacell (available at https://github.com/tanaylab/metacell, version 0.3.7, accessed in Feb. 2022. Other required packages will be automatically installed along with the metacell installation.).

2 Data Loading and Preprocessing

2.1 Data Loading

The example dataset, *Mnemiopsis leidyi*, includes 6144 cells and 21,623 genes/transcripts. To use the Scanpy or MetaCell for single-cell analysis, users must first import it and other necessary packages to the Python or R working environment, respectively.

2.1.1 Import Required Packages

To use Scanpy or MetaCell, users can import required packages in the Jupyter Notebook or RStudio with the following snippets in Boxes 1 or 2.

Box 1
Import the required packages in the Jupyter Notebook for Scanpy.

```
import pandas as pd
import os
import numpy as np
import anndata as ad
import scanpy as sc
import seaborn as sns
import matplotlib.pyplot as plt
```

Box 2
Import the package in RStudio for MetaCell.

```
library("metacell")
```

2.1.2 Data Loading

In Scanpy, the output *adata* of the loading function is an **anndata.AnnData** object,[2] which consists of three components: one is a unique molecular identifier (UMI) data matrix with shape *#cells* × *# genes* (*see* Fig. 5 for an illustration.); the others are two **pandas.DataFrame** objects,[3] which are, respectively, used to store the statistic information for genes (e.g., selected highly variable genes) and cells (e.g., total UMI counts in cells) during the future data analysis process. In our example, the count matrix is stored in multiple files, so we first read each data file as **pandas.DataFrame** object, where the index corresponds to the observations (or cell barcodes) and the column corresponds to variables (gene names),

[2] https://anndata.readthedocs.io/en/stable/anndata.AnnData.html

[3] https://pandas.pydata.org/docs/reference/api/pandas.DataFrame.html

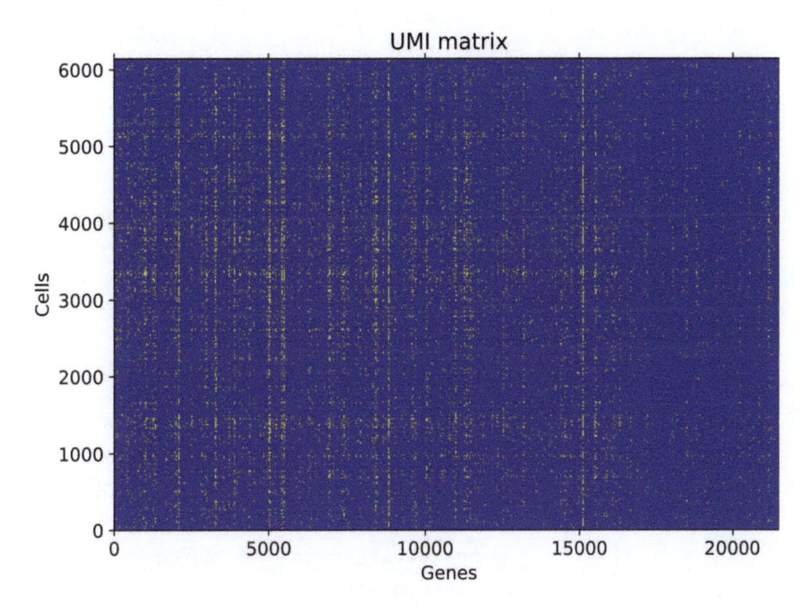

Fig. 5 Visualization of the imported UMI matrix. The *x*-axis represents gene IDs, and the *y*-axis represents cells. Brighter and warmer colors correspond to the larger expression counts, and the dark blue color corresponds to the zero value

and then combine them. Finally, we can build the **anndata. AnnData** object. Code snippets in Boxes 3, 4, 5, and 6 describe how to import data in Scanpy when the count matrices are stored in ".txt" files.

Box 3

Data loading with Scanpy. In our example, "data_dir" variable is the directory with batches of count matrices stored in ".txt" format.

```
for p,d,f in os.walk(f'{data_dir}'): # read each txt file individually and combine
them to one dataframe object
 df = pd.read_csv(f'{data_dir}/{f[0]}', sep='\t')
for file in f[1:]:
      df = pd.concat([df, pd.read_csv(f'{data_dir}/{file}', sep='\t')], axis=1)
df = df.T # transpose the dataframe
obs = pd.DataFrame(index=df.index) # use index of df for observations
var = pd.DataFrame(index=df.columns) # use columns of df for variable names
adata = ad.AnnData(df, obs=obs, var=var) # create AnnData object
```

Box 4

If the data is stored in "h5ad" format, we could use alternative simple version of code.

```
adata = sc.read_h5ad(f'/{data_dir}/file_name.h5ad',
# create the directory with the '.h5ad' file
```

Box 5

After importing the data, users must ensure the variable's uniqueness and observation names. The following operations will add suffix'-1','-2' for all duplicated items.

```
adata.var_names_make_unique()
adata.obs_names_make_unique()
```

Box 6

Visualize of the imported count matrix. The result is shown in Fig. 5.

```
plt.spy(adata.X, aspect=2.4, origin='lower', cmap='plasma') # plot the count matrix
plt.xlabel('Genes')
plt.ylabel('Cells')
plt.title('UMI matrix')
```

MetaCell provides a loading function that allows users to read multiple count matrix files (i.e., one file contains one count matrix) and automatically combines them. An **R List** object[4] (served as a hidden variable) named *.scdb* is initialized with the combined count matrix, and all the future generated results will be inserted into this list, and users can access the results based on their predefined names. Different from Scanpy, MetaCell assumes the input count matrix with shape *#genes* × *# cells*, which is the transpose of Scanpy's input.

Box 7 shows how to use MetaCell to import multiple count matrices. The loading function requires an additional user-defined batch table file, which stores the file names of the count matrices.

[4] https://cran.r-project.org/doc/manuals/r-release/R-intro.html#Lists

The count matrices files and the corresponding batch table file are provided in the supplementary materials for easy reproducibility.

Box 7
Data loading with MetaCell.

```
mcell_import_multi_mars("mne", # variable name in R List
 dataset_table_fn="Mnemiopsis_Batch.txt",# the batch table
 base_dir=data_dir, # the directory with batches of count
 matrices stored in '.txt' file
 force = T)
mat = scdb_mat("mne") # matrix object of count matrix
print(dim(mat@mat)) # check dimension of count matrix
scfigs_init("./figs/") # set base directory for figures
```

To verify the correctness of data loading, users can check whether the loaded count matrix has the shape 6144 × 21,623. Figure 5 provides a visualization of the imported dataset by Scanpy. It is worth noting that both Scanpy and MetaCell have several other reading functions designed for different data formats, e.g., 10× formatted hdf5 files, and users should choose the proper one based on the specific data format.

2.2 Quality Control

Because of the current technical variations and noises of single-cell gene expression patterns, data preprocessing for raw scRNA-seq is essential to the downstream analysis. The first step of data preprocessing is quality control, which filters out the low-quality or potentially unreliable transcriptomic profiles, e.g., dying cells, damaged cells, and doublets. For example, researchers commonly remove data from a cell if less than a certain number of transcripts have nonzero counts or remove a transcript if it is detected in less than a certain number of cells.

Scanpy provides a function *scanpy.pp.calculate_qc_metrics* to calculate multiple quality control metrics from the raw expression matrix, including the total UMI counts in cells, the mean gene expression counts, and the percentage of counts for a certain number of highest expressed genes in cells. The optimal thresholds for different metrics depend on specific species, tissues, or other technical factors. Typically, in practice, they are manually determined by inspecting the plots of the quality control metrics. Subheading 2.6

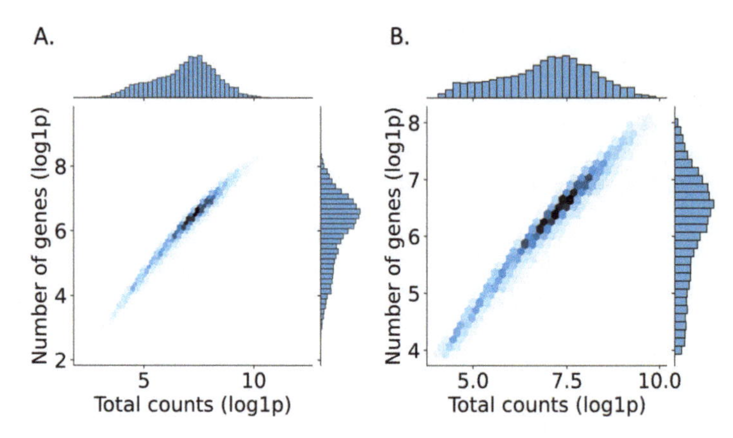

Fig. 6 Cell distribution of the example dataset before (**a**) and after quality control (**b**) using Scanpy. The *x* and *y* axes, respectively, represent the total counts of expressed genes and the number of unique genes with expression levels greater than 0 in each cell. Unreliable cells with smaller or larger total counts and expressed genes than the user-defined cutoffs were removed in the quality control step

investigates how different quality control thresholds impact the marker gene identification.

In general, we encourage users to first apply a relative "slack" filtering strategy with loose thresholds at the early stage of analysis and then gradually "tighten" the thresholds but keep monitoring the result changes to avoid losing the essential biological information. Figure 6 illustrates the difference in the sample distribution between the raw and filtered data, where we can find that all the cells and their expressed genes "located" on the long tail range over the distribution are removed, resulting in more reliable transcriptomic profiles. Scanpy also supports associated visualization functions, and Fig. 7 provides a violin plot for illustration. The corresponding codes are available in Boxes 8, 9, 10, and 11.

Box 8
Remove the ERCC genes in the example dataset in Scanpy.

```
cols = [col for col in adata.var.index if 'ERCC' not in col]
# select genes with ERCC prefix
adata = adata[:, cols]
```

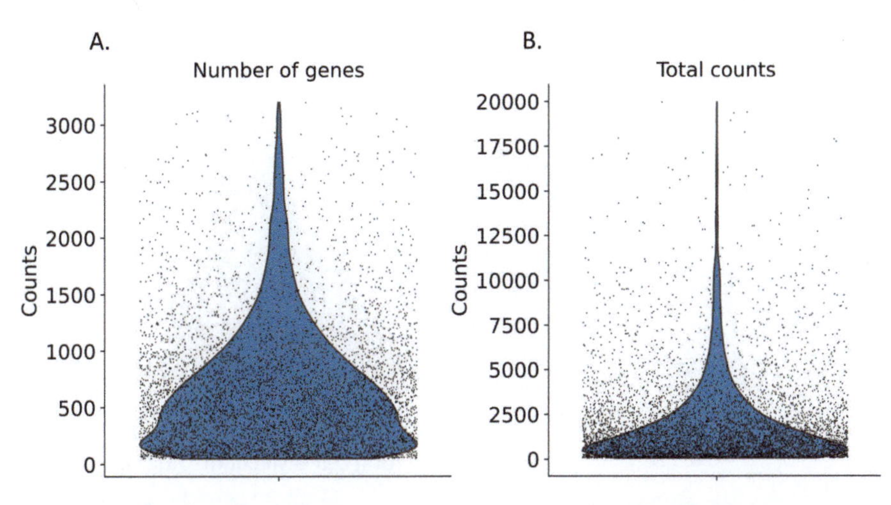

Fig. 7 Violin plot for cell distribution after quality control by Scanpy. Each dot denotes a cell, and two subplots correspond to two quality control metrics, i.e., the number of expressed unique genes (**a**) and total counts (**b**). Outlier cells, whose counts and expressed genes are smaller or larger than the user-defined cutoffs, have been removed in the quality control stage

Box 9

Scanpy quality control method 1: filtering outlier cells and genes with numbers or counts.

```
sc.pp.filter_cells(adata, min_genes=50) # filter out the cells with less than
50 expressed unique genes
sc.pp.filter_cells(adata, max_genes=3200) # filter out the cells with more than
3200 expressed unique genes
sc.pp.filter_genes(adata, min_cells=3) # filter out genes which are expressed in
less than 3 cells
```

Box 10

Scanpy quality control method 2: filter cells and genes based on quality control metrics.

```
sc.pp.calculate_qc_metrics(adata, inplace=True) # calculate quality control
metrics
adata = adata[adata.obs.n_genes_by_counts<=3200, :] # filter out the cells with
more than 3200 unique genes
adata = adata[adata.obs.total_counts<=20000, :] # filter out the cells with more
than 20000 counts of genes
```

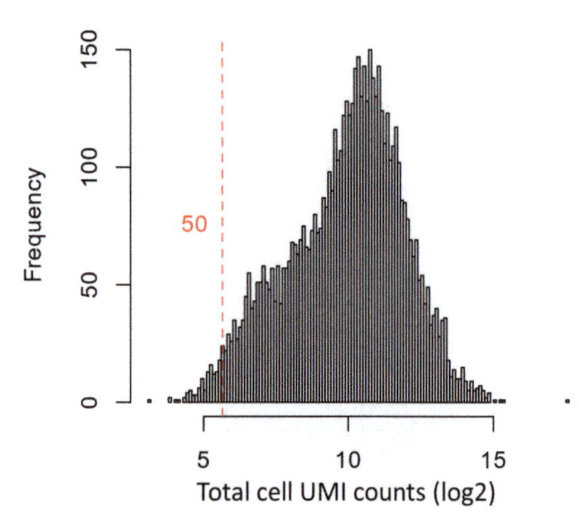

Fig. 8 Histogram of total UMI among cells generated by MetaCell. The red dash line represents a user-defined lower bound in quality control, and any cells below the bound will be removed

Box 11
Display the sample distribution after quality control in Scanpy.

```
sns.jointplot(data=adata.obs,
 x='log1p_total_counts', # logarithmic of total counts in cells
 y='log1p_n_genes_by_counts', # logarithmic of genes in cells
 kind='hex')
plt.xlabel('Total counts (log1p)')
plt.ylabel('Number of genes (log1p)')
sc.pl.violin(adata, ['n_genes_by_counts', 'total_counts'],
 jitter=0.4, multi_panel=True)
```

By contrast, MetaCell only supports the visualization of a histogram of total UMI among cells, shown in Fig. 8. Examples of filtering with basic quality control metrics are given in Boxes 12 and 13, providing hints for readers' future usage. After quality control, the example dataset now includes 5857 cells and 18,812 genes for Scanpy and 21,531 genes for MetaCell.

Box 12
Remove the ERCC genes in the example dataset in MetaCell.

```
erccs <- rownames(mat@mat)[grepl("ERCC",rownames(mat@mat))] # select genes with
ERCC prefix
mcell_mat_ignore_genes(new_mat_id="mne_noercc", # new variable name in R List
for filtered count matrix
 mat_id="mne",
 ig_genes=erccs) # remove genes with ERCC
mat <- scdb_mat("mne_noercc")
```

Box 13
Quality control in MetaCell.

```
mcell_plot_umis_per_cell("mne", min_umis_cutoff=50) # plot distribution of
cells with total UMI counts
gene_sizes <- colSums(as.matrix(mat@mat) != 0) # calculate number of genes in
each cell
cell_sizes <- colSums(as.matrix(mat@mat)) # calculate total counts in each cell
large_cells <- names(which(cell_sizes>20000)) # list of cells with more than
20000 counts
large_cells_gene <- names(which(gene_sizes>3200)) # list of cells with more than
3200 expressed genes
small_cells_gene <- names(which(gene_sizes<50)) # list of cells with less than
50 expressed genes
mcell_mat_ignore_cells("mne_filt", # new name in R List
 "mne",
 ig_cells=c(small_cells_gene, large_cells_gene,
 large_cells)) # remove outliers cells
```

Specifically for our example dataset, the authors use external RNA control consortium (ERCC) standards in dataset generation, so the corresponding genes with prefix ERCC need to be first removed before quality control. The code snippets in Boxes 8 and 12 provide solutions on how to remove them in Scanpy and MetaCell, respectively.

Comparing Fig. 6a with Fig. 8, we can find the ranges of total cell UMI counts are different, and this is because when calculating the UMI, Scanpy uses $\ln(x + 1)$ as the scales for both axes, while MetaCell uses $\log_2 x$.

2.3 Data Standardization

The amount of captured RNA per cell can vary significantly due to some biological (e.g., cell cycle state, differentiation, stress, etc.) or technical (e.g., capture, damage, sequencing depth, etc.) factors, which pose additional challenges to the comparison of the gene expressions between cells based on their count data [34].

The data standardization (or simply referred to as data normalization in some cases) technique is proposed to address this issue, e.g., by scaling count data of cell or transcript level with a size factor to obtain the correct relative expression profiles. The data standardization in Scanpy contains three steps: normalization, logarithmic transformation, and optional scaling.

The normalization method supported by Scanpy is the most straightforward algorithm named counters per millions (CPM), where the count of each gene is first divided by the total counts of all genes within its cell sample and then multiplied by 1×10^6, so that every cell has the same total count after normalization. However, a scRNA-seq dataset commonly consists of heterogeneous cell populations with different sizes and RNA molecule counts, which does not satisfy the original assumption of the above global scaling normalization method that all cells in the dataset contain the same number of RNA molecules. Therefore, more complex normalization methods, such as scran [44], SCnorm [45], etc., are usually favorable in practice, demonstrating more robust and fine-grained normalization capacities in many cases. However, unfortunately, they are not integrated into the current Scanpy toolbox, and readers who wish to understand the detailed comparison of complex normalization methods can refer to empirical surveys elsewhere [46, 47]. A logarithmic transformation usually follows the normalized expression matrices $\ln(x + 1)$, aiming to represent the expression difference in a canonical log-fold manner and reduce data skewness. Box 14 provides commands for the standardization process described above. Scanpy also provides a z-score scaling operation, which can scale the count of each gene individually to zero mean and unit variance. However, the magnitude of gene expression is biologically informative and does not necessarily be scaled equally. There is no consensus on whether to perform the scaling step over genes [32, 48]. An example of a scaling operation is described in Box 15.

Box 14
Data standardization with Scanpy.

```
sc.pp.normalize_total(adata, target_sum=1e4) # normalize
the data matrix to 10,000 reads per cell
sc.pp.log1p(adata) # logarithmic transformation
```

Box 15
Feature selection with Scanpy.

```
sc.pp.highly_variable_genes(adata,
 min_mean=0.0125,
 max_mean=3,
 min_disp=0.5) # select a set of HVGs with several cutoffs
sc.pl.highly_variable_genes(adata) # visualization of the
highly differential expressed genes
adata.raw = adata # keep a record of original data
adata = adata[:, adata.var.highly_variable] # only keep
the highly differential expressed genes
sc.pp.scale(adata, max_value=10) # scale data to zero
mean and unit variance
```

In MetaCell, data normalization is only performed on cells with high expression profiles, for which a random downsampling method without replacement is applied so that the total counts of each cell cannot exceed a certain threshold. The threshold value considers the data distribution of expression profiles and prior empirical knowledge. Additionally, MetaCell also supports the logarithmic transformation $\log(x + \epsilon)$ with an adjustable parameter ϵ. Different from Scanpy, MetaCell embeds data standardization in the graph construction step, which will be described in Subheading 3.

2.4 Batch Effect Correction

The batch effect occurs when the scRNA-seq data are collected from different sources, e.g., using distinct experimental procedures, devices, environments, or even the same protocols, yet implemented by different laboratories or at different time points for various samples.

Before integrating multiple batches, the correction step for batch effect is essential to adjust the nonbiological differences in expression level. Many efforts have been made to control such confounding effects. Accordingly, Scanpy supports several corresponding methods, such as ComBat [49], batch balanced K-nearest neighbors (BBKNN) [50], Harmony [51], matching mutual nearest neighbors (MNN) [52], and Scanorama [53].

ComBat [49] is one of the most pioneered works for batch effect correction, which fits a linear model, and its coefficient parameters are estimated by empirical Bayes. However, ComBat requires an assumption of prior normal distributions, and batch effect is estimated with all cells in batches, which may introduce extra confounders. To overcome the limitation, MNN reduces the batch effect by identifying possible mutual nearest neighbors among cells from different batches. Scanorama and BBKNN further optimize the idea of MNN with faster and less costly approaches, providing solutions for large-scale datasets. Harmony utilizes an iterative strategy, which clusters similar cells from batches while maximizing the diversity of batches. Readers are referred to a comprehensive review paper [54] for a detailed discussion and comparison of popular batch effect correction algorithms. In MetaCell, no explicit batch effect correction operation is applied in the preprocessing stage, but its graph construction strategy somewhat has a similar effect as MNN to normalize batch effects [20].

2.5 Feature Selection

A typical single-cell dataset may contain more than ten thousand genes or transcripts; for example, *Mnemiopsis leidyi* dataset has 21,531 genes. Although the quality control step can filter out some genes expressed in few cells, the remaining genes are still too many to be analyzed efficiently. Besides, most genes are expressed similarly between cell populations, playing less informative roles in cell-type clustering. Thus, using the feature selection step to filter out those genes and only keep the most "informative" ones is critical for efficient and accurate downstream analysis. Accordingly, Scanpy and MetaCell support several filtering metrics for the feature selection purpose, which can be summarized as two principles: (1) filtering out low average expressed genes and (2) keeping highly variable genes (HVGs) or the top differentially expressed genes among cells.

To select the HVGs, Scanpy provides two different types of methods. One is based on statistical dispersion [29, 55] measured by the variance-to-mean metric. Specifically, genes are grouped into discrete bins based on their average expression levels, and the dispersion measure (variance divided by mean) of the genes in each bin is z-normalized. Then, the HVGs are selected from each bin and compared to genes with similar average expression. The other type of method is based on the normalized variance [56], where a second-order polynomial predictor $f_\theta(\overline{u})$ is first used to fit the variance of each gene given the gene's mean expression value $\overline{u}$, and then, the count of gene i in cell j is standardized as:

$$z_{ij} = \frac{u_{ij} - \overline{u_i}}{\sqrt{f_\theta(\overline{u_i})}}$$

where u_{ij} is the raw counts of gene i in cell j, $\overline{u_i}$ is the mean expression value of gene i among cells, and θ is learnable parameters. The variance of each gene is calculated among cells with standardized counts, and genes with top variances are selected as the HVGs.

The feature selection step in Box 15 uses the minimum of dispersion and minimum and maximum of mean expression level as cutoff for filtration. After selection, the example dataset keeps the same number of cells (5857 cells), but the variables have been reduced to 5214. Figure 9 provides a visualization of selected HVGs, with large dispersion and significant mean expressions. In Subheading 2.6, we empirically evaluate the possible impacts of the different parameter settings on HVGs selection and provide some general suggestions.

In MetaCell, the HVGs are selected by jointly considering multiple metrics. Unliked the method in Scanpy, MetaCell defines different normalized variance as:

$$v'_i = \log_2\left(\frac{v_i}{\overline{u_i}}\right)$$

where $\overline{u}_i$ is the mean expression of gene i among cells and v_i is the corresponding variance. A fixed-size sliding window is then applied to the mean expression values of overall genes to select the similar

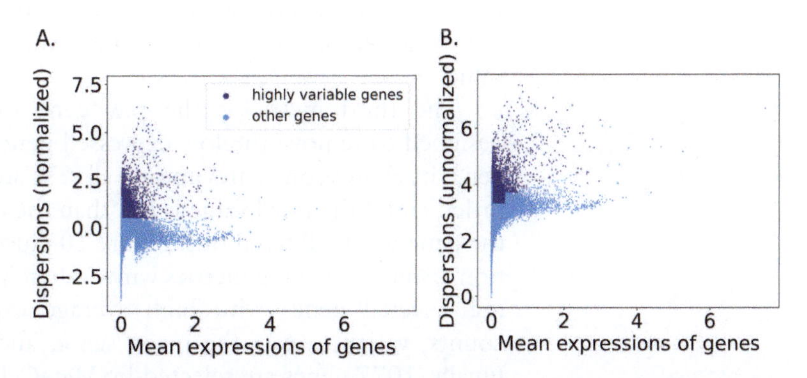

Fig. 9 Dispersion distributions of the selected highly variable genes over the mean expression level. Subplots (**a**) and (**b**) depict the normalized and unnormalized dispersion distributions calculated with Scanpy. The *x*-axis represents the mean expression values of genes among cells, and the *y*-axis represents their dispersions. The dark blue points are the selected genes and will be used in the following clustering steps

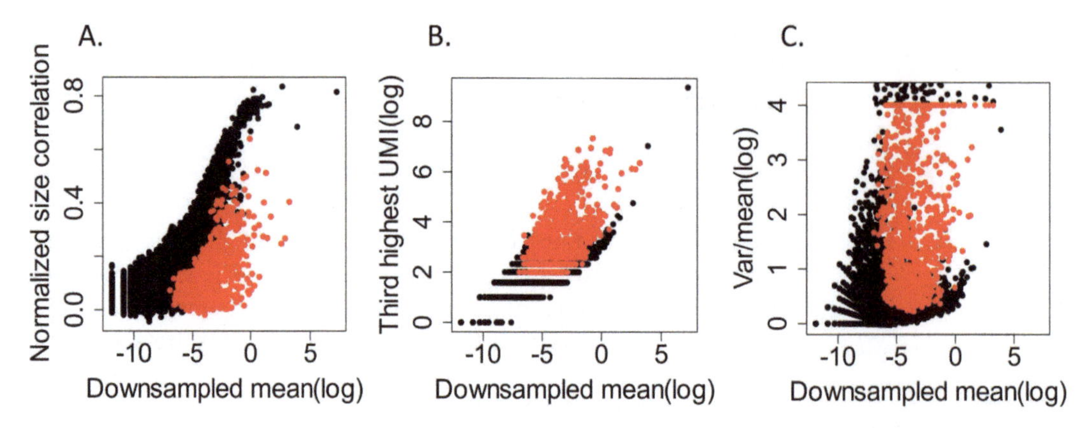

Fig. 10 Illustration of feature selection results using normalized size correlation (**a**), third highest UMI count (**b**), and normalized variance (**c**) in MetaCell. The *x*-axes of all three subplots represent the mean expression level of genes among cells, and the *y*-axes correspond to the different feature selection metrics. The red points represent the selected genes

expressed ones. A calibration term for every gene i, named variance empirical trend $v'_i(e)$, is set as the median value of previously calculated normalized variance for genes falling in the window centered on gene i. Then, the normalized variance is calibrated by $v'_i - v'_i(e)$, and a user-defined threshold is used to select for HVGs with large calibrated normalized variance.

The second metric is the size correlation measured by the Pearson correlation $r_i = r(\boldsymbol{u}_i, \boldsymbol{u}_{\text{total}})$ between expression values of gene i $\boldsymbol{u}_i = \{u_{i1}, \ldots, u_{iN}\}$ and a cell depth $\boldsymbol{u}^{\text{total}} = \{u_1^{\text{total}}, \ldots, u_N^{\text{total}}\}$, where u_j^{total} is the total count of the raw gene expressions in cell j. The calculated Pearson correlation value for each gene also needs to be calibrated, where a similar sliding window mechanism is applied over the sorted correlation values of genes and the median value is set a calibration term $r_i(e)$. Similarly, a user-defined threshold is used to select for HVGs with the small $r_i - r_i(e)$ Pearson correlation value.

The third metric is the raw gene expression level and is designed to remove the low expressed genes. Specifically, for each gene, its expression values among all cells are ranked in descending order, and if the third value is less than the user-defined threshold, the gene will be filtered out. Figure 10 illustrates the feature selection results over three metrics with code snippets in Box 16, implying selected genes with high average expressed counts, UMI counts, variance over the mean score, and low size correlation. Finally, 1077 genes are selected by MetaCell.

Box 16

Feature selection with MetaCell. Note that in MetaCell, the data standardization operation will be performed in the cell clustering stage, as described in Subheading 3.

```
mcell_add_gene_stat(gstat_id="mne", mat_id="mne_filt", force=T) # calculate
the feature selection metrics
mcell_gset_filter_multi(
 gset_id = "mne_feats", # new varible name in R List for selected HVGs
 gstat_id="mne",
 T_tot=200,
 T_top3=2,
 T_szcor=-0.1,
 T_niche=0.08,
 force_new=T) # gene selection with multiple cutoffs
mcell_plot_gstats(gstat_id="mne", gset_id="mne_feats") # visualization of the HVGs
```

2.6 Discussion on Parameters Selection for Preprocessing

This section discusses the possible impact on the scRNA data analysis results caused by different parameters in quality control and feature selection steps. Considering the diverse options and flexibility offered by Scanpy, the empirical evaluation in this section was conducted using the Scanpy package.

For the quality control (QC) evaluation, we focused on the cell outlier filtering operation *scanpy.pp.filter_cells* and tested critical criteria for removing "unreliable" cells (doublets, damaged cells with expressed minimum numbers of genes). Specifically, the example dataset was processed by four cell filtering functions following the same protocol but using different lower-bound cutoff values for the number of expressed genes, i.e., the parameter *min_genes* were extensively tuned from 0, 50, 100 to 200 for QC-strategies 1 to 4. Figure 11 illustrates the code snippets and the four resulting filtered cells and genes' counts, where the shape of "**adata**" object represents the *#cells* × *# genes*. Although with the increase of *min_genes* cutoff values, the kept cells and genes decreased accordingly, we did not find that different cutoff values yielded significant differences in the number of genes for this step. However, it is noted that they may introduce notable impacts on the selected HVGs even applying the same feature selection operation. From Fig. 11, we can observe that the count of the final HVGs varies a lot from 5225 to 3680 over different QC strategies.

For the feature selection (FS) step, we focused on the statistical dispersion-based method and tested three different settings. The first one, named FS-strategy-1, skipped the feature selection step directly, which means to set the two thresholds, including

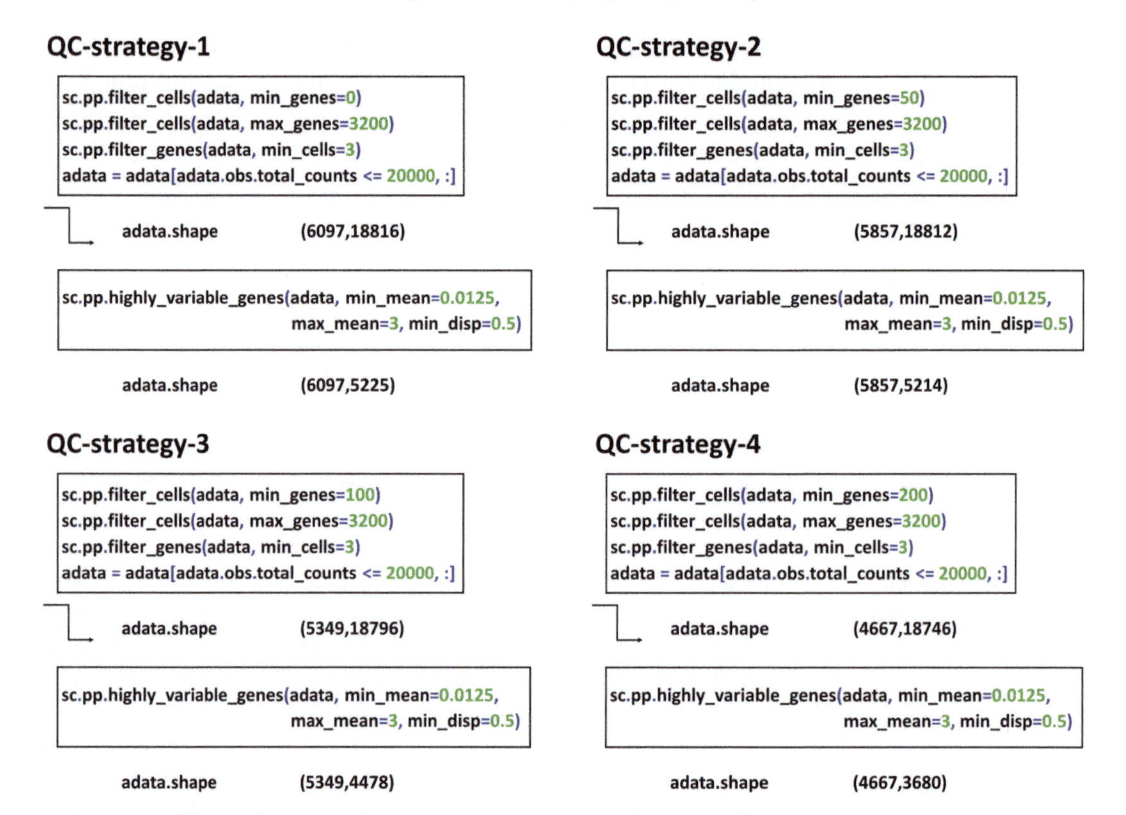

Fig. 11 Four quality control strategies with different cutoff values in Scanpy. The corresponding shape of data matrix is shown after each preprocessing step, and data standardization step is hidden here, because it does not affect the shape of the count matrix. The final selected HVGs vary a lot from 5225 to 3680, indicating the effect of cutoff values

min_disp, *min_mean* as minimum values, and *max_mean* threshold as a maximum value, so that all the genes were kept for the downstream clustering and marker gene selection. In the second (FS-strategy-2) and third (FS-strategy-3) settings, we tested the different lower-bound cutoff values *min_disp* for the normalized dispersions of genes, and the *min_disp* were, respectively, configured as 0.0 (the mean value of the normalized dispersions for all genes) and 0.5 (the default value used by Scanpy). The rest two parameters, i.e., *min_mean* and *max_mean*, the lower-bound and upper-bound cutoffs for the normalized gene expression levels, were fixed as 0.0125 and as 3.0. For a fair comparison, the inputs for all three FS-strategies were the same, 5857 cells and 18,812 variables, generated by the QC-strategy-2. From Fig. 12, we can observe that as the selection standards increase, the number of kept HVGs decreases significantly from 18,812 to 5214.

Data shape after QC (5857,18812)

FS-strategy-1

```
sc.pp.highly_variable_genes(adata, min_mean=-np.infty, max_mean=np.infty, min_disp=-np.infty)
```

adata.shape (5857,18812)

FS-strategy-2

```
sc.pp.highly_variable_genes(adata, min_mean=0.0125, max_mean=3, min_disp=0)
```

adata.shape (5857,8930)

FS-strategy-3

```
sc.pp.highly_variable_genes(adata, min_mean=0.0125, max_mean=3, min_disp=0.5)
```

adata.shape (5857,5214)

Fig. 12 Three feature selection strategies with different cutoff values in Scanpy. The quality control step is the same as QC-strategy-2, with data shape (5857,18,812). The final selected HVGs vary a lot from 18,812 to 5214, indicating the effect of cutoff values

To evaluate the selected HVGs, we applied the same clustering and marker gene selection methods configured with identical parameters on top of them and compared the final markers with the biologists' annotations. Figure 13 and Table 1 illustrate the intersection of markers between the Scanpy identified genes and the verified genes selected by the biologists. We can find that without removing any genes using the FS-strategy-1, Scanpy can identify as many as 14 biologists annotated marker genes. However, when increasing the *min_disp* criteria, the number of overlapping marker genes decreased, i.e., ten overlapping marker genes for FS-strategy-2 and seven for FS-strategy-3. The results suggest that when we excessively filter the data, we get a precedent when genes are not included in the analysis, but they are expressed. And we may erroneously conclude that these genes are not expressed. If we remove lower and upper values possibly, post-processing analysis does not show cluster-specific genes or important low-expressed genes as synaptic or some genes of receptors/channels. Low expressed genes are usually placed and seem like eliminations which intensify biological noise. But, in most cases, these genes form a molecular profile of cell-type expression.

In summary, to avoid omitting some essential biological information during the data preprocessing stage, we generally encourage users to start with a relatively "slack" filtering strategy with

FS-strategy-1: data shape (5857,18812)

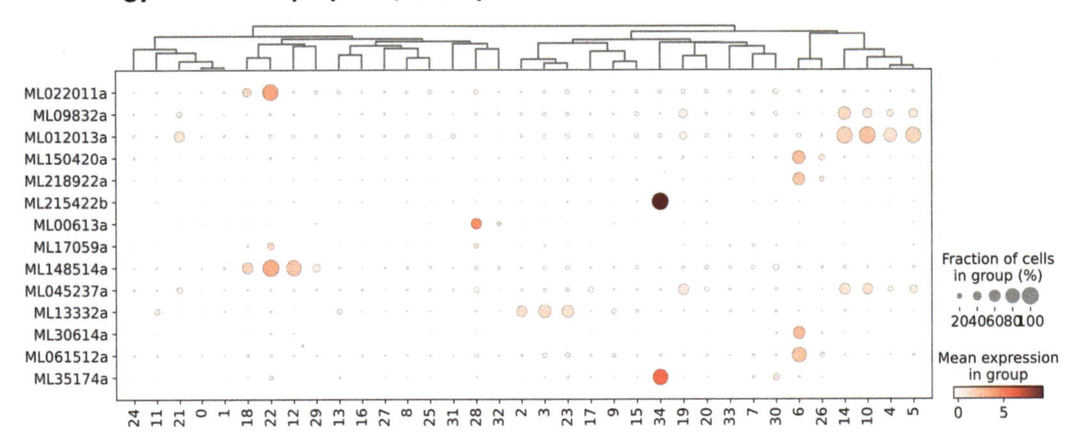

FS-strategy-2: data shape (5857,8930)

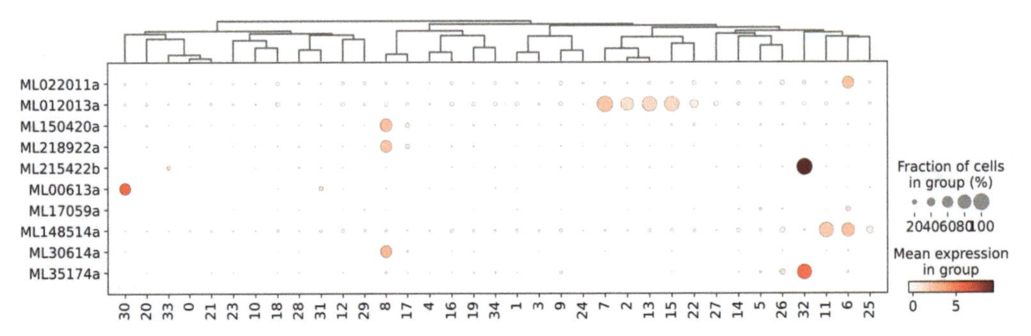

FS-strategy-3: data shape (5857,5214)

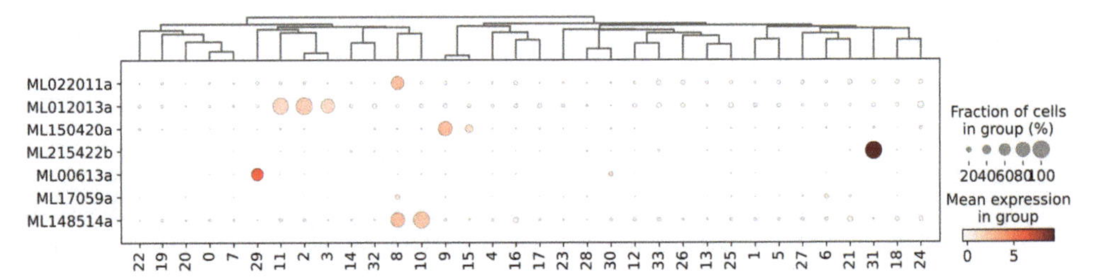

Fig. 13 Distribution of marker genes over Leiden clustering results. Each row represents a marker gene, which is verified by the domain experts and also identified by the Scanpy program. Three different feature selection strategies are applied to the example dataset. The size of the circle is proportional to the fraction of expressed genes in each cluster, and the color corresponds to the gene expression level, where a darker color means a higher expression

loose thresholds and then gradually "tighten" the selection criteria and closely monitor the results changes. Additionally, there is no one-size-fits-all standard for various scRNA datasets, but biological insights and domain knowledge always play critical roles in evaluating and selecting different strategies.

Table 1
Intersection of markers between the Scanpy identified genes with three feature selection strategies and selected of verified genes by the biologists

Gene ID	Annotation	FS-strategy-1	FS-strategy-2	FS-strategy-3
ML022011a	Myo II striated	+	+	+
ML09832a	Diaphanous	+	−	−
ML012013a	Peptidase C1	+	+	+
ML150420a	K_v ion channel	+	+	+
ML218922a	Innexin	+	+	−
ML215422b	Photoprotein	+	+	+
ML00613a	ShK toxin	+	+	+
ML17059a	Na_v ion channel	+	+	+
ML148514a	MLC	+	+	+
ML045237a	Cortactin	+	−	−
ML13332a	Immunoglobulin	+	−	−
ML30614a	Flagellar dynein	+	+	−
ML061512a	Amiloride-sensitive Na channel	+	−	−
ML35174a	K_v ion channel	+	+	−
ML022011a	Crystallin	+	+	+

Feature selection strategy 1 (FS-strategy-1) retains all annotated marker genes, while FS-strategy-2 and FS-strategy-3 maintain only ten and seven overlapping marker genes, respectively. This suggests a necessary trade-off when setting parameters in feature selection

Table 2 provides a comparison between Scanpy and MetaCell for the preprocessing stage. In short, Scanpy provides more choices for users to perform various quality control and batch effect correction steps, while MetaCell provides a more complex feature selection step by jointly considering several selection metrics.

3 Clustering

3.1 Dimensionality Reduction

Clustering scRNA-seq data is central for most analyses, which is essential to discovering novel cell types and their gene expression logics. The clustering usually starts with dimensionality reduction techniques to transform the raw expression data from a high-dimensional space to a low-dimensional space to reduce the computational cost for downstream analysis. Principal component analysis (PCA) [41] is widely used for the dimensionality reduction of single-cell data. An orthogonal linear transformation projects the data onto a lower-dimensional linear space, known as principal

Table 2
Comparison of preprocessing steps between Scanpy and Metacell

	Scanpy	MetaCell
Quality control	Provides user interfaces for several quality control metrics calculation [29] and visualization	Provides visualization of total UMI counts distribution, but the quality control metrics and the filtered-out genes and cells need to be calculated or given by users
Data standardization	Includes normalization, logarithmic transformation, and scaling	Includes normalization and logarithmic transformation
Feature selection	Genes can be selected by dispersion-based methods [28, 55] or normalized variance [56]	Genes are selected by jointly considering the normalized size correlation, third-highest UMI counts, normalized variance, total expression values, and niche statistics
Batch effect correction	Provides ComBat [49] for batch effect correction and supports external packages, e.g., BBKNN [50], Harmony [51], MNN [52], and Scanorama [53]	Does not include batch effect correction functions in preprocessing stage

Scanpy provides multiple choices for quality control metrics and batch effect correction approaches, which are convenient for users. MetaCell provides a more complex feature selection step, jointly considering several selection metrics

subspace, such that the variance of the projected data is maximized. In practice, PCA constructs the covariance matrix of the data and computes its eigenvectors. The eigenvector corresponding to the largest eigenvalue is used to define the first principal component that maximizes the variance of the projected data, and the second principal component defined by the eigenvector with the second largest eigenvalue is orthogonal to the first one and maximizes the variance in the subspace, and so on. Typically, a few principal components can reconstruct a large fraction of the variance of the original data with a small loss of information. For the single-cell data, thousands of selected genes after the data preprocessing step are usually projected to $10 \sim 50$ dimensions using the PCA method. Scanpy integrates the PCA algorithm, which is easy for users to call the function, whereas the MetaCell does not support any dimensionality reduction algorithm; thus, users are encouraged to leverage some other third-party packages to perform this operation.

3.2 Clustering Algorithms

Grouping cells based on the similarity of their gene expression profiles is critical to cell identification and other downstream tasks, such as cell-to-cell communication, gene regulatory networks, and so on. The similarity between two cells is typically measured by the distance metrics such as Euclidean distance in the reduced lower-dimensional space. Many different clustering algorithms have been proposed and applied to single-cell data analysis. Generally, they can be categorized into four different groups: (1) prototype-based clustering including *k*-means clustering [57], mixture of Gaussian model, and their variants [58, 59]; (2) hierarchical clustering, including agglomerative clustering with "bottom-up" manner [60, 61] and divisive clustering in the way of "top-down" [62]; (3) density-based clusterings, such as DBSCAN [63, 64]; and (4) community-detection-based clustering, including Leiden [65] and Louvain [66] algorithms. The reader is referred to [67–69] to get the overall picture of how clustering algorithms enable single-cell data analysis. This tutorial focuses on the community detection-based clustering methods, which have been broadly utilized by the single-cell community and well-supported by Scanpy and MetaCell packages.

Community detection-based clustering is a graph partitioning method, where clustering is performed on a graph representation of single-cell data. Typically, a *k*-nearest neighbor algorithm is utilized to build this initial graph, where each node corresponds to a cell, one node is allowed to connect to its *k* most similar cells, and the Euclidean distances between connected cells represent the edge weights. In Scanpy, a *k*-nearest neighbor (*k*-NN) graph is built using either the UMAP-based [43] or Gaussian kernels-based [70, 71] approaches. Louvain [66] is one of the most successful community detection-based clustering methods. Its objective is to maximize the modularity metric Q, which measures the density of links inside communities compared to links between communities.

$$Q = \frac{1}{2m} \sum_{ij} \left[A_{ij} - \frac{s_i s_j}{2m} \right] \delta(c_i, c_j), \delta(x, y) = \begin{cases} 1 & \text{if } x = y \\ 0 & \text{if } x \neq y \end{cases}$$

where A_{ij} denotes the edge weight between node i and j; s_i and s_j, respectively, represent the sum of edge weights linked to the node i and j; c_i and c_j represent their communities; and m is a normalization term denoting the sum of all edge weights in the graph. Before the optimization, each node is assigned to its own community. Louvain uses an iterative method containing two phases to maximize modularity greedily. In phase I, modularity is optimized by allowing only local changes to node-community memberships. Each node is moved into its neighboring communities, and the gain of the modularity is computed; then, the community with the most significant modularity increase is assigned as the node's new community. This operation is applied to all the nodes sequentially until no assignment can yield an increase in modularity. The identified communities are aggregated into super-nodes in phase II to build a new network. Specifically, super-nodes are connected if there is at least one edge between nodes of the corresponding communities. The weight of the edge between two super-nodes is the sum of weights from all edges between their corresponding communities. The second phase will terminate once the new network is created, which can be regarded as a hierarchical clustering result, and then the next round of phases I and II is rerun on the super-node network until the modularity no longer increases. The time complexity of the Louvain algorithm is $O(nlogn)$, where n is the number of nodes or cells in the graph. Due to the low time complexity and rapid convergence in practice, it can scale to sizeable single-cell datasets.

Leiden algorithm is a derived version of the Louvain algorithm proposed by Vincent Traag et al. [65]. They share the same objective, that is, to maximize modularity. The major difference is that the Leiden algorithm adds a new step to refine the proposed partition after the first phase to guarantee that the communities are well-connected when the optimization is applied iteratively. Overall, the Leiden algorithm is more robust and efficient than the Louvain algorithm. Scanpy provides fast implementations of both algorithms, and users can easily call the functions. It is worth noting that the most crucial parameter is the resolution, which directly and heavily affects the number of clusters in the final results. Usually, a relatively large resolution value tends to result in a fine-grained partitioning of cells, which can help users to understand the relationship among cells better. Then, based on the selected marker genes in the post-analysis step, users could merge several clusters with similar markers by gradually shrinking the resolution value and eventually obtain a calibrated clustering result. Although some computational methods have been proposed to help guide the number of neighbors k in k-NN graph construction and resolution determination, such as [57, 59, 72, 73], there is currently no consensus standard; thus, judgment with the researcher is still required. Some researchers empirically find that for datasets with 3000–5000 cells, a resolution value between 0.4 and 1.4 generally yields good clustering, and increased resolution is often required for larger datasets [74, 75]. Our dataset contains more than 6000 cells, and we manually set the resolution as 1.5 in our experiments. We encourage users to explore different parameter combinations and evaluate them based on the post-analysis results.

MetaCell proposes a new clustering algorithm, also a greedy community detection-based method. It first computes a cell-to-cell similarity matrix, based on which a balanced k-NN similarity graph is constructed by connecting pairs of cells representing reciprocally high-ranking neighbors. Unlike a possible highly nonsymmetric k-NN graph built directly from the similarity matrix, this balanced graph tends to have more balanced ingoing and outgoing degrees. Then, the graph is subsampled multiple times (e.g., n _ resamp times) to form n _ resamp sampled subgraph, and each time nodes in the original graph are sampled independently without replacement with a certain probability (p _ resamp), and the edges connecting among the sampled nodes are also added to the subgraph. A community detection algorithm partitions each subgraph;

initially, no cell in the subgraph has an assigned community. The graph partition process consists of two phases; in phase I, the algorithm iteratively samples the cells and their connected neighbors without any community and assigns them to a new community until every cell is in a community. Phase II is community adjustment, which optimizes the following objective function by iteratively reassigning cells to new communities until convergence:

$$w_{ij} = \frac{wi_{ij} wo_{ij}}{|M_j|^2}$$

where wi_{ij} and wo_{ij} denote the summation of edge weights linked from node i to community j and from community j to node i, respectively, and $|M_j|$ is a normalization term representing the size of community j. The objective design aims to assign node i to the community with the most reciprocal interactions, and the normalization term reduces the confound significant interactions introduced from a large community size. Once all the subgraph partitions are done, a co-occurrence matrix S^{boot} of each pair of cells is computed. The matrix element $s_{ij}^{\text{boot}} = c_{ij}/o_{ij}$, where c_{ij} represents the number of occurrences that the cell i and j are partitioned in the same community and o_{ij} is the number of occurrences that they are sampled into the same subgraph. Then, the matrix S^{boot} is used to construct a weighted and nondirected graph G^{boot}, and the final clustering results are generated by applying the above-described graph partitioning algorithm again on the resampled graph G^{boot}. MetaCell also provides a heuristic outlier cell filtering function based on clustering results to detect cells from experimental errors. Cells with one or more genes expressed far from other cells in the same cluster will be marked as outliers and filtered out. Overall, MetaCell with the bootstrapping steps can provide more robust clustering results but high computation time complexity, and the user-defined k value is crucial to the size distribution of the derived clusters, because it is repeatedly used for the clustering algorithm and thresholding graph edges. For the other parameters shown in the code section, *min_mc_size* determines the approximate minimum size for each cluster. n_resamp represents the iteration number of bootstrapping, which can significantly affect the algorithm's running time and encourage to be decreased for large datasets. α controls how harsh the edges in G^{boot} are filtered, where a lower α results in harsher filtering. In the paper, the authors suggested setting $n\_samp = 500$, $p\_samp = 0.75$, $\alpha = 2$, and the k value for G^{boot} was configured as 30. Similar to Scanpy, users are encouraged to try different parameter combinations and evaluate the clustering results with domain knowledge.

3.3 Visualization of Gene Expression in Low-Dimensional Space

Visualizing transcriptomic profiles in two- or three-dimensional space can provide an intuitionistic description of the dataset and be helpful in interpreting the clustering results. To obtain a precise visual representation, one needs to project the high-dimensional profiles into the low-dimensional space and preserve the cells' original underlying structural and distance information. The dimensionality reduction technique PCA introduced in Subheading 3.1 can also be utilized for this purpose. Still, due to the limitation of its linearity, in practice, nonlinear dimensionality reduction methods, such as t-distributed stochastic neighbor embedding (t-SNE) [42] and uniform manifold approximate and projection (UMAP) [43], are more widely adopted for the visualization of high-dimensional data. The basic idea of t-SNE is to use two probability distributions to represent neighborhood information in the high- and low-dimensional space and minimize the Kullback–Leibler (KL) divergence between the two distributions regarding the locations of points in the low-dimensional space. Intuitively, t-SNE aims to make the distances between projected points match the similarity matrix measured in the original high-dimensional space; the projected points with high similarity will attract each other and repulse the dissimilar points. The similarity matrices for the high- and low-dimensional data are measured by a Gaussian distribution and a Student's t-distribution. An important user-defined parameter perplexity describes the effective number of neighbors considered for each data point in the original space, and different perplexity values can capture different scales in the data.

Larger datasets usually require a larger perplexity, and values between 5 and 50 typically work well. Besides, because t-SNE has a non-convex objective function minimized by the gradient descent optimization, it is possible to yield different dimensionality reduction results in different runs. Hence, users are encouraged to make multiple runs with the same parameter setting and select the lowest value of the objective function as the final result. In practice, if the original dimension is very high, users are suggested first to run some other dimensionality reduction techniques, such as PCA, to reduce the dimension to a relatively small value, by which the informative dimensions can be effectively extracted from the potential noise, and then use t-SNE for the final reduction. This operation can also help to speed up the t-SNE computation for pairwise distance computation between cells.

UMAP proposed by McInnes et al. [43] is conceptually very similar to the t-SNE; it first builds a neighborhood-graph representation of data in the original space and then optimizes the layout of the graph in the low-dimensional space to be as similar as possible. Unlike the single perplexity value introduced in t-SNE, UMAP defines two critical parameters for tuning the results. One is the number of expected nearest neighbors, which can control the balance between local and global structure preservation; that is, a low value helps to preserve more local structure, and a high value pushes the algorithm to focus more on the global structure. The other parameter is the minimum distance, which specifies how tightly the algorithm packs the points into the low-dimensional space. Similar to t-SNE, UMAP is also a stochastic algorithm, and its results vary between runs. Thus, users are encouraged to run the algorithm multiple times and carefully adjust the hyperparameters. Overall, compared to t-SNE, UMAP often tends to preserve the global structure better, especially for more complex datasets, and it is more computationally efficient.

Both t-SNE and UMAP algorithms are integrated into the Scanpy toolbox so that users can directly call the corresponding functions, as shown in the following code section. The clustering results of the Leiden and Louvain algorithm with UMAP and t-SNE projection are shown in Figs. 14 and 15, respectively. Leiden

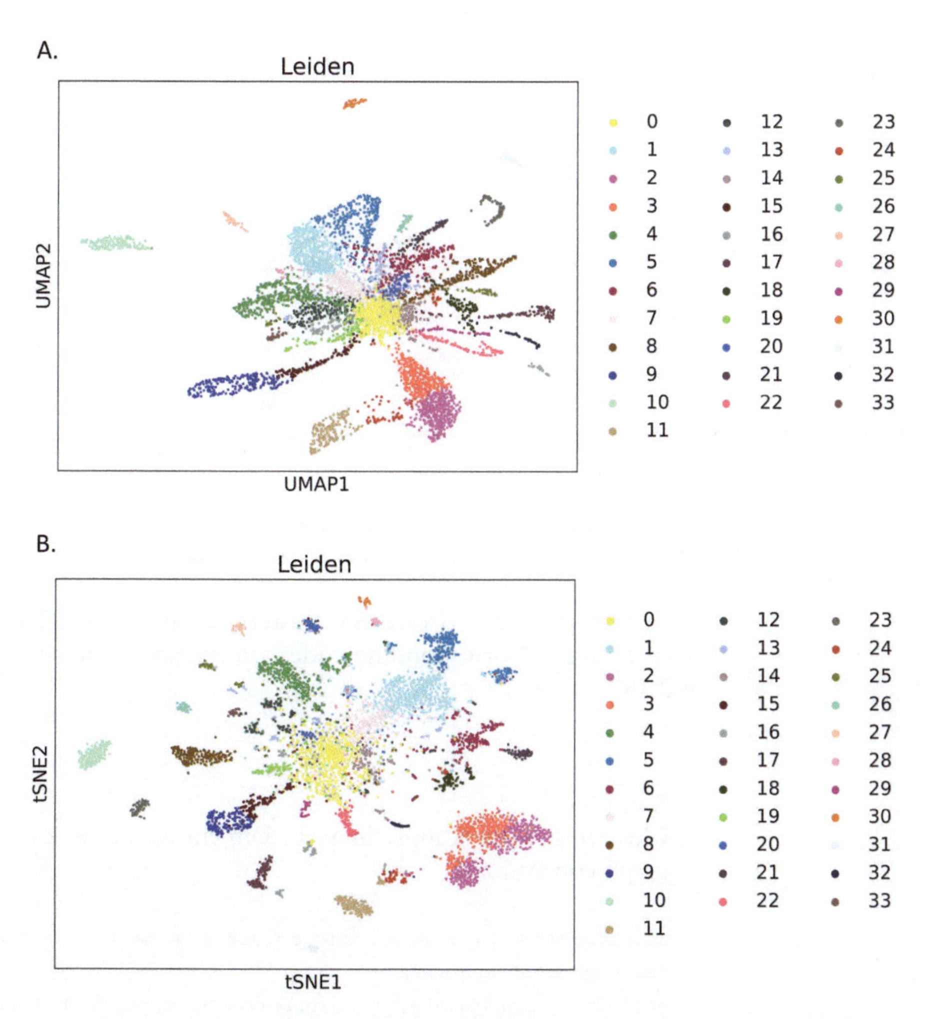

Fig. 14 Visualization of Leiden clustering results using UMAP (**a**) and t-SNE (**b**) 2-D projection. The result contains 34 clusters, and the right panel shows the cluster ids

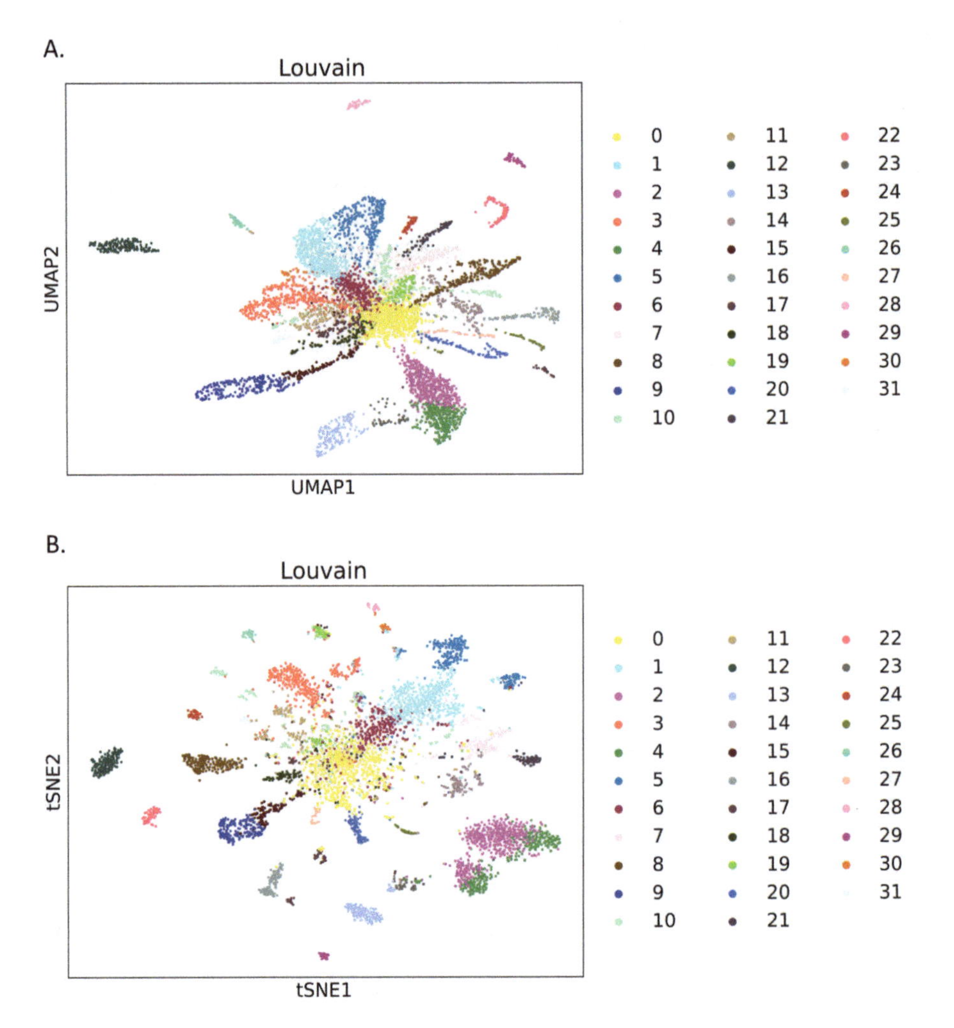

Fig. 15 Visualization of Louvain clustering results using UMAP (**a**) and t-SNE (**b**) 2-D projection. Thirty-two cell clusters are detected by the Louvain algorithm, and the right panel shows the cluster ids

algorithm generates overall 34 clusters compared to 32 clusters from Louvain. Corresponding codes are available in Boxes 17, 18, and 19.

Box 17
Clustering with Scanpy. Step 1: Dimension reduction and graph construction.

```
sc.tl.pca(adata, svd_solver='arpack') # perform princi-
pal component analysis
sc.pp.neighbors(adata, n_neighbors=10, n_pcs=25) # com-
puting the neighborhood graph
```

Box 18
Clustering with Scanpy. Step 2: Clustering with Leiden and Louvain.

```
sc.tl.leiden(adata, resolution = 1.5) # clustering with Leiden
 # resolution controls the number of clusters
sc.tl.louvain(adata, resolution = 1.5) # clustering with Louvain
 # resolution controls the number of clusters
```

Box 19
Clustering with Scanpy. Step 3: Clustering visualization.

```
sc.tl.umap(adata) # calculate UMAP 2-D projection
sc.tl.tsne(adata) # calculate TSNE 2-D projection

sc.pl.umap(adata,color='leiden') # plot UMAP 2-D projection for Leiden result
sc.pl.tsne(adata,color='leiden') # plot tSNE 2-D projection for Leiden result
sc.pl.umap(adata,color='louvain') # plot UMAP 2-D projection for Louvain result
sc.pl.tsne(adata,color='louvain') # plot tSNE 2-D projection for Louvain result
```

In MetaCell, the low-dimensional projection algorithm first generates a cluster similarity graph with the following calculation:

$$B = [b_{ml}] = \frac{k^2}{|M_m|\,|\,M_l\,|} \sum_{\{i \in M_m, j \in M_l\}} \lceil a_{ij}/C \rceil$$

where $|M_m|$ and $|\,M_l|$ represent the size of cluster m and l and a_{ij} is the edge weight from node i to node j on the balanced k-NN similarity graph. $C = \text{median}_i(|\,M_i|\,)$ is a scaling constant. b_{ml} represents the normalized interactions among cluster m and l. The adjacency matrix B is then symmetrized, and the top-D highest-scoring edges for each node are kept in the final similarity graph. Next, the force-directed graph drawing algorithm [76] is applied to illustrate the 2-D projection of the similarity graph. Figure 16 illustrates the MetaCell clustering results with 49 clusters and their corresponding balanced k-NN graphs. Each cell's position is determined by the average coordinates of some specific clusters, where the cell's neighbors on the balanced k-NN graph are assigned. The clustering pipeline of MetaCell is shown in Boxes 20, 21, 22, and 23.

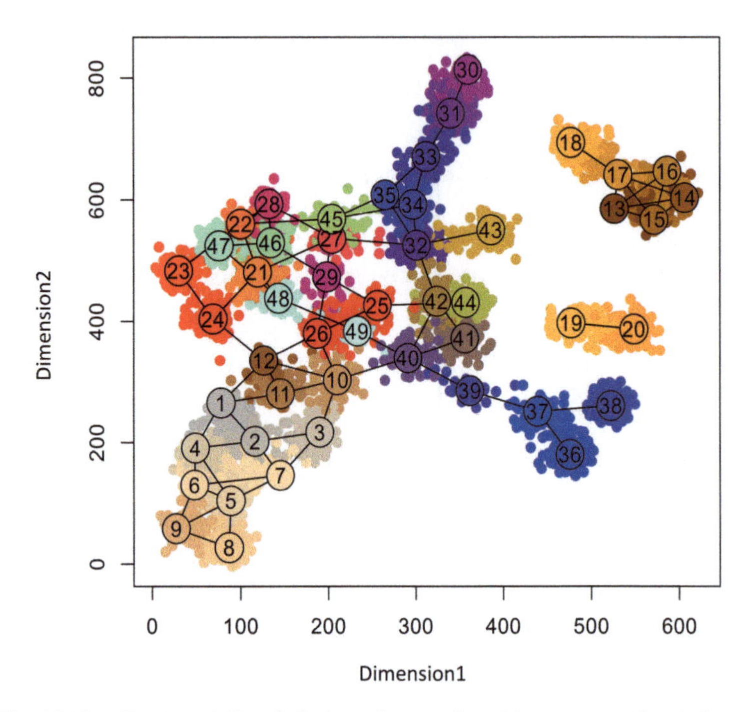

Fig. 16 Visualization of MetaCell clustering results with corresponding balanced *k*-NN graph. The 2-D projection of clusters is generated by the force-directed graph drawing algorithm in MetaCell

Box 20
Clustering with MetaCell. Step 1: Clustering construction.

```
mcell_add_cgraph_from_mat_bknn(
 mat_id="mne_filt", # use filtered dataset
 gset_id = "mne_feats", # use selected HVGs
 graph_id="mne_graph", # name in R List for balanced knn graph
 K=150, # number of neighbors in knn
 dsamp=T) # perform normalization before construct graph
```

Box 21
Clustering with MetaCell. Step 2: Clustering.

```
mcell_coclust_from_graph_resamp( # resample the balanced knn graph
 coc_id="mne_coc1000", # name of S^boot in R List
 graph_id="mne_graph", # name of knn graph in R List
 min_mc_size=30, # minimum samples in one cluster
 p_resamp=0.75, # fraction of cells sampled in one iteration
 n_resamp=1000) # iteration times
```

(continued)

```
mcell_mc_from_coclust_balanced( # generate final clustering results
 coc_id="mne_coc1000", # S^boot
 mat_id="mne_filt", # count matrix
 mc_id="mne_mc", # clustering result in R List
 K=30, # degrees in G^boot
 min_mc_size=20, # minimum samples in one cluster
 alpha=2) # the threshold for filtering edges in G^boot
```

Box 22
Clustering with MetaCell. Step 3: Outlier identification and filtering.

```
mcell_mc_split_filt( # filter out outlier cells
 new_mc_id="mne_mc_f", # new clustering result in R List
 mc_id="mne_mc",
 mat_id="mne_filt",
 T_lfc=3, # threshold for filtering
 plot_mats=F)
```

Box 23
Clustering with MetaCell. Step 4: Clustering visualization.

```
mcell_mc2d_force_knn(mc2d_id="mne_2dproj", # generate 2d projection of clusters
and cells
 mc_id="mne_mc_f",
 graph_id="mne_graph")
mcell_mc2d_plot(mc2d_id="mne_2dproj") # plot projection
```

Table 3 summarizes the main differences between Scanpy and MetaCell in the clustering stages and also lists their strengths and limitations. In short, the bootstrapping design in MetaCell makes the clustering results more robust than the other approaches; however, the downside is the increased running time, especially for the large dataset. By contrast, the Louvain and Leiden algorithms are more scalable and suitable for large datasets. The UMAP and t-SNE projections supported by Scanpy are more broadly used visualization techniques. Users could select the most suitable packages or methods according to the specific data and tasks.

Table 3
Comparison of the clustering stage between Scanpy and MetaCell

	Scanpy	MetaCell
Dimension reduction	Provides the principle component analysis [41] algorithm	Does not include a dimension reduction function
Graph construction	A k-nearest neighbor graph is built using either the UMAP-based [43] or Gaussian kernel-based [70, 71] approaches	Balanced directed *k*-nearest neighbor graph based on Pearson correlation similarity matrix for bootstrapping. Nondirected co-occurrence graph after bootstrapping for final clustering
Clustering algorithm	Supports Louvain and Leiden algorithms	Supports the clustering algorithm proposed by MetaCell
Strengths	The clustering algorithms are relatively efficient with low time complexities. Fewer parameters are required for users. Leiden provides more accurate clusters with better intra-cluster connectivity compared with Louvain	Provides robust and accurate clustering results and enables to filter out of the outlier clusters after clustering
Limitations	The clustering results are sensitive to the resolution parameter selection. It may not be accurate for small datasets	Time-consuming for large-scale datasets. The clustering results are sensitive to the setting of parameter *k*

4 Post-Analysis

This section describes several common post-analyses based on the obtained clustering results. The first one is identifying the biological heterogeneity and relations among cell groups to provide an interpretable topological analysis. Partition-based graph abstraction (PAGA) proposed by Wolf et al. [77] is a broadly used method in the single-cell community, which can generate a simple abstracted graph with the estimated connectivity of different manifold partitions at multiple resolutions. The nodes in the resulting PAGA graph correspond to the cell groups, and the edge weights represent the confidence in the presence of connectivity between the corresponding cell groups. The confidence of connectivity can be interpreted as a ratio of the inter-edge counts between two groups versus the expected value under random assignment. (*See* Appendix 8.2 for a detailed explanation.) With a much simpler, coarse-grained, and noise-reduced graph abstraction, PAGA is favorably helpful for researchers to find clearer pictures of the underlying biology compared to directly analyzing on the single-cell level. PAGA algorithm is supported by the Scanpy package

A.

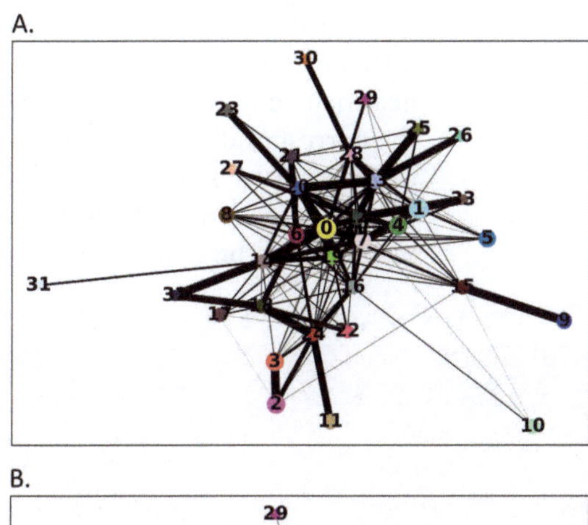

B.

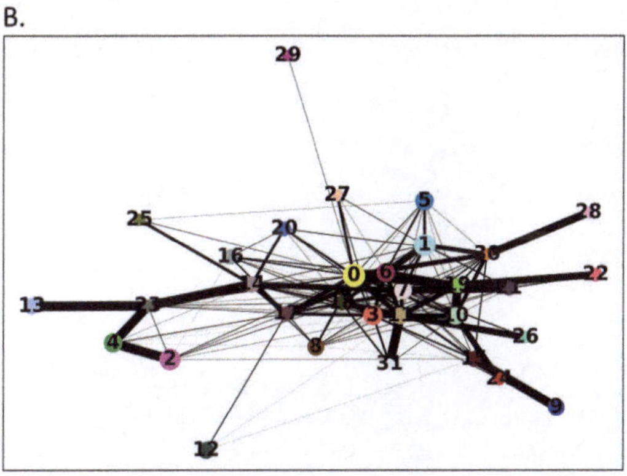

Fig. 17 Visualization of PAGA graphs for the Leiden (**a**) and Louvain (**b**) clustering results. The nodes represent the clusters' ids, and the width of an edge is proportional to the confidence of a connection. A thick edge means a strong relationship between two connected clusters

(as shown in Box 24), and the PAGA graphs generated for the Leiden and Louvain clustering results (presented in Subheading 3) are shown in Fig. 17. For MetaCell, the similarity graph analysis for clustering results has been described in the visualization Subheading 3.3.

Box 24
PAGA graph generation in Scanpy.

```
sc.tl.paga(adata, groups='leiden') # calculate PAGA based
on Leiden clusters
sc.pl.paga(adata, color='leiden') # plot PAGA graph based
on Leiden clusters
sc.tl.paga(adata, groups='louvain') # calculate PAGA
based on Louvain clusters
sc.pl.paga(adata, color='louvain') # plot PAGA graph
based on Louvain clusters
```

Another common way to analyze the relationship between different clusters is using a correlation heatmap. The color of a heatmap cell is proportional to the correlation coefficient between the corresponding two clusters, and the value of correlation ranges from −1.0 to 1.0. Figure 18 illustrates heatmaps associated with color bars generated by Scanpy functions in Box 25 for the clustering result of the Leiden and Louvain algorithms, where the brighter

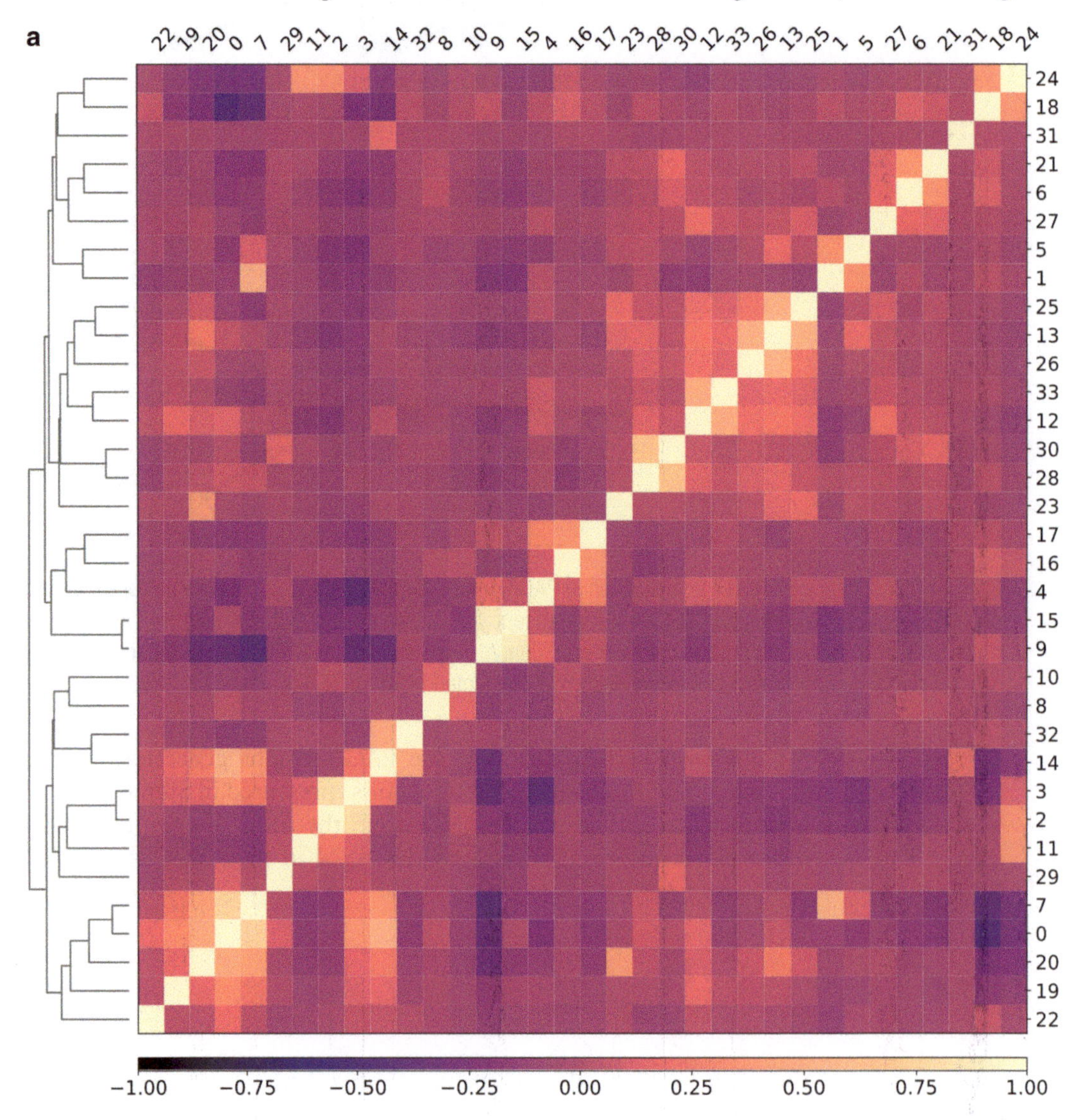

Fig. 18 Correlation heatmap for cell clusters generated by Leiden (**a**) and Louvain (**b**) clustering algorithms. The dendrogram is shown on the left side. Clusters with high correlation, e.g., clusters 15 and 9 in Leiden, are located in close branches in the dendrogram

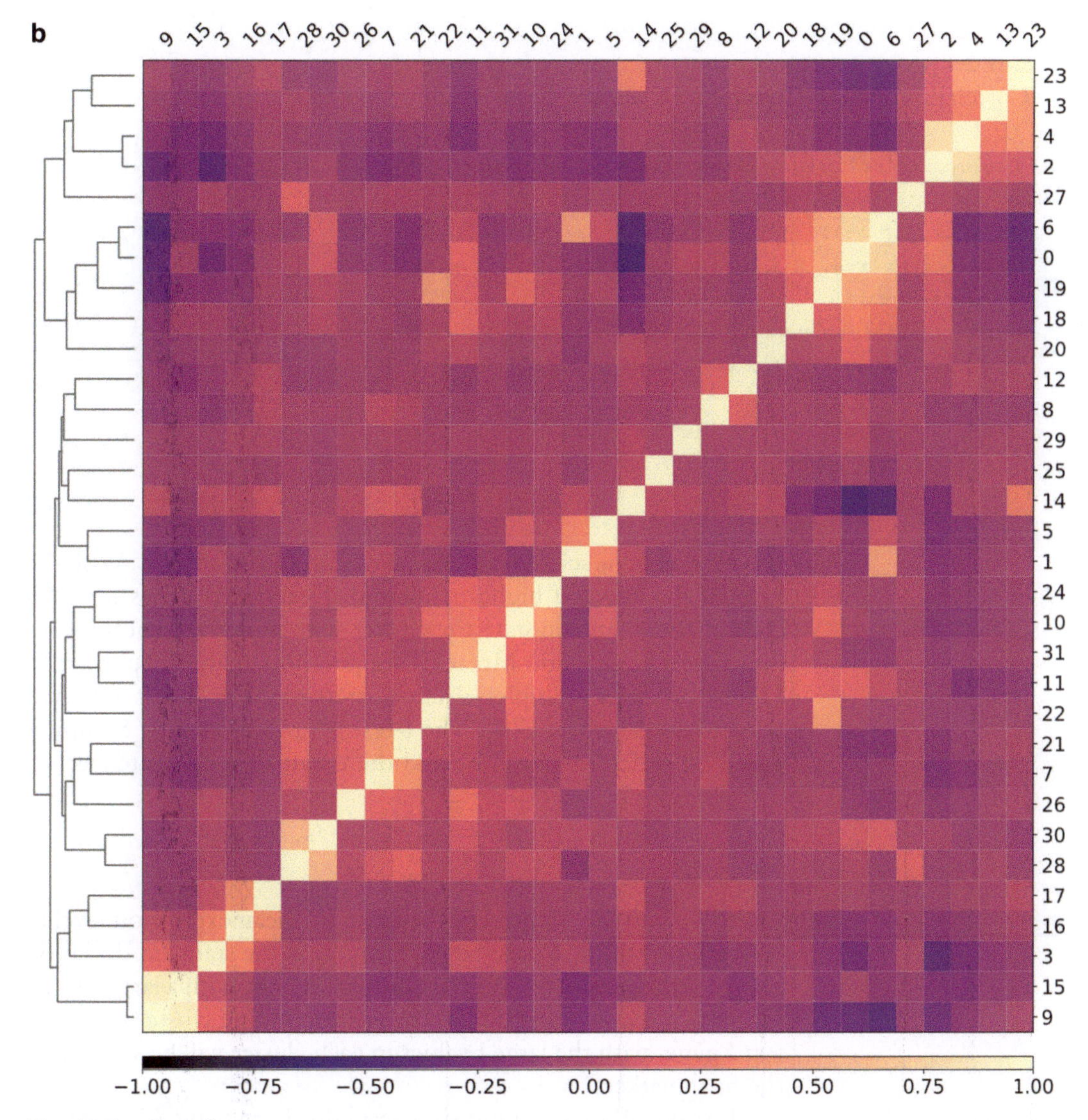

Fig. 18 (continued)

the color, the higher the correlation is. Values on the anti-diagonal of the matrix are always one, representing the correlation between a cluster and itself. Scanpy also supports attaching a dendrogram to the side of the heatmap, which is a tree-structured graph to visualize the result of a hierarchical clustering calculation. By highlighting the differences in the correlation between clusters, the heatmap makes the pattern more readable and provides clues for the underlying insight discovery.

Box 25
Correlation matrix generation in Scanpy.

```
sc.pl.correlation_matrix(adata,
 groupby='leiden', # result from Leiden
 figsize=(15,15),
 dendrogram=True, # show the dendrogram of clusters
 cmap='magma')
sc.pl.correlation_matrix(adata,
 groupby='louvain', # result from Louvain
 figsize=(15,15),
 dendrogram=True, # show the dendrogram of clusters
 cmap='magma')
```

Marker gene selection is also an essential post-analysis. With the help of the selected highest differential expressed genes for each cluster, users can annotate the cluster with more biological meanings. Both Scanpy and MetaCell support this analysis. Specifically, Scanpy provides multiple selection strategies, e.g., t-test and Wilcoxon rank-sum test [78]. The t-test approach used by Scanpy is based on Welch's t-test [79], and the t-statistic is calculated as:

$$t_{gi} = \frac{\overline{u}_{gi} - \overline{u}_{gi_{\text{res}}}}{\sqrt{s_{\overline{u}_{gi}}^2 + s_{\overline{u}_{gi_{\text{res}}}}^2}}, \qquad s_{\overline{u}_{gi}} = \frac{s_{gi}}{\sqrt{N_i}}$$

where $\overline{u}_{gi}$ and $\overline{u}_{gi_{\text{res}}}$ represent the gene g's mean expression level in cluster i and the rest part, respectively. s_{gi} is the standard deviation and N_i is the number of cells in cluster i. A gene with a large t value means it is highly expressed in the corresponding clusters, and the top k genes with the large t values in each cluster will be selected as the final markers.

For the Wilcoxon rank-sum test, the U-statistic U_{gi} is calculated as follow:

$$U_{gi} = \sum_{m \in M_i} \sum_{l \notin M_i} S(u_{gm}, u_{gl})$$

$$\text{where } S(u_{gm}, u_{gl}) = \begin{cases} 1, & \text{if } u_{gm} > u_{gl} \\ 1/2, & \text{if } u_{gm} = u_{gl} \\ 0, & \text{if } u_{gm} < u_{gl} \end{cases}$$

M_i represents cluster i. u_{gm} is the UMI count of gene g in cell m. U_{gi} will be high if gene g is specially highly expressed in cluster i.

Figure 19a–f shows the distributions of each cluster's top-5 highest expressed genes in Louvain and Leiden with t-test approach, which are generated by code snippet in Box 26.

Box 26
Marker gene selection in Scanpy.

```
sc.tl.rank_genes_groups(adata,
 groupby='leiden', # result from Leiden
 method='t-test') # use calculation in t-test

sc.pl.rank_genes_groups(adata,
 n_genes=25, # display top 25 highest differentially expressed genes
 sharey=False)

sc.pl.rank_genes_groups_dotplot(adata,
 groupby='leiden',
 n_genes=5, # top5 genes in each cluster are selected as markers
 swap_axes=True)

sc.tl.rank_genes_groups(adata,
 groupby='louvain', # result from Louvain
 method='t-test') # use calculation in t-test

sc.pl.rank_genes_groups(adata,
 n_genes=25, # display top 25 highest differentially expressed genes
 sharey=False)

sc.pl.rank_genes_groups_dotplot(adata,
 groupby='louvain',
 n_genes=5, # top5 genes in each cluster are selected as markers
 swap_axes=True)
```

Metacell uses a self-defined metric $\mathrm{lf}p_{gi}$ for marker gene selection:

$$p_{gi} = \exp\left[\left(\frac{1}{|M_i|}\sum_{j\in M_i}\log(1+u_{gj})\right)-1\right]/\left(\frac{1}{|M_i|}\sum_{j\in M_i}u_j\right)$$

$$\mathrm{lf}p_{gi} = \log\left(\left(p_{gi}+\epsilon\right)/\mathrm{median}_i\left(p_{gi}+\epsilon\right)\right)$$

where u_i is the cell depth and ϵ is a parameter to avoid division by zero. A large $\mathrm{lf}p_{gi}$ for gene g means its expression in cluster i is

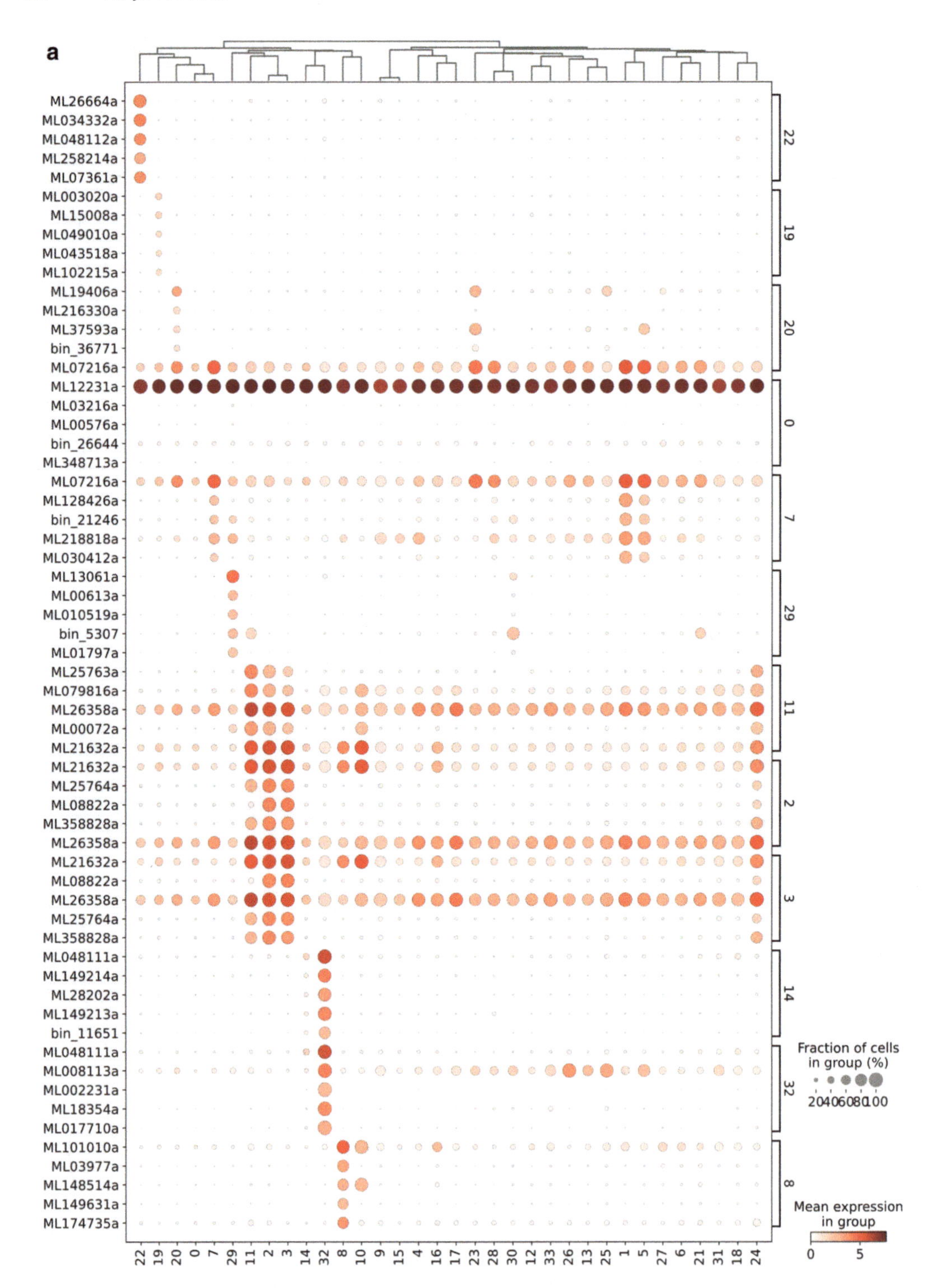

Fig. 19 Distribution of marker genes among clusters in Leiden (**a**–**c**) and Louvain (**d**–**f**). The marker genes are selected based on clusters' top-5 highest differential expressed genes. The *x*-axis represents the clusters from Leiden or Louvain, and the *y*-axis represents the selected marker genes with the corresponding clusters. The size of a circle is proportional to the fraction of expressed genes in each cluster, and the color corresponds to the gene expression level, where darker color means higher expression

Fig. 19 (continued)

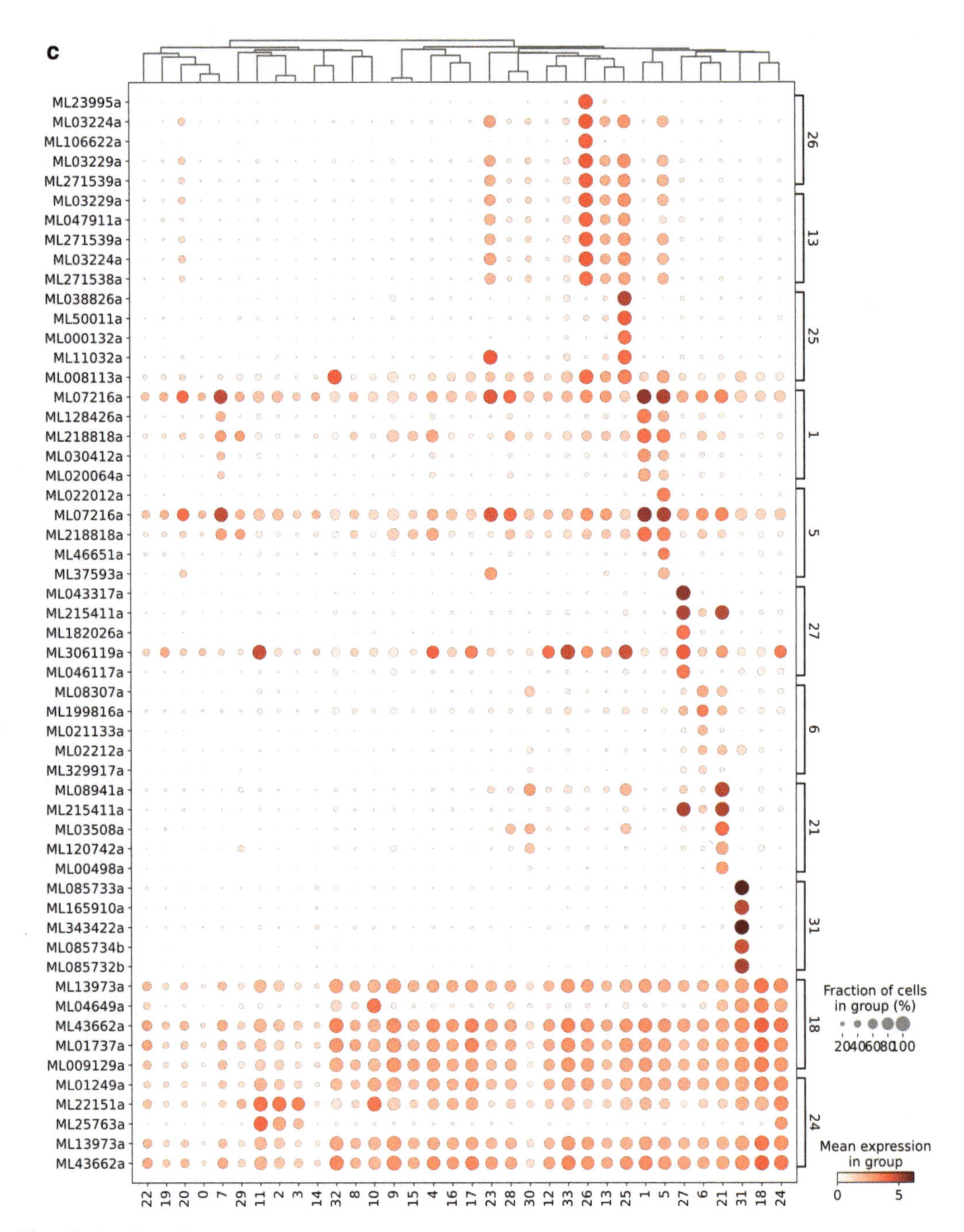

Fig. 19 (continued)

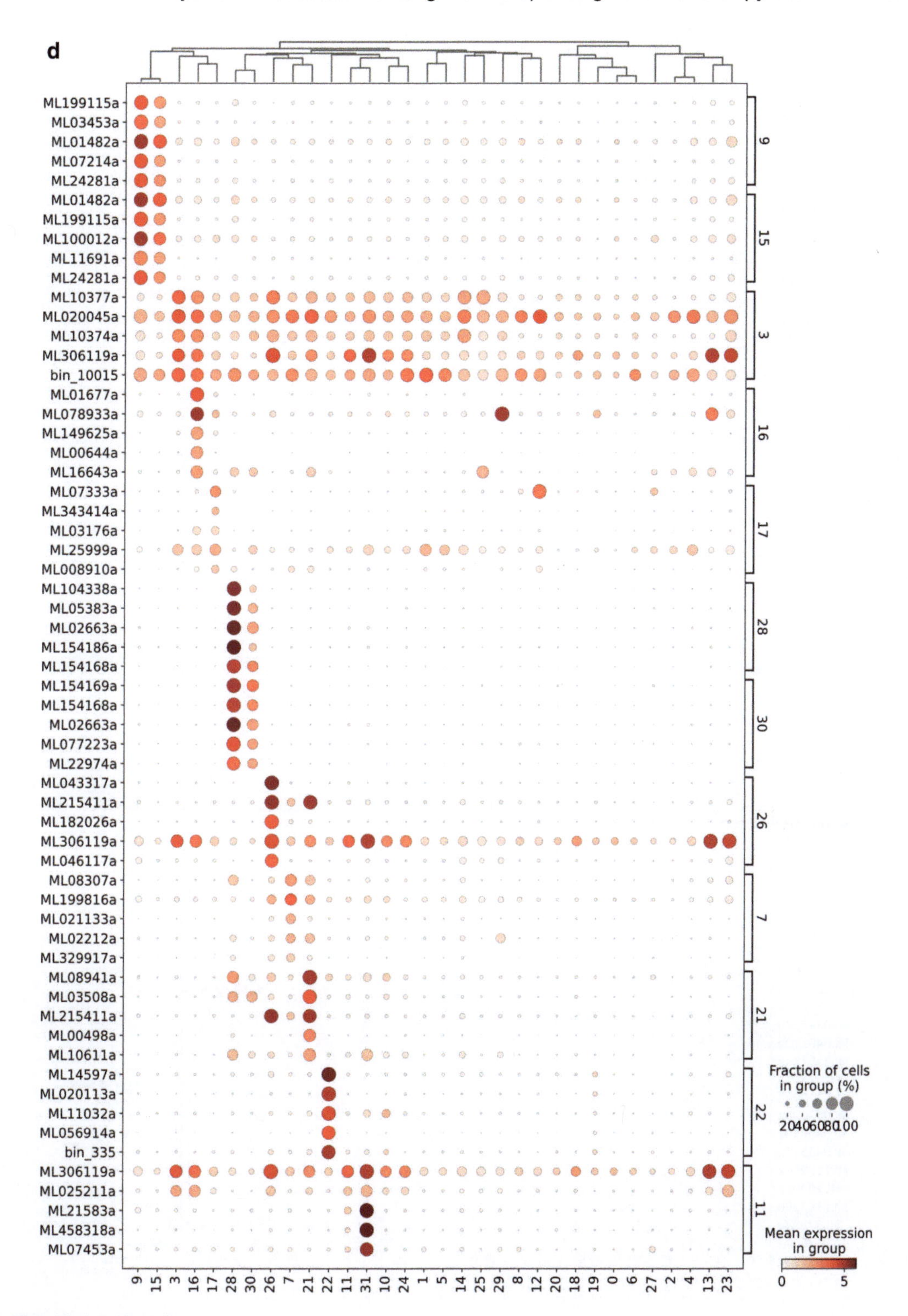

Fig. 19 (continued)

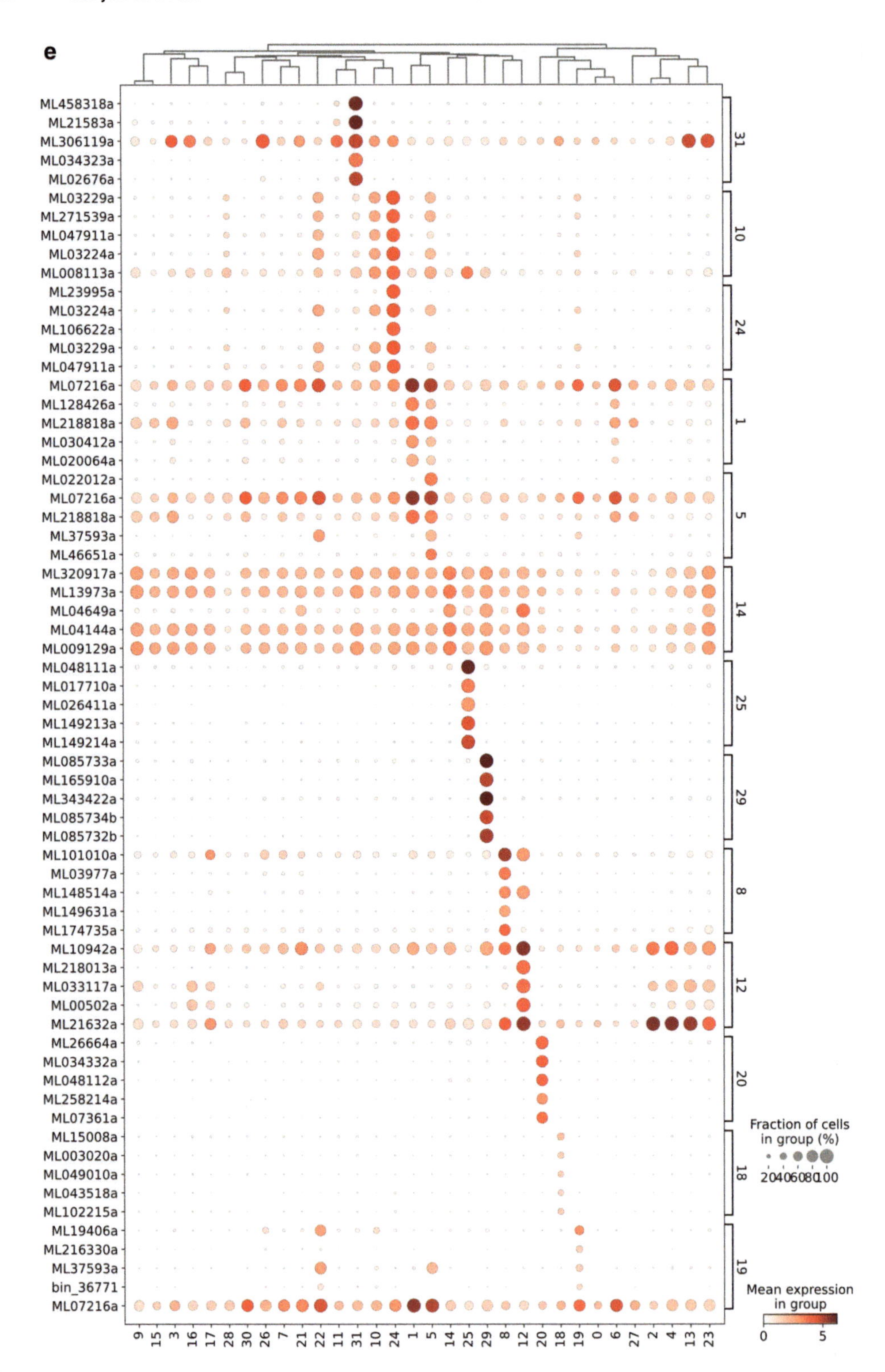

Fig. 19 (continued)

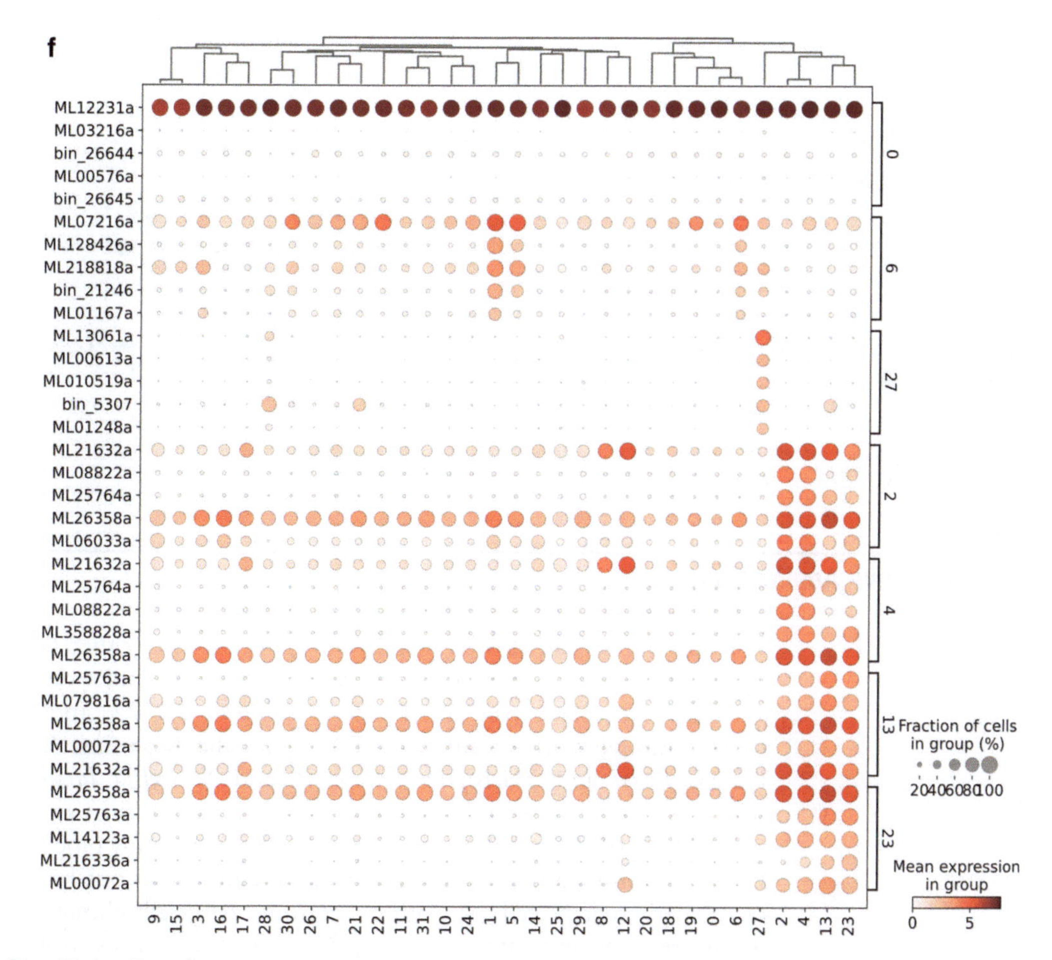

Fig. 19 (continued)

significant compared with its expressions in other clusters. Box 27 shows how to select marker genes in MetaCell, and Fig. 20a–c show the heatmap of marker genes distributions in MetaCell clusters. Supplementary material provides a full list of markers selected by MetaCell, and the top-5 highest differentially expressed genes selected by Scanpy for Leiden and Louvain clustering results. We can find that MetaCell selected 161 marker genes, overlapping with 90 markers out of 170 from Leiden, and 90 out of 160 from Louvain. Supplementary material also contains a list of the intersection of selected markers in three algorithms. Table 4 summarizes the available post-analysis functions supported by Scanpy and MetaCell. Users are encouraged to explore other useful tools to get interesting insights of identified clusters.

Box 27
Marker gene selection in MetaCell.

```
mc_f<- scdb_mc("mne_mc_f")
mc_f@colors <- colorRampPalette(c("darkgray", "burlywood1", "chocolate4","or-
ange", "red", "purple", "blue","darkgoldenrod3", "cyan"))(ncol(mc_f@mc_fp))
scdb_add_mc("mne_mc_f", mc_f)
mc_f <- scdb_mc("mne_mc_f")
mcell_gset_from_mc_markers(gset_id="mne_markers", # name in R List for markers
 mc_id="mne_mc_f")
mcell_mc_plot_marks(mc_id="mne_mc_f", gset_id="mne_markers",
mat_id="mne_filt", plot_cells=F)
```

5 Comparative Applications

The approaches can also be helpful for applications to other taxa, especially prebilaterian animals, where these tools are under development (e.g., placozoan and Porifera). Figures 21, 22, and 23 show some illustrative results of scRNA-data analyses using *Mnemiopsis leidyi* and *Trichoplax adhaerens* and *Spongilla lacustris* [80]. However, we would like to point out the need to be open-minded regarding default or suggesting parameters in any pipeline. We must consider that low-abundant genes and small but highly molecularly specific cell populations might be artificially eliminated or "missed" during data processing. As a result, in situ hybridization experiments might not correlate with identifying marker-gene-specific clusters.

Here, the first point is a selection of filtration parameters, which plays an essential role in all scRNA-seq pipelines, especially for clustering and gene implementation using UMAP or t-SNE visualization. Figures 21 and 22 illustrate this situation with different outcomes for different animals using the same pipelines.

Second, we present three different outcomes of scRNA-seq analysis of the very same dataset of the placozoan *Trichoplax adhaerens* [19] targeting 14 secretory peptides known to be cell-specific, as shown using single-molecule fluorescent in situ hybridization (smFISH), and physiologically relevant as signal molecules [81]. MetaCell and Scanpy demonstrate different results for clustering and cell-specific gene identification (Fig. 21). For example, the placozoan gene *elp* encodes a secretory molecule located in specific cells. Still, it was "missed" and appears not to be a cluster-specific marker using standard default parameters in Scanpy and MetaCell pipelines (Fig. 21a, c). However, without filtration, normalization, etc., Scanpy pipeline outcome nicely identifies specific

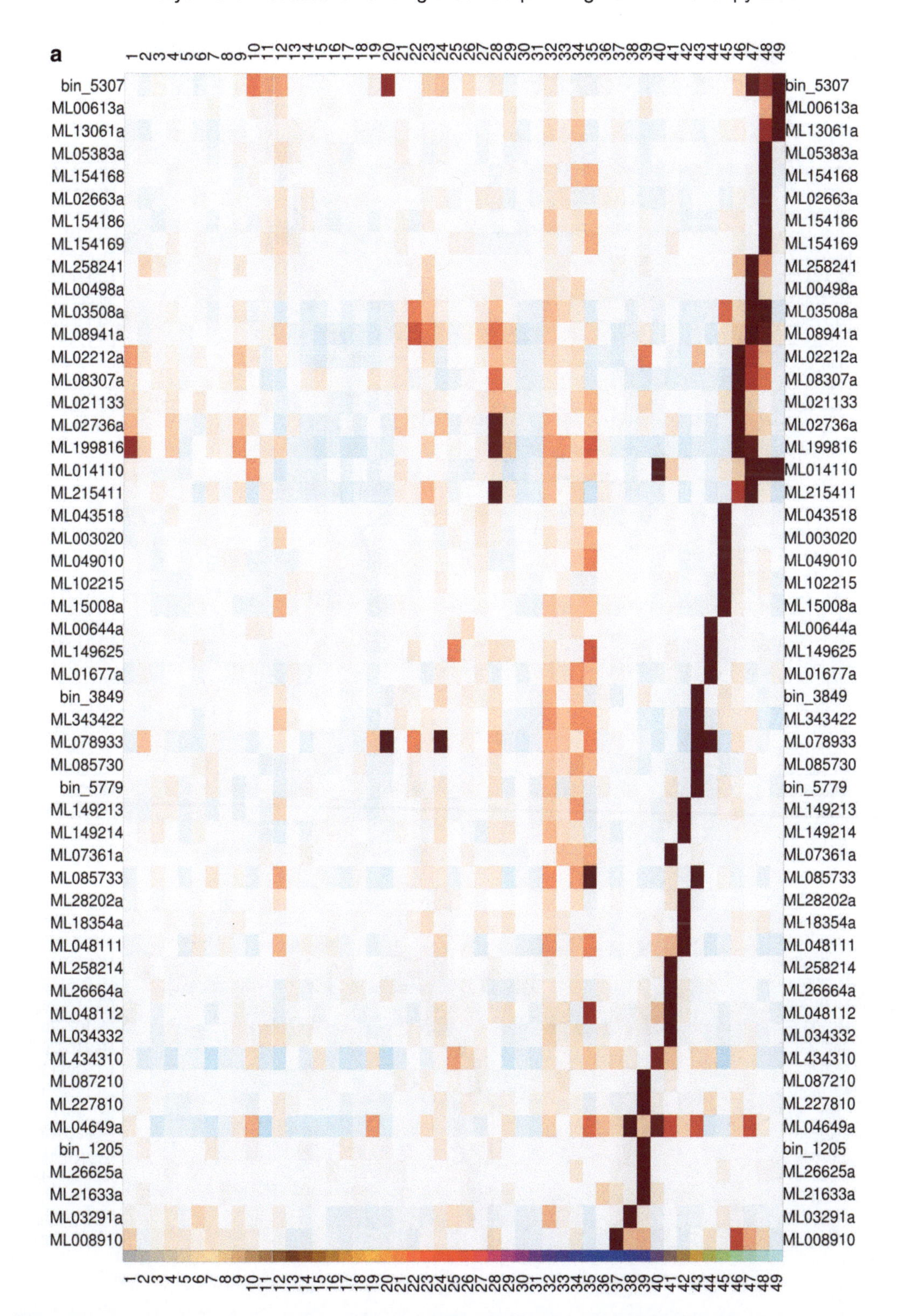

Fig. 20 Distribution of the marker genes over cell clusters generated by MetaCell (**a**–**c**). The *x*-axis represents different cell clusters, and the *y*-axis represents the selected marker genes. The color of the clusters is the same as that in Fig. 17

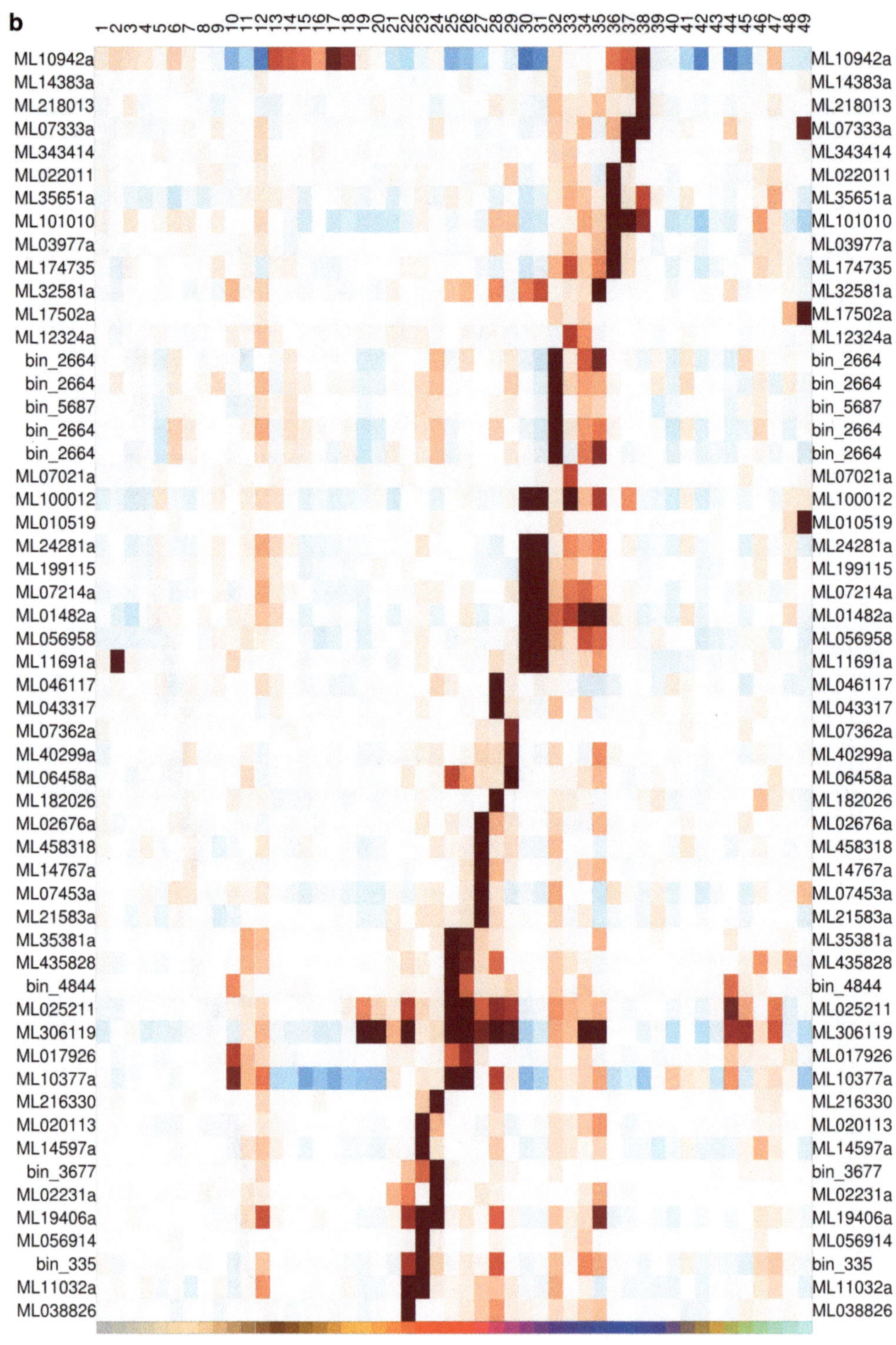

Fig. 20 (continued)

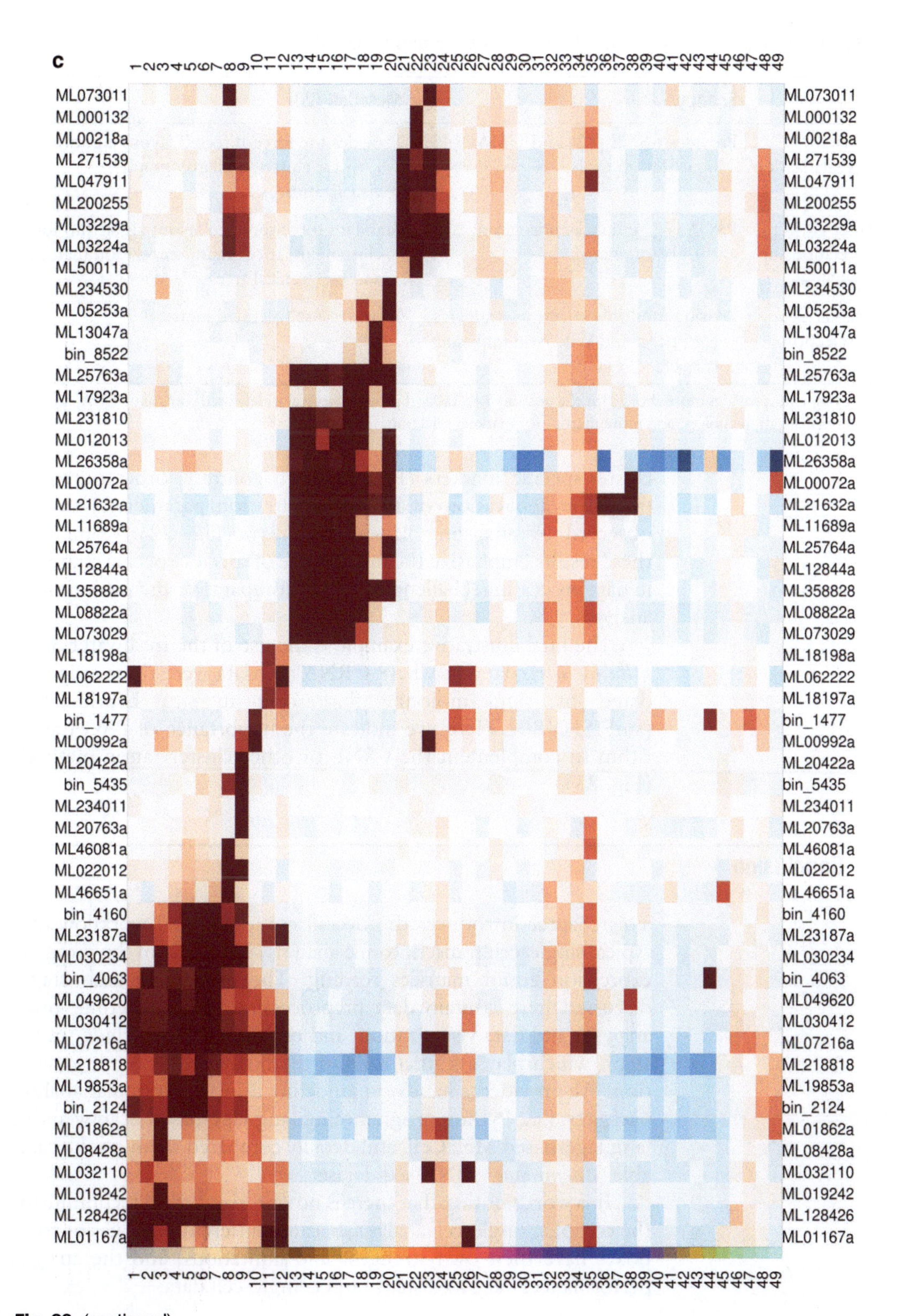

Fig. 20 (continued)

Table 4
Comparison of post-analysis tools between Scanpy and MetaCell

	Scanpy	MetaCell [20]
Cluster visualization	Provides t-SNE [42] and UMAP [43] visualization tools	Provides the force-directed layout algorithm [76] for cluster similarity graph visualization
Inter-cluster relationship analysis	PAGA [77] graph, dendrogram, and correlation matrix	Standard force-directed layout algorithm for cluster similarity graph, correlation matrix, and hierarchical trees
Marker gene selection	Provides multiple selection strategies, e.g., t-test and Wilcoxon rank-sum test, etc.	Provides a self-defined metric

Scanpy pipeline provides more choices for the post-analysis tasks; it provides several widely used tools like PAGA, multiple cluster visualization choices, and more metrics for markers selection

cluster-specific markers (Fig. 21b). In contrast, other outcomes from the application of the same filtration parameters) can be observed by analyzing *Mnemiopsis leidyi* (Fig. 22). Combined, these results emphasize the importance of species-specific strategies in data processing, challenging broad comparative and evolutionary analyses.

The final illustrative example is the use of the freshwater *Spongilla lacustris* dataset, where scRNA-seq data were rigorously validated by single-molecule fluorescent in situ hybridization (smFISH)[80]. Here, we indicate the importance of PAGA algorithm to complement the t-SNE or other clusterization methods (Fig. 23).

6 Conclusion

This chapter introduces the overall computational workflow of a typical single-cell transcriptomic analysis starting from the raw gene expression count matrices reading. The primary analysis stages discussed here contain data preprocessing, cell clustering, post-analysis, and data visualization, and each stage may include several steps, which play essential roles in the final analysis result generation. We provided a hands-on guideline using two popular analysis packages, i.e., Python programming language-based Scanpy and R language-based MetaCell, and detailly compared them on the valuable *Mnemiopsis* scRNA-seq dataset.

It is worth noting that there is no one-size-fits-all technique for the complex single-cell analysis, because different packages or toolboxes have their own strengths and limitations, and the analysis performances vary to distant target single-cell data.

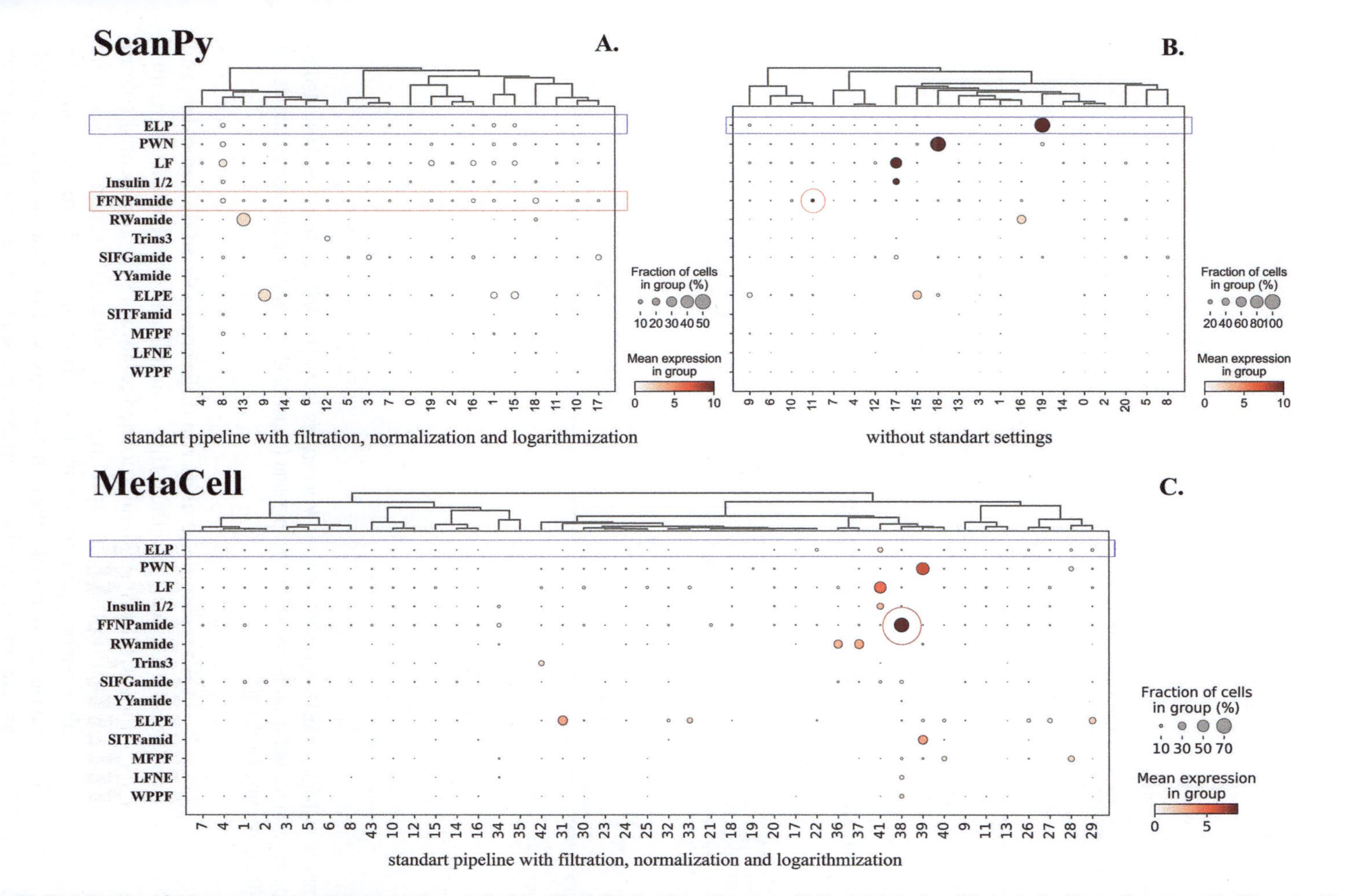

Fig. 21 Distribution of 14 secretory peptides experimentally identified *Trichoplax adhaerens* [81]: dotplots for different pipelines. Scanpy with (**A**) and without (**B**) default setting and MetaCell (**C**) pipeline for scRNA-seq analysis (default pipeline with filtration, normalization and logarithmization) [19]. Note that without filtration (**B**), one can more efficiently recognize rare genes associated with small cell populations and signaling [81]. Equally significant is a partial correlation across different clusterization techniques. However, MetaCell and Scanpy can identify the same cell populations (circle). *See* text for details

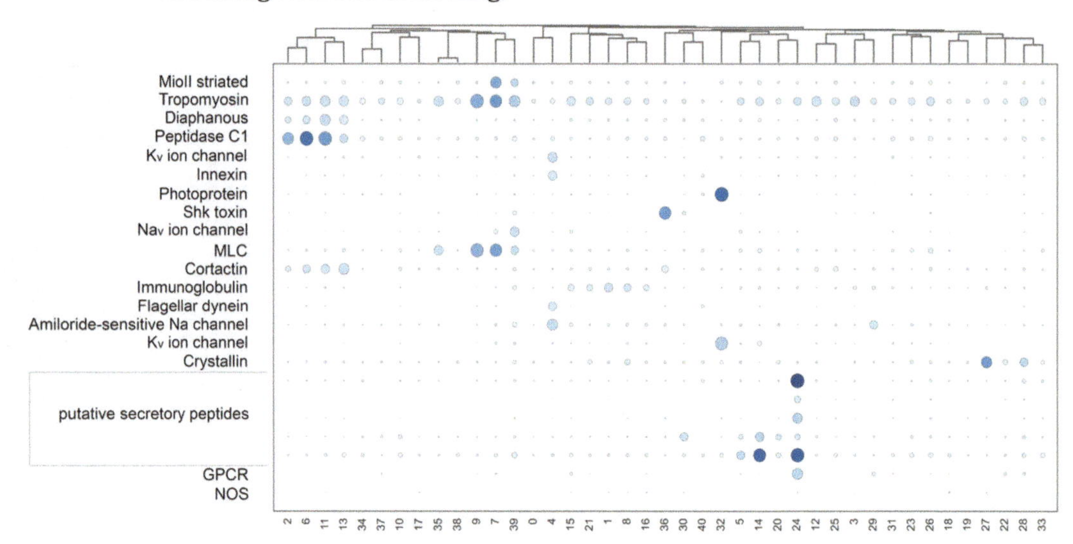

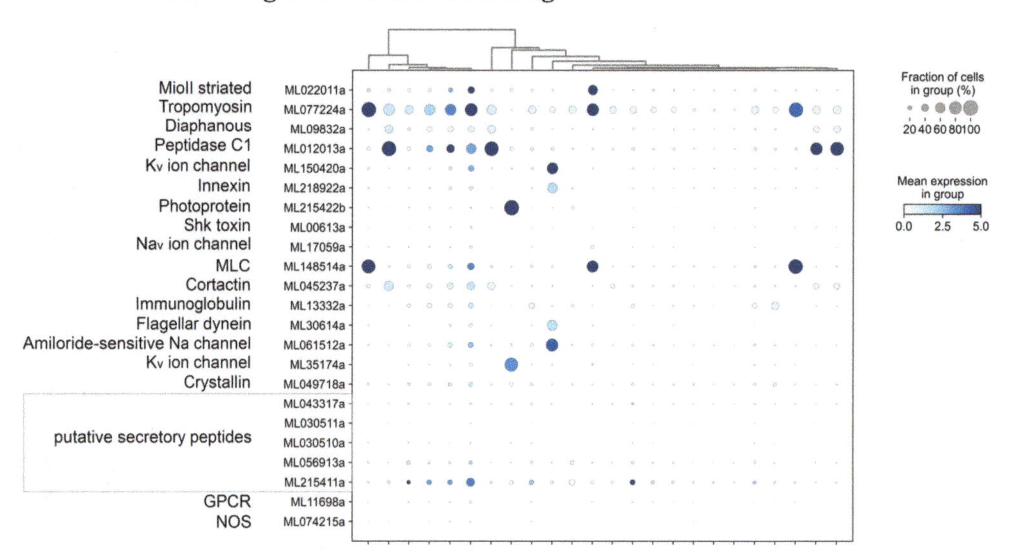

Fig. 22 Dotplots highlights cluster-specific genes of *Mnemiopsis leidyi* [19, 93], identified as illustrated by Scanpy pipeline with default parameters as the text pipeline (upper plot) and with customized setting as per reference [94]). *See* text for details

We recommend users run different tools on the same data and task to check whether they can obtain the consensus results. Biological insights and knowledge always play an essential role in evaluating and interpreting the obtained results. Users are also encouraged to attempt a hybrid approach to combine the favorable parts of different toolboxes. Take Scanpy and MetaCell as examples; for the preprocessing stage, Scanpy provides more quality control and batch effect correction choices than MetaCell.

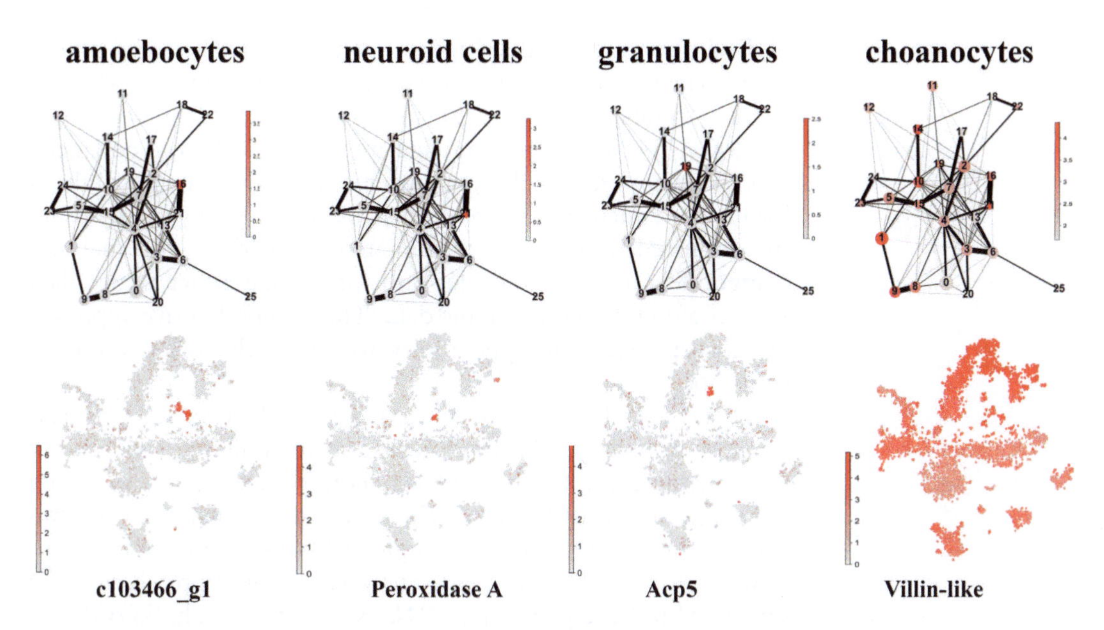

Fig. 23 Distribution of the marker genes of *Spongilla lacustris* [80] using t-SNE and PAGA graphs. Different cell types, such as amoebocytes, neuroid cells, and granulocytes, have remarkably different patterns compared to the most abundant marker genes (e.g., localized expression of villin-like gene) in choanocytes as the cell-type hallmark of sponges. In some cases, PAGA graphs are informative

Still, MetaCell supports more feature selection metrics, which could be integrated into the Scanpy pipeline. For the clustering stage, users could first use MetaCell's bootstrapping strategy to generate multiple subgraphs and apply other well-known clustering algorithms, such as Leiden supported by Scanpy, to partition each subgraph with good connectivity further.

Overall, we hope this tutorial can help the newcomers to the single-cell community become more familiar with the data analysis workflow quickly and possess the best implementation practices. It should be noted that with the rapid advance of the single-cell or even spatial transcriptomic field, more and more new comprehensive and robust analysis packages and algorithms will be proposed and evolved. Some specific techniques introduced here may lay behind the latest developments, but they should still be suitable for a strong baseline. More importantly, most new techniques share similar data processing logic and analysis [69, 82], and we believe that the conceptually similar computational workflow is not expected to change much in a short period. However, the broad evolutionary comparisons would require further development with the integration of multidisciplinary approaches.

Meanwhile, it is important to highlight that beyond the fundamental analysis tasks exhaustively described in this tutorial, scRNA-seq technologies and analysis have demonstrated immense applicability across a broad spectrum of fields. There are several successful applications and potential future trajectories for scRNA-seq data, coupled with sophisticated computational approaches in the realm of drug discovery and development, which hold substantial potential to enhance our understanding of disease biology and pharmacology significantly. Additionally, scRNA-seq data has been employed to train large language models for cell type and gene expression annotation. This advancement amplifies the possibilities for further enhancing compound virtual screening by jointly using the advanced deep learning-generated embeddings for chemical compounds and biomolecules and integrating other omics data sources along with phenotypic data. This comprehensive approach presents a significant opportunity to enrich the drug discovery landscape.

This tutorial can also benefit artificial intelligence (AI) applications by providing a robust foundation for building models and enabling hierarchical clustering of billions of cells across species. By following this tutorial, researchers and data scientists can gain insights into the intricacies of single-cell transcriptomics and learn how to effectively preprocess, analyze, and interpret large-scale single-cell RNA sequencing (scRNA-seq) datasets. This knowledge can then be applied to develop AI models that can accurately classify cell types, identify gene expression patterns, and uncover cell-to-cell interactions within complex biological systems. Moreover, the tutorial shed light on hierarchical clustering potentials, which can empower researchers to organize cells based on their gene expression profiles, facilitating the discovery of evolutionarily conserved or distinct cellular populations across multiple species. Ultimately, the prospective integration of single-cell transcriptomic analysis and AI techniques contributes to a deeper understanding of cellular heterogeneity and opens doors to novel discoveries in various fields, such as developmental biology, neuroscience, cancer research, and regenerative medicine.

In summary, we have to accept the following: different animals require different pipelines and parameters as well as combinations of multiple approaches with mandatory experimental validation. Currently, there is not a single well-validated platform allowing efficient cross-phyla comparisons and unbiased identification of homologous cell populations at larger evolutionary distances (although some approaches have been proposed and are under intensive development [80, 83–86]).

Acknowledgments

This work was supported in part by the Human Frontiers Science Program (RGP0060/2017) and National Science Foundation (IOS-1557923) grants to LLM. Research reported in this publication was also supported in part by the National Institute of Neurological Disorders and Stroke of the National Institutes of Health under Award Number R01NS114491 (to LLM). D.R. was supported by the Russian Science Foundation grant (23-14-00050).

The content is solely the authors' responsibility and does not necessarily represent the official views of the National Institutes of Health.

Appendix

Recommended Books

Pattern Recognition and Machine Learning [87]

Data Clustering: Algorithms and Applications [88]

Introduction to Machine Learning [40]

Introduction to Probability Models [89]

Python Programming: An Introduction to Computer Science [90]

Software for Data Analysis: Programming with R [91]

PAGA Graph

PAGA algorithm enables the similarity analysis between different partitions (or clusters) generated by the community-detection-based clustering methods. The edge weight or the connectivity in the page graph carries the essential information regarding the similarity. This section briefly introduces the edge weights computation of the PAGA graph. A partitioned directed graph is denotated as G containing e edges and n nodes, and each node corresponds to one cell. For the group i, there are overall e_i outgoing edges linked with n_i nodes in it. The target coarse-grained PAGA graph is represented as $G^* = (V^*, E^*)$, where $V^* = \{v_1^*, \ldots, v_M^*\}$ is a set of the M cell groups and $e^* \in E^*$ is a PAGA edge estimated by the PAGA algorithm. A random variable ϵ_{ij} is used to describe the number of edges connected from cell group i to cell group j in random connecting situation. $p(\epsilon_{ij})$ is calculated as:

$$p\left(e_{ij}^*|e_i, n_i, n_j\right) = \frac{\Omega_{\epsilon_{ij}|e_i,n_j,n}}{\Omega_{\text{total}|e_i,n}} = \frac{\binom{e_i}{\epsilon_{ij}} n_j^{\epsilon_{ij}} (n - n_j - 1)^{e_i - \epsilon_{ij}}}{(n-1)^{e_i}}$$
$$= \binom{e_i}{\epsilon_{ij}} \left(\frac{n_j}{n-1}\right)^{\epsilon_{ij}} \left(1 - \frac{n_j}{n-1}\right)^{e_i - \epsilon_{ij}}$$

The expression of $p(\epsilon_{ij}| e_i, n_i, n_j)$ is a binomial distribution with the expectation $\frac{e_i n_j}{n-1}$ and variance $\frac{e_i n_j (n - n_j - 1)}{(n-1)^2}$. A new variable $\epsilon = \epsilon_{ij} + \epsilon_{ji}$ is introduced to provide a symmetric metrics for the similarity of two clusters. Suppose the cluster size is large enough so that the binomial distributions can be approximated by Gaussian, then the distribution of ϵ can be approximated as:

$$p(\epsilon | e_i, e_j, n_i, n_j, n) \sim N\left(\epsilon | \frac{e_i n_j + e_j n_i}{n-1}, \frac{e_i n_j (n - n_j - 1) + e_j n_i (n - n_i - 1)}{(n-1)^2}\right)$$

Suppose the actual number of edges between cluster i and cluster j is $\epsilon_{ij}^{\mathrm{sym}}$ and the expected number of edges is $\widehat{\epsilon}_{ij} = \frac{e_i n_j + e_j n_i}{n-1}$; the edge weights w_{ij} is defined as:

$$w_{ij} = f(x) = \begin{cases} \frac{\epsilon_{ij}^{\mathrm{sym}}}{\widehat{\epsilon}_{ij}}, & \text{if } \epsilon_{ij}^{\mathrm{sym}} < \widehat{\epsilon}_{ij} \\ 1, & \text{otherwise} \end{cases}$$

If the number of actual edges is larger than the expected value, the connectivity will be set as 1, the upper bounder of the connectivity value. If the given partitioned graph is nondirected, one can convert it to a bi-directed graph by replacing a single edge with two independent edges pointing to the two linked nodes, respectively. Then, the same PAGA calculation strategy can be applied.

References

1. Moroz LL (2015) Biodiversity meets neuroscience: from the sequencing ship (Ship-Seq) to deciphering parallel evolution of neural systems in Omic's era. Integr Comp Biol 55(6): 1005–1017
2. Moroz LL (2018) NeuroSystematics and periodic system of neurons: model vs reference species at single-cell resolution. ACS Chem Neurosci 9(8):1884–1903
3. Hernandez-Nicaise M-L (1991) Ctenophora. In: Harrison FWFW, Westfall JA (eds) Microscopic anatomy of invertebrates: Placozoa, Porifera, Cnidaria, and Ctenophora. Wiley, New York, pp 359–418
4. Nielsen C (2012) Animal evolution: interrelationships of the living phyla. Oxford University Press, Oxford
5. Nielsen C (2019) Early animal evolution: a morphologist's view. R Soc Open Sci 6(7): 190638
6. Li Y et al (2021) Rooting the animal tree of life. Mol Biol Evol 38(10):4322–4333
7. Moroz LL (2012) Phylogenomics meets neuroscience: how many times might complex brains have evolved? Acta Biol Hung 63 (Suppl 2):3–19
8. Moroz LL et al (2014) The ctenophore genome and the evolutionary origins of neural systems. Nature 510(7503):109–114
9. Ryan JF et al (2013) The genome of the ctenophore *Mnemiopsis leidyi* and its implications for cell type evolution. Science 342(6164): 1242592
10. Schultz DT et al (2023) Ancient gene linkages support ctenophores as sister to other animals. Nature 618(7963):110–117
11. Whelan NV et al (2015) Error, signal, and the placement of Ctenophora sister to all other animals. Proc Natl Acad Sci U S A 112(18): 5773–5778
12. Whelan NV et al (2017) Ctenophore relationships and their placement as the sister group to all other animals. Nat Ecol Evol 1(11): 1737–1746
13. Moroz LL, Kohn AB (2015) Unbiased view of synaptic and neuronal gene complement in ctenophores: are there pan-neuronal and pan-synaptic genes across Metazoa? Integr Comp Biol 55(6):1028–1049
14. Moroz LL, Kohn AB (2016) Independent origins of neurons and synapses: insights from ctenophores. Philos Trans R Soc Lond Ser B Biol Sci 371(1685):20150041
15. Moroz LL, Romanova DY (2022) Alternative neural systems: what is a neuron? (Ctenophores, sponges and placozoans). Front Cell Dev Biol 10:1071961
16. Moroz LL, Romanova DY, Kohn AB (1821) Neural versus alternative integrative systems: molecular insights into origins of neurotransmitters. Philos Trans R Soc Lond Ser B Biol Sci 2021(376):20190762
17. Martindale MQ (2022) Emerging models: the "development" of the ctenophore *Mnemiopsis leidyi* and the cnidarian *Nematostella vectensis* as useful experimental models. Curr Top Dev Biol 147:93–120

18. Martindale MQ, Henry JQ (2015) Ctenophora. In: Wanninger A (ed) Evolutionary developmental biology of invertebrates 1: introduction, non-Bilateria, Acoelomorpha, Xenoturbellida, Chaetognatha. Springer Vienna, Vienna, pp 179–201
19. Sebe-Pedros A et al (2018) Early metazoan cell type diversity and the evolution of multicellular gene regulation. Nat Ecol Evol 2(7): 1176–1188
20. Baran Y et al (2019) MetaCell: analysis of single-cell RNA-seq data using K-nn graph partitions. Genome Biol 20(1):1–19
21. Sachkova MY et al (2021) Neuropeptide repertoire and 3D anatomy of the ctenophore nervous system. Curr Biol 31(23):5274–5285 e6
22. Hayakawa E et al (2022) Mass spectrometry of short peptides reveals common features of metazoan peptidergic neurons. Nat Ecol Evol 6(10):1438–1448
23. Moroz LL (2009) On the independent origins of complex brains and neurons. Brain Behav Evol 74(3):177–190
24. Zappia L, Theis FJ (2021) Over 1000 tools reveal trends in the single-cell RNA-seq analysis landscape. Genome Biol 22(1):1–18
25. Zappia L, Phipson B, Oshlack A (2018) Exploring the single-cell RNA-seq analysis landscape with the scRNA-tools database. PLoS Comput Biol 14(6):e1006245
26. Svensson V, da Veiga Beltrame E, Pachter L (2020) A curated database reveals trends in single-cell transcriptomics. Database 2020: baaa073
27. Wolf FA, Angerer P, Theis FJ (2018) SCANPY: large-scale single-cell gene expression data analysis. Genome Biol 19(1):1–5
28. Shannon P et al (2003) Cytoscape: a software environment for integrated models of biomolecular interaction networks. Genome Res 13(11):2498–2504
29. Satija R et al (2015) Spatial reconstruction of single-cell gene expression data. Nat Biotechnol 33(5):495–502
30. McCarthy DJ et al (2017) Scater: pre-processing, quality control, normalization and visualization of single-cell RNA-seq data in R. Bioinformatics 33(8):1179–1186
31. Jin S et al (2021) Inference and analysis of cell-cell communication using CellChat. Nat Commun 12(1):1–20
32. Luecken MD, Theis FJ (2019) Current best practices in single-cell RNA-seq analysis: a tutorial. Mol Syst Biol 15(6):e8746
33. Amezquita RA et al (2020) Orchestrating single-cell analysis with Bioconductor. Nat Methods 17(2):137–145
34. Andrews TS et al (2021) Tutorial: guidelines for the computational analysis of single-cell RNA sequencing data. Nat Protoc 16(1):1–9
35. Cao J et al (2019) The single-cell transcriptional landscape of mammalian organogenesis. Nature 566(7745):496–502
36. Ziegler CG et al (2020) SARS-CoV-2 receptor ACE2 is an interferon-stimulated gene in human airway epithelial cells and is detected in specific cell subsets across tissues. Cell 181(5):1016–1035. e19
37. Mathys H et al (2019) Single-cell transcriptomic analysis of Alzheimer's disease. Nature 570(7761):332–337
38. Bornstein C et al (2018) Single-cell mapping of the thymic stroma identifies IL-25-producing tuft epithelial cells. Nature 559(7715): 622–626
39. Giladi A et al (2018) Single-cell characterization of haematopoietic progenitors and their trajectories in homeostasis and perturbed haematopoiesis. Nat Cell Biol 20(7):836–846
40. Alpaydin E (2020) Introduction to machine learning. MIT press
41. Pedregosa F et al (2011) Scikit-learn: machine learning in Python. J Machine Learn Res 12: 2825–2830
42. Van der Maaten L, Hinton G (2008) Visualizing data using t-SNE. J Mach Learn Res 9(11): 2579
43. McInnes L, Healy J, Melville J (2018) Umap: uniform manifold approximation and projection for dimension reduction. arXiv preprint arXiv:180203426
44. Lun AT, Bach K, Marioni JC (2016) Pooling across cells to normalize single-cell RNA sequencing data with many zero counts. Genome Biol 17(1):1–14
45. Bacher R et al (2017) SCnorm: robust normalization of single-cell RNA-seq data. Nat Methods 14(6):584–586
46. Cole MB et al (2019) Performance assessment and selection of normalization procedures for single-cell RNA-Seq. Cell Syst 8(4): 315–328. e8
47. Lytal N, Ran D, An L (2020) Normalization methods on single-cell RNA-seq data: an empirical survey. Front Genet 11:41
48. Street K et al (2018) Slingshot: cell lineage and pseudotime inference for single-cell transcriptomics. BMC Genomics 19(1):1–16
49. Johnson WE, Li C, Rabinovic A (2007) Adjusting batch effects in microarray expression data using empirical Bayes methods. Biostatistics 8(1):118–127

50. Polański K et al (2020) BBKNN: fast batch alignment of single cell transcriptomes. Bioinformatics 36(3):964–965
51. Korsunsky I et al (2019) Fast, sensitive and accurate integration of single-cell data with harmony. Nat Methods 16(12):1289–1296
52. Haghverdi L et al (2018) Batch effects in single-cell RNA-sequencing data are corrected by matching mutual nearest neighbors. Nat Biotechnol 36(5):421–427
53. Hie B, Bryson B, Berger B (2019) Efficient integration of heterogeneous single-cell transcriptomes using Scanorama. Nat Biotechnol 37(6):685–691
54. Tran HTN et al (2020) A benchmark of batch-effect correction methods for single-cell RNA sequencing data. Genome Biol 21(1):1–32
55. Zheng GX et al (2017) Massively parallel digital transcriptional profiling of single cells. Nat Commun 8(1):1–12
56. Stuart T et al (2019) Comprehensive integration of single-cell data. Cell 177(7): 1888–1902. e21
57. Grün D et al (2015) Single-cell messenger RNA sequencing reveals rare intestinal cell types. Nature 525(7568):251–255
58. Wang B et al (2017) Visualization and analysis of single-cell RNA-seq data by kernel-based similarity learning. Nat Methods 14(4): 414–416
59. Kiselev VY et al (2017) SC3: consensus clustering of single-cell RNA-seq data. Nat Methods 14(5):483–486
60. Lin P, Troup M, Ho JW (2017) CIDR: ultrafast and accurate clustering through imputation for single-cell RNA-seq data. Genome Biol 18(1):1–11
61. Yau C (2016) pcaReduce: hierarchical clustering of single cell transcriptional profiles. BMC Bioinform 17(1):1–11
62. Zeisel A et al (2015) Cell types in the mouse cortex and hippocampus revealed by single-cell RNA-seq. Science 347(6226):1138–1142
63. Jiang L et al (2016) GiniClust: detecting rare cell types from single-cell gene expression data with Gini index. Genome Biol 17(1):1–13
64. Qiu X et al (2017) Reversed graph embedding resolves complex single-cell trajectories. Nat Methods 14(10):979–982
65. Traag VA, Waltman L, Van Eck NJ (2019) From Louvain to Leiden: guaranteeing well-connected communities. Sci Rep 9(1):1–12
66. Blondel VD et al (2008) Fast unfolding of communities in large networks. J Stat Mech Theory Exp 2008(10):P10008
67. Duò A, Robinson MD, Soneson C (2018) A systematic performance evaluation of clustering methods for single-cell RNA-seq data. F1000Research 7:1141
68. Zhang S et al (2020) Review of single-cell rna-seq data clustering for cell type identification and characterization. arXiv preprint arXiv:200101006
69. Liu B, Li Y, Zhang L (2022) Analysis and visualization of spatial transcriptomic data. Front Genet 12:785290
70. Coifman RR et al (2005) Geometric diffusions as a tool for harmonic analysis and structure definition of data: diffusion maps. Proc Natl Acad Sci 102(21):7426–7431
71. Haghverdi L et al (2016) Diffusion pseudotime robustly reconstructs lineage branching. Nat Methods 13(10):845–848
72. Xu C, Su Z (2015) Identification of cell types from single-cell transcriptomes using a novel clustering method. Bioinformatics 31(12): 1974–1980
73. Patterson-Cross RB, Levine AJ, Menon V (2021) Selecting single cell clustering parameter values using subsampling-based robustness metrics. BMC bioinform 22(1):1–13
74. Teaching team at the Harvard Chan Bioinformatics Core. Introduction to Single-cell RNA-seq. [cited 2022 04/10]; Available from: https://hbctraining.github.io/scRNA-seq/lessons/07_SC_clustering_cells_SCT.html
75. Paul Hoffman SL (2022) Seurat - guided clustering tutorial. [cited 2022 04/10]; Available from: https://satijalab.org/seurat/articles/pbmc3k_tutorial.html
76. Fruchterman TM, Reingold EM (1991) Graph drawing by force-directed placement. Softw Pract Exp 21(11):1129–1164
77. Wolf FA et al (2019) PAGA: graph abstraction reconciles clustering with trajectory inference through a topology preserving map of single cells. Genome Biol 20(1):1–9
78. Mann HB, Whitney DR (1947) On a test of whether one of two random variables is stochastically larger than the other. Ann Math Stat 18:50–60
79. Welch BL (1947) The generalization of 'STUDENT'S' problem when several different population variances are involved. Biometrika 34(1–2):28–35
80. Musser JM et al (2021) Profiling cellular diversity in sponges informs animal cell type and nervous system evolution. Science 374(6568): 717–723

81. Varoqueaux F et al (2018) High cell diversity and complex Peptidergic signaling Underlie Placozoan behavior. Curr Biol 28(21): 3495–3501 e2
82. Dries R et al (2021) Advances in spatial transcriptomic data analysis. Genome Res 31(10): 1706–1718
83. Tarashansky AJ et al (2021) Mapping single-cell atlases throughout Metazoa unravels cell type evolution. elife 10:e66747
84. Liu X, Shen Q, Zhang S (2023) Cross-species cell-type assignment from single-cell RNA-seq data by a heterogeneous graph neural network. Genome Res 33(1):96–111
85. Wang R et al (2023) Construction of a cross-species cell landscape at single-cell level. Nucleic Acids Res 51(2):501–516
86. Wang J et al (2021) Tracing cell-type evolution by cross-species comparison of cell atlases. Cell Rep 34(9):108803
87. Bishop CM, Nasrabadi NM (2006) Pattern recognition and machine learning, vol 4. Springer
88. Gan G, Ma C, Wu J (2020) Data clustering: theory, algorithms, and applications. SIAM
89. Ross SM (2014) Introduction to probability models. Academic press
90. Zelle JM (2004) Python programming: an introduction to computer science. Franklin, Beedle & Associates, Inc
91. Chambers JM (2008) Software for data analysis: programming with R, vol 2. Springer
92. Moroz LL (2023) Brief history of Ctenophora. Methods Mol Biol. in press
93. Burkhardt P, Jekely G (2021) Evolution of synapses and neurotransmitter systems: the divide-and-conquer model for early neural cell-type evolution. Curr Opin Neurobiol 71: 127–138
94. Moroz LL, Mukherjee K, Romanova DY (2023) Nitric oxide signaling in ctenophores. Front Neurosci 17:1125433

Chapter 18

DNA Methylation in Ctenophores

Emily C. Dabe, Andrea B. Kohn, and Leonid L. Moroz

Abstract

Epigenomic regulation and dynamic DNA methylation, in particular, are widespread mechanisms orchestrating the genome operation across time and species. Whole-genome bisulfite sequencing (WGBS) is currently the only method for unbiasedly capturing the presence of 5-methylcytosine (5-mC) DNA methylation patterns across an entire genome with single-nucleotide resolution. Bisulfite treatment converts unmethylated cytosines to uracils but leaves methylated cytosines intact, thereby creating a map of all methylated cytosines across a genome also known as a methylome. These epigenomic patterns of DNA methylation have been found to regulate gene expression and influence gene evolution rates between species. While protocols have been optimized for vertebrate methylome production, little adaptation has been done for invertebrates. Creating a methylome reference allows comparisons to be made between rates of transcription and epigenomic patterning in animals. Here we present a method of library construction for bisulfite sequencing optimized for non-bilateral metazoans such as the ctenophore, *Mnemiopsis leidyi*. We have improved upon our previously published method by including spike-in genomic DNA controls to measure methylation conversion rates. By pooling two bisulfite conversion reactions from the same individual, we also produced sequencing libraries that yielded a higher percentage of sequenced reads uniquely mapping to the reference genome. We successfully detected 5-mC in whole-animal methylomes at CpG, CHG, and CHH sites and visualized datasets using circos diagrams. The proof-of-concept tests were performed both under control conditions and following injury tests with changes in methylation patterns of genes encoding innexins, toxins and neuropeptides. Our approach can be easily adapted to produce epigenomes from other fragile marine animals.

Key words Ctenophora, Genome, DNA methylation, 5-mC, innexins, neuropeptides, *Mnemiopsis*

1 Introduction

A broad spectrum of epigenetic modifications control gene expression without altering the primary DNA sequence. Epigenomic regulation and dynamic DNA methylation, in particular, are widespread mechanisms orchestrating the genome operation across time and species [1–6]. Initial studies indicated that these mechanisms contribute to the development and specification of different ctenophore species (*Pleurobrachia*, *Mnemiopsis*, and *Beroe* [7, 8]), with the potential involvement of two types of genomic DNA methylation in ctenophores, 5-methyl cytosine (5-mC) and the

Leonid L. Moroz (ed.), *Ctenophores: Methods and Protocols*, Methods in Molecular Biology, vol. 2757,
https://doi.org/10.1007/978-1-0716-3642-8_18, © Springer Science+Business Media, LLC, part of Springer Nature 2024

unconventional form of methylation 6-methyl adenine (6-mA), including 6-mA genome modifications [7] in the aboral organ. Reversible 5-mC methylation and demethylation (by TET enzymes) were primarily detected during the early cleavage stages and in combs of *Pleurobrachia bachei* [8]. Notably, 5-mC methylation occurred both at the CpG sites and other locations within promoters, exons, and introns [7]. Thus, single-nucleotide resolution in mapping and quantifying DNA methylation is imperative. Bisulfite sequencing was the first technique developed to study epigenetic changes to DNA [9]. We sought to improve upon our previously published *Mnemiopsis leidyi* methylome to better study the evolution of epigenomic regulation in basal metazoans [7]. Our approach can also be easily adapted to produce epigenomes from other fragile marine animals.

Originally established in 1992, whole-genome bisulfite sequencing (BS-seq) is the only method currently available to measure DNA methylation at a single base resolution [9, 10]. This method's primary mechanism is converting all cytosine residues to uracil residues using bisulfite (HSO_3^-). The 5-methylcytosine (5-mC) residues are unaffected. Bisulfite reacts with unmethylated cytosine residues, which are converted to uracil in a three-step reaction: (1) a sulfonation reaction adds HSO_3^- to create cytosine sulfonate, (2) deamination creates uracil sulfonate, and finally, (3) alkali desulfonation finishes the conversion of cytosine to uracil [11]. Meanwhile, methylated cytosines are protected from this bisulfite conversion, so preserved cytosine residues in sequenced BS-seq samples are recognized as methylated sites [9]. Bisulfite conversion is a harsh process that causes DNA to become single-stranded and highly fragmented. Adding a "spike-in" at a low concentration with known cytosine methylation states to samples before the bisulfite conversion reaction allows for downstream assessment of how much degradation occurred and to calculate the percentage of unmethylated cytosine residues converted per sample [12]. This protocol provides an optimized method for adding methylation spike-in controls for *Mnemiopsis*.

Many protocols exist for bisulfite converting and creating sequencing libraries for mammalian or insect samples [10, 13, 14]; however, bisulfite conversion to genomic DNA from ctenophores and *Mnemiopsis*, in particular, poses additional challenges due to the propensity of isolated genomic DNA samples to rapidly degrade if not stored at deep freezer temperatures, ideally −80 °C. In this method, we compensate for the high rate of fragmentation or degradation caused by bisulfite conversion by performing multiple reactions per *M. leidyi* gDNA sample and combining these reactions during DNA methylation library construction.

2 Materials

2.1 Reagents

Library construction and quantification reagents.

1. Qubit® DNA assay kit (Cat # Q32850, Thermo Fisher Scientific).
2. RNase-free and DNase-free water (Cat # 10977-015, Thermo Fisher Scientific).
3. EZ methyl direct bisulfite (Cat # D5030, Zymo Research).
4. Qiagen genomic Tip 500/G (Cat # 10262) or 100/G (Cat # 10243).
5. Illumina TruSeq DNA methylation library kit 24 reactions Part # 15066014 Rev A.
6. Illumina TruSeq barcoded adapters (Cat # EGIDX81312, Epicentre/Illumina).
7. WiseGene 5-mC lambda spike-in control Cat # S001.
8. Promega D1501 lambda DNA, 250 μg.
9. 1.5 mL microcentrifuge tubes.
10. 0.2 or 0.5 mL microcentrifuge tubes.
11. FailSafe PCR Enzyme Epicentre, Catalog No. FSE51100.
12. Optional: TruSeq Index PCR Primers Illumina, Catalog No. EGIDX81312 (12 Indexes).
13. EZ DNA methyl direct 25 rxns Zymo Research, Catalog No. D5020.
14. AMPure XP System Beckman Coulter Genomics, Catalog No. A63880.

    Freshly prepared 80% (v/v) ethanol diluted with nuclease-free water.
15. Qubit broad-range dsDNA assay kit Q32850.
16. Qubit reader.
17. Qubit 0.5 mL tubes.
18. Qubit high-sensitivity dsDNA assay kit Q32851.
19. 2% agarose E-gel (fits 8 samples + 1 ladder) G401002.
20. Agilent D1000 DNA screen tapes (5067–5582).
21. Agilent D100 screen tape reagents (5067–5583, Agilent).
22. Qubit 0.5 mL assay tubes (Q32856, Thermo Fisher Scientific).
23. FailSafe PCR Enzyme Epicentre, Catalog No. FSE51100.
24. Agilent D1000 DNA screen tapes.

2.2 Equipment

1. Qubit® 2.0 Fluorometer (Cat # Q32866, Life Technologies).
2. Magnetic rack or stand for 1.5 mL tubes (Bangs Laboratories, Inc., Catalog No. LS001, MS002, MS003 or Life Technologies, Catalog No.12321D) or magnetic plate (Life Technologies, Catalog No.123310) for 1.5 mL microcentrifuge.
3. Thermal cycler, water bath, heating block, or another temperature-controlled device.
4. E-gel electronic gel reader or alternatively gel electrophoresis equipment and imager.
5. Vortex mixer.
6. NanoDrop UV-Vis Spectrophotometer Thermo Fisher.
7. Agilent TapeStation.

3 Methods

Day 1

Since genomic DNA was already collected (*see* Table 1), the protocol begins with EZ DNA methyl direct protocol bisulfite conversion step, skipping the proteinase K treatment step.

Bisulfite conversion of DNA

Genomic DNA preparation

1. For whole-animal ctenophore genomic DNA isolations, utilize Qiagen Genomic Tip 100/G or 500/G kits, depending on animal size. Quantify genomic DNA using Qubit BR dsDNA kit.
2. For each sample, if 20 μL of sample contains at least 100 ng but less than 500 ng of genomic DNA, *two* separate reactions for bisulfite conversion should be prepared for the sample which will be combined during the bisulfite library protocol before final PCR amplification. Failure to do this will result in final library concentrations being too low for sequencing without over-amplification during the final PCR step.

Preparation and dilution of 5-mC methylation spike-in control.

The goal is to add ~1 μL of methylation spike-in control to the gDNA sample before bisulfite conversion so that only 1% of starting material is spike-in control. *Note*: These instructions are assuming that 20 μL of your gDNA sample will only be around 100–200 ng. If higher than this range, dilution of spike-in control is not needed.

1. Perform a 1:5 dilution on the WiseGene 5-mC lambda spike-in control with nuclease-free water to create a 1.25 ng/μL final concentration.

Table 1
Genomic DNA isolations used for *Mnemiopsis leidyi* library construction

Final #	Species	Animal #	Tissue	Concentration (μg/μL)	Volume
737	*Mnemiopsis leidyi*	*Mnemiopsis*-2	Tip 500	0.173	700
739	*Mnemiopsis leidyi*	*Mnemiopsis*-3-2	Tip 500	0.0642	500
743	*Mnemiopsis leidyi*	*Mnemiopsis*-5-2	Tip 500	0.0874	500
773	*Mnemiopsis leidyi*	*Mnemiopsis*-1 2 h injury	Tip 100G	0.00828	500
774	*Mnemiopsis leidyi*	*Mnemiopsis*-2 2 h injury	Tip 100G	0.00568	500
775	*Mnemiopsis leidyi*	*Mnemiopsis*-3 2 h injury	Tip 100G	0.00444	500

2. Store dilution at +4 °C or on ice until ready to use.
3. Calculate total nanograms of ctenophore gDNA in 20 μL of sample, and add appropriate volume of lambda spike-in (not exceeding 2 μL).

When proceeding directly from isolated genomic DNA, we used 20 μL of sample isolated from the genomic tip. However, pilot testing revealed that this yielded bisulfite sequencing libraries with concentrations too low to be sequenced on NextSeq Illumina platforms. To compensate for this, and especially since these samples were whole-animal methylomes, we performed two separate bisulfite conversion reactions on 20 μL each per sample and combined the samples together during the final adapter cleanup step before PCR amplification. Since bisulfite conversion will generally cause breakages, no size selections were performed. Here we outline the Zymo methyl direct bisulfite kit and TruSeq DNA library construction kit used.

1. Add 21 μL of sample + spike-in control to 130 μL of CT Conversion Reagent solution in a PCR tube. Mix the sample, and then centrifuge briefly to ensure no droplets are in the cap or sides of the tube.
2. Proceed with the EZ DNA methyl direct protocol:

   Place the PCR tube(s) in a thermal cycler, and perform the following steps:

   (a) 98 °C for 8 min

   (b) 64 °C for 3.5 h

   (c) 4 °C storage for up to 20 h

   Note: The 4 °C storage step is optional.
3. Add 600 μL of M-Binding Buffer into a Zymo-Spin™ IC Column, and place the column into a provided collection tube.

   (a) Note: The capacity of the collection tube with the column inserted is 800 μL. Empty the collection tube whenever necessary to prevent contamination of the column contents by the flow-through.

4. Load the sample (from **Step 3**) into the Zymo-Spin™ IC Column containing the M-Binding Buffer. Close the cap, and mix by inverting the column several times.
5. Centrifuge at full speed (>10,000 × *g*) for 30 s. Discard the flow-through.
6. Add 100 μL of M-Wash Buffer to the column. Centrifuge at full speed for 30 s.
7. Add 200 μL of M-Desulfonation Buffer to the column, and let stand at room temperature (20–30 °C) for 15–20 min. After the incubation, centrifuge at full speed for 30 s.
8. Add 200 μL of M-Wash Buffer to the column. Centrifuge at full speed for 30 s.
   (a) Add another 200 μL of M-Wash Buffer, and centrifuge for an additional 30 s.
9. Place the column into a 1.5 mL microcentrifuge tube.
   (a) Add 11 μL of nuclease-free water directly to the column matrix. Centrifuge for 30 s at full speed to elute the DNA. Check columns for clogs, and centrifuge for an additional 30 s to get full elution.
   (b) Note: Even with the second spin, usually only ~9–9.5 μL of liquid will come through a column.
10. The DNA is ready for immediate library production or should be stored at or below −70 °C as ctenophore DNA is notorious for degradation at higher temperatures.

Day 2

Library preparation:

Two TruSeq DNA methylation library reactions will be needed to produce one ctenophore DNA methylation sequencing library to have sufficient concentration, sequence diversity, and coverage due to heavy gDNA fragmentation that occurs in *Mnemiopsis.*

Preparation:

1. Take all consumables for "Anneal the DNA Synthesis Primer" and "Synthesize DNA" sections of the protocol out of −20 °C storage and thaw on ice; keep all bisulfite converted samples at −80 °C until these reagents have completely thawed. Plus, vortex primers and buffers before creating master mixes. Do not pulse vortex mix enzymes.

2. While reagents are thawing, program the following programs into a thermocycler with a heated lid:

   Program 1 annealing primer.

   (a) 95 °C for 5 min in a thermal cycler with a heated lid

   Synthesize DNA program.

   (b) 25 °C, 5 min
   (c) 42 °C, 30 min
   (d) 37 °C, 2 min

   Manually pause thermocycler.

   (e) 37 °C for 10 min
   (f) 95 °C for 3 min
   (g) 25 °C for 2 min

   Terminal tagging program

   (a) 25 °C for 30 min
   (b) 95 °C for 3 min
   (c) Hold at 4 °C

3. Create enough master mix for all bisulfite reaction tubes to be processed for "Synthesize DNA." Pulse vortex the master mix before adding enzyme.
4. Prepare ice/cold water bath.
5. Take out AMPure beads, and place them at room temperature.

Producing DNA methylation library through tagmentation:

1. Proceed with TruSeq DNA methylation library protocol until the end of the "Tag DNA," and leave both sample reactions at 4 °C held in thermocycler or on ice until ready to proceed to post-tagmentation bead cleaning step.

Post-tagmentation bead cleaning alternative "Purify the Tagged DNA" protocol:

1. The DNA must be purified before PCR. Each reaction is now in a volume of 25 μL. Combine the two reactions from the same gDNA sample into a fresh 1.5 mL tube (50 μL).

   Illumina recommends using a 1.6× AMPure XP bead purification. Warm the AMPure XP beads to room temperature.

2. While the beads warm, prepare 800 μL of fresh 80% ethanol at room temperature for each sample. Vortex the AMPure XP beads until they are a homogenous suspension.

   (a) Caution: Ensure the AMPure XP beads are in a homogenous suspension before continuing.

3. Add 80 μL of the beads to each well of the plate or to each microcentrifuge tube containing ditagged DNA from "Tag the DNA."
4. Mix thoroughly by gently pipetting the entire volume of each well/tube for 10 times. If using microcentrifuge tubes, transfer each 130 μL volume to a separate 1.5 mL tube.
5. Incubate the plate/tubes at room temperature for 5 min.
6. Place the plate/tubes in a magnetic stand at room temperature for at least 5 min until the liquid appears clear.
7. Remove and discard the supernatant from each well/tube using a pipette (P200 or smaller). Some liquid can remain in each well/tube. Take care not to disturb the beads.
   (a) Note: To remove some residual liquid, use a smaller tipped P10 or P20 pipette.
8. With the plate/tubes remaining on the magnetic stand, slowly add 300–400 μL of 80% ethanol to each well/tube without disturbing the beads. Make sure that the beads are covered with 80% ethanol.
9. Incubate the plate/tubes at room temperature for at least 30 s, doing at least a quarter turn of the tube in the magnetic stand and a quarter turn back, then remove and discard all of the supernatants. Take care not to disturb the beads.
10. Repeat **Steps 9** and **10** one more time for a total of two 80% ethanol washes.
11. After the second wash, remove the ethanol by pipetting (as much as possible without disturbing the beads) with the plate/tubes still on the magnetic stand. Remove the plate/tubes, centrifuge for 10–30 s, and place the plate/tubes back on the magnetic stand for 1 min. Use a fine pipette tip to remove all the residual ethanol.
12. Let the wells/tubes air-dry for 3 min on the magnetic stand.
    (a) Note: Waving tubes with lids open for part of the incubation time is fine. If wafted smell of ethanol is still very strong after 3 min, let dry for 4 min total.
13. Add 25 μL of nuclease-free water to each well/tube, and remove the plate/tubes from the magnetic stand.
14. Thoroughly resuspend the beads by gently pipetting 20 times.
15. Incubate the plate/tubes at room temperature for 3 min.
16. Place the plate/tubes on the magnetic stand at room temperature for at least 5 min or until the liquid appears clear.
17. Carefully transfer 22.5 μL of the clear supernatant, which contains the ditagged DNA, from each well/tube to a new PCR

plate/tube. There should now be only one tube per library sample.

18. Place the plate/tubes on ice, and proceed to "Amplify the Library" and "Add an Index (Barcode)," or keep at −20 °C for longer-term storage.

Changes to the final PCR "Amplify library" protocol:

1. The recommended number of PCR cycles in the TruSeq DNA methylation protocol is ten cycles. This cycle number is based on the presumption that library construction begins with a starting material amount of 200–500 ng post-bisulfite treatment. However, the best pretreatment starting material for *Mnemiopsis* achieved is usually 100–200 ng per reaction before bisulfite conversion and when the two reactions are combined ~100 ng post-conversion.
2. Follow the PCR amplification protocol as described in the TruSeq DNA methylation library protocol for 14 cycles, and pause the thermocycler. Do not run the final thermocycler step (7 min at 68 °C). Take 1 μL of each library sample, and perform a 1:5 dilution. Prepare a 2% agarose E-gel on an E-gel reader. Load 2 μL of diluted library sample + 23 μL of nuclease-free water. Load 2 μL of diluted 50 bp ladder (1:40 dilution) into the ladder lane + 8 μL nuclease-free water. Fill all unloaded lanes with 25 μL of nuclease-free water. Run the 2% gel forward run program on the E-gel reader using the E-gel light button to check the run progress. Run until the reference ladder begins to separate (Fig. 1).
   (a) Note 1: If E-gel reader is unavailable, this can also be done by loading all 5 μL diluted library samples into a 2% agarose gel with appropriate loading dye.
   (b) Note 2: It is always advised to use TapeStation to visualize bisulfite sequencing libraries and obtain quantitative information about these libraries (Fig. 2).
3. Library bands should be both visible and bright. Continue with thermocycler library amplification for a total of 16–20 cycles.
4. Proceed to the final library bead cleaning step as instructed in the protocol. However, adjust elution volume to 30 μL as at least ~2–5 μL is lost since they cannot be recovered without also pipetting out residual magnetic beads, and ~5 μL may be needed for Qubit, TapeStation, and other concentration measurements pre-sequencing. This allows for ~10 μL to be used for sequencing and ~10 μL to be stored for backup/potential resequencing.

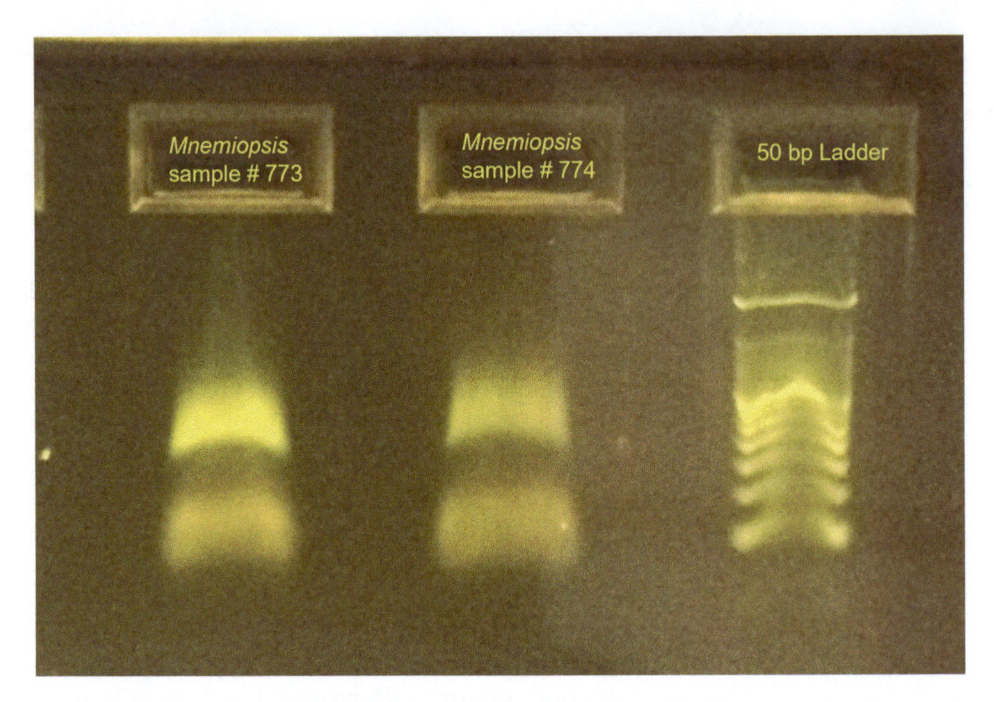

Fig. 1 Example 2% E-gel of amplified *Mnemiopsis leidyi* whole-animal bisulfite sequencing libraries. 2 μL of sample diluted 1:5 was loaded with a 50 bp ladder after 17 cycles of PCR (Lane 1, sample # 773; Lane 2, sample # 774; Lane 3, 50 bp ladder). The orange-colored band at the bottom of each sample lane is residual sequencing adapters and adapter dimers that will be cleaned out and removed during the final AMPure bead cleaning

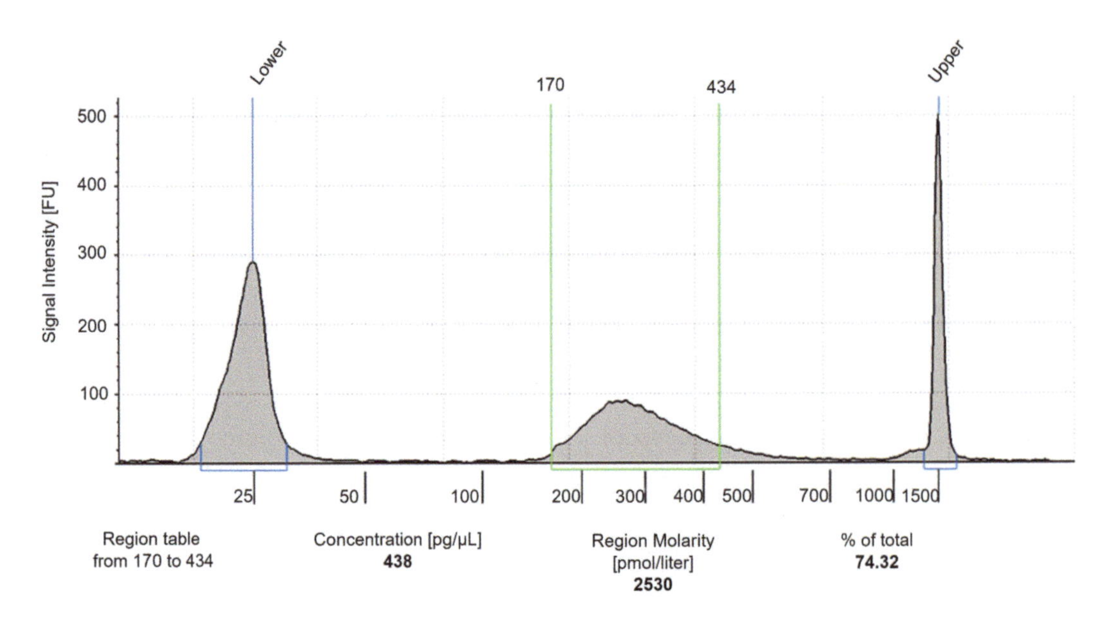

Fig. 2 Example TapeStation sample trace of *Mnemiopsis leidyi* whole-animal size distribution for bisulfite sequencing libraries. Control whole-animal bisulfite sequencing library (M37) diluted 1:5. The peak size and final concentration of this library are presented in Table 2

Table 2
Library sequencing statistics

Barcode	Barcode sequence	Amount (μL)	Sample name	Tissue	Library size	Library concentration nM	Number of reads	Number of bases
6	5′-GCCAAT-3′	10	*Mnemiopsis* an3 2 h injury	Whole animal	265	6.85	41,692,352	4,169,235,200
7	5′-CAGATC-3′	10	*Mnemiopsis* an1 2 h injury	Whole animal	261	3.255	33,478,600	3,347,860,000
8	5′-ACTTGA-3′	10	*Mnemiopsis* an2 2 h injury	Whole animal	270	3.535	35,043,424	3,504,342,400
6	5′-GCCAAT-3′	10	*Mnemiopsis*-737	Whole animal	246	4.8	71,595,718	7,159,571,800
7	5′-CAGATC-3′	10	*Mnemiopsis*-739	Whole animal	247	4.1	68,139,908	6,813,990,800
8	5′-ACTTGA-3′	10	*Mnemiopsis*-743	Whole animal	253	4.2	75,031,470	7,503,147,000

4 Downstream Bioinformatic Analysis

We used the *Mnemiopsis* Genome Project Portal as a key reference for mapping and annotation [15]. Our *Mnemiopsis* DNA methylation library sequencing reads were processed on the UF HiPerGator cloud computing system using the dmap2 methylation pipeline based on MOABS with the novel cscall methylation calling function [16]. The software can be downloaded using http://compbio.ufl.edu/software/cscall/.

5 Conclusions

The presented BS-seq library protocol described produced high-quality data for downstream analysis. Under two different states (control vs injury), we successfully detected 5-mC in whole-animal methylomes at CpG, CHG, and CHH sites with detection of changes in methylation of genes encoding gap junction proteins (innexins), pore-forming toxins and neuropeptides (Fig. 3). These new *Mnemiopsis* methylomes covered a higher percentage of the *Mnemiopsis leidyi* genome with a deeper depth of coverage than our previously published methylomes (Table 3). Differences were detected in the average gene body and average exonic DNA methylation in treatment and control groups. Although the overall gDNA methylation is lower than some previously reported methylation levels [7], the additional sequence coverage of these new samples also covered at least 300,000 additional CpG sites. Our methylation percentages are also consistent with levels reported in *Mnemiopsis* developmental larval methylomes [6]. Overall, these methods produced robust methylomes with novel insights into

Table 3
Coverage and percentages of CpG DNA methylation

Sample	Total CpGs in genome	Total CpGs called	% Sites called	Total methylated	Total methylated adjusted	% Methylated	% Methylated adjusted
Control	4,405,646	3,455,804	0.78	616,220	260,001	0.18	0.08
Injury	4,405,646	2,302,863	0.52	445,124	198,509	0.19	0.09

Mnemiopsis leidyi methylomes, the samples produced by these libraries had a higher percentage of genomic CpGs with sufficient sequencing coverage (78% of all CpGs for the control sample with 1,481,304 additional CpGs called than previous methylomes, and 52% injury with 328,363 additional CpGs called than previously reported) than our previously published methylome (45%). "Methylated sites" refer to any site where the number of called cytosines was greater than 0. However, when the presence of mCpG was adjusted to a binary based on false detection rates from spike-in controls, a lower overall rate of methylated CpG sites was detected than previously reported

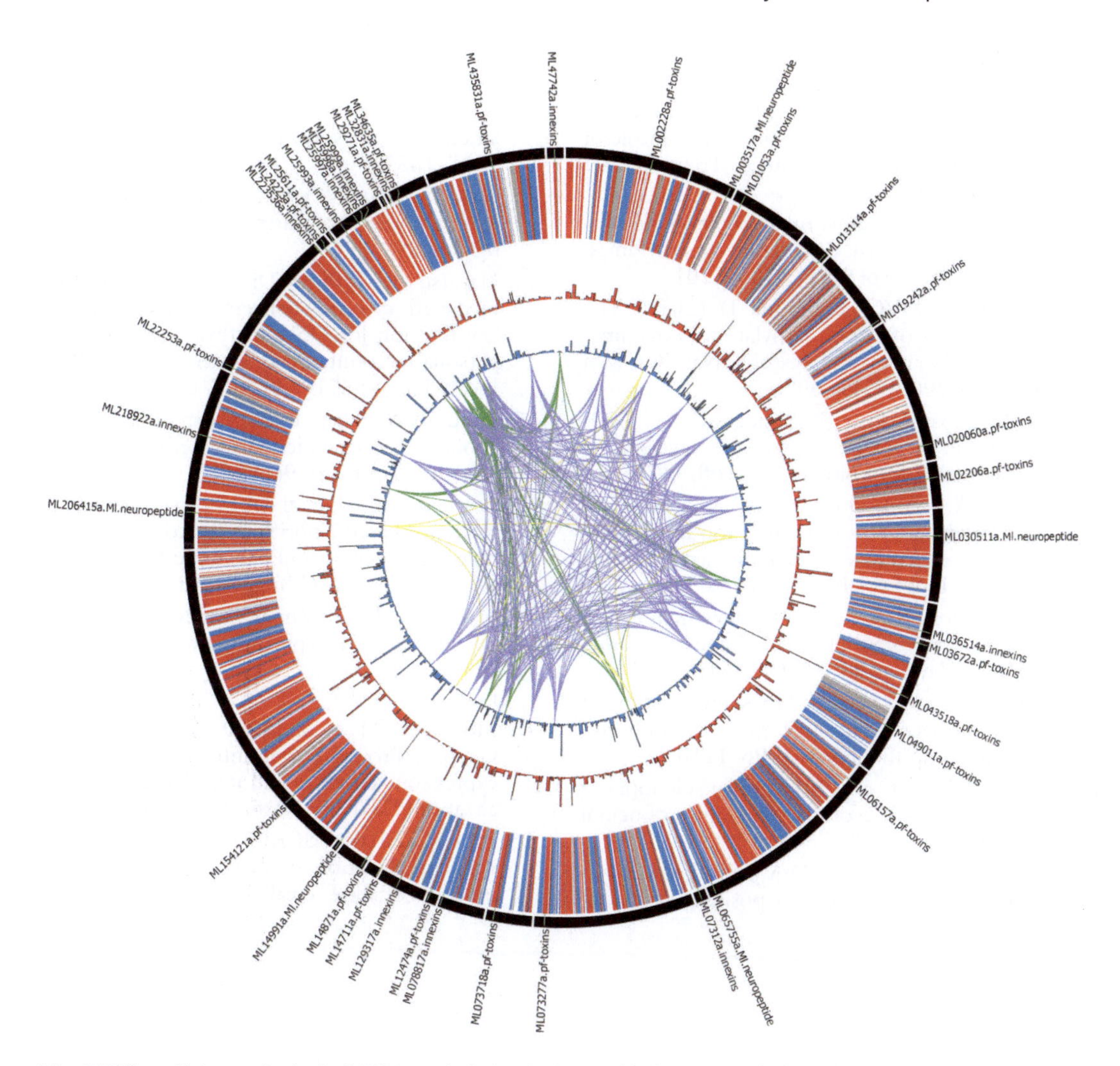

Fig. 3 Differential gene body CpG DNA methylation is detectable between whole-animal control and body-wall injury methylomes produced with the described technique. Circos genomic summary diagram for all genomic scaffolds that contain innexins (dark green linker lines), pore-forming toxins (dark purple linker lines), and neuropeptides (yellow linker lines). Genes of interest are labeled with their unique *Mnemiopsis* gene model identifiers and the gene type that is mirrored in the linker colors. The two inner barplot rings display the mean percentage of gene body methylation per gene for control (dark blue track) and 2 h injury methylomes (red track). The outer heat map plots whether the mean CpG methylation percentage—for just sites covered in both conditions—is higher in control (dark blue), 2 h post-injury (red), neither group (gray), or no overlapping cytosines were covered in the gene body between groups (white). Genes of interest are labeled with unique *Mnemiopsis* gene identifiers, and also gene groups are provided

the role of DNA methylation of innexins and neuropeptides in the regenerative abilities of *Mnemiopsis leidyi* (Fig. 3).

References

1. Hendrich B, Tweedie S (2003) The methyl-CpG binding domain and the evolving role of DNA methylation in animals. Trends Genet 19(5):269–277
2. Yi SV (2017) Insights into epigenome evolution from animal and plant methylomes. Genome Biol Evol 9(11):3189–3201
3. Greenberg MVC, Bourc'his D (2019) The diverse roles of DNA methylation in mammalian development and disease. Nat Rev Mol Cell Biol 20(10):590–607
4. de Mendoza A, Lister R, Bogdanovic O (2020) Evolution of DNA methylome diversity in eukaryotes. J Mol Biol 432(6):1687–1705
5. Zhang X, Jacobs D (2022) A broad survey of gene body and repeat methylation in Cnidaria reveals a complex evolutionary history. Genome Biol Evol 14(2):evab284
6. de Mendoza A et al (2019) Convergent evolution of a vertebrate-like methylome in a marine sponge. Nat Ecol Evol 3(10):1464–1473
7. Dabe EC et al (2015) DNA methylation in basal metazoans: insights from ctenophores. Integr Comp Biol 55(6):1096–1110
8. Moroz LL et al (2014) The ctenophore genome and the evolutionary origins of neural systems. Nature 510(7503):109–114
9. Frommer M et al (1992) A genomic sequencing protocol that yields a positive display of 5-methylcytosine residues in individual DNA strands. Proc Natl Acad Sci U S A 89(5): 1827–1831
10. Kurdyukov S, Bullock M (2016) DNA methylation analysis: choosing the right method. Biology (Basel) 5(1):3
11. Hayatsu H (2008) Discovery of bisulfite-mediated cytosine conversion to uracil, the key reaction for DNA methylation analysis--a personal account. Proc Jpn Acad Ser B Phys Biol Sci 84(8):321–330
12. Yong W-S, Hsu F-M, Chen P-Y (2016) Profiling genome-wide DNA methylation. Epigenetics Chromatin 9(1):26
13. Teschendorff AE et al (2017) A comparison of reference-based algorithms for correcting cell-type heterogeneity in Epigenome-Wide Association Studies. BMC Bioinformatics 18(1):105
14. Teschendorff AE, Zheng SC (2017) Cell-type deconvolution in epigenome-wide association studies: a review and recommendations. Epigenomics 9(5):757–768
15. Moreland RT et al (2020) The *Mnemiopsis* Genome Project Portal: integrating new gene expression resources and improving data visualization. Database (Oxford) 2020:baaa029
16. Riva A (2016) CSCALL: tools for genome-wide methylation analysis. http://compbio.ufl.edu/software/cscall/

Chapter 19

Ocean to Tree: Leveraging Single-Molecule RNA-Seq to Repair Genome Gene Models and Improve Phylogenomic Analysis of Gene and Species Evolution

Jan Hsiao, Lola Chenxi Deng, Leonid L. Moroz, Sreekanth H. Chalasani, and Eric Edsinger

Abstract

Understanding gene evolution across genomes and organisms, including ctenophores, can provide unexpected biological insights. It enables powerful integrative approaches that leverage sequence diversity to advance biomedicine. Sequencing and bioinformatic tools can be inexpensive and user-friendly, but numerous options and coding can intimidate new users. Distinct challenges exist in working with data from diverse species but may go unrecognized by researchers accustomed to gold-standard genomes. Here, we provide a high-level workflow and detailed pipeline to enable animal collection, single-molecule sequencing, and phylogenomic analysis of gene and species evolution. As a demonstration, we focus on (1) PacBio RNA-seq of the genome-sequenced ctenophore *Mnemiopsis leidyi*, (2) diversity and evolution of the mechanosensitive ion channel Piezo in genetic models and basal-branching animals, and (3) associated challenges and solutions to working with diverse species and genomes, including gene model updating and repair using single-molecule RNA-seq. We provide a Python Jupyter Notebook version of our pipeline (GitHub Repository: Ctenophore-Ocean-To-Tree-2023 https://github.com/000generic/Ctenophore-Ocean-To-Tree-2023) that can be run for free in the Google Colab cloud to replicate our findings or modified for specific or greater use. Our protocol enables users to design new sequencing projects in ctenophores, marine invertebrates, or other novel organisms. It provides a simple, comprehensive platform that can ease new user entry into running their evolutionary sequence analyses.

Key words Ctenophora, *Mnemiopsis*, De novo transcriptome, Single-molecule sequencing PacBio SMRT, Phylogenomics, Phylogenetic trees, Gene family tree, Mechanosensitive ion channels, Piezo

1 Introduction

Ctenophores are transparent gelatinous, almost alien-like, marine animals of enigmatic, if controversial, origin and biology [1–4]. They exhibit a number of unusual biological features, including rotational symmetry and fourfold structuring of the body plan (unique in animals) [1, 5–7], circumvention of physical constraints in cilia-based swimming that otherwise limit body size across

Leonid L. Moroz (ed.), *Ctenophores: Methods and Protocols*, Methods in Molecular Biology, vol. 2757,
https://doi.org/10.1007/978-1-0716-3642-8_19, © Springer Science+Business Media, LLC, part of Springer Nature 2024

kingdoms and do so by fusing cilia into massive comb plates of 100,000 cilia each [8–11], and a tripartite through-gut with anal pores at one end that make 'thru-ness' of the gut unobvious, a fact that some biologists largely forgot for over a century, and after its rediscovery and modern analysis, the gut's possible homology to tripartite through-guts in other animals remains unclear [12–14]. Moreover, common animal traits such as neurons, synapses, muscles, and mesoderm could potentially be a result of convergent evolution in ctenophores vs. other animals [15–17]. Finally, ctenophore genomes are unusual, even among marine invertebrate oddities, and are recognized as highly divergent in gene sequence and gene families in comparison with other animals [18, 19]. Overall, a deepened understanding of ctenophore diversity and evolution, from sequences to ecosystems, promises new insights into their extraordinary biology.

Initial genome publications for ctenophores were of draft assemblies for *Mnemiopsis leidyi* (*Mnemiopsis*) and *Pleurobrachia bachei* (*Pleurobrachia*) [18, 19]. More recent assembly of *Hormiphora californiensis* (*Hormiphora*) [20] and updated 3D genome assembly in *Pleurobrachia* [21] identified 13 chromosomes in both species and greatly improved available resources. Famously, the initial publications ignited controversy regarding phylogenetic placement of ctenophores in animal evolution [18, 19, 22–27]. The studies contradicted traditional views, indicating ctenophores are the basal-most branch in animals and calling into question conservation vs. convergence of basic animal features, including guts, muscles, and brains [7, 18, 19, 28, 29]. Subsequent studies and commentaries remained deeply divergent, often falling into ctenophore-first vs. sponge-first hypotheses [18, 19, 22–27, 30, 31].

Furthermore, it was shown that technical details regarding evolutionary models and available parameter space in different software packages may help account for dramatically different trees using similar data in different sequence-based studies [4]. Of note, our pipeline here uses maximum-likelihood tree building in IQTree2 and, as might be predicted for the tool [4] and as shown in Fig. 4, places ctenophores at the animal base. Until quite recently, it remained a subject of hot debates which of the two scenarios—ctenophore-first or sponge-first—is correct [2, 4]; although integrative analyses favored the ctenophore-first reconstructions. However, a novel approach using syntenic features across metazoan and unicellular outgroup genomes demonstrated ctenophore-first as correct [32].

Importantly, the divergent nature of ctenophore genes and genomes and their use in characterizing early animal evolution highlight fundamental challenges in sequence analysis when phylogenetic signal is limited [2, 33–35]. Work on ctenophores has led to new tools, approaches, and understanding in the phylogenomics field, including some of the methods used here [2, 33–35]. Whatever the evolutionary patterns of novelty, conservation, and convergence, phylogenomic study of ctenophores genes and genomes will be critical to understanding early animal evolution and basic principles of animal cell types and systems across phyla and in humans.

Biodiversity offers an incredibly rich potential for discovery of new genes and pathways that can be directly utilized or informatically leveraged in basic research and medicine. Sequencing and phylogenetic characterization of diverse genes in novel organisms have led to the discovery, advancement, and engineering of some of the most powerful genetic tools in science, including Taq for PCR [36], GFP for fluorescent imaging [37], channelrhodopsin for optogenetics [38], and CRISPR-Cas9 for gene editing [39]. Additional discovery and characterization of naturally occurring sequence diversity representing new or novel homologs in related species continue to advance each of these technologies and many others [40–43]. These tools and their advances underscore the importance of genome-scale sequencing of organismal diversity in closely and distantly related species and the importance of phylogenomic characterization of homologs across species; both are areas the protocol here enables for users.

Integrative experimental and phylogenomic approaches are commonly used in protein engineering and can be combined with additional increasingly powerful machine learning, deep learning, and other artificial intelligence (AI) methods. These tools exploit biodiversity and evolution's optimization of diverse sequences and functions through random, potentially near-comprehensive, explorations of sequence space. In this context, presence and distribution of existing biological or estimated ancestral phylogenetically related or unrelated neighboring sequences in sequence space can be highly informative [44], representing successful functional optimization in protein engineering and design by evolution. Leveraging natural sequence diversity at scale, the tools and approaches can massively collapse the sequence space that might otherwise have to be experimentally explored, highlighting a much smaller subset of potential variants. This can save by orders of magnitude the time and resources required, for instance, to rapidly identify or engineer a novel sequence and advance a genetic technology. This is especially true now that it is possible to predict the 3D structure of nearly all proteins based on sequence alignment and better infer potential function of motifs, domains, and regions for engineering [45–51]. Again, these areas in basic and biomedical research highlight the importance of genome-scale sequencing of organismal diversity in closely and distantly related species and the importance of phylogenomic characterization of homologs across species; both are areas the protocol here enables for users.

Phylogenomic analysis of gene family evolution includes three major steps:

1. Candidate homolog identification by searching reference sequences across species genomes and transcriptomes.
2. Multiple sequence alignment of reference and candidate sequences.
3. Gene tree generation based on aligned sequences [52].

Additional generation of a species tree by a similar process may also be needed, particularly when one or more species, or the collective set of species, have never been characterized. Phylogenomic analyses to produce gene and species trees are commonly undertaken in labs for the first time after newly acquired genome or transcriptome data sets arrive. Numerous tools and packages exist and can be run naively, often with reasonable accuracy and without detailed understanding of the command line or coding.

We recommend several such tools below. Still, their often extensive functionality can be overwhelming; some are not free, you are limited to specific subsets of tools developers happened to have packaged for a given task, and underlying steps and code may be inaccessible with little context. In contrast, cutting-edge phylogenomic and phylogenetic tools run at the command line are often freely available and part of a diverse universe of tools offering specific functionalities. These tools enable users to build powerful bioinformatic and phylogenomic pipelines targeted to their particular needs. They may readily scale but require proficiency in working at the command line and coding one or more languages that some researchers may not have yet acquired.

To facilitate sequencing and phylogenomic analysis by new users, we provide here a protocol that includes a high-level workflow of animal collection and single-molecule RNA-seq and a more detailed walk through phylogenomic analysis of species and gene evolution, including production of alignments, trees, and species-gene family heatmaps (Fig. 1).

As a workflow demonstration, we generated single-molecule (long-read) PacBio Sequel II RNA-seq data of an entire adult *Mnemiopsis leidyi* and produced a transcriptome. Our phylogenomic pipeline uses the PacBio transcriptome, which is predominantly accurate full-length transcripts, to improve gene models in the *Mnemiopsis* draft genome, and then performs species and gene evolutionary analysis of Piezo, a mechanosensitive ion channel [53], in basal-branching animals, including ctenophores *Mnemiopsis* and *Hormiphora*.

Finally, we provide a Python Jupyter Notebook that runs a fully fledged demonstration of the pipeline analysis. The notebook offers a simple means and user-friendly environment for nonexperts to run and adapt the pipeline. One can see the underlying code in

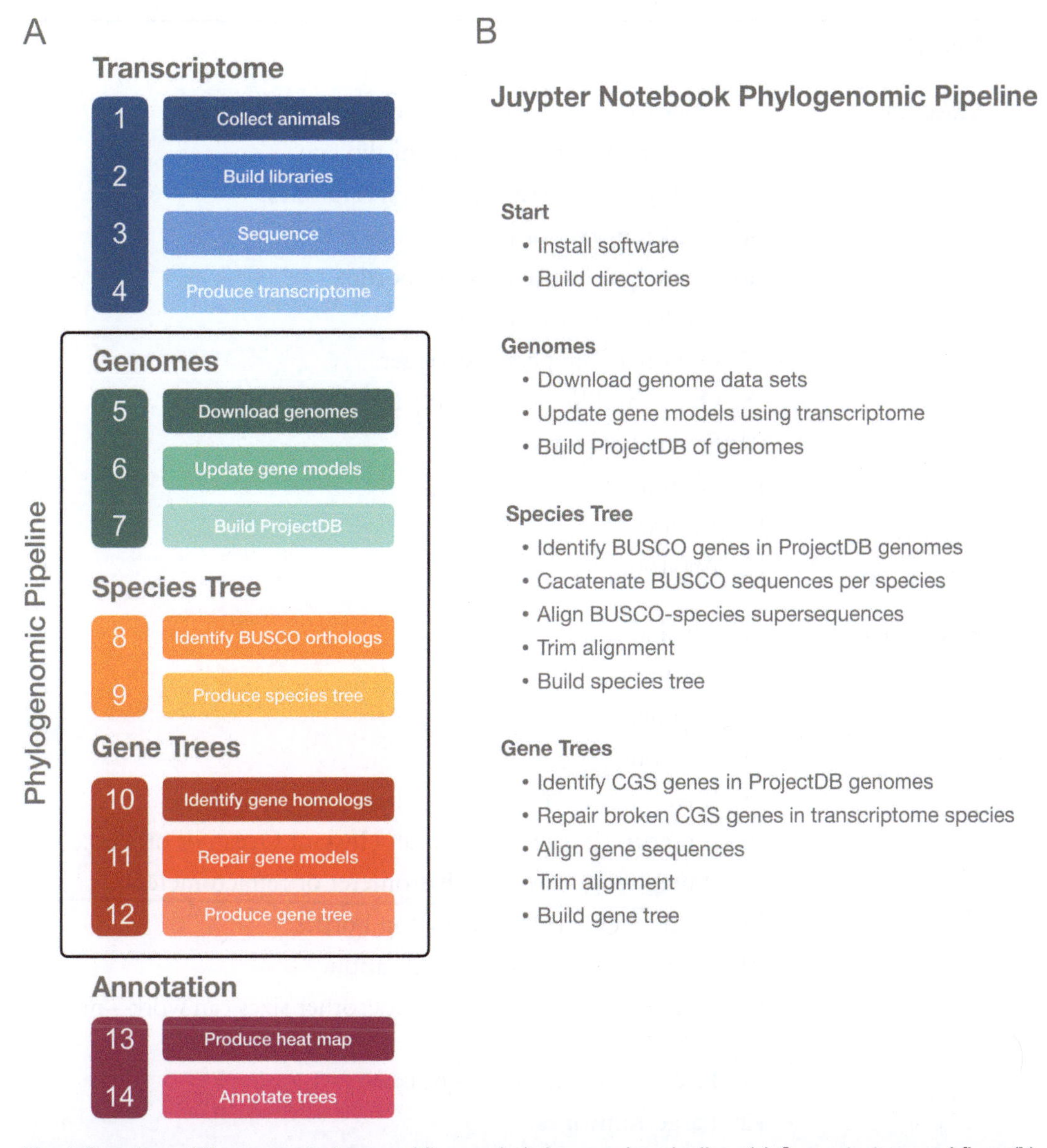

Fig. 1 Overview of the ocean-to-tree workflow and phylogenomics pipeline. (**a**) Ocean-to-tree workflow. (**b**) Outline of the phylogenomic pipeline that is run in the associated Jupyter Notebook that is provided at GitHub (https:/github.com/000generic/Ctenophore-Ocean-To-Tree-2023)

action and gain familiarity with the methods in the process of running the notebook. At the same time, the notebook provides full access to all scripts, allowing expert users to build off of our pipeline and advance their own specific projects. The notebook can be run for free in the Google Colab cloud, locally, or elsewhere.

2 Materials

2.1 Collection

1. Dock or seawall access to ocean or sea.
2. Large or aquarium dip net or dip cup.
3. 5–20 glass or plastic jars of varying sizes.
4. 5–20 L bucket or cooler.
5. Watertight insulated water bottle.
6. Cooler or Styrofoam box.
7. Optional: Temperature logger.
8. Optional: Blue ice or heat packs.
9. Small aquarium dip net.
10. Air pump.
11. Air line.
12. Air stone.

2.2 RNA Extraction

1. Small aquarium dip net.
2. Air pump.
3. Air line.
4. Air stone.
5. If collecting seawater: Bag filter.
6. If preparing seawater: Sea salt (Instant Ocean or others).
7. If preparing seawater: Hydrometer or refractometer.
8. Sterile filter that fits screw-top bottles.
9. Vacuum source for sterile filtration.
10. Five 1-L sterilized glass bottles (other sizes can work—need to hold 2.5–5 L total).
11. Five 1-L sterilized glass beakers.
12. Large Kimwipes.
13. Electric homogenizer (Polytron 1200 or others).
14. *Alternate: Glass mortar and pestle.*
15. RNA extraction kit (Qiagen RNeasy or others).
16. Surface decontamination solution (RNase Away or others).
17. RNase-free plasticware.
18. Low-binding DNA microcentrifuge tubes (DNA LoBind or others).
19. Fume hood.
20. Spectrophotometer (NanoDrop or others).
21. Fluorometer (Qubit or others).
22. Automated small-volume electrophoresis system (TapeStation, Bioanalyzer, or others).

2.3 Sequencing

1. Single-molecule sequencing system (PacBio Sequel II or others).
2. Single-molecule sequencing kit (PacBio Sequel II or others).
3. Spectrophotometer (NanoDrop or others).
4. Fluorometer (Qubit or others).
5. Automated small-volume electrophoresis system (TapeStation, Bioanalyzer, or others).

2.4 Phylogenomics

Python 3 programming language	https://tinyurl.com/yx2vt8mu
Python package installer: Anaconda	https://tinyurl.com/yrxzaase
Unix command line editor: Vim	https://tinyurl.com/4sxe3amu
Alternate: Emacs	https://tinyurl.com/yc8pjtdv
Lightweight editor: Atom	https://tinyurl.com/2p8ec7z7
Alternate: Sublime ($)	https://tinyurl.com/5yxekktu
Full IDE editor: Spyder	https://tinyurl.com/yeywt9pw
Alternate: PyCharm (free or $)	https://tinyurl.com/58kw5pv7
Alternate: Visual Studio Code	https://tinyurl.com/2p8f2xte
Pipeline run documentation: Jupyter Notebook	https://tinyurl.com/2w4trfku
Pipeline push-button automation: Snakemake	https://tinyurl.com/2cra9mxx
Alternate: YAML	https://tinyurl.com/ymr4c7ty
Pipeline standalone functionality: Singularity	https://tinyurl.com/jyy3sv7c
Computing: Cloud Google Colab (free or $)	https://tinyurl.com/489ttan7
Alternate: Google Cloud Life Sciences ($)	https://tinyurl.com/5e2wfvwn
Alternate: Cloud Amazon Web Services ($)	https://tinyurl.com/bdu4p6vj

(continued)

Alternate: Local research-grade machine or cluster ($)	NA
Alternate: Local laptop or desktop ($)	NA
Transcriptome QC: EvidentialGene	https://tinyurl.com/4mdkkrm9
Sequence searcher: Blast Suite	https://tinyurl.com/ycktxjsd
Single-gene gene family identifier: BUSCO	https://tinyurl.com/2p8mvjau
Sequence aligner: MAFFT	https://tinyurl.com/ve3cdzd2
Alignment trimmer: ClipKit	https://tinyurl.com/rvkyp4a7
Tree builder maximum likelihood-like: FastTree2	https://tinyurl.com/366ajd6t
Tree builder maximum likelihood: IQTree2	https://tinyurl.com/3mna5kte
Alternate: PhyloBayes (Bayesian Inference)	https://tinyurl.com/4kpbph6a
Alignment viewer: AlignmentViewer	https://tinyurl.com/2f7xscr7
Tree viewing: FigTree	https://tinyurl.com/5n8zay9z
Alternate: iTOL—Interactive Tree of Life (free or $)	https://tinyurl.com/4avjbunw
Spreadsheet data analysis: Google Sheets	https://tinyurl.com/mtdzdkf3
Graphics-friendly software: Google Slides	https://tinyurl.com/2p993bcn
Phylogenomic pipeline alternate: Geneious ($)	https://tinyurl.com/2bbr5y7u
Alternate: Galaxy (free or $)	https://tinyurl.com/4etvu7sv
Phylogenetic pipeline alternate: CIPRES (free or $)	https://tinyurl.com/2p8pc48v
Alternate: NGPhylogeny	https://tinyurl.com/2p88uss6
Alternate: PhyloToL	https://tinyurl.com/44etynyk

2.5 Source Data

Name	Type	Database ID	URL
Ichthyosporea *Sphaeroforma arctica*	Genome	NCBI RefSeq GCF_001186125.1	https://tinyurl.com/yrk8aeca
Filasterea *Capsaspora owczarzaki*	Genome	NCBI RefSeq GCF_000151315.2	https://tinyurl.com/yeu3dpsm
Choanoflagellata *Monosiga brevicollis*	Genome	NCBI RefSeq GCF_000002865.3	https://tinyurl.com/bdzh3huk
Porifera *Ephydatia muelleri*	Genome	EphyBase v1	https://tinyurl.com/2c3de66a
Porifera *Amphimedon queenslandica*	Genome	NCBI RefSeq GCF_000090795.1	https://tinyurl.com/yc69a5ae
Ctenophora *Mnemiopsis leidyi*	Genome	Ensembl Metazoa 51 MneLei_Aug2011	https://tinyurl.com/fmakzcz6
Ctenophora *Hormiphora californiensis*	Genome	GitHub Hormiphora Hcv1.av93	https://tinyurl.com/2p83cvpd
Cnidaria *Nematostella vectensis*	Genome	Stowers Institute NVEC200	https://tinyurl.com/2p893t24
Cnidaria *Morbakka virulenta*	Genome	OIST MG MOR05_r06	https://tinyurl.com/bd8jkvn6
Cnidaria *Rhopilema esculentum*	Genome	GigaDB 100720	https://tinyurl.com/2k8pxmcv
Cnidaria *Hydra vulgaris*	Genome	NIH NHGRI Hydra2.0	https://tinyurl.com/yckzbjew
Placozoa *Trichoplax adhaerens*	Genome	NCBI Genome GCF_000150275.1	https://tinyurl.com/mnw3z5cj
Chordata *Homo sapiens*	Genome	NCBI RefSeq GCF_000001405.39	https://tinyurl.com/4fd2c75v
Arthropoda *Drosophila melanogaster*	Genome	NCBI RefSeq GCF_000001215.4	https://tinyurl.com/mvbt9hw4

(continued)

Name	Type	Database ID	URL
Nematoda *Caenorhabditis elegans*	Genome	NCBI RefSeq GCA_000002985.3	https://tinyurl.com/2p8c6wnv
Ctenophora *Mnemiopsis leidyi*	Transcriptome	NCBI SRA SRR18002386	https://tinyurl.com/mxyy7jca
Chordata *Homo sapiens* Piezo 1	Gene	UniProt Q92508	https://tinyurl.com/ycknp27k
Chordata *Homo sapiens* Piezo 2	Gene	UniProt Q9H5I5	https://tinyurl.com/tpebj8fm
Arthropoda *Drosophila melanogaster* Piezo	Gene	UniProt M9MSG8	https://tinyurl.com/2p8evtrd
Arthropoda *Drosophila melanogaster* Piezo-like	Gene	UniProt A0A126GUQ2	https://tinyurl.com/253fvf54
Nematoda *Caenorhabditis elegans* Piezo	Gene	UniProt A0A061ACU2	https://tinyurl.com/4a7fsuz2

3 Methods

3.1 Collection

1. Ctenophore collections vary depending upon locations and conditions. Readily available genera include *Mnemiopsis*, *Pleurobrachia*, and *Beroe*. Capture one or more individuals of *Mnemiopsis leidyi* (Atlantic Ocean: University of Chicago Marine Biological Laboratory, Woods Hole, MA), *Pleurobrachia bachei* (Pacific Ocean: University of Washington Friday Harbor Laboratories, Friday Harbor, WA), other small-sized ctenophores, or other species found at a dock, seawall, or ocean using a dip net, dip cup, or small plankton net on a line (Fig. 2a–d) (*see* **Note 1**).

2. Maintain animals individually in small jars, ideally aerated, or in a 5–20 L bucket or cooler with fresh seawater during collection (Fig. 2b). When possible, replenish water periodically, particularly if air and water temperatures are very different. Avoid overloading a bucket with animals or plankton as water conditions can deteriorate quickly.

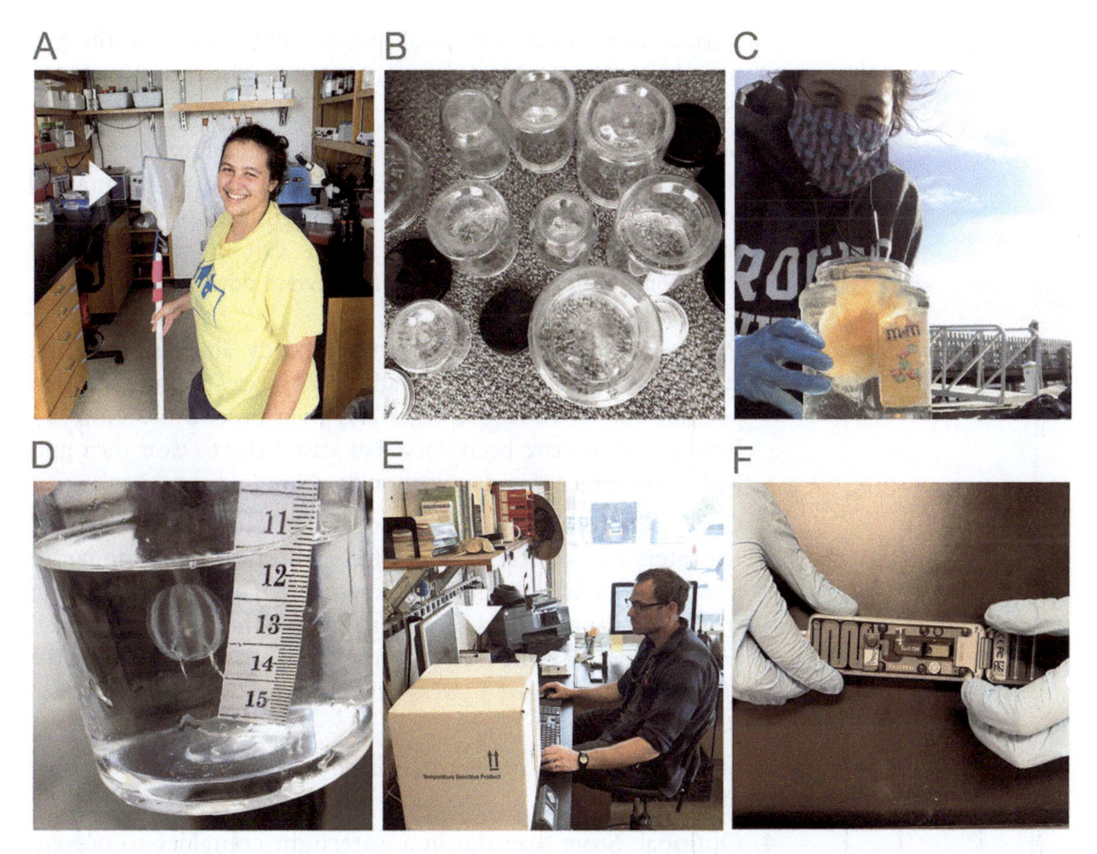

Fig. 2 Field collection, shipping, and sequencing of ctenophores, jellyfish, or other surface-dwelling marine organisms. (**a**) A medium-sized aquarium net duct-taped onto a piece of PVC piping (white arrow) can provide a simply effective tool for collecting surface-dwelling animals when walking along docks or seawalls at the ocean. (**b**) Individual ctenophores or jellyfish can be held in small jars after collection to help avoid damage. (**c**) Collecting ctenophores and jellyfish off docks in Bristol, RI, dock walker M. Cordeiro is pictured with the lion's mane jellyfish Cnidaria *Cyanea capillata*. (**d**) Ctenophora *Pleurobrachia pileus* collected by M. Cordeiro off docks in Bristol, RI. (**e**) Shipment of Ctenophora *Mnemiopsis leidyi* (boxed up and indicated by white arrow) by E. Edsinger and S. Bennet (pictured) from the University of Chicago Marine Biological Laboratory Marine Resources Center to the University of Florida Whitney Laboratory for Marine Bioscience Moroz Laboratory for single-cell sequencing. (**f**) Oxford Nanopore Technologies MinION and related sequencers provide inexpensive, user-friendly means of doing single-molecule genome and RNA-seq in a laboratory using a laptop and molecular biology tools and reagents

3. Transfer ctenophores to a watertight insulated bottle or cooler filled with fresh seawater using a small aquarium dip net (*see* **Note 2**). Aim for a density of five to ten animals golf ball or smaller in size in 1–2 L of seawater. Fill to brim to minimize sloshing that could damage animals.

4. Transport or ship animals to laboratory shortly after collection or else transport, maintain in aquaria with appropriate temperature and aeration, and ship later. Ship the insulated watertight container containing animals in a Styrofoam box (Fig. 2e). Blue ice for polar-to-cooler temperate species or heat packs for

warmer temperate-to-tropical species can be added to the box to help maintain appropriate temperatures for the animals (*see* **Note 3**).

5. Set up ctenophores with aeration upon arrival after transport or shipping. Polar-to-cooler temperate species can be maintained in a 4 °C cold room for 1–3 days without food. Similarly, warmer temperate-to-tropical species can be maintained at room or elevated temperatures for several days without food.

3.2 RNA Extraction

1. Process animals for RNA extraction shortly after arrival in the laboratory, or maintain with aeration at a species-appropriate temperature for up to several days, and then process (*see* **Note 4**). Ensure animals have been starved at least 1 day to clear their guts and avoid sequence contamination by prey.
2. Collect or prepare 5–10 L of seawater (*see* **Note 5**). If collecting seawater, bag or sterile filter after collection to minimize or remove organic and inorganic material. If preparing seawater from sea salt, match its salinity to that of the seawater animals were shipped in. The goal is to minimize osmotic shock to the animals and associated stress-related gene expression prior to RNA extraction.
3. Sterile filter 3–5 L of seawater using vacuum filtration or other methods.
4. Optional: Store seawater in a watertight container to prevent evaporation, or sterile-filtered seawater in sterilized 1 L or other sized glass or plastic bottles, for several days or weeks. Ideally, keep cold and in the dark to minimize growth by microorganisms for non-sterile water.
5. Equilibrate seawater and sterile seawater to the temperature of seawater animals were shipped in or to short-term culturing temperature, if animals will be maintained for a short time after arrival in the lab. Again, the goal is to minimize temperature shock to the animals and associated stress-related gene expression prior to RNA extraction.
6. Decontaminate lab bench and fume hood areas for work with RNA (*see* **Note 6**).
7. Prepare five beakers of temperature-equilibrated sterile filtered seawater at 500 mL to 1 L volumes.
8. Wash ctenophores one at a time in sterile seawater by gently collecting and transferring between beakers of sterile seawater using a clean aquarium dip net. Allow 30 s or longer in each beaker. It is possible to accumulate sterile seawater-washed animals in the final beaker.

9. Work on 1–3 large Kimwipes layered on one another on the decontaminated bench. Kimwipes can be replaced periodically as needed.
10. Transfer the animal onto scrunched up Kimwipes to dry briefly.
11. Transfer lightly dried animal to 50 mL tube containing an appropriate volume of RNA extraction buffer based on kit instructions (*see* **Note** 7).
12. Homogenize immediately in fume hood using electric homogenizer for 10–30 s (*see* **Note 8**).
13. Continue extraction according to kit protocol, and make the final elution into RNase-free water. Process the sample within a few days when possible, as samples will degrade over time in pure water.
14. Determine RNA purity using a spectrophotometer according to equipment instructions. Pure RNA has a 260/280 ratio of 2.0, and pure nucleic acid has a 260/230 ratio of 2.0–2.2. The lower values suggest contamination. Samples that are contaminated might be re-extracted per RNA extraction kit, RNA cleanup and concentration kit, or other protocols (*see* **Note 9**). RNA concentrations determined by spectrophotometry are typically less accurate than other methods.
15. Determine RNA concentration using a fluorometer according to equipment directions. General 2–10 μg is desirable, as PacBio RNA-seq libraries require 1 μg for preparation, though it is possible to use less.
16. Determine RNA quality using a small volume electrophoresis system according to equipment directions. RIN values greater than 8.0 and closer to 9 or 10 are needed to ensure full-length RNA molecules for full-length single-molecule sequencing (*see* **Note 10**).

3.3 Single-Molecule Sequencing

1. An overview of our sequencing strategy is provided (Fig. 2).
2. Prepare sequencing libraries for high-accuracy single-molecule RNA-seq according to sequencing technology kit directions (*see* **Note 11**). In our demonstration workflow, we used PacBio SMRT Sequel II single-molecule sequencing technology and sequencing library kit (https://www.pacb.com/). At this time (Feb 2022), it offers the highest available accuracy in single-molecule sequencing.
3. Perform sequencing of sequencing libraries according to sequencing technology equipment directions. In our demonstration workflow, we used PacBio Sequel II sequencing based on the technology's accuracy. Pictured is the incredibly small, inexpensive, and user-friendly Oxford Nanopore Technologies MinION sequencer (Fig. 2f).

4. Perform post-sequencing quality control and production of a final transcriptome using sequencing technology software. In our demonstration workflow, we used the PacBio SMRT Link software associated with the sequencing system.
5. Deposit reads and transcriptome in NCBI SRA, SRR, and TSA databases, respectively. Demonstration data sets for *Mnemiopsis leidyi* are available at NCBI (BioProject PRJNA806463; BioSample SAMN25884795; Reads SRR18002386; Transcriptome TSA) (*see* **Note 12**).

3.4 Genome Gene Model Updates

1. A demonstration of the following phylogenomic pipeline can be run using our provided Python Jupyter Notebook (GitHub Repository: Ctenophore-Ocean-To-Tree-2023 https://github.com/000generic/Ctenophore-Ocean-To-Tree-2023) in the Google Colab Free cloud or elsewhere. It requires no external input and produces Metazoa15 species (see below) and Piezo gene family alignments and trees.
2. Select computing and software options, and install all software in preparation of running the phylogenomic pipeline. *See* Subheading 2.4 for suggested options. Specific scripts and outside software used in the Jupyter Notebook version of our pipeline were selected or optimized in part to allow the pipeline to run on Google Colab Free (you will need a Google account; https://colab.research.google.com/). Thus, notebook specifics may not be the ideal option if adapting the pipeline to a specific project but can be a good place to start.
3. Download genome gene model gene sets (mRNA, CDS, and AA) and GFF or GTF files for each single-molecule sequenced species (*see* **Note 13**).
4. If isoforms are present in genome gene model gene sets (mRNA, CDS, and AA), collapse isoforms to the longest or best representative transcript. Processing can vary species to species and genome source to genome source. It can typically be done using GFF/GTF and/or fasta header information. Biopython (https://biopython.org/) tools or in-house Python scripts can be used to do this. In the case of the *Mnemiopsis leidyi* genome at Ensembl Metazoa (*see* Subheading 2.5 for link), no isoforms are present in the genome gene model gene set, so the downloaded data is ready to use for sequence repair.
5. Download or copy the single-molecule transcriptome data set (mRNA or CDS) to make it locally available, if it is not already. *See* Subheading 2.5 for link to our *Mnemiopsis* transcriptome.
6. Translate the transcriptome, and select the best transcript per "locus" independent of the genome using EvidentialGene [54] or other tools like CD-Hit [55] to produce a "T1"

transcriptome. EvidentialGene will produce many files, including mRNA (cdna), CDS (cds), and AA (aa) "okay" transcriptomes that include only EvidentialGene's designated best transcript per "locus" (*see* **Note 14**).

7. Build a Blastn database of T1 transcriptome mRNA using Blast + suite makeblastdb.
8. Blast genome CDS gene models against the T1 mRNA transcriptome database using Blastn (*see* **Note 15**).
9. Parse the blast report to identify T1 transcripts that have only a single gene model hit of e-value 0.0 (or can use other or additional Blast statistics and thresholds) (*see* **Notes 16** and **17**).
10. Update genome gene model mRNA, CDS, and AA to T1 transcriptome sequences. In our demonstration pipeline, this is referred to as the UPDATED genome for *Mnemiopsis.*

3.5 Project Database

1. An overview of our phylogenomics pipeline is provided (Fig. 1).
2. Download genome gene model gene sets (mRNA, CDS, and AA) and GFF or GTF files for species of interest (*see* **Notes 18** and **19**). For our demonstration pipeline, we focus on basal-branching animals (ctenophores, sponges, placozoans, and cnidarians) and unicellular outgroups, collectively referred to here as Metazoa15.
3. If isoforms are present in genome gene model gene sets (mRNA, CDS, and AA), collapse isoforms to the longest or the best representative transcript. Processing can vary species to species and genome source to genome source. It can typically be done using GFF/GTF and/or fasta header information. Biopython (https://biopython.org/) tools or in-house Python scripts can be used to do this.
4. Standardize file names and header information. To make visual interpretation of trees easier and provide phylogenetic context on an alignment or tree, our pipeline replaces header information with just the Phylum Genus species details and uses simple sequence identifiers (pdb0000000000) specific to the pipeline run. A map of source and pipeline header details is produced, so things can be mapped back and forth, if needed. There updated files are the pipeline's ProjectDB fastas.
5. Produce a Blastp database for each genome.

3.6 Species Tree Homologs

1. Run BUSCO and its latest Metazoa HMMs [56] on each genome to identify single-copy gene family orthologs to later use in generating a species tree.

2. Process the BUSCO gene fasta files produced by BUSCO Metazoa to reflect species names and ProjectDB identifiers and to provide sets of all BUSCO single-copy sequences per genome in a single file.

3.7 Species Alignment and Tree

1. Concatenate all BUSCO sequences per species genome. The order of sequences must be identical across species. For genes that are absent in a given genome, place holder sequence, such as 10 X's in a row, or simply no sequence, can be used.
2. Align the concatenated BUSCO sequences using MAFFT [57].
3. Trim the aligned sequences using ClipKit with the smartgap setting (*see* **Note 20**) [58].
4. Build a species tree using FastTree2 or IQTree2 for maximum-likelihood methods and/or PhyloBayes for Bayesian inference methods (*see* **Note 21**) [59–61].

3.8 Gene Family Homologs

1. Identify representative reference genes for the gene family of interest in the literature or elsewhere. These genes will be used to scan and identify homologs in Metazoa15 genomes. Ideally, sequences will be from species having high-quality well-annotated genomes. For our demonstration pipeline, we focus on three genetic models in neuroscience, human (*Homo sapiens*), fly (*Drosophila melanogaster*), and worm (*Caenorhabditis elegans*) (*see* **Note 22**). Critically, our pipeline uses a Reciprocal Top Family (RTF) method (detailed below). It requires all homologs of a gene in a genome be included as reference genes for the method to work correctly.
2. Collect CDS or AA fasta sequences for the selected reference genes from UniProt (protein sequences only; https://www.uniprot.org/), GenBank (https://www.ncbi.nlm.nih.gov/nuccore/), genome database websites, or other sources (*see* **Note 23**). This collection of reference sequences is referred to here as the reference gene set (RGS).
3. Blast RGS sequences against each Metazoa15 ProjectDB genome database (*see* **Note 24**).
4. Parse the blast reports and genome fastas to produce a single nonredundant fasta of all RGS hits in all Metazoa15 genomes, the All Hits Fasta.
5. Blast RGS sequences against just the RGS genomes (*see* **Note 25**).
6. Parse the blast report and genome fastas, and then update headers of identified RGS genes in each genome with the header used in the RGS fasta file.

7. Produce a blast database of the RGS-header updated RGS genomes combined.
8. Blast the All Hits Fasta sequences against the RGS genome database.
9. Parse the blast report, and produce a fasta file of all hits that any one of the RGS sequences as a top hit in the combined RGS genomes. This collection of sequences represents all identified potential homologs searched by RBF in the genomes. The sequences are referred to here as the candidate gene set (CGS).
10. Combine the RGS and CGS fasta sequences to produce a final gene set (FGS) fasta file. The FGS fasta will be used for subsequent phylogenetic characterization of the gene family.

3.9 Gene Model Repair

1. Align FGS sequences using MAFFT [57].
2. Trim the aligned sequences using ClipKit with the smartgap setting (*see* **Note 20**) [58].
3. Build a gene family tree using FastTree2 [59–61].
4. Examine the alignment and tree in viewing software, such as AlignmentViewer, FigTree or iTOL, and/or Geneious [62].
5. Focus in particular on branches having multiple homologs of cluster together for a single species and on the length of sequences in alignment relative to RGS sequences. Use this comparison to identify possible partial, expanded, or broken gene model artifacts. For the demonstration pipeline, note that there are four copies of Piezo from the *Mnemiopsis* draft genome clustered together on the tree but only a single copy from the *Mnemiopsis* PacBio transcriptome (Fig. 3a, b). There is also only a single homolog of Piezo in the *Hormiphora* genome (Fig. 3a, b). Based only on the tree, it would appear that there was an expansion of the Piezo gene family along the *Mnemiopsis* lineage within ctenophores and that only one of the four Piezo homologs in *Mnemiopsis* was detected in the PacBio transcriptome. However, in the alignment it is clear that all four copies from the draft genome are partial sequences and roughly line up in a 5′ to 3′ series relative to full-length RGS, *Hormiphora*, and *Mnemiopsis* PacBio sequences (Fig. 3a). Additional examination of genomic coordinates (GFF file) for the four gene models indicates they reside next to each other in the genome as neighbors. Based on this evidence, it appears there is a single copy of Piezo in *Mnemiopsis*, but the gene was broken into four gene models during the process of genome annotation for the draft genome. Thus, it seems reasonable to remove the four broken gene models from the FGS fasta and keep only the full-length PacBio sequence to represent the Piezo gene family in the *Mnemiopsis* genome.

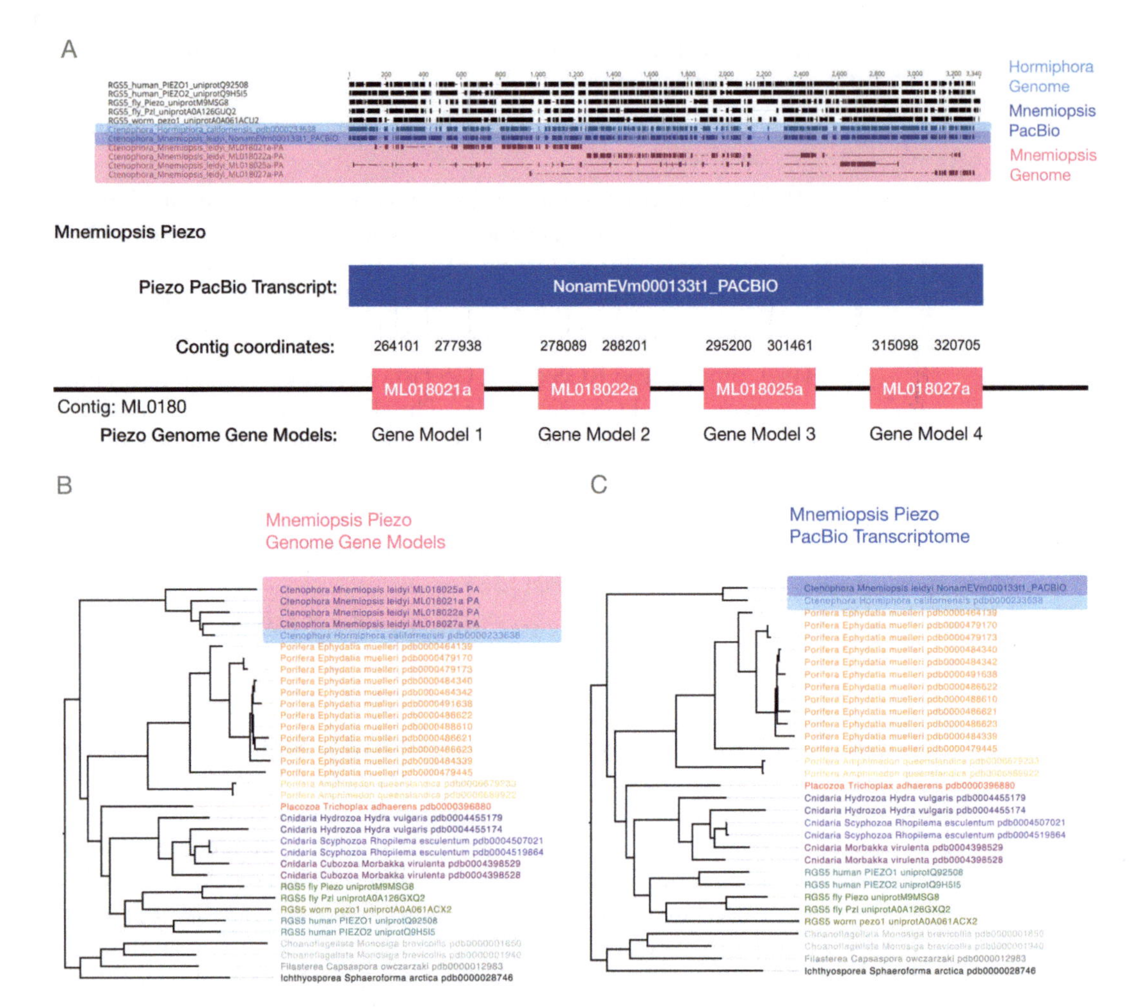

Fig. 3 Gene model repair in *Mnemiopsis* and evolution of the Piezo gene family tree in Metazoa15 genomes. (**a**) MAFFT alignment of Piezo reference sequences in human, fly, and worm and homologs identified in genomes of the ctenophores *Hormiphora* and *Mnemiopsis* and in the PacBio transcriptome of *Mnemiopsis*. *Note*: Multiple gene models appear to be partial sequences of a full-length sequence present as a single copy in the *Mnemiopsis* transcriptome and *Hormiphora* genome. (**b**) IQTree2 maximum-likelihood Piezo gene family tree for Metazoa15 genomes only. The *Mnemiopsis* lineage appears to have expanded the number of copies of Piezo in its genome. However, the alignment in 3A suggests Piezo in the *Mnemiopsis* genome is one gene broken up into four gene models. (**c**) IQTree2 maximum-likelihood Piezo gene family tree for Metazoa15 genomes but with *Mnemiopsis* Piezo gene models repaired by replacement with the *Mnemiopsis* transcriptome Piezo sequence. A single copy of Piezo appears present in the ancestor and is conserved in the *Mnemiopsis*–*Hormiphora* lineage in ctenophores

6. Remove partial, expanded, or broken gene models that can be represented instead by more accurate transcriptome sequences.
7. Remove any transcriptome sequences that are not being used as replacements for gene models exhibiting artifacts. In the case of our demonstration pipeline, FGS sequences have now been repaired for *Mnemiopsis* using the PacBio transcriptome, and the gene set is referred to here as REPAIRED.

3.10 Gene Family Alignment and Tree

1. Align the FGS REPAIRED sequences using MAFFT [57].
2. Trim the aligned sequences using ClipKit with the smartgap setting (*see* **Note 20**) [58].
3. Build a gene tree using FastTree2 or IQTree2 for maximum-likelihood methods and/or PhyloBayes for Bayesian Inference methods (*see* **Note 21**) [59–61].

3.11 Annotations and Heatmap

1. Visualize the species tree in tree viewing software such as FigTree, iTOL, or Geneious [61]. For the demonstration pipeline, this is the dot "fasttree." file for FastTree2 and the dot "treefile." for IQTree2.
2. Color annotate branches or taxon labels using the software, as preferred. Trees can also be rooted and branches rotated at this point.
3. Export the color annotated tree as a pdf or other scalable vector file (Fig. 4a).

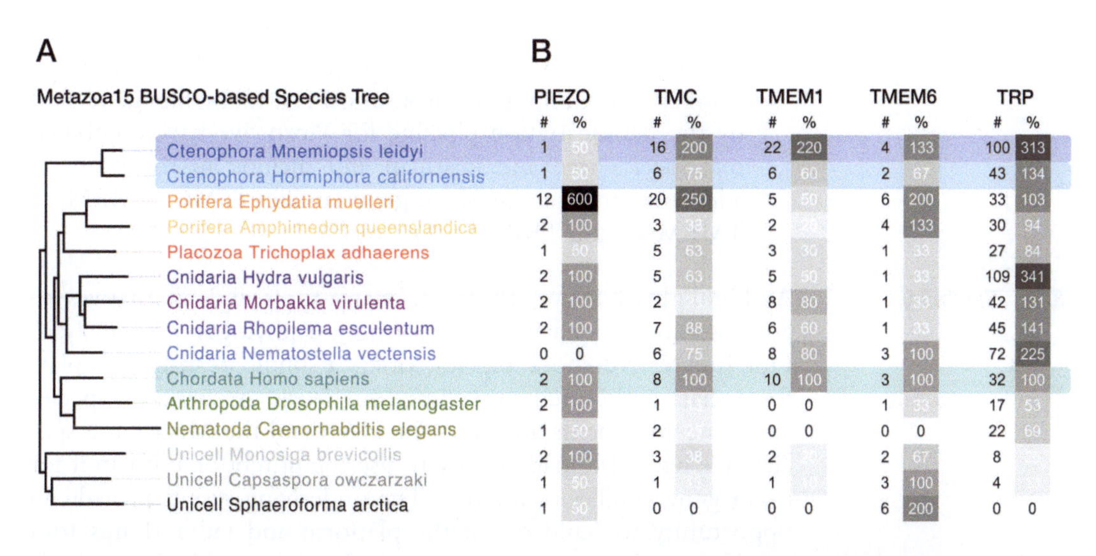

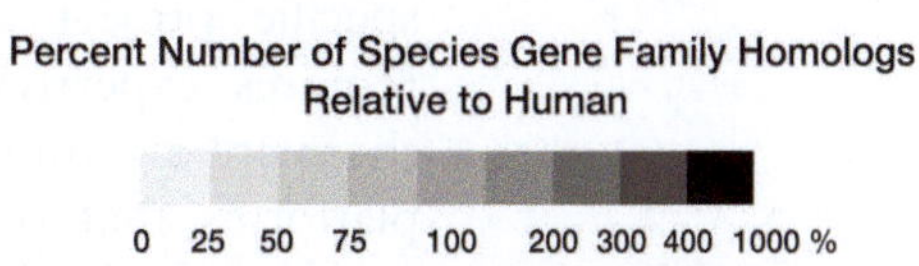

Fig. 4 BUSCO-based phylogenetic tree of Metazoa15 species evolution and associated heatmap of number and percentage relative to humans of homologs per gene family per species. (**a**) BUSCO Metazoa-based phylogenetic species tree using up to 982 genes per species in superalignment and IQTree2 maximum likelihood and C+60 model of evolution. *Note*: Ctenophores as the basal-most branch within animals is similar to recent studies that used similar methods, while the more traditional placement of sponge as the basal-most branch is commonly found using Bayesian methods. It remains unresolved which tree is correct in regard to ctenophore vs. sponge placement, but regardless, the situation highlights the importance of exploring and considering alternative methods and tools. (**b**) Heatmap of homologs per gene family per species

4. Import into spreadsheet software, such as Google Sheets, a file of traits per species you would like to map onto the species tree. The file should be structured with species names in one column and one or more columns of traits after it. For the demonstration pipeline, we include a file with counts of homologs per species genome.
5. Sort species names to match their vertical order on the species tree.
6. If useful, analyze the trait data to have additional features mapped onto the species tree.
7. If useful, generate thresholds, or use other methods to produce a heatmap that quantifies aspects of the traits based on a color scale that recolors each trait cell in the spreadsheet appropriately. This can also be done using more technical methods using tools such as Python, R, MatLab, or other programming language.
8. Export the heatmap as a pdf or other scalable vector file (Fig. 4b).
9. Combine the species tree and heatmap using any number of graphics-related software, including Google Slides, Keynote, PowerPoint, Gimp, Illustrator, or others. Using files generated by the demonstration pipeline for Piezo, we have combined our species tree and heatmap and include in the heatmap additional mechanosensitive gene families (TMC, TMEM16, TMEM63, and TRP; Fig. 4).

3.12 Summary

As highlighted above, there are many tools and alternatives to consider and work with in performing a phylogenomic analysis. Our Jupyter Notebook pipeline offers a simple push-button platform that can be used to replicate our work and generate species and gene alignments and trees for Piezo and the Metazoa15 species. We have also made it easy to use the notebook but run it for other gene families of interest. The underlying scripts provide an opportunity to build off of the platform and tailor things to a specific project and requirements. However, this increasingly requires expertise in coding and bioinformatics. Alternatively, there are a number of user-friendly desktop and online software platforms that offer extensive tools and options. In particular, ourselves and others commonly use or recommend Geneious or Galaxy for molecular biology, genomic, and phylogenomic tasks and NGPhylogeny or PhyloTol.

4 Notes

1. Ctenophores can also be collected in the surface or midwaters offshore from a boat using a plankton net or other collection device, particularly devices designed for delicate gelatinous animals.
2. A watertight bottle and Styrofoam container can also be used. The main thing is to maintain ambient seawater temperature for animals during shipping by immediately insulating the freshly collected seawater at ambient temperature.
3. For longer periods of transport or shipping, and particularly when there is a large difference in air and ideal seawater temperature, shipping boxes can include blue ice for polar to colder temperature species or heat packs for hotter temperate to tropical species. Test any shipping box for thermostability during shipping time by running a mock shipment in the lab with a temperature logger in the box. Adjust the amount of blue ice or heat packs, as needed. Also be sure to provide one or more pinhole-sized holes through any Styrofoam as otherwise oxygen will be used up by the heat packs. Travel times without aeration and with temperature control other than insulation and the addition of blue ice or heat packs can be 1–2 days without any issue.
4. Deep water and various other species can be delicate and difficult to maintain and might be processed immediately upon collection, when possible.
5. Preparing seawater from sea salt is relatively quick and easy but does require accurate measures of salinity. It is recommended to add around 80% of the sea salt, mix, and then gradually add the remaining salt while mixing and checking salinity. If time permits, the salts and seawater can be left to sit overnight, mixed, and salinity checked again the next day. Salinity can be measured with a hydrometer (inexpensive and less accurate) or refractometer (more expensive but highly accurate). For details on differences and use of each, see https://youtu.be/dQUSbruh7s4. Plunging a 1 L graduated cylinder to mix and dissolve sea salts provides highly efficient mixing.
6. Decontaminate lab bench, pipettes, tube holders, and other equipment using RNase Away or similar product that destroys RNAses. Periodically decontaminate or change gloves. Avoid breathing into tubes or solution bottles. This can be done by tilting them away from you when open. Do not leave tubes or bottles open. Open tube lids from sides without contacting the interior surface. Lids can be placed loosely on top, if helpful. Lids should be placed interior surface down on fresh Kimwipes when taken off.

7. Ctenophores are largely water, and kit recommendations for volume of homogenization buffer based on weight will often result in low final concentrations of RNA. Therefore, use 1–2× volumes for up to silver dollar-sized animals. It can be useful to test several volumes for a given species.
8. Homogenization can be done by any number of methods. For ctenophore preparations we have used variations of fresh tissue and electric or glass mortar and pestle homogenization and/or liquid nitrogen flash frozen material ground up in a ceramic mortar and pestle under liquid nitrogen and with the extraction buffer added and ground up before thawing. Avoid extended homogenization using an electric homogenizer, as buildup of heat and shearing can degrade samples.
9. There are two strategies to consider when faced with contaminated RNA at the end of extraction. If animals are not limited, you might simply redo the extraction with a new animal to see if it performs better. However some animals and samples seem to be highly resistant to producing clean samples after initial extraction. It is unclear why this is so, but we have even worked with companies making the latest kits, and they have had issues both in getting clean initial extractions. Ctenophores have been less problematic but it's worth noting. Alternatively, initially extracted contaminated samples can be cleaned up by any number of kits or older molecular biology methods. If the amount of RNA is limited, we find older methods can perform better in terms of minimizing loss of RNA or DNA after cleanup. If RNA is abundant, newer kits are quick and easy for cleanup. However, it is often the case that the amount of RNA after cleanup is greatly reduced after kit cleanup. It is unclear why this is so, but for both extraction and cleanup it may have something to do with interactions between the marine invertebrate samples and kits optimized for mammalian tissues. However, RIN values identification of ribosomal bands and assumes vertebrate sizes. However, many invertebrates have ribosomal bands that differ in size or even ribosomal molecules that separate under electrophoresis conditions and run at smaller sizes. A final evaluation of the molecular weight and distribution of the RNA smear, ribosomal bands, and RIN values might be used to make a final determination if the RNA quality should pass QC or not.
10. Ideally, three to five biological replicates will be sequenced. Here we use PacBio SMRT with Sequel II Chemistry, which offered at the time (Spring 2020) the highest single-molecule accuracy of any technology but with relatively high costs. Thus, our sequencing strategy was to use a single animal sequenced on two flow cells.

11. You will need to register with NCBI to submit data. Registering or updating your profile to link to an ORCID ID is a good idea. Data can be set to release at a later date during the submission process. It is useful to move data to NCBI once the initial processing is completed, as NCBI provides free archival storage; deposition in a public database like NCBI is generally required or encouraged in a publication, so taking care of the upload and processing now will streamline analyses and manuscript preparations later.
12. Ideally, genome gene models would be annotated after single-molecule sequencing; however, it is not always possible. Here, we provide a semi-automated method to repair candidate gene sequences after initial identification. Similar approaches could be more fully automated at genome-scale but can produce artifacts, and it might be worth considering re-annotating the genome.
13. EvidentialGene will collapse some paralogs and will fail to collapse some isoforms. These artifacts should be minimal but are an important caveat.
14. A PacBio Sequel II single-molecule transcript is likely to be a correct representation of an isoform of a gene. In contrast, genome gene models often integrate diverse data sets and methods and can be prone to artifacts, particularly in draft genome like that of *Mnemiopsis leidyi*, and can include partial, expanded, or broken gene models. For these reasons, our pipeline seeks to replace gene model sequences with transcriptome sequences when possible.
15. We have found that using genome gene model mRNA leads to many spurious Blastn hits that have surprisingly good statistics, including e-values of 0.0. Using CDS seems to greatly reduce these false positives. Potentially, the issue is related to untranslated regions in poorly called gene models, but we have not formally characterized things and do not fully understand what features of the mRNA and gene models are causing the false positive hits to arise when using mRNA for the gene models.
16. When there is only a single hit of e-value 0.0 (the best possible e-value score in Blast), the match of a gene between genome and transcriptome is clear, and the gene model can be updated to the transcriptome sequence with confidence. In cases where there are unassembled or unannotated paralogs, a transcriptome paralog could end up replacing a gene model ortholog sequence. This can be difficult to detect but should be rare and is a caveat to be aware of. Gene families with many sequence-similar paralogs will fail to be improved in these steps, as local alignments by Blastn will have highly similar or identical statistics. Importantly, BUSCO genes are generally single-gene

families, and we use them here for building a species tree. Because BUSCO genes generally have only a single gene per family, our gene model updating process will likely improve many BUSCO genes, particularly in draft genomes where gene model quality can vary a lot. Updating BUSCO gene models by the single-molecule transcriptome will improve the species tree, as gene models updated to their full sequence will perform better in alignment and tree building.

17. It is a good idea to include multiple single-cell outgroup species that are distantly related to one another but relatively close to animals for phylogenetic gene family trees of deep animal evolution. Losses of gene families do occur, and having distance can increase the odds of detecting ancient origins outside animals.

18. Metazoa15 genome datasets are from high-quality, published, and publicly available genomes: Ichthyosporea *Sphaeroforma arctica* [63], Filasterea *Capsaspora owczarzaki* [64], Choanoflagellata *Monosiga brevicollis* [65], Porifera *Ephydatia muelleri* [66], Porifera *Amphimedon queenslandica* [67], Ctenophora *Mnemiopsis leidyi* [18, 19], Ctenophora *Hormiphora californiensis* [20], Cnidaria *Nematostella vectensis* [68, 69], Cnidaria *Morbakka virulenta* [70], Cnidaria *Rhopilema esculentum* [71], Cnidaria *Hydra vulgaris* [72], Placozoa *Trichoplax adhaerens* [73], Chordata *Homo sapiens*, Arthropoda *Drosophila melanogaster*, and Nematoda *Caenorhabditis elegans*.

19. Other tools commonly used for trimming such as GBlocks [74] or TrimAl [75] can also be used; however, ClipKit provided superior trimming in our informal testing.

20. FastTree offers rapid fairly rigorous trees using maximum-likelihood-like methods. However, IQTree provides higher-quality trees, including better branch support and closer in matching to expectations. IQTree also includes ultrafast bootstraps which are statistically more rigorous and offer a declared threshold for evaluating branch support. FastTree jobs typically run seconds to hours on our machines, while the same data takes hours to days or even weeks for IQTree. IQTree is limited to the number of parameters it can explore in its model of evolution. PhyloBayes uses Bayesian inference to build trees and is not limited in parameters for its model of evolution. PhyloBayes typically takes longer than IQTree on a given data set, to the point we have killed jobs after several weeks in realizing they would take months to complete. All of the longer times are for large species trees. For most gene trees, run times will be very reasonable, in the seconds to days range in most cases.

21. There are many possible strategies and methods for identifying gene homologs in a species. Our demonstration pipeline uses what we refer to here as a blast-based reciprocal best family (RBF) strategy. It is lightweight (meaning it works well on smaller machines) and provides rapid accurate discovery of homologs. We like it because it performs well in regard to avoiding false-positive identification of homologs in genomes. In comparison with more sensitive and/or iterative detection methods, like HMMs and HMMer3 [76], Blast does miss some homologs in genomes that are highly sequence divergent from RGS sequences. More generally, most homolog detection methods have substantial overlap in true positive identification and varying levels of remote homolog detection and avoidance of false positives. In challenging cases, we find that one single method is rarely the best, so using multiple strategies can be a good idea, but it can also confound things without substantial gain. These are all things to consider or test for a given project. In regard to RBF, it is a variant of commonly used reciprocal best hit methods. Reference sequences are blasted against genomes of interest. Identified hit sequences are collected and blasted back against the genomes of the reference species. For RBF, we then keep all initial hits that have as a top hit at least one of the RGS sequences in at least one of the reference genomes. These filtered hits then form the candidate gene set. CGS and RGS sequences are subsequently combined and used for downstream phylogenetic analysis of the gene family.
22. Our demonstration pipeline is focused on deep evolution of Piezo in basal-branching animals and therefore requires the use of protein sequence, as phylogenetic signal is retained longer in protein sequences due to the redundant nature of codons. For more recent evolutionary comparisons, typically staring around 100 million years or less, DNA sequences can be considered and will be required for the most recent comparisons, as the amount of phylogenetic signal is increasingly limited (i.e., there are fewer and fewer sequence differences between genes of different species or individuals).
23. UniProt, GenBank, and other public databases are excellent sources for protein sequences that can be used to build reference gene sets. In addition, if human genes are of interest in building the reference gene set, the HUGO Genome Nomenclature Committee (HGNC) website (https://www.genenames.org/) provides curated gene groups that often represent entire gene families or superfamilies and can be useful to rapidly build reference gene sets. However, some gene groups are defined by function and include diverse gene families. In addition, a given gene can be in multiple gene groups, so care is required in selecting gene groups, but once

selected it can greatly facilitate the creation of a reference gene set for phylogenetic analysis.

24. Omissions of genes and gene families in genome gene model gene sets can occur, particularly in draft genomes, and we find that phylogenetic redundancy can be useful to compensate for these random losses. When possible, we include three distantly related species per major group, which in this case is three distantly related species within each phylum and for the unicellular outgroup.
25. RGS genes can potentially come from many sources, and their identifiers might not match those used for the same gene in their reference genome. Blasting RGS genes against the reference genomes enables identification of RGS genes in the reference genome and their associated reference genome identifiers. This is then used in later steps in sorting through initial CGS blast hits against the reference genomes. Importantly, for paralogs that are similar in sequence, or when there are highly conserved domains across homologs, it is possible to exceed the e-value sensitivity of BLAST, and RGS gene might technically be assigned the incorrect genome gene identifier internally. This is unlikely to impact the filtering of false positives based on initial CGS blast hits in the reference genomes; however, it is important to retain only genes outside RGS and then add the initial RGS sequences to the gene set prior to sequence alignment and tree building. In addition, RGS sequences on later gene family trees should be checked for sequence-identical or near-sequence-identical siblings that may represent overlooked RGS sequences that snuck through due to the limits of BLAST sensitivity.

Acknowledgments

We wish to thank the 2019 and 2020 field collectors at the University of Chicago Marine Biological Laboratory Marine Resources Center for providing animals, including S. Bennet, and similarly M. Cordeiro, as an undergraduate at Roger Williams University.
This work was supported in part by a National Institute of Mental Health of the National Institutes of Health award (MH119646) to S.C. and E.E., a National Institute of Neurological Disorders and Stroke of the National Institutes of Health award (R01NS114491) to L.L.M, a National Science Foundation award (IOS-1557923) to L.L.M, a Human Frontiers Science Program award (RGP0060/2017) to E.E. and L.L.M, Vetlesen Foundation funding, and a Connecticut Research Fund Grant (2018) award to E.E. The content is solely the authors' responsibility and does not necessarily represent official views of the funding agencies.

References

1. Hernandez-Nicaise M-L (1991) Ctenophora. In: Westfall FW, Harrison JA (eds) Microscopic anatomy of invertebrates: Placozoa, Porifera, Cnidaria, and Ctenophora. Wiley, pp 359–418
2. Redmond AK, McLysaght A (2021) Evidence for sponges as sister to all other animals from partitioned phylogenomics with mixture models and recoding. Nat Commun 12:1783. https://doi.org/10.1038/s41467-021-22074-7
3. Moroz LL, Romanova DY, Kohn AB (2021) Neural versus alternative integrative systems: molecular insights into origins of neurotransmitters. Philos Trans R Soc Lond Ser B Biol Sci 376:20190762. https://doi.org/10.1098/rstb.2019.0762
4. Li Y, Shen X-X, Evans B, Dunn CW, Rokas A (2021) Rooting the animal tree of life. Mol Biol Evol 38:4322–4333. https://doi.org/10.1093/molbev/msab170
5. Nielsen C (2012) Animal evolution: interrelationships of the living phyla. OUP, Oxford. Available: https://play.google.com/store/books/details?id=kr7HeXXq0o4C
6. Tamm SL (1982) Ctenophora. In: Shelton GAB (ed) Electrical conduction and behaviour in "simple" invertebrates. Clarendon Press, Oxford; New York. Available: https://www.worldcat.org/title/electrical-conduction-and-behaviour-in-simple-invertebrates/oclc/8894059
7. Nielsen C (2019) Early animal evolution: a morphologist's view. R Soc Open Sci 6: 190638. https://doi.org/10.1098/rsos.190638
8. Heimbichner Goebel WL, Colin SP, Costello JH, Gemmell BJ, Sutherland KR (2020) Scaling of ctenes and consequences for swimming performance in the ctenophore *Pleurobrachia bachei*. Invertebr Biol 139:e12297. https://doi.org/10.1111/ivb.12297
9. Omori T, Ito H, Ishikawa T (2020) Swimming microorganisms acquire optimal efficiency with multiple cilia. Proc Natl Acad Sci U S A 117: 30201–30207. https://doi.org/10.1073/pnas.2011146117
10. McDonald KA, Grünbaum D (2010) Swimming performance in early development and the "other" consequences of egg size for ciliated planktonic larvae. Integr Comp Biol 50: 589–605. https://doi.org/10.1093/icb/icq090
11. Tamm SL (2014) Cilia and the life of ctenophores. Invertebr Biol 133:1–46. https://doi.org/10.1111/ivb.12042
12. Dunn CW, Leys SP, Haddock SHD (2015) The hidden biology of sponges and ctenophores. Trends Ecol Evol 30:282–291. https://doi.org/10.1016/j.tree.2015.03.003
13. Presnell JS, Vandepas LE, Warren KJ, Swalla BJ, Amemiya CT, Browne WE (2016) The presence of a functionally tripartite through-gut in Ctenophora has implications for metazoan character trait evolution. Curr Biol 26: 2814–2820. https://doi.org/10.1016/j.cub.2016.08.019
14. Agassiz L (1850) Contributions to the natural history of the acalephæ of North America. Part I: on the naked-eyed medusæ of the shores of Massachusetts, in their perfect state of development. Mem Am Acad Arts Sci 4:221–316. https://doi.org/10.2307/25058163
15. Moroz LL (2014) The genealogy of genealogy of neurons. Commun Integr Biol 7:e993269. https://doi.org/10.4161/19420889.2014.993269
16. Moroz Leonid L, Kohn Andrea B (2016) Independent origins of neurons and synapses: insights from ctenophores. Philos Trans R Soc Lond Ser B Biol Sci 371:20150041. https://doi.org/10.1098/rstb.2015.0041
17. Moroz LL (2015) Convergent evolution of neural systems in ctenophores. J Exp Biol 218:598–611. https://doi.org/10.1242/jeb.110692
18. Moroz LL, Kocot KM, Citarella MR, Dosung S, Norekian TP, Povolotskaya IS et al (2014) The ctenophore genome and the evolutionary origins of neural systems. Nature 510:109–114. https://doi.org/10.1038/nature13400
19. Ryan JF, Pang K, Schnitzler CE, Nguyen A-D, Moreland RT, Simmons DK et al (2013) The genome of the ctenophore *Mnemiopsis leidyi* and its implications for cell type evolution. Science 342:1242592. https://doi.org/10.1126/science.1242592
20. Schultz DT, Francis WR, McBroome JD, Christianson LM, Haddock SHD, Green RE (2021) A chromosome-scale genome assembly and karyotype of the ctenophore *Hormiphora californensis*. G3 11:jkab302. https://doi.org/10.1093/g3journal/jkab302
21. Hoencamp C, Dudchenko O, Elbatsh AMO, Brahmachari S, Raaijmakers JA, van Schaik T et al (2021) 3D genomics across the tree of life reveals condensin II as a determinant of

architecture type. Science 372:984–989. https://doi.org/10.1126/science.abe2218

22. Whelan NV, Kocot KM, Moroz TP, Mukherjee K, Williams P, Paulay G et al (2017) Ctenophore relationships and their placement as the sister group to all other animals. Nat Ecol Evol 1:1737–1746. https://doi.org/10.1038/s41559-017-0331-3
23. Pisani D, Pett W, Dohrmann M, Feuda R, Rota-Stabelli O, Philippe H et al (2015) Genomic data do not support comb jellies as the sister group to all other animals. Proc Natl Acad Sci U S A 112:15402–15407. https://doi.org/10.1073/pnas.1518127112
24. Pisani D, Pett W, Dohrmann M, Feuda R, Rota-Stabelli O, Philippe H et al (2016) Reply to Halanych et al.: Ctenophore misplacement is corroborated by independent datasets. Proc Natl Acad Sci U S A 113:E948–E949. https://doi.org/10.1073/pnas.1525718113
25. Halanych KM, Whelan NV, Kocot KM, Kohn AB, Moroz LL (2016) Miscues misplace sponges. Proc Natl Acad Sci U S A 113: E946–E947. https://doi.org/10.1073/pnas.1525332113
26. Whelan NV, Kocot KM, Moroz LL, Halanych KM (2015) Error, signal, and the placement of Ctenophora sister to all other animals. Proc Natl Acad Sci U S A 112:5773–5778. https://doi.org/10.1073/pnas.1503453112
27. Feuda R, Dohrmann M, Pett W, Philippe H, Rota-Stabelli O, Lartillot N et al (2017) Improved modeling of compositional heterogeneity supports sponges as sister to all other animals. Curr Biol 27:3864–3870.e4. https://doi.org/10.1016/j.cub.2017.11.008
28. Jékely G, Paps J, Nielsen C (2015) The phylogenetic position of ctenophores and the origin (s) of nervous systems. EvoDevo 6:1. https://doi.org/10.1186/2041-9139-6-1
29. Rokas A (2013) Genetics. My oldest sister is a sea walnut? Science 342:1327–1329. https://doi.org/10.1126/science.1248424
30. Kapli P, Telford MJ (2020) Topology-dependent asymmetry in systematic errors affects phylogenetic placement of Ctenophora and Xenacoelomorpha. Sci Adv 6:eabc5162. https://doi.org/10.1126/sciadv.abc5162
31. Telford MJ, Moroz LL, Halanych KM (2016) Evolution: a sisterly dispute. Nature 529:286–287. https://doi.org/10.1038/529286a
32. Schultz DT, Haddock SHD, Bredeson JV, Green RE, Simakov O, Rokhsar DS (2023) Ancient gene linkages support ctenophores as sister to other animals. Nature 618:1–8. https://doi.org/10.1038/s41586-023-05936-6
33. Pandey A, Braun EL (2020) Phylogenetic analyses of sites in different protein structural environments result in distinct placements of the metazoan root. Biology 9:64. https://doi.org/10.3390/biology9040064
34. Hernandez AM, Ryan JF (2021) Six-state amino acid recoding is not an effective strategy to offset compositional heterogeneity and saturation in phylogenetic analyses. Syst Biol 70: 1200. https://doi.org/10.1093/sysbio/syab027
35. Natsidis P, Kapli P, Schiffer PH, Telford MJ (2020) Systematic errors in orthology inference: a bug or a feature for evolutionary analyses? Cold Spring Harbor Laboratory, p 2020.11.03.366625. https://doi.org/10.1101/2020.11.03.366625
36. Ishino S, Ishino Y (2014) DNA polymerases as useful reagents for biotechnology—the history of developmental research in the field. Front Microbiol 5:465. https://doi.org/10.3389/fmicb.2014.00465
37. Swaminathan S (2009) GFP: the green revolution. Nat Cell Biol 11:S20. https://doi.org/10.1038/ncb1953
38. Hegemann P, Nagel G (2013) From channelrhodopsins to optogenetics. EMBO Mol Med 5:173–176. https://doi.org/10.1002/emmm.201202387
39. Ishino Y, Krupovic M, Forterre P (2018) History of CRISPR-Cas from encounter with a mysterious repeated sequence to genome editing technology. J Bacteriol 200:e00580-17. https://doi.org/10.1128/JB.00580-17
40. Raghunathan G, Marx A (2019) Identification of Thermus aquaticus DNA polymerase variants with increased mismatch discrimination and reverse transcriptase activity from a smart enzyme mutant library. Sci Rep 9:590. https://doi.org/10.1038/s41598-018-37233-y
41. Nidhi S, Anand U, Oleksak P, Tripathi P, Lal JA, Thomas G et al (2021) Novel CRISPR-Cas systems: an updated review of the current achievements, applications, and future research perspectives. Int J Mol Sci 22:3327. https://doi.org/10.3390/ijms22073327
42. Lambert GG, Depernet H, Gotthard G, Schultz DT, Navizet I, Lambert T et al (2020) *Aequorea*'s secrets revealed: new fluorescent proteins with unique properties for bioimaging and biosensing. PLoS Biol 18: e3000936. https://doi.org/10.1371/journal.pbio.3000936
43. Zabelskii D, Alekseev A, Kovalev K, Rankovic V, Balandin T, Soloviov D et al (2020) Viral rhodopsins 1 are an unique family of light-gated cation channels. Nat Commun

11:5707. https://doi.org/10.1038/s41467-020-19457-7

44. Alley EC, Khimulya G, Biswas S, AlQuraishi M, Church GM (2019) Unified rational protein engineering with sequence-based deep representation learning. Nat Methods 16:1315–1322. https://doi.org/10.1038/s41592-019-0598-1
45. Varadi M, Anyango S, Deshpande M, Nair S, Natassia C, Yordanova G et al (2022) AlphaFold Protein Structure Database: massively expanding the structural coverage of protein-sequence space with high-accuracy models. Nucleic Acids Res 50:D439–D444. https://doi.org/10.1093/nar/gkab1061
46. Jumper J, Evans R, Pritzel A, Green T, Figurnov M, Ronneberger O et al (2021) Highly accurate protein structure prediction with AlphaFold. Nature 596:583–589. https://doi.org/10.1038/s41586-021-03819-2
47. Baek M, Baker D (2022) Deep learning and protein structure modeling. Nat Methods 19:13–14. https://doi.org/10.1038/s41592-021-01360-8
48. Humphreys IR, Pei J, Baek M, Krishnakumar A, Anishchenko I, Ovchinnikov S et al (2021) Computed structures of core eukaryotic protein complexes. Science 374:eabm4805. https://doi.org/10.1126/science.abm4805
49. Woodall NB, Weinberg Z, Park J, Busch F, Johnson RS, Feldbauer MJ et al (2021) De novo design of tyrosine and serine kinase-driven protein switches. Nat Struct Mol Biol 28:762–770. https://doi.org/10.1038/s41594-021-00649-8
50. Baek M, DiMaio F, Anishchenko I, Dauparas J, Ovchinnikov S, Lee GR et al (2021) Accurate prediction of protein structures and interactions using a three-track neural network. Science 373:871–876. https://doi.org/10.1126/science.abj8754
51. Anishchenko I, Pellock SJ, Chidyausiku TM, Ramelot TA, Ovchinnikov S, Hao J et al (2021) De novo protein design by deep network hallucination. Nature 600:547–552. https://doi.org/10.1038/s41586-021-04184-w
52. Kapli P, Yang Z, Telford MJ (2020) Phylogenetic tree building in the genomic age. Nat Rev Genet 21:428. https://doi.org/10.1038/s41576-020-0233-0
53. Lewis AH, Grandl J (2021) Piezo1 ion channels inherently function as independent mechanotransducers. Elife 10:e70988. https://doi.org/10.7554/eLife.70988
54. Gilbert DG (2019). Longest protein, longest transcript or most expression, for accurate gene reconstruction of transcriptomes? BioRxiv. https://doi.org/10.1101/829184
55. Fu L, Niu B, Zhu Z et al (2012) CD-HIT: accelerated for clustering the next-generation sequencing data. Bioinformatics 28(23):3150–3152. https://doi.org/10.1093/bioinformatics/bts565. Epub 2012 Oct 11. PMID: 23060610; PMCID: PMC3516142
56. Seppey M, Manni M, Zdobnov EM (2019) BUSCO: assessing genome assembly and annotation completeness. Methods Mol Biol 1962:227–245. https://doi.org/10.1007/978-1-4939-9173-0_14
57. Katoh K, Misawa K, Kuma K-I, Miyata T (2002) MAFFT: a novel method for rapid multiple sequence alignment based on fast Fourier transform. Nucleic Acids Res 30:3059–3066. Available: http://www.ncbi.nlm.nih.gov/pubmed/12136088
58. Steenwyk JL, Buida TJ 3rd, Li Y, Shen X-X, Rokas A (2020) ClipKIT: a multiple sequence alignment trimming software for accurate phylogenomic inference. PLoS Biol 18:e3001007. https://doi.org/10.1371/journal.pbio.3001007
59. Price MN, Dehal PS, Arkin AP (2010) FastTree 2--approximately maximum-likelihood trees for large alignments. PLoS One 5:e9490. https://doi.org/10.1371/journal.pone.0009490
60. Minh BQ, Schmidt HA, Chernomor O, Schrempf D, Woodhams MD, von Haeseler A et al (2020) IQ-TREE 2: new models and efficient methods for phylogenetic inference in the genomic era. Mol Biol Evol 37:1530–1534. https://doi.org/10.1093/molbev/msaa015
61. Lartillot N, Lepage T, Blanquart S (2009) PhyloBayes 3: a Bayesian software package for phylogenetic reconstruction and molecular dating. Bioinformatics 25:2286–2288. https://doi.org/10.1093/bioinformatics/btp368
62. Letunic I, Bork P (2021) Interactive Tree Of Life (iTOL) v5: an online tool for phylogenetic tree display and annotation. Nucleic Acids Res 49:W293. https://doi.org/10.1093/nar/gkab301
63. Dudin O, Ondracka A, Grau-Bové X, Haraldsen AA, Toyoda A, Suga H et al (2019) A unicellular relative of animals generates a layer of polarized cells by actomyosin-dependent cellularization. Elife 8:e49801. https://doi.org/10.7554/eLife.49801
64. Suga H, Chen Z, de Mendoza A, Sebé-Pedrós A, Brown MW, Kramer E et al

(2013) The *Capsaspora* genome reveals a complex unicellular prehistory of animals. Nat Commun 4:2325. https://doi.org/10.1038/ncomms3325

65. King N, Westbrook MJ, Young SL, Kuo A, Abedin M, Chapman J et al (2008) The genome of the choanoflagellate *Monosiga brevicollis* and the origin of metazoans. Nature 451:783–788. https://doi.org/10.1038/nature06617
66. Kenny NJ, Francis WR, Rivera-Vicéns RE, Juravel K, de Mendoza A, Díez-Vives C et al (2020) Tracing animal genomic evolution with the chromosomal-level assembly of the freshwater sponge *Ephydatia muelleri*. Nat Commun 11:1–11. https://doi.org/10.1038/s41467-020-17397-w
67. Srivastava M, Simakov O, Chapman J, Fahey B, Gauthier MEA, Mitros T et al (2010) The *Amphimedon queenslandica* genome and the evolution of animal complexity. Nature 466:720–726. https://doi.org/10.1038/nature09201
68. Zimmermann B, Robb SMC, Genikhovich G, Fropf WJ, Weilguny L, He S et al (2020) Sea anemone genomes reveal ancestral metazoan chromosomal macrosynteny. bioRxiv:2020.10.30.359448. https://doi.org/10.1101/2020.10.30.359448
69. Putnam NH, Srivastava M, Hellsten U, Dirks B, Chapman J, Salamov A et al (2007) Sea anemone genome reveals ancestral eumetazoan gene repertoire and genomic organization. Science 317:86–94. https://doi.org/10.1126/science.1139158
70. Khalturin K, Shinzato C, Khalturina M, Hamada M, Fujie M, Koyanagi R et al (2019) Medusozoan genomes inform the evolution of the jellyfish body plan. Nat Ecol Evol 3:811–822. https://doi.org/10.1038/s41559-019-0853-y
71. Li Y, Gao L, Pan Y, Tian M, Li Y, He C et al (2020) Chromosome-level reference genome of the jellyfish *Rhopilema esculentum*. Gigascience 9:giaa036. https://doi.org/10.1093/gigascience/giaa036
72. Chapman JA, Kirkness EF, Simakov O, Hampson SE, Mitros T, Weinmaier T et al (2010) The dynamic genome of *Hydra*. Nature 464:592–596. https://doi.org/10.1038/nature08830
73. Srivastava M, Begovic E, Chapman J, Putnam NH, Hellsten U, Kawashima T et al (2008) The *Trichoplax* genome and the nature of placozoans. Nature 454:955–960. https://doi.org/10.1038/nature07191
74. Talavera G, Castresana J (2007) Improvement of phylogenies after removing divergent and ambiguously aligned blocks from protein sequence alignments. Syst Biol 56:564–577. https://doi.org/10.1080/10635150701472164
75. Capella-Gutiérrez S, Silla-Martínez JM, Gabaldón T (2009) trimAl: a tool for automated alignment trimming in large-scale phylogenetic analyses. Bioinformatics 25:1972–1973. https://doi.org/10.1093/bioinformatics/btp348
76. Eddy SR (2009) A new generation of homology search tools based on probabilistic inference. Genome Inform 23:205–211. Available: https://www.ncbi.nlm.nih.gov/pubmed/20180275

Chapter 20

Parallel Evolution of Transcription Factors in Basal Metazoans

Krishanu Mukherjee and Leonid L. Moroz

Abstract

Transcription factors (TFs) play a pivotal role as regulators of gene expression, orchestrating the formation and maintenance of diverse animal body plans and innovations. However, the precise contributions of TFs and the underlying mechanisms driving the origin of basal metazoan body plans, particularly in ctenophores, remain elusive. Here, we present a comprehensive catalog of TFs in 2 ctenophore species, *Pleurobrachia bachei* and *Mnemiopsis leidyi*, revealing 428 and 418 TFs in their respective genomes. In contrast, morphologically simpler metazoans have a reduced TF representation compared to ctenophores, cnidarians, and bilaterians: the sponge *Amphimedon* encodes 277 TFs, and the placozoan *Trichoplax adhaerens* encodes 274 TFs. The emergence of complex ctenophore tissues and organs coincides with significant lineage-specific diversification of the zinc finger C2H2 (ZF-C2H2) and homeobox superfamilies of TFs. Notable, the lineages leading to *Amphimedon* and *Trichoplax* exhibit independent expansions of leucine zipper (BZIP) TFs. Some lineage-specific TFs may have evolved through the domestication of mobile elements, thereby supporting alternative mechanisms of parallel TF evolution and body plan diversification across the Metazoa.

Key words Ctenophora, Placozoa, Porifera, Genome, Transcription factors, *Pleurobrachia*, *Mnemiopsis*, *Amphimedon*, *Trichoplax*, Homeobox, BHLH, Neurons, Evolution of animal complexity, Transposons

1 Introduction

By specific DNA binding, transcription factors (TFs) evolved to directly interpret the genome, often acting as master regulators of gene expression. In humans alone, the genome encodes over 1600 TFs [1], controlling the expression of 20,000+ genes. Investigating the gain, loss, and lineage-specific origins of TFs during the transition from single-celled organisms to multicellular metazoans can provide valuable insights into the evolution of animal innovations. However, the origins of metazoan-specific genes are elusive, with some mechanisms emerging from comparative studies [2].

Leonid L. Moroz (ed.), *Ctenophores: Methods and Protocols*, Methods in Molecular Biology, vol. 2757,
https://doi.org/10.1007/978-1-0716-3642-8_20, © Springer Science+Business Media, LLC, part of Springer Nature 2024

The concept of gene duplication followed by functional diversification has been proposed as a major mechanism for the origin of novel genes and genome evolution [3]. According to this hypothesis, ancestral genes within the same animal lineage gave rise to duplicated copies through gene duplication events. This idea is supported by extensive evidence of gene conservation and the corresponding regulatory networks over large evolutionary distances [2], exemplified by the discovery of the highly conserved HOX gene cluster [4, 5]. However, the absence of canonical bilaterian master gene regulators, such as the HOX cluster, in the ctenophores genomes [6, 7] challenges tracing their origin and evolution.

Some TFs might evolve de novo or be a result of lateral gene (viral) transfer [2] or activity of mobile elements [8]. The birth of the majority of metazoan-specific TFs remains largely unexplored. Moreover, the exact number and annotation of TFs encoded in basal metazoan lineages remain incomplete. The challenges include limitations of gaps, errors, polymorphisms in genomes, and the limitations of homology detection algorithms for detecting precise orthologs. Meticulous, hand-curated annotation, especially across basal metazoan genomes, should be combined with systematic phylogenetic analyses of each class/family of genes to identify taxonomically restricted or novel TFs that lack close homologs.

Here, we comprehensively analyze major TF families encoded in two ctenophore genomes as representatives of the earliest branching animal lineage, sister to the rest of Metazoa [7, 9–11]. This survey is supported by complementary manual annotation of genomes from sponges and placozoans, representing the second and third basal branches of the animal tree of life, respectively. Our findings reveal that zinc finger, homeobox, and specific basic helix–loop–helix (BHLH) gene families of TFs have undergone lineage-specific expansions in ctenophores. Independent ctenophore lineage-specific expansions of homeobox genes Antennapedia (Antp) and the six protein families, as well as in leucine zipper (BZIP), are equally evident, contributing to the parallel evolution of animal complexity within comb jellies.

This annotation can be a valuable reference platform for further exploring the genomic basis of complex cellular and tissue phenotypes in ctenophores and other basal metazoan lineages.

2 Methods

1. **Sequence searches** were conducted using the local database. In cases where the online BLAST tool was unavailable, the standalone BLAST tool from NCBI (https://ftp.ncbi.nlm.nih.gov/blast/executables/LATEST/) was utilized. The local BLAST was installed and unpacked on a UNIX platform. To prepare the BLAST database specifically for the *Pleurobrachia*

bachei genome, the "makeblastdb" command was employed (./makeblastdb -in pleurobrachia.nt -dbtype nucl). The database was then queried using the DNA-binding domain of transcription factors, and the results were saved in a text file named "test.fasta." TBLASTn searches were performed against the genome database with the following command: (./tblastn -query test.fasta -db pleurobrachia.nt -evalue 5 -out output.txt). A cutoff range of 10^{-5} to 10^{-10} was employed for the standalone BLAST to identify potential homologs [12]. The DNA-binding domains were recursively subjected to BLAST analysis until no further homologs were obtained. Blast hits were manually scrutinized to determine their potential as genuine homologs. In cases where gene models (exomes) were unavailable, the genomic sequences surrounding the coding region were extracted, and homology-based gene prediction using hidden Markov models (HMMs) was performed using FGENESH+ (www.softberry.com).

2. **Sequence searches at the online database**: TBLASTn searches were conducted to retrieve *Mnemiopsis leidyi* transcription factors using the online BLAST server at https://research.nhgri.nih.gov/mnemiopsis/sequenceserver/. For *Amphimedon queenslandica* and *Trichoplax adhaerens*, a comprehensive set of transcription factors was obtained by performing TBLASTn searches at the Ensembl genome browser (https://metazoa.ensembl.org). To search against metazoan genomes, the complete sets of human and fruit fly transcription factors were downloaded from the AnimalTFDB3.0 online database (http://bioinfo.life.hust.edu.cn/AnimalTFDB/#!/species). The default e-value cutoff for the online BLAST was utilized to retrieve all potential homologs.
3. **Protein domain identification**: To identify the DNA-binding domain and other associated domains, searches were performed using the NCBI conserved domain database (CDD) [13] and the SMART [14, 15] online database. These databases were utilized to uncover conserved structural and functional domains relevant to DNA binding and transcription factor activity.
4. **Protein multiple-domain alignment**: The sequences were aligned using the MUSCLE alignment algorithm [15]. Alignment was performed either through an online tool available at EBI (https://www.ebi.ac.uk/Tools/msa/muscle/) or using the command-line approach in UNIX with the following command: "./muscle -in input.fasta -out output.muscle". This allowed for the alignment of the sequences to generate a comprehensive and accurate alignment for further analysis.

5. **Phylogeny reconstruction**: Maximum-likelihood (ML) trees were constructed using PhyML v3.0 [16, 17]. The most appropriate evolutionary model was determined using the AIC criterion as estimated by ProtTest [18]. ML phylogenies were generated using the JTT model, which accounts for rate heterogeneity, an estimated proportion of invariable sites, four rate categories, and an estimated alpha distribution parameter. This was achieved through the following command-line UNIX command: "./phyml -i input.phy -d aa -m JTT -v 0.0 -c 4 -a e -s 'BEST'". To optimize tree topology searches, both NNI (nearest-neighbor interchanges) and SPR (subtree pruning and regrafting) moves were employed [19]. Clade support was assessed using the SH-like approximate likelihood ratio test [20]. These analyses facilitated the construction of robust ML trees, capturing the evolutionary relationships among the studied transcription factors.
6. **Post-processing of tree file**: The tree file generated from the PhyML analysis was uploaded to an online tool (https://itol.embl.de/) for visualization. Once final adjustments were completed, the tree was exported in a vector-graphics format, such as SVG or EPS. The exported file was then imported into Adobe Illustrator CS6 version 16.0.0 for further refinement and the addition of labels and other annotations. Adobe Illustrator was utilized to create a polished and visually appealing representation of the final tree for presentation and publication purposes.
7. **Ortholog assessment**: To identify one-to-one orthologs across metazoan species, we employed the OrthoFinder tool [21]. Additionally, we employed a family-wise maximum-likelihood (ML) tree reconstruction approach to identify taxonomically restricted transcription factors (TFs). In addition to the four basal metazoan genomes, we incorporated the genomes of bilaterian organisms, specifically *Drosophila melanogaster* (fly) and *Homo sapiens* (humans), to assign the evolutionary conservative set of TF orthologs. Protein domain conservation served as a benchmark for evaluating the effectiveness of ortholog inference methods, ensuring robust and reliable identification of orthologous TFs across the metazoan lineage.

3 Illustrated Examples

3.1 Identification and Annotation of TFs in Basal Metazoan Lineages

We identify the nearly complete set of transcription factors (TFs) encoded in Ctenophora, Porifera, and Placozoa genomes. And representatives of each group show remarkable lineage-specific diversifications of different families of TFs, as summarized below.

Ctenophores *Pleurobrachia bachei* [6] and *Mnemiopsis leidyi* [7] encode 428 and 418 TFs, respectively (Fig. 1a, b), which exceed the previously reported number of 281 TFs in the *Mnemiopsis* genome [22]. We also found 277 TFs in *Amphimedon queenslandica* and 274 TFs in *Trichoplax adhaerens.*

The broader TF gene repertoire in ctenophores is likely associated with the greater complexity of cell types, tissues, and organs than in morphologically simpler sponges and placozoans. Still, both *Pleurobrachia* and *Mnemiopsis* have independent expansions of two major gene families: homeobox and zinc finger C2H2 (ZF-C2H2) (Fig. 1a–c). Approximately half of the total TFs encoded in ctenophore genomes are zinc finger TFs, which, unlike C2H2-ZF in tetrapods, lack KRAB and SCAN domains. The ZF-C2H2 family can be further classified into BED, THAP, and FLYWCH types derived from transposons or linked to transposon capture [23–25].

In *Amphimedon*, there is the overall expansion of different TFs: BZIP, CENPB [26, 27], FHY3 [28], HTH-Psq [29], and MADF [30] (Fig. 1a). Except for BZIP, all these TF families have transposon-derived origins or are associated with transposase capture. Similarly, the lineage-specific radiation of ZF-C2H2 in *Amphimedon* is attributed to expansions of ZF-BED and ZF-THAP (Fig. 1d), both of which have transposon origins [8].

In the third lineage leading to *Trichoplax*, the observed diversification of TFs in the genome primarily stems from BZIP, BHLH, CBFB, HMG/SOX, and specific homeobox genes (Fig. 1a, d).

In summary, these findings highlight the parallel evolution of TF families across all basal metazoan genomes, with transposons potentially playing a significant role [8].

3.2 Lineage-Specific Diversification of TFs and Their Evolutionary Implications

Phylogenetic reconstruction enables more precise identification of lineage-specific expansions within specific transcription factor [31] families across genomes. Figure 2 illustrates the expansion of the Antennapedia group in the ctenophore lineage. This group encompasses homeobox TF families, including ParaHox, EHGbox, and NK-like genes, which play crucial roles in developmental processes in bilaterians [32, 33]. The mechanisms underlying the independent expansion of Antennapedia and SIX homeobox genes in the ctenophore lineage remain unknown.

HOX genes determine the anterior–posterior body axis in bilaterians [34, 35]. Interestingly, ctenophores, sponges, and apparently placozoans do not encode recognizable HOX genes [6, 7, 36, 37], suggesting the origin of the HOX code in the common ancestor of Cnidaria+Bilateria.

Our analysis further reveals the lineage-specific expansion of posterior HOX genes, independently occurring in sea urchins and humans (Fig. 2). The posterior HOX gene cluster within vertebrates exhibits interesting expression patterns, including HOX10 to HOX13. For instance, HOX10 is involved in patterning the

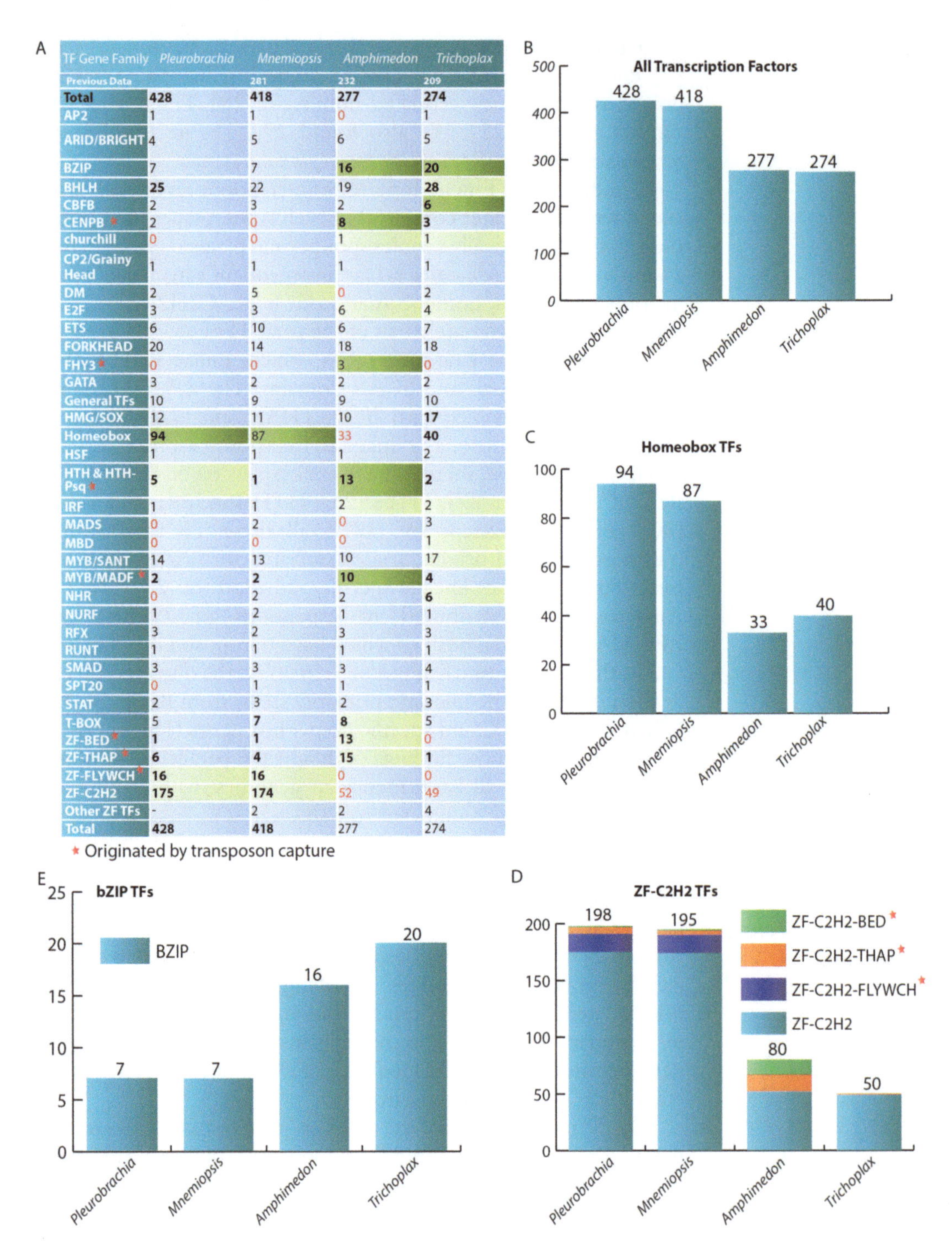

TF Gene Family	Pleurobrachia	Mnemiopsis	Amphimedon	Trichoplax
Previous Data		281	232	209
Total	**428**	**418**	**277**	**274**
AP2	1	1	0	1
ARID/BRIGHT	4	5	6	5
BZIP	7	7	**16**	**20**
BHLH	**25**	22	19	**28**
CBFB	2	3	2	**6**
CENPB *	2	0	**8**	**3**
churchill	0	0	1	1
CP2/Grainy Head	1	1	1	1
DM	2	5	0	2
E2F	3	3	6	4
ETS	6	10	6	7
FORKHEAD	20	14	18	18
FHY3 *	0	0	3	0
GATA	3	2	2	2
General TFs	10	9	9	10
HMG/SOX	12	11	10	**17**
Homeobox	**94**	87	33	**40**
HSF	1	1	1	2
HTH & HTH-Psq *	**5**	**1**	**13**	**2**
IRF	1	1	2	2
MADS	0	2	0	3
MBD	0	0	0	1
MYB/SANT	14	13	10	17
MYB/MADF *	**2**	**2**	**10**	**4**
NHR	0	2	2	**6**
NURF	1	2	1	1
RFX	3	2	3	3
RUNT	1	1	1	1
SMAD	3	3	3	4
SPT20	0	1	1	1
STAT	2	3	2	3
T-BOX	5	**7**	**8**	5
ZF-BED *	**1**	**1**	**13**	0
ZF-THAP *	**6**	**4**	**15**	**1**
ZF-FLYWCH *	**16**	**16**	0	0
ZF-C2H2	**175**	**174**	52	49
Other ZF TFs	-	2	2	4
Total	**428**	**418**	277	274

* Originated by transposon capture

Fig. 1 Diversity and independent expansion of transcription factors in the genomes of the three earliest branching metazoan lineages: Ctenophora (*Pleurobrachia bachei* and *Mnemiopsis leidyi*), Porifera (*Amphimedon queenslandica*), and Placozoa (*Trichoplax adhaerens*). (**a**) The metazoan genomes exhibit diverse transcription factors (TFs), with each lineage undergoing independent radiations highlighted in bold and distinct color-coding within the cell. The letter "red" signifies either the absence of TFs or a significant reduction in their numbers. (**b**) The bar diagram illustrates the total number of TFs encoded in the genomes from four basal metazoan lineages, with the corresponding numbers above each bar. Notably, the ctenophore

lumbar vertebra region, an evolutionary novelty associated with locomotion adaptations. HOX11 patterns the sacrum region, which is crucial for the male and female reproductive tracts, including the development of the oviduct, uterus, and cervix [38]. The observed correlation between innovation in the lumbar region and evolutionary novelties in the reproductive system might contribute to the shift from egg-laying to embryo implantation [38].

NK genes are essential for establishing neuronal identity along the vertebrate neural tube's dorsoventral (D–V) axis [39]. Similarly, the *sine oculis* homeobox (SIX) gene family plays a vital role in *Drosophila* retinal development and contributes to constructing multiple tissues and organs in vertebrates [40]. Additionally, vertebrate homeobox genes, apart from C2H2-ZF, are prone to transposase capture, indicating the need for further analysis to understand the involvement of transposons in the generation of these lineage-specific genes.

We support the ctenophore-first hypothesis, which places this phylum as the most basally branching clade, sister to other metazoans [6, 7, 9–11]. This topology of the metazoan phylogenetic tree helps to determine the evolutionary spectrum of gene expansion or gene loss events (see also details in [6]). If a TF homolog is present in Placozoa but absent in Ctenophora or Porifera, we inferred that this TF originated within the placozoan lineage. Similarly, if a TF homolog is present in the ctenophore lineage but absent in both Porifera and Placozoa, we inferred that this TF homolog originated within the Ctenophora lineage. The presence of the same TF homologs in Ctenophora and Cnidaria or Bilateria indicates that sponge and placozoan lineages secondarily lost these genes from the common metazoan ancestor.

In Fig. 3, the presence of BHLH gene families, including Achaete-scute, Dimm, Delilah, Fer, Hes 1/4, and Ahr, is depicted in the *Trichoplax* lineage while being absent from the other two more basal metazoan lineages (Ctenophora and Porifera). This evolutionary pattern suggests the origin of these transcription factors (TFs) in the common ancestor of Placozoa and Cnidaria, with their preservation in Bilateria. Notably, Achaete-scute, Dimm, and Delilah belong to the Group A BHLH gene, also known as proneuronal BHLH factors, which play a vital role in the early stages of

Fig. 1 (continued) genomes contain significantly more TFs than sponges (*Amphimedon*) and placozoans (*Trichoplax*). (**c**) The ctenophore genomes demonstrate the independent expansion of homeobox genes, indicating the diversification of this gene family within this lineage. (**d**) The represented basal metazoan genomes display an independent diversification of ZF-C2H2 TFs, highlighting their increased abundance and diversity within this group. (**e**) The *Amphimedon* and *Trichoplax* genomes exhibit independent expansions of leucine zipper (BZIP) TFs, signifying the specific diversification of BZIP TFs within these organisms

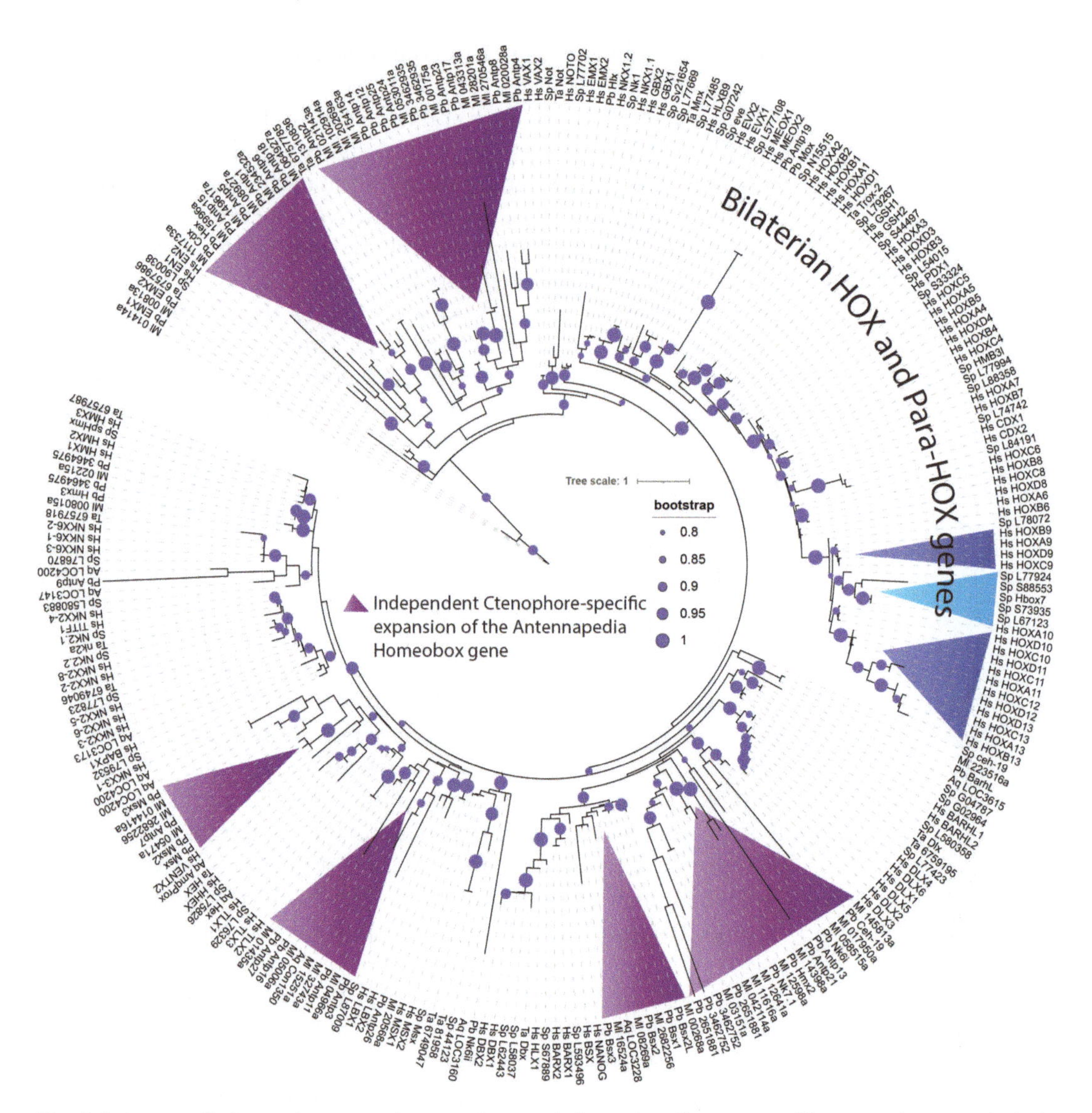

Fig. 2 Antennapedia homeobox genes have undergone independent lineage-specific expansions, particularly in the ctenophore lineage represented by *Pleurobrachia* and *Mnemiopsis*. Purple triangles denote these expanded genes and are absent from other basal metazoans, including sponge and placozoan lineages. Additionally, posterior HOX genes have independently expanded in some deuterostomes, such as sea urchins (cyan triangle) and humans (blue triangle). Abbreviations used in this and the other trees are Sci, *Sycon ciliatum*; Pb, *Pleurobrachia bachei*; ML, *Mnemiopsis leidyi*; Aq, *Amphimedon queenslandica*; Ta, *Trichoplax adhaerens*; Hs, *Homo sapiens*; Dm, *Drosophila melanogaster*; Sp, *Strongylocentrotus purpuratus*. (See text and notes for details)

nervous system development [41, 42]. Given that placozoans lack neurons [43–46], it would be intriguing to investigate whether the Achaete-scute gene in *Trichoplax* is expressed in specific secretory (neuroid-like) cells [47], providing an avenue for future studies of alternative integrative systems [48, 49].

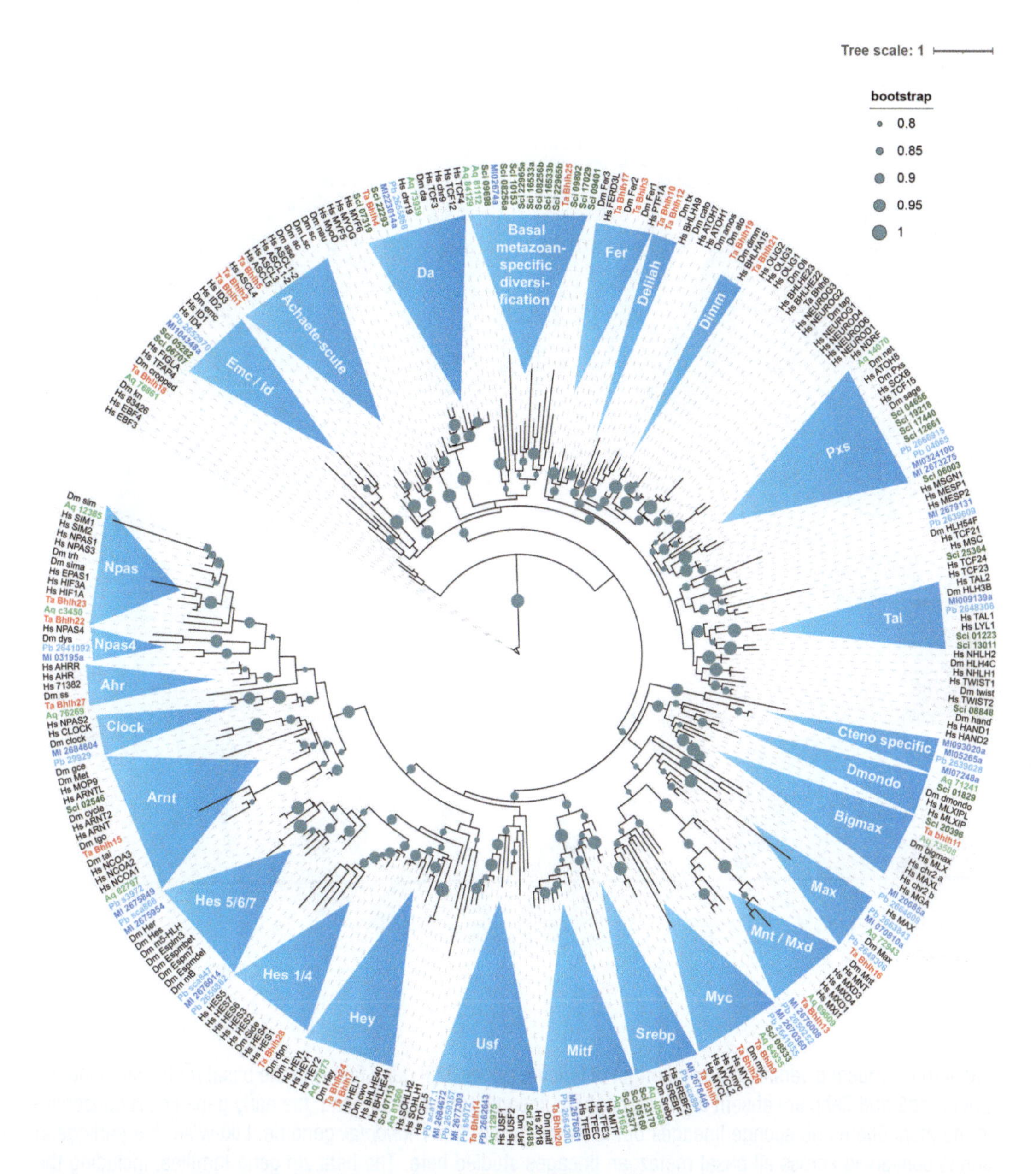

Fig. 3 Independent lineage-specific radiations of the basic helix–loop–helix (BHLH) genes. In the phylogenetic tree, the ctenophore species (*Pleurobrachia* and *Mnemiopsis*) are represented in blue color, Porifera (*Amphimedon*) is in green, and Placozoa (*Trichoplax*) is in red. The genes marked in black indicate their origin in bilaterians. Abbreviations: Sci, *Sycon ciliatum*; Pb, *Pleurobrachia bachei*; ML, *Mnemiopsis leidyi*; Aq, *Amphimedon queenslandica*; Ta, *Trichoplax adhaerens*; Hs, *Homo sapiens*; Dm, *Drosophila melanogaster*, Sp, *Strongylocentrotus purpuratus*. (See text and notes for details)

3.3 Lineage-Specific Expansion of Leucine Zipper (BZIP) TFs

Figure 4 illustrates the phylogenetic reconstruction of the BZIP TF families. Creb1, originating early in the common ancestor of all metazoans, is conserved across all animal lineages, including ctenophores. However, Creb2 and Cebp originated in the common ancestor of sponges (*Amphimedon*, Aq) and are preserved in

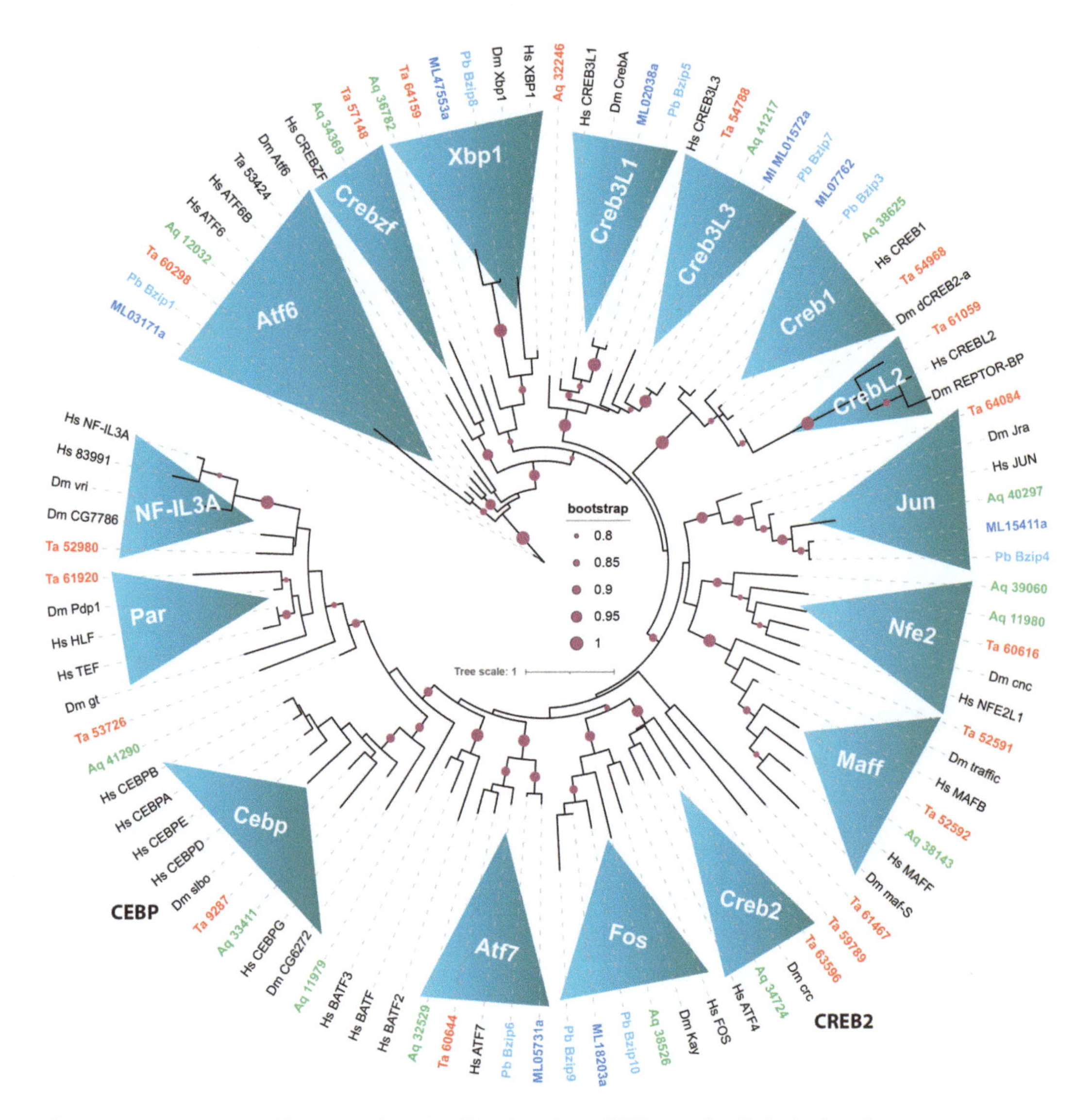

Fig. 4 Independent diversification and origin of leucine zipper BZIP gene family in the basal metazoan genome. Both Creb2 and Cebp are absent in the ctenophore lineage. On the other hand, the early gene Fos is conserved in the ctenophore and sponge lineages but is missing from the *Trichoplax* genome. Likewise, the early gene Jun is conserved across all basal metazoan lineages studied here. The beta zip gene families, including the canonical CrebL2, NFIL-3A, and Par, present in *Trichoplax* and are apparently absent from ctenophores and sponges. However, they are conserved in cnidarian and bilaterian lineages [77]. Abbreviations: Sci, *Sycon ciliatum*; Pb, *Pleurobrachia bachei*; ML, *Mnemiopsis leidyi*; Aq, *Amphimedon queenslandica*; Ta, *Trichoplax adhaerens*; Hs, *Homo sapiens*; Dm, *Drosophila melanogaster*; Sp, *Strongylocentrotus purpuratus*. (See text and notes for details)

Trichoplax, but they are absent in the ctenophore genomes (*Pleurobrachia* (Pb) and *Mnemiopsis* (Ml)). Our analysis also suggests that these genes are conserved in protostomes, lophotrochozoans, deuterostomes, and cnidarian genomes. This presents two possible scenarios. If the Porifera-first hypothesis is truthful [50, 51], the ctenophore lineage may have lost Creb2 and Cebp. Alternatively, and more likely, if the Ctenophora-first hypothesis is correct [9–11, 52], Creb2 and Cebp may have been primarily absent in stem ctenophores. Both cAMP response element-binding protein (Creb2) and CCAT enhancer-binding protein (Cebp) are involved in tissue-specific gene expression, proliferation, differentiation [53], learning [54], and long-term memory formation [55]. Thus, their absence in ctenophores raises intriguing questions regarding the mechanisms of neuroplasticity, stress response, and other cAMP-dependent processes in early-branching metazoans.

As described above, similar evolutionary distribution patterns for Creb2 and Cebp have been identified for particular beta zip subfamilies (BZIP). Specifically, Nfe2, Maff, and Crebzf are present in the *Amphimedon* and *Trichoplax* genomes as well as in cnidarians and bilaterians but absent in the *Pleurobrachia* and *Mnemiopsis* genomes, suggesting their 'later' origin (before the divergence of the sponge lineage). The functions of these BZIP TFs are currently unknown and require further analysis.

The beta zip subfamily CrebL2 presents in *Trichoplax* and is absent in all other prebilaterian metazoans but conserved among cnidarians and bilaterians. We interpret this phylogeny as a primary absence of this subfamily in ctenophores and sponges, originating in the common ancestor of the Placozoa+Cnidaria+Bilateria clade. The *Drosophila* ortholog of the human CrebL2 gene, named REPTOR-BP, functions downstream of mTORC1 and plays crucial roles in organismal metabolism, life span, and stress [56]. Additionally, human CrebL2 is a metabolic regulator in muscle and liver cells [57].

NF-IL3A also evolved in the common ancestor of Placozoa +Cnidaria+Bilateria. NF-IL3A is a human T-cell transcription factor transacting the interleukin-3 promoter [58]. The *Drosophila* ortholog of human NF-IL3A is vri, which controls cell growth and proliferation by regulating the actin skeleton [57]. Similarly, the canonical Par gene family, named after the *Drosophila* gene Pdp1 (PAR domain protein 1), which plays a crucial role in regulating muscle gene transcription, likely evolved in the common ancestor of *Trichoplax*, cnidarians, and bilaterians, with primary absence in sponges and ctenophores. The functions of many genes in early branched metazoans, including the examples mentioned above, remain largely unknown, and further studies on ctenophores and placozoans are crucial for shedding light on their evolutionary significance in Metazoa.

3.4 Conclusion and Future Directions

The lineage-specific expansion of transcription factors (TFs) is a widespread phenomenon observed in various TF families, including those that have potentially captured transposase domains independently. This hypothesis is one of many reasons in investigating the mechanisms underlying TF origins and evolution. Notably, zinc finger TFs exhibit significant lineage-specific expansions following the independent capture of transposase domains in the tetrapod lineage.

A similar trend is observed in the ctenophore lineage, where the FLYWCH transposon appears to be the primary contributor to the expansion of zinc finger TFs [8]. In contrast, the expansion of zinc finger TFs in the *Amphimedon* lineage seems to involve THAP and BED transposons but not FLYWCH, suggesting a preference for transposon recruitment in the independent diversification of zinc finger TFs in sponges and metazoans as a whole.

Further studies are required to understand the expansion of the homeobox Antennapedia and SIX genes in the ctenophore lineage. These genes play critical roles in the developmental control of numerous cell types associated with different organs and tissues. However, the functional understanding of transposon-domesticated FLYWCH genes is limited. In *C. elegans*, FLYWCH genes function as repressors of embryonic expression of microRNA genes [59], which regulate gene expression post-transcriptionally. Ctenophores do not possess recognized microRNAs [6], thus highlighting the need for future analysis of the radiation of FLYWCH genes in the ctenophore lineage.

Recent studies have demonstrated that FLYWCH1 in humans suppresses the nuclear beta-catenin pathway [59], which is responsible for cell fate decisions. Similarly, the BED zinc finger TF ZBED3 has been shown to modulate the Wnt/beta-catenin pathway by binding to Axin [60]. These findings underscore the potential regulatory roles of FLYWCH and BED zinc finger TFs in crucial signaling pathways.

Further investigations require understanding the functional significance of lineage-specific TF expansions in ctenophores and their involvement in their lineage-specific signaling pathways. This critical line of research would provide valuable insights into the evolutionary and regulatory mechanisms shaping the diversity of transcription factors and body plans across basal metazoans.

In summary, we propose the following hypotheses:

1. The lineage-specific diversification of TFs triggers the convergent evolution of diverse cell types in major basal metazoan clades, including events of independent origins of neurons, mesoderm, and muscles [47, 61–64].
2. Generating a broader spectrum of morphologically similar cell types might occur convergently across phyla.
3. Lineage-specific expansions of certain TF classes can be associated with the activity and domestication of transposase-captured genes.

Notably, most TFs described here, and subjects to lineage-specific origins and/or diversification, have no association with recognizable transposase-capture events, which needs further analyses of respective mechanisms (e.g., duplications, lateral gene transport, etc.). Notes below summarize various scenarios and examples of the evolution of TFs.

4 Notes

Illustrative examples for the role of mobile elements in the TF diversification and evolution.

1. Pax6, a homeobox gene, is a universal master control gene in bilaterian eye morphogenesis [65, 66]. The origin of the PAX gene can be attributed to the domestication of the transposable element Tc1/mariner, which is widely distributed across metazoans, certain unicellular eukaryotes, and plants [67]. The Pax family emerged through the fusion of a transposase domain with another gene, likely facilitated by exon-shuffling [68]. However, the precise timing and extent of transposase capture remain unclear due to the deep ancestry of the Pax genes. Additionally, it is currently unknown whether the capture of the transposase domain occurred once or multiple times during evolution.
2. Recent studies have provided evidence that the mobility of DNA transposons can facilitate exon-shuffling, allowing functional domains to be inserted into new genomic contexts. This phenomenon can lead to the formation of host-transposase fusion (HTF) genes through the splicing of these domains [69]. Cosby et al. demonstrated that transposase domains have been captured, primarily through alternate splicing, resulting in the generation of 106 distinct fusion proteins, mainly transcription factors. These events occurred independently around 106 times over approximately 350 million years during tetrapod evolution [69]. Remarkably, 77% of these HTF genes possess a DNA-binding domain, highlighting their role as transcription factors. These findings suggest that the fusion of host and transposase domains can give rise to novel regulatory genes, including deeply conserved and lineage-specific transcription factors.
3. The Krüppel-associated box zinc finger protein (KRAB-ZF) stands out as the largest and most prevalent family of zinc finger transcription factors (TFs), primarily found in tetrapod genomes but exhibiting a remarkable expansion in mammalian genomes [70]. Within the human TF repertoire of over 1600 genes, the KRAB domain-containing zinc finger family accounts for the highest abundance, encompassing around 423 TFs in the genome. Notably, a subset of KRAB-ZF proteins also possesses an additional SCAN domain. Interestingly, the KRAB and SCAN domains have been observed as

prominent targets for transposase fusion events [69]. Across the tetrapod phylogeny, the KRAB and SCAN domains have been involved in 32 and 19 independent fusion events, accounting for approximately 50% of such occurrences. Furthermore, 2 other transcription factor domains, C2H2-ZF and homeodomain, have undergone independent transposase fusion events 12 and 5 times, respectively, within the tetrapod lineage [69]. These analyses indicate that zinc finger and homeobox are among the most abundant transcription factor protein families, characterized by multiple independent instances of transposase capture throughout evolution.

4. See also Fig. 2 for independent diversification, origin, and losses of BHLH gene families in the basal metazoan genomes.

5. Notes about particular subfamilies of TFs relevant to their evolution.

   Achaete-scute—The Achaete-scute gene complex, which plays a crucial role in forming neural precursors and subsequent differentiation of specific neuronal lineages in *Drosophila* [71], is found in the *Trichoplax* genome. However, it is absent in the sequenced genomes of sponges (*Sycon* [72], *Amphimedon* (Aq)), and ctenophores (*Pleurobrachia* (Pb), and *Mnemiopsis* (Ml)). On the other hand, it is present in major bilaterian lineages. In addition to its role in neuronal development, Achaete-scute is also expressed in a cluster of mesodermal cells and specific cells in the gut [73], highlighting its involvement in multiple cell patterning processes.

   Dimm—The *Trichoplax* genome contains the BHLH gene Dimm, which is not found in the sequenced prebilaterian genomes of *Sycon* [74], *Amphimedon* (Aq), *Pleurobrachia* (Pb), and *Mnemiopsis* (Ml). However, it is present in bilaterians. Dimm is known for its role as a master regulator of secretory phenotypes in neuroendocrine cells. It is involved in the combinatorial code that regulates the terminal differentiation of peptidergic neurons [75].

   Delilah—The *Trichoplax* genome contains the Delilah gene, which is not found in the sequenced genomes of ctenophores *Pleurobrachia* and *Mnemiopsis* and sponges *Sycon* and *Amphimedon*. However, it is present in many bilaterians. Delilah is absent in vertebrates but has been found in the *Branchiostoma* genome. This gene also exhibits almost exclusive expression in insect apodemes, specialized structures involved in insect locomotion and attachment [76].

   Fer—The *Trichoplax* genome contains the Fer gene, which is not found in *Sycon*, *Amphimedon*, *Pleurobrachia*, and *Mnemiopsis* sequenced genomes. However, it is present in bilaterians. In *Drosophila*, Fer plays a crucial role in the development of a specific subset of circadian pacemaker neurons and dopaminergic neurons in the photocerebral anterior media [77]. It is also involved in the development of the photocerebral anterior lateral (PAL) clusters of the brain, which are essential for the survival of PAM cluster dopaminergic neurons during adulthood and oxidative stress response [12].

Dmondo—The Dmondo gene is found in sponges *Sycon* and *Amphimedon* but is absent from the sequenced genomes of *Pleurobrachia*, *Mnemiopsis*, and *Trichoplax.* Dmondo serves as a nutrient sensor, playing a role in sensing and responding to nutrient availability [52, 78].

Hes 1/4—The *Trichoplax* genome contains the Hes 1/4 gene, which is absent from the sequenced genomes of ctenophores (*Pleurobrachia* and *Mnemiopsis*) and sponges (*Sycon* and *Amphimedon*). However, it is present in bilaterians. Hes 1/4 functions as a repressor of differentiation by targeting genes involved in the Notch signaling pathway. This pathway regulates lineage specification decisions during the development of various tissues [78].

Hes 5/6/7—The Hes6 gene is found in the sequenced genomes of *Pleurobrachia* and *Mnemiopsis* but is absent from the sequenced genomes of *Sycon*, *Amphimedon*, and *Trichoplax.* It is present in protostomes but absent in the deuterostome genomes, including humans. Hes6 plays a role in promoting neuronal differentiation, contributing to the development of neurons [79].

Clock—The clock gene is present in the genomes of *Pleurobrachia* and *Mnemiopsis* but absent from the sequenced genomes of *Sycon*, *Amphimedon*, and *Trichoplax.* It is found in bilaterians. The clock gene is a key regulator of sleep, stress, learning, and memory, as studied extensively in mice [80].

Ahr—The Ahr gene is present in the *Trichoplax* genome but absent from the sequenced genomes of ctenophores *Pleurobrachia* and *Mnemiopsis* and sponges *Sycon* and *Amphimedon.* It is found in bilaterians. Ahr is involved in binding to a range of endogenous and exogenous chemicals, including TCDD, and contributes to immune responses and the regulation of development and pathology [81].

Acknowledgments

This work was supported in part by the Human Frontiers Science Program (RGP0060/2017) and National Science Foundation (IOS-1557923) grants to LLM. Research reported in this publication was also supported in part by the National Institute of Neurological Disorders and Stroke of the National Institutes of Health under Award Number R01NS114491 (to LLM). The content is solely the authors' responsibility and does not necessarily represent the official views of the National Institutes of Health.

References

1. Lambert SA et al (2018) The human transcription factors. Cell 175(2):598–599
2. de Mendoza A, Sebe-Pedros A (2019) Origin and evolution of eukaryotic transcription factors. Curr Opin Genet Dev 58–59:25–32
3. Ohno S, Wolf U, Atkin NB (1968) Evolution from fish to mammals by gene duplication. Hereditas 59(1):169–187
4. McGinnis W et al (1984) A conserved DNA sequence in homoeotic genes of the *Drosophila* Antennapedia and bithorax complexes. Nature 308(5958):428–433
5. Scott MP, Weiner AJ (1984) Structural relationships among genes that control development: sequence homology between the Antennapedia, Ultrabithorax, and fushi tarazu loci of *Drosophila*. Proc Natl Acad Sci U S A 81(13):4115–4119
6. Moroz LL et al (2014) The ctenophore genome and the evolutionary origins of neural systems. Nature 510(7503):109–114
7. Ryan JF et al (2013) The genome of the ctenophore *Mnemiopsis leidyi* and its implications for cell type evolution. Science 342(6164): 1242592
8. Mukherjee K, Moroz LL (2023) Transposon-derived transcription factors across metazoans. Front Cell Dev Biol 11:1113046
9. Schultz DT et al (2023) Ancient gene linkages support ctenophores as sister to other animals. Nature 618(7963):110–117
10. Whelan NV et al (2015) Error, signal, and the placement of Ctenophora sister to all other animals. Proc Natl Acad Sci U S A 112(18): 5773–5778
11. Whelan NV et al (2017) Ctenophore relationships and their placement as the sister group to all other animals. Nat Ecol Evol 1(11): 1737–1746
12. Tas D et al (2018) Parallel roles of transcription factors dFOXO and FER2 in the development and maintenance of dopaminergic neurons. PLoS Genet 14(3):e1007271
13. Marchler-Bauer A et al (2011) CDD: a conserved domain database for the functional annotation of proteins. Nucleic Acids Res 39 (Database issue):D225–D229
14. Letunic I, Bork P (2018) 20 years of the SMART protein domain annotation resource. Nucleic Acids Res 46(D1):D493–D496
15. Edgar RC (2004) MUSCLE: multiple sequence alignment with high accuracy and high throughput. Nucleic Acids Res 32(5): 1792–1797
16. Guindon S et al (2010) New algorithms and methods to estimate maximum-likelihood phylogenies: assessing the performance of PhyML 3.0. Syst Biol 59(3):307–321
17. Guindon S, Gascuel O (2003) A simple, fast, and accurate algorithm to estimate large phylogenies by maximum likelihood. Syst Biol 52(5):696–704
18. Abascal F, Zardoya R, Posada D (2005) ProtTest: selection of best-fit models of protein evolution. Bioinformatics 21(9):2104–2105
19. Hordijk W, Gascuel O (2005) Improving the efficiency of SPR moves in phylogenetic tree search methods based on maximum likelihood. Bioinformatics 21(24):4338–4347
20. Anisimova M et al (2011) Survey of branch support methods demonstrates accuracy, power, and robustness of fast likelihood-based approximation schemes. Syst Biol 60(5): 685–699
21. Emms DM, Kelly S (2019) OrthoFinder: phylogenetic orthology inference for comparative genomics. Genome Biol 20(1):238
22. Sebe-Pedros A et al (2018) Early metazoan cell type diversity and the evolution of multicellular gene regulation. Nat Ecol Evol 2(7): 1176–1188

23. Hayward A et al (2013) ZBED evolution: repeated utilization of DNA transposons as regulators of diverse host functions. PLoS One 8(3):e59940
24. Roussigne M et al (2003) The THAP domain: a novel protein motif with similarity to the DNA-binding domain of P element transposase. Trends Biochem Sci 28(2):66–69
25. Marquez CP, Pritham EJ (2010) Phantom, a new subclass of Mutator DNA transposons found in insect viruses and widely distributed in animals. Genetics 185(4):1507–1517
26. Mateo L, Gonzalez J (2014) Pogo-like transposases have been repeatedly domesticated into CENP-B-related proteins. Genome Biol Evol 6(8):2008–2016
27. Casola C, Hucks D, Feschotte C (2008) Convergent domestication of pogo-like transposases into centromere-binding proteins in fission yeast and mammals. Mol Biol Evol 25(1):29–41
28. Hudson ME, Lisch DR, Quail PH (2003) The FHY3 and FAR1 genes encode transposase-related proteins involved in regulation of gene expression by the phytochrome A-signaling pathway. Plant J 34(4):453–471
29. Siegmund T, Lehmann M (2002) The *Drosophila* Pipsqueak protein defines a new family of helix-turn-helix DNA-binding proteins. Dev Genes Evol 212(3):152–157
30. Kapitonov VV, Jurka J (2004) Harbinger transposons and an ancient HARBI1 gene derived from a transposase. DNA Cell Biol 23(5): 311–324
31. Lein ES et al (2007) Genome-wide atlas of gene expression in the adult mouse brain. Nature 445(7124):168–176
32. Beck F (2002) Homeobox genes in gut development. Gut 51(3):450–454
33. Harvey RP (1996) NK-2 homeobox genes and heart development. Dev Biol 178(2):203–216
34. McGinnis W, Krumlauf R (1992) Homeobox genes and axial patterning. Cell 68(2): 283–302
35. Hughes CL, Kaufman TC (2002) Hox genes and the evolution of the arthropod body plan. Evol Dev 4(6):459–499
36. Larroux C et al (2007) The NK homeobox gene cluster predates the origin of Hox genes. Curr Biol 17(8):706–710
37. Jakob W et al (2004) The Trox-2Hox/ ParaHox gene of *Trichoplax* (Placozoa) marks an epithelial boundary. Dev Genes Evol 214(4):170–175
38. Wellik DM, Capecchi MR (2003) Hox10 and Hox11 genes are required to globally pattern the mammalian skeleton. Science 301(5631): 363–367
39. McMahon AP (2000) Neural patterning: the role of Nkx genes in the ventral spinal cord. Genes Dev 14(18):2261–2264
40. Kumar JP (2009) The sine oculis homeobox (SIX) family of transcription factors as regulators of development and disease. Cell Mol Life Sci 66(4):565–583
41. Baker NE, Brown NL (2018) All in the family: proneural bHLH genes and neuronal diversity. Development 145(9)
42. Hartenstein V, Stollewerk A (2015) The evolution of early neurogenesis. Dev Cell 32(4): 390–407
43. Romanova DY et al (2021) Hidden cell diversity in Placozoa: ultrastructural insights from Hoilungia hongkongensis. Cell Tissue Res 385(3):623–637
44. Grell KG, Ruthmann A (1991) Placozoa. In: Harrison FW (ed) Microscopic anatomy of invertebrates. Wiley-Liss, New York, pp 13–27
45. Smith CL et al (2021) Microscopy studies of Placozoans. Methods Mol Biol 2219:99–118
46. Smith CL et al (2014) Novel cell types, neurosecretory cells, and body plan of the early-diverging metazoan *Trichoplax adhaerens*. Curr Biol 24(14):1565–1572
47. Moroz LL (2021) Multiple origins of neurons from secretory cells. Front Cell Dev Biol 9: 669087
48. Moroz LL, Romanova DY (2022) Alternative neural systems: what is a neuron? (Ctenophores, sponges and placozoans). Front Cell Dev Biol 10:1071961
49. Moroz LL, Romanova DY, Kohn AB (1821) Neural versus alternative integrative systems: molecular insights into origins of neurotransmitters. Philos Trans R Soc Lond Ser B Biol Sci 2021(376):20190762
50. Redmond AK, McLysaght A (2021) Evidence for sponges as sister to all other animals from partitioned phylogenomics with mixture models and recoding. Nat Commun 12(1):1783
51. Telford MJ, Moroz LL, Halanych KM (2016) Evolution: a sisterly dispute. Nature 529(7586):286–287
52. Li Y et al (2021) Rooting the animal tree of life. Mol Biol Evol 38(10):4322–4333
53. Tsukada J et al (2011) The CCAAT/enhancer (C/EBP) family of basic-leucine zipper (BZIP) transcription factors is a multifaceted highly-regulated system for gene regulation. Cytokine 54(1):6–19

54. Amar F et al (2021) Rapid ATF4 depletion resets synaptic responsiveness after cLTP. eNeuro 8(3):ENEURO.0239
55. Mirisis AA, Kopec AM, Carew TJ (2021) ELAV proteins bind and stabilize C/EBP mRNA in the induction of long-term memory in *Aplysia*. J Neurosci 41(5):947–959
56. Tiebe M et al (2015) REPTOR and REPTOR-BP regulate organismal metabolism and transcription downstream of TORC1. Dev Cell 33(3):272–284
57. Tiebe M et al (2019) Crebl2 regulates cell metabolism in muscle and liver cells. Sci Rep 9(1):19869
58. Zhang W et al (1995) Molecular cloning and characterization of NF-IL3A, a transcriptional activator of the human interleukin-3 promoter. Mol Cell Biol 15(11):6055–6063
59. Muhammad BA et al (2018) FLYWCH1, a novel suppressor of nuclear beta-catenin, regulates migration and morphology in colorectal cancer. Mol Cancer Res 16(12):1977–1990
60. Chen T et al (2009) Identification of zinc-finger BED domain-containing 3 (Zbed3) as a novel Axin-interacting protein that activates Wnt/beta-catenin signaling. J Biol Chem 284(11):6683–6689
61. Moroz LL (2009) On the independent origins of complex brains and neurons. Brain Behav Evol 74(3):177–190
62. Moroz LL (2014) The genealogy of genealogy of neurons. Commun Integr Biol 7(6): e993269
63. Moroz LL (2015) Biodiversity meets neuroscience: from the sequencing ship (Ship-Seq) to deciphering parallel evolution of neural systems in Omic's era. Integr Comp Biol 55(6): 1005–1017
64. Moroz LL, Kohn AB (2016) Independent origins of neurons and synapses: insights from ctenophores. Philos Trans R Soc Lond Ser B Biol Sci 371(1685):20150041
65. Hill RE et al (1991) Mouse small eye results from mutations in a paired-like homeobox-containing gene. Nature 354(6354):522–525
66. Ton CC et al (1991) Positional cloning and characterization of a paired box- and homeobox-containing gene from the aniridia region. Cell 67(6):1059–1074
67. Garcia-Fernandez J et al (1993) Infiltration of mariner elements. Nature 364(6433):109–110
68. Breitling R, Gerber JK (2000) Origin of the paired domain. Dev Genes Evol 210(12): 644–650
69. Cosby RL et al (2021) Recurrent evolution of vertebrate transcription factors by transposase capture. Science 371(6531)
70. Bellefroid EJ et al (1993) Clustered organization of homologous KRAB zinc-finger genes with enhanced expression in human T lymphoid cells. EMBO J 12(4):1363–1374
71. Bertrand N, Castro DS, Guillemot F (2002) Proneural genes and the specification of neural cell types. Nat Rev Neurosci 3(7):517–530
72. Hubbard T et al (2005) Ensembl 2005. Nucleic Acids Res 33(Database issue):D447–D453
73. Tepass U, Hartenstein V (1995) Neurogenic and proneural genes control cell fate specification in the *Drosophila* endoderm. Development 121(2):393–405
74. Fortunato SAV et al (2016) Conservation and divergence of bHLH genes in the calcisponge *Sycon ciliatum*. EvoDevo 7:23
75. Liu YT, Luo JN, Nassel DR (2016) The Drosophila transcription factor dimmed affects neuronal growth and differentiation in multiple ways depending on neuron type and developmental stage. Front Mol Neurosci 9
76. Armand P et al (1994) A novel basic helix-loop-helix protein is expressed in muscle attachment sites of the *Drosophila* epidermis. Mol Cell Biol 14(6):4145–4154
77. Pereira JF et al (2013) Boto, a class II transposon in *Moniliophthora perniciosa*, is the first representative of the PIF/harbinger superfamily in a phytopathogenic fungus. Microbiology 159(Pt 1):112–125
78. Iso T, Kedes L, Hamamori Y (2003) HES and HERP families: multiple effectors of the notch signaling pathway. J Cell Physiol 194(3): 237–255
79. Bae S et al (2000) The bHLH gene Hes6, an inhibitor of Hes1, promotes neuronal differentiation. Development 127(13):2933–2943
80. Bolsius YG et al (2021) The role of clock genes in sleep, stress and memory. Biochem Pharmacol 191:114493
81. Zhu K et al (2019) Aryl hydrocarbon receptor pathway: role, regulation and intervention in atherosclerosis therapy (review). Mol Med Rep 20(6):4763–4773

Chapter 21

Long-Term Culturing of Placozoans (*Trichoplax* and *Hoilungia*)

Daria Y. Romanova, Frédérique Varoqueaux, Michael Eitel, Masa-aki Yoshida, Mikhail A. Nikitin, and Leonid L. Moroz

Abstract

The phylum Placozoa remains one of the least explored among early-branching metazoan lineages. For over 130 years, this phylum had been represented by the single species *Trichoplax adhaerens*—an animal with the simplest known body plan (three cell layers without any organs) but complex behaviors. Recently, extensive sampling of placozoans across the globe and their subsequent genetic analysis have revealed incredible biodiversity with numerous cryptic species worldwide. However, only a few culture protocols are available to date, and all are for one species only. Here, we describe the breeding of four different species representing two placozoan genera: *Trichoplax adhaerens*, *Trichoplax* sp. H2, *Hoilungia* sp. H4, and *Hoilungia hongkongensis* originating from diverse biotopes. Our protocols allow to culture all species under comparable conditions. Next, we outlined various food sources and optimized strain-specific parameters enabling long-term culturing. These protocols can facilitate comparative analyses of placozoan biology and behaviors, which together will contribute to deciphering general principles of animal organization.

Key words *Trichoplax*, *Hoilungia*, *Polyplacotoma*, Placozoan, Early-branching metazoa, Evolution, Culture

1 Introduction

While studying the fauna of a seawater aquarium in Graz, Austria, in 1883, FE Schulze described a metazoan of deceivingly simple morphology—a flat, millimeter-wide disk-shaped multicellular organism devoid of organs or symmetry, which essentially grazed on the aquarium walls [47, 50]. Referring to this behavior and the numerous cilia observed on its surface, he named it *Trichoplax adhaerens* ("sticky hairy plate"). About 100 years later, upon further morphological and behavioral observations, Karl Grell coined the phylum Placozoa to classify this species among early animals [11]. Indeed, *Trichoplax* has only six currently recognized cell types (but *see* Varoqueaux et al. [66]), arranged in three layers, and is

Leonid L. Moroz (ed.), *Ctenophores: Methods and Protocols*, Methods in Molecular Biology, vol. 2757,
https://doi.org/10.1007/978-1-0716-3642-8_21, © Springer Science+Business Media, LLC, part of Springer Nature 2024

devoid of any organ [13, 15, 11, 19, 27, 57]. The morphological simplicity of placozoans contrasts with an array of complex behaviors, including components of social-like feeding patterns [10, 34, 51, 53, 58].

As one of the five early-branching metazoan lineages, placozoans serve as important models for understanding both metazoan evolution and integrative mechanisms of the animal organization (Srivastava et al. [59]) [9, 18]. However, we know very little about placozoan physiology [1, 3, 5, 26, 28–31, 40–42], mechanisms of their behaviors [2, 6, 8, 10, 21, 33, 49, 53, 54, 56, 66], and even less about their biological diversity. To date, placozoans have been collected across the world in marine aquaria and seawater systems, as well as the wild, mostly in warm temperate coastal waters ranging across different habitats such as mangroves, reefs, or harbors (*see*, e.g., [6, 32, 37, 55, 65]). Accordingly, they live in varied salinity (20–50 ppm), temperature (11–27 °C), depth (0–20 m), light, and pH conditions [8, 45, 48], where their population growth is subject to seasonal variations [20, 37, 43, 65].

Extensive sampling of placozoans across the globe [46] and their subsequent genetic analysis increasingly point to a phylum of morphologically similar animals with a sizeable cryptic biodiversity. Four genetically distinct, distantly related placozoan species, *Trichoplax adhaerens* [50], *Hoilungia hongkongensis* [9], *Polyplacotoma mediterranea* [35], and *Cladtertia collaboinventa*, nov. [61], and additionally 21 haplotypes have been described [8, 23, 60], while more than 100 cryptic species are expected to exist worldwide [46, 67]. Figure 1 shows phylogenetic relationships between known species of Placozoa.

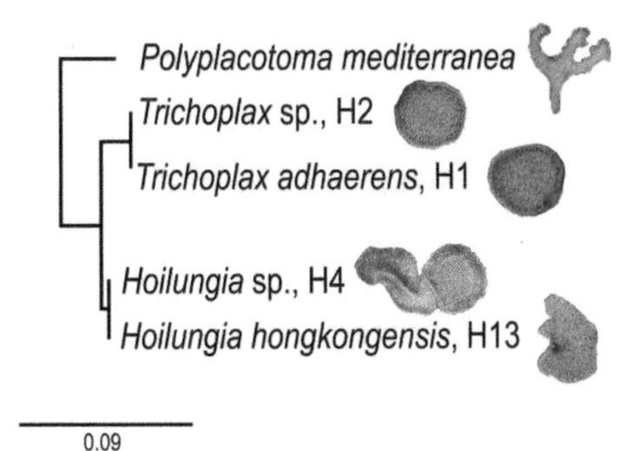

Fig. 1 Phylogenetic relationships between known species of Placozoa. *Trichoplax adhaerens* (Grell 1883), *Hoilungia hongkongensis* [9], and *Polyplacotoma mediterranea* [35] are the three formally described placozoan species. *Trichoplax* sp., H2, and *Hoilungia* sp., H4, are genetically closely related to *Trichoplax adhaerens* (H1) and *Hoilungia hongkongensis* (H13), respectively, while they might live in different marine habitats. The phylogenomic tree (based on mitochondrial-encoded proteins) and the *Polyplacotoma* picture have been modified from Osigus et al. [35]

Hence, the full characterization of several placozoan species needs to be done to assess their common and distinctive features. Currently, their ecological niches and exact diets are essentially unknown *in vivo*. Pearse and Voigt [37] have suggested that wild placozoan may be opportunistic grazers, scavenging on organic detritus, algae, and bacteria in biofilms.

The maintenance of placozoans in the lab has been scantily described [6, 8, 16, 55], although they have been maintained in "natural" aquaria [36] or Petri dishes supplemented with different food sources, such as *Cryptomonas* (e.g., [12, 44]), yeast extracts [65], red algae *Pyrenomonas helgolandii* [55], green algae (*Ulva* sp.; [52]), or a mix of green (*Nannochloropsis salina*) and red (*Rhodamonas salina*, *Pyrenomonas helgolandii*) algae [17, 57].

Heyland and his co-authors have proposed a detailed protocol defining optimized laboratory breeding conditions for *Trichoplax adhaerens*, primarily consisting of maintaining them in Petri dishes filled with ASW (35 ppm, pH = 8.0) changed bimonthly, feeding them on *Cryptomonas* sp., and placement in a climate chamber with constant light (12:8) and temperature (24 °C) conditions. To date, there have been no systematic comparisons across published culture protocols. The optimization of culture conditions is essential to characterize placozoan's biology and detecting species-specific behaviors reproducibly. It is also important to establish culture conditions that are close to those at the collection sites in the natural habitats of placozoans.

To compare different placozoan strains and species, we imported various haplotypes into the laboratory, where we monitored their survival in different salinity, temperature, light, and pH conditions. Here, we describe breeding conditions for four different species, representing two placozoan genera: *Trichoplax adhaerens* (16S haplotype H1), *Trichoplax* sp. H2 (16S haplotype H2), *Hoilungia* sp. H4 (16S haplotype H4, [38]), and *Hoilungia hongkongensis* (16S haplotype H13, [9]), all of which originate from different habitats (*see* details Subheading 2.1). We further refer to these species as "H1," "H2," "H4," and "H13" based on the mitochondrial 16S haplotype (for an overview of currently known haplotypes, *see* Eitel et al. [8] and Miyazawa et al. [23]).

Specifically, we propose three protocols for the culturing of placozoans:

1. A protocol for the culture of various algae needed as food sources for placozoans
2. An open/adaptive protocol that can be used as a starting point for culturing a majority of known placozoan strains, used here for haplotypes H1, H2, and H13
3. A specific protocol for breeding H4 in aquaria

In summary, these protocols can (i) support the long-term maintenance of different clonal lines in the laboratory; (ii) allow comparing species-specific development and behaviors under defined experimental manipulations, including applications of pharmacological agent, etc.; and (iii) facilitate future studies of placozoan biology from diverse representatives of this early-branching metazoan lineage.

2 Materials

2.1 Animals (Fig. 2)

1. *Trichoplax adhaerens* (16S haplotype H1, often referred to the type species described as *Trichoplax adherens* by Schulze [50]; the established laboratory clonal "Grell" strain is a lineage commonly referred to originate from the Red Sea close to Elat in Egypt, but more likely stems from an individual isolated near Villefranche in Southern France; [4]).

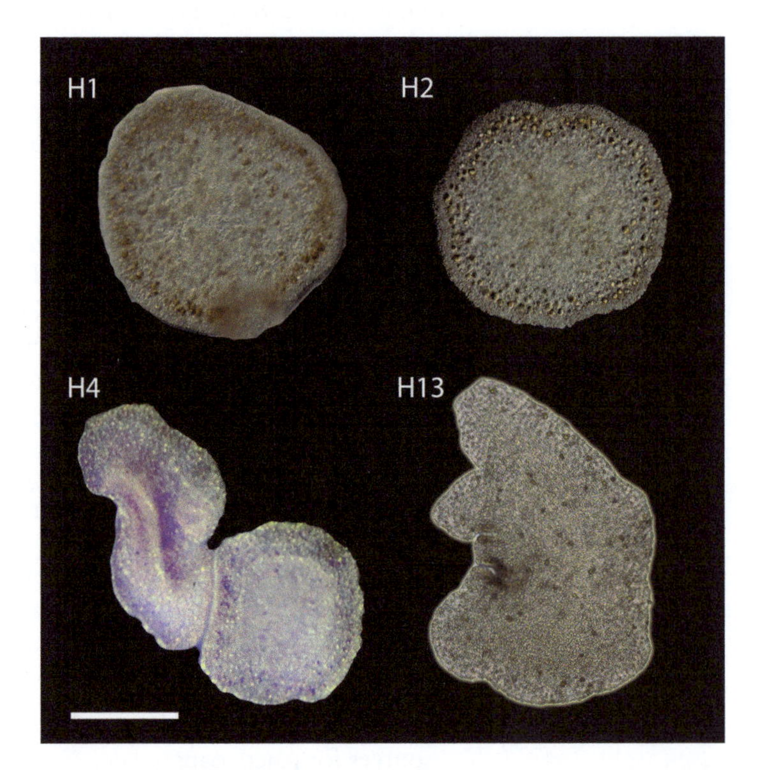

Fig. 2 Morphology of the four different haplotypes of Placozoa maintained in the laboratory. The four haplotypes, H1, H2, H4, and H13, likely represent distinct placozoan species (note: while H2 and H4 have not yet been formally described as distinct species, the analyses by Eitel et al. [9], might indicate this). Individual animals are similar in size (0.5–2 mm range). Note a pinkish pigmentation of H4: placozoans are partially transparent, yet they exhibit various color tones depending on the organisms they feed upon (see text). Scale bar: 200 μm

2. *Trichoplax* sp. H2 (16S haplotype H2; a clonal strain from a specimen isolated on the north coast of Bali) [8].
3. *Hoilungia* sp. H4 (16S haplotype H4; a clonal strain from an isolate of a marine aquarium of the Moscow Zoo, possibly originating from Indonesia; [38]).
4. *Hoilungia hongkongensis* (16S haplotype H13; clone from a mangrove stream in Hong Kong, SAR; [9]).

2.2 Algae (Fig. 3, See Note 1)

1. *Tetraselmis marina* (Cienkowski) R.E. Norris, Hori & Chihara, 1980 WoRMS Aphia ID 376158 (Chlorophyta) (Fig. 3e).
2. *Nannochloropsis salina* D.J.Hibberd, 1981 WoRMS Aphia ID 376044 (Ochrophyta) (Fig. 3c).
3. *Leptolyngbya ectocarpi* (Gomont) Anagnostidis & Komarek 1988 WoRMS Aphia ID 615645 (*Cyanobacteria*) (Fig. 3f).
4. *Spirulina versicolor* Cohn & Gomont 1892 WoRMS Aphia ID 495757 (*Cyanobacteria*) (Fig. 3g).
5. *Entomoneis paludosa* (*Entomoneis paludosa* (W. Smith) Reimer, 1975 WoRMS Aphia ID 163646 (Ochrophyta)) [39] (Fig. 3h).
6. *Rhodomonas salina* (Wislouch) D.R.A. Hill & R. Wetherbee, 1989 WoRMS Aphia ID 106316 (Cryptophyta) (Fig. 3d).

2.3 Equipment (See Note 2)

1. For algae cultures (Fig. 3a).
   - (a) Petri dishes (9 cm diameter), sterile.
   - (b) White LED ribbon (9.6 W/m, distance 10 cm).
2. For placozoan breeding (Fig. 4).
   - (a) Petri dishes (diameter ranging from 3 to 15 cm), sterile.
   - (b) Erlenmeyer flask (100 and 200 mL), sterile.
   - (c) Falcon® tubes, 50 mL.
   - (d) Pasteur pipettes, sterile.
   - (e) Pipetman® (for volumes ranging from 1 to 5000 μL).
   - (f) Non-stick, low-retention pipet tips, sterile (e.g., Clear-Line #713148).
   - (g) Aquarium (20–50 L) with a drainable false bottom and heat source (e.g., from Juwel or Eheim).
   - (h) Aragonite sand (e.g., Udeco Sea Coral, fraction 4–6 mm #UPC440156).
   - (i) Climate growth chamber, with or without light (e.g., KBW series, Binder, Germany).
   - (j) White LED ribbon (9.6 W/m, distance 100 cm).

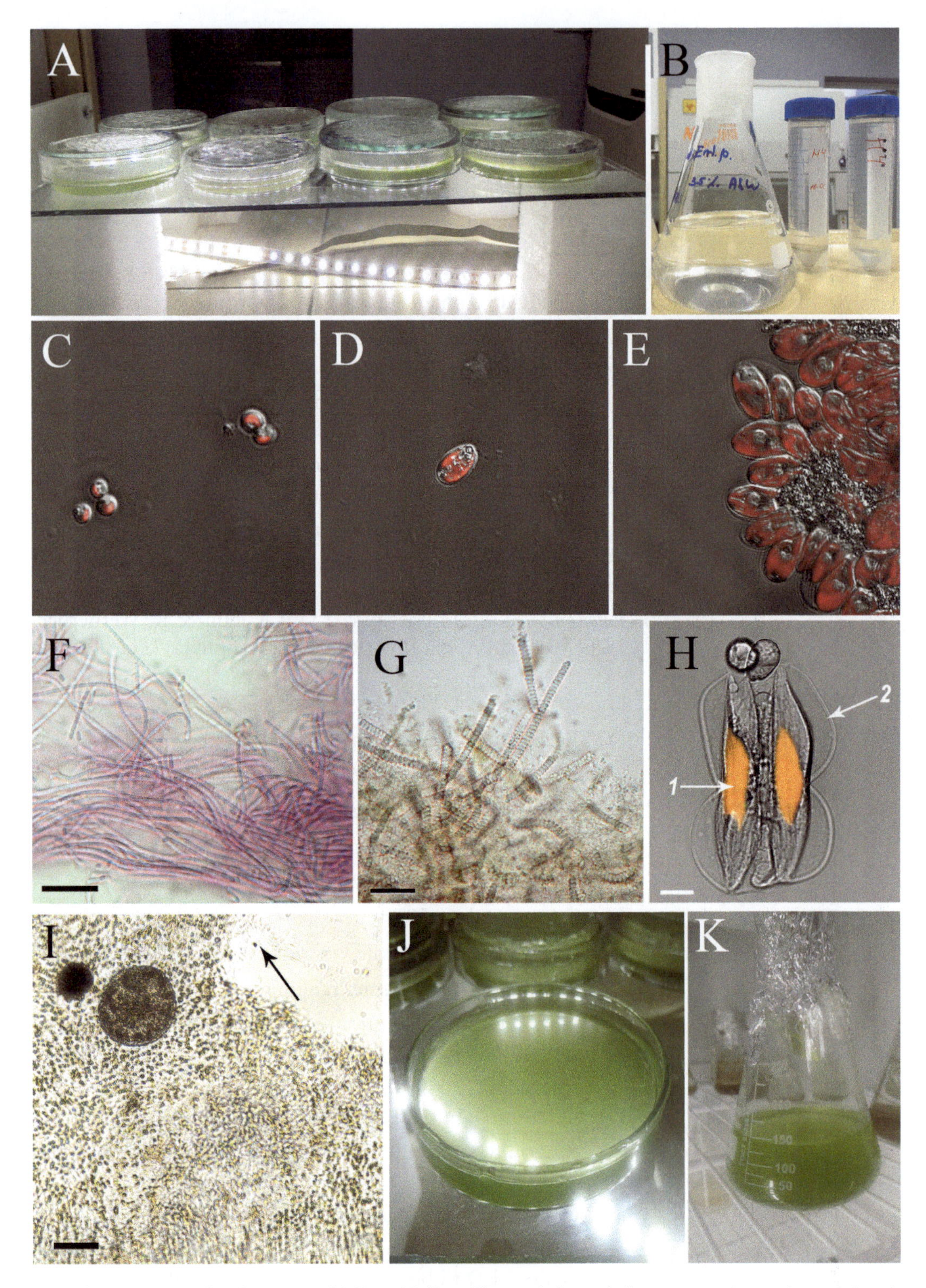

Fig. 3 Culturing algae for placozoans. (**a**) Algae biofilms (*Tetraselmis marina*) are grown at room temperature in Petri dishes. The dishes are illuminated from below with LED lights for 5–9 days before being used as a placozoan food source. (**b**) Placozoans can also be maintained temporarily in Erlenmeyer or Falcon tubes. (**c–h**) Different algal species frequently used to feed placozoans: (**c**) *Nannochloropsis salina*, (**d**) *Rhodomonas salina*,

3. Generic lab equipment.
   (a) Weighing scale (e.g., Adventure Pro, AV313C, Ohaus Corp., China).
   (b) Magnetic stirrer (e.g., Intelli-Stirrer MSH-300i, BIOSAN, Lithuania).
   (c) pH meter (e.g., Starter 2100, Ohaus Corp., China).
   (d) Refractometer.
   (e) Stereomicroscope (e.g., Stemi 305, Zeiss, Germany).
   (f) Inverted microscope (e.g., Eclipse Ts2R, Nikon, Japan).

2.4 Solutions

1. Artificial seawater (ASW; prepare and use within a week (*see* **Note 3**)).
   (a) Aquarium salt mix (e.g., Red Sea salt #222597521890, Red Sea, or Pro-Reef sea salt, #10524, Tropic Marin).
   (b) Freshly made distilled/Milli-Q water.
   (c) While stirring, add 38 g of salt to 1 L of distilled water.
   (d) Check salinity, and adjust if necessary, to reach a salinity of 35 ppm.
   (e) 0.22 μm filters and a sterile bottle (1 L).
2. Liquid medium for algae culture.

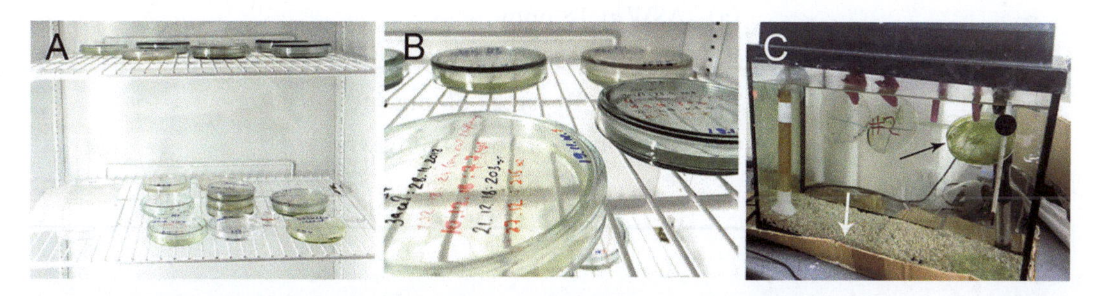

Fig. 4 Placozoan laboratory cultures. Illustrative examples of placozoan breeding in a climate growth chamber (**a**, **b**) and in the marine aquarium (**c**). (**a**, **b**) Animals (e.g., *Trichoplax* (H1, H2 haplotypes) and *Hoilungia hongkongensis* (H13 haplotype)) are placed in Petri dishes with algae in the artificial seawater (ASW, see methods) and maintained under defined temperature and light conditions. (**b**) Petri dishes can be marked with specific details of experiments. (**c**) Aquarium setup for culturing of *Hoilungia* sp.—the H4 haplotype. Petri dishes (black arrow) covered with the algal substrate are attached to the tank's walls to serve as a food source

Fig. 3 (continued) (**e**) *Tetraselmis marina*, (**f**) *Leptolyngbya ectocarpi*, (**g**) *Spirulina versicolor*, (**h**) *Entomoneis paludosa*. (1) chloroplast; (2) silica frustule. (**i**) Example of an algae biofilm (here with *Tetraselmis marina*) on which two small placozoans are crawling. The black arrow indicates the area without an algal substrate. (**j**) Example of a Petri dish with a ready-to-use thick algae substrate. (**k**) Density of *Tetraselmis marina* solution used for preparing Petri dishes. Scale bars: (**e**), (**f**), (**h**) = 10 μm; (**g**) = 20 μm; (**i**) = 100 μm

(a) Stock solutions (store at 4 °C).

(i) 1.75 M KNO_3.

(ii) 25 mM K_2HPO_4.

(iii) 25 mM Fe_2SO_4.

(iv) "A5" supplement solution for algae growth:

(1) 23,272 mg/L $FeC_6H_5O_7$.

(2) 1009.4 mg/L $MnSO_4$, $7H_2O$.

(3) 78 mg/L $CuSO_4$, $5H_2O$.

(4) 176 mg/L $ZnSO_4$, $7H_2O$.

(5) 50 mg/L Na_2MoO_4, $2H_2O$.

(6) 571 mg/L H_3BO_3.

(7) 95 mg/L $CoSO_4$, $7H_2O$.

(b) Working solution (store at 4 °C).

(i) In a small Erlenmeyer, add 200 μL KNO_3 (7 mM final), 20 μL K_2HPO_4 (0.1 mM final), and 20 μL mM $Fe_2(SO_4)_3$ (0.1 mM final) to 50 mL distilled water, and stir.

(ii) For optimal growth of *Tetraselmis marina* and cyanobacteria, also add 10 μL of the supplement solution for algae growth.

3. Liquid medium for diatoms (*Entomoneis paludosa*) culture.

(a) ASW at 18 ppm.

(b) Na_2SiO_3 (stock 15 g/L).

(c) Vitamin B_1 (stock 50 mg/mL).

(d) Vitamin B_{12} (stock 0.5 mg/mL).

3 Methods

Animals can be brought to the laboratory from different habitats. To ensure long-term viability, it is important to optimize their breeding conditions. We have noted in many cases that wild isolates, newly transferred to the unnatural laboratory environment, form a stable and healthy growing culture once they have survived a critical adjustment phase of a few weeks to several months. This often depends on the genetic background of the isolate, with H2 specimens adjusting very fast independent of their origin, while, e.g., haplotypes belonging to the 16S clade V (all *Hoilungia* spp.) tend to adapt slowly to a dish culture, if at all (as in the case of H4). We have successfully maintained H1, H2, and H13 in Petri dishes under controlled feeding and environmental conditions. H4 animals, on the contrary, thrive when kept in small aquaria. We describe both methods in detail below. Figure 2 shows the gross morphology of placozoans of the four different haplotypes maintained in our laboratories.

3.1 Culturing Algae and Cyanobacteria (5–7 Days)

H1, H2, and H13 grow well in the presence of green algae (e.g., *Tetraselmis marina*, Fig. 3e) in ASW. H4 requires further addition of cyanobacteria (e.g., *Spirulina versicolor* and *Leptolyngbya ectocarpi*, Fig. 3f, g). H4 culture is most stable, albeit slow-growing, on pure *Tetraselmis.* We found that optimal conditions are provided upon transfer of placozoans onto Petri dishes already covered with an algae mat (Fig. 3i, j), which can be grown 4–10 days beforehand. A well-attached algal substrate is formed under the constant illumination of the dishes by a white-LED light source placed underneath (Fig. 3a). We do not recommend illumination from above because it does not foster the concentration of algae at the bottom of Petri dishes.

1. Day 1: inoculation
   (a) Every Petri dish should be able to sustain a healthy population of placozoans. We recommend preparing novel Petri dishes every 7 days, on which fresh placozoans can be transferred to maintain or expand the cultures. Prepare sterile Petri dishes (9 or 20 cm diameter).
   (b) Prepare ASW (about 50 mL/9 cm and 500 mL for 20 cm diameter dishes, respectively).
   (c) To each 9 cm dish, add 40 mL ASW, and for algal growth, add 200 μL KNO_3 and 20 μL K_2HPO_4 and Fe_2SO_4.
   (d) Add also 10 μL A5 solution (see supplement solution for algae growth above) per dish for *Tetraselmis marina* and cyanobacteria.
   (e) Use a new tip for each Petri dish (*see* **Note 4**).
   (f) To produce a *Tetraselmis marina* algae mat, add about 2.5×10^5 cells/9 cm dish (i.e., about 1 mL of a dense culture; *see* Fig. 3k and **Note 5**).
   (g) To feed H4 placozoans, 1 mL of a dense culture mix of cyanobacteria (1:1 of *Leptolyngbya ectocarpi* and *Spirulina versicolor*, respectively) is added to Petri dishes already seeded with 1 mL of *Tetraselmis marina* (*see* **Note 5**).
   (h) Petri dishes can be maintained at room temperature (usually 22–24 °C) on a glass shelf under constant illumination by a white-LED ribbon placed underneath (*see* Fig. 3a).

2. Days 3, 5, 7, and 9: monitoring of algal growth
   (a) Every second day, dishes should be checked to monitor contamination (if any other organisms, e.g., ciliophorans, are observed, plates must be discarded).
   (b) When a dense algal biofilm of regular thickness covers the bottom of the dish, it can be used for culturing of placozoans (*see* Fig. 3i, j). This takes about 10–15 days for *Tetraselmis marina* and 8–10 days for cyanobacteria such as *Leptolyngbya ectocarpi* and *Spirulina versicolor*.
3. Biofilm-covered Petri dishes are washed three times for 2 min with ASW to remove algae in excess. Washing with ASW reduces nitrate and phosphate levels, which are toxic for placozoans.
4. The next day, the placozoans are added to the dish and placed in a climate growth chamber with a light/dark regime of 12: 12 h at 24 °C (Fig. 4a, b). However, the animals can also be cultured in darkness or ambient light illumination for several days.

3.2 Culturing Diatoms (*Entomoneis paludosa*)

We recommend using diatoms (Fig. 3h) as a food source when placozoans need to be transported in Falcon tubes (Fig. 3b), as they can form a biofilm on the walls of the tubes even in the absence of light. Healthy animals have been collected from such tubes after 5 days of travel or longer. Note that *Entomoneis* growth is best at lower salinity (18 ppm) than tolerated by most placozoans. However, it can survive for about a week in 35 ppm ASW with very low population growth rates.

1. Add 1 mL Na_2SiO_3, 0.25 mL vitamin B_1, and 2.5 mL vitamin B_{12} to 1 L bottle of ASW/Goldberg liquid medium (18 ppm, [39]).
2. Inoculate with 1 mL of diatoms (2.10^5/mL).
3. Keep at 17 °C in a climate growth chamber with a light regime of 6:18 (D:N) and illumination from top.
4. Renew ASW every 10–15 days.

3.3 Preparing Food Substrate for Placozoan Collection and Traveling

In several trials, placozoans shipped by plane (e.g., by express mail services) did not survive, and we often made special preparations for their transportation. We propose two options:

1. Prepare in advance *Tetraselmis marina* cultures in Falcon tubes as follows:
   (a) A sterile 50 mL Falcon tube inoculated by a sterile micropipette with *Tetraselmis marina* is filled with ASW (35 ppm), closed, and placed on its flank in an illuminated climatic chamber for 2–3 days so that a thin biofilm develops on its wall.

(b) On the day of travel, an excess of algae should be removed by careful replacement of the contents of the tube with fresh ASW while preserving the original biofilm. For example, the Falcon tube could be placed upright in a Petri dish and opened. A large volume of fresh ASW (prepare 1 L) could be poured on top, while overflowing water is collected in the underlying Petri dish and then discarded (*see* **Note 6**).

2. Alternately, diatoms (*Entomoneis paludosa*) can be used as a short-term food supply for placozoans (*see* the previous paragraph and **Note 7**).

   (a) Add 1 mL diatoms culture to a 50 mL Falcon tube filled with ASW (35 ppm).

In either case, 50 (medium to large) to a max 100 (small) animals are then transferred in the Falcon of choice, which is then filled with ASW and covered with parafilm to avoid the presence of air bubbles. The Falcon cap is screwed, and the tube is sealed with parafilm for transport.

If the animals are kept in Falcon tubes, it is necessary to change the ASW every 5–7 days and add 1 mL of algae.

3.4 Generic Protocol for Placozoan Breeding as Validated for H1, H2, and H13

To adapt fresh animals to laboratory culture conditions, we suggest transferring them in 50 mL Falcon tubes with 40 mL ASW (35 ppm) supplemented with 1–2 mL of dense algae suspension. Next, we usually transferred a few clonal animals onto an algae-biofilm-covered Petri dish (see preparation above), in ASW (35 ppm) at pH = 8.0 at 24 °C (Figs. 3i and 4a, b). The survival rate, size, and proliferation rate of the animals were monitored daily. If the animals do not proliferate, culture condition(s) are slightly modified, and a few animals are transferred anew to this slightly different environment(s). In order of priority/importance, several factors should be tested/adjusted: new food sources (green, red, or diatom algae, cyanobacteria), water with slightly different salinities (with 2 ppm increments, between 30 and 38 ppm), and finally, the ambient temperature (with 2 °C increments, between 20 and 28 °C). Light intensity and illumination profiles may also be modulated. In this context, it is helpful to collect information from the biotope the animals have been taken from. When a sufficient number of animals are available, several conditions should be tested in parallel.

When proper conditions have been established, 10–20 animals could be transferred to start a new culture plate, as illustrated in Fig. 4. Care should be taken that animals do not come into contact with air bubbles to dissociate. Instead, gentle flushing with a pipette will bring them to detach from their substrate and drift in the water, at which point they can be pipetted and transferred to the

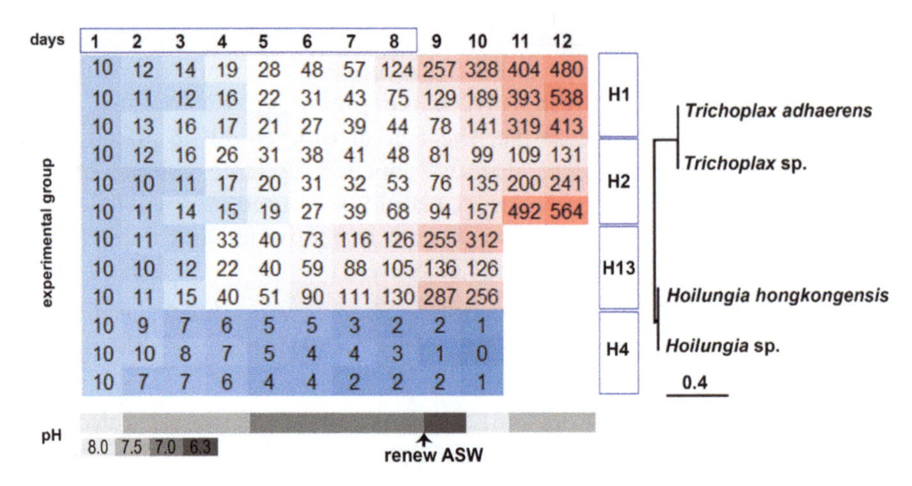

Fig. 5 Heatmap of the placozoan population growth and pH variations under defined culture conditions (artificial seawater (ASW) 35 ppm, starting pH = 8.0, 24 ± 2 °C, and daylight illumination). Two H1 or one H2 individual was added into a 9 cm Petri dish with *Tetraselmis marina* biofilm. Two-thirds of the water was replaced on Day 8. The numbers of individuals and the value of the pH in the media were assessed daily (at the same time of the day) for 10–12 days. H13 population growth was monitored under similar conditions except that ASW was changed daily, yielding a constant pH of 8.0. All experiments were made in triplicates

dish of choice. Whenever necessary, and particularly when animals have been collected from natural habitats, we moved a single animal to a Petri dish to obtain a clonal strain.

Dishes are maintained at a constant temperature of 24 ± 4 °C in a climate chamber illuminated from the top with natural or artificial (e.g., white neon or even LEDs, 12:12 D:N) light conditions, in ASW with a salinity of 35 ppm and a pH of 8.0.

We monitored H1, H2, and H13 population growth (Fig. 5) under controlled conditions (50 mL of ASW 35 ppm, pH = 8.0, 24 ± 2 °C, and daylight illumination) every second day for 2 weeks. It appears that under the conditions described above, all H1 (average number of animals per 9 cm dish Avg = 477 ± 71, $p < 0.05$), H2 (Avg = 312 ± 255, $p < 0.01$), and H13 (Avg = 232 ± 108, $p < 0.01$; Student's t-test, $n = 3$ for all haplotypes) population growth rates were exponential, validating the culture conditions. We noted that the pH value dropped in the dish within a week, which can be corrected by partial replacement of fresh ASW (Fig. 5), without yielding a decrease in the population size. During this interval, the size of H2 individuals decreased, a feature that is typical for culture dishes with a high population density. Thus, we suggest replacing ASW weekly and splitting cultures when reaching high population densities (~200 individuals per 9 cm dish and max 500 per 20 cm dish) to keep a healthy growing population.

Specifically, 2/3 of ASW is replaced every 7–10 days to maintain a healthy long-term culture. When algae are scarce in the dish and do not form a biofilm, a few milliliter of a suspension of *Tetraselmis marina* or *Nannochloropsis salina* are added to the culture dish (Fig. 3k). Mixed clonal cultures of algae can also be used as a food source (e.g., *Tetraselmis marina* can be mixed with cyanobacteria *Leptolyngbya ectocarpi* and *Spirulina versicolor*, or *Nannochloropsis salina* could be mixed with *Rhodomonas salina*) (*see* **Note 5**). Contamination with free-living bacteria should be monitored regularly under an inverted microscope, at least with each water change. If microbial contamination is detected, animals can still be collected and transferred to clean dishes. However, several washing steps (about six times) with fresh ASW should be performed to avoid cross-contamination before transferring specimens to a new dish.

Most placozoans are transparent, but their color varies along with the type of algae they feed on. We have observed that on *Tetraselmis marina*, they have light brown coloration; when they are fed with diatoms, they are medium brown, and they have a pinkish appearance upon addition of cyanobacteria (Fig. 2). Such pinkish coloration, due to the accumulation of phycobilins from cyanobacteria, red algae, and cryptophytes, may interfere with fluorescent/optical microscopy techniques, so feeding with green algae is recommended under certain conditions.

Under these long-term culture conditions, animals exhibit no drastic morphological changes over weeks to months other than the transient size growth before proliferation by binary fission.

3.5 H4 Breeding in the Aquarium

Our attempts to cultivate H4 in Petri dishes, similarly to H1, H2, and H13, were unsuccessful (Fig. 5). This has also been noticed earlier for various other haplotypes [6]. H4 animals were dying off after 2–4 weeks of cultivation in Petri dishes, indicating a higher sensitivity of this *Hoilungia* lineage to a fast accumulation of metabolic products in small water volumes. However, we have successfully cultivated H4 in larger water tanks using the protocol of Okshtein [34] with modifications. 20–50 L glass aquariums were well suited for this task (Fig. 4c). Aquaria should be equipped with a drainable plastic false bottom connected to an airlift filter and covered with coral sand. Illumination was solely provided by ambient light. A heater maintained water temperature at 28 °C, since at temperature below 27–28 °C, H4 did not reproduce vegetatively.

1. Setting up the aquarium (Fig. 4c).
   (a) Thoroughly wash all equipment with distilled water (aquarium, plastic false bottom, airlift, thermometers, heaters; *see* **Note 2**).
   (b) Boil coral sand for 30 min in distilled water, and wash it five times in distilled water.
   (c) Assemble a false bottom and airlift.
   (d) Add 15–20 mm coral sand.

(e) Fill up the tank with sterile ASW (35 ppm), and leave for 3–5 days with an active airlift in ambient light.

(f) Attach Petri dishes with algal biofilms (algae and cyanobacteria) to the walls of the aquarium (*see* Fig. 4c).

(g) Place Petri dishes with placozoans in the middle of the tank.

2. Maintenance of the aquarium.

(a) Replace 20–40% of ASW every 3 months.

(b) Algae and placozoans will inhabit aquarium walls within a few months. We essentially do not clean the aquarium walls because they bear a biofilm that is continuously consumed by placozoans (*see* **Note 8**).

(c) For experiments, freshly prepared algae plates can be placed in the aquarium and will be rapidly colonized by placozoans.

(d) When cultures decline (often after 12–24 months), it is best to set up a new tank. Animals that are present in the Petri dishes at the bottom of the tank or that can be pipetted from the aquarium walls are transferred in fresh dishes and washed four to six times with ASW (with 30–60 min between washings) to avoid the transfer of potentially harmful algae (*see* **Note 9**).

Since H4 was relatively difficult to maintain, we have isolated three microalgae species from the walls of an aquarium where these placozoans were initially discovered (shark tank in Moscow Zoo aquarium in a Moscow Zoo aquarium building). We identified the unicellular green algae *Tetraselmis marina* and two strains of filamentous cyanobacteria: motile *Spirulina* sp. and immobile *Leptolyngbya* sp. All three were used for H4 cultivation with different successes. The fastest growth was observed using *Leptolyngbya* sp., with duplication time as short as 24 h and animals that could grow up to 5 mm in diameter. Population growth on *Spirulina* sp. was slower and slowest on *Tetraselmis marina*. Neither strain of cyanobacteria provided monocultures sufficiently stable for long-term cultivation. In contrast, *Tetraselmis marina* provided very stable culture growth conditions. Hence, we often used a mix of both cyanobacteria strains with *Tetraselmis marina* to feed H4.

3.6 Toward Long-Term Cultures

In optimized culture conditions, individuals from all strains were usually 0.5–2 mm in diameter, flat, highly motile, and with rapid changes of body shapes (Figs. 2 and 6a–c). In dishes with low densities of algae, individual animals spread out and graze at the bottom of the dish (Fig. 6b, c). A few individuals have been seen "swimming" (staying at the water interface upside down, grazing at the water surface without sinking to the bottom; Fig. 6e). When

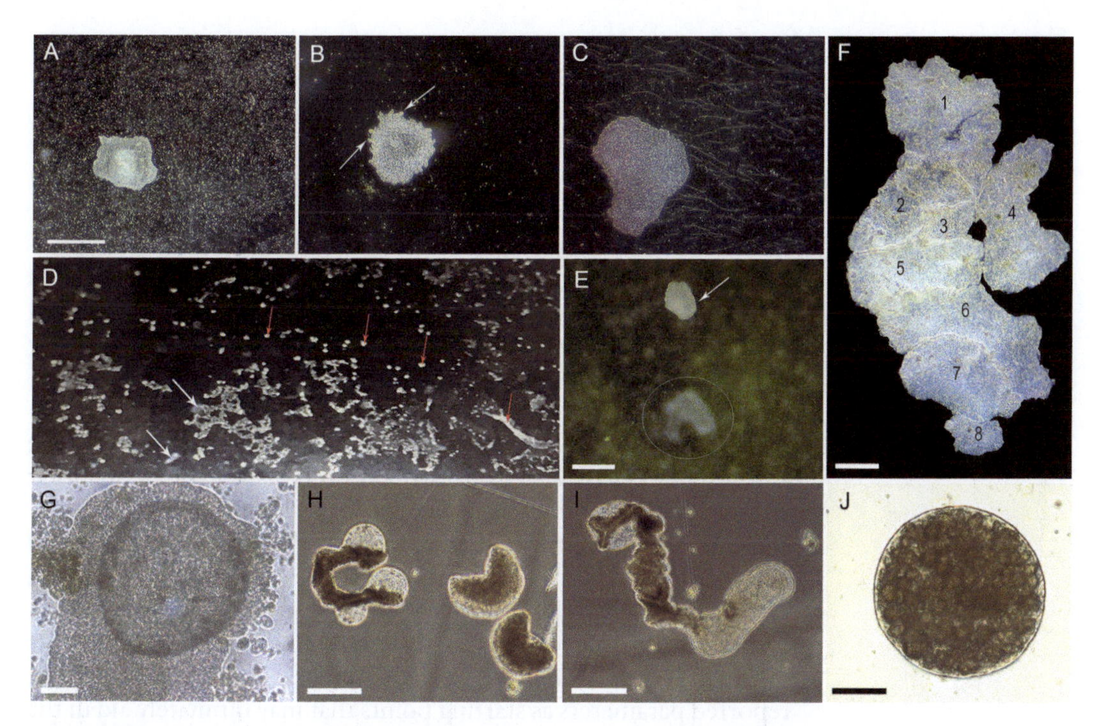

Fig. 6 Different states of placozoans in culture. H1, H2, H4, and H13 individuals maintained in the laboratory were about 0.5–2 mm in size; they most frequently graze at the bottom of the culture dishes (**a–c**, **f**). (**a**) H13 individual on a thick substrate of *Tetraselmis marina.* (**b**) H13 individual on a thin substrate of *Tetraselmis marina.* White arrows show invaginations of the marginal epithelium. (**c**) H13 individual on a substrate of cyanobacteria, conferring it a pinkish color. (**d**) When the placozoan population grows, individuals (in this case of haplotype H13) often migrate to the surface, where they remain isolated (red arrows) or can aggregate (white arrow). (**e**) H1 individual swimming/drifting in the dish; another animal at the bottom of the dish is visible but not in the focal plane. (**f**) On a thick algal substrate, animals often change their feeding behavior by forming aggregates (here from eight individuals of haplotype H1 numbered 1–8), which is sometimes named as "social behaviors." (**g–j**) Under long-term culturing conditions (>4–6 weeks in one Petri dish without transferring of placozoans), animals enter the so-called D-phase, a degenerative process in which animals sometimes elongate (**h**, **i**), start forming 'bubbles' (**j**) by lifting the upper epithelium (**g**), and sometimes start producing oocyte-like cells and structures (of uncertain nature) that are eventually free floating after being released by the degenerated mother animal. Scale bar: (**a**)–(**c**), (**e**), (**g**) = 200 μm; (**f**), (**h**), (**i**) = 500 μm; (**j**) = 50 μm

the density of animals is high, they crawl on the wall of the dish and reach the water surface (where they might aggregate; Fig. 6d). We have observed this frequently with *Hoilungia hongkongensis.* Single individuals can invariably be taken from the surface, from the interface, or from the bottom to set up new dishes and expand the cultures (*see* **Note 10**). When placozoans are grown on a thick biofilm (Fig. 6a, f), they tend to cluster to form large colonial-like "aggregates" moving in concert across the food mat, similar to social-like feeding behaviors [10].

Upon long-term culturing (more than 4 weeks) on a thin substrate of algae (with algae dispersed in the dish rather than forming a continuous layer at the bottom), without ASW change or addition of food, animals develop new morphological characters resembling (Fig. 6g–j) what has been reported as features of degeneration and sexual reproduction [7, 14]. In H1 and H2, we have occasionally observed the generation of oocyte-like cells and possible earlier developmental stages till about 64 cells (Fig. 6g, j). However, the exact nature of these structures are unknown.

Also, rarely, groups of small animals of different sizes detach from an individual (Figs. 3i and 6b). These may correspond to "swarmers," vegetative dispersal propagules described earlier [62–64].

3.7 Conclusion

We anticipate that the presented protocols will enable an effective culturing of placozoans from virtually every haplotype by adapting natural habitats to the laboratory, which includes the use of diverse algae (e.g., green, red, or diatom algae as Placozoa's food sources) and various laboratory conditions (e.g., diversity of dishes, flasks, temperature, photoperiod, etc.). Although we were focusing on four known placozoan species (H1, H2, H4, and H13, representing two genera *Trichoplax* and *Hoilungia*), we also consider the reported parameters as starting points that may ultimately aid in the development of more efficient recommendations to bring a broader placozoan diversity into the laboratory.

In summary, the current protocols can (a) support the long-term maintenance of different clonal lines in the laboratory; (b) allow comparing species-specific development and behaviors under defined experimental manipulations, including applications of pharmacological agents for analyses of cellular bases of behavior; and (c) facilitate future studies of placozoan biology from diverse representatives of this early-branching metazoan lineage such as single-cell sequencing (scRNA-seq) in the quest for homologous cell types, live imaging, and physiological studies across placozoan species [24, 25].

4 Notes

1. Most of our algae strains (*Tetraselmis marina*, *Leptolyngbya ectocarpi*, *Spirulina versicolor*, *Entomoneis paludosa*) have been collected by ourselves. Some (*Nannochloropsis salina*, *Rhodomonas salina*) have been ordered from SAG, Culture Collection of Algae, at Goettingen University, Goettingen, Germany. Many other nonprofit culture collections of algae exist (e.g., CCBA, University of Gdańsk, Gdańsk, Poland; UTEX, the University of Texas at Austin, Texas, USA).
2. Autoclave all glassware before use. Do not use any detergents or bleach to clean labware in contact with seawater and

placozoans. Wash off with sodium bicarbonate, and rinse several times with distilled water.

3. Depending on the type of salt mix used to prepare ASW, solutions tend to form a white deposit after 4–5 days (with the Red Sea salts more so than with Tropic Marin mix). Contamination by bacteria or ciliates may also spread during this time; therefore, always use freshly prepared solutions.
4. It is important to use new pipette tips to avoid the possible transfer of contaminants from one dish to another (bacteria, some species of *Ciliophora*, etc.).
5. A monoculture or a mixture of algae can be used as a food source. The initial volume of algae inoculated conditions the speed of biofilm formation. Algal biofilm usually forms within 7–10 days after the start of culturing when 400 μL of *Tetraselmis marina* is given per dish. If a faster biofilm formation is needed (e.g., 4–5 days), add 2 mL of *Tetraselmis marina* per dish. To maintain the H4 haplotype, a food mix is advisable, e.g., 1:1:1 of *Tetraselmis marina*, *Leptolyngbya ectocarpi*, and *Spirulina versicolor*, respectively. *Leptolyngbya* tends to form relatively dense, durable biofilms, which are difficult to dissociate. We used an inoculation loop to transfer cyanobacterial films to a new Petri dish. Also, *Leptolyngbya* is immobile and slow to expand across the Petri dish. It is recommended to place several pieces of *Leptolyngbya* biofilm across the Petri dish.
6. After transferring animals in Falcon tubes during traveling, we noted a decrease in the size of individuals; the placozoans can also be migrated to the top of tubes. So, we recommend using Falcon tubes only for transferring individual animals, but for no longer than 7 days and with refreshing ASW every 2 days (70% of total volume).
7. We have observed that placozoans transferred in a 50 mL Falcon filled ad hoc with ASW and *Entomoneis paludosa* kept growing and stayed at the bottom half of the Falcon, looking relatively healthy (average size 0.5–2 mm, animals can move vertically and fed on diatoms).
8. The clonal population can be obtained directly from the natural environment as original *T. adhaerens* was found and described from aquaria [50]. The Japanese placozoan "Shirahama" strain, 16S haplotype H15 [8, 22], was isolated from natural rocks introduced into an aquarium. The procedure became a valuable commodity in the aquarium industry and was often called "live rock." The natural rock harbors a wide variety of algae, sponges, and other invertebrates, including placozoans. We packed rocks in plastic bags with seawater and pure oxygen

in the ocean shore and transferred them in aquaria. After 2–4 weeks of cultivation under ambient light or aquarium lights, placozoans often appear on the wall of aquariums. We preferred rocks well encrusted with a variety of coralline algae. The animals can be maintained in the same setup as described above. From the same habitat as placozoans, holothurians can be kept in the same tank. These holothurians were suitable for maintaining the bottom sand in tanks, and they did not affect the placozoans mostly living on aquaria walls.

9. After several months of cultivation, algae and placozoans usually inhabit aquarium walls. After 12–24 months, aquarium placozoan culture might decline: the slow growth of animals and their numbers are reduced to not more 50–60 per dish. Sometimes it coincides with an infestation of an aquarium with various inedible microalgae, but it can also happen with no apparent reason. Water replacement and aquarium glass cleaning have limited effects. In this case, the best solution is to start a new aquarium. When transferring placozoans to a new aquarium, it is recommended to wash them with filtered artificial seawater four to six times (every 30–60 min). It is convenient to have 2–3 aquaria and reset them each 12–18 months at different times.

10. When individuals are very numerous (e.g., more than 400 per 9 cm Petri dish), some animals were seen swimming (*see* also Pearse and Voigt [37]). Whether this phenomenon is induced by culturing conditions or normal behavior of the animals at a very rare occurrence is not yet clear. However, when transferred to fresh Petri dishes with a new substrate, these animals showed standard locomotory and feeding patterns. We suggest that the observed swimming-like behavior is related to the unusually high density of animals in the dish and their ability to disseminate in search of food and new habitats broadly.

Acknowledgments

This work was supported in part by the Human Frontier Science Program (RGP0060/2017) and National Science Foundation (IOS-1557923) grants to L.L.M. Research reported in this publication was also supported in part by the National Institute of Neurological Disorders and Stroke of the National Institutes of Health under Award Number R01NS114491 (to L.L.M). D.R. was supported by the Russian Science Foundation grant (23-14-00050). The content is solely the authors' responsibility and does not necessarily represent the official views of the National Institutes of Health.

References

1. Aleoshin VV et al (2004) On the genetic uniformity of the genus *Trichoplax* (Placozoa). Russ J Genet 40(12):1423–1425
2. Armon S et al (2018) Ultra-fast contractions and emergent dynamics in a living active matter-the epithelium of the primitive animal *Trichoplax adhaerens*. Biophys J 114(3):649a
3. Belahbib H et al (2018) New genomic data and analyses challenge the traditional vision of animal epithelium evolution. BMC Genomics 19(1):393
4. Christen R et al (1991) An analysis of the origin of metazoans, using comparisons of partial sequences of the 28S RNA, reveals an early emergence of triploblasts. EMBO J 10(3): 499–503
5. DeSalle R, Schierwater B (2010) Key transitions in animal evolution. CRC Press
6. Eitel M, Schierwater B (2010) The phylogeography of the Placozoa suggests a taxon-rich phylum in tropical and subtropical waters. Mol Ecol 19(11):2315–2327
7. Eitel M et al (2011) New insights into placozoan sexual reproduction and development. PLoS One 6(5):e19639
8. Eitel M et al (2013) Global diversity of the Placozoa. PLoS One 8(4):e57131
9. Eitel M et al (2018) Comparative genomics and the nature of placozoan species. PLoS Biol 16(7):e2005359
10. Fortunato A, Aktipis A (2018) Social feeding behavior of *Trichoplax adhaerens*. Front Ecol Evol 7:19
11. Grell KG (1971) Trichoplax adhaerens F.E. Schulze und die Entstehung der Metazoen. Naturwiss Rundschau 24:160–161
12. Grell KG (1972) Eibildung und furchung von *Trichoplax adhaerens* F.E. Schulze (Placozoa). Zoomorphology 73(4):297–314
13. Grell KG, Benwitz G (1971) Die Ultrastruktur von *Trichoplax adhaerens* F.E. Schulze. Cytobiologie 4:216–240
14. Grell KG, Benwitz G (1974) Elektronenmikroskopische beobachtungen über das wachstum der eizelle und die bildung der "befruchtungsmembran" von *Trichoplax adhaerens* F.E. Schulze (Placozoa). Z Morphol Tiere 79:295–310
15. Grell KG, Ruthmann A (1991) Placozoa. In: Microscopic anatomy of invertebrates. Wiley-Liss, New York, pp 13–27
16. Heyland A et al (2014) *Trichoplax adhaerens*, an enigmatic basal metazoan with potential. Developmental Biology of the Sea Urchin and Other Marine Invertebrates. Methods Mol Biol 1128:45–61
17. Jackson AM, Buss LW (2009) Shiny spheres of placozoans (*Trichoplax*) function in anti-predator defense. Invertebr Biol 128:205–212
18. Laumer CE et al (2018) Support for a clade of Placozoa and Cnidaria in genes with minimal compositional bias. eLife 7:e36278
19. Malakhov VV (1990) Enigmatic groups of marine invertebrates: *Trichoplax*, Orthonectida, Dicyemida, Porifera. Moscow University Press, Moscow, p 144
20. Maruyama YK (2004) Occurrence in the field of a long-term, year-round, stable population of placozoans. Biol Bull 206(1):55–60
21. Mayorova TD et al (2018) Cells containing aragonite crystals mediate responses to gravity in *Trichoplax adhaerens* (Placozoa), an animal lacking neurons and synapses. PLoS One 13(1):e0190905
22. Miyazava H et al (2012) Mitochondrial genome of a Japanese Placozoa. Zool Sci 29: 223–228
23. Miyazawa H et al (2020) Mitochondrial genome evolution of placozoans: gene rearrangements and repeat expansions. Genome Biol Evol 13:evaa213
24. Moroz LL (2014) The genealogy of genealogy of neurons. Commun Integr Biol 7(6): e993269
25. Moroz LL (2018) NeuroSystematics and periodic system of neurons: model vs reference species at single-cell resolution. ACS Chem Neurosci 9:1884–1903
26. Moroz LL, Romanova DY (2021) Selective advantages of synapses in evolution. Front Cell Dev Biol 9:726563
27. Moroz LL, Romanova DY (2022) Alternative neural systems: what is a neuron? (Ctenophores, sponges and placozoans). Front Cell Dev Biol 10:1071961
28. Moroz LL et al (2020) Microchemical identification of enantiomers in early-branching animals: lineage-specific diversification in the usage of D-glutamate and D-aspartate. Biochem Biophys Res Commun 527(4):947–952
29. Moroz LL et al (2020) The diversification and lineage-specific expansion of nitric oxide sig-

naling in Placozoa: insights in the evolution of gaseous transmission. Sci Rep 10:13020
30. Moroz LL et al (2021) Neural versus alternative integrative systems: molecular insights into origins of neurotransmitters. Philos Trans R Soc B 376(1821):20190762
31. Moroz LL et al (2021) Evolution of glutamatergic signaling and synapses. Neuropharmacology 199:108740
32. Nakano H (2014) Survey of the Japanese coast reveals abundant placozoan populations in the Northern Pacific Ocean. Sci Rep 4:5356
33. Nikitin MA et al (2023) Amino acids integrate behaviors in nerveless placozoans. Front Neurosci 17:1125624
34. Okshtein IL (1988) New method of culturing of *Trichoplax* sp. (Placozoa). Zool Zhurnal 67: 923–926
35. Osigus HJ et al (2019) *Polyplacotoma mediterranea* is a new ramified placozoan species. Curr Biol 29(5):R148–R149
36. Pearse V (1989) Growth and behavior of *Trichoplax adhaerens*: first record of the phylum Placozoa in Hawaii. Pac Sci 43(2):117–121
37. Pearse V, Voigt O (2007) Field biology of placozoans (*Trichoplax*): distribution, diversity, biotic interactions. Integr Comp Biol 47(5): 677–692
38. Romanova DY (2019) Cell types diversity of H4 haplotype Placozoa sp. Mar Biol J 4(1): 81–90
39. Romanova DY et al (2017) Copper sulphate impact on growth and cell morphology of clonal strains of four benthic diatom species (Bacillariophyta) from the Black Sea. Mar Biol J 2(3):53–67
40. Romanova DY et al (2020) Glycine as a signaling molecule and chemoattractant in *Trichoplax* (Placozoa): insights into the early evolution of neurotransmitters. Neuroreport 31(6):490–497
41. Romanova DY et al (2020) Sodium action potentials in placozoa: insights into behavioral integration and evolution of nerveless animals. Biochem Biophys Res Commun 532(1): 120–126
42. Romanova DY et al (2021) Hidden cell diversity in Placozoa: ultrastructural insights from Hoilungia hongkongensis. Cell Tissue Res 385:623–637
43. Romanova DY et al (2022) Expanding of life strategies in Placozoa: insights from long-term culturing of *Trichoplax* and *Hoilungia*. Front Cell Dev Biol 10:823283
44. Ruthman A (1977) Cell differentiation, DNA content and chromosomes of *Trichoplax adhaerens* F.E. Schulze. - Cytibiologie. 15:58–64
45. Schierwater B (2005) My favorite animal, *Trichoplax adhaerens*. BioEssays 27:1294–1302
46. Schierwater B, De Salle R (2018) Placozoa. Curr Biol 28(3):R97–R98
47. Schierwater B et al (2009) Concatenated analysis sheds light on early metazoan evolution and fuels a modern "urmetazoon" hypothesis. PLoS Biol 7(1):e1000020
48. Schierwater B et al (2010) *Trichoplax* and Placozoa: one of the crucial keys to understanding metazoan evolution. In: Key transitions in animal evolution. CRC Press, pp 289–326
49. Schleicherová D et al (2017) The most primitive metazoan animals, the placozoans, show high sensitivity to increasing ocean temperatures and acidities. Ecol Evol 7(3):895–904
50. Schulze FE (1883) *Trichoplax adhaerens*, nov. gen., nov. spec. Zool Anz 6:92–97
51. Seravin LN (1989) Orientation of invertebrates in three-dimensional space: 4. Reaction of turning from the dorsal side to the ventral one. Zool Zhurnal 68:18–28. (in Russ)
52. Seravin LN, Gerasimova ZP (1998) Characteristics of the fine structure of *Trichoplax adhaerens*, feeding on solid plant substrates. Tsitologiya 30:1188–1193
53. Seravin LN, Gudkov AV (2005) *Trichoplax adhaerens* (Placozoa)—odno iz samykh primitivnykh mnogokletochnykh zhivotnykh. Tessa, St. Petersburg. 69 p. (in Russ)
54. Seravin LN, Karpenko AA (1987) Orientation of invertebrates in 3-dimentional space. 2. Turbellaria. Zool Zhurnal 66(9):1285–1292
55. Signorovitch AY et al (2006) Caribbean placozoan phylogeography. Biol Bull 211(2): 149–156
56. Smith CL, Mayorova TD (2019) Insights into the evolution of digestive systems from studies of *Trichoplax adhaerens*. Cell Tissue Res 377: 353–367
57. Smith CL et al (2014) Novel cell types, neurosecretory cells, and body plan of the early-diverging metazoan *Trichoplax adhaerens*. Curr Biol 24(14):1565–1572
58. Smith CL et al (2015) Coordinated feeding behavior in *Trichoplax*, an animal without synapses. PLoS One 10(9):e0136098
59. Srivastava M, et al. (2008) The Trichoplax genome and the nature of placozoans. Nature 454.7207:955–960

60. Telford MJ, Moroz LL, Halanych KM (2016) Evolution: a sisterly dispute. Nature 529(7586):286
61. Tessler M et al (2022) Phylogenomics and the first higher taxonomy of Placozoa, an ancient and enigmatic animal phylum. Front Ecol Evol 10:1016357
62. Thiemann M, Ruthmann A (1988) *Trichoplax adhaerens* F.E. Schulze (Placozoa): the formation of swarmers. Z Naturforsch C 43(11–12): 955–957
63. Thiemann M, Ruthmann A (1990) Spherical form of *Trichoplax adhaerens*. Zoomorphology 110(1):37–45
64. Thiemann M, Ruthmann A (1991) Alternative modes of sexual reproduction in *Trichoplax adhaerens* (Placozoa). Zoomorphology 110(3):165–174
65. Ueda T, Koya S, Maruyama YK (1999) Dynamic patterns in the locomotion and feeding behaviors by the placozoan *Trichoplax adhaerens*. Biosystems 54(1–2):65–70
66. Varoqueaux F et al (2018) High cell diversity and complex peptidergic signaling underlie placozoan behavior. Curr Biol 28.21:3495–3501
67. Voigt O et al (2004) Placozoa–no longer a phylum of one. Curr Biol 14(22):R944–R945

Chapter 22

Bioinformatic Prohormone Discovery in Basal Metazoans: Insights from *Trichoplax*

Mikhail A. Nikitin, Daria Y. Romanova, and Leonid L. Moroz

Abstract

Experimental discovery of neuropeptides and peptide hormones is a long and tedious task. Mining the genomic and transcriptomic sequence data with robust secretory peptide prediction tools can significantly facilitate subsequent experiments. We describe the application of various in silico neuropeptide discovery methods for the placozoan *Trichopax adhaerens* as an illustrated example and a powerful experimental paradigm for cellular and evolutionary biology. In total, 33 placozoan (neuro)peptide-like hormone precursors were found using homology-based BLAST search and repeat-based and comparative evolutionary methods. Some of the discovered precursors are homologous to insulins and RFamide precursors from Cnidaria and other animal phyla.

Key words Placozoa, Neuropeptides, Secretory molecules, Behavior, Basal metazoans, Signaling, Prohormone processing, Cnidaria, Ctenophora, Nervous system evolution

1 Introduction

The origin of the neural system and signal molecules, including neurotransmitters, is a major and long-standing evolutionary question [1–4]. Recent advances in this field have significantly benefited from the ever-decreasing cost of whole-genome sequencing. Sequenced genomes are available now for the representatives of all four basal metazoan phyla: Porifera [5], Ctenophora [6, 7], Placozoa [8, 9], Cnidaria [10–13], and numerous Bilateria. All systematic comparative and evolutionary neurobiology studies require the identification of neurotransmitters, especially secretory signaling peptides, which contribute to the majority of neurotransmitter diversity across species [14]. Representatives of early-branching metazoan lineages are vital in such evolutionary reconstructions.

Some peptidomics data are available for *Pleurobrachia bachei* and *Mnemiopsis leidyi* as one initial reference set (Ctenophora) and several species of Cnidaria [6, 15, 16] as the second set. The

Leonid L. Moroz (ed.), *Ctenophores: Methods and Protocols*, Methods in Molecular Biology, vol. 2757,
https://doi.org/10.1007/978-1-0716-3642-8_22, © Springer Science+Business Media, LLC, part of Springer Nature 2024

absence of peptidomics data for Placozoa and Porifera (two basal metazoan phyla lacking neurons) hampers attempts to trace the possible origin of ancestral peptide transmitters in the common ancestor of cnidarians and bilaterians.

Bioinformatic prediction of regulatory peptides/neuropeptides can facilitate their subsequent experimental discovery. For example, expression data on prohormone genes may pinpoint the exact body region and life cycle stage of producing a particular peptide. Extensive literature exists on the various prohormone and peptide bioinformatic predictions [17–20] and references therein.

Prohormone/peptide prediction methods are based on three conceptual ideas:

1. Homology-based search includes analysis of sequence similarity between species and model/reference organisms with extensive peptidomics data;
2. Non-homology-based search using signatures of prohormone sequences: secretory pathway signal peptide, prohormone convertase cleavage sites, and often multiple copies of a mature peptide within a single precursor (paracopies) flanked by cleavage sites [20–22];
3. Evolutionary methods are based on amino acid substitution patterns along a prohormone sequence, where mature peptides and cleavage sites are more conservative than discarded parts of the prohormone sequence [17, 18].

These three approaches can be combined. However, homology-based prohormone/regulatory peptide prediction in basal metazoans (except Cnidaria) is complicated by the considerable phylogenetic distances between these phyla and bilaterian model species and the need for extensive experimental characterization of active sites in signaling peptides (*Nematostella, Hydra, Aplysia,* arthropods, and chordates). Such large phylogenetic distances suggest that homologous peptide families evolved beyond recognition. Also, early-branching metazoan lineages have a great diversity of species-specific peptide families that do not present in Bilateria and Cnidaria.

Evolutionary methods were successfully used to predict the neuropeptides of the parasitic flatworms [17], where phylogenetic distance is also a significant issue. Nevertheless, they require data from several species within the analyzed phyla. (Neuro)Peptide predictions perform best only at certain phylogenetic distances between studied species (e.g., different orders of mammals and species of *Caenorhabditis*).

Multiple genomes were published only recently for Placozoa [9, 23]. All extant placozoans are descendants of relatively recent common ancestors, and the divergence between them and their genetic distances are similar to those of a single order or family of mammals [9].

In detecting unique metazoan peptides, we have to resort to methods based on shared prohormone properties: signal peptide, convertase cleavage sites, and presence of peptide paracopies flanked by these cleavage sites. These structural features are crucial for prohormone processing. Leading signal peptides are universally cleaved. Then prohormone convertase (PC1/3 and PC2 in humans) cuts prohormone at specific sites, typically two basic amino acids (KK, RR, KR, or RK) [24–26]. Carboxypeptidase E (CPE) removes trailing basic amino acids [25]. If glycine is at the C-end of the resulting peptide, it is processed by peptidylglycine alpha-amidating monooxygenase (PAM) [25, 27], leaving an amidated C-end of the mature peptide. Search for these enzymes in *Trichoplax* genome will be described below.

Bioinformatic prediction of prohormones/peptides is also critically dependent on the quality of genome assembly and its annotation. If a prohormone gene is missed in a given annotation, only homology-based BLAST search could recover it from transcriptome or genome sequences. If some of prohormone's exons are missing from annotation, signal peptide, convertase cleavage site, or some peptide paracopies could be lost, hampering non-homology-based search methods. This chapter outlines a practical protocol to search for putative secretory peptides in the placozoan, *Trichoplax adhaerens*, as an illustrated example and a powerful experimental paradigm for cellular and evolutionary biology [28–30].

2 Materials

2.1 Genomic Data of Target Species

Predicting prohormones and mature peptides require high-quality sequence data from species of interest. DNA sequences in the form of assembled chromosomes or genomic scaffolds, transcriptome (assembled mRNAs or expressed sequence tags), and predicted protein sequences are used throughout the workflow. Here, we used the following:

- The *Trichoplax adhaerens* genomic assembly and annotation v.1.0 [8] available from JGI website (https://mycocosm.jgi.doe.gov/Triad1/Triad1.home.html).
- Improved annotation by Kamm et al. [23]. (https://genomevolution.org/coge/GenomeView.pl?gid=31909); we must use two different annotations instead of one latest because they have different errors in the prediction of 5′-exons and translation start sites, and the combination of two annotations helps to mitigate these errors.
- For comparative search, we used *Hoilungia hongkongensis* assembly and annotation by Eitel et al. [9] (https://bitbucket.org/molpalmuc/hoilungia-genome/src/master/).

2.2 Reference Prohormone Protein Sequences from Other Species

A comprehensive collection of neuropeptides from various species is required for homology-based neuropeptides prediction. Peptides from closely related species are best suited for this analysis, but none of the model species are related to placozoans. Therefore, we could use the (neuro)peptide precursors from the broad range of taxa, concentrating on the model species with more data available.

- Cnidaria: *Nematostella vectensis* [31]
- Mollusca: *Aplysia californica, Lottia gigantea* [32]
- Nematoda: *Caenorhabditis elegans.* WormBase [33, 34]
- Arthropoda: *D. melanogaster*, FlyBase [35]; *Tribolium castaneum* [36]
- Echinodermata: *Strongylocentrotus purpuratus* [37]
- Chordata: *Homo sapiens* [38], *Gallus gallus* [39]

2.3 Bioinformatic Tools

Our search strategies, homology-based or not, will invariably require multiple searches of homologous sequences, often within custom databases. NCBI BLAST is the software of choice. Web BLAST will be used for quest and tracking in large databases (nr), while stand-alone BLAST is more convenient for small custom databases. Specialized software is required to predict export pathway signal peptide and convertase cleavage sites, namely, SignalP 4.0 and NeuroPred [40, 41].

- BLAST webserver https://blast.ncbi.nlm.nih.gov/BLAST.cgi
- Local BLAST download page ftp://ftp.ncbi.nlm.nih.gov/blast/executables/blast+/LATEST/ (on most popular Linux platforms installation from the repository is available and preferred)
- SignalP 4.0 web server http://www.cbs.dtu.dk/services/SignalP-4.0/
- NeuroPred web server http://stagbeetle.animal.uiuc.edu

3 Methods

3.1 Homology-Based Prediction

3.1.1 Create a List of Prohormones from Model Species

As illustrated examples, we included *Nematostella vectensis* [31], *Aplysia californica*, *Lottia gigantea* [32], *Caenorhabditis* [33, 34], *Drosophila* [35], *Tribolium* [36], *Strongylocentrotus* [37], *Homo* [38], and *Gallus* [39].

3.1.2 Perform Search

We compile the full list of prohormones and neuropeptide precursors from the species listed above: metazoa_masterlist.fasta, containing 340 proteins. This list is available in Table 1. Then we use it as a query to search the predicted proteins of *Trichoplax*:

Table 1
List of neuropeptide and peptide hormone precursors used for homology-based search

GenBank accession	Name	Species
NP_000930.1	pro-opiomelanocortin preproprotein	*Homo sapiens*
CAG46792.1	ADM	*Homo sapiens*
NP_001129.1	agouti-related protein precursor	*Homo sapiens*
NP_000406.1	islet amyloid polypeptide precursor	*Homo sapiens*
P01160.1	ANF HUMAN RecName: Full=Atrial natriuretic factor	*Homo sapiens*
AAF25815.1	AF179680 1 apelin	*Homo sapiens*
AAA58403.1	calcitonin	*Homo sapiens*
EAW95694.1	cocaine- and amphetamine-regulated transcript	*Homo sapiens*
NP_000720.1	cholecystokinin preproprotein	*Homo sapiens*
CAA34070.1	cgrp propeptide	*Homo sapiens*
NP_077720.1	natriuretic peptide precursor C precursor	*Homo sapiens*
NP_001293.2	cortistatin preproprotein	*Homo sapiens*
NP_000747.1	corticoliberin preproprotein	*Homo sapiens*
NP_077722.1	proenkephalin-B preproprotein	*Homo sapiens*
AAA52339.1	endothelin-1	*Homo sapiens*
NP_001947.1	endothelin-2 preproprotein	*Homo sapiens*
NP_996917.1	endothelin-3 isoform 1 preproprotein	*Homo sapiens*
NP_001129162.1	proenkephalin-A	*Homo sapiens*
NP_000796.1	gastrin preproprotein	*Homo sapiens*
CAA01907.1	galanin	*Homo sapiens*
NP_002045.1	glucagon preproprotein	*Homo sapiens*
NP_066567.1	growth hormone releasing hormone preproprotein	*Homo sapiens*
P07492.1	GRP	*Homo sapiens*
AAA35915.1	guanylin	*Homo sapiens*
NP_001076580.1	gonadotropin-releasing hormone 1 preproprotein	*Homo sapiens*
AAA63214.1	melanin concentrating hormone	*Homo sapiens*
AAA59860.1	motilin	*Homo sapiens*
NP_006672.1	neuromedin-U precursor	*Homo sapiens*
NP_006174.1	neurotensin/neuromedin N preproprotein	*Homo sapiens*
NP_004549.1	neurturin preproprotein	*Homo sapiens*
NP_006219.1	nociceptin precursor	*Homo sapiens*

(continued)

Table 1
(continued)

GenBank accession	Name	Species
AAA59944.1	neuropeptide Y	*Homo sapiens*
NP_001515.1	orexin precursor	*Homo sapiens*
NP_000906.1	oxytocin-neurophysin 1 preproprotein	*Homo sapiens*
AAB21470.1	pituitary adenylate cyclase activating polypeptide	*Homo sapiens*
NP_002713.1	pancreatic prohormone preproprotein	*Homo sapiens*
AAA61289.1	preprovasoactive intestinal peptide	*Homo sapiens*
NP_056977.1	prolactin releasing hormone precursor	*Homo sapiens*
NP_000306.1	parathyroid hormone preproprotein	*Homo sapiens*
NP_945317.1	parathyroid hormone-related protein isoform 1 preproprotein	*Homo sapiens*
NP_004151.2	peptide YY precursor	*Homo sapiens*
NP_068739.1	secretin preproprotein	*Homo sapiens*
AAH32625.1	Somatostatin	*Homo sapiens*
EAW76738.1	tachykinin, precursor 1 (substance K, substance P, neurokinin 1, neurokinin 2, neuromedin L, neurokinin alpha, neuropeptide K, neuropeptide gamma), isoform CRA e	*Homo sapiens*
AAB29732.1	prostate secretory protein	*Homo sapiens*
CAA53472.1	TEGT	*Homo sapiens*
NP_009048.1	prothyroliberin	*Homo sapiens*
AAA61291.1	vasopressin/neurophysin	*Homo sapiens*
AAO66192.1	gonadotropin-releasing hormone	*Ciona intestinalis*
AAP06794.1	preprogonadotropin-releasing hormone 1	*Ciona intestinalis*
AAP06796.1	preprogonadotropin-releasing hormone 2	*Ciona intestinalis*
BAD52219.1	tachykinin precursor	*Ciona intestinalis*
BAG72193.1	vasopressin/oxytocin-related peptide	*Ciona intestinalis*
BAI63095.1	calcitonin	*Ciona intestinalis*
NP_001123204.1	insulin-like 1	*Ciona intestinalis*
NP_001123345.1	insulin-like 2	*Ciona intestinalis*
NP_001123344.1	insulin-like 3 protein	*Ciona intestinalis*
NP_001027711.1	cionin precursor	*Ciona intestinalis*
XP_001176287.1	PREDICTED: hypothetical protein	*Strongylocentrotus purpuratus*

(continued)

Table 1 (continued)

GenBank accession	Name	Species
XP_785647.1	PREDICTED: similar to pedal peptide precursor protein	*Strongylocentrotus purpuratus*
XP_001175944.1	PREDICTED: hypothetical protein	*Strongylocentrotus purpuratus*
XP_799858.1	PREDICTED: hypothetical protein	*Strongylocentrotus purpuratus*
XP_001175555.1	PREDICTED: hypothetical protein	*Strongylocentrotus purpuratus*
XP_001199000.1	PREDICTED: similar to arginine/serine-rich splicing factor 4	*Strongylocentrotus purpuratus*
ABU55286.1	GnRH-like tetrapeptide	*Strongylocentrotus purpuratus*
AAC26833.1	thymosin beta	*Strongylocentrotus purpuratus*
XP_001175484.1	PREDICTED: hypothetical protein	*Strongylocentrotus purpuratus*
XP_001186882.1	PREDICTED: similar to LFRFa precursor	*Strongylocentrotus purpuratus*
XP_001175691.1	PREDICTED: hypothetical protein	*Strongylocentrotus purpuratus*
Q6T5C1	Pituitary glycoprotein hormone alpha subunit	*Gallus gallus*
P08998.2	Somatotropin	*Gallus gallus*
Q1KNA7	GHRH/PACAP-like	*Gallus gallus*
Q2ACD0	Mesotocin-neurophysin I	*Gallus gallus*
P14676.2	Prolactin	*Gallus gallus*
Q7T2S6	[Pro2]-somatostatin	*Gallus gallus*
Q6Q273	Prepro-urotensinII-relatedpeptide	*Gallus gallus*
NP_523514.1	diuretic_hormone_31,_isoform_A	*Drosophila_melanogaster*
NP_524489.2	allatostatin	*Drosophila_melanogaster*
NP_523542.1	allatostatin_C,_isoform_A	*Drosophila_melanogaster*
NP_650983.1	bursicon	*Drosophila_melanogaster*
NP_524552.1	capability	*Drosophila_melanogaster*
NP_524350.1	corazonin	*Drosophila_melanogaster*
NP_651083.2	cardioacceleratory_peptide	*Drosophila_melanogaster*

(continued)

Table 1
(continued)

GenBank accession	Name	Species
NP_649922.2	diuretic_hormone,_isoform_A	*Drosophila_melanogaster*
NP_523514.1	diuretic_hormone_31,_isoform_A	*Drosophila_melanogaster*
NP_524699.1	ecdysis_triggering_hormone	*Drosophila_melanogaster*
NP_524386.1	eclosion_hormone	*Drosophila_melanogaster*
NP_523669.2	FMRFamide-related	*Drosophila_melanogaster*
NP_611937.2	CG4681	*Drosophila_melanogaster*
NP_001163293.1	ion_transport_peptide,_isoform_E	*Drosophila_melanogaster*
NP_524893.2	leucokinin	*Drosophila_melanogaster*
NP_648971.1	myoinhibiting_peptide_precursor	*Drosophila_melanogaster*
NP_536772.1	dromyosuppressin	*Drosophila_melanogaster*
NP_536741.1	neuropeptide_F	*Drosophila_melanogaster*
NP_611993.1	Neuropeptide-like_precursor_1,_isoform_A	*Drosophila_melanogaster*
NP_524329.1	hugin	*Drosophila_melanogaster*
NP_524517.1	Pigment-dispersing_factor	*Drosophila_melanogaster*
NP_608537.2	prothoracicotropic_hormone	*Drosophila_melanogaster*
NP_523605.1	short_neuropeptide_F_precursor,_isoform_A	*Drosophila_melanogaster*
NP_524845.2	drosulfakinin	*Drosophila_melanogaster*
NP_650141.2	tachykinin	*Drosophila_melanogaster*
NP_507840.1	abnormal_DAuer_Formation_family_member_(daf-28)	*Caenorhabditis_elegans*
NP_501926.1	INSulin_related_family_member_(ins-1)	*Caenorhabditis_elegans*
NP_495194.1	INSulin_related_family_member_(ins-2)	*Caenorhabditis_elegans*
NP_495195.1	INSulin_related_family_member_(ins-3)	*Caenorhabditis_elegans*
NP_495196.1	INSulin_related_family_member_(ins-4)	*Caenorhabditis_elegans*
NP_495197.3	INSulin_related_family_member_(ins-5)	*Caenorhabditis_elegans*
NP_495198.1	INSulin_related_family_member_(ins-6)	*Caenorhabditis_elegans*
NP_501717.1	INSulin_related_family_member_(ins-7)	*Caenorhabditis_elegans*
NP_001023609.1	INSulin_related_family_member_(ins-8)	*Caenorhabditis_elegans*
NP_001024362.1	INSulin_related_family_member_(ins-9)	*Caenorhabditis_elegans*
NP_001024125.1	INSulin_related_family_member_(ins-10)	*Caenorhabditis_elegans*
NP_495071.1	INSulin_related_family_member_(ins-11)	*Caenorhabditis_elegans*
NP_001021963.1	INSulin_related_family_member_(ins-12)	*Caenorhabditis_elegans*

(continued)

Table 1
(continued)

GenBank accession	Name	Species
NP_001021962.1	INSulin_related_family_member_(ins-13)	*Caenorhabditis_elegans*
NP_001022153.1	INSulin_related_family_member_(ins-14)	*Caenorhabditis_elegans*
NP_001022154.1	INSulin_related_family_member_(ins-15)	*Caenorhabditis_elegans*
NP_001022841.1	INSulin_related_family_member_(ins-16)	*Caenorhabditis_elegans*
NP_497911.1	INSulin_related_family_member_(ins-17)	*Caenorhabditis_elegans*
NP_492231.1	INSulin_related_family_member_(ins-18)	*Caenorhabditis_elegans*
NP_001022339.1	INSulin_related_family_member_(ins-19)	*Caenorhabditis_elegans*
NP_001022529.1	INSulin_related_family_member_(ins-20)	*Caenorhabditis_elegans*
NP_499222.2	INSulin_related_family_member_(ins-21)	*Caenorhabditis_elegans*
NP_499223.1	INSulin_related_family_member_(ins-22)	*Caenorhabditis_elegans*
NP_499224.1	INSulin_related_family_member_(ins-23)	*Caenorhabditis_elegans*
NP_493443.2	INSulin_related_family_member_(ins-24)	*Caenorhabditis_elegans*
NP_001021849.1	INSulin_related_family_member_(ins-25)	*Caenorhabditis_elegans*
NP_493445.1	INSulin_related_family_member_(ins-26)	*Caenorhabditis_elegans*
NP_001021847.1	INSulin_related_family_member_(ins-27)	*Caenorhabditis_elegans*
NP_001021850.1	INSulin_related_family_member_(ins-28)	*Caenorhabditis_elegans*
NP_001021846.1	INSulin_related_family_member_(ins-29)	*Caenorhabditis_elegans*
NP_493444.1	INSulin_related_family_member_(ins-30)	*Caenorhabditis_elegans*
NP_494454.1	INSulin_related_family_member_(ins-31)	*Caenorhabditis_elegans*
NP_494655.1	INSulin_related_family_member_(ins-32)	*Caenorhabditis_elegans*
NP_494655.1	INSulin_related_family_member_(ins-32)	*Caenorhabditis_elegans*
NP_502702.2	INSulin_related_family_member_(ins-34)	*Caenorhabditis_elegans*
NP_507926.1	INSulin_related_family_member_(ins-35)	*Caenorhabditis_elegans*
NP_001021786.1	INSulin_related_family_member_(ins-36)	*Caenorhabditis_elegans*
NP_496902.1	INSulin_related_family_member_(ins-37)	*Caenorhabditis_elegans*
NP_001021964.1	INSulin_related_family_member_(ins-38)	*Caenorhabditis_elegans*
NP_508247.1	INSulin_related_family_member_(ins-39)	*Caenorhabditis_elegans*
NP_501592.1	FMRF-Like_Peptide_family_member_(flp-1)	*Caenorhabditis_elegans*
NP_001024946.1	FMRF-Like_Peptide_family_member_(flp-2)	*Caenorhabditis_elegans*
NP_509694.1	FMRF-Like_Peptide_family_member_(flp-3)	*Caenorhabditis_elegans*
NP_496173.1	FMRF-Like_Peptide_family_member_(flp-4)	*Caenorhabditis_elegans*

(continued)

Table 1
(continued)

GenBank accession	Name	Species
NP_509445.1	FMRF-Like_Peptide_family_member_(flp-5)	*Caenorhabditis_elegans*
NP_505444.1	FMRF-Like_Peptide_family_member_(flp-6)	*Caenorhabditis_elegans*
NP_508985.1	FMRF-Like_Peptide_family_member_(flp-7)	*Caenorhabditis_elegans*
NP_741934.1	FMRF-Like_Peptide_family_member_(flp-8)	*Caenorhabditis_elegans*
NP_502436.2	FMRF-Like_Peptide_family_member_(flp-9)	*Caenorhabditis_elegans*
NP_501306.1	FMRF-Like_Peptide_family_member_(flp-10)	*Caenorhabditis_elegans*
NP_001024754.1	FMRF-Like_Peptide_family_member_(flp-11)	*Caenorhabditis_elegans*
NP_508789.1	FMRF-Like_Peptide_family_member_(flp-12)	*Caenorhabditis_elegans*
NP_501255.1	FMRF-Like_Peptide_family_member_(flp-13)	*Caenorhabditis_elegans*
NP_499682.2	FMRF-Like_Peptide_family_member_(flp-14)	*Caenorhabditis_elegans*
NP_499820.1	FMRF-Like_Peptide_family_member_(flp-15)	*Caenorhabditis_elegans*
NP_001022091.1	FMRF-Like_Peptide_family_member_(flp-16)	*Caenorhabditis_elegans*
NP_503051.1	FMRF-Like_Peptide_family_member_(flp-17)	*Caenorhabditis_elegans*
NP_508514.2	FMRF-Like_Peptide_family_member_(flp-18)	*Caenorhabditis_elegans*
NP_509776.1	FMRF-Like_Peptide_family_member_(flp-19)	*Caenorhabditis_elegans*
NP_509574.2	FMRF-Like_Peptide_family_member_(flp-20)	*Caenorhabditis_elegans*
NP_505011.2	FMRF-Like_Peptide_family_member_(flp-21)	*Caenorhabditis_elegans*
NP_492344.2	FMRF-Like_Peptide_family_member_(flp-22)	*Caenorhabditis_elegans*
NP_498907.2	FMRF-Like_Peptide_family_member_(flp-23)	*Caenorhabditis_elegans*
NP_497243.3	FMRF-Like_Peptide_family_member_(flp-24)	*Caenorhabditis_elegans*
NP_001022665.1	FMRF-Like_Peptide_family_member_(flp-25)	*Caenorhabditis_elegans*
NP_741827.1	FMRF-Like_Peptide_family_member_(flp-26)	*Caenorhabditis_elegans*
NP_495111.1	FMRF-Like_Peptide_family_member_(flp-27)	*Caenorhabditis_elegans*
NP_001024947.1	FMRF-Like_Peptide_family_member_(flp-28)	*Caenorhabditis_elegans*
NP_510551.1	FMRF-Like_Peptide_family_member_(flp-32)	*Caenorhabditis_elegans*
NP_871818.1	FMRF-Like_Peptide_family_member_(flp-33)	*Caenorhabditis_elegans*
NP_503365.1	hypothetical_protein_R09A1.5	*Caenorhabditis_elegans*
NP_508640.1	Neuropeptide-Like_Protein_family_member_(nlp-1)	*Caenorhabditis_elegans*
NP_508426.1	Neuropeptide-Like_Protein_family_member_(nlp-2)	*Caenorhabditis_elegans*
NP_510187.1	Neuropeptide-Like_Protein_family_member_(nlp-3)	*Caenorhabditis_elegans*
NP_492750.1	Neuropeptide-Like_Protein_family_member_(nlp-4)	*Caenorhabditis_elegans*

(continued)

Table 1
(continued)

GenBank accession	Name	Species
NP_495735.1	Neuropeptide-Like_Protein_family_member_(nlp-5)	*Caenorhabditis_elegans*
NP_510850.2	Neuropeptide-Like_Protein_family_member_(nlp-6)	*Caenorhabditis_elegans*
NP_509456.1	Neuropeptide-Like_Protein_family_member_(nlp-7)	*Caenorhabditis_elegans*
NP_492158.1	Neuropeptide-Like_Protein_family_member_(nlp-8)	*Caenorhabditis_elegans*
NP_504168.1	Neuropeptide-Like_Protein_family_member_(nlp-9)	*Caenorhabditis_elegans*
NP_497795.1	Neuropeptide-Like_Protein_family_member_(nlp-10)	*Caenorhabditis_elegans*
NP_496091.1	Neuropeptide-Like_Protein_family_member_(nlp-11)	*Caenorhabditis_elegans*
NP_490908.1	Neuropeptide-Like_Protein_family_member_(nlp-12)	*Caenorhabditis_elegans*
NP_504170.1	Neuropeptide-Like_Protein_family_member_(nlp-13)	*Caenorhabditis_elegans*
NP_509508.1	Neuropeptide-Like_Protein_family_member_(nlp-14)	*Caenorhabditis_elegans*
NP_493278.2	Neuropeptide-Like_Protein_family_member_(nlp-15)	*Caenorhabditis_elegans*
NP_001023384.1	Neuropeptide-Like_Protein_family_member_(nlp-16)	*Caenorhabditis_elegans*
NP_502603.1	Neuropeptide-Like_Protein_family_member_(nlp-17)	*Caenorhabditis_elegans*
NP_496365.2	Neuropeptide-Like_Protein_family_member_(nlp-18)	*Caenorhabditis_elegans*
NP_741898.1	Neuropeptide-Like_Protein_family_member_(nlp-19)	*Caenorhabditis_elegans*
NP_501244.1	Neuropeptide-Like_Protein_family_member_(nlp-20)	*Caenorhabditis_elegans*
NP_499467.1	Neuropeptide-Like_Protein_family_member_(nlp-21)	*Caenorhabditis_elegans*
NP_508424.1	Neuropeptide-Like_Protein_family_member_(nlp-22)	*Caenorhabditis_elegans*
NP_508425.1	Neuropeptide-Like_Protein_family_member_(nlp-23)	*Caenorhabditis_elegans*
NP_505946.2	Neuropeptide-Like_Protein_family_member_(nlp-24)	*Caenorhabditis_elegans*
NP_507801.1	Neuropeptide-Like_Protein_family_member_(nlp-25)	*Caenorhabditis_elegans*
NP_507802.1	Neuropeptide-Like_Protein_family_member_(nlp-26)	*Caenorhabditis_elegans*
NP_504111.1	Neuropeptide-Like_Protein_family_member_(nlp-27)	*Caenorhabditis_elegans*
NP_504110.1	Neuropeptide-Like_Protein_family_member_(nlp-28)	*Caenorhabditis_elegans*
NP_504109.1	Neuropeptide-Like_Protein_family_member_(nlp-29)	*Caenorhabditis_elegans*
NP_504108.1	Neuropeptide-Like_Protein_family_member_(nlp-30)	*Caenorhabditis_elegans*
NP_504107.1	Neuropeptide-Like_Protein_family_member_(nlp-31)	*Caenorhabditis_elegans*
NP_497204.1	Neuropeptide-Like_Protein_family_member_(nlp-32)	*Caenorhabditis_elegans*
NP_505834.1	Neuropeptide-Like_Protein_family_member_(nlp-33)	*Caenorhabditis_elegans*
NP_741520.1	Neuropeptide-Like_Protein_family_member_(nlp-34)	*Caenorhabditis_elegans*
NP_501687.1	Neuropeptide-Like_Protein_family_member_(nlp-35)	*Caenorhabditis_elegans*

(continued)

Table 1 (continued)

GenBank accession	Name	Species
NP_499088.1	Neuropeptide-Like_Protein_family_member_(nlp-36)	*Caenorhabditis_elegans*
NP_508397.2	Neuropeptide-Like_Protein_family_member_(nlp-37)	*Caenorhabditis_elegans*
NP_740928.1	Neuropeptide-Like_Protein_family_member_(nlp-38)	*Caenorhabditis_elegans*
NP_493163.1	Neuropeptide-Like_Protein_family_member_(nlp-39)	*Caenorhabditis_elegans*
NP_490661.1	Neuropeptide-Like_Protein_family_member_(nlp-40)	*Caenorhabditis_elegans*
NP_001021917.1	Neuropeptide-Like_Protein_family_member_(nlp-41)	*Caenorhabditis_elegans*
NP_507687.1	Neuropeptide-Like_Protein_family_member_(nlp-42)	*Caenorhabditis_elegans*
ABB58739.1 A	AKH-I preprohormone	*Tribolium castaneum*
EFA12888.1	AKH like peptide	*Tribolium castaneum*
EFA04704.1	adipokinetic hormone 2	*Tribolium castaneum*
NP_001137202.1	allatostatin type b precursor	*Tribolium castaneum*
EFA09152.1	allatostatin-C	*Tribolium castaneum*
ACJ38497.1	allatotropin preprohormone isoform 2	*Tribolium castaneum*
EEZ99367.1	diuretic hormone 31 like protein	*Tribolium castaneum*
XP_008192828.1	PREDICTED: uncharacterized protein LOC103312856 isoform X1	*Tribolium castaneum*
NP_001280497.1	similar to crustacean cardioactive peptide precursor	*Tribolium castaneum*
NP_001165744.1	ecdysis triggering hormone preproprotein	*Tribolium castaneum*
XP_008191572.1	PREDICTED: FMRFamide-related neuropeptides	*Tribolium castaneum*
EFA12055.1	myosuppressin	*Tribolium castaneum*
XP_008190594.1	PREDICTED: uncharacterized protein LOC103312231	*Tribolium castaneum*
EFA11568.1	pyrokinin	*Tribolium castaneum*
DAA34847.1	TPA_inf: short neuropeptide F	*Tribolium castaneum*
XP_001814498.1	PREDICTED: SIFamide-related peptide	*Tribolium castaneum*
XP_008194373.1	PREDICTED: drosulfakinins	*Tribolium castaneum*
XP_008196871.1	PREDICTED: tachykinins	*Tribolium castaneum*
NP_001107779.1	bursicon precursor	*Tribolium castaneum*
NP_001107780.1	bursicon beta precursor	*Tribolium castaneum*
XP_969164.1	PREDICTED: eclosion hormone	*Tribolium castaneum*
XP_008190383.1	PREDICTED: uncharacterized protein LOC103312180 isoform X1	*Tribolium castaneum*

(continued)

Table 1 (continued)

GenBank accession	Name	Species
NP_001164244.1	glycoprotein hormone alpha 2 precursor	*Tribolium castaneum*
EFA10760.1	glycoprotein hormone beta 5	*Tribolium castaneum*
EFA02918.1	insulin-like peptide 1	*Tribolium castaneum*
XP_001814181.1	PREDICTED: LIRP	*Tribolium castaneum*
EEZ99258.1	insulin-like peptide	*Tribolium castaneum*
EFA04588.1	neuroparsin-like protein	*Tribolium castaneum*
EEZ99381.1	prothoracicotropic hormone	*Tribolium castaneum*
ABN79655.1	arginine vasopressin-like peptide	*Tribolium castaneum*
XP_008195863.1	PREDICTED: uncharacterized protein LOC103313694	*Tribolium castaneum*
NP_001164303.1	ADFb like protein precursor	*Tribolium castaneum*
NP_001164302.1	anti-diuretic peptide precursor	*Tribolium castaneum*
EFA07533.1	ADFb like protein	*Tribolium castaneum*
XP_008195861.1	PREDICTED: uncharacterized protein LOC103313692	*Tribolium castaneum*
NP_001164096.1	diuretic hormone 37 like protein	*Tribolium castaneum*
XP_008196925.1	PREDICTED: prohormone-2-like	*Tribolium castaneum*
EFA07415.1	brain peptide ITGQGNRIF-like protein	*Tribolium castaneum*
EFA09268.1	neuropeptide-like precursor 1	*Tribolium castaneum*
P07712.1	Abdominal ganglion neuropeptides L5–67 precursor [Contains: Luqin; Luqin-B; Luqin-C; Proline-rich mature peptide (PRMP)]	*Aplysia californica*
Q17043.1	Aplysianin A precursor	*Aplysia californica*
O96910.1	Attractin precursor	*Aplysia californica*
Q00676.1	Bradykinin-like neuropeptide (LUQ-1)	*Aplysia californica*
P20481.2	Buccalin precursor (buccalin precursor)	*Aplysia californica*
Q25461.1	Central nervous system APGWamide	*Aplysia californica*
C0HK25.1	capsulin	*Aplysia californica*
Q8T112.1	Cerebrin prohormone precursor	*Aplysia californica*
P01360.2	ELH Atrial gland peptide A precursor (ELH-18)	*Aplysia californica*
P01361.1	ELH egg-laying hormone-related precursor [Pro 25] B	*Aplysia californica*
P01362.2	ELH precursor [Contains: Beta-bag cell peptide (Beta-BCP)	*Aplysia californica*

(continued)

Table 1
(continued)

GenBank accession	Name	Species
Q95P23.1	Enterin	*Aplysia californica*
Q8I817.1	Enticin precursor	*Aplysia californica*
Q8ISH7.1	feeding circuit activating peptide precursor	*Aplysia californica*
P08021.3	FMRFamide neuropeptide precursor	*Aplysia californica*
NP_001191526.1	Seductin/Hypothetical protein 2	*Aplysia californica*
AAN83920.1	hypothetical protein 3	*Aplysia californica*
Q9NDE7.1	insulin precursor	*Aplysia californica*
P06518.1	L11 neuropeptide precursor	*Aplysia californica*
Q9NDE8.1	MIP-related peptide precursor	*Aplysia californica*
P15513.2	myomodulin []	*Aplysia californica*
Q2VF17.1	myomodulin gene 2 neuropeptide precursor	*Aplysia californica*
Q8T0Y7.1	Neuropeptide CP2 precursor	*Aplysia californica*
Q27441.1	Neuropeptide Y	*Aplysia californica*
Q86MA7.1	PRQFVamide precursor protein	*Aplysia californica*
NP_001191482.1	preprogonadotropin-releasing hormone-like protein	*Aplysia californica*
NP_001191460.1	putative pheromone	*Aplysia californica*
XP_005096950.1	R15–1 neuroactive peptide precursor	*Aplysia californica*
P12285.1	R15–2 Neuroactive polyprotein precursor	*Aplysia californica*
P01364.1	R3–14 neuropeptide precursor	*Aplysia californica*
P29233.1	Sensorin-A precursor	*Aplysia californica*
Q7Z0T3.1	Temptin precursor	*Aplysia californica*
P09892.1	Small cardioactive peptides precursor	*Aplysia californica*
NP_001191589.1	achatin-like neuropeptide precursor [Aplysia californica]	*Aplysia californica*
XP_005099829.1	PREDICTED: uncharacterized protein LOC101854331 [Aplysia californica]	*Aplysia californica*
NP_001191416.1	Lys-conopressin preprohormone precursor [Aplysia californica]	*Aplysia californica*
NP_001191641.1	glycoprotein hormone alpha subunit [Aplysia californica]	*Aplysia californica*
NP_001191597.1	glycoprotein hormone beta subunit [Aplysia californica]	*Aplysia californica*
NP_001191590.1	fulicin-like neuropeptide precursor [Aplysia californica]	*Aplysia californica*
NP_001191621.1	insulin-like 7 precursor [Aplysia californica]	*Aplysia californica*

(continued)

Table 1 (continued)

GenBank accession	Name	Species
NP_001191624.1	insulin-like 0 precursor [Aplysia californica]	*Aplysia californica*
NP_001191503.1	insulin-like peptide precursor II precursor [Aplysia californica]	*Aplysia californica*
XP_005090025.1	PREDICTED: uncharacterized protein LOC101847921 [Aplysia californica]	*Aplysia californica*
XP_005104578.1	PREDICTED: major royal jelly protein 1-like [Aplysia californica]	*Aplysia californica*
NP_001191604.1	putative toxin I precursor [Aplysia californica]	*Aplysia californica*
NP_001191605.1	putative toxin II precursor [Aplysia californica]	*Aplysia californica*
NP_001191585.1	pedal peptide-1 precursor [Aplysia californica]	*Aplysia californica*
NP_001191623.1	pedal peptide 2 precursor [Aplysia californica]	*Aplysia californica*
NP_001191625.1	pedal peptide 3 precursor [Aplysia californica]	*Aplysia californica*
NP_001191626.1	pedal peptide 4 [Aplysia californica]	*Aplysia californica*
NP_001191654.1	pleurin precursor [Aplysia californica]	*Aplysia californica*
NP_001191584.1	schistosomin-like precursor [Aplysia californica]	*Aplysia californica*
NP_001191611.1	PTSP-like peptide neurotransmitter precursor [Aplysia californica]	*Aplysia californica*
NP_001191586.1	whitnin precursor [Aplysia californica]	*Aplysia californica*
NP_001191429.1	FRFamide precursor protein precursor [Aplysia californica]	*Aplysia californica*
XP_005098207.1	PREDICTED: FMRFamide-related neuropeptides-like [Aplysia californica]	*Aplysia californica*
XP_005099491.1	PREDICTED: myomodulin neuropeptides 1-like [Aplysia californica]	*Aplysia californica*
XP_005109599.1	PREDICTED: uncharacterized protein LOC101856383 isoform X1 [Aplysia californica]	*Aplysia californica*
XP_005094943.1	PREDICTED: uncharacterized protein LOC101845765 isoform X3 [Aplysia californica]	*Aplysia californica*
XP_005097583.1	PREDICTED: uncharacterized protein LOC101845620 [Aplysia californica]	*Aplysia californica*
XP_005106030.1	PREDICTED: uncharacterized protein LOC101859590 isoform X1 [Aplysia californica]	*Aplysia californica*
XP_005112794.1	PREDICTED: uncharacterized protein LOC101859220 [Aplysia californica]	*Aplysia californica*
NP_001191666.1	allatotropin-like peptide precursor [Aplysia californica]	*Aplysia californica*

(continued)

Table 1 (continued)

GenBank accession	Name	Species
NP_001191666.1	allatotropin-like peptide precursor [Aplysia californica]	*Aplysia californica*
XP_005103388.1	PREDICTED: uncharacterized protein LOC101863916 [Aplysia californica]	*Aplysia californica*
XP_005103388.1	PREDICTED: uncharacterized protein LOC101863916 [Aplysia californica]	*Aplysia californica*
NP_001268793.1	adipokinetic prohormone type 1-like precursor [Aplysia californica]	*Aplysia californica*
XP_005101218.2	PREDICTED: uncharacterized protein LOC101859417 [Aplysia californica]	*Aplysia californica*
XP_005105032.1	PREDICTED: ecdysteroid-regulated 16 kDa protein-like [Aplysia californica]	*Aplysia californica*
ATD50219.1	Leucokinin	*Aplysia californica*
XP_005096263.1	PREDICTED: uncharacterized protein LOC101857558 isoform X1 [Aplysia californica]	*Aplysia californica*
XP_012944053.1	PREDICTED: uncharacterized protein LOC101859757 [Aplysia californica]	*Aplysia californica*
XP_005094684.1	PREDICTED: uncharacterized protein LOC101848188 [Aplysia californica]	*Aplysia californica*
XP_005089090.1	PREDICTED: uncharacterized protein LOC101864107 [Aplysia californica]	*Aplysia californica*
XP_001638801.1	Antho-Rfamide	*Nematostella vectensis*
XP_001618682.1	RFaP-1	*Nematostella vectensis*
XP_001622334.1	RFaP-1_precursor	*Nematostella vectensis*
XP_001635550.1	RFap-2_precursor	*Nematostella vectensis*
XP_001634202.1	Amidating_Enzyme_1	*Nematostella vectensis*
XP_001626495.1	Amidating_Enzyme_2	*Nematostella vectensis*
XP_001624069.1	Antho-Rlamide-like	*Nematostella vectensis*
XP_001638094.1	Antho-RNamide-like	*Nematostella vectensis*
XP_001640991.1	Antho-RPamide-like	*Nematostella vectensis*
XP_032225345.1	Antho-RWamide-like	*Nematostella vectensis*
XP_001626456.1	LW-amide_precursor	*Nematostella vectensis*
XP_001636030.1	Galanin-like_precursor_1	*Nematostella vectensis*
XP_001625905.1	Galanin-like_precursor_2	*Nematostella vectensis*
XP_001622181.1	Galanin-like_precursor_3	*Nematostella vectensis*

(continued)

Table 1
(continued)

GenBank accession	Name	Species
XP_001638793.1	Tachykinin-precursor_1	*Nematostella vectensis*
XP_001636630.1	Tachykinin-precursor_2	*Nematostella vectensis*
XP_001625405.1	GnRH-like_precursor	*Nematostella vectensis*
XP_001641986.1	Vasopressin-like_precursor_1	*Nematostella vectensis*
XP_001633670.1	Vasopressin-like_precursor_2	*Nematostella vectensis*
XP_001618986.1	alpha-MSH-like_precursor_1	*Nematostella vectensis*
XP_001641703.1	alpha-MSH-like_precursor_2	*Nematostella vectensis*
XP_001639325.1	Insulin-like_1	*Nematostella vectensis*
XP_001632788.1	Insulin-like_2	*Nematostella vectensis*

```
blastp -db triadv2_all -evalue 0.1 -num_threads
2 -num_descriptions 5 -num_alignments 5 -out
Metazoa_neuropep-vs-TriadV2_blastout_e01.txt -query ../
metazoa_masterlist.fasta
```

```
blastp -db triadv2_all -evalue 0.1 -num_threads 2 -outfmt
6 -max_target_seqs 5 -out Metazoa_neuropep-vs-TriadV2_-
blastout_e01.tsv -query ../metazoa_masterlist.fasta
```

These two BLAST runs differ only in output format: first is human-readable plain text output for visual inspection of alignments. The second is the tabular format for sorting and filtering in MS Excel.

3.1.3 Filter BLAST Results and Find Potential Prohormones

We import neuropep-vs-TriadV2_blastout_e01.tsv in Excel, sort by subject name, and see that 33 *Trichoplax* proteins were found. We extract sequences of these proteins from the Fasta file with all TriadV2 proteins and analyze them using web versions of SignalP, TMHMM, NeuroPred, and NCBI BLAST versus NR database. Twenty-two out of 33 protein hits are unlikely peptide precursors because of one or more of the following reasons:

- Presence of transmembrane regions (e.g., TriadITZ_011332)
- Size over 500 amino acids but only single hits to peptide precursor (TriadITZ_006195, TriadITZ_010221, TriadITZ_003114, TriadITZ_008509, and others)

```
TriadITZ_004971-RA DQPPRWamide precursor, similar to Nematostella
antho-RFamide
MLTNRFIIWILFLGITTAQNVAKGKAQIGNHKSVFLKNEATRPERDQPPRWGRDQPTRWGRDQPPR
WGRDQPPRWGRDQPSRWGRDQPPRWGRDQPPRWGRDQPPRWGRDQPPRWGRDQPPRWGRDQPPRWG
RDQPPRWGRDQPPRWGRDQPPRWGRDQPPRWGRDQPPRWGRDQPPRWGGDQLPEMEKNHAPPRWGR
DQYSWWNQEQYPSRWGREYSTPDNTAEKLLDSLTHQSENAKKNNFQEINSDSNSGNESAVHRLFSN
KLKNQKAKSDSNKLMNSFSGSESISRPREKSLKRSETLDNMRIDLI

TriadITZ_009688-RA SITFamide precursor, similar to Aplysia
FIRFamide-related-like
MATNSIFLSYTILMIAVVVATTPGNARFISSAKRHTLREQPTFIPLGFGRRNLEDIKVSRQSHNSD
VLDRRNSESIQQGMPSIMFGKRNSESTQQGIPSITFGKRNSESTQQGIPSITFGKRNSASTQQGIP
SITFGKRNSESIQQGIPSITFGKRNGKSIQQGIPSITFGKRNSESIRQNIPPIMFGKRDSKRMEFR
MPTLTFSNRDNEPKLYGLPSINLHKRDAWTKQTIVPRDNLLARQLYNYGPEMYTMPEDIAYLTSMT
EPSSLDYQLSFAQSTPDSKAFGKYNLPIWNLDSTGLASGGFVDVPTVYLESGNEIPMAMLKRRSSE
RFTKDMSTSYDQKKKLFLGLPRFGDQDRSAKSALKDELRLINRKREYKQPPPIIYEGK

TriadITZ_006520-RA SAKIPME precursor, similar to C.elegans
FMRFamide-related flp-7
MKISNAVIVLTVAIALVTNSADSRYLYDNSIKTPKSKHSIKDFADETDRLSYLLMVSTSNGNIKRL
DKTYMERPDFISMERSNKISKRSLTKTAMKKSDDIPMERSSDISFERSKTIPMERHIPFEKSENIP
MERSAKIPMERSAKIPMERSSNIPFERSAKIPMERSNNIPFERSYKIPFERSAKIP
```

Fig. 1 Predicted precursors for *Trichoplax adhaerens* short regulatory peptides. Signal peptides are highlighted in yellow. The recognition and cleavage sites are highlighted in gray. Glycine residues that are likely to be converted into C-terminal amides are highlighted in dark green. Mature peptides are highlighted in cyan

- Absence of cleavage sites in the region of similarity with neuropeptide precursor (e.g., TriadITZ_003380)
- Absence of conserved cysteines in the region of similarity with insulin precursor (TriadITZ_003512 and TriadITZ_009919)
- High similarity to proteins with other functions (e.g., TriadITZ_000575—bacterial DNA alkylation repair protein)

Seven of the 11 remaining proteins display all characteristics of prohormone/neuropeptide precursors: N-end signal peptide and multiple paracopies interspersed with prohormone convertase cleavage sites (Fig. 1, Table 2). We briefly describe them:

- Precursor TriadITZ_004971: 14 paracopies of DQPPRWamide peptides.
- TriadITZ_009688: 5 paracopies of NSESTQQGIPSITFamide peptides and 3 others.
- TriadITZ_004974: 14 identical copies of DGQFFNPamide peptides, 4 slightly different paracopies, and 9 other peptides.
- TriadITZ_005098: 5 paracopies of xYDDYYYx and three other tyrosine-rich peptides.
- TriadITZ_000026: 5 identical copies of QEPGISLFNE peptides, 3 slightly different paracopies, monobasic cleavage sites RQ are predicted only using mollusk model of convertase specificity [42].

Table 2
Full list of preprohormones discovered in *Trichoplax adhaerens* using three different prediction methods, and their homologs in *Hoilungia hongkongensis*

#	Gene ID (TriadV2 annotation)	Major mature peptide product	#Paracopies	Detected by: BLAST	Detected by: Paracopies	Detected by: Comparative	Ortholog in H13	Previous predictions
1	TriadITZ_004971-RA	DQPPRWamide	14	Nematostella Antho-Rfamide	+	–	braker1_g04334.t1	Nikitin, 2014
2	TriadITZ_009688-RA	NSESTQQGIPSITFamide	5	Drosophila FIRFamide-related	+	+	braker1_g03415.t1	Nikitin, 2014
3	TriadITZ_008272-RA	EDQANLKSIFGamide	36	–	+	–	braker1_g04983.t1	Nikitin, 2014
4	TriadITZ_004974-RA	DGQFFNPamide	14	Aplysia Myomodulin and PRQFVamide	+	+	braker1_g04338.t1	Nikitin, 2014
5	TriadITZ_007434-RA	QGVLVDVPWN	8	–	+	+	–	Nikitin, 2014
6	TriadITZ_005098-RA	GYDDYYYamide	5	Drosophila FMRFamide-related	+	–	braker1_g05846.t1	Nikitin, 2014
7	TriadITZ_000027-RA	DDSQGYPLF	11	–	+	+	–	Nikitin, 2014
8	TriadITZ_004972-RA	DDQQNKPYNGWPPF	13	–	+	–	braker1_g04333.t1	Nikitin, 2014
9	TriadITZ_000026-RA	QEPGISLFNE	5	C.elegans FMRF-related Flp-18	–	+	–	Nikitin, 2014
10	TriadITZ_009692-RA	QDYPFFamide	3	–	+	–	braker1_g03419.t1	Smith et al., 2014
11	TriadITZ_005528-RA	MFPF	17	–	–	–	braker1_g05652.t1	

(continued)

Table 2
(continued)

#	Gene ID (TriadV2 annotation)	Major mature peptide product	#Paracopies	Detected by			Ortholog in H13	Previous predictions
				BLAST	Paracopies	Comparative		
								Varoqueaux et al., 2018
12	TriadITZ_004633-RA	nxKAQKWW	2	–	+	+	braker1_g10045	
13	TriadITZ_006520-RA	SAKIPME	10	C.elegans FMRF-related flp-7	+	–	braker1_g00251.t1	
14	TriadITZ_008559-RA	SFE(F/L)PE	6	–	+	–	–	Varoqueaux et al., 2018
15	TriadITZ_011042-RA	AVVV, IL	5	–	+	–	–	
16	TriadITZ_004550-RA	FIT(F/L)	3	–	+	–	–	
17	TriadITZ_004552-RA	F(F/Y)Y	5	–	+	–	–	
18	TriadITZ_004786-RA	FYI/FLY	4	–	+	–	–	
19	EDV26034.1	LYY	7	–	+	–	–	
20	TriadITZ_004547-RA	TrIns-1	–	C.elegans Ins-18	–	+	braker1_g03237.t1	Nikitin, 2014
21	TriadITZ_006796-RA	TrIns-2	–	C.elegans Ins-18	–	–	braker1_g07431.t1	Nikitin, 2014

22	GR413862.1	TrIns-3		C.elegans Ins-18	–	–	–	Nikitin, 2014
23	TriadITZ_006795-RA	TrIns-4	–	C.elegans Ins-18	–	+	braker1_g07432.t1	Nikitin, 2014
24	TriadITZ_006796-RA	TrIns-5	–	C.elegans Ins-18	–	–	–	
25	TriadITZ_004694-RA	TrIns-6	–	C.elegans Ins-18	–	–	–	
26	TriadITZ_000151-RA	CISVGNNSKECVGISGALIFFK	–	–	–	+	braker1_g08556.t1	
27	TriadITZ_001762-RA	QFERTRSRLWRCTNPW	–	–	–	+	braker1_g03129.t1	
28	TriadITZ_004547-RA	LYCLPEVFEFFINNCPTS	–	–	–	+	braker1_g03237.t1	
29	TriadITZ_007599-RA	HWWVLamide	–	–	–	+	braker1_g09095.t1	
30	TriadITZ_008271-RA	SADRTQPALDETFSP	–	–	–	+	braker1_g04984.t1	
31	TriadITZ_009601-RA	NKSKCGRFGSFWLMCILG	–	–	–	+	braker1_g03345.t1	
32	TriadITZ_009689-RA	AREDQPPPYLYPRYamide	–	–	–	+	braker1_g03416.t1	
33	TriadITZ_000618-RA	ENELENWDPLWPF	–	–	–	+	braker1_g03657.t1	

- TriadITZ_011314: highly similar to TriadITZ_000026, possibly alleles of the same gene. We exclude this from subsequent gene counts.
- TriadITZ_006520: 3 identical and 7 diverged paracopies of SAKIPME, monobasic cleavage sites ERS are predicted only using the molluscan model of convertase specificity [42].

Another protein, TriadITZ_004547, is a placozoan member of the insulin family with a signal peptide, two cleavage sites, and six conserved cysteine residues.

Three more proteins are similar to small secreted proteins with functions unrelated to neuroendocrine signaling. These are TriadITZ_010167 (similar to *Aplysia* pheromone carrier temptin), TriadITZ_001732, and TriadITZ_002192 (similar to ependymin-related proteins from a wide range of invertebrates—see the phylogenetic tree in Fig. 5).

It is worth noting that insulin TriadITZ_004547 was found on the margin of BLAST sensitivity (e-value = 0.0003). We repeat BLAST search with a relaxed e-value cutoff of 1 and using only TriadITZ_004547 as query and find three more insulins: TriadITZ_004694, 006795, and 006796. TriadITZ_006796 is a chimera of two insulins, possibly due to misprediction. Therefore, *Trichoplax* annotation TriadV2 contains five insulins.

TriadITZ_004547, 006795, and 3′ half of 006796 have canonical insulin structure with A and B chains, C-peptide in between, and six conserved cysteines (Fig. 2). 5′ half of TriadITZ_006796 have an unusual arrangement of Cys residues (3 in A-chain and 3 in B-chain) and lack a C-peptide (Figs. 2 and 3). TriadITZ_004694 has 8 Cys residues (3 in A-chain and 5 in B-chain) and no C-peptide.

```
TriadITZ_004547-RA (insulin 1)
MVRLNPCLVIIALLLVSHSASIICQNDDDDDSKERKLYCLPEVFEFFINNCPTSRRKRFIIDSKSS
TSTHKQHDVLMRTLRDANHDCCNPPGCTRDEILAYC
TriadITZ_006796-RA 3′ part of chimeric protein (insulin 2)
MKSQVFAIIALVVVAIVIVDATGTQQQFCAPDAFDWLINNCPTKRHVAVNGLRFRRSTLRLRHLYQ
NIARDLHTDCCGQPGCTEETVIQNYC
TriadITZ_006795-RA (insulin 4)
MNSKTMVIFAALLLALAVMQAESSDIVLKNDAKTSICTPEAFEWLINHCPTRSAARRAIRRLERGD
SEGAYNDIGMRRILRSIGDCCGVPGCSDSELLTYC
TriadITZ_006796-RA 5′ part of chimeric protein (insulin 5)
MGSRQITLIALILLAFIIIDVNCSEPRQFCAPDCFDWLVENCPPFNPARSAIGLRHSFHHIRNIYR
HVARDLHADCCEQPGCTEETVINYYC
TriadITZ_004694-RA (insulin 6)
MKAFTLFAVSLTVFLLVNITTTSACIGGIPKEQLCVPAAFDWLIDNCPTRSVRWDRFHHYHAARHS
VYLRNLHNECCTPEGCCPEYAMQYC
```

Fig. 2 Predicted precursors for *Trichoplax adhaerens* insulins. Signal peptides are highlighted in yellow. The recognition and cleavage sites are highlighted in gray. Cysteine residues are in red. Mature peptides are highlighted in cyan. TrIns3 is not shown because it is missed in both genome annotations and was found only in the transcriptome [43]

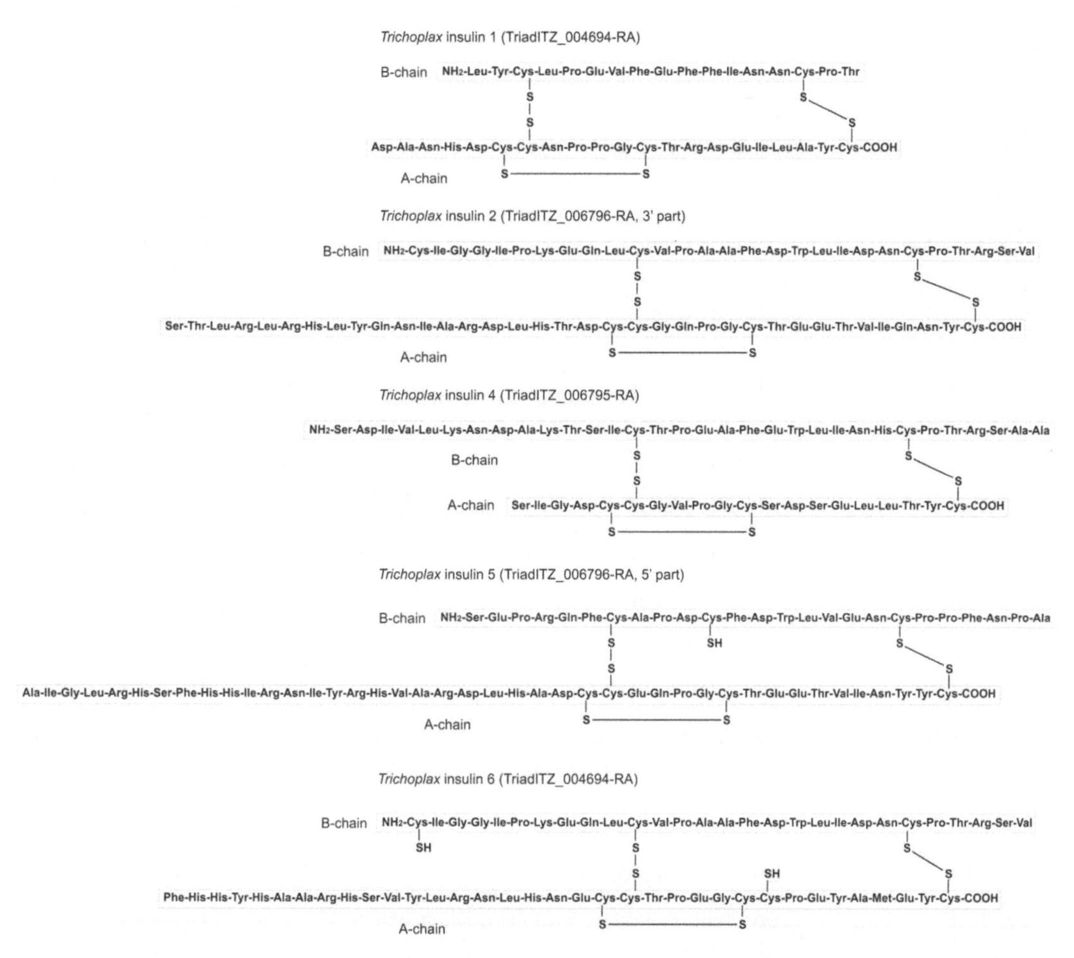

Fig. 3 Putative structures of mature *Trichoplax adhaerens* insulins. TrIns3 is not shown because it is missed in both genome annotations and was found only in the transcriptome [35]

We compiled an alignment of insulins found in *Trichoplax* (including TrIns3 found in the transcriptome in [43] but missed in both genome annotations) with all other insulins from our set of prohormones and added some more insulins from Cnidaria and Mollusca (complete list in Table 3). On the phylogenetic tree (Fig. 4), all placozoan insulins form a monophyletic clade sister to the insulins of *Hydra* and other cnidarians.

Similarly, we used TriadITZ_010167 (similar to temptin), TriadITZ_001732, and TriadITZ_002192 (similar to ependymin-related proteins) as queries to find other members of these families among the *Trichoplax* proteins. We have found 9 ependymins in total for *Trichoplax adhaerens* (Table 2). On the phylogenetic tree, they all (with 10 ependymins from *Hoilungia hongkongensis*) form isolated clade without affinities to any of ependymin-related proteins from various animal phyla (Fig. 5).

Table 3
List of proteins used for phylogenetic tree reconstructions

Accession or gene ID	Protein name	Species
Prohormone convertases		
braker1_g01213.t1	predicted protein	*Hoilungia hongkongensis*
braker1_g01217.t1	predicted protein	*Hoilungia hongkongensis*
braker1_g08092.t1	predicted protein	*Hoilungia hongkongensis*
ML07022a	predicted protein	*Mnemiopsis leydi*
ML11682a	predicted protein	*Mnemiopsis leydi*
ML200226a	predicted protein	*Mnemiopsis leydi*
ML20261a	predicted protein	*Mnemiopsis leydi*
ML207912a	predicted protein	*Mnemiopsis leydi*
ML279823a	predicted protein	*Mnemiopsis leydi*
NP_001191568.1	FUR protein precursor	*Aplysia californica*
P09958.2	Furin	*Homo sapiens*
P16519.2	Neuroendocrine convertase 2	*Homo sapiens*
P29120.2	Neuroendocrine convertase 1	*Homo sapiens*
P29122.1	Proprotein convertase subtilisin/kexin type 6	*Homo sapiens*
Q16549.2	Proprotein convertase subtilisin/kexin type 7	*Homo sapiens*
Q6UW60.2	Proprotein convertase subtilisin/kexin type 4	*Homo sapiens*
Q92824.4	Proprotein convertase subtilisin/kexin type 5	*Homo sapiens*
sb3462183	predicted protein	*Pleurobrachia bachei*
sb3464607	predicted protein	*Pleurobrachia bachei*
sb3466502	predicted protein	*Pleurobrachia bachei*
scpid25436	predicted protein	*Sycon ciliatum*
scpid28547	predicted protein	*Sycon ciliatum*
scpid34651	predicted protein	*Sycon ciliatum*
scpid35388	predicted protein	*Sycon ciliatum*
scpid44026	predicted protein	*Sycon ciliatum*
scpid74752	predicted protein	*Sycon ciliatum*

(continued)

Table 3
(continued)

Accession or gene ID	Protein name	Species
XP_001639811.2	neuroendocrine convertase 1	*Nematostella vectensis*
XP_001742149.1	protein MONBRDRAFT 30798	*Monosiga brevicollis MX1*
XP_001743348.1	protein MONBRDRAFT 14515, partial	*Monosiga brevicollis MX1*
TriadITZ_003258-RA	Trichoplax adhaerens predicted protein	*Trichoplax adhaerens*
TriadITZ_003262-RA	Trichoplax adhaerens predicted protein	*Trichoplax adhaerens*
TriadITZ_008375-RA	Trichoplax adhaerens predicted protein	*Trichoplax adhaerens*
XP_002166571.1	PREDICTED: furin-like protease kpc1	*Hydra vulgaris*
XP_004994196.1	prohormone convertase 1	*Salpingoeca rosetta*
XP_004998978.1	protease PC6 isoform A	*Salpingoeca rosetta*
XP_005098392.1	PREDICTED: proprotein convertase subtilisin/kexin type 7-like	*Aplysia californica*
XP_009048793.1	hypothetical protein LOTGIDRAFT 112524	*Lottia gigantea*
XP_009065921.1	hypothetical protein LOTGIDRAFT 81964, partial	*Lottia gigantea*
XP_011404106.1	PREDICTED: PC3-like endoprotease variant B	*Amphimedon queenslandica*
XP_011405480.2	PREDICTED: furin1-like	*Amphimedon queenslandica*
XP_011407660.1	PREDICTED: neuroendocrine convertase 1-like	*Amphimedon queenslandica*
XP_011450807.2	neuroendocrine convertase 1-like isoform X1	*Crassostrea gigas*
XP_013095075.1	PREDICTED: neuroendocrine convertase 1-like isoform X2	*Biomphalaria glabrata*
XP_019852355.1	PREDICTED: neuroendocrine convertase 1-like	*Amphimedon queenslandica*
XP_019853479.1	PREDICTED: neuroendocrine convertase 1-like	*Amphimedon queenslandica*
XP_019855384.1	PREDICTED: proprotein convertase subtilisin/kexin type 7-like	*Amphimedon queenslandica*
XP_019855895.1	PREDICTED: proprotein convertase subtilisin/kexin type 4-like	*Amphimedon queenslandica*

(continued)

Table 3
(continued)

Accession or gene ID	Protein name	Species
XP_021372747.1	neuroendocrine convertase 1-like	*Mizuhopecten yessoensis*
XP_022785006.1	neuroendocrine convertase 1-like	*Stylophora pistillata*
XP_022803926.1	PC3-like endoprotease variant B	*Stylophora pistillata*
XP_027053636.1	proprotein convertase subtilisin/kexin type 6-like	*Pocillopora damicornis*
XP_029186554.1	PC3-like endoprotease variant B	*Acropora millepora*
XP_029194159.1	proprotein convertase subtilisin/kexin type 5-like isoform X1	*Acropora millepora*
XP_029206484.1	neuroendocrine convertase 1-like	*Acropora millepora*
XP_029634454.1	proprotein convertase subtilisin/kexin type 4-like isoform X2	*Octopus vulgaris*
XP_029636186.1	proprotein convertase subtilisin/kexin type 7-like	*Octopus vulgaris*
XP_031553653.1	PC3-like endoprotease variant B	*Actinia tenebrosa*
XP_031571001.1	proprotein convertase subtilisin/kexin type 4-like isoform X2	*Actinia tenebrosa*
XP_031573618.1	proprotein convertase subtilisin/kexin type 6-like	*Actinia tenebrosa*
XP_034302009.1	proprotein convertase subtilisin/kexin type 7	*Crassostrea gigas*
XP_034325707.1	furin-like protease kpc1 isoform X1	*Crassostrea gigas*
Amidating monnoxygenases		
AFP99909.1	peptidylglycine alpha-amidating monooxygenase 1	*Acropora millepora*
AFP99911.1	peptidyl alpha-hydroxylating monooxygenase	*Acropora millepora*
braker1_g02349.t1	predicted protein	*Hoilungia hongkongensis*
braker1_g04121.t1	predicted protein	*Hoilungia hongkongensis*
EDO34395.1	predicted protein	*Nematostella vectensis*
EDO34601.1	predicted protein	*Nematostella vectensis*

(continued)

Table 3
(continued)

Accession or gene ID	Protein name	Species
EDO42139.1	predicted protein	*Nematostella vectensis*
ML00116a	predicted protein	*Mnemiopsis leydi*
ML02634a	predicted protein	*Mnemiopsis leydi*
P19021.2	Peptidylglycine alpha-amidating monooxygenase	*Homo sapiens*
sb3461822	predicted protein	*Pleurobrachia bachei*
sb3462501	predicted protein	*Pleurobrachia bachei*
scpid29202	predicted protein	*Sycon ciliatum*
scpid34778	predicted protein	*Sycon ciliatum*
scpid51609	predicted protein	*Sycon ciliatum*
TriadITZ_002979-RA	predicted protein	*Trichoplax adhaerens*
TriadITZ_003133-RA	predicted protein	*Trichoplax adhaerens*
XP_002160828.2	PREDICTED: peptidylglycine alpha-amidating monooxygenase-like, partial	*Hydra vulgaris*
XP_003388865.3	PREDICTED: peptidylglycine alpha-amidating monooxygenase-like	*Amphimedon queenslandica*
XP_011417387.2	peptidylglycine alpha-amidating monooxygenase isoform X1	*Crassostrea gigas*
XP_012944541.1	PREDICTED: probable peptidylglycine alphahydroxylating monooxygenase Y71G12B.4	*Aplysia californica*
XP_014771980.1	PREDICTED: peptidylglycine alpha-amidating monooxygenase B-like	*Octopus bimaculoides*
XP_020916085.1	peptidylglycine alpha-amidating monooxygenase B	*Exaiptasia pallida*
XP_020916094.1	peptidylglycine alpha-amidating monooxygenase isoform X1	*Exaiptasia pallida*
XP_021342878.1	peptidylglycine alpha-amidating monooxygenase-like	*Mizuhopecten yessoensis*
XP_022792700.1	peptidylglycine alpha-amidating monooxygenase B-like isoform X1	*Stylophora pistillata*
XP_027044794.1	peptidylglycine alpha-amidating monooxygenase-like isoform X1	*Pocillopora damicornis*
XP_029202965.1	peptidylglycine alpha-amidating monooxygenase B-like	*Acropora millepora*

(continued)

Table 3
(continued)

Accession or gene ID	Protein name	Species
XP_031553478.1	peptidylglycine alpha-amidating monooxygenase B-like isoform X1	*Actinia tenebrosa*
XP_031572438.1	peptidylglycine alpha-amidating monooxygenase-like	*Actinia tenebrosa*
XP_032239436.1	peptidylglycine alpha-amidating monooxygenase A	*Nematostella vectensis*
XP_033749526.1	peptidylglycine alpha-amidating monooxygenase-like isoform X1	*Pecten maximus*
Carboxypeptidases		
AAC47412.1	carboxypeptidase E	*Aplysia californica*
braker1_g00958	predicted protein	*Hoilungia hongkongensis*
braker1_g01018	predicted protein	*Hoilungia hongkongensis*
braker1_g03126	predicted protein	*Hoilungia hongkongensis*
braker1_g08971	predicted protein	*Hoilungia hongkongensis*
braker1_g08973	predicted protein	*Hoilungia hongkongensis*
ML02203a	predicted protein	*Mnemiopsis leydi*
ML214323a	predicted protein	*Mnemiopsis leydi*
NP_001191551.1	carboxypeptidase D precursor	*Aplysia californica*
O75976.2	Carboxypeptidase D	*Homo sapiens*
P14384.2	Carboxypeptidase M	*Homo sapiens*
P15169.1	Carboxypeptidase N	*Homo sapiens*
P16870.1	Carboxypeptidase E	*Homo sapiens*
Q66K79.2	Carboxypeptidase Z	*Homo sapiens*
XP_002117289.1	predicted protein	*Trichoplax adhaerens*
sb3461983	predicted protein	*Pleurobrachia bachei*
sb3464191	predicted protein	*Pleurobrachia bachei*
scpid13267_Sycon	predicted protein	*Sycon ciliatum*
scpid23739_Sycon	predicted protein	*Sycon ciliatum*

(continued)

Table 3
(continued)

Accession or gene ID	Protein name	Species
scpid70932_Sycon	predicted protein	*Sycon ciliatum*
XP_001624056.2	carboxypeptidase D	*Nematostella vectensis*
XP_001633785.2	carboxypeptidase D	*Nematostella vectensis*
XP_001638948.2	carboxypeptidase D	*Nematostella vectensis*
XP_001639021.2	carboxypeptidase D	*Nematostella vectensis*
XP_001747863.1	uncharacterized protein MONBRDRAFT 27454, partial	*Monosiga brevicollis MX1*
XP_002108045.1	predicted protein	*Trichoplax adhaerens*
XP_002112076.1	predicted protein	*Trichoplax adhaerens*
XP_002112103.1	predicted protein	*Trichoplax adhaerens*
XP_002114978.1	predicted protein	*Trichoplax adhaerens*
XP_002117288.1	predicted protein	*Trichoplax adhaerens*
XP_002117835.1	predicted protein	*Trichoplax adhaerens*
XP_002158333.2	carboxypeptidase D-like	*Hydra vulgaris*
XP_002169920.3	carboxypeptidase D-like	*Hydra vulgaris*
XP_003384134.1	PREDICTED: carboxypeptidase D-like	*Amphimedon queenslandica*
XP_004211138.2	carboxypeptidase D-like	*Hydra vulgaris*
XP_004991969.1	carboxypeptidase H	*Salpingoeca rosetta*
XP_009051142.1	hypothetical protein LOTGIDRAFT 226642	*Lottia gigantea*
XP_011428565.2	carboxypeptidase E	*Crassostrea gigas*
XP_012555394.1	carboxypeptidase D-like isoform X1	*Hydra vulgaris*
XP_012557491.1	carboxypeptidase D-like	*Hydra vulgaris*
XP_012943223.1	carboxypeptidase D-like	*Aplysia californica*
XP_012945414.1	carboxypeptidase M-like	*Aplysia californica*

(continued)

Table 3 (continued)

Accession or gene ID	Protein name	Species
XP_029180622.1	carboxypeptidase D-like	*Acropora millepora*
XP_029190811.1	carboxypeptidase D-like	*Acropora millepora*
XP_029191340.1	carboxypeptidase D-like	*Acropora millepora*
XP_029195890.1	carboxypeptidase D-like	*Acropora millepora*
XP_029195891.1	carboxypeptidase D-like	*Acropora millepora*
XP_029212572.1	carboxypeptidase D-like	*Acropora millepora*
XP_032219153.1	carboxypeptidase D	*Nematostella vectensis*
XP_032238532.1	carboxypeptidase D	*Nematostella vectensis*
XP_034299322.1	carboxypeptidase D	*Crassostrea gigas*
XP_034318111.1	carboxypeptidase D	*Crassostrea gigas*
Insulins		
ARS73224.1	insulin 4 partial	*Deroceras reticulatum*
ABU82759.1	insulin-like peptide 1 precursor	*Tritonia diomedea*
ANC47992.1	insulin-like protein 1 partial	*Conus monile*
AQS80518.1	insulin-like 2A precursor	*Charonia tritonis*
XP_022289108.1	conIns Im2-like	*Crassostrea virginica*
XP_029634273.1	conIns Im1-like	*Octopus vulgaris*
XP_021355405.1	Insulin-like	*Mizuhopecten yessoensis*
P01308.1	insulin	*Homo sapiens*
P01344.1	Insulin-like growth factor II	*Homo sapiens*
P05019.1	Insulin-like growth factor I	*Homo sapiens*
AAF50205.1	Insulin-like peptide 1	*Drosophila melanogaster*
AAF50202.1	Insulin-like peptide 4	*Drosophila melanogaster*

(continued)

Table 3
(continued)

Accession or gene ID	Protein name	Species
NP_648360.2	Insulin-like peptide 3	*Drosophila melanogaster*
EFA02918.1	insulin-like peptide	*Tribolium castaneum*
EFA02796.1	insulin-like peptide	*Triboilum castaneum*
EEZ99258.1	insulin-like peptide	*Triboilum castaneum*
AAF50204.1	Insulin-like peptide 2	*Drosophila melanogaster*
XP_001632788.2	Con-Ins G1c	*Nematostella vectensis*
XP_031565084.1	conIns T1-like isoform X1	*Actinia tenebrosa*
XP_029199002.1	uncharacterized protein LOC114963903 isoform X1	*Acropora millepora*
XP_020629342.1	Bombyxin-related peptide B-like isoform X1	*Orbicella faveolata*
XP_032220293.1	uncharacterized protein LOC5519428	*Nematostella vectensis*
XP_031572344.1	uncharacterized protein LOC116306422 isoform X1	*Actinia tenebrosa*
XP_028516297.1	uncharacterized protein LOC110243949	*Exaiptasia pallida*
XP_022786957.1	molluscan insulin-related peptide 1-like	*Stylophora pistillata*
NP_001267870.1	uncharacterized LOC100205786 precursor	*Hydra vulgaris*
A0A0B5ADU4.1	Con-Ins T1	*Conus tulipa*
NP_001267872.1	insulin-like peptide 2 precursor	*Hydra vulgaris*
NP_001267878.1	uncharacterized LOC100202208 precursor	*Hydra vulgaris*
TriadITZ_004694-RA	predicted protein	*Trichoplax adhaerens*
TriadITZ_006796-RA	predicted protein	*Trichoplax adhaerens*
H13_braker1_g07431.t1	predicted protein	*Hoilungia hongkongensis*
TriadITZ_006796-RA_2	predicted protein	*Trichoplax adhaerens*
braker1_g07432.t1	predicted protein	*Hoilungia hongkongensis*

(continued)

Table 3
(continued)

Accession or gene ID	Protein name	Species
TriadITZ_006795RA	predicted protein	*Trichoplax adhaerens*
H1_TrIns3_isoform_a	predicted protein	*Trichoplax adhaerens*
H1_TrIns3_isoform_b	predicted protein	*Trichoplax adhaerens*
braker1_g03237.t1	predicted protein	*Hoilungia hongkongensis*
TriadITZ_004547-RA	predicted protein	*Trichoplax adhaerens*
XP_975459.1	similar to Ilp7 CG13317-PA	*Tribolium castaneum*
Q9Y581.2	insulin-like peptide INSL6 precursor	*Homo sapiens*
P04808.1	relaxin	*Homo sapiens*
Ependymin-like proteins		
AAK26441.1	ependyminrelated protein-1	*Homo sapiens*
braker1_g01593.t1	predicted protein	*Hoilungia hongkongensis*
braker1_g01615.t1	predicted protein	*Hoilungia hongkongensis*
braker1_g01629.t1	predicted protein	*Hoilungia hongkongensis*
braker1_g03807.t1	predicted protein	*Hoilungia hongkongensis*
braker1_g03808.t1	predicted protein	*Hoilungia hongkongensis*
braker1_g03809.t1	predicted protein	*Hoilungia hongkongensis*
braker1_g03810.t1	predicted protein	*Hoilungia hongkongensis*
braker1_g03811.t1	predicted protein	*Hoilungia hongkongensis*
braker1_g03812.t1	predicted protein	*Hoilungia hongkongensis*
braker1_g03814.t1	predicted protein	*Hoilungia hongkongensis*

(continued)

Table 3 (continued)

Accession or gene ID	Protein name	Species
NP_001164689.1	Ependymin-like protein precursor	*Saccoglossus kowalevskii*
NP_001191622.1	ependymin-related protein precursor	*Aplysia californica*
scpid101243	Ependymin	*Sycon ciliatum*
scpid75046	Mammalian ependymin-related protein 1	*Sycon ciliatum*
scpid81382	Mammalian ependymin-related protein 1	*Sycon ciliatum*
scpid81637	Mammalian ependymin-related protein 1	*Sycon ciliatum*
scpid83742	Mammalian ependymin-related protein 1	*Sycon ciliatum*
scpid86964	Mammalian ependymin-related protein 1	*Sycon ciliatum*
scpid92940	Mammalian ependymin-related protein 1	*Sycon ciliatum*
scpid94850	Mammalian ependymin-related protein 1	*Sycon ciliatum*
TriadITZ_001728-RA	predicted protein	*Trichoplax adhaerens*
TriadITZ_001729-RA	predicted protein	*Trichoplax adhaerens*
TriadITZ_001730-RA	predicted protein	*Trichoplax adhaerens*
TriadITZ_001732-RA	predicted protein	*Trichoplax adhaerens*
TriadITZ_001734-RA	predicted protein	*Trichoplax adhaerens*
TriadITZ_002192-RA	predicted protein	*Trichoplax adhaerens*
TriadITZ_002208-RA	predicted protein	*Trichoplax adhaerens*
TriadITZ_011126-RA	predicted protein	*Trichoplax adhaerens*
TriadITZ_011128-RA	predicted protein	*Trichoplax adhaerens*
XP_005099896.1	PREDICTED: mammalian ependymin-related protein 1-like	*Aplysia californica*
XP_005099911.1	PREDICTED: mammalian ependymin-related protein 1-like	*Aplysia californica*
XP_011413899.2	mammalian ependymin-related protein 1-like	*Crassostrea gigas*
XP_011413900.2	mammalian ependymin-related protein 1	*Crassostrea gigas*
XP_011413901.2	mammalian ependymin-related protein 1	*Crassostrea gigas*

(continued)

Table 3 (continued)

Accession or gene ID	Protein name	Species
XP_011452850.1	mammalian ependymin-related protein 1-like	*Crassostrea gigas*
XP_012934740.1	PREDICTED: mammalian ependymin-related protein 1-like	*Aplysia californica*
XP_012939903.1	PREDICTED: mammalian ependymin-related protein 1-like	*Aplysia californica*
XP_012942722.1	PREDICTED: mammalian ependymin-related protein 1-like isoform X1	*Aplysia californica*
XP_020604257.1	mammalian ependymin-related protein 1-like	*Orbicella faveolata*
XP_020604379.1	mammalian ependymin-related protein 1-like, partial	*Orbicella faveolata*
XP_022329571.1	mammalian ependymin-related protein 1-like	*Crassostrea virginica*
XP_022794351.1	mammalian ependymin-related protein 1-like	*Stylophora pistillata*
XP_030829649.1	uncharacterized protein LOC581361	*Strongylocentrotus purpuratus*
XP_031564114.1	mammalian ependymin-related protein 1-like	*Actinia tenebrosa*
XP_031564115.1	mammalian ependymin-related protein 1-like	*Actinia tenebrosa*
XP_031564127.1	mammalian ependymin-related protein 1-like	*Actinia tenebrosa*
XP_034303021.1	mammalian ependymin-related protein 1	*Crassostrea gigas*
XP_034326212.1	mammalian ependymin-related protein 1-like	*Crassostrea gigas*

3.2 Search for Prohormone-Processing Enzymes

Several enzymes must process all predicted peptide precursors to produce mature peptides. These enzymes include prohormone convertases, carboxypeptidases, and peptidylglycine alpha-amidating monooxygenases. We should check if these enzymes are present in *Trichoplax* before using the convertase recognition sites as criteria for non-homology-based peptide predictions. We will use the following human enzymes as BLAST queries:

- Neuroendocrine converatases PC1/3 (NP_000430), PC2 (NP_002585), and furin (NP_002560)
- Carboxypeptidases D and E (O75976.2, P16870.1)
- Amidating monooxygenase P19021.2

BLAST search of *Trichoplax adhaerens* proteins (annotation TriadV2) returns three putative convertases, seven carboxypeptidases, and two monooxygenases. *Hoilungia hongkongensis* have an inventory of neuropeptide-processing enzymes broadly similar to *T. adhaerens*; the only difference is fewer carboxypeptidases (5 vs.

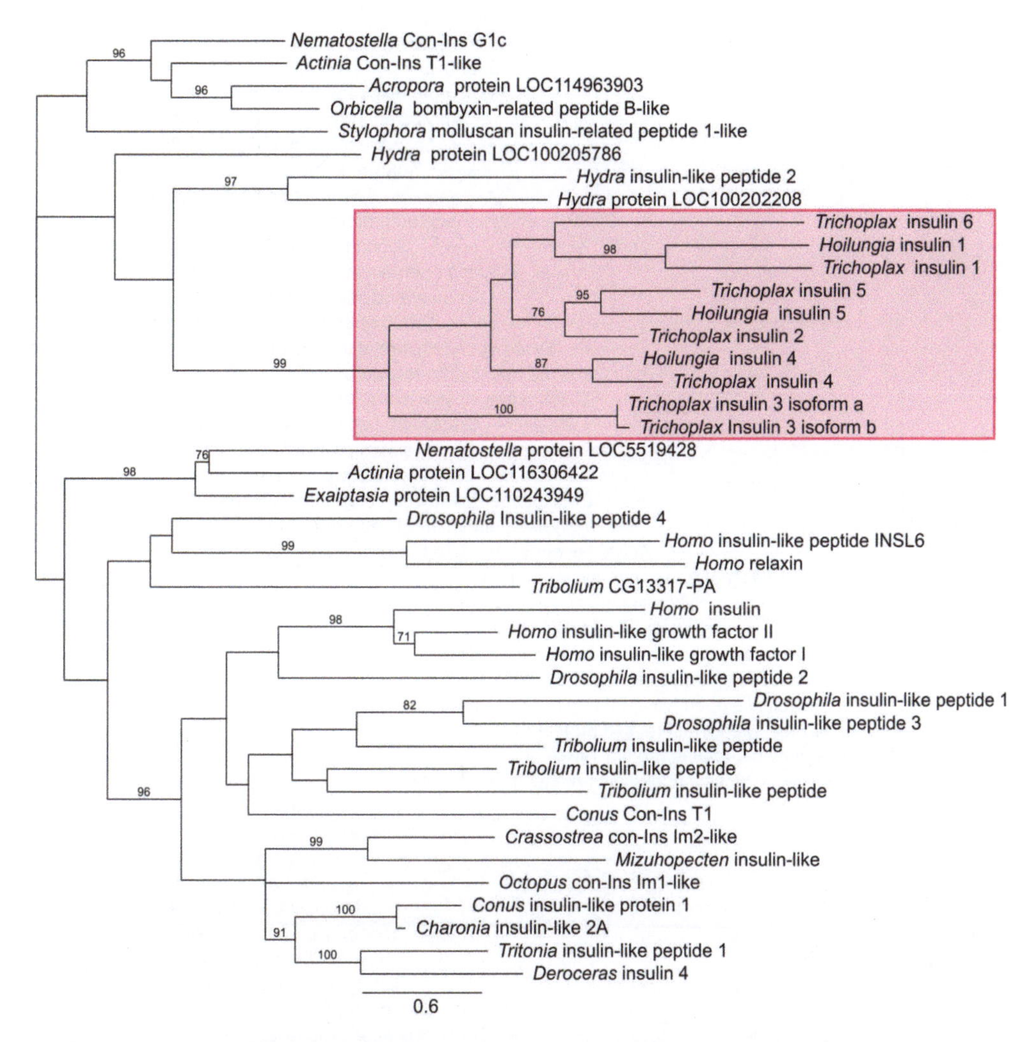

Fig. 4 Maximum likelihood phylogenetic tree of animal insulins. Placozoan insulins are highlighted in the pink box. Protein sequences were aligned in MAFFT Online (https:/mafft.cbrc.jp/alignment/server/) [36], and the phylogenetic tree was built using IQTree [37] with 1000 replicas of ultrafast bootstrap approximation. Bootstrap support values are listed near the tree nodes

7). Using a similar BLAST search, we retrieve homologous enzymes from representatives of cnidarians, mollusks, sponges (*Amphimedon queenslandica* and *Sycon ciliatum*), and ctenophores (*Pleurobrachia bachei* and *Mnemiopsis leidyi*). All found proteins were aligned in MAFFT [44], and phylogenetic trees were built using IQTree [45].

On the phylogenetic tree of prohormone convertases (Fig. 6), three placozoan convertases are grouped with distinct subtypes of convertases from Bilateria. One placozoan convertase (TriadITZ_003262-RA) is within the neuroendocrine-like convertases, including human PC1/3 and PC2, second

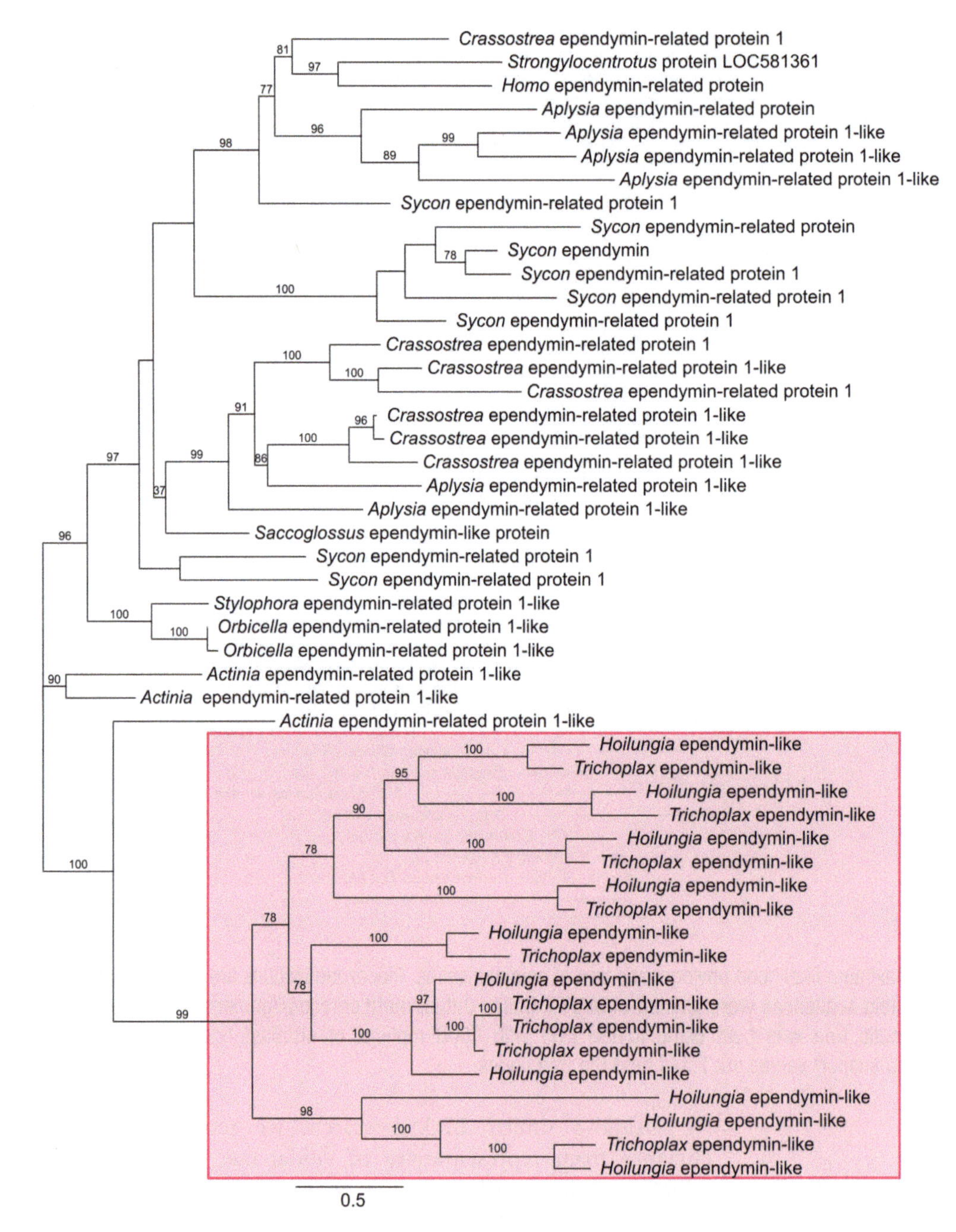

Fig. 5 Maximum likelihood phylogenetic tree of animal ependymins and ependymin-like proteins. Placozoan proteins are highlighted in the pink box. Protein sequences were aligned in MAFFT Online (https:/mafft.cbrc.jp/alignment/server/) [36], and the phylogenetic tree was built using IQTree [37] with 1000 replicas of ultrafast bootstrap approximation. Bootstrap support values are listed near the tree nodes

(TriadITZ_008375-RA)—with PCSK7-like proprotein convertases, and third (TriadITZ_003258-RA)—with PCSK6-like proprotein convertases.

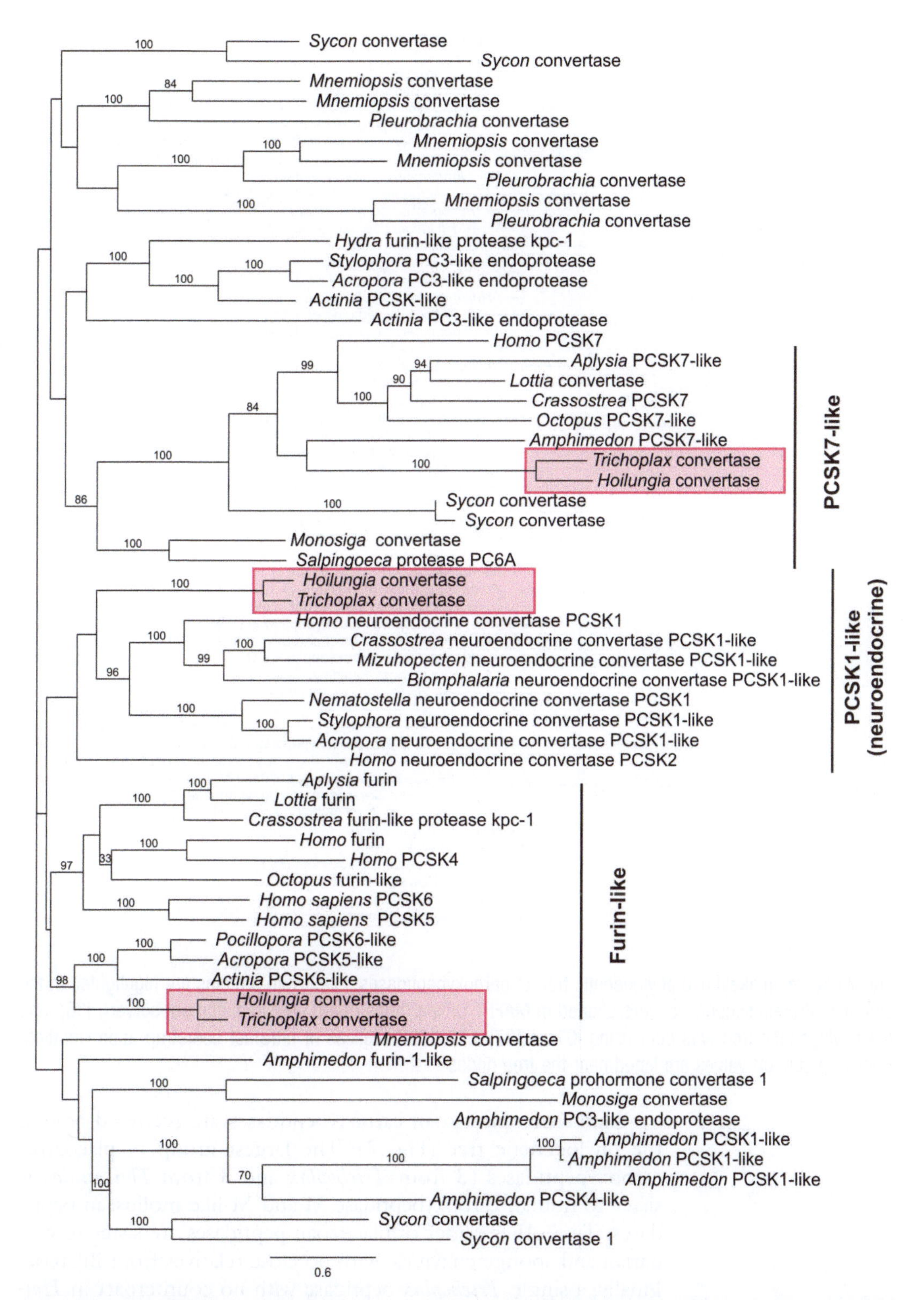

Fig. 6 Maximum likelihood phylogenetic tree of prohormone convertases and related enzymes. Placozoan proteins are highlighted in the pink box. Protein sequences were aligned in MAFFT Online (https:/mafft.cbrc.jp/alignment/server/) [36], and the phylogenetic tree was built using IQTree [37] with 1000 replicas of ultrafast bootstrap approximation. Bootstrap support values are listed near the tree nodes

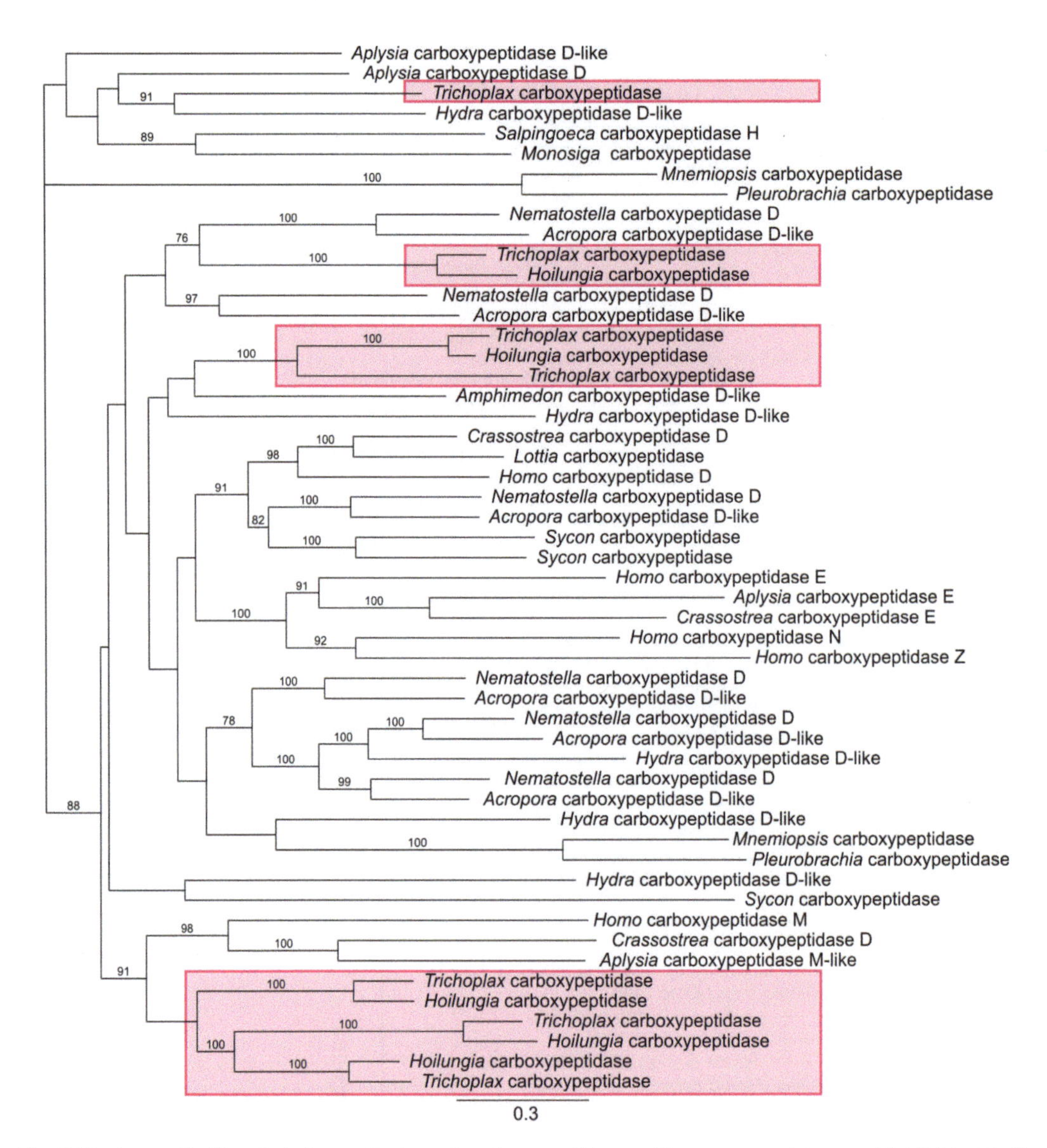

Fig. 7 Maximum likelihood phylogenetic tree of carboxypeptidases. Placozoan proteins are highlighted in the pink box. Protein sequences were aligned in MAFFT Online (https:/mafft.cbrc.jp/alignment/server/) [36], and the phylogenetic tree was built using IQTree [37] with 1000 replicas of ultrafast bootstrap approximation. Bootstrap support values are listed near the tree nodes

Numerous placozoan carboxypeptidases are scattered among the phylogenetic tree (Fig. 7). The largest group of placozoan carboxypeptidases (3 from *Trichoplax* and 3 from *Hoilungia*) is sister to human carboxypeptidase M and M-like molluscan peptidases. Two other clades of placozoan peptidases are sister to cnidarian and sponge enzymes, with no close relatives from Bilateria. Finally, a single *Trichoplax* peptidase with no counterpart in *Hoilungia* has grouped with *Hydra* and *Aplysia* peptidases annotated as D-like but very distant from human carboxypeptidase D.

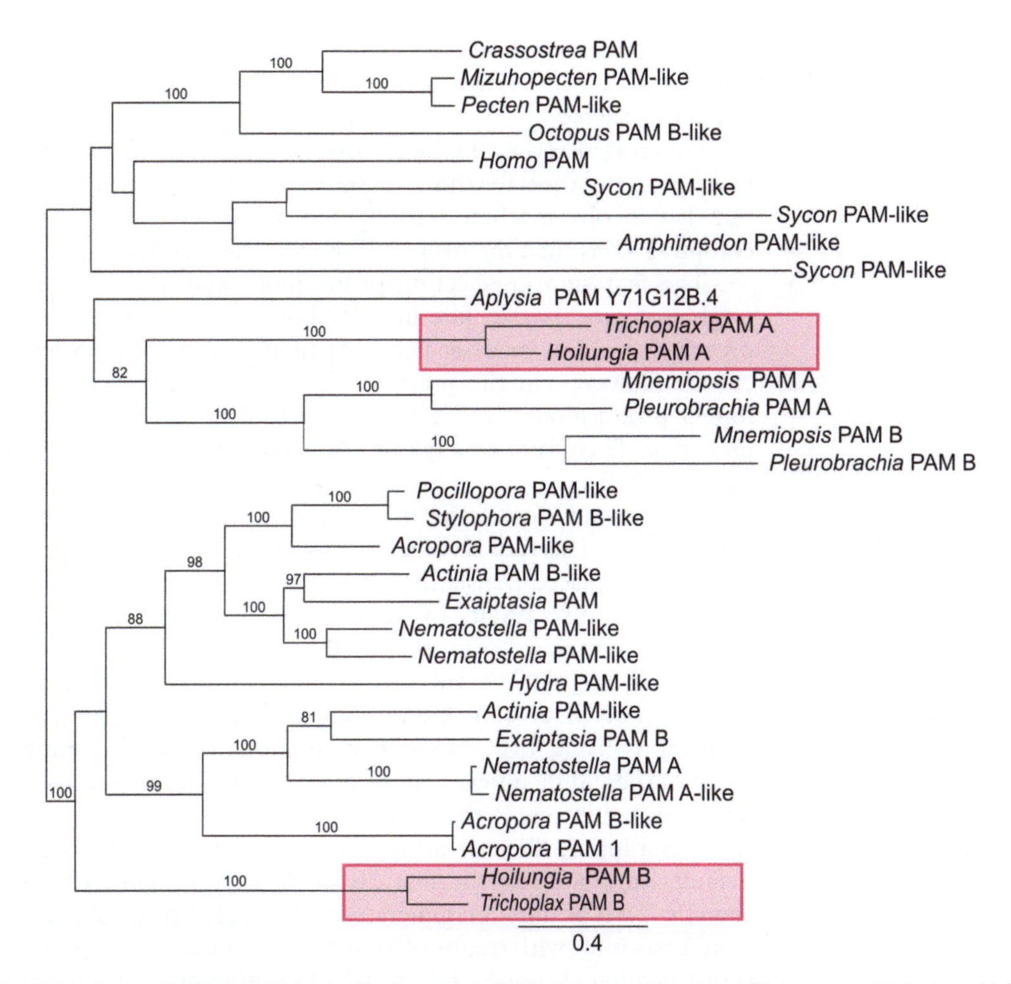

Fig. 8 Maximum likelihood phylogenetic tree of amidating monooxygenases. Placozoan proteins are highlighted in the pink box. Protein sequences were aligned in MAFFT Online (https:/mafft.cbrc.jp/alignment/server/) [36], and the phylogenetic tree was built using IQTree [37] with 1000 replicas of ultrafast bootstrap approximation. Bootstrap support values are listed near the tree nodes

Two placozoan-amidating monooxygenases bear little similarity to each other. In the phylogenetic tree (Fig. 8), one of the placozoan monooxygenases (TriadITZ_003133-RA) is sister to a large group of cnidarian monooxygenases, while other (TriadITZ_002979-RA) grouped with Ctenophora sequences.

From these datasets, we can conclude that placozoans possess the full set of prohormone-processing enzymes, including convertases, carboxypeptidases, and amidating monooxygenases.

3.3 Non-Homology-Based Predictions

3.3.1 Construction of Filtered Secretome Database

To detect prohormones without detectable similarity to sequences from other phyla, we have to use the general hallmarks of neuropeptide precursors: the presence of N-end signal peptide, convertase cleavage, and multiple paracopies of mature peptides between these sites. The protocol for this type of analysis has been developed for *Drosophila melanogaster* [21]. However, the reproduction of this analysis is complicated because scripts for this protocol are not publicly available.

We created several Python scripts for this analysis. There are more powerful tools for this analysis, for example, NeuroPID [46], but they involve machine learning and require considerably more computer skills than the protocols used in this chapter.

The first step in selecting prohormone candidates from the total proteome is size filtering. Prohormones are usually short proteins; previous analysis [17, 21] used 250 or 300 aa length cutoff. However, in the previous chapter, we have found much longer prohormones in *Trichoplax*, up to 649 aa for SIFGamide precursor. Therefore, we suggest using a more conservative length cutoff of 700 aa.

The second step is the prediction of a signal peptide using SignalP [40]. This step is vulnerable to less efficient protein prediction, as a misplaced translation start can lead to both false negative and false positive results of SignalP. Therefore, we can pass all *Trichoplax* proteins shorter than 700 aa in the first round through the SignalP analysis. We will then repeat downstream analysis using a total <700 aa proteome and compare our results. Comparing predictions using two different annotations could also mitigate some of these difficulties. Length filtering and SignalP analysis are implemented in one Python script signalp_fasta.py. It accepts protein Fasta input, filters it using defined length cutoff (700 aa by default), and subjects the first 70 aa of each protein to SignalP analysis. Two outfiles are generated: CSV table listing all proteins from Fasta infile with results of SignalP (D-score and most probable signal peptide cleavage site), and Fasta outfile with secreted proteins (D-score > 0.5) after removal of signal peptides. We have 588 proteins in outfile for annotation Triad1 [8] and 691 for annotation TriadV2 [23].

This list of secreted proteins should be further refined by removing proteins with known functions not related to secretory peptide signaling. This is done using BLAST search in SwissProt database: a manually curated collection of experimentally confirmed proteins. We perform local BLAST search versus downloaded SwissProt database:

```
blastp -db swissprot -evalue 10E-10 -out
triadv2_secretome691_vs_swissprot.tsv -num_threads 2 -outfmt
'6 qaccver saccver stitle pident length mismatch gapopen qstart
qend sstart send evalue bitscore' -max_target_seqs 2 -query
triadv2_secretome691.fasta
```

E-value is set to 10^{-10} to exclude spurious hits. The output format is set to tabular. We will need the titles of proteins from BLAST database, which is not written to tabular BLAST out by default. Therefore, option -outfmt '6 qaccver saccver stitle pident length mismatch gapopen qstart qend sstart send evalue bitscore' adds field "subject title" to the third column of output (*see* **Note 1**).

When BLAST versus Swissprot is finished, its tabular output is used for filtering the secretome. Script blastfilter.py does this job. It reads Fasta file with secreted proteins and writes to outfile only those proteins that either (1) do not have BLAST hits in SwissProt database or (2) have BLAST hits whose title contains words "neuropeptide," "hormone," or "uncharacterized" (*see* **Note 2**):

```
blastfilter.py triadv2_secretome691.fasta
triadv2_secretome691_vs_swissprot.tsv triadv2_secretome691.
filtered.fasta
```

This command leaves 260 and 299 proteins in the filtered Fasta file for annotations Triad1 and TriadV2, respectively. We can split them into short peptides and use them to search for paracopies.

3.3.2 Prediction of Propeptide Cleavage and Search for Paracopies

Script cleave_propeptide.py does this. It finds the most common prohormone cleavage sites recognized by convertases (RR, RK, KR, KK, and RxGR, *see* **Note 3**), cut protein sequence at these sites, and filters resulting peptides by length. Only peptides longer than 5 and shorter than 30 amino acids (including C-end basic residues which are usually cleaved by carboxypeptidase) are saved. Peptide sequences are saved as separate sequences in Fasta file. The start and end positions of a peptide on its precursor are appended to each peptide name.

```
cleave_propeptide.py triadv2_secretome691.filtered.fasta
triadv2_sec691_peptides.fasta
```

At this point, we have 463 and 476 peptides from 144 and 161 precursors for annotations V1 and V2, respectively. Other precursors did not have any peptides of appropriate length. Now we are ready to make BLAST database from filtered secretome and search it using predicted short peptides as queries:

```
makeblastdb -dbtype prot -in triadv2_secretome691.filtered.
fasta -out triadv2_sec691     blastp -task blastp-short -db
triadv2_sec691 -evalue 1 -out triadv2-691_peps_vs_secre-
tome.tsv -num_threads 2 -outfmt 6 -max_target_seqs
1 -comp_based_stats 0 -query triadv2_sec691_peptides.fasta
```

BLAST is run with parameters optimized for short queries (-task blastp-short), tabular output (-outfmt 6), and no more than one subject sequence for each query (-max_target_seqs 1). Also, we disable the composition-based calculations of E-value (-comp_based_stats 0), which otherwise will increase E-value for 1–2 orders for compositionally biased sequences. Most short peptides are compositionally biased and will be missed unless we disable composition-based statistics. If several paracopies of predicted peptides exist within precursor protein, then several BLAST hits for this peptide query will be reported. Otherwise, we could see only one BLAST hit with 100% similarity (to query peptide's own position within precursor). Script blastpep.py will filter the tabular BLAST output, leaving only queries with 2 or more hits.

```
cleave_propeptide.py triadv2_secretome691.filtered.fasta
triadv2_sec691_peptides.fasta
```

```
blastpep.py triadv2-691_peps_vs_secretome.tsv triadv2-
691_peps_vs_secr_filtered.tsv
```

Now we have the table triadv2-691_peps_vs_secr_filtered.tsv with peptides present in multiple paracopies. All of them are produced from 13 precursors. Manual inspection of each precursor sequence is required to verify it as a short peptide precursor. NeuroPred prediction of cleavage sites is useful here.

Three of 16 proteins in the table triadv2-691_peps_vs_secr_filtered.tsv (TriadITZ_001490, TriadITZ_002201, and TriadITZ_011126) have only two potential paracopies with a similarity of less than 50%, and only one of them is surrounded by cleavage sites. These are unlikely neuropeptide precursors. Four precursors (DQPPRWamide, SITFamide, FFNPamide, SAKIPME) were already found by homology, other 9 are putative new precursors (Table 2, Fig. 3), which we briefly describe here:

- Precursor TriadITZ_000027: 11 paracopies of DDSQGPLF peptides
- TriadITZ_007434: 8 paracopies of QGVLVDVPWN peptides
- TriadITZ_004550: 3 paracopies of FIT(F/L) peptides
- TriadITZ_004552: 5 paracopies of FYY/FFY peptides
- TriadITZ_004633: 2 paracopies of FSYYxxxxQxPEQQKKW-Wamide or (dependent on convertase specificity) FSYYxxxxQx-PEQQ, and 2 peptides with C-end KAQKWW
- TriadITZ_004786: 4 paracopies of FYI/FLY peptides

- TriadITZ_008272: 36 paracopies of EDQANLKSIFGamide peptide
- TriadITZ_008559: 6 paracopies of SFE(L/F)PE peptides
- TriadITZ_011042: 5 copies of AVVV peptide

In the analysis of Triad1 annotation, we could see 15 peptide precursors. Most of them are the same as in TriadV2. Some precursors in Triad1 are fragmented into two genes; for example, Triad1_63920 and 56030 are parts of TriadITZ_004633, and Triad1_55947 with 55948 are isoforms of TriadITZ_004552. Therefore, in Triad1, we can find (partial) copies of 10 out of 14 TriadV2 precursors, plus 2 additional:

- Triad1_56359: 13 paracopies of DDQQNKPYNGWPPF peptides. Present in TriadV2 as TriadITZ_004972 but was not detected because of misprediction of the translation start and undetected signal peptide.
- Triad1_63870: 7 copies of LYY peptides, completely absent in TriadV2 annotation, but can be found in chromosome using TblastN.

On the other hand, 3 precursors were found only in TriadV2 (TriadITZ_004971, 006520, 008272). This outcome was caused by their complete absence in Triad1 annotation (004971, 006520) or misprediction of the translation start causing failure to detect signal peptide (TriadITZ_008272).

3.3.3 Repeat of Analysis Without Signal Peptide Prediction

We can see that out of 16 peptide precursors found in two versions of *Trichoplax* genome annotation using multiple paracopies detection, 4 had issues with SignalP analysis due to mispredicted 5′ end. Therefore, we will repeat our analysis skipping the SignalP step, and hope to find some more peptide precursors.

Without filtering by SignalP, sequence lists grow ~tenfold: 2500 proteins after filtering TriadV2 through SwissProt, 5083 peptides for 1621 potential precursors after cleave_propeptide.py, and 48 possible precursors with at least 2 paracopies after blastpep.py. Sixteen of 48 were already found in Triad1 or TriadV2 using SignalP. Either other 30 proteins do not have sufficiently similar paracopies, or strong cleavage sites do not flank them according to NeuroPred predictions. There are only two new good precursors. One protein is endomorphin-like peptide precursor TriadITZ_009692 described from transcriptome by Smith et al. [25], absent in Triad1 and lacking a signal peptide in TriadV2. Second is GYDDYYYamide precursor TriadITZ_005098, found in the previous chapter by homology with *Drosophila* FMRFamide-related peptide precursor. Therefore, despite some annotation errors leading to signal peptide misprediction, discarding the SignalP step result only in marginally more discoveries at the cost of tenfold more manual analysis.

3.4 Evolutionary-Based Comparative Predictions

To compare predicted *T. adhaerens* peptide precursors with their counterparts from *Hoilungia*, we created BLAST database from the *Hoilungia hongkongensis* (haplotype H13) proteome:

```
makeblastdb -dbtype prot -out Hhon_prot -in
Hhon_BRAKER1_proteins.fasta
```

and search it using *T. adhaerens* length-filtered secreted proteins as queries:

```
blastp -db Hhon -evalue 10 -outfmt "6 qseqid qstart qend qlen
qseq sseqid sstart send slen sseq evalue" -max_target_seqs 5 -out
Triadv2_vs_Hhon.tsv -query TriadV2_secretome691.filtered.
fasta
```

Output format specification "-outfmt "6 qseqid qstart qend qlen qseq sseqid sstart send slen sseq evalue" is required by downstream analysis script PeptFischer.py. This script requires infile name to be entered in dialog mode, so we launch PeptFischer.py and enter infile name Triadv2_vs_Hhon.tsv (*see* **Note 4**). Two outfiles are created: Triadv2_vs_Hhon_out1.txt is the TSV table, and Triadv2_vs_Hhon_out2.txt is the human-readable text file similar to plain text BLAST output.

We further analyzed Triadv2_vs_Hhon_out1.txt in a spreadsheet editor such as MS Excel or LibreOffice Calc. It contains 724 BLAST hits. First filtering step by the segment length between dibasic sites (longin) in the range 371 hits passed 6–30 aa. After the second filtering by p-value below 0.1, only 93 hits to 51 proteins remain.

We extract sequences of these proteins from the Fasta file with all TriadV2 proteins and analyze them using web versions of TMHMM, NeuroPred, and NCBI BLAST versus NR database. Thirty-four out of 51 protein hits are unlikely peptide precursors because of one or more of the following reasons:

1. Presence of transmembrane regions (e.g., TriadITZ_001029)
2. Size over 300 amino acids but only a single hit to peptide precursor (e.g., TriadITZ_007584)
3. Cleavage sites do not align between *Trichoplax* and *Hoilungia* proteins (e.g., TriadITZ_003380)
4. High similarity to proteins with known functions (e.g., TriadITZ_006082—riboflavin-binding protein)

Among 17 remaining proteins, 6 TriadV2 peptide precursors were detected in previous chapters (SITFamide, FFNPamide, PWN, GYPLF, LFNE, and KAQKWW precursors). Nine other proteins are likely new precursors with a single copy of peptides (Fig. 4, Table 2). Two more new peptide precursors are missed in TriadV2 annotation but can be found in Triad1 (EDV23381.1 and EDV25674.1). Finally, 2 proteins (TriadITZ_008298, TriadITZ_002547) have all hallmarks of peptide precursors, but their pattern of Cys residues shows similarities to folded secreted proteins: WAP family protease inhibitors (TriadITZ_008298) and cystine knot proteins (growth factors and toxins, TriadITZ_002547).

3.5 Discussion

We have successfully predicted 33 secretory peptide products for basal metazoan *Trichoplax adhaerens*. This selection is a significant increase compared to prior knowledge: Nikitin [43] predicted 13 precursors (4 insulins and 9 small peptide precursors), Senatore et al. [47] had found an opioid peptide, and Varoqueaux et al. [48]—two precursors for peptides MFPF and ELPE.

It is important that we use three different methods. As illustrated in Fig. 9, there is little overlap between peptides found using different methods.

Most of the small peptides found do not have recognizable homologs outside Placozoa. A few exceptions are RWamide (similar to anthozoan LWamides and RFamides [49, 50]) and FFNPamide (similar to *Aplysia* myomodulin [51, 52]). No homologs of short peptide precursors were identified in sponges or ctenophores.

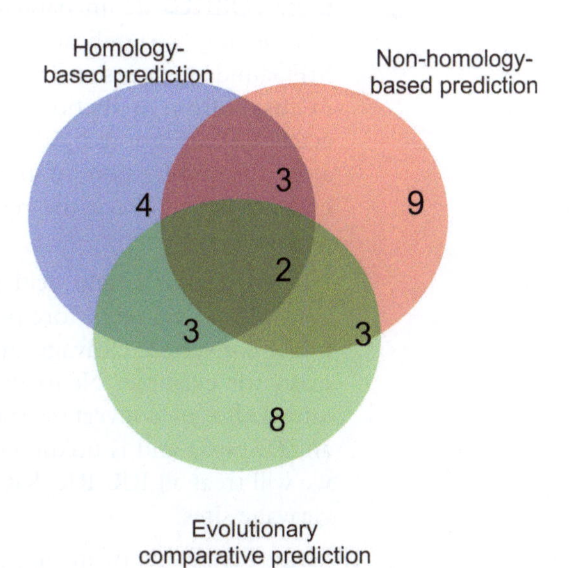

Fig. 9 Venn diagram showing the intersection between sets of *Trichoplax adhaerens* peptide precursors identified with three different methods

On the other hand, insulins and ependymin-related proteins have recognizable homologs in Cnidaria and Bilateria [53] (Figs. 4 and 5). We were unable to find any insulins in sponges and ctenophores. Ependymin-related proteins were not detected in ctenophores but are present in demosponge *Amphimedon* and calcareous sponge *Sycon ciliatum* [54], suggesting their origin before the divergence of Placozoa-Cnidaria-Bilateria.

The full complement of propeptide-processing enzymes (convertase, carboxypeptidase, amidating monooxygenase) was found in all four basal metazoan phyla: Placozoa, Ctenophora, Cnidaria, and Porifera (Table 4) and can be a helpful reference for future studies and peptidomics work. In summary, we think that this list of newly predicted peptides in *Trichoplax* will be further expanded with additional microanalytical and single-cell omic approaches.

4 Notes

1. BLAST offers several tabular output formats, including CSV (comma-separated, -outfmt 10) and TSV (tab-separated, -outfmt 6). In our case, Swissprot titles often contain commas that break down CSV columns; therefore, TSV must be used.
2. Liu's analysis [21] used only "neuropeptide" and "hormone" as stop words. Since then, high-throughput proteomics led to the discovery of many proteins that are experimentally verified and present in SwissProt, but without known functions. One of these (Q8IIG7.2, uncharacterized protein from *Plasmodium falciparum*) is significantly similar (E-value $3.5*10^{-23}$) with SIFGamide precursor TriadITZ_008272 and will cause its exclusion from predictions unless hits with the word "uncharacterized" are ignored. Additional issues like this will likely arise after the publication of this chapter; therefore, readers are encouraged to check the results of BLAST against SwissProt manually.
3. Not all dibasic amino acid sites are subject to prohormone convertase cleavage. More precise tools for predicting prohormone convertase cleavage sites using Hidden Markov Models exist, for example, NeuroPred [41]. Still, they require prior knowledge of convertase specificity, which differs across the animal phyla and is unknown for basal metazoans. Therefore, we will treat all RR, RK, KR, KK, and RxGR sites as potential cleavage sites.
4. Script PeptFisher.py in supplemental data of [17] is written in DOS character encoding. To run this script on Linux or Mac systems, it must be converted with dos2unix program.

Table 4
List of prohormone-processing enzymes in Placozoa, Ctenophora, Porifera, and Cnidaria

Phylum	Species	Gene ID
Convertases		
Placozoa	Hoilungia hongkongensis	braker1_g01213.t1
Placozoa	Hoilungia hongkongensis	braker1_g01217.t1
Placozoa	Hoilungia hongkongensis	braker1_g08092.t1
Placozoa	Trichoplax adhaerens	TriadITZ_003258-RA
Placozoa	Trichoplax adhaerens	TriadITZ_003262-RA
Placozoa	Trichoplax adhaerens	TriadITZ_008375-RA
Ctenophora	Mnemiopsis leydi	ML07022a
Ctenophora	Mnemiopsis leydi	ML11682a
Ctenophora	Mnemiopsis leydi	ML200226a
Ctenophora	Mnemiopsis leydi	ML20261a
Ctenophora	Mnemiopsis leydi	ML207912a
Ctenophora	Mnemiopsis leydi	ML279823a
Ctenophora	Pleurobrachia bachei	sb3462183
Ctenophora	Pleurobrachia bachei	sb3464607
Ctenophora	Pleurobrachia bachei	sb3466502
Porifera	Sycon ciliatum	scpid25436
Porifera	Sycon ciliatum	scpid28547
Porifera	Sycon ciliatum	scpid34651
Porifera	Sycon ciliatum	scpid35388
Porifera	Sycon ciliatum	scpid44026
Porifera	Sycon ciliatum	scpid74752
Porifera	Amphimedon queenslandica	XP_011404106.1
Porifera	Amphimedon queenslandica	XP_011405480.2
Porifera	Amphimedon queenslandica	XP_011407660.1
Porifera	Amphimedon queenslandica	XP_019852355.1
Porifera	Amphimedon queenslandica	XP_019853479.1
Porifera	Amphimedon queenslandica	XP_019855384.1
Porifera	Amphimedon queenslandica	XP_019855895.1
Cnidaria	Nematostella vectensis	XP_001639811.2
Cnidaria	Hydra vulgaris	XP_002166571.1

(continued)

Table 4 (continued)

Phylum	Species	Gene ID
Carboxypeptidases		
Placozoa	Hoilungia hongkongensis	braker1_g00958
Placozoa	Hoilungia hongkongensis	braker1_g01018
Placozoa	Hoilungia hongkongensis	braker1_g03126
Placozoa	Hoilungia hongkongensis	braker1_g08971
Placozoa	Hoilungia hongkongensis	braker1_g08973
Placozoa	Trichoplax adhaerens	XP_002117289.1
Placozoa	Trichoplax adhaerens	XP_002108045.1
Placozoa	Trichoplax adhaerens	XP_002112076.1
Placozoa	Trichoplax adhaerens	XP_002112103.1
Placozoa	Trichoplax adhaerens	XP_002114978.1
Placozoa	Trichoplax adhaerens	XP_002117288.1
Placozoa	Trichoplax adhaerens	XP_002117835.1
Ctenophora	Mnemiopsis leydi	ML02203a
Ctenophora	Mnemiopsis leydi	ML214323a
Ctenophora	Pleurobrachia bachei	sb3461983
Ctenophora	Pleurobrachia bachei	sb3464191
Porifera	Sycon ciliatum	scpid13267_Sycon
Porifera	Sycon ciliatum	scpid23739_Sycon
Porifera	Sycon ciliatum	scpid70932_Sycon
Porifera	Amphimedon queenslandica	XP_003384134.1
Cnidaria	Nematostella vectensis	XP_001624056.2
Cnidaria	Nematostella vectensis	XP_001633785.2
Cnidaria	Nematostella vectensis	XP_001638948.2
Cnidaria	Nematostella vectensis	XP_001639021.2
Cnidaria	Nematostella vectensis	XP_032219153.1
Cnidaria	Nematostella vectensis	XP_032238532.1
Cnidaria	Hydra vulgaris	XP_002158333.2
Cnidaria	Hydra vulgaris	XP_002169920.3
Cnidaria	Hydra vulgaris	XP_004211138.2
Cnidaria	Hydra vulgaris	XP_012555394.1
Cnidaria	Hydra vulgaris	XP_012557491.1

(continued)

Table 4 (continued)

Phylum	Species	Gene ID
Amidating monooxygenases		
Placozoa	Hoilungia hongkongensis	braker1_g02349.t1
Placozoa	Hoilungia hongkongensis	braker1_g04121.t1
Placozoa	Trichoplax adhaerens	TriadITZ_002979-RA
Placozoa	Trichoplax adhaerens	TriadITZ_003133-RA
Ctenophora	Mnemiopsis leydi	ML00116a
Ctenophora	Mnemiopsis leydi	ML02634a
Ctenophora	Pleurobrachia bachei	sb3461822
Ctenophora	Pleurobrachia bachei	sb3462501
Porifera	Sycon ciliatum	scpid29202
Porifera	Sycon ciliatum	scpid34778
Porifera	Sycon ciliatum	scpid51609
Porifera	Amphimedon queenslandica	XP_003388865.3
Cnidaria	Nematostella vectensis	EDO34395.1
Cnidaria	Nematostella vectensis	EDO34601.1
Cnidaria	Nematostella vectensis	EDO42139.1
Cnidaria	Hydra vulgaris	XP_002160828.2

Acknowledgments

This work was supported by Russian Foundation for Basic Research #18–29-13014mk grant. This work was also supported in part by the Human Frontiers Science Program (RGP0060/2017) and National Science Foundation (IOS-1557923) grants to L.L.M. Research reported in this publication was also supported in part by the National Institute of Neurological Disorders and Stroke of the National Institutes of Health under Award Number R01NS114491 (to L.L.M). D.R. and M.A.N. were supported by the Russian Science Foundation grant (23-14-00050). The content is solely the authors' responsibility and does not necessarily represent the official views of the National Institutes of Health.

References

1. Moroz LL, Romanova DY, Kohn AB (2021) Neural versus alternative integrative systems: molecular insights into origins of neurotransmitters. Philos Trans R Soc Lond Ser B Biol Sci 376:20190762
2. Martinez P, Sprecher SG (2020) Of circuits and brains: the origin and diversification of neural architectures. Front Ecol Evol 8:82
3. Jékely G (2021) The chemical brain hypothesis for the origin of nervous systems. Philos Trans R Soc Lond Ser B Biol Sci 376:20190761
4. Arendt D (2021) Elementary nervous systems. Philos Trans R Soc Lond Ser B Biol Sci 376: 20200347
5. Srivastava M, Simakov O, Chapman J et al (2010) The *Amphimedon queenslandica* genome and the evolution of animal complexity. Nature 466:720–726
6. Moroz LL, Kocot KM, Citarella MR et al (2014) The ctenophore genome and the evolutionary origins of neural systems. Nature 510(7503):109–114
7. Ryan JF, Pang K, Schnitzler CE et al (2013) The genome of the ctenophore *Mnemiopsis leidyi* and its implications for cell type evolution. Science 342:1242592
8. Srivastava M, Begovic E, Chapman J et al (2008) The *Trichoplax* genome and the nature of placozoans. Nature 454:955–960
9. Eitel M, Francis WR, Varoqueaux F et al (2018) Comparative genomics and the nature of placozoan species. PLoS Biol 16:e2005359
10. Putnam NH, Srivastava M, Hellsten U et al (2007) Sea anemone genome reveals ancestral eumetazoan gene repertoire and genomic organization. Science 317:86–94
11. Leclère L, Horin C, Chevalier S et al (2019) The genome of the jellyfish *Clytia hemisphaerica* and the evolution of the cnidarian life-cycle. Nat Ecol Evol 3:801–810
12. Khalturin K, Shinzato C, Khalturina M et al (2019) Medusozoan genomes inform the evolution of the jellyfish body plan. Nat Ecol Evol 3:811–822
13. Voolstra CR, Li Y, Liew YJ et al (2017) Comparative analysis of the genomes of *Stylophora pistillata* and *Acropora digitifera* provides evidence for extensive differences between species of corals. Sci Rep 7:17583
14. Veenstra JA (2011) Neuropeptide evolution: neurohormones and neuropeptides predicted from the genomes of *Capitella teleta* and *Helobdella robusta*. Gen Comp Endocrinol 171:160–175
15. Hayakawa E, Watanabe H, Menschaert G et al (2019) A combined strategy of neuropeptide prediction and tandem mass spectrometry identifies evolutionarily conserved ancient neuropeptides in the sea anemone *Nematostella vectensis*. PLoS One 14:e0215185
16. Sachkova MY, Nordmann E-L, Soto-Àngel JJ et al (2021) Neuropeptide repertoire and 3D anatomy of the ctenophore nervous system. Curr Biol 31:5274–5285.e6
17. Koziol U, Koziol M, Preza M et al (2016) De novo discovery of neuropeptides in the genomes of parasitic flatworms using a novel comparative approach. Int J Parasitol 46:709–721
18. Toporik A, Borukhov I, Apatoff A et al (2014) Computational identification of natural peptides based on analysis of molecular evolution. Bioinformatics 30:2137–2141
19. Clynen E, Liu F, Husson SJ et al (2010) Bioinformatic approaches to the identification of novel neuropeptide precursors. In: Soloviev M (ed) Peptidomics: methods and protocols. Humana Press, Totowa, pp 357–374
20. Karsenty S, Rappoport N, Ofer D et al (2014) NeuroPID: a classifier of neuropeptide precursors. Nucleic Acids Res 42:W182–W186
21. Liu F, Baggerman G, D'Hertog W et al (2006) In silico identification of new secretory peptide genes in *Drosophila melanogaster*. Mol Cell Proteomics 5:510–522
22. McVeigh P, Mair GR, Atkinson L et al (2009) Discovery of multiple neuropeptide families in the phylum Platyhelminthes. Int J Parasitol 39: 1243–1252
23. Kamm K, Osigus H-J, Stadler PF et al (2018) *Trichoplax* genomes reveal profound admixture and suggest stable wild populations without bisexual reproduction. Sci Rep 8:11168
24. Jansen E, Ayoubi TAY, Meulemans SMP et al (1995) Neuroendocrine-specific expression of the human prohormone convertase 1 Gene hormonal regulation of transcription through distinct cAMP responsive elements. J Biol Chem 270:15391–15397
25. Fricker LD (2005) Neuropeptide-processing enzymes: applications for drug discovery. AAPS J 7:E449–E455
26. Seidah NG, Prat A (2012) The biology and therapeutic targeting of the proprotein convertases. Nat Rev Drug Discov 11:367–383
27. Van Bael S, Watteyne J, Boonen K et al (2018) Mass spectrometric evidence for neuropeptide-amidating enzymes in *Caenorhabditis elegans*. J Biol Chem 293:6052–6063

28. Schierwater B, Osigus H-J, Bergmann T et al (2021) The enigmatic Placozoa part 1: exploring evolutionary controversies and poor ecological knowledge. BioEssays 43:2100080
29. Schierwater B, Osigus H-J, Bergmann T et al (2021) The enigmatic Placozoa part 2: exploring evolutionary controversies and promising questions on earth and in space. BioEssays 43: 2100083
30. Fortunato A, Fleming A, Aktipis A et al (2021) Upregulation of DNA repair genes and cell extrusion underpin the remarkable radiation resistance of *Trichoplax adhaerens*. PLoS Biol 19:e3001471
31. Anctil M (2009) Chemical transmission in the sea anemone *Nematostella vectensis*: a genomic perspective. Comp Biochem Physiol Part D Genomics Proteomics 4:268–289
32. Veenstra JA (2010) Neurohormones and neuropeptides encoded by the genome of *Lottia gigantea*, with reference to other mollusks and insects. Gen Comp Endocrinol 167:86–103
33. Husson SJ, Mertens I, Janssen T et al (2007) Neuropeptidergic signaling in the nematode *Caenorhabditis elegans*. Prog Neurobiol 82: 33–55
34. Li C (2008) Neuropeptides. WormBook 25:1–36
35. Clynen E, Reumer A, Baggerman G et al (2010) Neuropeptide biology in *Drosophila*. Adv Exp Med Biol 692:192–210
36. Li B, Predel R, Neupert S et al (2008) Genomics, transcriptomics, and peptidomics of neuropeptides and protein hormones in the red flour beetle *Tribolium castaneum*. Genome Res 18:113–122
37. Menschaert G, Vandekerckhove TTM, Baggerman G et al (2010) A hybrid, de novo based, genome-wide database search approach applied to the sea urchin neuropeptidome. J Proteome Res 9:990–996
38. Sonmez K, Zaveri NT, Kerman IA et al (2009) Evolutionary sequence modeling for discovery of peptide hormones. PLoS Comput Biol 5: e1000258
39. Delfino KR, Southey BR, Sweedler JV et al (2010) Genome-wide census and expression profiling of chicken neuropeptide and prohormone convertase genes. Neuropeptides 44:31–44
40. Petersen TN, Brunak S, Heijne G von et al (2011) SignalP 4.0: discriminating signal peptides from transmembrane regions. Nat Methods 8:785–786
41. Southey BR, Amare A, Zimmerman TA et al (2006) NeuroPred: a tool to predict cleavage sites in neuropeptide precursors and provide the masses of the resulting peptides. Nucleic Acids Res 34:W267–W272
42. Hummon AB, Hummon NP, Corbin RW et al (2003) From precursor to final peptides: a statistical sequence-based approach to predicting prohormone processing. J Proteome Res 2: 650–656
43. Nikitin M (2014) Bioinformatic prediction of *Trichoplax adhaerens* regulatory peptides. Gen Comp Endocrinol 212:145–155
44. Katoh K, Standley DM (2013) MAFFT multiple sequence alignment software version 7: improvements in performance and usability. Mol Biol Evol 30:772–780
45. Trifinopoulos J, Nguyen L-T, von Haeseler A et al (2016) W-IQ-TREE: a fast online phylogenetic tool for maximum likelihood analysis. Nucleic Acids Res 44:W232–W235
46. Ofer D, Linial M (2014) NeuroPID: a predictor for identifying neuropeptide precursors from metazoan proteomes. Bioinformatics 30: 931–940
47. Senatore A, Reese TS, Smith CL (2017) Neuropeptidergic integration of behavior in *Trichoplax adhaerens*, an animal without synapses. J Exp Biol 220:3381–3390
48. Varoqueaux F, Williams EA, Grandemange S et al (2018) High cell diversity and complex peptidergic signaling underlie placozoan behavior. Curr Biol 28:3495–3501.e2
49. Gajewski M, Leitz T, Schloßherr J et al (1996) LWamides from Cnidaria constitute a novel family of neuropeptides with morphogenetic activity. Rouxs Arch Dev Biol 205:232–242
50. Pernet V, Anctil M, Grimmelikhuijzen CJP (2004) Antho-RFamide-containing neurons in the primitive nervous system of the anthozoan *Renilla koellikeri*. J Comp Neurol 472: 208–220
51. Cropper EC, Tenenbaum R, Kolks MA et al (1987) Myomodulin: a bioactive neuropeptide present in an identified cholinergic buccal motor neuron of *Aplysia*. Proc Natl Acad Sci U S A 84:5483–5486
52. Miller M, Beushausen S, Vitek A et al (1993) The myomodulin-related neuropeptides: characterization of a gene encoding a family of peptide cotransmitters in *Aplysia*. J Neurosci 13:3358–3367
53. Chan SJ, Steiner DF (2000) Insulin through the ages: phylogeny of a growth promoting and metabolic regulatory hormone. Am Zool 40: 213–222
54. McDougall C, Hammond MJ, Dailey SC et al (2018) The evolution of ependymin-related proteins. BMC Evol Biol 18:182

INDEX

Leonid L. Moroz (ed.), *Ctenophores: Methods and Protocols*, Methods in Molecular Biology, vol. 2757,
https://doi.org/10.1007/978-1-0716-3642-8, © Springer Science+Business Media, LLC, part of Springer Nature 2024

E

G

H

I

L

M

N

O

MIX
Papier aus verantwortungsvollen Quellen
Paper from responsible sources
FSC® C105338

Printed by Libri Plureos GmbH
in Hamburg, Germany